Texts in Computer Science

For further volumes:
www.springer.com/series/3191

Zoran Majkić

Big Data Integration Theory

Theory and Methods of Database Mappings, Programming Languages, and Semantics

Springer

Zoran Majkić
ISRST
Tallahassee, FL, USA

Series Editors
David Gries
Department of Computer Science
Cornell University
Ithaca, NY, USA

Fred B. Schneider
Department of Computer Science
Cornell University
Ithaca, NY, USA

ISSN 1868-0941 ISSN 1868-095X (electronic)
Texts in Computer Science
ISBN 978-3-319-35539-9 ISBN 978-3-319-04156-8 (eBook)
DOI 10.1007/978-3-319-04156-8
Springer Cham Heidelberg New York Dordrecht London

Printed on acid-free paper

Springer is part of Springer Science+Business Media (www.springer.com)

Preface

Big data is a popular term used to describe the exponential growth, availability and use of information, both structured and unstructured. Much has been written on the big data trend and how it can serve as the basis for innovation, differentiation and growth.

According to International Data Corporation (IDC) (one of the premier global providers of market intelligence, advisory services, and events for the information technology, telecommunications and consumer technology markets), it is imperative that organizations and IT leaders focus on the ever-increasing volume, variety and velocity of information that forms big data. From Internet sources, available to all riders, here I briefly cite most of them:

- **Volume**. Many factors contribute to the increase in data volume—transaction-based data stored through the years, text data constantly streaming in from social media, increasing amounts of sensor data being collected, etc. In the past, excessive data volume created a storage issue. But with today's decreasing storage costs, other issues emerge, including how to determine relevance amidst the large volumes of data and how to create value from data that is relevant.
- **Variety**. Data today comes in all types of formats—from traditional databases to hierarchical data stores created by end users and OLAP systems, to text documents, email, meter-collected data, video, audio, stock ticker data and financial transactions.
- **Velocity**. According to Gartner, velocity means both how fast data is being produced and how fast the data must be processed to meet demand. Reacting quickly enough to deal with velocity is a challenge to most organizations.
- **Variability**. In addition to the increasing velocities and varieties of data, data flows can be highly inconsistent with periodic peaks. Daily, seasonal and event-triggered peak data loads can be challenging to manage—especially with social media involved.
- **Complexity**. When you deal with huge volumes of data, it comes from multiple sources. It is quite an undertaking to link, match, cleanse and transform data across systems. However, it is necessary to connect and correlate relationships, hierarchies and multiple data linkages or your data can quickly spiral out

of control. Data governance can help you determine how disparate data relates to common definitions and how to systematically integrate structured and unstructured data assets to produce high-quality information that is useful, appropriate and up-to-date.

Technologies today not only support the collection and storage of large amounts of data, they provide the ability to understand and take advantage of its full value, which helps organizations run more efficiently and profitably.

We can consider a Relational Database (RDB) as an unifying framework in which we can integrate all commercial databases and database structures or also unstructured data wrapped from different sources and used as relational tables. Thus, from the theoretical point of view, we can chose RDB as a general framework for data integration and resolve some of the issues above, namely volume, variety, variability and velocity, by using the existing Database Management System (DBMS) technologies.

Moreover, simpler forms of integration between different databases can be efficiently resolved by Data Federation technologies used for DBMS today.

More often, emergent problems related to the complexity (the necessity to connect and correlate relationships) in the systematic integration of data over hundreds and hundreds of databases need not only to consider more complex schema database mappings, but also an evolutionary graphical interface for a user in order to facilitate the management of such huge and complex systems.

Such results are possible only under a clear theoretical and *algebraic* framework (similar to the algebraic framework for RDB) which extends the standard RDB with more powerful features in order to manage the complex schema mappings (with, for example, merging and matching of databases, etc.). More work about Data Integration is given in pure logical framework (as in RDB where we use a subset of the First Order Logic (FOL)). However, unlike with the pure RDB logic, here we have to deal with a kind of Second Order Logic based on the tuple-generating dependencies (tgds). Consequently, we need to consider an 'algebraization' of this subclass of the Second Order Logic and to translate the declarative specifications of logic-based mapping between schemas into the algebraic graph-based framework (sketches) and, ultimately, to provide denotational and operational semantics of data integration inside a universal algebraic framework: the *category theory*.

The kind of algebraization used here is different from the Lindenbaum method (used, for example, to define Heyting algebras for the propositional intuitionistic logic (in Sect. 1.2), or used to obtain cylindric algebras for the FOL), in order to support the compositional properties of the inter-schema mapping.

In this framework, especially because of Big Data, we need to theoretically consider both the inductive and coinductive principles for databases and infinite databases as well. In this semantic framework of Big-Data integration, we have to investigate the properties of the basic **DB** category both with its topological properties.

Integration across heterogeneous data resources—some that might be considered "big data" and others not—presents formidable logistic as well as analytic challenges, but many researchers argue that such integrations are likely to represent the most promising new frontiers in science [2, 5, 6, 10, 11]. This monograph is a synthesis of my personal research in this field that I developed from 2002 to 2013: this work presents a complete formal framework for these new frontiers in science.

Since the late 1960s, there has been considerable progress in understanding the algebraic semantics of logic and type theory, particularly because of the development of categorical analysis of most of the structures of interest to logicians. Although there have been other algebraic approaches to logic, none has been as far reaching in its aims and in its results as the categorical approach. From a fairly modest beginning, categorical logic has matured very nicely in the past four decades.

Categorical logic is a branch of category theory within mathematics, adjacent to mathematical logic but more notable for its connections to theoretical computer science [4]. In broad terms, categorical logic represents both syntax and semantics by a category, and an interpretation by a functor. The categorical framework provides a rich conceptual background for logical and type-theoretic constructions. The subject has been recognizable in these terms since around 1970.

This monograph presents a categorical logic (denotational semantics) for database schema mapping *based on views* in a very general framework for database-integration/exchange and peer-to-peer. The base database category **DB** (instead of traditional **Set** category), with objects instance-databases and with morphisms (mappings which are not simple functions) between them, is used at an *instance level* as a proper semantic domain for a database mappings based on a set of complex query computations.

The higher logical *schema level* of mappings between databases, usually written in some high expressive logical language (ex. [3, 7], GLAV (LAV and GAV), tuple generating dependency) can then be translated functorially into this base "computation" category.

Different from the Data-exchange settings, we are not interested in the 'minimal' instance B of the target schema $\mathcal{B}$, of a schema mapping $\mathcal{M}_{AB} : \mathcal{A} \to \mathcal{B}$. In our more general framework, we do not intend to determine 'exactly', 'canonically' or 'completely' the instance of the target schema $\mathcal{B}$ (for a fixed instance A of the source schema $\mathcal{A}$) just because our setting is more general and the target database is *only partially* determined by the source database $\mathcal{A}$. Another part of this target database can be, for example, determined by another database $\mathcal{C}$ (that is not any of the 'intermediate' databases between $\mathcal{A}$ and $\mathcal{B}$), or by the software programs which update the information contained in this target database $\mathcal{B}$. In other words, the Data-exchange (and Data integration) settings are only special particular simpler cases of this *general framework* of database mappings where each database can be mapped from other sources, or maps its own information into other targets, and is to be locally updated as well.

The new approach based on the behavioral point of view for databases is assumed, and behavioral equivalences for databases and their mappings are established. The introduction of observations, which are computations without side-effects, defines the fundamental (from Universal algebra) monad endofunctor T, which is also the closure operator for objects and for morphisms such that the database lattice $\langle Ob_{DB}, \preceq \rangle$ is an algebraic (complete and compact) lattice, where

Ob_{DB} is a set of all objects (instance-database) of **DB** category and "$\preceq$" is a pre-order relation between them. The join and meet operators of this database lattice are Merging and Matching database operators, respectively.

The resulting 2-category **DB** is symmetric (also a mapping is represented as an object (i.e., instance-database)) and hence the mappings between mappings are a 1-cell morphisms for all higher meta-theories. Moreover, each mapping is a homomorphism from a Kleisli monadic T-coalgebra into the cofree monadic T-coalgebra. The database category **DB** has nice properties: it is equal to its dual, complete and cocomplete, locally small and locally finitely presentable, and monoidal biclosed V-category enriched over itself. The monad derived from the endofunctor T is an enriched monad.

Generally, database mappings are not simply programs from values (i.e., relations) into computations (i.e., views) but an equivalence of computations because each mapping between any two databases A and B is symmetric and provides a duality property to the **DB** category. The denotational semantics of database mappings is given by morphisms of the Kleisli category $\mathbf{DB}_T$ which may be "internalized" in **DB** category as "computations". Special attention is devoted to a number of practical examples: query definition, query rewriting in the Database-integration environment, P2P mappings and their equivalent GAV translation.

The book is intended to be accessible to readers who have specialist knowledge of theoretical computer science or advanced mathematics (category theory), so it attempts to treat the important database mapping issues accurately and in depth. The book exposes the author's original work on database mappings, its programming language (algebras) and denotational and operational semantics. The analysis method is constructed as a combination of technics from a kind of Second Order Logic, data modeling, (co)algebras and functorial categorial semantics.

Two primary audiences exist in the academic domain. First, the book can be used as a text for a graduate course in the Big Data Integration theory and methods within a database engineering methods curriculum, perhaps complementing another course on (co)algebras and category theory. This would be of interest to teachers of computer science, database programming languages and applications in category theory. Second, researches may be interested in methods of computer science used in databases and logics, and the original contributions: a category theory applied to the databases. The secondary audience I have in mind is the IT software engineers and, generally, people who work in the development of the database tools: the graph-based categorial formal framework for Big Data Integration is helpful in order to develop new graphic tools in Big Data Integration. In this book, a new approach to the database concepts developed from an observational equivalence based on views is presented. The main intuitive result of the obtained basic database category **DB**, more appropriate than the category **Set** used for categorial Lawvere's theories, is to have the possibility of making synthetic representations of database mappings and queries over databases in a graphical form, such that all mapping (and query) arrows can be composed in order to obtain the complex database mapping diagrams. For example, for the P2P systems or the mappings between databases in complex data warehouses. Formally, it is possible to develop a graphic (sketch-based) tool for a meta-mapping description of complex (and partial) mappings in various contexts with a formal mathematical background. A part of this book has been presented to several audiences at various conferences and seminars.

Dependencies Between the Chapters

After the introduction, the book is divided into three parts. The first part is composed of Chaps. 2, 3 and 4, which is a nucleus of this theory with a number of practical examples. The second part, composed of Chaps. 5, 6 and 7, is dedicated to computational properties of the **DB** category, compared to the extensions of the Codd's SPRJU relational algebra Σ_R and Structured Query Language (SQL). It is demonstrated that the **DB** category, as a denotational semantics model for the schema mappings, is computationally equivalent to the Σ_{RE} relational algebra which is a complete extension of Σ_R with all update operations for the relations. Chapter 6 is then dedicated to define the abstract computational machine, the categorial RDB machine, able to support all **DB** computations by SQL embedding. The final sections are dedicated to categorial semantics for the database transactions in time-sharing DBMS. Based on the results in Chaps. 5 and 6, the final chapter of the second part, Chap. 7, then presents full operational semantics for database mappings (programs).

The third part, composed of Chaps. 8 and 9, is dedicated to more advanced theoretical issues about **DB** category: matching and merging operators (tensors) for databases, universal algebra considerations and algebraic lattice of the databases. It is demonstrated that the **DB** category is not a Cartesian Closed Category (CCC) and hence it is not an elementary topos. It is demonstrated that **DB** is monoidal biclosed, finitely complete and cocomplete, locally small and locally finitely presentable category with hom-objects ("exponentiations") and a subobject classifier.

Thus, **DB** is a weak monoidal topos and hence it does not correspond to propositional intuitionistic logic (as an elementary, or "standard" topos) but to one intermediate superintuitionistic logic with strictly more theorems than intuitionistic logic but less then the propositional logic. In fact, as in intuitionistic logic, it does not hold the excluded middle $\phi \vee \neg\phi$, rather the *weak* excluded middle $\neg\phi \vee \neg\neg\phi$ is valid.

Detailed Plan

1. Chapter 1 is a formal and short introduction to different topics and concepts: logics, (co)algebras, databases, schema mappings and category theory, in order to render this monograph more self-contained; this material will be widely used in the rest of this book. It is important also due to the fact that usually database experts do a lot with logics and relational algebras, but much less with programming languages (their denotational and operational semantics) and still much less with categorial semantics. For the experts in programming languages and category theory, having more information on FOL and its extensions used for the database theory will be useful.

2. In Chap. 2, the formal logical framework for the schema mappings is defined, based on the second-order tuple generating dependencies (SOtgds), with existentially quantified functional symbols. Each tgd is a material implication from the conjunctive formula (with relational symbols of a source schema, preceded with negation as well) into a particular relational symbol of the target schema. It provides a number of algorithms which transform these logical formulae into the algebraic structure based on the theory of R-operads. The schema database integrity constraints are transformed in a similar way so that both the schema mappings and schema integrity-constraints are formally represented by R-operads. Then the compositional properties are explored, in order to represent a database mapping system as a graph where the nodes are the database schemas and the arrows are the schema mappings or the integrity-constraints for schemas. This representation is used to define the database mapping sketches (small categories), based on the fact that each schema has an identity arrow (mapping) and that the mapping-arrows satisfy the associative low for the composition of them.

   The algebraic theory of R-operads, presented in Sect. 2.4, represents these algebras in a non-standard way (with carrier set and the signature) because it is oriented to express the compositional properties, useful when formalizing the algebraic properties for a composition of database mappings and defining a categorial semantics for them. The standard algebraic characterization of R-operads, as a kind of relational algebras, will be presented in Chap. 4 in order to understand the relationship with the extensions of the Select–Project–Rename–Join–Union (SPRJU) Codd's relational algebras. Each Tarski's interpretation of logical formulae (SOtgds), used to specify the database mappings, results in the instance-database mappings composed of a set of particular functions between the source instance-database and the target instance-database. Thus, an interpretation of a database-mapping system may be formally represented as a functor from the sketch category (schema database graph) into a category where an object is an instance-database (i.e., a set of relational tables) and an arrow is a set of mapping functions. Section 2.5 is dedicated to the particular property for such a category, namely the duality property (based on category symmetry).

   Thus, at the end of this chapter, we obtain a complete algebraization of the Second Order Logic based on SOtgds used for the logical (declarative) specification of schema mappings. The sketch category (a graph of schema mappings) represents the syntax of the database-programming language. A functor α, derived from a specific Tarski's interpretation of the logical schema database mapping system, represents the semantics of this programming language whose objects are instance-databases and a mapping is a set of functions between them. The formal *denotational semantics* of this programming language will be provided by a database category **DB** in Chap. 3, while the *operational semantics* of this programming language will be presented in Chap. 7.
3. Chapter 3 provides the basic results of this theory, including the definition of the **DB** category as a denotational semantics for the schema database mappings. The objects of this category are the instance-databases (composed of the relational tables and an empty relation $\bot$) and every arrow is just a set of functions

(mapping-interpretations defined in Sect. 2.4.1 of Chap. 2) from the set of relations of the source object (a source database) into a particular relation of the target object (a target database). The power-view endofunctor $T : \mathbf{DB} \to \mathbf{DB}$ is an extension of the power-view operation for a database to morphisms as well. For a given database instance A, TA is the set of all views (which can be obtained by SPRJU statements) of this database A. The Data Federation and Data Separation operators for the databases and a partial ordering and the strong (behavioral) and weak equivalences for the databases are introduced.

4. In Chap. 4, the *categorial functorial semantics of database mappings* is defined also for the database integrity constraints. In Sect. 4.2, we present the applications of this theory to data integration/exchange systems with an example for query-rewriting in GAV data integration system with (foreign) key integrity constraints, based on a coalgebra semantics. In the final section, a fixpoint operator for an infinite canonical solution in data integration/exchange systems is defined.

   With this chapter we conclude the first part of this book. It is intended for all readers because it contains the principal results of the data integration theory, with a minimal introduction of necessary concepts in schema mappings based on SOtgds, their algebraization resulting in the **DB** category and the categorial semantics based on functors. A number of applications are given in order to obtain a clear view of these introduced concepts, especially for database experts who have not worked with categorial semantics.
5. In Chap. 5, we consider the extensions of Codd's SPRJU relational algebra Σ_R and their relationships with the internal algebra of the **DB** category. Then we show that the computational power of the **DB** category (used as denotational semantics for database-mapping programs) is equivalent to the Σ_{RE} relational algebra, which extends the Σ_R algebra with all update operations for relations, and which is implemented as SQL statements in the software programming. We introduce an "action" category **RA** where each tree-term of the Σ_{RE} relational algebra is equivalently represented by a single path term (an arrow in this category) which, applied to its source object, returns its target object. The arrows of this action category will be represented as the Application Plans in the abstract categorical RDB machines, in Chap. 6, during the executions of the embedded SQL statements.
6. Chapter 6 is a continuation of Chap. 5 and is dedicated to computation systems and categorial RDB machines able to support all computations in the **DB** category (by translations of the arrows of the action category **RA**, represented by the Application Plans of the RDB machine, into the morphisms of the database category **DB**). The embedding of SQL into general purpose programs, synchronization process for execution of SQL statements as morphisms in the **DB** category, and transaction recovery are presented in a unifying categorial framework. In particular, we consider the concurrent categorial RDB machines able to support the time-shared "parallel" execution of several user programs.
7. Chapter 7 provides a complete framework of the operational semantics for database-mapping programs, based on final coalgebraic semantics (dual of the initial algebraic semantics introduced in Chap. 5 for the syntax monads (programming languages), and completed in this chapter) of the database-mapping

programs. We introduce an observational comonad for the final coalgebra operational semantics and explain the duality for the database mapping programs: specification versus solution. The relationship between initial algebras (denotational semantics) and final coalgebras (operational semantics) and their semantic adequateness is then presented in the last Sect. 7.5.

Thus, Chaps. 5, 6 and 7 present the second part of this book, dedicated to the syntax (specification) and semantics (solutions) of database-mapping programs.

8. The last part of this book begins with Chap. 8. In this chapter, we analyze advanced features of the **DB** category: matching and merging operators (tensors) for databases, present universal algebra considerations and algebraic lattice of the databases. It is demonstrated that the **DB** category is not a Cartesian Closed Category (CCC) and hence it is not an elementary topos, so that its computational capabilities are strictly inferior to those of typed λ-calculus (as more precisely demonstrated in Chap. 5). It is demonstrated that **DB** is a V-category enriched over itself. Finally, we present the inductive principle for objects and the coinductive principle for arrows in the **DB** category, and demonstrate that its "computation" Kleisly category is embedded into the **DB** category by a faithful forgetful functor.
9. Chapter 9 considers the topological properties of the **DB** category: in the first group of sections, we show the Database metric space, its Subobject classifier, and demonstrate that **DB** is a *weak monoidal topos*. It is proven that **DB** is monoidal biclosed, finitely complete and cocomplete, locally small and locally finitely presentable category with hom-objects ("exponentiations") and a subobject classifier. It is well known that the intuitionistic logic is a logic of an elementary (standard) topos. However, we obtain that **DB** is not an elementary but a weak monoidal topos. Consequently, in the second group of sections, we investigate which kind of logic corresponds to the **DB** weak monoidal topos. We obtain that in the specific case when the universe of database values is a finite set (thus, without Skolem constants which are introduced by existentially quantified functions in the SOtgds) this logic corresponds to the standard propositional logic. This is the case when the database-mapping system is completely specified by the FOL. However, in the case when we deal with incomplete information and hence we obtain the SOtgds with existentially quantified Skolem functions and our universe must include the infinite set of distinct Skolem constants (for recursive schema-mapping or schema integrity constraints), our logic is then an intermediate or superintuitionistic logic in which the weak excluded middle formula $\neg\phi \vee \neg\neg\phi$ is valid. Thus, this weak monoidal topos of **DB** has more theorems than intuitionistic logic but less than the standard propositional logic.

Acknowledgements

Although it contains mostly original material (some of it has not been published before), this book is largely based on the previous work of other people, especially in the framework of the First Order Logic, and could not have been written without that work. I would like to acknowledge the contribution of these authors with a number of references to their important research papers. Also, many of the ideas contained are the result of personal interaction between the author and a number of his colleagues and friends. It would be impossible to acknowledge each of these contributions individually, but I would like to thank all the people who read all or parts of the manuscript and made useful comments and criticisms: I warmly thank Maurizio Lenzerini with whom I carried out most of my research work on data integration [1, 8, 9]. I warmly thank Giuseppe Longo who introduced me to category theory and Sergei Soloviev who supported me while writing my PhD thesis. I warmly thank Eugenio Moggi and Giuseppe Rosolini for their invitation to a seminar at DISI Computer Science, University of Genova, Italy, December 2003, and for a useful discussion that have offered me the opportunity to make some corrections and to improve an earlier version of this work. Also, I thank all the colleagues that I have been working with in several data integration projects, in particular Andrea Calì and Domenico Lembo. Thanks are also due to the various audiences who endured my seminars during the period when these ideas where being developed and who provided valuable feedback and occasionally asked hard questions.

Notational Conventions

We use logical symbols both in our formal languages and in the metalanguage. The notation slightly differs: in classical propositional and FOL, we use $\wedge_c, \vee_c, \Rightarrow_c, \neg_c$ (and $\exists, \forall$ for the FOL quantifiers), while in intuitionistic, $\wedge, \vee, \Rightarrow, \neg$ (in Heyting algebra, we use $\rightharpoonup$ for the relative-pseudocomplement instead of $\Rightarrow$, but for different carrier sets of Heyting algebras we will use the special symbols as well). As the metasymbol of conjunction, the symbol & will be used.

In our terminology, we distinguish functions (graphs of functions) and maps. A (graph of) *function* from X to Y is a binary relation $F \subseteq X \times Y$ (subset of the Cartesian product of the sets X and Y) with domain A satisfying the functionality condition $(x, y) \in F \& (x, z) \in F$ implies $y = z$, and the triple $\langle f, X, Y\rangle$ is then called a *map* (or morphism in a category) from A to B, denoted by $f : X \to Y$ as well. The composition of functions is denoted by $g \cdot f$, so that $(g \cdot f)(x) = g(f(x))$, while the composition of mappings (in category) by $g \circ f$. $\mathcal{N}$ denotes the set of natural numbers.

We will use the symbol $:=$ (or $=_{\mathrm{def}}$, or simply $=$) for definitions and, more often, $\triangleq$ as well. For the equality we will use the standard symbol $=$, while for different equivalence relations we will employ the symbols $\simeq, \approx, \equiv$, etc. In what follows, 'iff' means 'if and only if'. Here is some other set-theoretic notation:

- $\mathcal{P}(X)$ denotes the power set of a set X, and $X^n \triangleq \overbrace{X \times \cdots \times X}^{n}$ the n-ary Cartesian product, and Y^X denotes the set of all functions from X to Y;
- We use $\subseteq$ for inclusion, $\subset$ for proper inclusion (we use $\leq, \preceq, \sqsubseteq$ for partial orders), and $X \subseteq_\omega Y$ denotes that X is a *finite* subset of an infinite set Y;
- R^{-1} is the converse of a binary relation $R \subseteq X \times Y$ and $\overline{R}$ is the complement of R (equal to $(X \times Y) \backslash R$, where $\backslash$ is the set-difference operation);

- id_X is the identity map on a set X. $|X|$ denotes the cardinality of a set (or sequence) X;
- For a set of elements $x_1, \ldots, x_n \in X$, we denote by $\mathbf{x}$ the sequence (or *tuple*) $\langle x_1, \ldots, x_n \rangle$, and if $n = 1$ simply by x_1, while for $n = 0$ the empty tuple $\langle\rangle$. An n-ary relation R, with $n = ar(R) \geq 1$, is a set (also empty) of tuples $\mathbf{x}_i$ with $|\mathbf{x}_i| = n$, with $\langle\rangle \in R$ (the empty tuple is a tuple of every relation);
- By $\pi_{\mathbf{K}}(R)$, where $\mathbf{K} = [i_1, \ldots, i_n]$ is a sequence of indexes with $n = |\mathbf{K}| \geq 1$, we denote the projection of R with columns defined by ordering in $\mathbf{K}$. If $|\mathbf{K}| = 1$, we write simply $\pi_i(R)$;
- Given two sequences $\mathbf{x}$ and $\mathbf{y}$, we write $\mathbf{x} \subseteq \mathbf{y}$ if every element in the list $\mathbf{x}$ is an element in $\mathbf{y}$ (not necessarily in the same position) as well, and by $\mathbf{x}\&\mathbf{y}$ their concatenation; $(\mathbf{x}, \mathbf{y})$ denotes a tuple $\mathbf{x}\&\mathbf{y}$ composed of variables in $\mathbf{x}$ and $\mathbf{y}$, while $\langle \mathbf{x}, \mathbf{y} \rangle$ is a tuple of two tuples $\mathbf{x}$ and $\mathbf{y}$.

A *relational symbol* (a predicate letter in FOL) r and its extension (relation table) R will be called often shortly as "relation" where it is clear from the context. If R is the extension of a relational symbol r, we write $R = \|r\|$.

References

1. D. Beneventano, M. Lenzerini, F. Mandreoli, Z. Majkić, Techniques for query reformulation, query merging, and information reconciliation—part A. Semantic webs and agents in integrated economies, D3.2.A, IST-2001-34825 (2003)
2. M. Bohlouli, F. Schulz, L. Angelis, D. Pahor, I. Brandic, D. Atlan, R. Tate, Towards an integrated platform for Big Data analysis, in *Integration of Practice-Oriented Knowledge Technology: Trends and Prospectives* (Springer, Berlin, 2013), pp. 47–56
3. R. Fagin, P.G. Kolaitis, R.J. Miller, L. Popa, DATA exchange: semantics and query answering, in *Proc. of the 9th Int. Conf. on Database Theory (ICDT 2003)* (2003), pp. 207–224
4. B. Jacobs, *Categorical Logic and Type Theory*. Studies in Logic and the Foundation of Mathematics, vol. 141 (Elsevier, Amsterdam, 1999)
5. M. Jones, M. Schildhauer, O. Reichman, S. Bowers, The new bioinformatics: integrating ecological data from the gene to the biosphere. Annu. Rev. Ecol. Evol. Syst. **37**(1), 519–544 (2006)
6. A. Labrinidis, H.V. Jagadish, Challenges and opportunities with Big Data. Proc. VLDB **5**(12), 2032–2033 (2012)
7. M. Lenzerini, Data integration: a theoretical perspective, in *Proc. of the 21st ACM SIGACT SIGMOD SIGART Symp. on Principles of Database Systems (PODS 2002)* (2002), pp. 233–246
8. M. Lenzerini, Z. Majkić, First release of the system prototype for query management. Semantic webs and agents in integrated economies, D3.3, IST-2001-34825 (2003)
9. M. Lenzerini, Z. Majkić, General framework for query reformulation. Semantic webs and agents in integrated economies, D3.1, IST-2001-34825, February (2003)

10. T. Rabl, S.G. Villamor, M. Sadoghi, V.M. Mulero, H.A. Jacobsen, S.M. Mankovski, Solving Big Data challenges for enterprise application performance management. Proc. VLDB **5**(12), 1724–1735 (2012)
11. S. Shekhar, V. Gunturi, M. Evans, K. Yang, Spatial Big- Data challenges intersecting mobility and cloud computing, in *Proceedings of the Eleventh ACM International Workshop on Data Engineering for Wireless and Mobile Access* (2012), pp. 1–6

Tallahassee, USA Zoran Majkić

Contents

Introduction and Technical Preliminaries 1

1.1 Historical Background

Database schema mappings are so important in information integration that many mapping formalisms have been proposed for different tasks. A schema mapping is a high-level *declarative specification* of the relationship between two schemas, it specifies how data structured under one schema, called source schema, is to be converted into data structured under possibly different schema, called the target schema. It the last decade, schema mappings have been fundamental components for both data exchange and data integration. In this work, we will consider the declarative schema mappings between relational databases. A widely used formalism for specifying relational-to-relational schema mappings is that of tuple generating dependencies (tgds). In the terminology of data integration, tgds are equivalent to global-and-local-as-view (GLAV) assertions. Using a language that is based on tgds for specifying (or 'programming') database schema mappings has several advantages over lower-level languages, such as XSLT scripts of Java programs, in that it is declarative and it has been widely used in the formal study of the semantics of data exchange and data integration. Declarative schema mapping formalisms have been used to provide formal semantics for data exchange [21], data integration [38], peer data management [25, 29], pay-as-you-go integration systems [72], and model management operators [5].

Indeed, the use of higher-level declarative language for 'programming' schema mappings is similar to the goal of model management [4, 63]. One of the goals in model management is to reduce programming effort by allowing a user to manipulate higher-level abstractions, called models, and mappings between models (in this case, models and mappings between models are database schemas and mappings between schemas). The goal of model management is to provide an algebra for explicitly manipulating schemas and mappings between them. A whole area of model management has focused on such issues as mapping composition [22, 43, 67] and mapping inversion [20, 23].

Z. Majkić, *Big Data Integration Theory*, Texts in Computer Science,
DOI 10.1007/978-3-319-04156-8_1,

Our approach is very close to the model management approach, however, with denotational semantics based on the theory of category sketches. In fact, here we define a composition of database schema mappings and two principal algebraic operators for DBMS composition of database schemas (data separation and data federation), which are used for composition of complex schema mappings graphs. At the instance database level, we define also the matching and merging algebraic operators for databases and the perfect inverse mappings.

Most of the work in the data integration/exchange and peer-to-peer (P2P) framework is based on a logical point of view (particularly for the integrity constraints, in order to define the right models for certain answers) in a 'local' mode (source-to-target database) where proper attention to the general 'global' problem of the *compositions* of complex partial mappings which possibly involve a high number of databases has not been given. Today, this 'global' approach cannot be avoided because of the necessity of P2P open-ended networks of heterogenous databases. The aim of this work is a *definition of a* ***DB*** *category for the database mappings* which has to be more suitable than a generic **Set** domain category since the databases are more complex structures w.r.t. the sets and the mappings between them are so complex that they cannot be represented by a *single* function (which is one arrow in **Set**). Why do we need an enriched categorical semantic domain for the databases? We will try, before an exhaustive analysis of the problem presented in next two chapters, to give a partial answer to this question.

- This work is an attempt to give a proper solution for a general problem of complex database-mappings and for the high-level algebra operators of the databases (merging, matching, etc.), by preserving the traditional common practice logical language for schema database mapping definitions.
- The schema mapping specifications are not the integral parts of the standard relational-database theory (used to define a database schema with its integrity constraints); they are the *programs* and we need an enriched denotational semantics context that is able to formally express these programs (derived by the mappings between the databases).
- Let us consider, for example, the P2P systems or the mappings in a complex data warehouse. We would like to have a synthetic graphical representations of the database mappings and queries, and to be able to develop a graphical tool for the meta-mapping descriptions of complex (and partial) mappings in various contexts, with a formal mathematical background.

Only a limited amount of research has been reported in the literature [2, 14, 22, 42, 43, 62, 67] that addressed the general problem presented in this book. One of these works uses category theory [2]. However, it is too restrictive: institutions can only be applied to the simple *inclusion* mappings between databases.

A lot of work has been done for a sketch-based and fibrational formulation of denotational semantics for databases [16, 31, 32, 37, 70]. But all these works are using the elements of an ER-scheme of a database, such as relations, attributes, etc., as the objects of a sketch category but not the *whole databases* as a single object. Hence we need a framework of inter-database mappings. The main difference between the previous categorial approaches to databases and this one is the *level* of abstraction used for the prime objects of the theory.

Another difference is methodological. In fact, the logics for relational databases are based on different kinds of First-Order Logic (FOL) sublanguages as, for example, Description Logic, Relational Database logic, DATALOG, etc. Consequently, the previous work on categorical semantics for the database theory strictly follows an earlier well-developed research for categorial FOL on the predicates with types (many-sorted FOL) where each attribute of a given predicate has a particular sort with a given set of values (domain). Thus, the *fibred semantics* for predicates is assumed for such a typed logic, where other basic operations as negation, conjunction and FOL quantifiers (that are algebraically connected with the Galois connection of their types, traduced by left and right adjunction of their functors in categorical translation) are defined *algebraically* in such a fibrational formulation. This algebraic method, applied in order to translate the FOL in a categorical language, is successively and directly applied to the database theory seen as a sublanguage of the FOL. Consequently, there are no particularly important new results, from the previous developed for the FOL, in this simple translation of DB-theory into a categorical framework. No new particular *base category* is defined for databases (different from **Set**), as it happened in the cases, for example, of the Cartesian Closed Categories (CCC) for typed λ-calculus, Bicartesian Closed Poset Categories for Heyting algebras, or the elementary topos (with the subobject classifier diagrams) for the intuitionistic logic [26, 71]. Basically, all previously works use the **Set** category as the base denotational semantics category, without considering the question if such a topos is also a *necessary* requirement for the database-mapping theory.

This manuscript, which is a result of more than ten years of my personal but not always continuative research, begins with my initial collaboration with Maurizio Lenzerini [3, 39, 40] and from the start its methodological approach was *coalgebraic*, that is, based on an observational point of view for the databases. Such a coalgebraic approach was previously adopted in 2003 for logic programming [48] and for the databases with preorders [47] and here it is briefly exposed.

In our case, we are working with Relational Databases (RDB), and consequently with Structured Query Language (SQL), which is an extension of Codd's "Select–Project–Join+Union" (SPJRU) relational algebra [1, 13]. We assume *a view* of a database A as an observation on this database, presented as a relation (a set of tuples) obtained by a query $q(\mathbf{x})$ (SPJRU term with a list of free variables in $\mathbf{x}$), where $\mathbf{x}$ is a list of attributes of this view. Let $\mathcal{L}_A$ be the set of all such queries over A and $\mathcal{L}_A/_\approx$ be the quotient term algebra obtained by introducing the equivalence relation $\approx$ such that $q(\mathbf{x}) \approx q'(\mathbf{x})$ if both queries return with the same relation (view). Thus, a view can be equivalently considered as a *term* of this quotient-term algebra $\mathcal{L}_A/_\approx$ with a carrier set of relations in A and a finite arity of their SPRJU operators whose computation returns a set of tuples of this view. If this query is a finite term of this algebra then it is called a "finitary view" (a finitary view can have an infinite number of tuples as well).

In this coalgebraic methodological approach to databases, we consider a database instance A of a given database schema $\mathcal{A}$ (i.e., the set of relations that satisfy all integrity constraints of a given database schema) as a black box and any view (the response to a given query) is considered as an *observation*. Thus, in this framework we do not consider a categorical semantic for the free syntax algebra of a given query language, but only the resulting observations and the query-answering system of this database (an Abstract Object Type (AOT), that is, the *coalgebra* presented in Sect. 2.4.2). Consequently, all algebraic aspects of the query language are encapsulated in the single power-view operator T, such that for a given database instance A (first object in our base database category) the object TA is the set of all possible views of this database A that can be obtained from a given query language $\mathcal{L}_A/_\approx$.

A functorial translation of database schema inter-mappings (a small graph category) into the database category **DB**, defined in Sect. 3.2, is fundamentally based on a functor that represents a given *model* of this database schema inter-mappings theory. This functor maps a data schema of a given database into a single object of the **DB** category, that is, a database instance A of this database schema $\mathcal{A}$ (a *model* of this database schema, composed of a set of relations that satisfy the schema's integrity constraints).

The morphisms in the **DB** category are not simple functions as in the **Set** category. Thus, the category **DB** is not necessarily an elementary (standard) topos and, consequently, we investigate its structural properties. In fact, it was shown in [15] that if we want to progress to more expressive sketches w.r.t. the original Ehresmann's sketches for diagrams with limits and coproducts, by eliminating non-database objects as, for example, Cartesian products of attributes or powerset objects, we need *more expressive arrows* for sketch categories (diagram predicates in [15] that are analog to the approach of Makkai in [60]). As we progress to a more abstract vision in which objects are the whole databases, following the approach of Makkai, we obtain more complex arrows in this new basic **DB** category for databases in which objects are just the database instances (each object is a set of relations that compose this database instance). Such arrows are not just simple functions as in the case of the **Set** category but complex trees (i.e., operads) of view-based mappings: each arrow is equivalent to the *sets of functions*. In this way, while Ehresmann's approach prefers to deal with a few fixed diagram properties (commutativity, (co)limitness), we enjoy the possibility of setting a full relational-algebra signature of diagram properties.

This work is an attempt to provide a proper algebraic solution for these problems while preserving the traditional common practice logical language for the schema database mapping definitions: thus we develop a number of algorithms to translate the logical into algebraic mappings.

The *instance level* base database category **DB** has been introduced for the first time in [45] and it was also used in [46]. Historically, in the first draft of this category, we tried to consider its limits and colimits as candidates for the matching and merging type operations on database instances, but after some problems with this interpretation for the coproducts, kindly indicated to me by Giuseppe Rosolini, after my presentation of this initial draft at DISI Computer Science, University of Genova, Italy, December 2003, I realized that it needs additional investigation in order to understand which kind of categorical operators has to be used for matching and merging database objects in the **DB** category. However, I could not finish this work immediately after the visiting seminar at DISI because I received an important invitation to work at College Park University, MD, USA, on some algebraic problems in temporal probabilistic logic and databases. Only after 2007, I was again able to consider these problems of the **DB** category and to conclude this work. Different properties of this **DB** category were presented in a number of previously published papers, in initial versions, [53–57] as well, and it has been demonstrated that this category is a *weak monoidal topos*. The fundamental power view-operator T has been defined in [52]. The Kleisli category and the semantics of morphisms in the **DB** category, based on the monad (endofunctor T) have been presented in [59]. The semantics for merging and matching database operators based on complete database lattice, as in [7], were defined as well and presented in a number of papers cited above. But in this book, the new material represents more than 700 percent w.r.t. previously published research.

In what follows, in this chapter we will present only some basic technical notions for algebras, database theory and the extensions of the first-order logic language (FOL) for database theory and category theory that will be used in the rest of this work. These are very short introductions and more advanced notions can be found in given references.

This work is not fully self-contained; it needs a good background in Relational Database theory, Relational algebra and First Order Logic. This very short introduction is enough for the database readers inexperienced in category theory but interested in understanding the first two parts of this book (Chaps. 2 through 7) where basic properties of the introduced **DB** category and Categorical semantics for schema database mappings based on views, with a number of more interesting applications, are presented.

The third part of this book is dedicated to more complex categorical analysis of the (topological) properties of this new base **DB** category for databases and their mappings, and it requires a good background in the Universal algebra and Category theory.

1.2 Introduction to Lattices, Algebras and Intuitionistic Logics

Lattices are the posets (partially ordered sets) such that for all their elements a and b, the set $\{a, b\}$ has both a join (lub—least upper bound) and a meet (glb—greatest lower bound)) with a partial order $\leq$ (reflexive, transitive and anti-symmetric). A *bounded* lattice has the greatest (top) and least (bottom) element, denoted by convention as 1 and 0. Finite meets in a poset will be written as $1, \wedge$ and finite joins as $0, \vee$. By $(W, \leq, \wedge, \vee, 0, 1)$ we denote a bounded lattice iff for every $a, b, c \in W$ the following equations are valid:

1. $a \vee a = a$, $a \wedge a = a$, (idempotency laws)
2. $a \vee b = b \vee a$, $a \wedge b = b \wedge a$, (commutativity laws)
3. $a \vee (b \vee c) = (a \vee b) \vee c$, $a \wedge (b \wedge c) = (a \vee b) \wedge c$, (associativity laws)
4. $a \vee 0 = a$, $a \wedge 1 = a$,
5. $a \vee (b \wedge a) = a$, $a \wedge (b \vee a) = a$. (absorption laws)

It is distributive if it satisfies the distributivity laws:
6. $a \vee (b \wedge c) = (a \vee b) \wedge (a \vee c)$, $a \wedge (b \vee c) = (a \wedge b) \vee (a \wedge c)$.
A lattice W is *complete* if each (also infinite) subset $S \subseteq W$ (or, $S \in \mathcal{P}(W)$ where $\mathcal{P}$ is the powerset symbol, with the empty set $\emptyset \in \mathcal{P}(W)$) has the least upper bound (lub, supremum) denoted by $\bigvee S \in W$. When S has only two elements, the supremum corresponds to the join operator '$\vee$'. Each finite bounded lattice is a complete lattice. Each subset S has the greatest lower bound (glb, infimum) denoted by $\bigwedge S \in W$, given as $\bigvee\{a \in W \mid \forall b \in S.a \leq b\}$. A complete lattice is bounded and has the bottom element $0 = \bigvee \emptyset = \bigwedge W \in W$ and the top element $1 = \bigwedge \emptyset = \bigvee W \in W$. An element $a \in W$ is *compact* iff whenever $\bigwedge S$ exists and $a \leq \bigwedge S$ for $S \subseteq W$ then $a \leq \bigwedge S'$ for some finite $S' \subseteq W$. W is *compactly generated* iff every element in W is a supremum of compact elements. A lattice W is *algebraic* if it is complete and compactly generated.

A function $l : W \to Y$ between the posets W, Y is *monotone* if $a \leq a'$ implies $l(a) \leq l(a')$ for all $a, a' \in W$. Such a function $l : W \to Y$ is said to have a right (or upper) adjoint if there is a function $r : Y \to W$ in the reverse direction such that $l(a) \leq b$ iff $a \leq r(b)$ for all $a \in W, b \in Y$. Such a situation forms a Galois connection and will often be denoted by $l \dashv r$. Then l is called a left (or lover) adjoint of r. If W, Y are complete lattices (posets) then $l : W \to Y$ has a right adjoint iff l preserves all joins (it is *additive*, i.e., $l(a \vee b) = l(a) \vee l(b)$ and $l(0_W) = 0_Y$ where $0_W, 0_Y$ are bottom elements in complete lattices W and Y, respectively). The right adjoint is then $r(b) = \bigvee\{c \in W \mid l(c) \leq b\}$. Similarly, a monotone function $r : Y \to W$ is a right adjoint (it is *multiplicative*, i.e., has a left adjoint) iff r preserves all meets; the left adjoint is then $l(a) = \bigwedge\{c \in Y \mid a \leq r(c)\}$.

Each monotone function $l : W \to Y$ on a complete lattice (poset) W has both the *least* fixed point (Knaster–Tarski) $\mu l \in W$ and *greatest* fixed point $\nu l \in W$. These can be described explicitly as: $\mu l = \bigwedge\{a \in W \mid l(a) \leq a\}$ and $\nu l = \bigvee\{a \in W \mid a \leq l(a)\}$.

In what follows, we write $b < a$ iff ($b \leq a$ and not $a \leq b$) and we denote by $a \bowtie b$ two unrelated elements in W (so that not ($a \leq b$ or $b \leq a$)). An element in a lattice $c \neq 0$ is a *join-irreducible* element iff $c = a \vee b$ implies $c = a$ or $c = b$ for any $a, b \in W$. An element in a lattice $a \in W$ is an *atom* iff $a > 0$ and $\nexists b$ such that $a > b > 0$.

A lower set (down closed) is any subset Y of a given poset $(W, \leq)$ such that, for all elements a and b, if $a \leq b$ and $b \in Y$ then $a \in Y$.

A *Heyting algebra* is a distributive bounded lattice W with finite meets and joins such that for each element $a \in W$, the function $(\_) \wedge a : W \to W$ has a right adjoint $a \rightharpoonup (\_)$, also called an algebraic implication. An equivalent definition can be given by considering a bonded distributive lattice such that for all a and b in W there is a greatest element c in W, denoted by $a \rightharpoonup b$, such that $c \wedge a \leq b$, i.e., $a \rightharpoonup b = \bigvee\{c \in W \mid c \wedge a \leq b\}$ (relative pseudo-complement). We say that a lattice is *relatively pseudo-complemented* (r.p.c.) lattice if $a \rightharpoonup b$ exists for every a and b in W. Thus, a Heyting algebra is, by definition, an r.p.c. lattice that has 0.

Formally, a distributive bounded lattice $(W, \leq, \wedge, \vee, 0, 1)$ is a Heyting algebra iff there is a binary operation $\rightharpoonup$ on W such that for every $a, b, c \in W$:

7. $a \rightharpoonup a = 1$,
8. $a \wedge (a \rightharpoonup b) = a \wedge b, b \wedge (a \rightharpoonup b) = b$,
9. $a \rightharpoonup (b \wedge c) = (a \rightharpoonup b) \wedge (a \rightharpoonup c)$.

In a Heyting algebra, we can define the negation $\neg a$ as a pseudo-complement $a \rightharpoonup 0$. Then, $a \leq \neg\neg a$. Each *finite distributive* lattice is a Heyting algebra. A complete Heyting algebra is a Heyting algebra $\mathbf{H} = (W, \leq, \wedge, \vee, \rightharpoonup, \neg, 0, 1)$ which is complete as a poset. A complete distributive lattice is thus a complete Heyting algebra iff the following *infinite distributivity* holds [69]:

10. $a \wedge \bigvee_{i \in I} b_i = \bigvee_{i \in I}(a \wedge b_i)$ for every $a, b_i \in W, i \in I$.

The negation and implication operators can be represented as the following monotone functions: $\neg : W \rightarrow W^{OP}$ and $\rightharpoonup: W \times W^{OP} \rightarrow W^{OP}$, where W^{OP} is the lattice with inverse partial ordering and $\wedge^{OP} = \vee$, $\vee^{OP} = \wedge$.

The following facts are valid in any $\mathbf{H}$:

(H1) $a \leq b$ iff $a \rightharpoonup b = 1$, $(a \rightharpoonup b) \wedge (a \rightharpoonup \neg b) = \neg a$;
(H2) $\neg 0 = 0^{OP} = 1, \neg(a \vee b) = \neg a \vee^{OP} \neg b = \neg a \wedge \neg b$; (additive negation)

with the following weakening of classical propositional logic:

(H3) $\neg a \vee b \leq a \rightharpoonup b$, $a \leq \neg\neg a$, $\neg a = \neg\neg\neg a$;
(H4) $a \wedge \neg a = 0$, $a \vee \neg a \leq \neg\neg(a \vee \neg a) = 1$; (weakening excluded-middle)
(H5) $\neg a \vee \neg b \leq \neg(a \wedge b)$; (weakening of De Morgan laws)

Notice that since negation $\neg : W \rightarrow W^{OP}$ is a monotonic and additive operator, it is also a *modal* algebraic negation operator. The smallest complete distributive lattice is denoted by $\mathbf{2} = \{0, 1\}$ with two classic values, false and true, respectively. It is also a complemented Heyting algebra and hence it is Boolean.

From the point of view of Universal algebra, given a signature Σ with a set of functional symbols $o_i \in \Sigma$ with arity $ar : \Sigma \rightarrow \mathcal{N}$, an algebra (or algebraic structure) $\mathbf{A} = (W, \Sigma_{\mathbf{A}})$ is a carrier set W together with a collection $\Sigma_{\mathbf{A}}$ of operations on W with an arity $n \geq 0$. An n-ary operator (functional symbol) $o_i \in \Sigma, ar(o_i) = n$ on W will be named an n-ary operation (a function) $\widehat{o_i} : W^n \rightarrow W$ in $\Sigma_{\mathbf{A}}$ that takes n elements of W and returns a single element of W. Thus, a 0-ary operator (or nullary operation) can be simply represented as an element of W, or a constant, often denoted by a letter like a (thus, all 0-ary operations are included as constants into the carrier set of an algebra). An algebra $\mathbf{A}$ is *finite* if the carrier set W is finite; it is *finitary* if each operator in $\Sigma_{\mathbf{A}}$ has a finite arity. For example, a lattice is an algebra with signature $\Sigma_L = \{\wedge, \vee, 0, 1\}$, where $\wedge$ and $\vee$ are binary operations (meet and join operations, respectively), while 0, 1 are two nullary operators (the constants). The equational semantics of a given algebra is a set of equations E between the terms (or expressions) of this algebra (for example, a distributive lattice is defined by the first six equations above).

Given two algebras $\mathbf{A} = (W, \Sigma_{\mathbf{A}})$, $\mathbf{A}' = (W', \Sigma_{\mathbf{A}})$ of the same *type* (with the same signature Σ and set of equations), a map $h : W \rightarrow W'$ is called a *homomorphism* if for each n-ary operation $\widehat{o_i} \in \Sigma_{\mathbf{A}}$ and $a_1, \ldots, a_n \in W$, $h(\widehat{o_i}(a_1, \ldots, a_n)) = \widehat{o_i}'(h(a_1), \ldots, h(a_n))$. A homomorphism h is called an *isomorphism* if h is a bijection between respective carrier sets; it is called a monomorphism (or embedding) if h is an injective function from W into W'. An algebra $\mathbf{A}'$ is called a homomorphic image of $\mathbf{A}$ if there exists a homomorphism from $\mathbf{A}$ onto $\mathbf{A}'$. An algebra $\mathbf{A}'$ is a *subalgebra* of $\mathbf{A}$ if $W' \subseteq W$, the nullary operators are equal, and the other operators of $\mathbf{A}'$ are the restrictions of operators of $\mathbf{A}$ to W'.

A *subuniverse* of $\mathbf{A}$ is a subset W' of W which is closed under the operators of $\mathbf{A}$, i.e., for any n-ary operation $\widehat{o_i} \in \Sigma_{\mathbf{A}}$ and $a_1, \ldots, a_n \in W'$, $\widehat{o_i}(a_1, \ldots, a_n) \in W'$. Thus, if $\mathbf{A}'$ is a subalgebra of $\mathbf{A}$, then W' is a subuniverse of $\mathbf{A}$. The empty set may be a subuniverse, but it is not the underlying carrier set of any subalgebra. If $\mathbf{A}$ has nullary operators (constants) then every subuniverse contains them as well.

Given an algebra $\mathbf{A}$, *Sub*($\mathbf{A}$) denotes the set of subuniverses of $\mathbf{A}$, which is an algebraic lattice. For $Y \subseteq W$ we say that Y *generates* $\mathbf{A}$ (or Y is a set of generators of $\mathbf{A}$) if $W = Sg(Y) \triangleq \bigcap\{Z \mid Y \subseteq Z \text{ and } Z \text{ is a subuniverse of } \mathbf{A}\}$. Sg is an algebraic closure operator on W: for any $Y \subseteq W$, let $F(Y) = Y \cup \{\widehat{o_i}(b_1, \ldots, b_k) | o_i \in \Sigma_{\mathbf{A}}$ and $b_1, \ldots, b_k \in Y\}$, with $F^0(Y) = Y$, $F^{n+1}(Y) = F(F^n(Y)), n \geq 0$, so that for a finitary $\mathbf{A}$, $Y \subseteq F(Y) \subseteq F^2(Y) \subseteq \cdots$, and, consequently, $Sg(Y) = Y \cup F(Y) \cup F^2(Y) \cup \cdots$, and from this it follows that if $a \in Sg(Y)$ then $a \in F^n(Y)$ for some $n < \omega$; hence for some *finite* $Z \subseteq Y$, $a \in F^n(Z)$, thus $a \in Sg(Z)$, i.e., Sg is an algebraic closure operator.

The algebra $\mathbf{A}$ is *finitely generated* if it has a finite set of generators.

Let X be a set of variables. We denote by $\mathcal{T}X$ the set of terms with variables $x_1, x_2, \ldots$ in X of a type Σ of algebras, defined recursively by:

- All variables and constants (nullary functional symbols) are in $\mathcal{T}X$;
- If $o_i \in \Sigma$, $n = ar(o_i) \geq 1$, and $t_1, \ldots, t_n \in \mathcal{T}X$ then $o_i(t_1, \ldots, t_n) \in \mathcal{T}X$.

If $X = \emptyset$, then $\mathcal{T}\emptyset$ denotes the set of ground terms. Given a class K of algebras of the same type (signature Σ), the term algebra $(\mathcal{T}X, \Sigma)$ is a free algebra with the universal (initial algebra) property: for every algebra $\mathbf{A} = (W, \Sigma) \in K$ and map $f : X \to W$, there is a unique homomorphism $f_\# : \mathcal{T}X \to W$ that extends f to all terms (more in Sect. 5.1.1). Given a term $t(x_1, \ldots, x_n)$ over X and given an algebra $\mathbf{A} = (W, \Sigma_{\mathbf{A}})$ of type Σ, we define a mapping $\widehat{t} : W^n \to W$ by: (i) if t is a variable x_i then $\widehat{t}(a_1, \ldots, a_n) = a_i$ is the ith projection map; (ii) if t is of the form $o_i(t_1(x_1, \ldots, x_n), \ldots, t_k(x_1, \ldots, x_n))$, where $o_i \in \Sigma$, then $\widehat{t}(a_1, \ldots, a_n) = \widehat{o_i}(\widehat{t_1}(a_1, \ldots, a_n), \ldots, \widehat{t_k}(a_1, \ldots, a_n))$. Thus, $\widehat{t}$ is the *term function* on $\mathbf{A}$ corresponding to term t. For any subset $Y \subseteq W$,

$$Sg(Y) = \{\widehat{t}(a_1, \ldots, a_n) \mid t \text{ is } n\text{-ary term of type } \Sigma, n < \omega, \text{ and } a_1, \ldots, a_n \in Y\}.$$

The product of two algebras of the same type $\mathbf{A}$ and $\mathbf{A}'$ is the algebra $\mathbf{A} \times \mathbf{A}' = (W \times W', \Sigma_\times)$ such that for any n-ary operator $\widehat{o_{i,2}} \in \Sigma_\times$ and $(a_1, b_1), \ldots, (a_n, b_n) \in W \times W'$, $n \geq 1$, $\widehat{o_{i,2}}((a_1, b_1), \ldots, (a_n, b_n)) = (\widehat{o_i}(a_1, \ldots, a_n), \widehat{o_i'}(b_1, \ldots, b_n))$. In what follows, if there is no ambiguity, we will write $o_i(a_1, \ldots, a_n)$ for $\widehat{o_i}(a_1, \ldots, a_n)$ as well, and Σ for $\Sigma_{\mathbf{A}}$ of any algebra $\mathbf{A}$ of this type Σ.

Given a Σ-algebra $\mathbf{A}$ with the carrier W, we say that an equivalence relation Q on W agrees with the n-ary operation $o_i \in \Sigma$ if for n-tuples $(a_1, \ldots, a_n), (b_1, \ldots, b_n) \in W^n$ we have $(o_i(a_1, \ldots, a_n), o_i(b_1, \ldots, b_n)) \in Q$ whenever $(a_i, b_i) \in Q$ for $i = 1, \ldots, n$. We say that an equivalence relation Q on a Σ-algebra $\mathbf{A}$ is a *congruence* on $\mathbf{A}$ if it agrees with every operation in Σ. If $\mathbf{A}$ is a Σ-algebra and Q a congruence on $\mathbf{A}$ then there exists a unique Σ-algebra on the quotient set W/Q of the carrier W of $\mathbf{A}$ such that the natural mapping $W \rightarrow W/Q$ (which maps each element $a \in W$ into its equivalence class $[a] \in W/Q$) is a homomorphism. We will denote such an algebra as $\mathbf{A}/Q = (W/Q, \Sigma)$ and will call it a *quotient algebra* of an algebra $\mathbf{A}$ by the congruence Q, such that for each its k-ary operation $\widehat{o}'_i$, we have $\widehat{o}'_i([a_1], \dots, [a_k]) = [\widehat{o}_i(a_1, \dots, a_k)]$.

Let K be a class of algebras of the same type. We say that K is a *variety* if K is closed under homomorphic images, subalgebras and products. Each variety can be seen as a *category* with objects being the algebras and arrows the homomorphisms between them.

The fundamental Birkhoff's theorem in Universal algebra demonstrates that a class of algebras forms a variety iff it is equationally definable. For example, the class of all Heyting algebras (which are definable by the set E of nine equations above), denoted by $\mathcal{HA} = (\Sigma_H, E)$, is a variety. Arend Heyting produced an axiomatic system of propositional logic which was claimed to generate as theorems precisely those sentences that are valid according to the *intuitionistic* conception of truth. Its axioms are all axioms of the Classical Propositional Logic (CPL) having a set of propositional symbols $p, q, \ldots \in PR$ and the following axioms (ϕ, ψ, φ denote arbitrary propositional formulae):

1. $\phi \Rightarrow (\phi \wedge \phi)$;
2. $(\phi \wedge \psi) \Rightarrow (\psi \wedge \phi)$;
3. $(\phi \Rightarrow \psi) \Rightarrow ((\phi \wedge \varphi) \Rightarrow (\psi \wedge \varphi))$;
4. $((\phi \Rightarrow \psi) \wedge (\psi \Rightarrow \varphi)) \Rightarrow (\phi \Rightarrow \varphi)$;
5. $\psi \Rightarrow (\phi \Rightarrow \psi)$;
6. $(\phi \wedge (\phi \Rightarrow \psi)) \Rightarrow \psi$;
7. $\phi \Rightarrow (\phi \vee \psi)$;
8. $(\phi \vee \psi) \Rightarrow (\psi \vee \phi)$;
9. $((\phi \Rightarrow \varphi) \wedge (\psi \Rightarrow \varphi)) \Rightarrow ((\phi \vee \psi) \Rightarrow \varphi)$;
10. $\neg\phi \Rightarrow (\phi \Rightarrow \psi)$;
11. $((\phi \Rightarrow \psi) \wedge (\phi \Rightarrow \neg\psi)) \Rightarrow \neg\phi$;
12. $\phi \vee \neg\phi$;

except for the 12th axiom of "excluded middle", which, according to constructivist attitude (L.E.J. Brouwer) has to be replaced by "either I have constructively demonstrated ϕ, or I have constructively demonstrated that ϕ is false", equivalent to modal formula $\Box\phi \vee \Box\neg\phi$, where $\Box$ is a "necessity" universal modal operator in S4 modal logic (with transitive and symmetric accessibility relation between the possible worlds in Kripke semantics, i.e., where this relation is a partial ordering $\leq$). In the same constructivist attitude, $\neg\neg\phi \Rightarrow \phi$ is not valid (different from CLP). According to Brouwer, to say that ϕ is not true means only that I have not at this time constructed ϕ, which is not the same as saying ϕ is false.

In fact, in Intuitionistic Logic (IL), $\phi \Rightarrow \psi$ is equivalent to $\Box(\phi \Rightarrow_c \psi)$, that is, to $\Box(\neg_c\phi \vee \psi)$ where '$\Rightarrow_c$' is classical logical implication and '$\neg_c$' is classical negation and $\neg\phi$ is equivalent to $\Box\neg_c\phi$. Thus, in IL, the conjunction and disjunction are that of CPL, and only the implication and negation are modal versions of classical versions of the implication and negation, respectively.

Each theorem ϕ obtained from the axioms (1 through 11), and by Modus Ponens (MP) and Substitutions inference rules, is denoted by '$\vdash_{\mathrm{IL}} \phi$'. We denote by **IPC** the set of all theorems of IL, that is, $\mathbf{IPC} = \{\phi | \vdash_{\mathrm{IL}} \phi\}$ (the set of formulae closed under MP and substitution) and, analogously, by **CPC** the set of all theorems of CPL.

We introduce an *intermediate logic* (a consistent superintuitionistic logic) such that a set $\mathbf{L}$ of its theorems (closed under MP and substitution) satisfies $\mathbf{IPC} \subseteq \mathbf{L} \subseteq \mathbf{CPC}$. For every intermediate logic $\mathbf{L}$ and a formula ϕ, $\mathbf{L} + \phi$ denotes the smallest intermediate logic containing $\mathbf{L} \cup \{\phi\}$. Then we obtain

$$\mathbf{CPC} = \mathbf{IPC} + (\phi \vee \neg\phi) = \mathbf{IPC} + (\neg\neg\phi \Rightarrow \phi).$$

The topological aspects of intuitionistic logic (IL) were discovered independently by Alfred Tarski [75] and Marshall Stone [74]. They have shown that the open sets of a topological space form an "algebra of sets" in which there are operations satisfying laws corresponding to the axioms of IL.

In 1965, Saul Kripke published a new formal semantics for IL in which IL-formulae are interpreted as upward-closed hereditary subsets of a partial ordering $(W, \leq)$. More formally, we introduce an *intuitionistic Kripke frame* as a pair $\mathfrak{F} = (W, R)$, where $W \neq \emptyset$ and R is a binary relation on a set W, exactly a partial order (a reflexive, transitive and anti-symmetric relation). Then we define a subset $S \subseteq W$, called an *upset* of $\mathfrak{F}$ if for every $a, b \in W$, $a \in W$ and $(a, b) \in R$ imply $b \in S$. Here we will briefly present this semantics, but with a semantics equivalent (dual) to it, based on downward-closed hereditary subsets of W, where a subset $S \subseteq W$ is hereditary in W if we have:

$$\text{whenever} \quad a \in S \text{ and } b \leq a, \quad \text{then} \quad b \in S$$

(that is, when S is an upset for the binary relation R equal to $\leq^{-1}$). The collection of hereditary subsets of W will be denoted by $H(W)$. A valuation $V : PR \rightarrow H(W)$, for a set PR of propositional symbols of IL, is a mapping which assigns to each propositional symbol $p \in PR$ a hereditary subset $V(p) \in H(W)$. A Kripke model on the set of "possible worlds" in W is a pair $\mathcal{M} = (\mathfrak{F}, V)$ where V is a valuation and $\mathfrak{F} = (W, R)$ a Kripke frame with $R = \leq^{-1}$ (i.e., $R = \geq$) such that $V(p)$ is the set of all possible worlds in which p is true. The requirement that $V(p)$ be downward hereditary formalizes (according to Kripke) the "persistence in time of truth", satisfying the condition: $a \in V(p)$ and $(a, b) \in R$ implies $b \in V(p)$.

We now extend the notion of truth at a particular possible world $a \in W$ to all IL formulae, by introducing the expression $\mathcal{M} \models_a \phi$, to be read "a formula ϕ is true in $\mathcal{M}$ at a", defined inductively as follows:

1. $\mathcal{M} \models_a p$ iff $a \in V(p)$;
2. $\mathcal{M} \models_a \phi \wedge \psi$ iff $\mathcal{M} \models_a \phi$ and $\mathcal{M} \models_a \psi$;
3. $\mathcal{M} \models_a \phi \vee \psi$ iff $\mathcal{M} \models_a \phi$ or $\mathcal{M} \models_a \psi$;
4. $\mathcal{M} \models_a \phi \Rightarrow \psi$ iff $\forall b$ ($a \geq b$ implies ($\mathcal{M} \models_b \phi$ implies $\mathcal{M} \models_b \psi$));
5. $\mathcal{M} \models_a \neg\phi$ iff $\forall b$ ($a \geq b$ implies not $\mathcal{M} \models_b \phi$).

In fact, for the binary accessibility relation R on X equal to the binary partial ordering '$\geq$' which is reflexive and transitive, we obtain the S4 modal framework with the universal modal quantifier $\square$, defined by
6. $\mathcal{M} \models_a \square\phi$ iff $\forall b$ ($a \geq b$ implies $\mathcal{M} \models_b \phi$),
so that from 4. and 5. we obtain for the inuitionistic implication and negation that $\Rightarrow$ is equal to $\square \Rightarrow_c$ and $\neg$ is equal to $\square\neg_c$, where $\Rightarrow_c$ and $\neg_c$ are the classical (standard) propositional implication and negation, respectively. Thus, we extend V to any given formula ϕ by $V(\phi) = \{a \mid \mathcal{M} \models_a \phi\} \in H(W)$ and say that ϕ is true in $\mathcal{M}$ if $V(\phi) = W$. Consequently, the complex algebra of the truth values is a Heyting algebra:
7. $\mathbf{H}(W) = (H(W), \subseteq, \cap, \cup, \Rightarrow_h, \neg_h, \emptyset, W)$,
where $\Rightarrow_h$ is a relative pseudo-complement in $H(W)$ and $\neg_h$ is a pseudo-complement such that for any hereditary subset $S \in H(W)$, $\neg_h(S) = S \Rightarrow_h \emptyset$ (the empty set $\emptyset$ is the bottom element in the truth-value lattice $(H(W), \subseteq, \cap, \cup)$, corresponding to falsity; thus, a formula ϕ is false in $\mathcal{M}$ if $V(\phi) = \emptyset$).

Let ϕ be a propositional formula, $\mathfrak{F}$ be a Kripke frame, $\mathcal{M}$ be a model on $\mathfrak{F}$, and K be a class of Kripke frames, then:

(a) We say that ϕ is *true* in $\mathcal{M}$, and write $\mathcal{M} \models \phi$, if $\mathcal{M} \models_a \phi$ for every $a \in W$ (i.e., if $V(\phi) = W$);
(b) We say that ϕ is *valid* in the frame $\mathfrak{F}$, and write $\mathfrak{F} \models \phi$, if $V(\phi) = W$ for every valuation V on $\mathfrak{F}$. We denote by $Log(\mathfrak{F}) = \{\phi \mid \mathfrak{F} \models \phi\}$ the set of all formulae that are valid in $\mathfrak{F}$;
(c) We say that ϕ is *valid* in K, and write K $\models \phi$, if $\mathfrak{F} \models \phi$ for every $\mathfrak{F} \in$ K.

Analogously, let $\mathbf{H} = (W, \leq, \wedge, \vee, \rightharpoonup, \neg, 0, 1)$ be a Heyting algebra. A function $v : PR \to W$ is called a valuation into this Heyting algebra. We extend the valuation from PR to all propositional formulae via the recursive definition:

$$v(\phi \wedge \psi) = v(\phi) \wedge v(\psi), \qquad v(\phi \vee \psi) = v(\phi) \vee v(\psi),$$
$$v(\phi \Rightarrow \psi) = v(\phi) \rightharpoonup v(\psi).$$

A formula ϕ is *true* in $\mathbf{H}$ under v if $v(\phi) = 1$; ϕ is *valid* into $\mathbf{H}$ if ϕ is true for every valuation in $\mathbf{H}$, denoted by $\mathbf{H} \models \phi$. Algebraic completeness means that a formula ϕ is **HA**-valid iff it is valid in *every* Heyting algebra:
8. ϕ is **HA**-valid iff $\vdash_{\mathrm{IL}} \phi$.
The "soundness" part of 8. consists in showing that the axioms 1–11 are **HA**-valid and that Modus Ponens inference preserves this property (in fact, if for a given valuation v, $v(\phi) = v(\phi \Rightarrow \psi) = 1$ then $v(\phi) \leq v(\psi)$ so $v(\psi) = 1$).

The completeness of IL w.r.t. **HA**-validity can be shown by the Lindenbaum–Tarski algebra method by establishing the equivalence relation $\sim_{\mathrm{IL}}$ for the IL-formulae (**IPC**), as follows:
9. $\phi \sim_{\mathrm{IL}} \psi$ iff $\vdash_{\mathrm{IL}} \phi \Rightarrow \psi$ and $\vdash_{\mathrm{IL}} \psi \Rightarrow \phi$ (i.e., iff $\vdash_{\mathrm{IL}} \phi \Leftrightarrow \psi$).
The Lindenbaum algebra for IL is then the quotient Heyting algebra

$$\mathbf{H}_{\mathrm{IL}} = (\mathbf{IPC}_{\sim_{\mathrm{IL}}}, \sqsubseteq, \sqcap, \sqcup, \rightharpoonup, \neg)$$

where for any two equivalence classes $[\phi], [\psi] \in \mathbf{IPC}_{\sim_{\mathrm{IL}}}$, $[\phi] \sqsubseteq [\psi]$ iff $\vdash_{\mathrm{IL}} \phi \Rightarrow \psi$, with

$$[\phi] \sqcap [\psi] \triangleq [\phi \wedge \psi], \qquad [\phi] \sqcup [\psi] \triangleq [\phi \vee \psi],$$

$$[\phi] \rightharpoonup [\psi] \triangleq [\phi \Rightarrow \psi], \qquad \neg[\phi] \triangleq [\neg\phi].$$

Then the valuation $v(\phi) = [\phi]$ can be used to show $\vdash_{\mathrm{IL}} \phi$ iff $\mathbf{H}_{\mathrm{IL}} \models \phi$, hence any **HA**-valid sentence will be $\mathbf{H}_{\mathrm{IL}}$-valid and so an IL-theorem.

We can extend the algebraic semantics of **IPC** to all intermediate logics. With every intermediate logic $\mathbf{L} \supseteq \mathbf{IPC}$ we associate the class $\mathbf{V_L}$ of Heyting algebras in which all the theorems of **L** are valid. It is well known that $\mathbf{V_L}$ is a variety. For example, $\mathbf{V_{IPC}} = \mathcal{HA}$ denotes the variety of all Heyting algebras. For every variety $\mathbf{V} \subseteq \mathcal{HA}$, let $\mathbf{L_V}$ be the logic of all formulae valid in **V**, so that, for example, $\mathbf{L}_{\mathcal{HA}} = \mathbf{IPC}$. The Lindenbaum–Tarski construction shows that every intermediate logic is complete w.r.t. its algebraic semantics. In fact, it was shown that every intermediate logic **L** (an extension of **IPC**) is sound and complete w.r.t. the variety $\mathbf{V_L}$.

1.3 Introduction to First-Order Logic (FOL)

We will shortly introduce the syntax of the First-order Logic language $\mathcal{L}$, as an extension of the propositional logic, and its semantics based on Tarski's interpretations:

Definition 1 The syntax of the First-order Logic (FOL) language $\mathcal{L}$ is as follows:

- Logical operators $(\wedge, \neg, \exists)$ over the bounded lattice of truth values $\mathbf{2} = \{0, 1\}$, 0 for falsity and 1 for truth;
- Predicate letters $r_1, r_2, \ldots$ with a given finite arity $k_i = ar(r_i) \geq 1$, $i = 1, 2, \ldots$ in $\mathbb{R}$;
- Functional letters $f_1, f_2, \ldots$ with a given arity $k_i = ar(f_i) \geq 0$ in $\mathbb{F}$ (language constants $\overline{0}, \overline{1}, \ldots, \overline{c}, \overline{d}, \ldots$ are considered as a particular case of nullary functional letters);
- Variables $x, y, z, \ldots$ in X, and punctuation symbols (comma, parenthesis);
- A set PR, with *truth* $r_\emptyset \in PR \cap \mathbb{R}$, of propositional letters (nullary predicates);
- The following simultaneous inductive definition of *term* and *formulae*:
  1. All variables and constants are terms. All propositional letters are formulae.
  2. If $t_1, \ldots, t_k$ are terms and $f_i \in \mathbb{F}$ is a k-ary functional symbol then $f_i(t_1, \ldots, t_k)$ is a term, while $r_i(t_1, \ldots, t_k)$ is a formula for a k-ary predicate letter $r_i \in \mathbb{R}$.
  3. If ϕ and ψ are formulae then $(\phi \wedge \psi)$, $\neg\phi$, and $(\exists x_i)\phi$ for $x_i \in X$ are formulae.

An interpretation (Tarski) I_T consists of a nonempty domain $\mathcal{U}$ and a mapping that assigns to any predicate letter $r_i \in \mathbb{R}$ with $k = ar(r_i) \geq 1$, a relation $\|r_i\| = I_T(r_i) \subseteq \mathcal{U}^k$, to any k-ary functional letter $f_i \in \mathbb{F}$ a function $I_T(f_i) : \mathcal{U}^k \to \mathcal{U}$, to each individual constant $\overline{c} \in F$ one given element $I_T(\overline{c}) \in \mathcal{U}$, with $I_T(\overline{0}) = 0$, $I_T(\overline{1}) = 1$ for natural numbers $\mathcal{N} = \{0, 1, 2, \ldots\}$, and to any propositional letter $p \in PR$ one truth value $I_T(p) \in \mathbf{2} = \{0, 1\} \subseteq \mathcal{N}$. We assume the countable infinite set of Skolem constants (marked null values) $SK = \{\omega_0, \omega_1, \ldots\}$ to be a subset of the universe $\mathcal{U}$.

Notice that, when $\mathbb{R}$, $\mathbb{F}$ and X are empty, this definition reduces to the Classical Propositional Logic CPL, where I_T is its valuation.

In a formula $(\exists x)\phi$, the formula ϕ is called "action field" for the quantifier $(\exists x)$. A variable y in a formula ψ is called a bounded variable iff it is the variable of a quantifier $(\exists y)$ in ψ, or it is in the action field of a quantifier $(\exists y)$ in the formula ψ. A variable x is free in ψ if it is not bounded. The universal quantifier is defined by $\forall = \neg\exists\neg$.

Disjunction $\phi \vee \psi$ and implication $\phi \Rightarrow \psi$ are expressed by $\neg(\neg\phi \wedge \neg\psi)$ and $\neg\phi \vee \psi$, respectively. In FOL with the identity $\doteq$, the formula $(\exists_1 x)\phi(x)$ denotes the formula $(\exists x)\phi(x) \wedge (\forall x)(\forall y)(\phi(x) \wedge \phi(y) \Rightarrow (x \doteq y))$. We use the built-in binary identity relational symbol (predicate) $r_{\top}$, with $r_{\top}(x, y)$ for $x \doteq y$, as well.

We can introduce the sorts in order to be able to assign each variable x_i to a sort $S_i \subseteq \mathcal{U}$ where $\mathcal{U}$ is a given domain for the FOL (for example, for natural numbers, for reals, for dates, etc., as used for some attributes in database relations). An assignment $g : X \to \mathcal{U}$ for variables in X is applied only to free variables in terms and formulae. If we use sorts for variables then for each sorted variable $x_i \in X$ an assignment g must satisfy the auxiliary condition $g(x_i) \in S_i$. Such an assignment $g \in \mathcal{U}^X$ can be recursively and uniquely extended into the assignment $g^* : \mathcal{T}X \to \mathcal{U}$, where $\mathcal{T}X$ denotes the set of all terms with variables in X, by

1. $g^*(t) = g(x) \in \mathcal{U}$ if the term t is a variable $x \in X$.
2. $g^*(t) = I_T(\overline{c}) \in \mathcal{U}$ if the term t is a constant $\overline{c} \in F$.
3. If a term t is $f_i(t_1, \ldots, t_k)$, where $f_i \in \mathbb{F}$ is a k-ary functional symbol and $t_1, \ldots, t_k$ are terms, then $g^*(f_i(t_1, \ldots, t_k)) = I_T(f_i)(g^*(t_1), \ldots, g^*(t_k))$.

We denote by t/g (or ϕ/g) the ground term (or formula) without free variables, obtained by assignment g from a term t (or a formula ϕ), and by $\phi[x/t]$ the formula obtained by uniformly replacing x by a term t in ϕ. A *sentence* is a formula having no free variables. A Herbrand base of a logic $\mathcal{L}$ is defined by $H = \{r_i(t_1, \ldots, t_k) \mid r_i \in \mathbb{R}$ and $t_1, \ldots, t_k$ are ground terms$\}$. We define the satisfaction for the logical formulae in $\mathcal{L}$ and a given assignment $g : X \to \mathcal{U}$ inductively, as follows:

- If a formula ϕ is an atomic formula $r_i(t_1, \ldots, t_k)$, then this assignment g satisfies ϕ iff $(g^*(t_1), \ldots, g^*(t_k)) \in I_T(r_i)$;
- If a formula ϕ is a propositional letter, then g satisfies ϕ iff $I_T(\phi) = 1$;
- g satisfies $\neg\phi$ iff it does not satisfy ϕ;
- g satisfies $\phi \wedge \psi$ iff g satisfies ϕ and g satisfies ψ;
- g satisfies $(\exists x_i)\phi$ iff there exists an assignment $g' \in \mathcal{U}^X$ that may differ from g only for the variable $x_i \in X$, and g' satisfies ϕ.

A formula ϕ is `true` for a given interpretation I_T iff ϕ is satisfied by every assignment $g \in \mathcal{U}^X$. A formula ϕ is `valid` (i.e., tautology) iff ϕ is true for every Tarski's interpretation $I_T \in \mathfrak{I}_T$ (for example, $r_\emptyset$ and, for each propositional letter $p \in PR$, $p \Rightarrow p$ are valid). An interpretation I_T is a `model` of a set of formulae Γ iff every formula $\phi \in \Gamma$ is true in this interpretation. We denote by FOL(Γ) the FOL with a set of assumptions Γ, and by $\mathfrak{I}_T(\Gamma)$ the subset of Tarski's interpretations that are models of Γ, with $\mathfrak{I}_T(\emptyset) = \mathfrak{I}_T$. A formula ϕ is said to be a *logical consequence* of Γ, denoted by $\Gamma \Vdash \phi$, iff ϕ is true in all interpretations in $\mathfrak{I}_T(\Gamma)$. Thus, $\Vdash \phi$ iff ϕ is a tautology.

The basic set of axioms of the FOL are that of the propositional logic CPL with two additional axioms:

(A1) $(\forall x)(\phi \Rightarrow \psi) \Rightarrow (\phi \Rightarrow (\forall x)\psi)$ (x does not occur in ϕ and it is not bounded in ψ), and

(A2) $(\forall x)\phi \Rightarrow \phi[x/t]$ (neither x nor any variable in t is bounded in ϕ).

For the FOL with identity, we need the *proper* axiom

(A3) $x_1 \doteq x_2 \Rightarrow (x_1 \doteq x_3 \Rightarrow x_2 \doteq x_3)$.

We denote by $R_=$ the Tarski's interpretation of identity $\doteq$, that is, $R_= = \|r_\top\| = I_T(r_\top)$ is the *built-in* identity relation (equal for any Tarski's interpretation), with, for example, $\langle 0, 0\rangle, \langle 1, 1\rangle \in R_=$.

The inference rules are Modus Ponens and generalization (G) "if ϕ is a theorem and x is not bounded in ϕ, then $(\forall x)\phi$ is a theorem".

In what follows, any open-sentence, a formula ϕ with nonempty tuple of free variables $\mathbf{x} = \langle x_1, \ldots, x_m\rangle$ will be called an m-ary *virtual predicate*, denoted also by $\phi(x_1, \ldots, x_m)$ or by $\phi(\mathbf{x})$. This definition contains the precise method of establishing the *ordering* of variables in this tuple. The method that will be adopted here is the ordering of appearance, from left to right, of free variables in ϕ. This method of composing the tuple of free variables is a unique and canonical way of defining the virtual predicate from a given formula. The FOL is considered as an extensional logic because two open-sentences with the same tuple of variables $\phi(x_1, \ldots, x_m)$ and $\psi(x_1, \ldots, x_m)$ are `equal` iff they have the *same extension* in a given interpretation I_T, that is, iff $I_T^*(\phi(x_1, \ldots, x_m)) = I_T^*(\psi(x_1, \ldots, x_m))$, where I_T^* is the unique extension of I_T to all formulae, as follows:

1. For a (closed) sentence ϕ/g, $I_T^*(\phi/g) = 1$ iff g satisfies ϕ, as recursively defined above.
2. For an open-sentence ϕ with the tuple of free variables $\langle x_1, \ldots, x_m\rangle$,

$$I_T^*\big(\phi(x_1, \ldots, x_m)\big) =_{\text{def}} \big\{\big\langle g(x_1), \ldots, g(x_m)\big\rangle \mid g \in \mathcal{U}^X \text{ and } I_T^*(\phi/g) = 1\big\}.$$

It is easy to verify that for a formula ϕ with the tuple of free variables $\langle x_1, \ldots, x_m\rangle$,

$$I_T^*\big(\phi(x_1, \ldots, x_m)/g\big) = 1 \quad \text{iff} \quad \big\langle g(x_1), \ldots, g(x_m)\big\rangle \in I_T^*\big(\phi(x_1, \ldots, x_m)\big).$$

1.3.1 Extensions of the FOL for Database Theory

One of the most important issues of mathematical logic is that our understanding of mathematical phenomena is enriched by elevating the languages we use to describe mathematical structures to objects of explicit study. It is this aspect of logic which is most prominent in model theory which deals with the relation between a formal language and its interpretations. The specialization of model theory to finite structures should find manifold applications in computer science, particularly in the framework of specifying programs to query databases: phenomena whose understanding requires close attention to the interaction between language and structure. Beginning with connection to automata theory, the *finite model* theory has developed through a range of applications to problems in graph theory, database and complexity theory and artificial intelligence.

Remark First of all, we will use the FOL extended by a number of binary *built-in predicates*, necessary for composition of queries, as $\doteq, \neq, <$, etc., that can be used for compositions of database queries, without using logical negation operator $\neg$. For example, $\neg(x \doteq y)$ will be expressed by $x \neq y$, $x \leq y$ by $(x < y) \vee (x \doteq y)$, $\neg(x > y)$ by $(x < y) \vee (x \doteq y)$, $\neg(x \leq y)$ by $x > y$, etc. These built-in predicates have the equal prefixed extension for a given FOL domain $\mathcal{U}$, so do not depend on a particular Tarski's interpretation I_T in Definition 1.

Notice that we will use the symbol $\doteq$ formally for FOL formulae, while informally we will use the common symbol for equality $=$ in all other metalanguage cases.

First-order logic (FOL) corresponds to relational calculus, existential second-order logic (∃SOL: they start with existential second-order quantifiers, followed by a first-order formula) to the complexity class NP [18] (existential second-order quantifiers correspond to the guessing stage of an NP algorithm, and the remaining first-order formula corresponds to the polynomial time verification of an NP algorithm), and second-order logic with quantifiers ranging over sets (of positions) describes regular languages, as $(aa)*$, for example. It can be shown that the transitive closure in the database theory is not expressible in FOL. Such inexpressibility results have traditionally been a core theme of the finite model theory [17, 28, 76].

Let us consider the reachability query: can we get from x to y for a given binary relation r, by considering the following list of queries:

$$q_0(x, y) = r(x, y), \; q_1(x, y) = \exists z \big(r(x, z_1) \wedge r(z_1, y)\big),$$

$$\ldots$$

$$q_n(x, y) = \exists z_1 \ldots z_n \big(r(x, z_1) \wedge r(z_1, z_2) \wedge \cdots \wedge r(z_n, y)\big),$$

that is, one wants to compute the transitive closure of R. The problem with this is that we do not know in advance what n is supposed to be. So the query that we need to write is

$$\bigvee_{n \in \mathcal{N}} q_n$$

where $\mathcal{N}$ is the set of natural numbers. But it is not an FOL formula. The inability of FOL to express some important queries motivated a lot of research on extensions of FOL that can do queries such as transitive closure or cardinality comparisons (as in SQL that can count). Such extensions, for example,

- *Fixed point logics* (fragment of second-order logic). We can extend FOL to express properties that algorithmically require recursion. Such extensions have fixed point operators as the least, inflationary, and partial fixed point operators.

  The resulting fixed point logics, in the presence of a linear order, capture complexity classes PTIME (for least and inflationary fixed points) and PSPACE (for partial fixed points). A well-known database query language that adds fixed points in FOL is DATALOG. By adding the transitive closure to FOL, over order structures, it captures nondeterministic logarithmic space.

Fixed point logics can be embedded into a logic which uses infinitary connectives but has a restriction that every formula mentions finitely many variables.

- *Counting logics* that are important for database theory. For example [41], in SQL one can write a query that finds all pairs of managers x and y who have the same number of people reporting to them (Reports_To relation stores pairs (x, y) where x is an employee and y is his/her immediate manager):

  Select R_1.manager, R_2.manager
  from Reports_To R_1, Reports_To R_2
  where (select count (Reports_To.employee)
  from Reports_To
  where Reports_To.manager $=$ R_1.manager)
  $=$ (select count (Reports_To.employee)
  from Reports_To
  where Reports_To.manager $=$ R_2.manager)

  In general, we add mechanisms for counting, such as counting terms, counting quantifiers, or certain generalized quantifiers. Usually with this counting power, these extended languages remain local, as FOL. We can apply these results in the database setting, by considering a standard feature of many query-languages, namely aggregate functions.

Interesting extensions of FOL by a number of second-order features are *monadic* second-order quantifiers (MSO). Such quantifiers can range over particular *subsets* of the universe (in monadic extensions, we can use the quantification $\exists X$ where X is a subset of the universe, differently from FOL where X is an element of the universe). We can consider two particular restrictions:

1. An $\exists$MSO formula starts with a sequence of existential second-order quantifiers, which is followed by an FOL formula.
2. An $\forall$MSO formula starts with a sequence of universal second-order quantifiers, which is followed by an FOL formula.

For example, $\exists$MSO and $\forall$MSO are different for graphs. For strings MSO collapses to $\exists$MSO and captures exactly the *regular languages* [6]. If we restrict attention to FOL over strings then it captures exactly the star-free languages.

MSO can be used over *trees* (if we view the XML documents as trees, such queries choose certain nodes from trees) and tree automata, for example, for monadic DATALOG. Furthermore, monadic DATALOG can be evaluated in time linear both in the size of the program and the size of the string [27].

1.4 Basic Database Concepts

The database mappings, for a given logical language (we assume the FOL language in Definition 1), are usually defined at a schema level as follows:

- A *database schema* is a pair $\mathcal{A} = (S_A, \Sigma_A)$ where S_A is a countable set of relational symbols (predicates in FOL) $r \in \mathbb{R}$ with finite arity $n = ar(r) \geq 1$ ($ar : \mathbb{R} \rightarrow \mathcal{N}$), disjoint from a countable infinite set **att** of attributes (a domain of $a \in$ **att** is a nonempty finite subset $dom(a)$ of a countable set of individual symbols **dom**, with $\mathcal{U} =$ **dom** $\cup\, SK$). For any $r \in \mathbb{R}$, the sort of r, denoted by tuple $\mathbf{a} = atr(r) = \langle atr_r(1), \ldots, atr_r(n)\rangle$ where all $a_i = atr_r(m) \in \mathbf{att}$, $1 \leq m \leq n$, must be distinct: if we use two equal domains for different attributes then we denote them by $a_i(1), \ldots, a_i(k)$ (a_i equals to $a_i(0)$). Each index ("column") i, $1 \leq i \leq ar(r)$, has a distinct column name $nr_r(i) \in SN$ where SN is the set of names with $nr(r) = \langle nr_r(1), \ldots, nr_r(n)\rangle$. A relation $r \in \mathbb{R}$ can be used as an atom $r(\mathbf{x})$ of FOL with variables in $\mathbf{x}$ assigned to its columns, so that Σ_A denotes a set of sentences (FOL formulae without free variables) called *integrity constraints* of the sorted FOL with sorts in **att**. We denote the empty schema by $\mathcal{A}_\emptyset = (\{r_\emptyset\}, \emptyset)$, where $r_\emptyset$ is the relation with empty set of attributes (*truth* propositional letter in FOL, Definition 1), and we denote the set of all database schemas for a given (also infinite) set $\mathbb{R}$ by $\mathbb{S}$.

- An *instance-database* of a nonempty schema $\mathcal{A}$ is given by $A = (\mathcal{A}, I_T) = \{R = \|r\| = I_T(r) \mid r \in S_A\}$ where I_T is a Tarski's FOL interpretation in Definition 1 which satisfies *all* integrity constraints in Σ_A and maps a relational symbol $r \in S_A$ into an n-ary relation $R = \|r\| \in A$. Thus, an instance-database A is a set of n-ary relations, managed by relational database systems (DBMSs). Let A and $A' = (\mathcal{A}, I'_T)$ be two instances of $\mathcal{A}$, then a function $h : A \rightarrow A'$ is a *homomorphism* from A into A' if for every k-ary relational symbol $r \in S_A$ and every tuple $\langle v_1, \ldots, v_k\rangle$ of this k-ary relation in A, $\langle h(v_1), \ldots, h(v_k)\rangle$ is a tuple of the same symbol r in A'. If A is an instance-database and ϕ is a sentence then we write $A \models \phi$ to mean that A satisfies ϕ. If Σ is a set of sentences then we write $A \models \Sigma$ to mean that $A \models \phi$ for every sentence $\phi \in \Sigma$. Thus the set of all instances of $\mathcal{A}$ is defined by $Inst(\mathcal{A}) = \{A \mid A \models \Sigma_A\}$. We denote the set of all values in A by $val(A) \subseteq \mathcal{U}$. Then 'atomic database' $\mathfrak{J}_A = \{\{\langle v_i\rangle\} \mid v_i \in val(A)\}$ is infinite iff $SK \subseteq val(A)$. Note that for each $a \in atr(r)$, a subset $dom(a) \subseteq \mathbf{dom}$ is finite, and any introduction of Skolem constants is ordered $\omega_0, \omega_1, \ldots$.
- We consider a rule-based *conjunctive query* over a database schema $\mathcal{A}$ as an expression $q(\mathbf{x}) \longleftarrow r_1(\mathbf{u}_1), \ldots, r_n(\mathbf{u}_n)$, with finite $n \geq 0$, r_i are the relational symbols (at least one) in $\mathcal{A}$ or the built-in predicates (e.g., $\leq, =$, etc.), q is a relational symbol not in $\mathcal{A}$ and $\mathbf{u}_i$ are free tuples (i.e., one may use either variables or constants). Recall that if $\mathbf{v} = (v_1, \ldots, v_m)$ then $r(\mathbf{v})$ is a shorthand for $r(v_1, \ldots, v_m)$. Finally, each variable occurring in $\mathbf{x}$ is a *distinguished* variable that must also occur at least once in $\mathbf{u}_1, \ldots, \mathbf{u}_n$. Rule-based conjunctive queries (called rules) are composed of a subexpression $r_1(\mathbf{u}_1), \ldots, r_n(\mathbf{u}_n)$ that is the *body*, and the *head* of this rule $q(\mathbf{x})$. The *Yes/No* conjunctive queries are the rules with an empty head. If we can find values for the variables of the rule, such that the body is logically satisfied, then we can deduce the head-fact. This concept is captured by a notion of "valuation". The deduced head-facts of a conjunctive query $q(\mathbf{x})$ defined over an instance A (for a given Tarski's interpretation I_T of schema $\mathcal{A}$) are equal to $\|q(x_1, \ldots, x_k)\|_A = \{\langle v_1, \ldots, v_k\rangle \in \mathcal{U}^k \mid A \models \exists\mathbf{y}(r_1(\mathbf{u}_1) \wedge \cdots \wedge r_n(\mathbf{u}_n))[x_i/v_i]_{1\leq i\leq k}\} = I^*_T(\exists\mathbf{y}(r_1(\mathbf{u}_1) \wedge \cdots \wedge r_n(\mathbf{u}_n)))$, where $\mathbf{y}$ is a set of variables which are not in the head of query. We recall that the conjunctive queries are monotonic and satisfiable, and that a (Boolean) query is a class of instances that is closed under isomorphism [12]. Each conjunctive query corresponds to a "select–project–join" term $t(\mathbf{x})$ of SPRJU algebra obtained from the formula $\exists\mathbf{y}(r_1(\mathbf{u}_1) \wedge \cdots \wedge r_n(\mathbf{u}_n))$, as explained in Sect. 5.1.

- We consider a finitary *view* as a union of a finite set S of conjunctive queries with the same head $q(\mathbf{x})$ over a schema $\mathcal{A}$, and from the equivalent algebraic point of view, it is a "select–project–join+union" (SPJRU) finite-length term $t(\mathbf{x})$ which corresponds to a union of the terms of conjunctive queries in S. In what follows, we will use the same notation for an FOL formula $q(\mathbf{x})$ and its equivalent algebraic SPJRU expression $t(\mathbf{x})$. A materialized view of an instance-database A is an n-ary relation $R = \bigcup_{q(\mathbf{x})\in S} \|q(\mathbf{x})\|_A$. Notice that a finitary view can also have an infinite number of tuples. We denote the set of all finitary materialized views that can be obtained from an instance A by TA.
- Given two autonomous instance-databases A and B, we can make a *federation* of them, in order to be able to *compute* the queries with relations of both autonomous instance-databases. A federated database system is a type of meta-database management system (DBMS) which transparently integrates multiple autonomous database systems into a single federated database. The constituent databases are interconnected via a computer network, and may be geographically decentralized. Since the constituent database systems remain autonomous, a federated database system is a contrastable alternative to the task of *merging* together several disparate databases. A federated database, or virtual database, is a fully-integrated, logical composite of all constituent databases in a federated database system. McLeod and Heimbigner [61] were among the first to define a federated database system. Among other surveys, Sheth and Larsen [73] define a Federated Database as a collection of cooperating component systems which are autonomous and are possibly heterogeneous.

We consider the views as a universal property for databases: they are the possible observations of the information contained in an instance-database. We can use them in order to establish an equivalence relation between databases. Database category **DB**, which will be introduced in Chap. 3, is at the *instance level*, i.e., any object in **DB** is an instance-database. The connection between a *schema level* and this category is based on the *interpretation* functors. Thus, each rule-based conjunctive query at the schema level over a schema $\mathcal{A}$ will be translated (by an interpretation functor) in a morphism in **DB**, from an instance-database A (a model of the schema $\mathcal{A}$) into the instance-database TA (composed by all materialized views of A).

1.4.1 Basic Theory about Database Observations: Idempotent Power-View Operator

We will introduce a class of coalgebras for database query-answering systems for a given instance-database A of a schema $\mathcal{A}$ in Sect. 2.4.2. They will be presented in an algebraic style by providing a co-signature. In particular, the sorts include a single "hidden sort", corresponding to the carrier of a coalgebra, and other "visible" sorts for the inputs and outputs with a given fixed interpretation. Visible sorts will be interpreted as the sets without any algebraic structure defined on them. For us, the coalgebraic terms, built by operations (destructors), are interpreted by the basic *observations* which one can make on the states of a coalgebra.

Input sorts for a given instance-database A is a countable set $\mathcal{L}_A$ of the union of a finite set S of conjunctive *finite-length* queries $q(\mathbf{x})$ (with the same head with a finite tuple of variables $\mathbf{x}$) so that $R = ev_A(q(\mathbf{x})) = \bigcup_{q(\mathbf{x})\in S} \|q(\mathbf{x})\|_A$ is the relation (a materialized view) obtained by applying this query to A.

Each query (FOL formula introduced in Sect. 1.4) has an equivalent finite-length *algebraic term* of the SPJRU algebra (or equivalent to it, SPCU algebra, Chaps. 4.5, 5.4 in [1]) as shortly introduced in the previous section, and hence the power view-operator T can be defined by the initial SPRJU algebra of ground terms (see Sect. 5.1.1). We define this fundamental *idempotent power-view* operator T, with the domain and codomain equal to the set of all instance-databases, such that for any instance-database A, the object $TA = T(A)$ denotes a database composed of the set of *all views* of A. The object TA, for a given instance-database A, corresponds to the carrier of the quotient-term Lindenbaum algebra $\mathcal{L}_A/_{\approx}$, i.e., the set of the *equivalence classes* of queries (such a query is equivalent to a term in $\mathcal{T}_P X$ of an SRRJU relational algebra Σ_R, formally given in Definition 31 of Sect. 5.1, with the select, project, join and union operators, with relational symbols of a database schema $\mathcal{A}$). More precisely, TA is "generated" from A by this quotient-term algebra $\mathcal{L}_A/_{\approx}$ and a given evaluation of queries in $\mathcal{L}_A$, $ev_A : \mathcal{L}_A \twoheadrightarrow TA$, which is surjective function. From the factorization theorem, there is a unique bijection $is_A : \mathcal{L}_A/_{\approx} \rightarrow TA$ such that

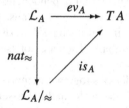

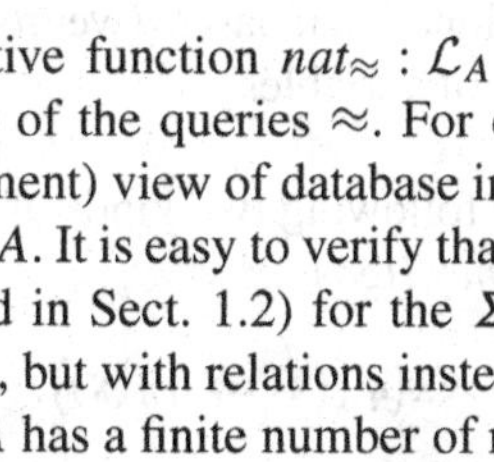

commutes, where the surjective function $nat_{\approx} : \mathcal{L}_A \rightarrow \mathcal{L}_A/_{\approx}$ is a natural representation for the equivalence of the queries $\approx$. For every object A, $A \subseteq TA$ and $TA = T(TA)$, i.e., each (element) view of database instance TA is also an element (view) of a database instance A. It is easy to verify that T corresponds to the closure operator Sg on A (introduced in Sect. 1.2) for the Σ_R relational "select–project–join+union" (SPJRU) algebra, but with relations instead of variables (i.e., relational symbols). Notice that when A has a finite number of relations, but at least one relation with an infinite number of tuples, then TA has an infinite number of relations (i.e., views of A) and hence can be an infinite object.

1.4.2 Introduction to Schema Mappings

The problem of sharing data from multiple sources has recently received significant attention, and a succession of different architectures has been proposed, beginning with federated databases [61, 73], followed by data integration systems [9, 10, 38], data exchange systems [20, 21, 25] and Peer-to-Peer (P2P)) data management systems [11, 24, 29, 34, 49, 50].

A lot of research has been focused on the development of logic languages for semantic mapping between data sources and mediated schemas [7, 14, 30, 38, 45, 64], and algorithms that use mappings to answer queries in data sharing systems [3, 10, 39, 40, 42, 46, 51, 58, 77].

We consider that a *mapping* between two database schemas $\mathcal{A} = (S_A, \Sigma_A)$ and $\mathcal{B} = (S_B, \Sigma_B)$ is expressed by an union of "conjunctive queries with the same head". Such mappings are called "view-based mappings", defined by a set of FOL sentences

$$\{\forall \mathbf{x}_i (q_{Ai}(\mathbf{x}_i) \Rightarrow q_{Bi}(\mathbf{y}_i)) \mid \text{with } \mathbf{y}_i \subseteq \mathbf{x}_i, 1 \leq i \leq n\},$$

where $\Rightarrow$ is the logical implication between these conjunctive queries $q_{Ai}(\mathbf{x}_i)$ and $q_{Bi}(\mathbf{x}_i)$, over the databases $\mathcal{A}$ and $\mathcal{B}$, respectively.

Schema mappings are often specified by the source-to-target tuple-generating dependencies (tgds), used to formalize a data exchange [21], and in the data integration scenarios under a name "GLAV assertions" [9, 38]. A tgd is a logical sentence (FOL formula without free variables) which says that if some tuples satisfying certain equalities exist in the relation, then some other tuples (possibly with some unknown values) must also exist in another specified relation.

An equality-generating dependency (egd) is a logical sentence which says that if some tuples satisfying certain equalities exist in the relation, then some values in these tuples must be equal. Functional dependencies are egds of a special form, for example, primary-key integrity constraints. Thus, egds are only used for the specification of integrity constraints of a single database schema, which define the set of possible models of this database. They are not used for inter-schema database mappings.

These two classes of dependencies together comprise the *embedded implication dependencies* (EID) [19] which seem to include essentially all of the naturally-occurring constraints on relational databases (we recall that the bold symbols $\mathbf{x}, \mathbf{y}, \ldots$ denote a nonempty list of variables):

Definition 2 We introduce the following two kinds of EIDs [19]:

1. A *tuple-generating dependency* (*tgd*)

$$\forall \mathbf{x}(q_A(\mathbf{x}) \Rightarrow q_B(\mathbf{x})),$$

where $q_A(\mathbf{x})$ is an existentially quantified formula $\exists \mathbf{y} \phi_A(\mathbf{x}, \mathbf{y})$ and $q_B(\mathbf{x})$ is an existentially quantified formula $\exists \mathbf{z} \psi_A(\mathbf{x}, \mathbf{z})$, and where the formulae $\phi_A(\mathbf{x}, \mathbf{y})$ and $\psi_A(\mathbf{x}, \mathbf{z})$ are conjunctions of atomic formulae (conjunctive queries) over the given database schemas. We assume the safety condition, that is, that every *distinguished* variable in $\mathbf{x}$ appears in q_A.

We will consider also the class of *weakly-full* tgds for which query answering is decidable, i.e., when $q_B(\mathbf{x})$ has no existentially quantified variables, and if each $y_i \in \mathbf{y}$ appears at most once in $\phi_A(\mathbf{x}, \mathbf{y})$.

2. An *equality-generating dependency* (egd)

$$\forall \mathbf{x}(q_A(\mathbf{x}) \Rightarrow (\mathbf{y} \doteq \mathbf{z})),$$

where $q_A(\mathbf{x})$ is a conjunction of atomic formulae over a given database schema, and $\mathbf{y} = \langle y_1, \dots, y_k \rangle, \mathbf{z} = \langle z_1, \dots, z_k \rangle$ are among the variables in $\mathbf{x}$, and $\mathbf{y} \doteq \mathbf{z}$ is a shorthand for the formula $(y_1 \doteq z_1) \wedge \dots \wedge (y_k \doteq z_k)$ with the built-in binary identity predicate $\doteq$ of the FOL.

Note that a tgd $\forall \mathbf{x}(\exists \mathbf{y} \phi_A(\mathbf{x}, \mathbf{y}) \Rightarrow \exists \mathbf{z} \psi_A(\mathbf{x}, \mathbf{z}))$ is logically equivalent to the formula $\forall \mathbf{x} \forall \mathbf{y}(\phi_A(\mathbf{x}, \mathbf{y}) \Rightarrow \exists \mathbf{z} \psi_A(\mathbf{x}, \mathbf{z}))$, i.e., to $\forall \mathbf{x}_1(\phi_A(\mathbf{x}_1) \Rightarrow \exists \mathbf{z} \psi_A(\mathbf{x}, \mathbf{z}))$ with the set of distinguished variables $\mathbf{x} \subseteq \mathbf{x}_1$.

We will use for the integrity constraints Σ_A of a database schema $\mathcal{A}$ both tgds and egds, while for the inter-schema mappings, between a schema $\mathcal{A} = (S_A, \Sigma_A)$ and a schema $\mathcal{B} = (S_B, \Sigma_B)$, only the tgds $\forall \mathbf{x}(q_A(\mathbf{x}) \Rightarrow q_B(\mathbf{x}))$, as follows:

Definition 3 An elementary schema mapping is a triple $(\mathcal{A}, \mathcal{B}, \mathcal{M})$ where $\mathcal{A}$ and $\mathcal{B}$ are schemas with no relational symbol in common and $\mathcal{M}$ is a set of tgds $\forall \mathbf{x}(q_A(\mathbf{x}) \Rightarrow q_B(\mathbf{x}))$, such that $q_A(\mathbf{x})$ is a conjunctive query with conjuncts equal to relational symbols in S_A or to a formula with built-in relational symbols, $\doteq, <, >, \dots$), while $q_B(\mathbf{x})$ is a conjunctive query with relational symbols in S_B.

An instance of $\mathcal{M}$ is an instance pair (A, B) (where A is an instance of $\mathcal{A}$ and B is an instance of $\mathcal{B}$) that satisfies every tgds in $\mathcal{M}$, denoted by $(A, B) \models \mathcal{M}_{AB}$. We write *Inst*($\mathcal{M}$) to denote all instances (A, B) of $\mathcal{M}$.

Notice that the formula with built-in predicates, in the left side of implication of a tgd, can be expressed by only two logical connectives, conjunction and negation, from the fact that implication and disjunction can be reduced to equivalent formulae with these two logical connectives.

Recall that in data exchange terminology, B is a solution for A under $\mathcal{M}$ if $(A, B) \in$ *Inst*($\mathcal{M}$), and that an instance of $\mathcal{M}$ satisfies all FOL formulae in $\Sigma_A \cup \Sigma_B \cup \mathcal{M}$.

For a given set of FOL formulas S, we denote by $\bigwedge S$ the conjunction of all formulae in the set S.

Lemma 1 *For any given Tarski's interpretation I_T that is a model of the schemas $\mathcal{A} = (S_A, \Sigma_A)$ and $\mathcal{B} = (S_B, \Sigma_B)$ and of the set of tgds in the mapping $\mathcal{M}$, that is, when $I_T^*(\bigwedge \Sigma_A) = I_T^*(\bigwedge \Sigma_B) = I_T^*(\bigwedge \mathcal{M}) = 1$, one has $(\{I_T(r) \mid r \in S_A\}, \{I_T(r) \mid r \in S_B\}) \in$ Inst($\mathcal{M}$).*

Proof Due to the fact that $I_T^*(\bigwedge \Sigma_A) = 1$ means that all integrity constraints of $\mathcal{A}$ are satisfied, one has that $A = \{I_T(r) \mid r \in S_A\}$ is an instance (model) of $\mathcal{A}$. The same holds for the schema $\mathcal{B}$, so that $B = \{I_T(r) \mid r \in S_B\}$ is an instance of $\mathcal{B}$.

From the fact that $I_T^*(\bigwedge \mathcal{M}) = 1$, each tgd $\phi \in \mathcal{M}_{AB}$ is satisfied, i.e., $I_T^*(\phi) = 1$, and hence $(A, B) \models \mathcal{M}$, i.e., $(A, B) \in$ *Inst*($\mathcal{M}$). □

The formulae (tgds) in the set $\mathcal{M}$ express the constraints that an instance (A, B) over the schemas $\mathcal{A}$ and $\mathcal{B}$ must satisfy. We assume that the satisfaction relation between formulae and instances is preserved under isomorphism, which means that if an instance satisfies a formula then every isomorphic instance also satisfies that formula.

This is a mild condition that is true for all standard logical formalisms, such as first-order logic, second-order logic, fixed-point logics, and infinitary logics.

Thus, such formulae represent the queries in the sense of Chandra and Harel [12]. An immediate consequence of this property is that $Inst(\mathcal{M})$ is closed under isomorphism.

Remark 1 Differently from [22, 43, 64], each formula in $\mathcal{M}$ contains the relational symbols of both source and target schema (the integrity constraints are contained in their schemas), in order to represent an inter-schema mapping graphically as a graph edge $\mathcal{M}_{AB} : \mathcal{A} \to \mathcal{B}$ as in standard mathematical denotation of a mapping.

The problem of computing semantic mappings [5, 22, 23, 43, 63, 67], given a semantic mapping $\mathcal{M}_{AB}$ between data schemas $\mathcal{A}$ and $\mathcal{B}$, and $\mathcal{M}_{BC}$ between $\mathcal{B}$ and $\mathcal{C}$, generally was to answer if it is possible to generate a direct semantic mapping $\mathcal{M}_{AC}$ (possibly in the same logic language formalism) between $\mathcal{A}$ and $\mathcal{C}$ that is "equivalent" to the original mappings. Here "equivalent" means that for any query in a given class of queries Q and for any instance of data sources, using the direct mapping yields exactly the same answer that would be obtained by the two original mappings [43].

The semantics of the composition of the schema mappings proposed by Madhavan and Halevy [43] was a significant first step. However, it suffers from certain drawbacks that are caused by the fact that this semantics is given relative to a class of queries. In this setting, the set of formulae specifying a composition $\mathcal{M}_{AC}$ of $\mathcal{M}_{AB}$ and $\mathcal{M}_{BC}$ relative to a class Q of queries need not be unique up to logical equivalence, even when the class Q of queries is held fixed.

It was shown [22] that this semantics is rather fragile because a schema mapping $\mathcal{M}_{AC}$ may be a composition of $\mathcal{M}_{AB}$ and $\mathcal{M}_{BC}$ when Q is the class of conjunctive queries, but may fail to be a composition of these two (inter-)schema mappings when Q is the class of conjunctive queries with inequalities.

Using these results, a proper theory of composition of schema mappings, based on Second-Order tgds, will be developed in Chap. 2.

1.5 Basic Category Theory

Category theory is an area of mathematics that examines in an abstract way the properties of particular mathematical concepts, by formalizing them as collections of objects and arrows (also called morphisms, although this term also has a specific, non-category-theoretic sense), where these collections satisfy certain basic conditions. Many significant areas of mathematics can be formalized as categories, and the use of category theory allows many intricate and subtle mathematical results in these fields to be stated, and proved, in a much simpler way than without the use of categories.

The most accessible example of a category is the category **Set** of sets, where the objects are sets and the arrows are functions from one set to another. However, it is important to note that the objects of a category need not be sets or the arrows functions; any way of formalizing a mathematical concept such that it meets the basic conditions on the behavior of objects and arrows is a valid category, and all the results of category theory will apply to it.

Categories were first introduced by Samuel Eilenberg and Saunders Mac Lane in 1942–45 [36], in connection with algebraic topology. The study of categories is an attempt to axiomatically capture what is commonly found in various classes of related mathematical structures by relating them to the structure-preserving functions between them. A systematic study of category theory then allows us to prove general results about any of these types of mathematical structures from the axioms of a category.

A category **C** consists of the following three mathematical entities:

- A class $Ob_{\mathbf{C}}$ whose elements are called objects;
- A class $Mor_{\mathbf{C}}$ whose elements are called morphisms or maps, or arrows. Each morphism f has a unique source object A and target object B. The expression (arrow) $f : A \rightarrow B$ would be verbally stated as "f is a morphism from A to B". The expression $hom(A, B)$—alternatively expressed as $hom_{\mathbf{C}}(A, B)$, or $mor(A, B)$, or $\mathbf{C}(A, B)$—denotes the hom-class of all morphisms from A to B. If for all A, B this hom-class is a set, this category **C** is called *locally small*.
- A binary operation $\circ$, called composition of morphisms, such that for any three objects A, B, and C, we have a function $\circ : hom(A, B) \times hom(B, C) \rightarrow hom(A, C)$. Each morphism $f \in hom(A, B)$ has a domain $A = dom(f)$ and codomain $B = cod(f)$ which are objects.

  The composition of $f : A \rightarrow B$ and $g : B \rightarrow C$ is written as $g \circ f$ and is governed by two axioms:
  - *Associativity*. If $f : A \rightarrow B, g : B \rightarrow C$ and $h : C \rightarrow D$ then $h \circ (g \circ f) = (h \circ g) \circ f$, and
  - *Identity*. For every object A, there exists a morphism $id_A : A \rightarrow A$ called the identity morphism for A, such that for every morphism $f : A \rightarrow B$, we have $id_B \circ f = f \circ id_A = f$.

From these axioms, it can be proved that there is exactly one identity morphism for every object. Because of that we can identify each object with its identity morphism. Relations among morphisms are often depicted using commutative diagrams, with nodes representing objects and arrows representing morphisms, for example, $g \circ f = h$ is graphically represented by the following commutative diagram:

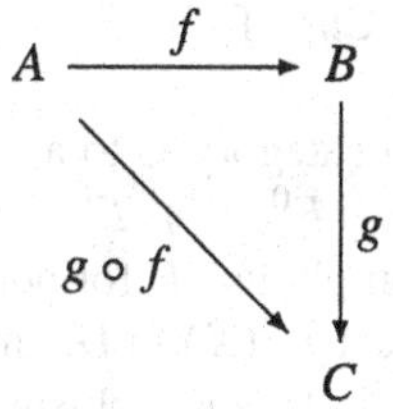

Examples:

- Category **Set** where the objects are sets and the arrows are functions from one set to another.
- A preorder is a set X together with a binary relation $\leq$ which is reflexive (i.e., $x \leq x$ for all $x \in X$), and transitive (i.e., $x \leq y$ and $y \leq z$ imply $x \leq z$ for all $x, y, z \in X$). This can be seen as a category, with set of objects X and for every pair of objects (x, y) such that $x \leq y$, exactly one arrow $x \to y$.
- Any set X can be seen as a discrete category, with a set of objects X and only with the identity morphisms.

Properties of morphisms:

Morphisms can have any of the following properties. A morphism $f : A \to B$ is a

- *Monomorphism* (or monic) if $f \circ g_1 = f \circ g_2$ implies $g_1 = g_2$ for all morphisms $g_1, g_2 : X \to A$. It is denoted by $f : A \hookrightarrow B$.
- *Epimorphism* (or epic) if $g_1 \circ f = g_2 \circ f$ implies $g_1 = g_2$ for all morphisms $g_1, g_2 : B \to X$. It is denoted by $f : A \twoheadrightarrow B$. This epic is called *split* if there is $g : B \to A$ such that $f \circ g = id_B$ (then f is a *retraction* of g).
- *Isomorphism* if there exists a morphism $g : B \to A$ such that $f \circ g = id_B$ and $g \circ f = id_A$. It is denoted by $f : A \simeq B$. Often we denote simply by $A \simeq B$ the isomorphism of two objects. If f is split epic and monic, f is isomorphism.

For example, in **Set** the epimorphisms are surjective functions, the monomorphisms are injective functions, the isomorphisms are bijective functions.

The definition of epic is *dual* to the definition of monic. With $\mathbf{C}^{OP}$ we denote the opposite category of $\mathbf{C}$, where all arrows are reversed. Thus, f is monic in the category $\mathbf{C}$ iff f^{OP} is epic in $\mathbf{C}^{OP}$, and vice versa. In general, given a property P of an object, arrow, diagram, etc., we can associate with P the dual property P^{OP}: The object or arrow has a property P in $\mathbf{C}$ iff it has P^{OP} in $\mathbf{C}^{OP}$.

For example, if $g \circ f$ is monic then f is monic. From this, by duality, if $f^{OP} \circ g^{OP}$ is epic then f^{OP} is epic.

An object X is called *terminal* if for any object Y there is exactly one morphism from it into X in the category. Dually, one object X is called *initial* if for any object Y there is exactly one morphism from X to Y in the category. An object X is called a *zero* object if it is both terminal and initial. For example, in **Set** the empty set is initial, while any singleton set is (up to isomorphism) a terminal object.

Given two categories $\mathbf{C}$ and $\mathbf{D}$, we can define the *product category* $\mathbf{C} \times \mathbf{D}$ which has as objects pairs $(X, Y) \in Ob_C \times Ob_D$, and as arrows $(X, Y) \to (X', Y')$ pairs (f, g) with $f : X \to X'$ in $\mathbf{C}$, and $g : Y \to Y'$ in $\mathbf{D}$.

Functors are structure-preserving maps between categories. They can be thought of as morphisms in the category **Cat** of all (small) categories as objects, and all functors as morphisms

A (covariant) functor F from a category $\mathbf{C}$ to a category $\mathbf{D}$, written $F : \mathbf{C} \to \mathbf{D}$, consists of a pair of functions $F = (F^0, F^1)$, $F^0 : Ob_{\mathbf{C}} \to Ob_{\mathbf{D}}$ and $F^1 : Mor_{\mathbf{C}} \to Mor_{\mathbf{D}}$ (in what follows, we will simply use F for both functions as well):

- For each object X in $\mathbf{C}$, an object $F^0(X)$ in $\mathbf{D}$; and
- For each morphism $f : X \to Y$ in $\mathbf{C}$, a morphism $F^1(f) : F^0(X) \to F^0(Y)$,

such that the following two properties hold:

- For every object X in $\mathbf{C}$, $F^1(id_X) = id_{F(X)}$;
- For all morphisms $f: X \to Y$ and $g: Y \to Z$, $F^1(g \circ f) = F^1(g) \circ F^1(f)$.

A contravariant functor $F: \mathbf{C} \to \mathbf{D}$ is like a covariant functor, except that it "turns morphisms around" ("reverses all the arrows"). More specifically, every morphism $f: X \to Y$ in $\mathbf{C}$ must be assigned to a morphism $F^1(f): F^0(Y) \to F^0(X)$ in $\mathbf{D}$. In other words, a contravariant functor is a covariant functor from the opposite category $\mathbf{C}^{\mathrm{op}}$ to $\mathbf{D}$. Opposite category $\mathbf{C}^{\mathrm{op}}$ has the same objects as the category $\mathbf{C}$. Given a composition of arrows $f \circ g$ in $\mathbf{C}$, its opposite arrow in $\mathbf{C}^{\mathrm{op}}$ is $(f \circ g)^{OP} = g^{OP} \circ f^{OP}$.

Every functor preserves epi-, mono- and isomorphisms.

A locally small category $\mathbf{C}$, for any object X, has a *representable functor* $F_X = \mathbf{C}(X, \_): \mathbf{C} \to \mathbf{Set}$, such that for any object Y, $F_X(Y) = \mathbf{C}(X, Y)$, and any arrow $f: Y \to Y'$ gives by composition a function $F_X(f): \mathbf{C}(X, Y) \to \mathbf{C}(X, Y')$.

A functor $F: \mathbf{C} \to \mathbf{D}$ is called *full* if for every two objects A and B of $\mathbf{C}$, $F: \mathbf{C}(A, B) \to \mathbf{D}(F(A), F(B))$ is a surjection. A functor F is called *faithful* if this map is always injective. A functor F is called *endofunctor* if $\mathbf{D} = \mathbf{C}$.

A functor F *reflects* a property P if whenever the F-image of something (object, arrow, etc.) has P, then that something has P. For example, the faithful functor reflects epimorphisms and monomorphisms.

It is well known that each endofunctor $F: \mathbf{C} \to \mathbf{C}$ defines the algebras and coalgebras (the left and right commutative diagrams)

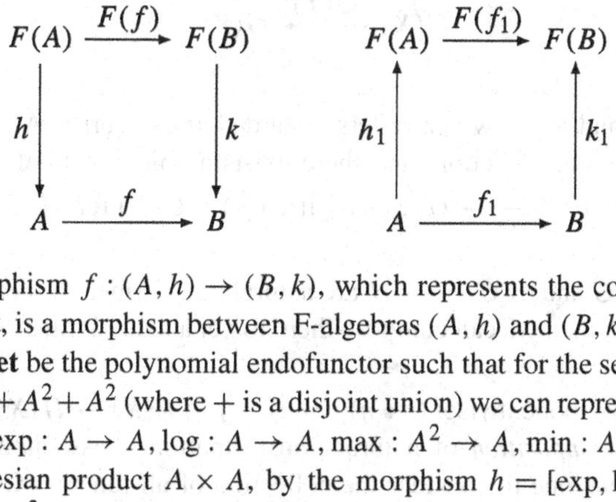

so that the morphism $f: (A, h) \to (B, k)$, which represents the commutative diagram on the left, is a morphism between F-algebras (A, h) and (B, k). For example, let $F: \mathbf{Set} \to \mathbf{Set}$ be the polynomial endofunctor such that for the set A of all reals, $F(A) = A + A + A^2 + A^2$ (where $+$ is a disjoint union) we can represent the algebra for operators $\{\exp: A \to A, \log: A \to A, \max: A^2 \to A, \min: A^2 \to A\}$, where A^2 is the Cartesian product $A \times A$, by the morphism $h = [\exp, \log, \max, \min]: (A + A + A^2 + A^2) \to A$ in $\mathbf{Set}$, that is, by the F-algebra (A, h).

Analogously, the morphism $f_1: (A, h_1) \to (B, k_1)$, which represents the commutative diagram on the right, is a morphism between F-coalgebras (A, h_1) and (B, k_1).

Let $G: \mathbf{D} \to \mathbf{C}$ be a functor, and C an object of $\mathbf{C}$. A *universal arrow from C to G* is a pair (D, g) where D is an object of $\mathbf{D}$ and $g: C \to G(D)$ an morphism in $\mathbf{C}$ such that, for any object D' of $\mathbf{D}$ and morphism $f: C \to G(D')$, there exists a unique morphism $\widehat{f}: D \to D'$ in $\mathbf{D}$ such that $f = G(\widehat{f}) \circ g$. Diagrammatically,

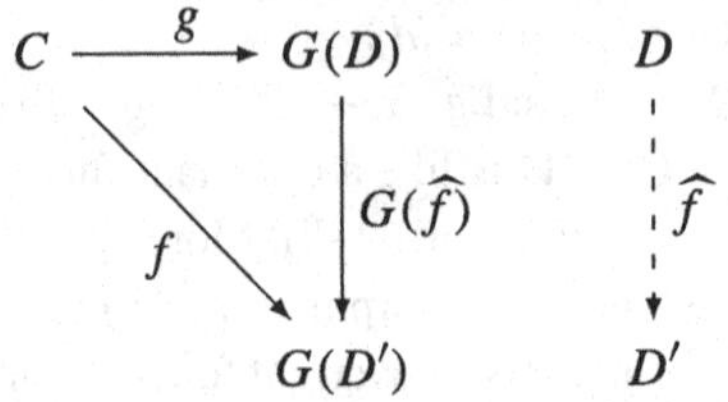

A *natural transformation* is a relation between two functors. Functors often describe "natural constructions" and natural transformations then describe "natural homomorphisms" between two such constructions. Sometimes two quite different constructions yield "the same" result; this is expressed by a natural isomorphism between the two functors.

If F and G are (covariant) functors between the categories $\mathbf{C}$ and $\mathbf{D}$, then a natural transformation η from F to G associates to every object X in $\mathbf{C}$ a morphism $\eta_X : F(X) \to G(X)$ in $\mathbf{D}$ such that for every morphism $f : X \to Y$ in $\mathbf{C}$, we have $\eta_Y \circ F(f) = G(f) \circ \eta_X$.

This means that the following diagram is commutative:

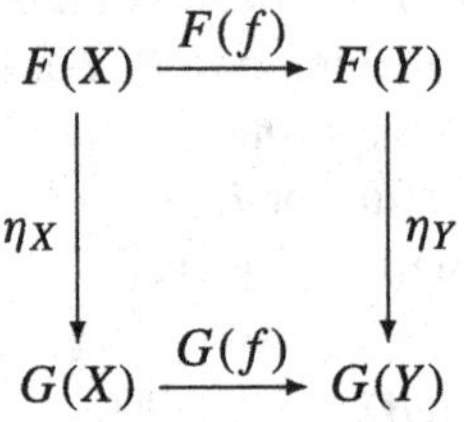

Example Let P and Q be two preorders, regarded as categories. A functor F from P to Q is a monotone function, and there exists a unique natural transformation between two such, $\eta : F \stackrel{\cdot}{\longrightarrow} G$, exactly if $F(X) \leq G(X)$ for all X in P.

If $\eta : F \stackrel{\cdot}{\longrightarrow} G$ and $\varepsilon : G \stackrel{\cdot}{\longrightarrow} H$ are natural transformations between functors $F, G, H : \mathbf{C} \to \mathbf{D}$, then we can compose them to get a natural transformation $\varepsilon \bullet \eta : F \stackrel{\cdot}{\longrightarrow} H$.

This is done componentwise: $(\varepsilon \bullet \eta)_X = \varepsilon_X \circ \eta_X : F(X) \to H(X)$.

This *vertical composition* of natural transformation is associative and has an identity, and allows one to consider the collection of all functors $\mathbf{C} \to \mathbf{D}$ itself as a category $\mathbf{D}^{\mathbf{C}}$ of all functors from $\mathbf{C}$ to $\mathbf{D}$.

There is also a *horizontal composition* of natural transformations, given the functors $F, G : \mathbf{B} \to \mathbf{C}$, $F', G' : \mathbf{C} \to \mathbf{D}$, and two natural transformations $\eta : F \stackrel{\cdot}{\longrightarrow} G$ and $\varepsilon : F' \stackrel{\cdot}{\longrightarrow} G'$, defined by $\varepsilon \circ \eta : F' \circ F \stackrel{\cdot}{\longrightarrow} G' \circ G$, between the composed functors $F' \circ F, G' \circ G : \mathbf{B} \to \mathbf{D}$.

The two functors F and G are called naturally isomorphic if there exists a natural transformation η from F to G such that η_X is an isomorphism for every object X in $\mathbf{C}$.

The Yoneda lemma is one of the most famous basic results of category theory. It describes representable functors in functor categories. The Yoneda embedding for any given category **C** is given by the covariant functor $H : \mathbf{C} \to \mathbf{Set}^{\mathbf{C}^{OP}}$ such that, for any object X in **C**, we obtain the contravariant hom-functor $h_X = \mathbf{C}^{OP}(X, \_) : \mathbf{C}^{OP} \to \mathbf{Set}$.

Consequently, for any object Y' in $\mathbf{C}^{OP}$, $h_X(Y_1) = \mathbf{C}^{OP}(X, Y_1)$, while for any arrow $f^{OP} : Y_1 \to Y_2$ in $\mathbf{C}^{OP}$, we obtain the arrow (function) in **Set**, $h_X(f^{OP}) : \mathbf{C}^{OP}(X, Y_1) \to \mathbf{C}^{OP}(X, Y_2)$ is a composition with f^{OP}, i.e., for any $g^{OP} : X \to Y_1$ in $\mathbf{C}^{OP}(X, Y_1)$ we obtain the arrow $h_X(f^{OP})(g^{OP}) = f^{OP} \circ g^{OP} = (g \circ f)^{OP} : X \to Y_2$, such that the following diagram commutes in $\mathbf{C}^{OP}$ (or, dually in **C**):

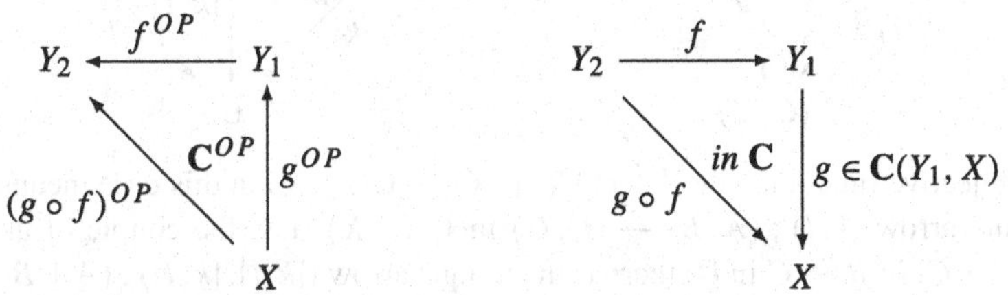

In the case when $f^{OP} : Y_1 \to Y_2$ in $\mathbf{C}^{OP}$ is a *universal arrow*, the function $h_X(f^{OP}) : \mathbf{C}^{OP}(X, Y_1) \to \mathbf{C}^{OP}(X, Y_2)$ (an arrow in **Set**) is a bijection.

Let us consider two significant cases of Yoneda embedding for categorial *products* and *coproducts* (without the necessity to introduce the more general case of limits and colimits):

1. The universal arrows for the *products* $f^{OP} = (\pi_1, \pi_2) : (A \times B, A \times B) \twoheadrightarrow (A, B)$ in $\mathbf{C}^{OP} = \mathbf{D} \times \mathbf{D}$, and a given object $X = (C, C) \in \mathbf{C}^{OP}$, where $\pi_1 : A \times B \twoheadrightarrow A$, $\pi_2 : A \times B \twoheadrightarrow B$ are two epimorphisms in **D** corresponding to the first and second projections, respectively. Thus we obtain the following case for the commutative diagram above in $\mathbf{C}^{OP} = \mathbf{D} \times \mathbf{D}$ (and its corresponding commutative diagram in **D** on the right), where $Y_2 = (A, B)$ and $Y_1 = (A \times B, A \times B)$, $g^{OP} = (\langle k, l \rangle, \langle k, l \rangle) : (C, C) \to (A \times B, A \times B)$, with $k : C \to A, l : C \to B$ two arrows in **D**:

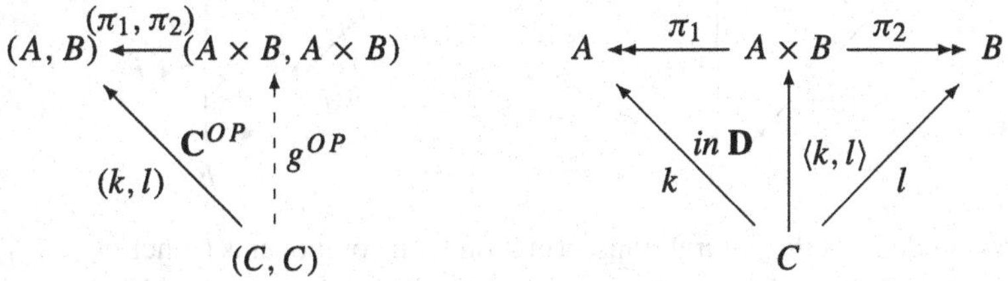

The bijective function $h_X(f^{OP}) : \mathbf{C}^{OP}(X, Y_1) \to \mathbf{C}^{OP}(X, Y_2)$ in this case means that for any arrow $(k, l) : (C, C) \to (A, B)$ in $\mathbf{C}^{OP}(X, Y_2)$ (i.e., the couple of arrows $k : C \to A$, $l : C \to B$ in **D**) there is the unique arrow $(\langle k, l \rangle, \langle k, l \rangle) : (C, C) \to (A \times B, A \times B)$ in $\mathbf{C}^{OP}(X, Y_1)$ (i.e., the unique arrow $\langle k, l \rangle : C \to A \times B$ in **D**), such that the diagrams above commute.

2. Universal arrows for the *coproducts* $f = (in_1, in_2) : (A, B) \hookrightarrow (A+B, A+B)$ in $\mathbf{C} = \mathbf{D} \times \mathbf{D}$, and a given object $X = (C, C) \in \mathbf{C}$, where $in_1 : A \hookrightarrow A+B$, $in_2 : B \hookrightarrow A+B$ are two monomorphisms in $\mathbf{D}$ corresponding to the first and second injections, respectively. Thus we obtain the following case for the commutative diagram above in $\mathbf{C} = \mathbf{D} \times \mathbf{D}$ (and its corresponding commutative diagram in $\mathbf{D}$ on the right), where $Y_2 = (A, B)$ and $Y_1 = (A+B, A+B)$, $g = ([k, l], [k, l]) : (A+B, A+B) \to (C, C)$, with $k : A \to C$, $l : B \to C$ two arrows in $\mathbf{D}$:

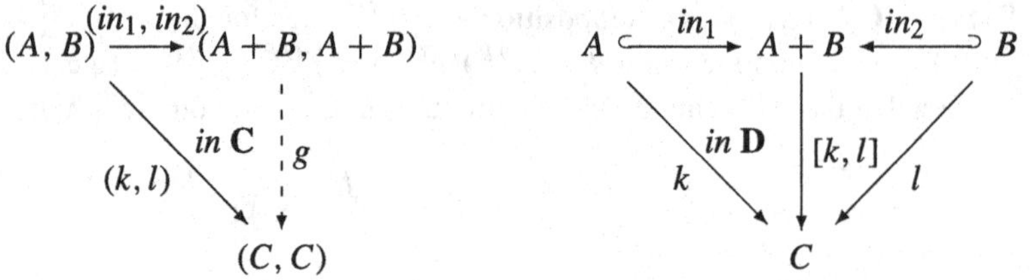

The bijective function $h_X(f^{OP}) : \mathbf{C}(Y_1, X) \to \mathbf{C}(Y_2, X)$ in this case means that for any arrow $(k, l) : (A, B) \to (C, C)$ in $\mathbf{C}(Y_2, X)$ (i.e., the couple of arrows $k : A \to C$, $l : B \to C$ in $\mathbf{D}$) there is the unique arrow $([k, l], [k, l]) : (A+B, A+B) \to (C, C)$ in $\mathbf{C}(Y_1, X)$ (i.e., the unique arrow $[k, l] : A+B \to C$ in $\mathbf{D}$), such that the diagrams above commute.

For example, in **Set** the object $A \times B$ is the Cartesian product of the set A and set B, while $A + B$ is the disjoint union of the set A and set B.

Products and coproducts are the particular cases of adjunctions. Given two functors $F : \mathbf{C} \to \mathbf{D}$, $G : \mathbf{D} \to \mathbf{C}$, we say that F is *left adjoint* to G, or G is *right adjoint* to F, is there are the natural isomorphism: $\theta : \mathbf{C}(\_, G(\_)) \stackrel{\cdot}{\longrightarrow} \mathbf{D}(F(\_), \_)$, where $\mathbf{C}(\_, G(\_)) : \mathbf{C}^{OP} \times \mathbf{D} \to \mathbf{Set}$ is the result of composing the bivariant hom-functor $\mathbf{C}(\_, \_)$ with $Id_{C^{OP}} \times G$, and $\mathbf{D}(F(\_), \_)$ is similar.

Equivalently, they are adjoint if there are two natural transformations, *counit* $\varepsilon : FG \stackrel{\cdot}{\longrightarrow} Id_D$ and *unit* $\eta : Id_C \stackrel{\cdot}{\longrightarrow} GF$, such that the following diagrams of natural transformations commute:

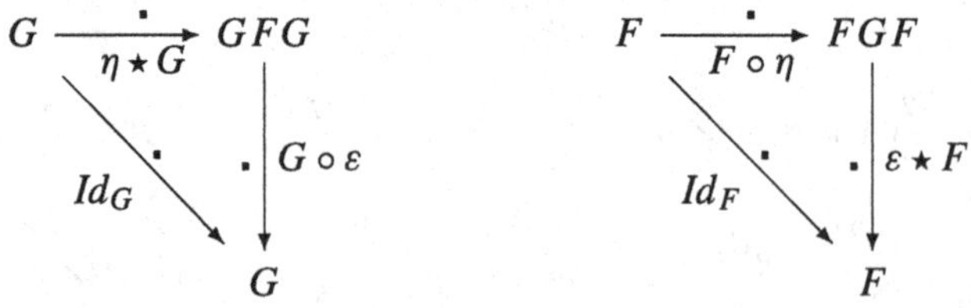

where $\eta \star G$ denotes the natural transformation with components (functions) $\eta_{G(X)} : G(X) \to GFG(X)$ and $G \circ \varepsilon$ denotes the natural transformation with components $G(\varepsilon_X) : GFG(X) \to G(X)$, for each object X in $\mathbf{D}$.

This adjunction is denoted by tuple $(F, G, \varepsilon, \eta)$, with $\eta_C : C \to G(F(C))$ a universal arrow for each object C. In fact, for any object D' in $\mathbf{D}$ and morphism $f : C \to G(D')$ there exists a unique morphism $\widehat{f} : D \to D'$, where $D = F(C)$, such that the following two adjoint diagrams commute:

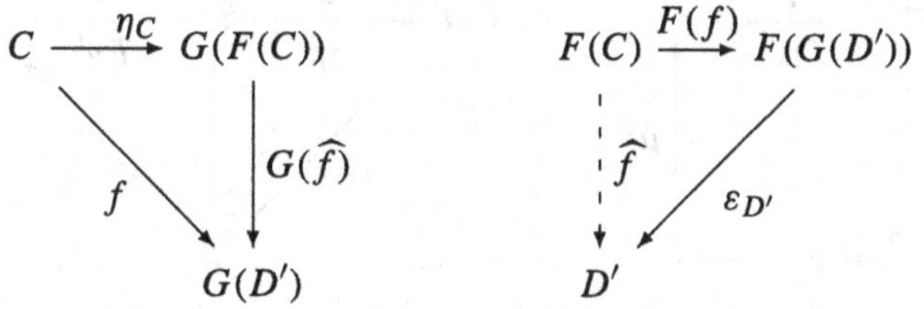

so that η_C is a universal arrow from C into G and, dually, $\varepsilon_{D'}$ is a couniversal arrow from F into D'. Consequently, in an given adjunction, the unit η generates a universal arrow for each object in **C** and counit ε generates a couniversal arrow for each object in **D**. In the case when **C** and **D** are poset categories, f and $\widehat{f}$ define this adjunction as a Galois connection, denoted by $F \dashv G$.

Two categories **C** and **D** are *equivalent* if they are adjoint with the unit and counit whose components are all isomorphisms.

Given an adjunction $(F, G, \varepsilon, \eta)$ let us look at the endofunctor $T = GF : \mathbf{C} \rightarrow \mathbf{C}$. We have a natural transformation $\eta : id_{\mathbf{C}} \stackrel{\cdot}{\longrightarrow} T$ and a natural transformation $\mu : T^2 \stackrel{\cdot}{\longrightarrow} T$ with components for each object C, $\mu_C = G(\varepsilon_{F(C)}) : T^2C \rightarrow TC$ (Here T^2 denotes the composition TT and T^3 the composition TTT). Furthermore, the equalities (commutative diagrams of natural transformations)

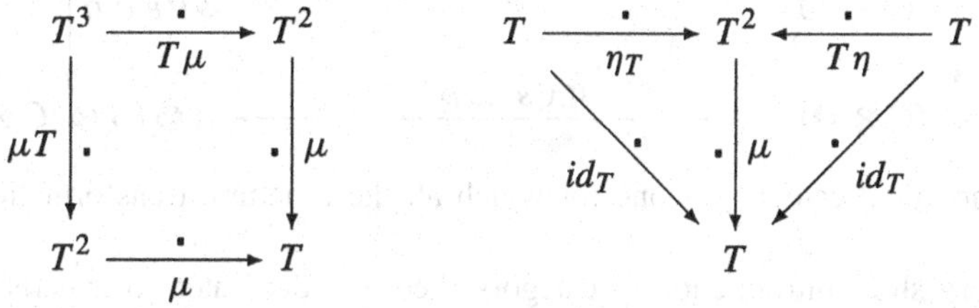

hold. Here $(T\mu)_C = T(\mu_C) : T^3C \rightarrow TC$ and $(\mu T)_C = \mu_{TC} : T^3C \rightarrow TC$ (similarly for ηT and $T\eta$).

A triple (T, η, μ) satisfying these equalities is called a *monad*.

The notion of a monad is one of the most general mathematical notions. For instance, *every* algebraic theory, that is, every set of operations satisfying equational laws, can be seen as a monad (which is also a *monoid* in a category of endofunctors of a given category: the "operation" μ being the associative multiplication of this monoid and η its unit).

Thus, monoid laws of the monad do subsume all possible algebraic laws.

We will use monads [33, 35, 36] for giving *denotational semantics to database mappings*, and more specifically as a way of modeling computational/collection types [8, 65, 66, 68].

A dual structure to a monad is a comonad (T, η^C, μ^C) such that all arrows in the commutative monad's diagrams above are inverted, so that we obtain:

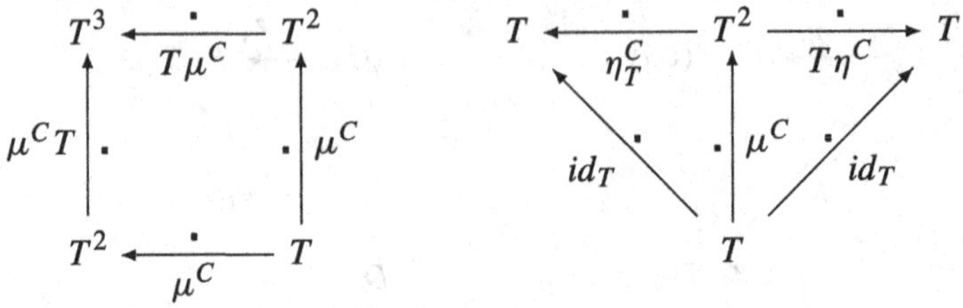

A *monoidal category* is a structure $(\mathbf{C}, \otimes, I, \alpha, \beta, \gamma)$ where $\mathbf{C}$ is a category, $\otimes : \mathbf{C} \times \mathbf{C} \to \mathbf{C}$ is a functor (*tensor*), I is a distinguished object of $\mathbf{C}$ (unit), α, β, γ are natural transformations (*structural isos* for associativity, left and right identity) with components: $\alpha_{A,B,C} : (A \otimes B) \otimes C \simeq A \otimes (B \otimes C)$, $\beta_A : I \otimes A \simeq A$, $\gamma_A : A \otimes I \simeq A$, such that the following diagrams commute (coherence conditions)

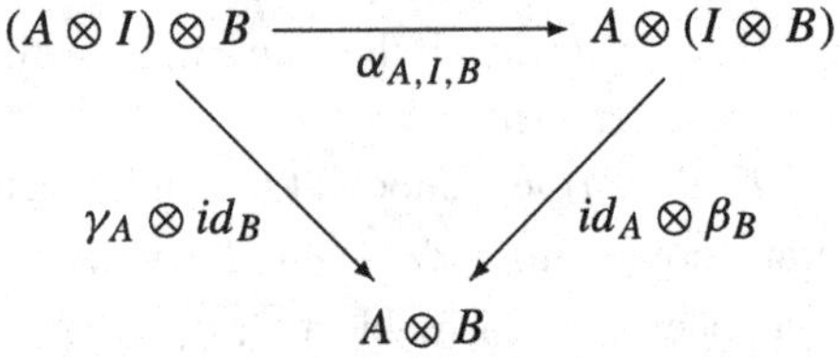

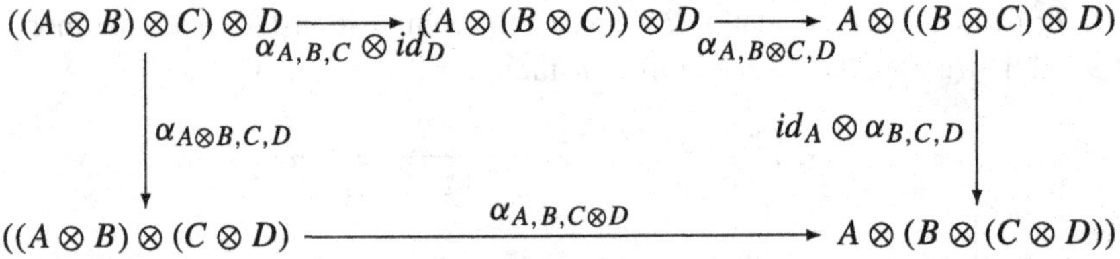

A *strict* monoidal category is one for which all these natural transformations are identities.

This very short introduction to category theory is dedicated to experts in DB theory which did not work previously with the concepts of the category theory. Chapters 8 and 9 require a more deep expertise in the category theory.

1.5.1 Categorial Symmetry

The concept of categorial symmetry was introduced in my PhD thesis [44] with a number of its applications. Here I will only introduce some of the fundamental properties of this particular class of categories, which will be considered in the database category **DB** (one of the principal issues of this manuscript).

It is well known that all categories can be defined by only their morphisms (arrows) because each object is equivalently represented by its identity morphism. Fundamentally, a symmetric category has a nice property of a kind of (incomplete) duality between objects and arrows. In fact, there is an operator '$*$' for the composition of objects, analogous to the standard composition of arrows '$\circ$'. However, differently form $\circ$ that is not commutative (preserves the sequential composition of the arrows), the composition $*$ of objects can be a (not necessarily) commutative binary operator as well because the objects have no directionality.

The objects obtained from arrows will be called *conceptualized* objects. For example, let us consider the polynomial function $f(x) = (x^2 - 3x + 1)$, with the domain being the closed interval of reals from 0 to 1, and codomain $\mathbb{R}$ of all reals. It is an arrow $f : [0, 1] \to \mathbb{R}$ in the category **Set**. The conceptualized object obtained from this arrow, denoted by $\widetilde{f}$, is the *set* (hence an object in **Set**) equal to the graph of this polynomial function. That is, $\widetilde{f} = \{(x, f(x)) \mid x \in [0, 1]\} \in Ob_{Set}$.

Let us now formally introduce the following definitions [44] for the categorial symmetry:

Definition 4 (CATEGORIAL SYMMETRY) Let $\mathbf{C}$ be a category with a function $B_T : Mor_C \longrightarrow Ob_C$ such that for each identity arrow $id_A : A \to A$, $B_T(id_A) \simeq A$ (the "representability" principle), and with an associative composition operator for objects $*$ such that, for any composition $g \circ f \in Mor_C$ of arrows, $B_T(g) * B_T(f) = B_T(g \circ f)$.

The *conceptualized* object $B_T(f)$ of an arrow $f : A \longrightarrow B$ in $\mathbf{C}$ will be denoted by $\widetilde{f}$.

Remark This symmetry property allows us to consider all properties of an arrow as properties of objects and their compositions. For a given category $\mathbf{C}$, its "arrow category" is denoted by $\mathbf{C} \downarrow \mathbf{C}$. Objects of this arrow category are the triples $\langle A, B, f\rangle$ where $f : A \to B$ is a morphism in $\mathbf{C}$, and each morphism in this arrow category is a couple of two morphisms $(k_1; k_2) : \langle A, B, f\rangle \to \langle C, D, g\rangle$ such that $k_2 \circ f = g \circ k_1 : A \to D$.

Let us introduce, for a category $\mathbf{C}$ and its arrow category $\mathbf{C} \downarrow \mathbf{C}$, an encapsulation operator $J : Mor_C \longrightarrow Ob_{C\downarrow C}$ and its inverse ψ, such that for any arrow $f : A \longrightarrow B$, $J(f) = \langle A, B, f\rangle$ is its corresponding object in $\mathbf{C} \downarrow \mathbf{C}$ and $\psi(\langle A, B, f\rangle) = f$.

We denote the first and the second comma functorial projections by $F_{st}, S_{nd} : (\mathbf{C} \downarrow \mathbf{C}) \longrightarrow \mathbf{C}$ such that for any arrow $(k_1; k_2) : \langle A, B, f\rangle \to \langle A', B', g\rangle$ in $\mathbf{C} \downarrow \mathbf{C}$ (i.e., when $k_2 \circ f = g \circ k_1$ in C) we have that $F^0_{st}(\langle A, B, f\rangle) = A$, $F^1_{st}(k_1; k_2) = k_1$, $S^0_{nd}(\langle A, B, f\rangle) = B$ and $S^1_{nd}(k_1; k_2) = k_2$. It is easy to extend the operator ψ into a natural transformation $\psi : F_{st} \stackrel{\bullet}{\longrightarrow} S_{nd}$ such that its component for an object $J(f)$ in $\mathbf{C} \downarrow \mathbf{C}$ is the arrow $\psi_{J(f)} = \psi(J(f)) = f$. We denote the diagonal functor by $\blacktriangle : \mathbf{C} \longrightarrow (\mathbf{C} \downarrow \mathbf{C})$ such that, for any object A in a category $\mathbf{C}$, $\blacktriangle^0(A) = \langle A, A, id_A\rangle$.

An important subset of symmetric categories are the Conceptually Closed and Extended symmetric categories, as follows:

Definition 5 [44] *A conceptually closed* category is a symmetric category $\mathbf{C}$ with a functor $T_e = (T^0_e, T^1_e) : (\mathbf{C} \downarrow \mathbf{C}) \longrightarrow \mathbf{C}$ such that $T^0_e = B_T\psi$, i.e., $B_T = T^0_e J$, with a natural isomorphism $\varphi : T_e \circ \blacktriangle \simeq I_C$, where I_C is an identity functor for $\mathbf{C}$.

$\mathbf{C}$ is an *extended symmetric* category if $\tau^{-1} \bullet \tau = \psi : F_{st} \stackrel{\bullet}{\longrightarrow} S_{nd}$, for vertical composition of natural transformations $\tau : F_{st} \stackrel{\bullet}{\longrightarrow} T_e$ and $\tau^{-1} : T_e \stackrel{\bullet}{\longrightarrow} S_{nd}$.

It is easy to verify also that in extended symmetric categories the following is valid:

$$\tau = \left(T_e^1\left(\tau_I F_{st}^0; \psi\right)\right) \bullet \left(\varphi^{-1} F_{st}^0\right), \qquad \tau^{-1} = \left(\varphi^{-1} S_{nd}^0\right) \bullet \left(T_e^1\left(\psi; \tau_I S_{nd}^0\right)\right),$$

where $\tau_I : I_C \longrightarrow I_C$ is an identity natural transformation (for any object A in **C**, $\tau_I(A) = id_A$).

Example 1 The **Set** is an extended symmetric category: given any function $f : A \longrightarrow B$, the conceptualized object of this function is the graph of this function (which is a set), $\widetilde{f} = B_T(f) \triangleq \{(x, f(x)) \mid x \in A\}$.

The composition of objects $*$ is defined as an associative composition of binary relations (graphs of functions), $B_T(g \circ f) = \{(x, (g \circ f)(x)) \mid x \in A\} = \{(y, g(y)) \mid y \in B\} \circ \{(x, f(x)) \mid x \in A\} = B_T(g) * B_T(f)$.

Set is also conceptually closed by the functor T_e such that for any object $J(f) = \langle A, B, f\rangle$, $T_e^0(J(f)) \triangleq B_T(f) = \{(x, f(x)) \mid x \in A\}$, and for any arrow $(k_1; k_2) : J(f) \to J(g)$, the component T_e^1 is defined by:

$$\text{for any } \left(x, f(x)\right) \in T_e^0\left(J(f)\right), \quad T_e^1(k_1; k_2)\left(x, f(x)\right) = \left(k_1(x), k_2\left(f(x)\right)\right).$$

It is easy to verify the compositional property for T_e^1, and that $T_e^1(id_A; id_B) = id_{T_e^0(J(f))}$. For example, **Set** is also an extended symmetric category such that for any object $J(f) = \langle A, B, f\rangle$ in **Set** $\downarrow$ **Set**, $\tau(J(f)) : A \twoheadrightarrow B_T(f)$ is an epimorphism such that for any $x \in A$, $\tau(J(f))(x) = (x, f(x))$, while $\tau^{-1}(J(f)) : B_T(f) \hookrightarrow B$ is the second projection, such that for any $(x, f(x)) \in B_T(f)$, $\tau^{-1}(J(f))(x, f(x)) = f(x)$. Thus, each arrow in **Set** is a composition of an epimorphism (surjective function) and a monomorphism (injective function).

The categorial symmetry can be better understood by considering a category **C** as a two-sorted algebra $Alg_{\mathbf{C}} = ((Ob_C, Mor_C), \Sigma_C)$ (two-sorted carrier set is composed of its objects and its morphisms) with the signature $\Sigma_{\mathbf{C}} = \{dom, cod, id, \circ\} \cup \{o_i \mid 0 \le i \le n\}$, where

1. $dom, cod : Mor_C \to Ob_C$ are two operations such that for any morphism $f : A \to B$, $dom(f) = A$, $cod(f) = B$;
2. $id : Ob_C \to Mor_C$ is the operations such that for any object A, $id(A) = id_A : A \to A$ is the identity arrow for this object;
3. $\circ : Mor_C^2 \to Mor_C$ is a *partial* function, such that for any two morphisms f, g, $f \circ g$ is defined if $dom(f) = cod(g)$, with the following set of equations:
   - (associativity) $(f \circ g) \circ h = f \circ (g \circ h)$, if $dom(g) = cod(h)$ and $dom(f) = cod(g)$,
   - (identity) $f \circ id(dom(f)) = f$ and $id(cod(f)) \circ f = f$;
4. For each operator $o_i \in \Sigma_C$, the composition of objects $o_i : Ob_C^{ar(o_i)} \to Ob_C$ (as, for example, the product, coproduct, etc.), or the composition of morphisms $o_i : Mor_C^{ar(o_i)} \to Mor_C$ (as, for example, the product, coproduct, the various pairings (structural operations) $(,)$, $\langle,\rangle$, $[,]$, etc.).

Consequently, the categorial symmetry can be done by the following corollary:

Corollary 1 *A category* $\mathbf{C}$ *is symmetric if for its algebra* $Alg_{\mathbf{C}} = ((Ob_C, Mor_C), \Sigma_{\mathbf{C}})$ *we have the additional binary partial operator* $* : Ob_C^2 \rightarrow Ob_C$ *and unary operator* $B_T \in \Sigma_{\mathbf{C}}$ *which is a homomorphism* $B_T : (Mor_C, \circ) \rightarrow (Ob_C, *)$ *such that for every object* $A \in Ob_C$ *there exists the isomorphism,* $B_T(id(A)) \simeq A$ *(representability principle).*

Notice that the partial operations can be equivalently represented by a set of standard (total) operations. For example, the partial operation '$\circ$' can be equivalently represented by the set of total operations,

$$\{\circ_{A,B,C} : \mathbf{C}(A, B) \times \mathbf{C}(B, C) \rightarrow \mathbf{C}(A, C) \mid A, B, C \in OB_C\}.$$

References

1. S. Abiteboul, R. Hull, V. Vianu, *Foundations of Databases* (Addison Wesley Publ. Co., Reading, 1995)
2. S. Alagić, P. Bernstein, A model theory for generic schema management, in *DBPL 2001*. LNCS, vol. 2397 (Springer, Berlin, 2002), pp. 228–246
3. D. Beneventano, M. Lenzerini, F. Mandreoli, Z. Majkić, Techniques for query reformulation, query merging, and information reconciliation—part A. Semantic webs and agents in integrated economies, D3.2.A, IST-2001-34825 (2003)
4. P.A. Bernstein, Applying model management to classical meta data problems, in *CIDR* (2003), pp. 209–220
5. P. Bernstein, A. Halevy, R. Pottinger, A vision for management of complex models. SIGMOD Rec. **29**(4), 55–63 (2000)
6. J.R. Büchi, Weak second-order arithmetic and finite automata. Z. Math. Log. Grundl. Math. **6**, 66–92 (1960)
7. P. Buneman, S. Davidson, A. Kosky, Theoretical aspects of schema merging, in *EDBT 1992* (1992), pp. 152–167
8. P. Buneman, S. Naqui, V. Tanen, L. Wong, Principles of programming with complex objects and collection types. Theor. Comput. Sci. **149**(1), 3–48 (1995)
9. A. Calì, D. Calvanese, G. Giacomo, M. Lenzerini, Data integration under integrity constraints, in *Proc. of the 14th Conf. on Advanced Information Systems Engineering (CAiSE 2002)* (2002), pp. 262–279
10. A. Calì, D. Calvanese, G. Giacomo, M. Lenzerini, Reasoning in data integration systems: why LAV and GAV are siblings, in *Proceedings of the 14th International Symposium on Methodologies for Intelligent Systems, ISMIS 2003*. Lecture Notes in Computer Science, vol. 2871 (Springer, Berlin, 2003), pp. 282–289
11. D. Calvanese, G. Giacomo, D. Lembo, M. Lenzerini, R. Rosati, Inconsistency tollerance in P2P data integration: an epistemic approach, in *Proc. 10th Int. Workshop on Database Programming Language* (2005)
12. A.K. Chandra, D. Harel, Structure and complexity of relational queries. J. Comput. Syst. Sci. **25**(1), 99–128 (1982)
13. E.F. Codd, A relational model of data for large shared data banks. Commun. ACM **13**(6), 377–387 (1970)
14. S. Davidson, P. Buneman, A. Kosky, Semantics of database transformations, in *Semantics in Databases*, ed. by B. Thalheim, L. Libkin. LNCS, vol. 1358 (1998), pp. 55–91
15. Z. Diskin, Generalized sketches as an algebraic graph-based framework for semantic modeling and database design. Laboratory for Database Design, FIS/LDBD-97-03, May (1997)

16. Z. Diskin, B. Cadish, Algebraic graph-based approach to management of multibase systems: schema integration via sketches and equations, in *Proc. 2nd Int. Workshop on Next Generation Information Technologies and Systems (NGITS'95)*, Naharia, Israel (1995), pp. 69–79
17. H.D. Ebbinghaus, J. Flum, *Finite Model Theory* (Springer, Berlin, 1995)
18. R. Fagin, Generalized first-order spectra and polynomial-time recognizable sets, in *Complexity of Computation*, ed. by R.M. Karp. SIAM-AMS Proceedings, vol. 7 (1974), pp. 43–73
19. R. Fagin, Horn clauses and database dependencies. J. ACM **29**, 952–985 (1982)
20. R. Fagin, Inverting schema mappings. ACM Trans. Database Syst. **32**(4), 25–78 (2005)
21. R. Fagin, P.G. Kolaitis, R.J. Miller, L. Popa, DATA exchange: semantics and query answering, in *Proc. of the 9th Int. Conf. on Database Theory (ICDT 2003)* (2003), pp. 207–224
22. R. Fagin, P.G. Kolaitis, L. Popa, W. Tan, Composing schema mappings: second-order dependencies to the rescue. ACM Trans. Database Syst. **30**(4), 994–1055 (2005)
23. R. Fagin, P.G. Kolaitis, L. Popa, W. Tan, Quasi-inverses of schema mappings. ACM Trans. Database Syst. **33**(2), 1–52 (2008)
24. E. Franconi, G. Kuper, A. Lopatenko, L. Serafini, A robust logical and computational characterization of peer-to- peer data systems. Technical Report DIT-03-051, University of Trento, Italy, September (2003)
25. A. Fuxman, P.G. Kolaitis, R. Miller, W.C. Tan, Peer data exchange. ACM Trans. Database Syst. **31**(4), 1454–1498 (2006)
26. R. Goldblat, *Topoi: The Categorial Analysis of Logic* (North-Holland, Amsterdam, 1979)
27. G. Gotlob, C. Koch, Monadic datalog and the expressive power of languages for web information extraction. J. ACM **51**, 74–113 (2004)
28. E. Grädel, Ph. Kolaitis, L. Libkin, M. Marx, J. Spencer, M.Y. Vardi, Y. Venema, S. Weinstein, *Finite Model Theory and Its Applications* (Springer, Berlin, 2004)
29. A.Y. Halevy, Z.G. Ives, J. Madhavan, P. Mork, D. Suciu, I. Tatarinov, The Piazza peer data management system. IEEE Trans. Knowl. Data Eng. **16**(7), 787–798 (2004)
30. A.Y. Halevy, D. Suiu, Z.G. Ives, I. Tatarinov, Schema mediation for large-scale semantic data sharing. VLDB J. **14**(1), 68–83 (2005)
31. A. Islam, W. Phoa, Categorical models of relational databases I: fibrational formulation, schema integration, in *Theoretical Aspects of Comp. Sci.* Lect. Notes Comp. Sci., vol. 789 (Springer, Berlin, 1994), pp. 618–641
32. M. Johnson, R. Rosebrugh, Entity-relationship models and sketches. J. Theory Appl. Categ. (2000)
33. G.M. Kelly, A.J. Power, Adjunctions whose counits are coequalizers, and presentations of finitary enriched monads. J. Pure Appl. Algebra **89**, 163–179 (1993)
34. A. Kementsietsidis, M. Arenas, R.J. Miller, Mapping data in peer-to-peer systems: semantics and algorithmic issues, in *Proc. of SIGMOD* (2003)
35. J. Lambek, P. Scott, *Introduction to Higher Order Categorial Logic* (Cambridge University Press, Cambridge, 1986)
36. S.M. Lane, *Categories for the Working Mathematician* (Springer, Berlin, 1971)
37. S.K. Lellahi, N. Spyratos, Toward a categorical database model supporting structured objects and inheritance, in *Int. Workshop in Information Systems 90'*, October, Kiev, USSR (1990)
38. M. Lenzerini, Data integration: a theoretical perspective, in *Proc. of the 21st ACM SIGACT SIGMOD SIGART Symp. on Principles of Database Systems (PODS 2002)* (2002), pp. 233–246
39. M. Lenzerini, Z. Majkić, First release of the system prototype for query management. Semantic webs and agents in integrated economies, D3.3, IST-2001-34825 (2003)
40. M. Lenzerini, Z. Majkić, General framework for query reformulation. Semantic webs and agents in integrated economies, D3.1, IST-2001-34825, February (2003)
41. L. Libkin, *Elements of Finite Model Theory*. EACTS Series, vol. XIV (Springer, Berlin, 2004)
42. J. Madhavan, P.A. Bernstein, P. Domingos, A.Y. Halevy, Representing and reasoning about mappings between domain models, in *AAAI/IAAI* (2002), pp. 80–86
43. J. Madhavn, A. Halevy, Composing mappings among data sources, in *Proc. of VLDB*, (2003), pp. 572–583

44. Z. Majkić, Categories: symmetry, n-dimensional levels and applications. PhD Thesis, University "La Sapienza", Roma, Italy (1998)
45. Z. Majkić, The category-theoretic semantics for database mappings. Technical Report 14-03, University 'La Sapienza', Roma, Italy (2003)
46. Z. Majkić, Fixpoint semantics for query answering in data integration systems, in *AGP03—8. th Joint Conference on Declarative Programming*, Reggio Calabria (2003), pp. 135–146
47. Z. Majkić, Modal query language for databases with partial orders, in *International Conference on Ontologies, Databases and Applications of SEmantics (ODBASE'04)*, Larnaca, Cyprus. LNCS, vol. 3290 (2004), pp. 768–782. ISSN 0302-9743
48. Z. Majkić, Coalgebraic semantics for logic programming, in *18th Workshop on (Constraint) Logic Programming, W(C)LP 2004*, March 04–06, Berlin, Germany (2004)
49. Z. Majkić, Weakly-coupled ontology integration of P2P database systems, in *1st Int. Workshop on Peer-to-Peer Knowledge Management (P2PKM)*, August 22, Boston, USA (2004)
50. Z. Majkić, Weakly-coupled P2P system with a network repository, in *6th Workshop on Distributed Data and Structures (WDAS'04)*, July 5–7, Lausanne, Switzerland (2004)
51. Z. Majkić, Flexible intentional query-answering for RDF peer-to- peer systems, in *7th International Conference on Flexible Query Answering Systems (FQAS 2006)*, 7–10 June, Milano, Italy (2006)
52. Z. Majkić, Abstract database category based on relational-query observations, in *International Conference on Theoretical and Mathematical Foundations of Computer Science (TMFCS-08)*, Orlando FL, USA, July 7–9 (2008)
53. Z. Majkić, Algebraic operators for matching and merging of relational databases, in *International Conference in Artificial Intelligence and Pattern Recognition (AIPR-09)*, Orlando FL, USA, July 13–16 (2009)
54. Z. Majkić, Induction principle in relational database category, in *Int. Conference on Theoretical and Mathematical Foundations of Computer Science (TMFCS-09)*, Orlando FL, USA, July 13–16 (2009)
55. Z. Majkić, Data base mappings and monads: (co)induction. arXiv:1102.4769v1 (2011), 22 February, pp. 1–31
56. Z. Majkić, Matching, merging and structural properties of data base category. arXiv:1102.2395v1 (2011), 11 February, pp. 1–27
57. Z. Majkić, DB category: denotational semantics for view-based database mappings. arXiv:1103.0217v1 (2011), 24 February, pp. 1–40
58. Z. Majkić, B. Prasad, Soft query-answering computing in P2P systems with epistemically independent peers, in *Book on Soft Computing Applications in Industry, STUDFUZZ 226* (Springer, Berlin, 2008), pp. 331–356
59. Z. Majkić, B. Prasad, Kleisli category for database mappings. Int.l J. Intell. Inf. Database Syst. **4**(5), 509–527 (2010)
60. M. Makkai, Generalized sketches as a framework for completeness theorems. Technical report, McGill University (1994)
61. D. McLeod, D. Heimbigner, A federated architecture for information management. ACM Trans. Inf. Syst. **3**(3), 253–278 (1985)
62. S. Melnik, E. Rahm, P.A. Bernstein, Rondo: a programming platform for generic model management, in *SIGMOD 2003*, June 9–12, San Diego, CA (2003)
63. S. Melnik, P.A. Bernstein, A. Halevy, E. Rahm, Supporting executable mapping in model management, in *SIGMOD* (2005), pp. 167–178
64. R. Miller, L. Haas, M. Hernandez, Schema mapping as query discovery, in *Proc. VLDB* (2000), pp. 77–88
65. E. Moggi, Computational lambda-calculus and monads, in *Proc. of the 4th IEEE Symp. on Logic in Computer Science (LICS'89)* (1989), pp. 14–23
66. E. Moggi, Notions of computation and monads. Inf. Comput. **93**(1), 55–92 (1991)
67. A. Nash, P. Bernstein, S. Melnik, Composition of mappings given by embedded dependencies, in *Proc. of PODS*, (2005), pp. 172–183

68. G.D. Plotkin, A.J. Power, Adequacy for algebraic effects, in *Proc. FOSSACS 2001*. LNCS, vol. 2030 (2001), pp. 1–24
69. H. Rasiowa, R. Sikorski, *The Mathematics of Metamathematics*, 3rd edn. (PWN-Polisch Scientific Publishers, Warsaw, 1970)
70. R. Rosebrugh, R.J. Wood, Relational databases and indexed categories, in *Proc. Int. Category Theory Meeting*. CMS Conference Proceedings, vol. 13 (1992), pp. 391–407
71. G. Rosolini, Continuity and effectiveness in Topoi. Ph.D dissertation, Oxford University. Also preprint, Department of Computer Science, Carnegie-Mellon University (1986)
72. M.A.V. Salles, J.P. Dittrich, S.K. Karakashian, O.R. Girard, L. Blunschi, iTrial: pay-as-you-go information integration in dataspaces, in *Proc. of VLDB*, Vienna, Austria (2007), pp. 663–674
73. A. Sheth, J. Larsen, Federated database systems for maniging distributed, heterogenuous and autonomous databases. ACM Comput. Surv. **22**(3), 183–236 (1990)
74. M.H. Stone, Topological representation of distributive lattices and Brouwerian logics. Čas. Pěst. Math. Fys. **67**, 1–25 (1937)
75. A. Tarski, Der aussagenkalkül und die topologie. Fundam. Math. **31**, 103–134 (1938)
76. V. Vianu, Databases and finite model theory, in *Descriptive Complexity and Finite Models, Proc. of DIMACS Workshop* (AMS, Providence, 1997), pp. 97–148
77. I. Zaihrayeu, Query answering in peer-to-peer database networks. Technical Report DIT-03-012, University of Trento, Italy (2003)

2 Composition of Schema Mappings: Syntax and Semantics

2.1 Schema Mappings: Second-Order tgds (SOtgds)

Based on previous considerations and weak points in the schema-mapping developments presented in the introduction (Sect. 1.4.2), a different semantics for composition that is valid for every class of queries has been proposed in [5]. The suggested semantics does not carry along a class of queries as a parameter, and the authors have shown that the set of formulae defining a composition is unique up to logical equivalence.

On the negative side, however, they demonstrated that the composition of a finite set of tgds with a finite set of *full* tgds (a tgd is full if no existentially quantified variables occur in the tgd) may not be definable by any set (finite or infinite) of tgds. Moreover, the composition of a finite set of tgds with a finite set of tgds is not definable in least-fixed point logics.

Based on these negative results, a class of existential second-order formulae with function symbols (introduced as a result of Skolemization of the existentially quantified variables) and equalities, the so-called second-order tgds (SOtgds), has been introduced in [5] as follows:

Definition 6 [5] Let $\mathcal{A}$ be a source schema and $\mathcal{B}$ a target schema. A second-order tuple-generating dependency (SOtgd) is a formula of the form:

$$\exists \mathbf{f}\big(\big(\forall \mathbf{x}_1(\phi_1 \Rightarrow \psi_1)\big) \wedge \cdots \wedge \big(\forall \mathbf{x}_n(\phi_n \Rightarrow \psi_n)\big)\big),$$

where

1. Each member of the tuple $\mathbf{f}$ is a functional symbol;
2. Each ϕ_i is a conjunction of:
   - Atomic formulae of the form $r_A(y_1, \ldots, y_k)$, where $r_A \in S_A$ is a k-ary relational symbol of schema $\mathcal{A}$ and $y_1, \ldots, y_k$ are variables in $\mathbf{x}_i$, not necessarily distinct;
   - The formulae with conjunction and negation connectives and with built-in predicate's atoms of the form $t \odot t'$, $\odot \in \{\doteq, <, >, \ldots\}$, where t and t' are the terms based on $\mathbf{x}_i$, $\mathbf{f}$ and constants.

Z. Majkić, *Big Data Integration Theory*, Texts in Computer Science,
DOI 10.1007/978-3-319-04156-8_2,

3. Each ψ_i is a conjunction of atomic formulae $r_B(t_1, \ldots, t_m)$ where $r_B \in S_B$ is an m-ary relational symbol of schema $\mathcal{B}$ and $t_1, \ldots, t_m$ are terms based on $\mathbf{x}_i$, $\mathbf{f}$ and constants.
4. Each variable in $\mathbf{x}_i$ appears in some atomic formula of ϕ_i.

Notice that each constant $\overline{a}$ in an atom on the left-hand side of implications must be substituted by a new fresh variable y_i and by adding a conjunct $(y_i = \overline{a})$ on the left-hand side of this implication, so that such atoms will have only the variables (condition 2 above). For the empty set of tgds, we will use the SOtgd tautology $r_\emptyset \Rightarrow r_\emptyset$. The forth condition is a "safety" condition, analogous to that made for the (first-order) tgds. It is easy to see that every tgd is equivalent to one SOtgd without equalities. For example, let σ be the tgd $\forall x_1 \ldots \forall x_m (\phi_A(x_1, \ldots, x_m) \Rightarrow \exists y_1 \ldots \exists y_n \psi_B(x_1, \ldots, x_m, y_1, \ldots, y_n))$. It is logically equivalent to the following SOtgd without equalities, which is obtained by Skolemizing existential quantifiers in σ:

$$\exists f_1 \ldots \exists f_n \big(\forall x_1 \ldots \forall x_m \big(\phi_A(x_1, \ldots, x_m) \Rightarrow \\ \psi_B\big(x_1, \ldots, x_m, f_1(x_1, \ldots, x_m), \ldots, f_n(x_1, \ldots, x_m)\big)\big)\big).$$

Given a finite set S of tgds of an inter-schema mapping in Definition 3, we can find a *single* SOtgd that is equivalent to S by taking, for each tgd σ in S, a conjunct of the SOtgd to capture σ as described above (we use disjoint sets of function symbols in each conjunct, as before). Based on these observations, we can define the algorithm for transformation of a given set of tgds into an single SOtgd:

Transformation algorithm *TgdsToSOtgd*(S)
Input. A set S of tgds from a schema $\mathcal{A}$ into a schema $\mathcal{B}$,

$$S = \big\{\forall \mathbf{x}_1 \big(\phi_{A,1}(\mathbf{x}_1) \Rightarrow \exists \mathbf{z}_1 \psi_{B,1}(\mathbf{y}_1, \mathbf{z}_1)\big), \ldots, \forall \mathbf{x}_n \big(\phi_{A,n}(\mathbf{x}_n) \Rightarrow \mathbf{z}_n \psi_{B,n}(\mathbf{y}_n, \mathbf{z}_n)\big)\big\},$$

where $\mathbf{y}_i \subseteq \mathbf{x}_i$ and possibly $\mathbf{z}_i$ the empty set, for $1 \leq i \leq n$.
Output. A SOtgd

- (*Create the set of implications from tgds*)
  Initialize S_{AB} to be the empty set.
  Put each of the n implications of tgds in S, $\phi_{A,i}(\mathbf{x}_i) \Rightarrow \exists \mathbf{z}_i \psi_{B,i}(\mathbf{y}_i, \mathbf{z}_i)$ in S_{AB}.
- (*Eliminate the existential quantifiers by Skolemization*)
  Repeat the following as far as possible:
  For each implication χ in S_{AB} of the form $\phi_{A,i}(\mathbf{x}) \Rightarrow \exists \mathbf{z} \psi_{B,i}(\mathbf{y}, \mathbf{z})$ with $\mathbf{z} = \langle z_1, \ldots, z_k \rangle$, $k \geq 1$, eliminate this χ from S_{AB} and add the following implication in S_{AB} (by introducing the set of new functional symbols $f_{i,j}$, for $1 \leq j \leq k$):

$$\big(\phi_{A,i}(\mathbf{x}) \wedge \big(z_1 \doteq f_{i,1}(\mathbf{x})\big) \wedge \cdots \wedge \big(z_k \doteq f_{i,k}(\mathbf{x})\big)\big) \Rightarrow \psi_{B,i}(\mathbf{y}, \mathbf{z}).$$

- (*Remove the variables originally quantified*)
  For each implication χ_i in S_{AB} constructed in the previous step, perform an elimination of the variables z_i from this implication as far as possible: Select an equality $z_j \doteq f_{i,j}(\mathbf{x})$ that was generated in the previous step. Remove the

equality $z_j \doteq f_{i,j}(\mathbf{x})$ from χ_i and replace every remaining occurrence of z_j in χ by $f_{i,j}(\mathbf{x})$.

- (*Construct* SOtgd)
If $S_{AB} = \{\chi_1, \ldots, \chi_n\}$ is the set where $\chi_1, \ldots, \chi_n$ are all the implications from the previous step, then SOtgd is the formula $\exists \mathbf{f}(\forall \mathbf{x}_1 \chi_1 \wedge \cdots \wedge \forall \mathbf{x}_n \chi_n)$, $\mathbf{f}$ is the tuple of all function symbols that appear in any of the implications in S_{AB}, and the variables in $\mathbf{x}_i$ are all the variables found in the implication χ_i, for $1 \leq i \leq n$.
Return the SOtgd $\exists \mathbf{f}(\forall \mathbf{x}_1 \chi_1 \wedge \cdots \wedge \forall \mathbf{x}_n \chi_n)$.

Every SOtgd is equivalent to an SOtgd in a *normal form*, where the right-hand sides (i.e., the formulae ψ_i) are atomic formulae rather than conjunctions of atomic formulae. For example, $\exists f \forall x(r(x) \Rightarrow (r_1(x, f(x)) \wedge r_2(f(x), x)))$ is logically equivalent to $\exists f(\forall x(r(x) \Rightarrow r_1(x, f(x))) \wedge \forall x(r(x) \Rightarrow r_2(f(x), x)))$. This is unlike the situation for (first-order) tgds (with existential quantifiers on the right-hand sides of implications), where we would lose expressive power if we required that the right-hand sides consist only of atomic formulae and not conjunctions of atomic formulae.

It was shown that the composition of SOtgds is also definable by an SOtgd (with algorithm given in [5]), and that the computing of certain answers of conjunctive queries in data exchange settings specified by SOtgds can be done in *polynomial* time in the size of source database instance. The class of SOtgds is obtained from (first-order) tgds by Skolemizing the existential first-order quantifiers into existential quantified functional symbols. This process gives rise to a class of existential second-order formulae with no equalities, but having equalities is not sufficient to expressively define the composition of finite sets of (first-order) tgds.

In the study of data exchange [5], one usually assumes an open-world assumption (OWA) semantics, making it possible to extend instances of target schema. The closed-world assumption (CWA) semantics was considered in [6] as an alternative that only moves 'as much data as needed' from the source instance into the target instance to satisfy constraints of a inter-schema mapping. It extends instances with nulls as well (incomplete information), and their language of CQ-SkSTDs slightly extends the syntax of SOtgds and is closed under composition. It was shown [6] that the Skolemized constraints are closed under composition not only under OWA but also under the CWA. If only conjunctive queries are used in mappings then, under both OWA and CWA, the composition problem is generally NP-complete (a model-checking for SOtgds, i.e., determining whether a given instance over the source and target schema satisfies a SOtgd can be NP-complete, in contrast with the first-order case, where such model-checking can be done in polynomial time). It is valid for composition of inter-schema mappings that mix open and closed attributes in mapping tgds.

There is subtle issue about the choice of universe in the semantics of SOtgds. In [5], the universe has been taken to be a countably infinite set of elements that included *active domain*. We recall that for a given instance $(A, B) \in Inst(\mathcal{M}_{AB})$ of an inter-schema mapping $\mathcal{M}_{AB} : \mathcal{A} \rightarrow \mathcal{B}$ the active domain consists of those values that appear in A and B, so that the second-order variables (here functional symbols that can be semantically represented by a binary relation of the function-graph) are interpreted over relations on the active domain.

Since SOtgds have existentially quantified function symbols, one needs sufficiently many elements in the universe in order to interpret these function symbols by values that are not contained in the source database instance as well. In [5], it was shown that as long as we take the universe to be finite but sufficiently large, the semantics of SOtgds remains unchanged from an infinite universe semantics. The most natural choice for the universe, of an instance (A, C) of a composition $\mathcal{M}_{AC} : \mathcal{A} \rightarrow \mathcal{C}$ (obtained from mappings $\mathcal{M}_{AB} : \mathcal{A} \rightarrow \mathcal{B}$ and $\mathcal{M}_{BC} : \mathcal{B} \rightarrow \mathcal{C}$ with a "intermediate" database $\mathcal{B}$), seems to be the active domain. But for the semantics of composition of mappings in [5] the authors did not use this simple choice when considering the necessity in the composition of instances (A, B) and (B, C) to take on values in the missing middle instance B for quantified functions in SOtgds.

The approach for what is the *proper* semantics of a given SOtgd mapping between $\mathcal{A}$ and $\mathcal{B}$ in our setting is more general than in the OWA data-exchange setting used in [5]: in our case, we consider that a more adequate semantics of a mapping from $\mathcal{A}$ into $\mathcal{B}$ has to be a *strict mapping* semantics, which considers the part of information that is mapped (only) from $\mathcal{A}$ into $\mathcal{C}$, without the information taken from another intermediate databases (as $\mathcal{B}$ in this case), and we need to derive it formally from the SOtgd.

In fact, in the former approach to the semantics of composition of mappings, the authors are using a non-well-defined "extended" semantics for mapping instances $(\mathcal{U}, A, C)$ instead of (A, C), where $\mathcal{U}$ is an unspecified but "sufficiently large" universe, "fixed and understandable from the context" [5], that includes the active domain A and C.

Remark In what follows, we introduce the characteristic functions for predicates (i.e., relations) during the composition of mappings and hence assume that the set of the classic truth logical values $\mathbf{2} = \{0, 1\}$ is a subset of the universe $\mathcal{U} = \mathbf{dom} \cup SK$.

With the syntax choice for SOtgds in [5], the authors encapsulate the necessary information contained in the intermediate database $\mathcal{B}$ into the semantics of the quantified functional symbols. In the interpreted functions of the composed mapping SOtgd, the information is mixed from both database instances A and B. Consequently, we will not consider two mappings between a given source and target databases equal if they are logically equivalent (as in data-exchange setting presented in [5]), but only if the *strict semantics* of the SOtgds of these two mappings are equal. Let us clarify these different approaches to the semantics of inter-schema mappings:

Example 2 Let us consider the following four schemas $\mathcal{A}, \mathcal{B}, \mathcal{C}$ and $\mathcal{D}$ (an extension of Example 2.3 in [5]). Schema $\mathcal{A}$ consists of a single binary relational symbol `Takes` that associates student names with courses they take, i.e., $S_A = \{\texttt{Takes}(x_n, x_c)\}$, $\Sigma_A = \emptyset$. Schema $\mathcal{B}$ consists of one binary relational symbol

Takes1 that is intended to provide a copy of Takes, and of an additional binary relational symbol Student that associates each student name with a student id, i.e.,

$$S_B = \{\texttt{Takes1}(x_n, x_c), \texttt{Student}(x_n, y)\}, \qquad \Sigma_B = \emptyset.$$

Schema $\mathcal{D}$ consists of one binary relational symbol Enrolment that associates student ids with the courses the student takes, and of an additional binary relational symbol Teaching that associates professor names with the courses, i.e., $S_D = \{\texttt{Enrolment}(y, x_c), \texttt{Teaching}(x_p, x_c)\}$, $\Sigma_D = \emptyset$. Schema $\mathcal{C}$ consists of one ternary relational symbol Learning that associates student name with courses he/she takes and professor names, and of an additional unary relational symbol Professor with professor names, i.e., $S_C = \{\texttt{Learning}(x_n, x_c, x_p), \texttt{Professor}(x_p)\}$, and with integrity constraints

$$\begin{aligned}\Sigma_C = \{&\forall x_p(\texttt{Professor}(x_p) \Rightarrow \exists x_n \exists x_c \texttt{Learning}(x_n, x_c, x_p)),\\ &\forall x_n \forall x_c \forall y \forall z((\texttt{Learning}(x_n, x_c, z_1) \wedge \texttt{Learning}(x_n, x_c, z_2))\\ &\Rightarrow z_1 = z_2)\},\end{aligned}$$

where the second constraint defines the tuple (x_n, x_c) as a key of the relation Learning.

Let us consider now the following schema mappings $\mathcal{M}_{AB} : \mathcal{A} \to \mathcal{B}$, $\mathcal{M}_{BD} : \mathcal{B} \to \mathcal{D}$, $\mathcal{M}_{AC} : \mathcal{A} \to \mathcal{C}$ and $\mathcal{M}_{CD} : \mathcal{C} \to \mathcal{D}$, where

$$\begin{aligned}\mathcal{M}_{AB} = \{&\forall x_n \forall x_c(\texttt{Takes}(x_n, x_c) \Rightarrow \texttt{Takes1}(x_n, x_c)),\\ &\forall x_n \exists y \forall x_c(\texttt{Takes}(x_n, x_c) \Rightarrow \texttt{Student}(x_n, y))\},\end{aligned}$$

that is, by Skolemization of $\exists y$,

$$\begin{aligned}\mathcal{M}_{AB} = \{&\forall x_n \forall x_c(\texttt{Takes}(x_n, x_c) \Rightarrow \texttt{Takes1}(x_n, x_c)),\\ &\forall x_n \forall x_c(\texttt{Takes}(x_n, x_c) \Rightarrow \texttt{Student}(x_n, f_1(x_n)))\},\\ \mathcal{M}_{BD} = \{&\forall x_n \forall x_c \forall y((\texttt{Takes1}(x_n, x_c) \wedge \texttt{Student}(x_n, y))\\ &\Rightarrow \texttt{Enrolment}(y, x_c))\},\\ \mathcal{M}_{AC} = \{&\forall x_n \forall x_c \exists x_p(\texttt{Takes}(x_n, x_c) \Rightarrow \texttt{Learning}(x_n, x_c, x_p))\},\end{aligned}$$

that is, by Skolemization of $\exists x_p$,

$$\begin{aligned}\mathcal{M}_{AC} &= \{\forall x_n \forall x_c(\texttt{Takes}(x_n, x_c) \Rightarrow \texttt{Learning}(x_n, x_c, f_2(x_n, x_c)))\},\\ \mathcal{M}_{CD} &= \{\forall x_n \forall x_c \forall x_p(\texttt{Learning}(x_n, x_c, x_p) \Rightarrow \texttt{Teaching}(x_p, x_c))\}.\end{aligned}$$

Then the composition $\mathcal{M}_{AD} : \mathcal{A} \to \mathcal{D}$, obtained by the composition (with the algorithm in [5]) of $\mathcal{M}_{AB}$ and $\mathcal{M}_{BD}$, is equal to the SOtgd:

(i) $\exists f_1(\forall x_n \forall x_c(\texttt{Takes}(x_n, x_c) \Rightarrow \texttt{Enrolment}(f_1(x_n), x_c)))$.

Analogously, the composition $\mathcal{M}'_{AD} : \mathcal{A} \to \mathcal{D}$, obtained by the composition of $\mathcal{M}_{AC}$ and $\mathcal{M}_{CD}$, can be equivalently represented by the SOtgd:
(ii) $\exists f_2(\forall x_n \forall x_c(\mathtt{Takes}(x_n, x_c) \Rightarrow \mathtt{Teaching}(f_2(x_n, x_c), x_c)))$.
Clearly, by using the semantics of inter-schema mappings defined in [5], we obtain that these two composed mappings $\mathcal{M}_{AD}$ and $\mathcal{M}'_{AD}$ are different.

As we mentioned, both SOtgds (i) and (ii) are not strict mappings from $\mathcal{A}$ into $\mathcal{D}$. Since f_1 in (i) encapsulates the data from $\mathcal{B}$ as well, it substantially remains a mapping from $\mathcal{A}$ and $\mathcal{B}$ into D. Analogously, f_2 in SOtgd in (ii) encapsulates the data from $\mathcal{C}$ as well, so it substantially remains a mapping from $\mathcal{A}$ and $\mathcal{C}$ into $\mathcal{D}$. The differences from the intermediate databases $\mathcal{B}$ and $\mathcal{C}$ explain why these two SOtgds are different (i.e., are not logically equivalent).

In spite of that, the strict information that is transferred (only) from $\mathcal{A}$ into $\mathcal{D}$ is equal for SOtgds (i) and (ii) and corresponds for each instance A of the schema $\mathcal{A}$ to the projection of its relation $\mathtt{Takes}(x_n, x_c)$ over the attribute x_c (see the complete proof in Examples 6 and 15). Consequently, in our semantics for composed mappings, these two composed mappings have to be equal (differently from the data exchange framework presented in [5]) and hence we need a new formal definition for this strict inter-schema mapping composition.

Consequently, the main difference between the data-exchange framework and our more general approach is the following: In a data-exchange framework, the mapping from a source to a target database determines, for a given source database instance, the target database instance, so that two equal mappings "produce" two equal solutions for the target database. From this point of view, two equal inter-schema mappings have to be logically equivalent [5].

In our more general framework, each database in a schema mapping system is relatively independent and can be consistently modified locally, by requiring only that after this local modification of this database all other instances of the correlated database schemas in this mapping system have to be (minimally) updated in order to obtain a new instance of this mapping system where all inter-schema mappings are satisfied again (as explained in Chap. 7 dedicated to operational semantics for database mappings).

Consequently, we do not require the strong determination of the target instance by a given mapping, but only that this target instance *satisfies* the mappings as well, and hence this general framework satisfies the OWA. Hence in given composed mappings from a source schema to a given target schema, in order to verify the equality of two different mappings between these two databases, we are not interested in contributions of the intermediate databases to the target database, but only in the strict information contribution from a given source database-instance into the target database-instance. Such an information contribution of the source database can be only filtered (thus decremented) by intermediate databases and not incremented. This strict contribution can be represented only at the instance-level semantics of a composed mapping for a given instance of the source schema and hence the *equality of two mappings* in our framework can be considered only at this *instance-level*. What is common for both semantic approaches to composition of schema mappings is that they cannot be obtained by the first-order logic formulae. From the logical point of view, we can consider the representation of inter-schema mappings and their compositions by SOtgds also in our approach, but with different meaning assigning to them.

In our case, the two equal compositions are not necessarily logically equivalent formulae (which requires the consideration of all intermediate databases involved in this composition). We will use SOtgds to define a mathematical framework from which we will be able to determine the strict meaning of composed mappings at an instance-level, as a quantity of information (set of views) specified by mapping to be transmitted from the active domain of the given source instance into the target instance.

Moreover, in order to pass to a categorical logic and its functorial semantics, we will use an algebraic representation based on the theory of operads both at schema (e.e., database sketches) and at instance database levels. We will show that this algebraic representation based on operads is semantically equivalent to the representation of the logical mappings based on SOtgds. Consequently, we will develop the algorithms for the transformation of SOtgds into the mapping operads and the algorithms for determination of the strict semantics at an instance database level. We will show that the egds can be represented as particular SOtgds and mapping operads so that the whole set of integrity constraints over a given schema can be equivalently represented by a schema mapping as well. The algorithm for composition of SOtgds given in [5] for the data-exchange setting has to be generalized for this more general semantics of schema mappings that we intend to use.

2.2 Transformation of Schema Integrity Constraints into SOtgds

Codd initially defined two sets of constraints but, in his second version of the relational model, he came up with the following integrity constraints:

- *Entity integrity*. The entity integrity constraint states that no primary key value can be null. This is because the primary key value is used to identify individual tuples in a relation. Having null value for the primary key (PK) implies that we cannot identify some tuples. This also specifies that there may not be any duplicate entries in the primary key column key row.
- *Referential Integrity*. The referential integrity constraint is specified between two relations and is used to maintain the consistency among tuples in the two relations. Informally, the referential integrity constraint states that a tuple in one relation that refers to another relation must refer to an existing tuple in that relation. It is a rule that maintains consistency among the rows of the two relations.
- *Domain Integrity*. The domain integrity states that every element from a relation should respect the type and restrictions of its corresponding attribute. A type can have variable length which needs to be respected. Restrictions could be the range of values that the element can have, the default value if none is provided, and if the element can be NULL.

- User Defined Integrity. (For example, $Age \geq 18 \wedge Age \leq 60$.) A business rule is a statement that defines or constrains some aspect of the business. It is intended to assert business structure or to control or influence the behavior of the business.

A typical example of the entity integrity is the primary-key integrity constraint for relational database relations, and it can be expressed in the FOL by egds. A typical example of the referential integrity is the foreign-key integrity constraint (FK) between relations in a give database, an it can be expressed in the FOL by tgds.

The domain integrity for a given attribute $a \in \mathbf{att}$ is defined by $dom(a) \subset \mathcal{U}$ in the introduction (Sect. 1.4).

Each user-defined integrity of a given relation r with attribute-variables in $\mathbf{x}$, expressed by a formula $\psi(\mathbf{x})$ by using the build-in predicates $\doteq, \neq, <, \ldots$ (Extensions of FOL, Sect. 1.3.1), can be defined as a tgd $r(\mathbf{x}) \Rightarrow \psi(\mathbf{x})$.

Consequently, we need only to consider two kinds of transformations, from the egds into a SOtgd and from the tgds into a SOtgd.

Such transformations of the integrity constraints into an SOtgd have to define a unique schema mapping $\mathcal{M} : \mathcal{A} \rightarrow \mathcal{A}_\top$ between a schema database $\mathcal{A}$ into a distinct target schema $\mathcal{A}_\top$, in a way such that a mapping has no significant compositions with another "real" inter-schema mappings of a given database mapping system.

Moreover, these obtained SOtgd (in $\mathcal{M}$) from the integrity constraints of a given schema, have to be only "logical", that is, with no transfer of information from the source database $\mathcal{A}$ into the target database $\mathcal{A}_\top$ (differently from the inter-schema mappings in Definition 3).

As we will see, it will be provided by the fact that the right-hand side of each implication in the obtained SOtgd (in $\mathcal{M}$) will have the false ground atom $r_\top(\overline{0}, \overline{1})$, where the relational symbol $r_\top$ (introduced in Sect. 1.3 as a binary built-in predicate for the FOL identity $\doteq$) is the unique relational symbol of the schema $\mathcal{A}_\top = (\{r_\top\}, \emptyset)$.

2.2.1 Transformation of Tuple-Generating Constraints into SOtgds

A normalized (with a simple atom on the right-hand side of implication) integrity constraint (tgd) $\forall \mathbf{x}(\phi_A(\mathbf{x}) \Rightarrow r(\mathbf{t})) \in \Sigma_A$ of a given schema database $\mathcal{A} = (S_A, \Sigma_A)$ (in Definition 2) has on the right-hand side of implication the relational symbol $r \in \mathcal{A}$ of this database schema. Thus, in order to satisfy the previously considered requirements for a "logical" representation of such a tgd, we have to transform such an implication by keeping in mind the following considerations:

Each normalized tgd $\forall \mathbf{x}(\phi_A(\mathbf{x}) \Rightarrow r(\mathbf{t}))$ is satisfied if $(\phi_A(\mathbf{x}) \wedge \neg \mathbf{r}(\mathbf{t}))$ is a *falsity* (false for every assignment of the values to variables in $\mathbf{x}$). Consequently, we will represent this tgd by the formula $\forall \mathbf{x}((\phi_A(\mathbf{x}) \wedge \neg r(\mathbf{t})) \Rightarrow r_\top(\overline{0}, \overline{1}))$, with the built-in identity relational symbol $r_\top$ used for the FOL identity $\doteq$, as follows:

Lemma 2 *Any normalized tgd constraint of a schema*

$$\mathcal{A} = (S_A, \Sigma_A), \ \forall \mathbf{x}\big(\phi_A(\mathbf{x}) \Rightarrow r(\mathbf{t})\big) \in \Sigma_A^{\text{egd}} \subseteq \Sigma_A,$$

where $\mathbf{t}$ *is a tuple of terms with variables in* $\mathbf{x}$ *and* $r \in S_A$, *is logically equivalent to the FOL sentence* $\forall \mathbf{x}((\phi_A(\mathbf{x}) \wedge \neg r(\mathbf{t})) \Rightarrow r_\top(\overline{0}, \overline{1}))$.

Proof For any assignment g, the formula $r_\top(\overline{0}, \overline{1})$ cannot be satisfied because $\langle g(\overline{0}), g(\overline{1})\rangle = \langle 0, 1\rangle \notin R_= = I_T(r_\top)$, so that $\phi_A(\mathbf{x}) \Rightarrow r(\mathbf{t})$ is logically equivalent to the formula $\phi_A(\mathbf{x}) \Rightarrow (r(\mathbf{t}) \vee r_\top(\overline{0}, \overline{1}))$, i.e., to the formula $\neg\phi_A(\mathbf{x}) \vee r(\mathbf{t}) \vee r_\top(\overline{0}, \overline{1})$, or equivalently, to the formula $\neg(\phi_A(\mathbf{x}) \wedge \neg r(\mathbf{t})) \vee r_\top(\overline{0}, \overline{1})$, and hence to the formula $(\phi_A(\mathbf{x}) \wedge \neg r(\mathbf{t})) \Rightarrow r_\top(\overline{0}, \overline{1})$. □

Based on these considerations, we can define the following algorithm for transformation of a given set of tgds into a single SOtgd:

Transformation algorithm *TgdsToConSOtgd*$(\Sigma_A^{\mathrm{tgd}})$
Input. A set Σ_A^{egd} of tgds of a given schema $\mathcal{A}$,

$$\Sigma_A^{\mathrm{tgd}} = \{\forall \mathbf{x}_1(\phi_{A,1}(\mathbf{x}_1) \Rightarrow \exists \mathbf{z}_1 \psi_{B,1}(\mathbf{y}_1, \mathbf{z}_1)), \ldots, \forall \mathbf{x}_n(\phi_{A,n}(\mathbf{x}_n) \Rightarrow \mathbf{z}_n \psi_{B,n}(\mathbf{y}_n, \mathbf{z}_n))\},$$

where $\mathbf{y}_i \subseteq \mathbf{x}_i$ and possibly $\mathbf{z}_i$ is the empty set, for $1 \le i \le n$.
Output. An SOtgd

1. (*Create the set of implications from tgds*)
   Initialize S_{AB} to be the empty set.
   Put each of the n implications of tgds in Σ_A^{tgd}, $\phi_{A,i}(\mathbf{x}_i) \Rightarrow \exists \mathbf{z}_i \psi_{B,i}(\mathbf{y}_i, \mathbf{z}_i)$, in S_{AB}.
2. (*Eliminate the existential quantifiers by Skolemization*)
   Repeat the following as far as possible:
   For each implication χ in S_{AB} of the form $\phi_{A,i}(\mathbf{x}) \Rightarrow \exists \mathbf{z}\psi_{B,i}(\mathbf{y}, \mathbf{z})$ with $\mathbf{z} = \langle z_1, \ldots, z_k\rangle$, $k \ge 1$, eliminate this χ from S_{AB} and add the following implication in S_{AB} (by introducing the set of new functional symbols $f_{i,j}$, for $1 \le j \le k$):
   $$(\phi_{A,i}(\mathbf{x}) \wedge (z_1 \doteq f_{i,1}(\mathbf{x})) \wedge \cdots \wedge (z_k \doteq f_{i,k}(\mathbf{x}))) \Rightarrow \psi_{B,i}(\mathbf{y}, \mathbf{z}).$$
3. (*Remove the variables originally quantified*)
   For each implication χ_i in S_{AB}, constructed in the previous step, perform the elimination of the variables z_i from this implication as far as possible: Select an equality $z_j \doteq f_{i,j}(\mathbf{x})$ that was generated in the previous step. Remove the equality $z_j \doteq f_{i,j}(\mathbf{x})$ from χ_i and replace every remaining occurrence of z_j in χ by $f_{i,j}(\mathbf{x})$.
4. (*Normalize the tgds*)
   Each implication χ in S_{AB} has the form $\phi_{A,i}(\mathbf{x}_i) \Rightarrow \bigwedge_1^k r_j(\mathbf{t}_j)$ where every member of $\mathbf{x}_i$ is a universally quantified variable, and each $\mathbf{t}_j$, for $1 \le j \le k$, is a sequence (a tuple) of terms over $\mathbf{x}_i$. We then replace each such implication χ in S_{AB} with k implications $\phi_{A,i}(\mathbf{x}_i) \Rightarrow r_1(\mathbf{t}_1), \ldots, \phi_{A,i}(\mathbf{x}_i) \Rightarrow r_k(\mathbf{t}_k)$.
5. (*Convert the tgds*)
   Remove each implication χ in S_{AB} of the form $\phi_{A,i}(\mathbf{x}_i) \Rightarrow r_j(\mathbf{t}_j)$ from S_{AB} and add the implication $(\phi_{A,i}(\mathbf{x}_i) \wedge \neg r_j(\mathbf{t}_j)) \Rightarrow r_\top(\overline{0}, \overline{1})$ in S_{AB}.

6. (*Construct* SOtgd)
If $S_{AB} = \{\chi_1, \ldots, \chi_m\}$ is the set where $\chi_1, \ldots, \chi_m$ are all the implications from the previous step then SOtgd is the formula $\exists \mathbf{f}(\forall \mathbf{x}_1 \chi_1 \wedge \cdots \wedge \forall \mathbf{x}_m \chi_m)$, where $\mathbf{f}$ is the tuple of all function symbols that appear in any of the implications in S_{AB}, and the variables in $\mathbf{x}_i$ are all the variables found in the implication χ_i, for $1 \leq i \leq m$.
Return the SOtgd $\exists \mathbf{f}(\forall \mathbf{x}_1 \chi_1 \wedge \cdots \wedge \forall \mathbf{x}_m \chi_m)$.

Let us consider the well known foreign-key integrity constraints:

Example 3 Let us consider Example 2 with relation $\texttt{Student}(x_n, y)$ where x_n is the primary-key of this relation (student's id) in the schema $\mathcal{B}$ and the relation $\texttt{Takes1}(x_n, x_c)$ where x_n is the foreign-key of this relation.

Thus, the foreign-key integrity constraint can be given by a single tgd in $\Sigma_B^{\text{tgd}} = \{\forall x_n, x_c(\texttt{Takes1}(x_n, x_c) \Rightarrow \exists y \texttt{Student}(x_n, y))\}$. Consequently, with this input Σ_B^{tgd}, we have the following steps of the algorithm:
1. We obtain $S_{AB} = \{\texttt{Takes1}(x_n, x_c) \Rightarrow \exists y \texttt{Student}(x_n, y)\}$.
2. We obtain

$$S_{AB} = \{(\texttt{Takes1}(x_n, x_c) \wedge (y = f_1(x_n, x_c)) \Rightarrow \texttt{Student}(x_n, y)\}.$$

3. We obtain $S_{AB} = \{\texttt{Takes1}(x_n, x_c) \Rightarrow \texttt{Student}(x_n, f_1(x_n, x_c))\}$.
4. Non applicable.
5. We obtain

$$S_{AB} = \{(\texttt{Takes1}(x_n, x_c) \wedge \neg \texttt{Student}(x_n, f_1(x_n, x_c))) \Rightarrow r_\top(\bar{0}, \bar{1})\}.$$

6. **Return**

$$\exists f_1 (\forall x_n, x_c((\texttt{Takes1}(x_n, x_c) \wedge \neg \texttt{Student}(x_n, f_1(x_n, x_c))) \Rightarrow r_\top(\bar{0}, \bar{1}))).$$

Consequently, the output of this algorithm is the SOtgd

$$\begin{aligned} &TgdsToConSOtgd(\{\forall x_n, x_c(\texttt{Takes1}(x_n, x_c) \Rightarrow \exists y \texttt{Student}(x_n, y))\}) = \\ &\exists f_1 (\forall x_n, x_c((\texttt{Takes1}(x_n, x_c) \wedge \neg \texttt{Student}(x_n, f_1(x_n, x_c))) \Rightarrow r_\top(\bar{0}, \bar{1}))). \end{aligned}$$

2.2.2 Transformation of Equality-Generating Constraints into SOtgds

An integrity constraint (egd) $\forall \mathbf{x}(\phi_A(\mathbf{x}) \Rightarrow (\mathbf{y} \doteq \mathbf{z})) \in \Sigma_A$ of a given schema database $\mathcal{A} = (S_A, \Sigma_A)$ (in Definition 2) has no relation from the target database on the right-hand side of the implication. It can be directly represented by SOtgds if we consider the equality '$\doteq$' as a built-in binary predicate with relational symbol $r_\top$ such that for any two given tuples $\mathbf{y} = \langle y_1, \ldots, y_k \rangle$ and $\mathbf{z} = \langle z_1, \ldots, z_k \rangle$ the formula $\mathbf{y} \doteq \mathbf{z}$ is an abbreviation for the formula $r_\top(y_1, z_1) \wedge \cdots \wedge r_\top(y_k, z_k)$.

However, in order to unify this presentation with tgd's transformation in the previous section, we will have in mind the following considerations:

Each egd is logically equivalent to the sentence $\forall \mathbf{x} \neg(\phi_A(\mathbf{x}) \wedge \neg(\mathbf{y} \doteq \mathbf{z}))$ which is satisfied if $\phi_A(\mathbf{x}) \wedge \neg(\mathbf{y} \doteq \mathbf{z})$ is a *falsity* (i.e., false for every assignment of the values to variables in $\mathbf{x}$).

Consequently, we will represent this egd by the formula

$$\forall \mathbf{x}\big((\phi_A(\mathbf{x}) \wedge (\mathbf{y} \neq \mathbf{z})) \Rightarrow r_\top(\overline{0}, \overline{1})\big),$$

where $\mathbf{y} \neq \mathbf{z}$ is equivalent to $\neg(\mathbf{y} \doteq \mathbf{z})$, i.e., an abbreviation for the formula $(y_1 \neq z_1) \vee \cdots \vee (y_k \neq z_k)$, as follows:

Lemma 3 *Any egd* $\forall \mathbf{x}(\phi_A(\mathbf{x}) \Rightarrow (\mathbf{y} \doteq \mathbf{z})) \in \Sigma_A^{\mathrm{egd}} \subseteq \Sigma_A$ *of a given schema database* $\mathcal{A} = (S_A, \Sigma_A)$, *where* $\mathbf{y} = \langle x_{j_1}, \ldots, x_{j_k}\rangle \subseteq \mathbf{x}$ *and* $\mathbf{z} = \langle x_{l_1}, \ldots, x_{l_k}\rangle \subseteq \mathbf{x}$ *such that* $j_i \neq l_i$ *for* $1 \leq i \leq k$, *is logically equivalent to the FOL formula*:

$$\forall \mathbf{x}\big((\phi_A(\mathbf{x}) \wedge (\mathbf{y} \neq \mathbf{z})) \Rightarrow r_\top(\overline{0}, \overline{1})\big),$$

where $\mathbf{y} \neq \mathbf{z}$ *is an abbreviation for the formula* $(x_{j_1} \neq x_{l_1}) \vee \cdots \vee (x_{j_k} \neq x_{l_k})$.

Proof For any assignment g, the atom $r_\top(\overline{0}, \overline{1})$ cannot be satisfied because

$$\langle g(\overline{0}), g(\overline{1})\rangle = \langle 0, 1\rangle \notin R_= = I_T(r_\top),$$

so that $\phi_A(\mathbf{x}) \Rightarrow (\mathbf{y} = \mathbf{z})$ is logically equivalent to the formula $\phi_A(\mathbf{x}) \Rightarrow ((\mathbf{y} \doteq \mathbf{z}) \vee r_\top(\overline{0}, \overline{1}))$, i.e., to the formula $\neg\phi_A(\mathbf{x}) \vee (\mathbf{y} \doteq \mathbf{z}) \vee r_\top(\overline{0}, \overline{1})$, or equivalently, to the formula $\neg(\phi_A(\mathbf{x}) \wedge (\mathbf{y} \neq \mathbf{z})) \vee r_\top(\overline{0}, \overline{1})$, and hence to the formula $(\phi_A(\mathbf{x}) \wedge (\mathbf{y} \neq \mathbf{z})) \Rightarrow r_\top(\overline{0}, \overline{1})$. □

This formula does not transfer any information, differently from tgds, as it is expected because on the right side of the implication we have the ground atom without variables.

Based on these considerations, we can define the algorithm for transformation of a given set of egds into a single SOtgd:

Transformation algorithm *EgdsToSOtgd*(Σ_A^{egd})
Input. A set of egds of a given schema $\mathcal{A}$,

$$\Sigma_A^{\mathrm{egd}} = \big\{\forall \mathbf{x}_1\big(\phi_{A,1}(\mathbf{x}_1) \Rightarrow (\mathbf{y}_1 \doteq \mathbf{z}_1)\big), \ldots, \forall \mathbf{x}_n\big(\phi_{A,n}(\mathbf{x}_n) \Rightarrow (\mathbf{y}_n \doteq \mathbf{z}_n)\big)\big\}.$$

Output. A SOtgd

1. (*Create the set of implications from egds*)
   Initialize S_{AB} to be the empty set.
   Put each of the n implications of tgds in Σ_A^{egd}, $\phi_{A,i}(\mathbf{x}_i) \Rightarrow (\mathbf{y}_i \doteq \mathbf{z}_i)$, in S_{AB}.
2. (*Convert the egds*)
   Remove each implication χ in S_{AB} of the form $\phi_{A,i}(\mathbf{x}_i) \Rightarrow (\mathbf{y}_i \doteq \mathbf{z}_i)$ from S_{AB} and add the implication $(\phi_{A,i}(\mathbf{x}_i) \wedge (\mathbf{y}_i \neq \mathbf{z}_i)) \Rightarrow r_\top(\overline{0}, \overline{1})$ in S_{AB}.

3. (*Construct* SOtgd)
   If $S_{AB} = \{\chi_1, \dots, \chi_n\}$ is the set, where $\chi_1, \dots, \chi_n$ are all the implications from the previous step, then SOtgd is the formula $\forall \mathbf{x}_1 \chi_1 \wedge \cdots \wedge \forall \mathbf{x}_n \chi_n$, where the variables in $\mathbf{x}_i$ are all the variables found in the implication χ_i, for $1 \leq i \leq n$.
   Return the SOtgd $\forall \mathbf{x}_1 \chi_1 \wedge \cdots \wedge \forall \mathbf{x}_n \chi_n$.

Example 4 Let us consider Example 2 with the relation $\texttt{Student}(x_1, x_2)$ in the schema $\mathcal{B}$ and introduce the primary-key integrity constraint for the attribute x_1 (i.e., the student's id) of this relation $\texttt{Student}$, expressed by the following egd:

$$\forall x_1 \forall x_2 \forall x_3 \big((\texttt{Student}(x_1, x_2) \wedge \texttt{Student}(x_1, x_3)) \Rightarrow (x_2 \doteq x_3) \big),$$

i.e., the egd $\forall \mathbf{x}(\phi_A(\mathbf{x}) \Rightarrow (\mathbf{y} \doteq \mathbf{z}))$ with $\mathbf{x} = (x_1, x_2, x_3)$, $\mathbf{y} = x_2$, $\mathbf{z} = x_3$ where $\phi_A(\mathbf{x})$ is the conjunctive formula $\texttt{Student}(x_1, x_2) \wedge \texttt{Student}(x_1, x_3)$.

Then,

$$\begin{aligned} &\mathit{EgdsToSOtgd}\big(\big\{\forall x_1 \forall x_2 \forall x_3 \big((\texttt{Student}(x_1, x_2) \\ &\qquad \wedge \texttt{Student}(x_1, x_3)\big) \Rightarrow (x_2 \doteq x_3)\big\}\big) \\ &\qquad = \big(\forall x_1 \forall x_2 \forall x_3 \big((\texttt{Student}(x_1, x_2) \\ &\qquad \wedge \texttt{Student}(x_1, x_3) \wedge (x_2 \neq x_3)\big) \Rightarrow r_\top(\overline{0}, \overline{1})\big). \end{aligned}$$

2.3 New Algorithm for General Composition of SOtgds

First of all, we have to specify what is an *atomic* (or basic) schema mapping in our framework, so that we can use them in the process of composition of other (composed) mappings. We will use only these atomic mappings in order to define a database mapping graphs $G = (V_G, E_G)$ where the vertices (or nodes) in V_G are the database schemas and the directed edges in E_G are the atomic mappings. The mappings obtained by the composition of two consecutive directed edges of this mapping graph are called *composed* mappings and hence, in what follows, we will define a new algorithm for this composition.

Definition 7 An *atomic mapping* is:
1. An inter-schema mapping $\mathcal{M}_{AB} = \{\Phi\} : \mathcal{A} \to \mathcal{B}$ where an SOtgd Φ is obtained by the algorithm *TgdsToSOtgd* for a given set S of tgds from a source schema $\mathcal{A}$ to the target database schema $\mathcal{B}$;
2. An integrity-constraints mapping $\mathcal{M}_{AA_\top} = \{\Phi \wedge \Psi\} : \mathcal{A} \to \mathcal{A}_\top$ where Φ is obtained by the algorithm *EgdsToSOtgd* for the set of egds Σ_A^{egd} and Ψ is obtained by the algorithm *TgdToConSOtgd* for the set Σ_A^{tgd} of tgd-constraints of a schema $\mathcal{A}$.

We can use the atomic mappings in order to compose another composed mappings. The algorithm for composition of SOtgds

$$\exists \mathbf{f}\big((\forall \mathbf{x}_1(\phi_{A,1} \Rightarrow \psi_{B,1})) \wedge \cdots \wedge (\forall \mathbf{x}_n(\phi_{A,n} \Rightarrow \psi_{B,n}))\big) \in \mathcal{M}_{AB} : \mathcal{A} \to \mathcal{B}$$

and

$$\exists \mathbf{g}\big((\forall \mathbf{y}_1(\phi_{B,1} \Rightarrow \psi_{D,1})) \wedge \cdots \wedge (\forall \mathbf{y}_m(\phi_{B,m} \Rightarrow \psi_{D,m}))\big) \in \mathcal{M}_{BD} : \mathcal{B} \to \mathcal{D},$$

presented in [5] for the data-exchange settings, assumes that all relational symbols in $\{\phi_{B,1}, \dots, \phi_{B,m}\}$ are contained in $\{\psi_{B,1}, \dots, \psi_{B,n}\}$ as well, so that a composition of these two SOtgds does not introduce new functional symbols. Differently from this setting in [5], it often happens that in our general framework, where a target database $\mathcal{B}$ is not completely determined by the mappings from the source database $\mathcal{A}$, the target database $\mathcal{B}$ can have some relational symbols that are not used in the mappings $\mathcal{M}_{AB} : \mathcal{A} \to \mathcal{B}$.

Consequently, if these 'independent' (w.r.t. mapping $\mathcal{M}_{AB}$) relational symbols in $\mathcal{B}$ are used in a mapping $\mathcal{M}_{BD} : \mathcal{B} \to \mathcal{D}$ then we will need to introduce new characteristic-functional symbols for them (by considering that they are the predicates in FOL) during the composition of the mappings $\mathcal{M}_{AB}$ and $\mathcal{M}_{BD}$ into a composed mapping $\mathcal{M}_{AD} = \mathcal{M}_{BD} \circ \mathcal{M}_{AB} : \mathcal{A} \to \mathcal{D}$ because they are not the relational symbols in the source schema $\mathcal{A}$.

Hence, in our more general setting, we need a more general algorithm than that presented in [5]. Note that, differently from the data-exchange setting, the mapping $\mathcal{M}_{AB}$ has SOtgds where each relational symbol in $\psi_{B,i}$ is in $\mathcal{B}$, and the mapping $\mathcal{M}_{BD}$ has SOtgds where each relational symbol in $\psi_{D,i}$ is in $\mathcal{D}$.

Notice that this is, from a logical point of view, a conservative extension of the algorithm in [5] because we only substitute some atoms with logically equivalent equations composed by characteristic functions of the relational symbols of these substituted atoms. This point is important as we will see in the next subsection in order to demonstrate the associativity property of the composition of schema mappings.

In fact, the algorithm in [5] is extended by the new step 4, while steps 1 and 3 are identical (and step 2 is slightly modified) to those presented in [5], as follows:

New algorithm Compose $(\mathcal{M}_{AB}, \mathcal{M}_{BD})$
Input. Two schema mappings $\mathcal{M}_{AB} : \mathcal{A} \to \mathcal{B}$ and $\mathcal{M}_{BD} : \mathcal{B} \to \mathcal{D}$ given by a set of SOtgds. We assumed by Definition 1 of FOL that a "sufficiently large" universe $\mathcal{U}$ contains the set $\mathbf{2} = \{0, 1\}$ of classic truth values for the constants $\overline{0}$ and $\overline{1}$.
Output. A schema mapping $\mathcal{M}_{AD} = \mathcal{M}_{BD} \circ \mathcal{M}_{AB} : \mathcal{A} \to \mathcal{B}$, which is the composition of $\mathcal{M}_{AB}$ and $\mathcal{M}_{BD}$, represented by an SOtgd.

1. (*Normalize the SOtgds in* $\mathcal{M}_{AB}$ *and* $\mathcal{M}_{BD}$)
   Rename the functional symbols so that the functional symbols which appear in $\mathcal{M}_{AB}$ are all distinct from the functional symbols which appear in $\mathcal{M}_{BD}$. For notational convenience, we shall refer to variables in $\mathcal{M}_{AB}$ as x's, possibly with

subscripts, and the variables in $\mathcal{M}_{BD}$ as y's, possibly with subscripts. Initialize S_{AB} and S_{BD} to be the empty sets. Assume that the SOtgd in $\mathcal{M}_{AB}$ is

$$\exists \mathbf{f}\big(\big(\forall \mathbf{x}_1(\phi_{A,1} \Rightarrow \psi_{B,1})\big) \wedge \cdots \wedge \big(\forall \mathbf{x}_n(\phi_{A,n} \Rightarrow \psi_{B,n})\big)\big).$$

For $1 \leq i \leq n$, put the n implications $\phi_{A,i} \Rightarrow \psi_{B,i}$ into S_{AB} if these implications are not ground formulas. We do likewise for $\mathcal{M}_{BD}$ and S_{BD}. If $S_{AB} = \emptyset$ or $S_{BD} = \emptyset$ then set $S_{BD} = \{r_\emptyset \Rightarrow r_\emptyset\}$ and go to step 5.

Each implication χ in S_{AB} has the form $\phi_{A,i}(\mathbf{x}_i) \Rightarrow \bigwedge_{1 \leq j \leq k} r_j(\mathbf{t}_j)$ where every member of $\mathbf{x}_i$ is a universally quantified variable, and each $\mathbf{t}_j$, for $1 \leq j \leq k$, is a sequence (tuple) of terms over $\mathbf{x}_i$. We then replace each such implication χ in S_{AB} with k implications $\phi_{A,i}(\mathbf{x}_i) \Rightarrow r_1(\mathbf{t}_1), \ldots, \phi_{A,i}(\mathbf{x}_i) \Rightarrow r_k(\mathbf{t}_k)$.

In all implications in S_{BD} replace all equations $(f_{r_i}(\mathbf{t}_i) = 1)$ with $r_i \in \mathcal{A}$ on the left-hand sides by the atoms $r_i(\mathbf{t}_i)$.

2. (*Compose S_{AB} with S_{BD}*)
Repeat the following as far as possible:
For each implication χ in S_{BD} of the form $\varphi \Rightarrow \psi_D$ where there is an atom $r(\mathbf{y})$ in φ, we perform the following steps to (possibly) replace $r(\mathbf{y})$ with atoms over $\mathcal{A}$. (The part of formula φ composed of built-in predicates $\doteq, \neq, >, <, \ldots$ is left unchanged). Let $\phi_1 \Rightarrow r(\mathbf{t}_1), \ldots, \phi_l \Rightarrow r(\mathbf{t}_l)$ be all the implications in S_{AB} whose right-hand side is an atom with the relational symbol r. For each such implication $\phi_i \Rightarrow r(\mathbf{t}_i)$, rename the variables in this implication so that they do not overlap with the variables in χ. (In fact, every time we compose this implication, we take a fresh copy of the implication, with new variables.) Let θ_i be the conjunction of the equalities between the variables in $r(\mathbf{y})$ and the corresponding terms in $r(\mathbf{t}_i)$, position by position. For example, the conjunction of equalities, position by position, between $r(y_1, y_2, y_3)$ and $r(x_1, f_2(x_2), f_1(x_3))$ is $(y_1 \doteq x_1) \wedge (y_2 \doteq f_2(x_2)) \wedge (y_3 \doteq f_1(x_3))$. Observe that every equality that is generated has the form $y \doteq t$ where y is a variable in $\mathcal{M}_{BD}$ and t is a term based on variables in $\mathcal{M}_{BD}$ and on functions in the tuple $\mathbf{f}$. Remove χ from S_{BD} and add l implications to S_{BD} as follows: replace $r(\mathbf{y})$ in χ with $\phi_i \wedge \theta_i$ and add the resulting implication to S_{BD}, for $1 \leq i \leq l$.
3. (*Remove the variables originally in $\mathcal{M}_{BD}$*)
For each implication χ constructed in the previous step, perform the elimination of the variables y_i from $\mathcal{M}_{BD}$ as far as possible: Select an equality $y_i \doteq t$ that was generated in the previous step (thus y_i is a variable in $\mathcal{M}_{BD}$ and t is a term based on variables in $\mathcal{M}_{AB}$ and on $\mathbf{f}$). Remove the equality $y_i \doteq t$ from χ and replace every remaining occurrence of y_i in χ by t.
4. (*Remove remained atoms with relational symbols in $\mathcal{B}$ that are not in $\mathcal{A}$*)
For the implications in S_{BD} repeat the following, in order to eliminate all relational symbols (on the left-hand side of implications) that are not in $\mathcal{A}$:
For each χ in S_{BD} such that its left-hand side is equal to a conjunction $\phi_1(\mathbf{x})$ (that contains only the relational symbols in $\mathcal{A}$ and the equations with functional symbols) and $\phi_2(\mathbf{x}', y_1, \ldots, y_k)$ (with $\mathbf{x}'$ a subset of $\mathbf{x}$) that contains only the atoms with relational symbols that are not in $\mathcal{A}$, we modify ϕ_2 from the left-hand side

of χ by replacing each atom $r_i(\mathbf{x}_i, \mathbf{y}_i)$ in ϕ_2 (where $\mathbf{x}_i \subseteq \mathbf{x}'$ and $\mathbf{y}_i \subseteq \{y_1, ..y_k\}$) with the equation $(f_{r_i}(\mathbf{x}_i, \mathbf{y}_i) \doteq \overline{1})$ by introducing its characteristic function f_{r_i} (i.e., for a given Tarski's interpretation I_T in Definition 1, $I_T(f_{r_i})(\mathbf{c}) = 1$ if $\mathbf{c}$ is a tuple of the interpreted relation $I_T(r_i)$; 0 otherwise).

5. (*Construct* $\mathcal{M}_{AD}$)
If $S_{BD} = \{\chi_1, \ldots, \chi_j\}$ is the set, where $\chi_1, \ldots, \chi_j$ are all the implications from the previous step, then $\mathcal{M}_{AD}$ is a singleton set composed of an SOtgd element

$$\exists \mathbf{g}(\forall \mathbf{z}_1 \chi_1 \wedge \cdots \wedge \forall \mathbf{z}_j \chi_j),$$

where $\mathbf{g}$ is the tuple of all function symbols that appear in any of the implications in S_{BD}, and where the variables in $\mathbf{z}_i$ are all the variables found in the implication χ_i, for $1 \leq i \leq j$.
Return $\mathcal{M}_{AD} = \{\exists \mathbf{g}(\forall \mathbf{z}_1 \chi_1 \wedge \cdots \wedge \forall \mathbf{z}_j \chi_j)\} : \mathcal{A} \rightarrow \mathcal{D}$.

The assumption that $\mathbf{2} \subset \mathcal{U}$ is another generalization of the previous algorithm in [5]; it is necessary for the introduction of characteristic functions in the step 4 of this new algorithm.

Note that we also do not obtain the empty composition if there is no direct mapping from the relations in $\mathcal{A}$ into $\mathcal{D}$. For example, let us consider the database

$$\mathcal{A} = (S_A, \emptyset) \quad \text{with } S_A = \{\texttt{Takes}(x_n, x_c)\},$$
$$\mathcal{B} = (S_B, \emptyset) \quad \text{with } S_B = \{\texttt{Takes1}(x_n, x_c), \texttt{Professor}(x_p)\},$$
$$\mathcal{D} = (S_D, \emptyset) \quad \text{with } S_D = \{\texttt{Professor1}(x_p)\}$$

and with mappings

$$\mathcal{M}_{AB} = \{\forall x_n \forall x_c (\texttt{Takes}(x_n, x_c) \Rightarrow \texttt{Takes1}(x_n, x_c))\} : \mathcal{A} \rightarrow \mathcal{B}$$

and

$$\mathcal{M}_{BD} = \{\forall x_p (\texttt{Professor}(x_p) \Rightarrow \texttt{Professor1}(x_p))\} : \mathcal{B} \rightarrow \mathcal{D}.$$

Then,

$$\begin{aligned}\mathcal{M}_{AD} &= \mathcal{M}_{BD} \circ \mathcal{M}_{AB} \\ &= \{\exists f_{Professor}(\forall x_p ((f_{Professor}(x_p) = \overline{1}) \Rightarrow \texttt{Professor1}(x_p)))\} : \mathcal{A} \rightarrow \mathcal{D}.\end{aligned}$$

Example 5 Let us consider Example 2.3 in [5] (also discussed in our Example 2) for the composition of SOtgds $\mathcal{M}'_{AD} = \mathcal{M}_{BD} \circ \mathcal{M}_{AB}$, where

$$\begin{aligned}\mathcal{M}_{AB} = \{&\forall x_n \forall x_c (\texttt{Takes}(x_n, x_c) \Rightarrow \texttt{Takes1}(x_n, x_c)),\\ &\forall x_n \exists y \forall x_c (\texttt{Takes}(x_n, x_c) \Rightarrow \texttt{Student}(x_n, y))\},\end{aligned}$$

or expressed equivalently by SOtgd

$$\mathcal{M}_{AB} = \{\exists f_1 (\forall x_n \forall x_c (\texttt{Takes}(x_n, x_c) \Rightarrow \texttt{Takes1}(x_n, x_c)) \\ \wedge \forall x_n \forall x_c (\texttt{Takes}(x_n, x_c) \Rightarrow \texttt{Student}(x_n, f_1(x_n))))\},$$

and

$$\mathcal{M}_{BD} = \{\forall x_n \forall x_c \forall y ((\texttt{Takes1}(x_n, x_c) \\ \wedge \texttt{Student}(x_n, y)) \Rightarrow \texttt{Enrolment}(y, x_c))\}.$$

1. We rename the variables as required by the algorithm and obtain

$$\mathcal{M}_{AB} = \{\exists f_1 (\forall x_1 \forall x_2 (\texttt{Takes}(x_1, x_2) \Rightarrow \texttt{Takes1}(x_1, x_2)) \\ \wedge \forall x_1 \forall x_2 (\texttt{Takes}(x_1, x_2) \Rightarrow \texttt{Student}(x_1, f_1(x_1))))\}$$

$$\mathcal{M}_{BD} = \{\forall y_1 \forall y_2 \forall y_3 ((\texttt{Takes1}(y_1, y_2) \\ \wedge \texttt{Student}(y_1, y_3)) \Rightarrow \texttt{Enrolment}(y_3, y_2))\}.$$

Then we obtain

$$S_{AB} = \{\texttt{Takes}(x_1, x_2) \Rightarrow \texttt{Takes1}(x_1, x_2), \\ \texttt{Takes}(x_1, x_2) \Rightarrow \texttt{Student}(x_1, f_1(x_1))\},$$

$$S_{BD} = \{(\texttt{Takes1}(y_1, y_2) \wedge \texttt{Student}(y_1, y_3)) \Rightarrow \texttt{Enrolment}(y_3, y_2)\}.$$

2. We obtain

$$S_{BD} = \{((\texttt{Takes}(x_1, x_2) \wedge (y_1 \doteq x_1) \wedge (y_2 \doteq x_2)) \\ \wedge (\texttt{Takes}(x_1, x_2) \wedge (y_1 \doteq x_1) \wedge (y_3 \doteq f_1(x_1)))) \\ \Rightarrow \texttt{Enrolment}(y_3, y_2)\}.$$

3. We obtain $S_{BD} = \{\texttt{Takes}(x_1, x_2) \Rightarrow \texttt{Enrolment}(f_1(x_1), x_2)\}$.
4. Non applicable.
5. **Return**

$$\mathcal{M}'_{AD} = \{\exists f_1 (\forall x_1 \forall x_2 (\texttt{Takes}(x_1, x_2) \Rightarrow \texttt{Enrolment}(f_1(x_1), x_2)))\}.$$

Let us explain the difference of this more general algorithm w.r.t. the original algorithm presented in [5]:

Example 6 Let us consider the composition of SOtgds $\mathcal{M}'_{AD} = \mathcal{M}_{BD} \circ \mathcal{M}_{AB}$ where $\mathcal{M}_{AB}$ in Example 5 above is reduced to

$$\mathcal{M}_{AB} = \{\forall x_n \forall x_c (\texttt{Takes}(x_n, x_c) \Rightarrow \texttt{Takes1}(x_n, x_c))\}$$

while $\mathcal{M}_{BD}$ remains unchanged.

1. We rename the variables as required by the algorithm and obtain

$$\mathcal{M}_{AB} = \{\forall x_1 \forall x_2 (\texttt{Takes}(x_1, x_2) \Rightarrow \texttt{Takes1}(x_1, x_2))\},$$
$$\mathcal{M}_{BD} = \{\forall y_1 \forall y_2 \forall y_3 ((\texttt{Takes1}(y_1, y_2) \wedge \texttt{Student}(y_1, y_3)) \Rightarrow \texttt{Enrolment}(y_3, y_2))\}.$$

Hence,

$$S_{AB} = \{\texttt{Takes}(x_1, x_2) \Rightarrow \texttt{Takes1}(x_1, x_2)\},$$
$$S_{BD} = \{(\texttt{Takes1}(y_1, y_2) \wedge \texttt{Student}(y_1, y_3)) \Rightarrow \texttt{Enrolment}(y_3, y_2)\}.$$

2. We obtain

$$S_{BD} = \{((\texttt{Takes}(x_1, x_2) \wedge (y_1 \doteq x_1) \wedge (y_2 \doteq x_2)) \wedge \texttt{Student}(y_1, y_3)) \Rightarrow \texttt{Enrolment}(y_3, y_2)\}.$$

3. We obtain

$$S_{BD} = \{(\texttt{Takes}(x_1, x_2) \wedge \texttt{Student}(x_1, y_3)) \Rightarrow \texttt{Enrolment}(y_3, x_2)\}.$$

4. The left-hand side of the implication in S_{BD} contains the atom $\texttt{Student}(x_1, y_3)$ where `Student` is a relational symbol in $\mathcal{B}$, and hence we replace it by the equation $(h_{Student}(x_1, y_3) \doteq \overline{1})$. Thus we obtain

$$S_{BD} = \{(\texttt{Takes}(x_1, x_2) \wedge (h_{Student}(x_1, y_3) \doteq \overline{1})) \Rightarrow \texttt{Enrolment}(y_3, x_2)\}.$$

5. **Return**

$$M'_{AD} = \{\exists h_{Student} (\forall x_1 \forall x_2 \forall y_3 ((\texttt{Takes}(x_1, x_2) \wedge (h_{Student}(x_1, y_3) \doteq \overline{1})) \Rightarrow \texttt{Enrolment}(y_3, x_2)))\}.$$

Note that in the two cases, for

$$\mathcal{M}_{AB} = \{\exists f_1 (\forall x_n \forall x_c (\texttt{Takes}(x_n, x_c) \Rightarrow \texttt{Takes1}(x_n, x_c)) \wedge \forall x_n \forall x_c (\texttt{Takes}(x_n, x_c) \Rightarrow \texttt{Student}(x_n, f_1(x_n)))\}$$

and for $\mathcal{M}_{AB} = \{\forall x_n \forall x_c (\texttt{Takes}(x_n, x_c) \Rightarrow \texttt{Takes1}(x_n, x_c))\}$, we obtained different SOtgds of the resulting composition. In fact, in the second case, we did not express the fact that the second argument in `Student` depends only on its first argument, as in the first case. So, if we wanted to obtain equivalent functions for both f_1 and $h_{Student}$, in the second case we would need to introduce the key integrity constraint for the relation `Student` in the schema $\mathcal{B}$ given by the egd (from Example 4)

$$\forall z_1 \forall z_2 \forall z_3 ((\texttt{Student}(z_1, z_2) \wedge \texttt{Student}(z_1, z_3)) \Rightarrow (z_2 \doteq z_3))$$

so that for any instance $\mathcal{D}$ that satisfies also the mapping $\mathcal{M}'_{AD}$, $h_{Student}(x_1, y_3) \doteq 1$ is equivalent to $y_3 \doteq f_1(x_1)$.

2.3.1 Categorial Properties for the Schema Mappings

Now we will show that this new algorithm for composition of schema mappings expressed by SOtgds can be used as the composition of morphisms in a category where the objects are database schemas and morphisms (the arrows in such a category) are the mappings. We recall that, as it was presented in [5], the space of instances $Inst(\mathcal{M}_{AB})$ of a mapping $\mathcal{M}_{AB} : \mathcal{A} \to \mathcal{B}$ is a binary relation between instance-databases of $\mathcal{A}$ and target instance-databases of $\mathcal{B}$. Consequently, the schema mapping $\mathcal{M}_{AC}$ is a composition of two schema mappings $\mathcal{M}_{AB}$ and $\mathcal{M}_{BC}$, denoted by $\mathcal{M}_{BC} \circ \mathcal{M}_{AB}$, if the space of instances of $\mathcal{M}_{AC}$ is the set-theoretic composition of the spaces of instances of $\mathcal{M}_{AB}$ and $\mathcal{M}_{BC}$. The advantage of this approach is that the set of formulae defining a composition $\mathcal{M}_{AC}$ of $\mathcal{M}_{AB}$ and $\mathcal{M}_{BC}$ *is unique* up to logical equivalence. Consequently,

$$(Comp) \quad \mathcal{M}_{AC} = \mathcal{M}_{BC} \circ \mathcal{M}_{AB} \quad \text{if } Inst(\mathcal{M}_{AC}) = Inst(\mathcal{M}_{AB}) * Inst(\mathcal{M}_{BC}),$$

where $*$ denotes the composition of binary relations.

Lemma 4 *The composition of two schema mappings $\mathcal{M}_{AB}$ and $\mathcal{M}_{BC}$, presented by the new algorithm as $\mathcal{M}_{BC} \circ \mathcal{M}_{AB} \triangleq Compose(\mathcal{M}_{AB}, \mathcal{M}_{BC})$, is associative.*

Proof From the fact that the new algorithm is a conservative extension of the algorithm in [5] (see the proof of Proposition 4.1 in [5] for more details) such that in the new step 4 we only replace the remaining atoms (that are removed in this step) by their characteristic functions, the property (Comp) above is still valid. Consequently, $\mathcal{M}_{AD} = \mathcal{M}_{CD} \circ (\mathcal{M}_{BC} \circ \mathcal{M}_{AB}) = \mathcal{M}_{CD} \circ \mathcal{M}_{AC}$ if

$$\begin{aligned} Inst(\mathcal{M}_{AD}) &= Inst(\mathcal{M}_{AC}) * Inst(\mathcal{M}_{CD}) \\ &= \big(Inst(\mathcal{M}_{AB}) * Inst(\mathcal{M}_{BC})\big) * Inst(\mathcal{M}_{CD}) \quad \text{(by (Comp))} \\ &= Inst(\mathcal{M}_{AB}) * Inst(\mathcal{M}_{BC}) * Inst(\mathcal{M}_{CD}). \quad \text{(by associativity of } *) \end{aligned}$$

Analogously, $\mathcal{M}'_{AD} = (\mathcal{M}_{CD} \circ \mathcal{M}_{BC}) \circ \mathcal{M}_{AB} = \mathcal{M}_{BD} \circ \mathcal{M}_{AB}$ if

$$\begin{aligned} Inst\big(\mathcal{M}'_{AD}\big) &= Inst(\mathcal{M}_{AB}) * Inst(\mathcal{M}_{BD}) \\ &= Inst(\mathcal{M}_{AB}) * \big(Inst(\mathcal{M}_{BC}) * Inst(\mathcal{M}_{CD})\big) \quad \text{(by (Comp))} \\ &= Inst(\mathcal{M}_{AB}) * Inst(\mathcal{M}_{BC}) * Inst(\mathcal{M}_{CD}). \quad \text{(by associativity of } *) \end{aligned}$$

Consequently, $Inst(\mathcal{M}'_{AD}) = Inst(\mathcal{M}_{AD})$, that is, $\mathcal{M}'_{AD} = \mathcal{M}_{AD}$, i.e.,

$$(\mathcal{M}_{CD} \circ \mathcal{M}_{BC}) \circ \mathcal{M}_{AB} = \mathcal{M}_{CD} \circ (\mathcal{M}_{BC} \circ \mathcal{M}_{AB}),$$

or in other words, the operation of composition $\circ$ for the new algorithm *Compose* is associative. □

Let us show that we are able to define the identity mappings for any schema $\mathcal{A}$, such that composition with another arrow is equal to this arrow:

Lemma 5 *For a schema $\mathcal{A} = (S_A, \Sigma_A)$ we define its identity mapping as*

$$\mathcal{M}_{AA} = Id_A = \left\{ \bigwedge \{ \forall \mathbf{x}_i (r_i(\mathbf{x}_i) \Rightarrow r_i(\mathbf{x}_i)) \mid r_i \in S_A \} \right\} : \mathcal{A} \to \mathcal{A}.$$

Consequently, for any two mappings $\mathcal{M}_{BA} : \mathcal{B} \to \mathcal{A}$ and $\mathcal{M}_{AB} : \mathcal{A} \to \mathcal{B}$,

$$\mathcal{M}_{AB} \circ Id_A = Compose(\mathcal{M}_{AB}, Id_A) = \mathcal{M}_{AB}$$

and $Id_A \circ \mathcal{M}_{BA} = Compose(Id_A \circ \mathcal{M}_{BA}) = \mathcal{M}_{BA}$.

Proof Let us verify that $Compose(\mathcal{M}_{AB}, Id_A) = \mathcal{M}_{AB}$. In step 2, each $r_i(\mathbf{y}_i)$ on the left-hand side of the implication in S_{BD} is replaced by the conjunction $r_i(\mathbf{x}_i) \wedge (\mathbf{y}_i \doteq \mathbf{x}_i)$. Consequently, in step 3, we replace all y's on the right-hand side of implications with x's and eliminate the equations $\mathbf{y}_i \doteq \mathbf{x}_i$ from the left-hand side of these implications, so that we obtain exactly the SOtgd of the mapping $\mathcal{M}_{AB}$.

Let us verify that $Compose(Id_A \circ \mathcal{M}_{BA}) = \mathcal{M}_{BA}$. In step 2, each $r_i(\mathbf{y}_i)$ on the left-hand side of the implication in S_{BD} such that there is an implication $\phi_i(\mathbf{y}_i) \Rightarrow r_i(\mathbf{t}_i)$ in S_{AB} is replaced by the conjunction $\phi_i(\mathbf{y}_i) \wedge (\mathbf{y}_i \doteq \mathbf{t}_i)$. Consequently, in step 3, we replace all y's on the right-hand side of implications with terms in $\mathbf{t}_i$ and we eliminate the equations $\mathbf{y}_i \doteq \mathbf{t}_i$ from the left-hand side of these implications, so that we obtain exactly the SOtgd of the mapping $\mathcal{M}_{BA}$. □

Based on these two lemmas, we have the following important corollary:

Corollary 2 *The database schemas and the composition of their schema mappings can be represented by a sketch category.*

Proof It is a direct result of Lemma 4 which demonstrates the associativity for composition of the mappings (i.e., morphisms of a sketch category) and Lemma 5 that demonstrates that for each schema (i.e., object of a sketch category) there is an identity mapping. □

Note that the identity SOtgds are not tuple-generating, that is, they do not insert new tuples in the target database. From the fact that each implication is of the form $r_i(\mathbf{x}_i) \Rightarrow r_i(\mathbf{x}_i)$ and hence if for a tuple of values $\mathbf{c}$, $r_i(\mathbf{c})$ is true, then $\mathbf{c}$ is already a tuple in the relation r_i. Thus, this implication is true without inserting a new tuple in the relation r_i.

From the logical point of view, the SOtgd of an identity mapping is a tautology and, consequently, its conjunction with another SOtgd is equivalent logically to this SOtgd. Consequently, based on this general algorithm, we can represent the logic-syntax of (composed) schema mappings by SOtgds. However, we need a new semantics for the mappings in order to define the (strict) equality between two mappings, different from the logical equivalence [5] used in data-exchange settings. The data-exchange setting is only a particular case of our more general setting. Thus, in our setting two logically equivalent schema-mappings are always equal at instance-level as well. However, we can have equal mappings (at instance level) which are not logically equivalent at schema level as it was presented by Example 2. Thus, in what follows, we will investigate this equality of mappings at the instance-level between instance-databases, and the first step is to pass from logic to algebras and hence to functorial semantics of database mappings.

2.4 Logic versus Algebra: Categorification by Operads

We consider that logical syntax for composition of schema mappings, represented by SOtgds introduced in [5], is well suited for consideration of the logical syntax and semantics of schema mappings, and we explained how SOtgds hide the relations of 'intermediate' databases by substituting them with existentially quantified characteristic functions of such relations.

Another way, more closed to the functorial categorial representation, is to assume an algebraic point of view for the representation of atomic and composed mappings. It is possible because each conjunctive formula (a view) used on the left-hand side of the implication forms of tgds can be equivalently represented by a term of SPJRU-relational algebra. Any elementary (or atomic) mapping may be represented by a set of these algebraic terms, while the composition of atomic mappings may be represented algebraically by the term-trees, or formally, by using the algebraic theory of operads.

In what follows, we will use this algebraic point of view based on the operads, which is a more convenient than a standard Lindenbaum-like translation of the second-order tgds logic into the algebraic categorial setting.

In what follows, we will work with the typed operads, first developed for homotopy theory [1, 3, 7], having a set $\mathbb{R}$ of types (each finitary relation symbol is a type), or "R-operads" for short. The basic idea of an R-operad O is that, given types $r_1, \ldots, r_k, r \in \mathbb{R}$, there is a set $O(r_1, \ldots, r_k, r)$ of abstract k-ary "operations" with inputs of type $r_1, \ldots, r_k$ and output of type r. In short, an operad describes a family of composable operations with multiple inputs and one output, satisfying several intuitive properties like associativity of composition and permutability of the inputs. The main difference from the *universal algebra* approach where R would be a carrier set and the operad's operations would compose the signature of such an algebra, here we emphasize just the "operations" by treating them as a carrier set so that the symbol of their composition '$\cdot$' is just an operator of the R-operads signature.

Categorification is a process of finding category-theoretic analog of set-theoretic concepts by introducing 'n-categories', i.e., algebraic structures having objects, morphisms between objects, 2-morphisms between morphisms, and so on up to n-morphisms. In [2], the existence of a close relation between n-categories and the

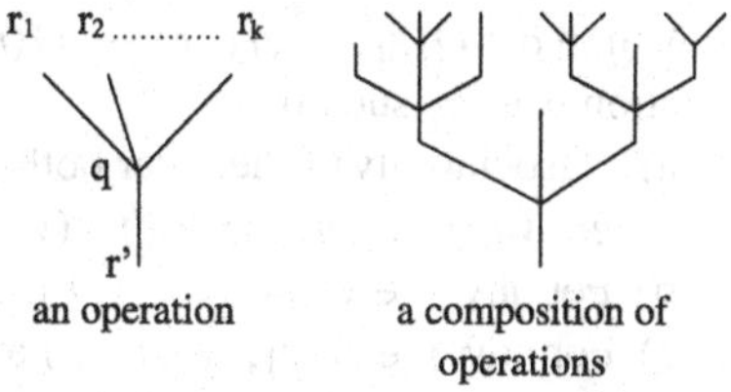

Fig. 2.1 Operations of an R-operad

homotopy theory is demonstrated, based on theory of operads. As we will see in our case of **DB** category used for denotational semantics of database mappings, this **DB** category is an n-category as well, but with a particular property that each arrow can be equivalently represented by an object (i.e., a categorial symmetry) and hence the higher-order arrows are represented just by ordinary 1-arrows in **DB**. Basically, the work presented in this book is a categorification of the logic theory of the schema database mappings.

Remark In our case, the input types are the relational symbols of the source database of a given schema mapping (represented logically by an SOtgd) while the output type is a relational symbol of the target database. Each single abstract "operation" is derived from a logical implication, whose left-hand side is a conjunctive query over the source database schema, into a particular relation of the target database schema. Thus, it represents an SPJRU term of the relational algebra corresponding to the left-hand side (a conjunctive query, possibly with negations (in the case of integrity constraints), over the source schema) of this implication, labeled by a node q on the left tree (an operation) in Fig. 2.1. In what follows, we will provide an algorithm *MakeOperads* that transforms a given SOtgd (of a given mapping from a source to a target schema database) into the set of such abstract operations with input types in the source schema and output types in the target schema.

We can visualize such an operation as a tree with only one node (a relational symbol of the target schema). In an operad, we can obtain new operations from old ones by composing them; it can be visualized in terms of trees in Fig. 2.1. We can obtain the new operators from old ones by permuting arguments, and there is a unary "identity" operation of each type. Finally, we insist on a few plausible axioms: the identity operations act as identities for composition, permuting arguments is compatible with composition, and *composition is associative*. Thus, formally, we have the following:

Definition 8 For any set $\mathbb{R}$ of relational symbols with finite arity (the predicate letters in Definition 1) an R-operad O consists of:

1. A set $O(r_1, \dots, r_k, r)$ for any $r_1, \dots, r_k, r \in \mathbb{R}$ and an element $1_r \in O(r, r)$ for any $r \in \mathbb{R}$. We denote the empty type by $r_\emptyset$ (the truth propositional letter in FOL, Definition 1), with $ar(r_\emptyset) = 0$ and $1_{r_\emptyset} \in O(r_\emptyset, r_\emptyset)$.
2. An element $f \cdot (g_1, \dots, g_k) \in O(r_{11}, \dots, r_{1i_1}, \dots, r_{k1}, \dots, r_{ki_k}, r)$ for any $f \in O(r_1, \dots, r_k, r)$ and any $g_1 \in O(r_{11}, \dots, r_{1i_1}, r_1), \dots, g_k \in O(r_{k1}, \dots, r_{ki_k}, r_k)$.

3. A map $\sigma : O(r_1, \dots, r_k, r) \to O(r_{\sigma(1)}, \dots, r_{\sigma(k)}, r)$, $f \longmapsto f\sigma$ for any permutation $\sigma \in \mathbb{R}_k$ such that:
   (a) (associativity) Whenever both sides make sense, $f \cdot (g_1 \cdot (h_{11}, \dots, h_{1i_1}), \dots, g_k \cdot (h_{k1}, \dots, h_{ki_k})) = (f \cdot (g_1, \dots, g_k)) \cdot (h_{11}, \dots, h_{1i_1}, \dots, h_{k1}, \dots, h_{ki_k})$;
   (b) For any $f \in O(r_1, \dots, r_k, r)$, $f = 1_r \cdot f = f \cdot (1_{r_1}, \dots, 1_{r_k})$;
   (c) For any $f \in O(r_1, \dots, r_k, r)$ and $\sigma, \sigma_1 \in \mathbb{R}_k$, $f(\sigma\sigma_1) = (f\sigma)\sigma_1$;
   (d) For any $f \in O(r_1, \dots, r_k, r)$, $\sigma \in \mathbb{R}_k$ and $g_1 \in O(r_{11}, \dots, r_{1i_1}, r_1), \dots, g_k \in O(r_{k1}, \dots, r_{ki_k}, r_k)$, $(f\sigma) \cdot (g_{\sigma(1)}, \dots, g_{\sigma(k)}) = (f \cdot (g_1, \dots, g_k))\rho(\sigma)$ where $\rho : \mathbb{R}_k \longrightarrow \mathbb{R}_{i_1 + \dots + i_k}$ is the obvious homomorphism;
   (e) For any $f \in O(r_1, \dots, r_k, r)$, $g_1 \in O(r_{11}, \dots, r_{1i_1}, r_1), \dots, g_k \in O(r_{k1}, \dots, r_{ki_k}, r_k)$ and $\sigma_1 \in \mathbb{R}_{i_1}, \dots, \sigma_k \in \mathbb{R}_{i_k}$,

$$\big(f \cdot (g_1\sigma_1, \dots, g_k\sigma_k)\big) = \big(f \cdot (g_1, \dots, g_k)\big)\varrho_1(\sigma_1, \dots, \sigma_k),$$

   where $\varrho_1 : \mathbb{R}_{i_1} \times \dots \times \mathbb{R}_{i_k} \longrightarrow \mathbb{R}_{i_1 + \dots + i_k}$ is the obvious homomorphism.

Thus, an R-operad O can be considered as an algebra where the variables (a carrier set) are the typed operators $f \in O(r_1, \dots, r_k, r)$ and the signature is composed of the set of basic operations used for composition of the elements of this carrier set:

1. The binary associative composition operation (a partial function) '$\cdot$' such that

$$\cdot\big(f, (g_1, \dots, g_k)\big) = f \cdot (g_1, \dots, g_k),$$

   if f is a k-ary typed operator.
2. The left and right identity (partial) operations $1_r \cdot \_$, $\_ \cdot (1_{r_1}, \dots, 1_{r_1})$ such that for a k-ary typed operator f,

$$1_r \cdot \_(f) = 1_r \cdot f = f, \qquad \_ \cdot (1_{r_1}, \dots, 1_{r_1})(f) = f \cdot (1_{r_1}, \dots, 1_{r_1}) = f.$$

3. The permutation (partial) operations,

$$\sigma : O(r_1, \dots, r_k, r) \to O(r_{\sigma(1)}, \dots, r_{\sigma(k)}, r),$$

   such that for each $f \in O(r_1, \dots, r_k, r)$, $\sigma(f) = f\sigma$.

Thus, all operations in the signature of this R-operads algebra are the *partial* functions (defined only for the subsets of elements of the carrier set of this algebra).

It is clear that in this formal syntax of R-operads algebra, the relational symbols are left out of its syntax; thus, the terms of this algebra do not contain the relational symbols. Consequently, it appears fundamentally different from the relational algebras (as, for example, the basic SPRJU Codd's algebra and its extensions, examined in Chap. 5). However, in Chap. 5, where we will consider the extensions of the Codd's relational algebra, we will take a different point of view of the R-algebras of operads by considering its carrier set (the variables) equal to the set $\mathbb{R}$ of relational symbols (considered only as implicit "types" for operads). Hence, each $f \in O(r_1, \dots, r_k, r)$ will be a composed operation of such a relational Σ_α algebra, and we will define its complete signature Σ_α. Then we will demonstrate in Sect. 5.4, Theorem 9 and Corollary 19, that such a Σ_α algebra (where $f \in O(r_1,\dots,r_k,r)$ are considered not as a carrier-set but as the set of composed operations in this algebra) is computationally equivalent to the largest extension of the Codd's relational algebra (which contains all update operations for the relations).

We consider an R-algebra in Definition 8 in the way where not the relations but the operads $f \in O(r_1,\dots,r_k,r)$ constitute its carrier-set and hence it is similar to the consideration of a category as a kind of algebra (see the introduction, Sect. 1.5.1) where the carrier set is composed of objects and arrows with the basic operation just the associative composition '$\circ$' of the arrows, which is a partial function, as is the associative operation '$\cdot$' for the operads. This analogy (both with a composition with the identity elements) is a natural justification for why we are using the interpretation of R-operads algebra in Definition 8 as a basis for the definition of the **DB** category for database-mappings.

Let us define the algorithm that transform an SOtgd into a set of abstract operads:

Operads algorithm *MakeOperads*($\mathcal{M}_{AB}$)
Input. A schema mapping $\mathcal{M}_{AB} : \mathcal{A} \to \mathcal{B}$ given by an SOtgd

$$\exists \mathbf{f}\big((\forall \mathbf{x}_1(\phi_{A,1} \Rightarrow \psi_{B,1})) \wedge \dots \wedge (\forall \mathbf{x}_n(\phi_{A,n} \Rightarrow \psi_{B,n}))\big).$$

Output. Set of abstract operad's operations from $\mathcal{A}$ into $\mathcal{B}$

1. (*Normalize the SOtgd*)
   Initialize S to be the empty set $\emptyset$. Let $\mathcal{M}_{AB}$ be the singleton set

   $$\big\{\exists \mathbf{f}\big((\forall \mathbf{x}_1(\phi_{A,1} \Rightarrow \psi_{B,1})) \wedge \dots \wedge (\forall \mathbf{x}_n(\phi_{A,n} \Rightarrow \psi_{B,n}))\big)\big\}.$$

   For $1 \le i \le n$, put the implication $\phi_{A,i} \Rightarrow \psi_{B,i}$ into S if this implication is not a ground formula. If S is empty go to step 4.

   Each implication χ in S has the form $\phi_{A,i}(\mathbf{x}_i) \Rightarrow \bigwedge_{1\le j\le k} r_j(\mathbf{t}_j)$ where every member of $\mathbf{x}_i$ is a universally quantified variable and each $\mathbf{t}_j$, for $1 \le j \le k$, is a sequence (tuple) of terms over $\mathbf{x}_i$. We then replace each such implication χ in S with k implications $\phi_{A,i}(\mathbf{x}_i) \Rightarrow r_1(\mathbf{t}_1),\dots,\phi_{A,i}(\mathbf{x}_i) \Rightarrow r_k(\mathbf{t}_k)$.
2. (*Reinsert the hidden relations*)
   In each implication $\phi_{A,i}(\mathbf{x}_i) \Rightarrow r_k(\mathbf{t}_k)$ in S (obtained in the previous step), on the left-hand side of this implication replace each equation $(f_r(\mathbf{y}_i) \doteq \overline{1})$, where $\mathbf{y}_i \subseteq \mathbf{x}_i$ and f_r is the characteristic function of a hidden relational symbol $r \notin \mathcal{A}$ (introduced by the algorithm *Compose*), by the atom $r(\mathbf{y}_i)$.
3. (*Transformation into operad's operations*)
   Consider $S = \{\chi_1,\dots,\chi_m\}$ with χ_i being a view-based query mapping (implication from conjunctive query over $\mathcal{A}$ into a relation of $\mathcal{B}$), $q_{Ai}(\mathbf{x}_i) \Rightarrow r'(\mathbf{t}_i)$, where $S_q = (r_{i_1,1},\dots,r_{i_k,k})$ is the ordered list of relational symbols as they appear (from left to right) in the conjunctive query $q_{Ai}(\mathbf{x}_i)$. Then replace each such implication χ_i in S with the operad's operations $q_i \in O(r_{i_1,1},\dots,r_{i_k,k},r')$ where $(r_{i_1,1},\dots,r_{i_k,k}) = S_q$ and q_i is the expression obtained from $q_{Ai}(\mathbf{x}_i)$ by replacing

each relational symbol $r_{i_j,j}$ by the place symbol $(\_)_j$ (while replacing $\neg r_{i_j,j}$ by the place symbol $\neg(\_)_j$) and by replacing r' by unlabeled $(\_)$.

We will represent each operation q_i by a composition of two operations $v_i \cdot q_{A,i}$, where $q_{A,i} \in O(r_{i_1,1}, \ldots, r_{i_k,k}, r_q)$ is the same expression as q_i above where r_q is a relational symbol of the same type as r', and $v_i \in O(r_q, r')$ is the expression $(\_)_1(\mathbf{y}_i) \Rightarrow (\_)(\mathbf{y}_i)$ where $\mathbf{y}_i$ is the tuple of variables of the atoms $r'(\mathbf{y}_i)$ and $r_q(\mathbf{y}_i)$.

In the simplest case when χ_i is a tautology $r(\mathbf{x}_i) \Rightarrow r(\mathbf{x}_i)$, χ_i is replaced in S by $q_i = q_{A,i} = v_i = 1_r \in O(r, r)$ (that is, by the identity mapping expression $(\_)_1(\mathbf{x}_i) \Rightarrow (\_)(\mathbf{x}_i)$ if $\mathbf{x}_i$ is not empty; and by the expression $(\_)_1 \Rightarrow (\_)$ for the identity operation $1_{r_\emptyset} \in O(r_\emptyset, r_\emptyset)$, otherwise).

4. (*Construct operad's operations*)
Add the 'empty operation' operad $1_{r_\emptyset} \in O(r_\emptyset, r_\emptyset)$ in S, represented by the mapping expression $(\_)_1 \Rightarrow (\_)$.
Return The set of abstract operad's operations denoted by an (mapping) arrow $\mathbf{M}_{AB} = S : \mathcal{A} \to \mathcal{B}$.

In what follows, we will use the term 'mapping-operad' from such a set of abstract operad operations. The reason why we insert the identity 'empty operad's operation' $1_{r_\emptyset} \in O(r_\emptyset, r_\emptyset)$ in $\mathbf{M}_{AB} = S$ is that, as we will see, we can have the Tarski's interpretations of database mappings where we do not transfer any information from the source to the target database (i.e., with the empty information flux of such a mapping). Consequently, $1_{r_\emptyset}$ is always an element of the set of operad's operations obtained from a given SOtgd. As we will see in the algorithm of decomposition of SOtgds, we will have cases where the mappings are generated with SOtgd equal to the tautology $r_\emptyset \Rightarrow r_\emptyset$.

Example 7 In the most trivial case when $\mathcal{M}_{AB} = \{r_\emptyset \Rightarrow r_\emptyset\}$ (that is, when there is no effective mapping between the schemas $\mathcal{A}$ and $\mathcal{B}$) where $r_\emptyset$ is a nullary predicate symbol (i.e., the *truth* propositional letter in FOL, Definition 1), so that SOtgd is a banal tautology $r_\emptyset \Rightarrow r_\emptyset$ and, consequently, $\mathcal{M}_{AB}$ is always satisfied, we obtain that *MakeOperads*$(\mathcal{M}_{AB}) = \{1_{r_\emptyset}\}$. The particular cases for such trivial mappings are the following:

$$\mathcal{M}_{AA_\emptyset} = \{r_\emptyset \Rightarrow r_\emptyset\} : \mathcal{A} \to \mathcal{A}_\emptyset,$$
$$\mathcal{M}_{A_\emptyset A} = \{r_\emptyset \Rightarrow r_\emptyset\} : \mathcal{A}_\emptyset \to \mathcal{A}, \text{ and}$$
$$Id_{A_\emptyset} = \{r_\emptyset \Rightarrow r_\emptyset\} : \mathcal{A}_\emptyset \to \mathcal{A}_\emptyset, \text{ the identity mapping,}$$

where $\mathcal{A}_\emptyset = (\{r_\emptyset\}, \emptyset)$ is the empty database schema. It is easy to verify that

$$\mathit{MakeOperads}\big(\{r_\emptyset \Rightarrow r_\emptyset\}\big) = \{1_{r_\emptyset}\}.$$

Obviously, we cannot have the Tarski's interpretations for the empty schema $\mathcal{A}_\emptyset$, and we will need to define its interpretation in an ad hoc way in Sect. 2.4.1 dedicated to R-algebras for the operads.

There is an obvious inverse transformation of a given mapping-operad (a set of k-ary operations) $\mathbf{M}_{AB} = \{q_1, \ldots, q_n, 1_{r_\emptyset}\} : \mathcal{A} \to \mathcal{B}$, into the SOtgd of this mapping.

Inverse operads algorithm *InverseOperads*($\mathbf{M}_{AB}$)
Input. A mapping-operad $\mathbf{M}_{AB} = \{q_1, \ldots, q_n, 1_{r_\emptyset}\} : \mathcal{A} \to \mathcal{B}$.
Output. Mapping $\mathcal{M}_{AB}$ from $\mathcal{A}$ into $\mathcal{B}$ represented by the SOtgd.

1. (*Transform operads into logical formulae*)
Initialize S to be the empty set $\emptyset$.
For each k-ary mapping operad's operation $q_i \in O(r_{i,1}, \ldots, r_{i,k}, r')$, for $1 \le i \le n$, such that $q_i \neq 1_{r_\emptyset}$, add in S the logical sentence $\forall \mathbf{x}_i \phi_i$ where $\mathbf{x}_i$ is the tuple of all variables in the expression of the operad's operation q_i and ϕ_i is the logical formula where each labeled place symbol $(\_\,)_m$, for $1 \le m \le k$, is replaced by the relational symbol $r_{i,k}$ and each unlabeled place symbol $(\_\,)$ is replaced by $r' \in \mathcal{B}$.
2. (*Elimination of relational symbols not in* $\mathcal{A}$)
In each formula $\forall \mathbf{x}_i \phi_i \in S$ where ϕ is an implication $\phi_{A,i}(\mathbf{x}_i) \Rightarrow r'(\mathbf{t}_i)$, on the left-hand side of this implication replace each atom $r(\mathbf{y}_i)$ with $r \notin \mathcal{A}$, where $\mathbf{y}_i \subseteq \mathbf{x}_i$, by the equation $(f_r(\mathbf{y}_i) \doteq \bar{1})$, where f_r is the characteristic function of a hidden relational symbol $r \notin \mathcal{A}$.
3. (*Construct* SOtgd)
The SOtgd is a formula Φ equal to $r_\emptyset \Rightarrow r_\emptyset$ if S is empty; $\exists \mathbf{f}(\forall \mathbf{x}_1 \phi_1 \wedge \cdots \wedge \forall \mathbf{x}_n \phi_n)$ otherwise, where $\mathbf{f}$ is the tuple of all functional symbols in the formulae in S.
Return the logical mapping $\mathcal{M}_{AB} = \{\Phi\} : \mathcal{A} \to \mathcal{B}$.

Consequently, the algebraic formalism based on the mapping-operads is equivalent to the logical formalism based on SOtgds.

Example 8 Let us consider the following example (corresponding to Example 4.3 in [5]) that explains in detail the transformation of the logic into R-algebras, that is, the transformation of logical formulae used to define a schema mapping into the set of abstract operad's operations and their R-algebras.

Schema $\mathcal{A} = (S_A, \emptyset)$ consists of a unary relation `EmpAcme` that represents the employees of Acme, a unary relation `EmpAjax` that represents the employees of Ajax, and a unary relation `Local` that represents employees that work in the local office of their company. Schema $\mathcal{B} = (S_B, \emptyset)$ consists of a unary relation `Emp` that represents all employees, a unary relation `Local1` that is intended to be a copy of `Local`, and a unary relation `Over65` that is intended to represent people over age 65. Schema $\mathcal{C} = (S_C, \emptyset)$ consists of a binary relation `Office` that associates employees with office numbers and a unary relation `CanRetire` that represents employees eligible for retirement. Consider now the following schema mappings:

$$\mathcal{M}_{AB} = \{\forall x_e(\texttt{EmpAcme}(x_e) \Rightarrow \texttt{Emp}(x_e)) \\ \wedge \forall x_e(\texttt{EmpAjax}(x_e) \Rightarrow \texttt{Emp}(x_e)) \\ \wedge \forall x_p(\texttt{Local}(x_p) \Rightarrow \texttt{Local1}(x_p))\},$$

with $\mathbf{M}_{AB} = MakeOperads(\mathcal{M}_{AB}) = \{q_1^A, q_2^A, q_3^A, 1_{r_\emptyset}\}$ where:

1. The operation $q_1^A \in O(\texttt{EmpAcme}, \texttt{Emp})$ is the expression $(\_)_1(x_e) \Rightarrow (\_)(x_e)$;
2. The operation $q_2^A \in O(\texttt{EmpAjax}, \texttt{Emp})$ is the expression $(\_)_1(x_e) \Rightarrow (\_)(x_e)$;
3. The operation $q_3^A \in O(\texttt{Local}, \texttt{Local1})$ is the expression $(\_)_1(x_p) \Rightarrow (\_)(x_p)$; and

$$\mathcal{M}_{BC} = \{\exists f_1 (\forall x_e((\texttt{Emp}(x_e) \wedge \texttt{Local1}(x_e)) \Rightarrow \texttt{Office}(x_e, f_1(x_e)))$$
$$\wedge \forall x_e((\texttt{Emp}(x_e) \wedge \texttt{Over65}(x_e)) \Rightarrow \texttt{CanRetire}(x_e)))\},$$

with $\mathbf{M}_{BC} = MakeOperads(\mathcal{M}_{BC}) = \{q_1^B, q_2^B, 1_{r_\emptyset}\}$ where:

4. The operation $q_1^B \in O(\texttt{Emp}, \texttt{Local1}, \texttt{Office})$ is the expression $((\_)_1(x_e) \wedge (\_)_2(x_e)) \Rightarrow (\_)(x_e, f_1(x_e))$;
5. The operation $q_2^B \in O(\texttt{Emp}, \texttt{Over65}, \texttt{CanRetire})$ is the expression $((\_)_1(x_e) \wedge (\_)_2(x_e)) \Rightarrow (\_)(x_e)$.

Then, by applying the new algorithm for composition, we obtain the composed mapping $\mathcal{M}_{AC} = Compose(\mathcal{M}_{AB}, \mathcal{M}_{BC})$ equal to

$$\mathcal{M}_{AC} = \{\exists f_1 \exists f_2 \exists f_{Over65}($$
$$\forall x_e((\texttt{EmpAcme}(x_e) \wedge \texttt{Local}(x_e)) \Rightarrow \texttt{Office}(x_e, f_1(x_e))$$
$$\wedge \forall x_e((\texttt{EmpAjax}(x_e) \wedge \texttt{Local}(x_e)) \Rightarrow \texttt{Office}(x_e, f_2(x_e)))$$
$$\wedge \forall x_e((\texttt{EmpAcme}(x_e) \wedge (f_{Over65}(x_e) \doteq \overline{1})) \Rightarrow \texttt{CanRetire}(x_e))$$
$$\wedge \forall x_e((\texttt{EmpAjax}(x_e) \wedge (f_{Over65}(x_e) \doteq \overline{1})) \Rightarrow \texttt{CanRetire}(x_e))))\}.$$

Then, by a transformation into abstract operad's operations, we obtain

$$\mathbf{M}_{AC} = MakeOperads(\mathcal{M}_{AC}) = \{q_1, q_2, q_3, q_4, 1_{r_\emptyset}\}$$

where:

6. The operation $q_1 \in O(\texttt{EmpAcme}, \texttt{Local}, \texttt{Office})$ is the expression
$$((\_)_1(x_e) \wedge (\_)_2(x_e)) \Rightarrow (\_)(x_e, f_1(x_e));$$
7. The operation $q_2 \in O(\texttt{EmpAjax}, \texttt{Local}, \texttt{Office})$ is the expression
$$((\_)_1(x_e) \wedge (\_)_2(x_e)) \Rightarrow (\_)(x_e, f_2(x_e));$$
8. The operation $q_3 \in O(\texttt{EmpAcme}, \texttt{Over65}, \texttt{CanRetire})$ is the expression
$$((\_)_1(x_e) \wedge (\_)_2(x_e)) \Rightarrow (\_)(x_e);$$
9. The operation $q_4 \in O(\texttt{EmpAjax}, \texttt{Over65}, \texttt{CanRetire})$ is the expression
$$((\_)_1(x_e) \wedge (\_)_2(x_e)) \Rightarrow (\_)(x_e).$$

Note that in operad's operations q_3 and q_4, we have the relational symbol `Over65` that is not in the source schema $\mathcal{A}$ (it was a hidden relation in the SOtgd). Consequently, the operad's operations generally can have the relational symbols of the schemas that are different from the source and target schemas. Let us consider the case when no relational symbols of the source schema are used on the left-hand side of the tgd's implications:

Example 9 Let us consider the previous Example 8 where the mapping $\mathcal{M}_{AB}$ is reduced to $\mathcal{M}_{AB} = \{\forall x_p(\texttt{Local}(x_p) \Rightarrow \texttt{Local1}(x_p))\}$ with

$$\mathbf{M}_{AB} = \mathit{MakeOperads}(\mathcal{M}_{AB}) = \{q_1^A, 1_{r\emptyset}\},$$

where the operation $q_1^A \in O(\texttt{Local}, \texttt{Local1})$ is the expression $(\_\ )_1(x_e) \Rightarrow (\_\ )(x_e)$.

Then, by applying the new algorithm for composition, we obtain the composed mapping $\mathcal{M}_{AC} = \mathit{Compose}(\mathcal{M}_{AB}, \mathcal{M}_{BC})$ equal to

$$\begin{aligned}\mathcal{M}_{AC} = \{&\exists f_1 \exists f_{Emp} \exists f_{Over65}(\\ &\forall x_e(((f_{Emp}(x_e) \doteq \overline{1}) \wedge \texttt{Local}(x_e)) \Rightarrow \texttt{Office}(x_e, f_1(x_e)))\\ &\wedge \forall x_e(((f_{Emp}(x_e) \doteq \overline{1}) \wedge (f_{Over65}(x_e) \doteq \overline{1})) \Rightarrow \texttt{CanRetire}(x_e))))\}.\end{aligned}$$

Note that in the second implication of the composed SOtgd, on the left-hand side of this implication, we have no any relational symbol from $\mathcal{A}$, so that in step 2 of the algorithm *MakeOperad* we will introduce the relational symbols `Emp` and `Over65` in this implication.

Then, by transformation into operads, we obtain

$$\mathbf{M}_{AC} = \mathit{MakeOperads}(\mathcal{M}_{AC}) = \{q_1, q_2, 1_{r\emptyset}\}$$

where:

1. The operation $q_1 \in O(\texttt{Emp}, \texttt{Local}, \texttt{Office})$ is the expression

$$((\_\ )_1(x_e) \wedge (\_\ )_2(x_e)) \Rightarrow (\_\ )(x_e, f_1(x_e));$$

2. The operation $q_2 \in O(\texttt{Emp}, \texttt{Over65}, \texttt{CanRetire})$ is the expression

$$((\_\ )_1(x_e) \wedge (\_\ )_2(x_e)) \Rightarrow (\_\ )(x_e).$$

Note that the operation q_2 means that we do not transfer any tuple from $\mathcal{A}$ into $\mathcal{C}$ (but only the tuples from $\mathcal{B}$ into $\mathcal{C}$). Consequently, the information flux transmitted from $\mathcal{A}$ into $\mathcal{C}$ by this operation has to be empty, as we will show in the next example.

Based on the equivalence of representation of schema mappings by *SOtgd* and by mapping-operads, we can introduce the composition of mapping operads analogously to the composition of schema mappings:

Definition 9 For any given schema mapping $\mathcal{M}_{AB} : \mathcal{A} \to \mathcal{B}$ and an *atomic* schema mapping $\mathcal{M}_{BC} : \mathcal{B} \to \mathcal{C}$, we can define the corresponding mapping-operads

$$\mathbf{M}_{AB} = MakeOperads(\mathcal{M}_{AB}) = \{q_1^A, \dots, q_n^A, 1_{r_\emptyset}\} : \mathcal{A} \to \mathcal{B},$$
$$\mathbf{M}_{BC} = MakeOperads(\mathcal{M}_{BC}) = \{q_1^B, \dots, q_m^B, 1_{r_\emptyset}\} : \mathcal{B} \to \mathcal{C},$$

and their composition

$$\mathbf{M}_{BC} \circ \mathbf{M}_{AB}$$
$$\triangleq \{1_{r_\emptyset}\} \bigcup_{1 \le i \le m} \{q_i = q_i^B \cdot (q_{i1}^A, \dots, q_{ij}^A) \mid q_i^B \in O(r_{B,i1}, \dots, r_{B,ij}, r_{C,i}) \in \mathbf{M}_{BC}$$
$$\text{and } (q_{ik}^A \in O(r_{k1}, \dots, r_{kl}, r_{B,ik}) \text{ if } q_{ik}^A \in \mathbf{M}_{AB};\ q_{ik}^A = 1_{r_{B,ik}} \text{ otherwise })_{1 \le k \le j}\}.$$

Note that each abstract operation q_i in the composed operad mapping $\mathbf{M}_{AC} = \mathbf{M}_{BC} \circ \mathbf{M}_{AB}$ is represented by the operation composition $q^B \cdot (q_1^A, \dots, q_n^A)$ where $q^B \in \mathbf{M}_{BC}$ and each q_j^A, for $1 \le j \le n$, is an operation in $\mathbf{M}_{AB}$ or an identity operation for relations in $\mathcal{B}$. Let us show that the transformation of the SOtgd of a given mapping into operads is well defined and that the properties of the mapping-operads obtained by the algorithm *MakeOperad* satisfy the general properties of operads in Definition 8.

Proposition 1 *The transformation by the algorithm MakeOperads of SOtgds of the schema mappings into the mapping-operads is well defined and satisfies the general properties of operads in Definition* 8.

Proof Let us show the following mapping-operads properties required by Definition 8:

1. There exist the identity mappings that are transformed into the identity operad's operations for each relational symbol (as required by point 3 in Definition 8).
   In fact, for any relational symbol r we can define a database schema $\mathcal{A} = (\{r\}, \emptyset)$ with only this relation and its identity mapping $Id_{\mathcal{A}} : \mathcal{A} \to \mathcal{A}$, where $Id = \forall \mathbf{x}(r(\mathbf{x}) \Rightarrow r(\mathbf{x}))$. Then, the identity operad's operation $1_r \in O(r, r)$ is defined by $MakeOperad(Id_{\mathcal{A}}) = \{(\_)_1 \Rightarrow (\_), 1_{r_\emptyset}\} = \{1_r, 1_{r_\emptyset}\}$.
2. Let us define the composition of mapping-operads '$\cdot$' so that it satisfies the properties in point 2 of Definition 8, namely for any $f \in O(r_1, \dots, r_k, r)$ and any $g_1 \in O(r_{11}, \dots, r_{1i_1}, r_1), \dots, g_k \in O(r_{k1}, \dots, r_{ki_k}, r_k)$, an element $f \cdot (g_1, \dots, g_k) \in O(r_{11}, \dots, r_{1i_1}, \dots, r_{k1}, \dots, r_{ki_k}, r)$.
   Define the database schemas $\mathcal{A} = (S_A, \emptyset)$ with $S_A = \bigcup_{1 \le j \le k} \{r_{j1}, \dots, r_{ji_j}\}$, $\mathcal{B} = (S_B, \emptyset)$ with $S_B = \{r_1, \dots, r_k\}$, and $\mathcal{C} = (\{r\}, \emptyset)$, with the mappings $\mathbf{M}_{BC} = \{f, 1_{r_\emptyset}\} : \mathcal{B} \to \mathcal{C}$ and $\mathbf{M}_{AB} = \{g_1, \dots, g_k, 1_{r_\emptyset}\} : \mathcal{A} \to \mathcal{B}$.
   Hence, the operad's operation composition '$\cdot$', based on Definition 9, is defined by $\{f \cdot (g_1, \dots, g_k), 1_{r\emptyset}\} = \mathbf{M}_{BC} \circ \mathbf{M}_{AB}$.

3. The permutation in a given operad's operation $f \in O(r_1, \ldots, r_k, r)$ of its relational symbols in $\{r_1, \ldots, r_k\}$ is possible on the left-hand side of implication of the expression $e \Rightarrow (\_)$ because the atoms of these relational symbols are connected by the logical conjunction $\wedge$ which is a commutative operation. Thus, all properties for the permutations in point 4 of Definition 8 are satisfied for the mapping-operads as well. □

Corollary 3 *The database schemas and the composition of their operad-mappings can be represented by a sketch category.*

Proof It is a direct result of Corollary 2, the equality of these two representations, and from Proposition 1. From the fact that for each schema $\mathcal{A} = (S_A, \Sigma_A)$ there is its identity mapping-operad $\mathbf{Id}_A : \mathcal{A} \to \mathcal{A}$ such that $\mathbf{Id}_A = \{1_r | r \in S_A\} \cup \{1_{r_\emptyset}\}$, and that the composition of mapping-operads is associative, we conclude that the set of schemas (the objects) and the set of mapping-operads (the morphisms) can define a sketch category. The associativity of composition is a consequence of the associativity of the operads-composition '·' defined in point 4(a) of Definition 8. Thus, $\mathbf{M}_{CD} \circ (\mathbf{M}_{BC} \circ \mathbf{M}_{AB}) = (\mathbf{M}_{CD} \circ \mathbf{M}_{BC}) \circ \mathbf{M}_{AB}$. □

2.4.1 R-Algebras, Tarski's Interpretations and Instance-Database Mappings

The theory of typed operads represents the syntax of the database mapping, translated from the schema-database logic into the algebraic framework. The semantic part of the operad's algebra theory, corresponding to the semantics of FOL based on Tarski's interpretations, is represented by the R-algebras which are particular interpretations of the operads.

Let us now define the R-algebra of a database mapping-operad based on homotopy theory [1, 3, 7], where its abstract operations are represented by actual functions:

Definition 10 For a given universe of values $\mathcal{U}$ and R-operad O, an R-algebra α consists of (here '\' denotes the set-difference):

1. A set $\alpha(r)$ for any $r \in \mathbb{R}$, which is a set of tuples (relation), with the empty relation $\alpha(r_\emptyset) = \bot = \{\langle\rangle\}$, unary universe-relation $\alpha(r_\infty) = \{\langle d \rangle \mid d \in \mathcal{U}\}$, and binary relation $\alpha(r_\top) = R_=$ for the "equality" type $r_\top$. The α^* is the extension of α to a list of symbols $\alpha^*(\{r_1, \ldots, r_k\}) \triangleq \{\alpha(r_1), \ldots, \alpha(r_k), \alpha(r_\emptyset)\}$.
2. A mapping function $\alpha(q_i) : R_1 \times \cdots \times R_k \longrightarrow \alpha(r)$ for any $q_i \in O(r_1, \ldots, r_k, r)$, where for each $1 \le i \le k$, $R_i = \mathcal{U}^{ar(r_i)} \backslash \alpha(r_i)$ if the place symbol $(\_)_i \in q_i$ is preceded by negation operator $\neg$; $\alpha(r_i)$ otherwise.

   Consequently, $\alpha^*(\{q_1, \ldots, q_k\}) \triangleq \{\alpha(q_1), \ldots, \alpha(q_k)\}$, and

   (a) For any $r \in \mathbb{R}$, $\alpha(1_r)$ acts as an identity function on relation $\alpha(r)$. The $q_\bot = \alpha(1_{r_\emptyset}) : \bot \to \bot$ is the empty function (with $q_\bot(\langle\rangle) = \langle\rangle$ for the empty tuple $\langle\rangle$) with the empty graph).

(b) For the associative composition '·',

$$\alpha\big(q \cdot (q_1, \ldots, q_k)\big) = \alpha(q)\big(\alpha(q_1) \times \cdots \times \alpha(q_k)\big)$$

(c) For any $q \in O(r_1, \ldots, r_k, r)$ and a permutation $\sigma \in \mathbb{R}_k$, $\alpha(q\sigma) = \alpha(q)\sigma$, where σ acts on the function $\alpha(q)$ on the right by permuting its arguments.

3. We introduce two functions ∂_0 and ∂_1 such that for any $\alpha(q), q \in O(r_1, \ldots, r_k, r)$, $\partial_0(q) = \{r_1, \ldots, r_k\}$, $\partial_0(\alpha(q)) = \alpha^*(\partial_0(q)) = \{\alpha(r_1), \ldots, \alpha(r_k)\}$, $\partial_1(q) = \{r\}$, and $\partial_1(\alpha(q)) = \alpha^*(\partial_1(q)) = \{\alpha(r)\}$.

Remark In what follows, an instance database $A = \alpha^*(S_A)$ of a schema $\mathcal{A} = (S_A, \Sigma_A)$ will be denoted by $\alpha^*(\mathcal{A})$ as well. The empty schema is denoted by $\mathcal{A}_\emptyset = (\{r_\emptyset\}, \emptyset)$ so that for any α, from point 1 of this definition, $\bot^0 = \alpha^*(\mathcal{A}_\emptyset) = \alpha^*(S_{A_\emptyset}) = \{\alpha(r_\emptyset)\} = \{\bot\}$ is the empty instance-database.

Note that this empty database is the zero object of the **DB** category defined in the next chapter, and this is the reason that in the computation of α^* in point 1 of this definition we added $\alpha(r_\emptyset)$ as well, so that we will have for each schema $\mathcal{A} = (S_A, \Sigma_A)$ that $\alpha(r_\emptyset) = \bot \in \alpha^*(S_A) = \alpha^*(\mathcal{A})$.

Consequently, we can think of an operad as a simple sort of theory, used to define a schema mapping between databases, and its R-algebras as models of this theory used to define the mappings between instance-databases, where an R-algebra α is considered as an interpretation of relational symbols of a given database schema. For the empty operation $1_{r_\emptyset} \in \mathcal{M}_{AB}$ of a mapping $\mathcal{M}_{AB}$, we have, by Definition 10, that its interpretation is prefixed by the empty function $q_\bot : \bot \to \bot$, while for the rest of operad's operations in $\mathcal{M}_{AB}$ we have to define their formal semantics. What we need is to specify the subsets of R-algebras that contain well defined *mapping-interpretations* for the operad's operations obtained from schema mappings by the *MakeOperads* algorithm. They are particular extensions of Tarski's interpretations of the database mappings (when we eliminate the existential quantifiers over functions from SOtgds of a given schema database mapping system then we obtain the FOL with its Tarski's interpretations I_T, and their extensions I_T^* to all formulae introduced in Sect. 1.3) as follows:

Definition 11 Let $\phi_{Ai}(\mathbf{x}) \Rightarrow r_B(\mathbf{t})$ be an implication χ in a normalized SOtgd $\exists \mathbf{f}(\Psi)$ (where Ψ is an FOL formula) of the mapping $\mathcal{M}_{AB}$, let $\mathbf{t}$ be a tuple of terms with variables in $\mathbf{x} = \langle x_1, \ldots, x_m \rangle$, and $q_i \in$ *MakeOperads*$(\mathcal{M}_{AB})$ be the operad's operation of this implication obtained by *MakeOperads* algorithm, equal to the expression $(e \Rightarrow (\_\ )(\mathbf{t})) \in O(r_1, \ldots, r_k, r_B)$, where $q_i = v_i \cdot q_{A,i}$ with $q_{A,i} \in O(r_1, \ldots, r_k, r_q)$ and $v_i \in O(r_q, r_B)$ such that for a new relational symbol r_q, $ar(r_q) = ar(r_B) \geq 1$.

Let S be an empty set and $e[(\_\ )_n / r_n]_{1 \leq n \leq k}$ be the formula obtained from expression e where each place-symbol $(\_\ )_n$ is substituted by the relational symbol r_n for $1 \leq n \leq k$. Then do the following as far as possible: For each two relational symbols r_j, r_n in the formula $e[(\_\ )_n / r_n]_{1 \leq n \leq k}$ such that the j_hth *free variable* (which is not an argument of a functional symbol) in the atom $r_j(\mathbf{t}_j)$ is equal to the n_hth free variable in the atom $r_n(\mathbf{t}_n)$ (both atoms in $e[(\_)_n/r_n]_{1\le n\le k}$), we insert the set $\{(j_h, j), (n_h, n)\}$ as one element of S. At the end, S is the set of sets that contain the pairs of mutually equal free variables.

An R-algebra α is a *mapping-interpretation* of $\mathcal{M}_{AB} : \mathcal{A} \to \mathcal{B}$ if it is an extension of a Tarski's interpretation I_T (in Definition 1, of all predicate and functional symbols in FOL formula Ψ, with I_T^* being its extension to all formulae), and if for each $q_i \in MakeOperads(\mathcal{M}_{AB})$ it satisfies the following:

1. For each relational symbol $r_i \neq r_\emptyset$ in $\mathcal{A}$ or $\mathcal{B}$, $\alpha(r_i) = I_T(r_i)$.
2. We obtain a function $f = \alpha(q_{A,i}) : R_1 \times \cdots \times R_k \to \alpha(r_q)$, where for each $1 \le i \le k$, $R_i = \mathcal{U}^{ar(r_i)} \backslash \alpha(r_i)$ if the place symbol $(\_)_i \in q_i$ is preceded by negation operator $\neg$; $\alpha(r_i)$ otherwise, such that for every $\mathbf{d}_i \in R_i$:

$$f\big(\langle \mathbf{d}_1, \ldots, \mathbf{d}_k\rangle\big) = g^*(\mathbf{t}) = \big\langle g^*(t_1), \ldots, g^*(t_{ar(r_B)})\big\rangle$$

if $\bigwedge\{\pi_{j_h}(\mathbf{d}_j) = \pi_{n_h}(\mathbf{d}_n) \mid \{(j_h, j), (n_h, n)\} \in S\}$ is true and the assignment g satisfies the formula $e[(\_)_n/r_n]_{1\le n\le k}$; $\langle\rangle$ (empty tuple) otherwise, where the assignment $g : \{x_1, \ldots, x_m\} \to \mathcal{U}$ is defined by the tuple of values $\langle g(x_1), \ldots, g(x_m)\rangle = Cmp(S, \langle \mathbf{d}_1, \ldots, \mathbf{d}_k\rangle)$, and its extension g^* to all terms is such that for any term $f_i(t_1, \ldots, t_n)$:

$$g^*\big(f_i(t_1, \ldots, t_n)\big) = I_T(f_i)\big(g^*(t_1), \ldots, g^*(t_n)\big) \text{ if } n \ge 1; \quad I_T(f_i) \text{ otherwise.}$$

The algorithm *Cmp* (compacting the list of tuples by eliminating the duplicates defined in S) is defined as follows:

Input. A set S of joined (equal) variables defined above, and a list of tuples $\langle \mathbf{d}_1, \ldots, \mathbf{d}_k\rangle$.

Initialize $\mathbf{d}$ to $\mathbf{d}_1$. Repeat consecutively the following, for $j = 2, \ldots, k$:

Let $\mathbf{d}_j$ by a tuple of values $\langle v_1, \ldots, v_{j_n}\rangle$, then for $i = 1, \ldots, j_n$ repeat consecutively the following:

$\mathbf{d} = \mathbf{d}\ \& v_i$ if there does not exist an element $\{(j_h, j), (n_h, n)\}$ in S such that $j \le n$; $\mathbf{d}$, otherwise.

(The operation of concatenation '&' appends the value v_i at the end of tuple $\mathbf{d}$)

Output. The tuple $Cmp(S, \langle \mathbf{d}_1, \ldots, \mathbf{d}_k\rangle) = \mathbf{d}$.

3. $\alpha(r_q)$ is equal to the image of the function f in point 2 above.
4. The function $h = \alpha(v_i) : \alpha(r_q) \to \alpha(r_B)$ such that for each $\mathbf{b} \in \alpha(r_q)$, $h(\mathbf{b}) = \mathbf{b}$ if $\mathbf{b} \in \alpha(r_B)$; empty tuple $\langle\rangle$ otherwise.

Note that the formulae $\phi_{Ai}(\mathbf{x})$ and expression $e[(\_)_n/r_n]_{1\le n\le k}$ are logically equivalent, with the only difference that the atoms with characteristic functions $f_r(\mathbf{t}) \doteq \overline{1}$ in the first formula are substituted by the atoms $r(\mathbf{t})$, based on the fact that the assignment g satisfies $r(\mathbf{t})$ iff $g^*(f_r(\mathbf{t})) = \overline{f}_r(g^*(\mathbf{t})) = 1$ (and for every assignment $g(\overline{1}) = 1$), where $\overline{f}_r : \mathcal{U}^{ar(r)} \to \{0, 1\}$ is the characteristic function of a relation $\alpha(r)$ such that for each tuple $\mathbf{c} \in \mathcal{U}^{ar(r)}$, $\overline{f}_r(\mathbf{c}) = 1$ if $\mathbf{c} \in \alpha(r)$; 0 otherwise.

Example 10 Let us show how we construct the set S and the compacting of tuples given by Definition 11 above:

Let us consider an operad $q_i \in MakeOperads(\mathcal{M}_{AB})$, obtained from a normalized implication $\phi_{Ai}(\mathbf{x}) \Rightarrow r_B(\mathbf{t})$ in $\mathcal{M}_{AB}$, $((y \doteq f_1(x,z)) \wedge r_1(x,y,z) \wedge r_2(v,x,w) \wedge (f_{r_3}(y,z,w',w) \doteq \overline{1})) \Rightarrow r_B(x,z,w,f_2(v,z))$, so that q_i is equal to the expression $(e \Rightarrow (\_)(\mathbf{t})) \in O(r_1,r_2,r_3,r_B)$, where $\mathbf{x} = \langle x,y,z,v,w,w'\rangle$ (the ordering of variables in the atoms (with database relational symbols) from left to right), $\mathbf{t} = \langle x,z,w,f_2(v,z)\rangle$, and the expression e equal to $(y \doteq f_1(x,z)) \wedge (\_\ )_1(\mathbf{t}_1) \wedge (\_\ )_2(\mathbf{t}_2) \wedge (\_\ )_3(\mathbf{t}_3)$, with $\mathbf{t}_1 = \langle x,y,z\rangle$, $\mathbf{t}_2 = \langle v,x,w\rangle$ and $\mathbf{t}_3 = \langle y,z,w',w\rangle$.

Consequently, we obtain

$$S = \{\{(1,1),(2,2)\},\{(2,1),(1,3)\},\{(3,1),(2,3)\},\{(3,2),(4,3)\}\}$$

that are the positions of duplicates (or joined variables) of x, y, z, and w, respectively.

Thus, for given tuples $\mathbf{d}_1 = \langle a_1,a_2,a_3\rangle \in \alpha(r_1)$, $\mathbf{d}_2 = \langle b_1,b_2,b_3\rangle \in \alpha(r_2)$ and $\mathbf{d}_3 = \langle c_1,c_2,c_3,c_4\rangle \in \alpha(r_3)$, the statement

$$\bigwedge\{\pi_{j_h}(\mathbf{d}_j) = \pi_{n_h}(\mathbf{d}_n) \mid \{(j_h,j),(n_h,n)\} \in S\}$$

is equal to

$$(\pi_1(\mathbf{d}_1) = \pi_2(\mathbf{d}_2)) \wedge (\pi_2(\mathbf{d}_1) = \pi_1(\mathbf{d}_3)) \wedge (\pi_3(\mathbf{d}_1) = \pi_2(\mathbf{d}_3)) \wedge (\pi_3(\mathbf{d}_2) = \pi_4(\mathbf{d}_3)),$$

which is true when $a_1 = b_2$, $a_2 = c_1$, $a_3 = c_2$, and $b_3 = c_4$.

The compacting of these tuples is equal to

$$\mathbf{d} = Cmp(S, \langle \mathbf{d}_1,\mathbf{d}_2,\mathbf{d}_3\rangle) = \langle a_1,a_2,a_3,b_1,b_3,c_3\rangle,$$

with the assignment to variables $[x/a_1]$, $[y/a_2]$, $[z/a_3]$, $[v,b_1]$, $[w/b_3]$, and $[w'/c_3]$.

That is, $\mathbf{d} = \mathbf{x}[x/a_1, y/a_2, z/a_3, v/b_1, w/b_3, w'/c_3]$ is obtained by this assignment g to the tuple of variables $\mathbf{x}$, so that the sentence $e[(\_\ )_n/r_n]_{1\le n\le k}/g$ is well defined and equal to

$$(a_2 = I_T(f_1)(a_1,a_3)) \wedge r_1(a_1,a_2,a_3) \wedge r_2(b_1,a_1,b_3) \wedge r_3(a_2,a_3,c_3,b_3),$$

that is, to

$$(a_2 = I_T(f_1)(a_1,a_3)) \wedge r_1(\mathbf{d}_1) \wedge r_2(\mathbf{d}_2) \wedge r_3(\mathbf{d}_3),$$

and if this formula is satisfied by such an assignment g, i.e., $I_T^*(e[(\_\ )_n/r_n]_{1\le n\le k}/g) = 1$, then

$$\begin{aligned} f(\langle \mathbf{d}_1,\mathbf{d}_2,\mathbf{d}_3\rangle) &= g^*(\mathbf{t}) = \langle g(x), g(z), g(w), g^*(f_2(v,z))\rangle \\ &= \langle a_1, a_3, b_3, I_T(f_2)(b_1,a_3)\rangle, \end{aligned}$$

for a given Tarski's interpretation I_T, where I_T^* is the extension of I_T to all FOL formulae, as defined in Sect. 1.3.

If $\mathcal{M}_{AB}$ is satisfied by the mapping-interpretation α, this value of $f(\langle \mathbf{d}_1, \mathbf{d}_2, \mathbf{d}_3\rangle)$ corresponds to the truth of the normalized implication in the SOtgd of $\mathcal{M}_{AB}$, $\phi_{Ai}(\mathbf{x}) \Rightarrow r_B(\mathbf{t})$ for the assignment g derived by substitution $[\mathbf{x}/\mathbf{d}]$, when $\phi_{Ai}(\mathbf{x})/g$ is true. Hence, $r_B(\mathbf{t})/g$ is equal to $r_B(\langle a_1, a_3, b_3, I_T(f_2)(b_1, a_3)\rangle)$, i.e., to $r_B(f(\langle \mathbf{d}_1, \mathbf{d}_2, \mathbf{d}_3\rangle))$ and has to be true as well (i.e., $I_T^*(r_B(f(\langle \mathbf{d}_1, \mathbf{d}_2, \mathbf{d}_3\rangle))) = 1$ or, equivalently, $f(\langle \mathbf{d}_1, \mathbf{d}_2, \mathbf{d}_3\rangle) \in \alpha(r_B) = I_T(r_B)$).

Consequently, if $\mathcal{M}_{AB}$ is satisfied by a mapping-interpretation α (and hence $\alpha(v_i)$ is an injection function with $\alpha(r_q) \subseteq \alpha(r_B)$) then $f(\langle \mathbf{d}_1, \mathbf{d}_2, \mathbf{d}_3\rangle) \in \|r_B\|_{\alpha^*(\mathcal{B})}$, so that the function $f = \alpha(q_{A,i})$ represents the transferring of the tuples in relations of the source instance databases into the target instance database $B = \alpha^*(\mathcal{B})$, according to the SOtgd Φ of the mapping $\mathcal{M}_{AB} = \{\Phi\} : \mathcal{A} \to \mathcal{B}$.

In this way, for a given R-algebra α which satisfies the conditions for the mapping-interpretations in Definition 11, we translate a logical representation of database mappings, based on SOtgds, into an algebraic representation based on relations of the instance databases and the functions obtained from mapping-operads.

It is easy to verify that for a *query mapping* $\phi_{Ai}(\mathbf{x}) \Rightarrow r_B(\mathbf{t})$, a mapping-interpretation α is an R-algebra such that the relation $\alpha(r_q)$ is just equal to the image of the function $\alpha(q_{A,i})$. The mapping-interpretation of v_i is the transfer of information of this computed query into the relation $\alpha(r_B)$ of the database $\mathcal{B}$.

When α satisfies this query mapping $\phi_{Ai}(\mathbf{x}) \Rightarrow r_B(\mathbf{t})$, then $\alpha(r_q) \subseteq \alpha(r_B)$ (with proof in Proposition 4) and, consequently, the function $\alpha(v_i)$ is an injection, i.e., the *inclusion* of $\alpha(r_q)$ into $\alpha(r_B)$.

Proposition 2 *Let $\phi_{Ai}(\mathbf{x}) \Rightarrow r_B(\mathbf{t})$ be an implication in the normalized SOtgd $\exists \mathbf{f}\Psi$ of a mapping $\mathcal{M}_{AB} : \mathcal{A} \to \mathcal{B}$, where $\mathbf{t}$ is a tuple of terms with variables in $\mathbf{x} = \langle x_1, \dots, x_m\rangle$. Let $q_i \in MakeOperads(\mathcal{M}_{AB})$ be the operad's operation of this implication obtained by MakeOperads algorithm, equal to the expression $(e \Rightarrow (\_)(\mathbf{t})) \in O(r_1, \dots, r_k, r_B)$, and let S be the set of sets that contain the pairs of mutually equal (joined) free variables in q_i as specified in Definition* 11.

Then, for a given Tarski's interpretation of all FOL formulae in Ψ and, extended from it, a mapping-interpretation α in Definition 11 *(such that for each $1 \le i \le k$, $R_i = \mathcal{U}^{ar(r_i)} \backslash I_T(r_i)$ if the place symbol $(\_)_i \in q_i$ is preceded by negation operator $\neg$; $I_T(r_i)$ otherwise), the following is true:*

If for every tuple $\langle \mathbf{d}_1, \dots, \mathbf{d}_k\rangle \in R_1 \times \dots \times R_k$ such that $\bigwedge\{\pi_{j_h}(\mathbf{d}_j) = \pi_{n_h}(\mathbf{d}_n) \mid \{(j_h, j), (n_h, n)\} \in S\}$ is true, we have that $\alpha(q_i)(\langle \mathbf{d}_1, \dots, \mathbf{d}_k\rangle) \in I_T(r_B)$, then $I_T^*(\forall \mathbf{x}(\phi_{Ai}(\mathbf{x}) \Rightarrow r_B(\mathbf{t}))) = 1$, and vice versa.

Proof We have to show that for every $\langle \mathbf{d}_1, \dots, \mathbf{d}_k\rangle \in R_1 \times \dots \times R_k$ such that $\bigwedge\{\pi_{j_h}(\mathbf{d}_j) = \pi_{n_h}(\mathbf{d}_n) \mid \{(j_h, j), (n_h, n)\} \in S\}$ is true, with $\mathbf{d} = Cmp(S, \langle \mathbf{d}_1, \dots, \mathbf{d}_k\rangle)$ and the assignment g derived from the substitution $[\mathbf{x}/\mathbf{d}]$, $I_T^*(\phi_{Ai}(\mathbf{x})/g \Rightarrow r_B(\mathbf{t})/g) = 1$.

If $I_T^*(\phi_{Ai}(\mathbf{x})/g) = 0$ then it is satisfied. Thus, we have to consider only the following cases:

(a) $I_T^*(\phi_{Ai}(\mathbf{x})/g) = 1$, that is, when $I_T^*(e[(\_)_n/r_n]_{1\leq n\leq k}/g) = 1$ (see the comments after Definition 11).

From (a), the statement $\bigwedge\{\pi_{j_h}(\mathbf{d}_j) = \pi_{n_h}(\mathbf{d}_n) \mid \{(j_h, j), (n_h, n)\} \in S\}$ is true and, by Definition 11, $g^*(\mathbf{t}) = \alpha(q_i)(\langle \mathbf{d}_1, \ldots, \mathbf{d}_k\rangle) \in I_T(r_B)$ (from assumption of this proposition), that is, equivalently;

(b) $I_T^*(r_B(g^*(\mathbf{t}))) = I_T^*(r_B(\mathbf{t})/g) = 1$.

Thus, from (a) and (b), we obtain $I_T^*((\phi_{Ai}(\mathbf{x}) \Rightarrow r_B(\mathbf{t}))/g) = 1$, and hence from the fact that it holds for every $\langle \mathbf{d}_1, \ldots, \mathbf{d}_k\rangle \in R_1 \times \cdots \times R_k$, we obtain by generalization that $I_T^*(\forall\mathbf{x}(\phi_{Ai}(\mathbf{x}) \Rightarrow r_B(\mathbf{t}))) = 1$.

Vice versa, if $I_T^*(\forall\mathbf{x}(\phi_{Ai}(\mathbf{x}) \Rightarrow r_B(\mathbf{t}))) = 1$, then for each $\langle \mathbf{d}_1, \ldots, \mathbf{d}_k\rangle \in R_1 \times \cdots \times R_k$ such that $\bigwedge\{\pi_{j_h}(\mathbf{d}_j) = \pi_{n_h}(\mathbf{d}_n) \mid \{(j_h, j), (n_h, n)\} \in S\}$ is true, with $\mathbf{d} = Cmp(S, \langle \mathbf{d}_1, \ldots, \mathbf{d}_k\rangle)$ and the assignment g derived from the substitution $[\mathbf{x}/\mathbf{d}]$, $I_T^*((\phi_{Ai}(\mathbf{x}) \Rightarrow r_B(\mathbf{t}))/g) = 1$. Hence, if $I_T^*(\phi_{Ai}(\mathbf{x})/g) = 1$, that is, if $I_T^*(e[(\_)_n/r_n]_{1\leq n\leq k}/g) = 1$, then $I_T^*(r_B(\mathbf{t})/g) = I_T^*(r_B(g^*(\mathbf{t}))) = 1$, i.e., $g^*(\mathbf{t}) \in I_T(r_B)$. Moreover, from Definition 11, $g^*(\mathbf{t}) = \alpha(q_i)(\langle \mathbf{d}_1, \ldots, \mathbf{d}_k\rangle)$ and hence $\alpha(q_i)(\langle \mathbf{d}_1, \ldots, \mathbf{d}_k\rangle) \in I_T(r_B)$. □

Let us explain the mapping-interpretations by the following example:

Example 11 Let us consider Example 5.2 in [5] for the composition of SOtgds $\mathcal{M}_{AD} = \mathcal{M}_{BD} \circ \mathcal{M}_{AB}$, where $\mathcal{A} = (S_A, \emptyset)$ and $S_A = \{\texttt{Emp}(x_e)\}$ with a single unary relational symbol of employees, $\mathcal{B} = (S_B, \emptyset)$ and $S_B = \{\texttt{Emp1}(x_e), \texttt{Mgr}(x_e, x_m)\}$ with Emp1 intended as a copy of Emp and Mgr a binary relational symbol that associates each employee with a manager, and $\mathcal{D} = (S_D, \emptyset)$ and $S_D = \{\texttt{SelfMgr}(x_e)\}$ with a single unary relational symbol that is intended to store employees who are their own manager, and with mappings

$$\mathcal{M}_{AB} = \{\forall x_e(\texttt{Emp}(x_e) \Rightarrow \texttt{Emp1}(x_e)), \forall x_e(\texttt{Emp}(x_e) \Rightarrow \exists x_m \texttt{Mgr}(x_e, x_m))\},$$

or equivalently, by the following SOtgd:

$$\mathcal{M}_{AB} = \{\exists f_1(\forall x_e(\texttt{Emp}(x_e) \Rightarrow \texttt{Emp1}(x_e)) \wedge \forall x_e(\texttt{Emp}(x_e) \Rightarrow \texttt{Mgr}(x_e, f_1(x_e))))\},$$
$$\mathcal{M}_{BD} = \{\forall x_e(\texttt{Mgr}(x_e, x_e) \Rightarrow \texttt{SelfMgr}(x_e))\}.$$

Consequently, from the new algorithm for composition we obtain

$$\mathcal{M}_{AD} = \{\exists f_1(\forall x_e((\texttt{Emp}(x_e) \wedge (x_e \doteq f_1(x_e))) \Rightarrow \texttt{SelfMgr}(x_e)))\}.$$

Therefore,

$$\mathbf{M}_{AB} = MakeOperads(\mathcal{M}_{AB}) = \{q_1, 1_{r_\emptyset}\}$$

where $q_1 \in O(\texttt{Emp}, \texttt{SelfMgr})$ is an abstract operation represented by the expression $((\_)_1(x_e) \wedge (x_e \doteq f_1(x_e))) \Rightarrow (\_)(x_e)$ and by a composition $q_1 = v_1 \cdot q_{A,1}$ where $q_{A,1} \in O(\texttt{Emp}, r_q)$, $v_1 \in O(r_q, \texttt{SelfMgr})$, with a new relational symbol r_q for the relation obtained by the query $\texttt{Emp}(x_e) \wedge (x_e \doteq f_1(x_e))$.

Then, for a mapping-interpretation (an R-algebra) α such that it is a model (i.e., satisfies all the constraints in Σ_A) of the source schema $\mathcal{A} = (S_A, \Sigma_A)$ and defines its database instance $A = \alpha^*(S_A) = \{\alpha(r_i) \mid r_i \in S_A\}$ and, analogously, a model of $\mathcal{B}$ with $B = \alpha^*(S_B)$ such that it satisfies also the schema mapping $\mathcal{M}_{AB}$. Consequently, $(A, B) \in Inst(\mathcal{M}_{AB})$ and α satisfies the SOtgd of the mapping $\mathcal{M}_{AB}$ by fixing Tarski's interpretation for the functional symbol f_1 in this SOtgd (denoted by $I_T(f_1)$). We obtain the *function* $\alpha(q_1) = \alpha(v_1)(\alpha(q_{A,1})) : \alpha(\mathtt{Emp}) \rightarrow \alpha(\mathtt{SelfMgr})$, such that for any tuple $\langle a\rangle \in \alpha(\mathtt{Emp})$:

$$\alpha(q_{A,1})(\langle a\rangle) = \langle a\rangle \quad \text{if } a = I_T(f_1)(a); \qquad \langle\rangle \quad \text{otherwise.}$$

Relation $\alpha(r_q)$ is equal to the image of $\alpha(q_{A,1})$, so that for any $\langle a\rangle \in \alpha(r_q)$ we have

$$\alpha(v_1)(\langle a\rangle) = \langle a\rangle \quad \text{if } \langle a\rangle \in \alpha(\mathtt{SelfMgr}); \qquad \langle\rangle \quad \text{otherwise.}$$

Hence, the function $\alpha(v_1)$ is an injection for the inclusion $\alpha(r_q) \subseteq \alpha(\mathtt{SelfMgr})$.

We consider that every relation has the empty tuple as well; an empty relation in this case is considered as a relation that has only the empty tuple.

Remark Note that in the case when $(B, D) \in Inst(\mathcal{M}_{BD})$ as well (i.e., when the schema mapping $\mathcal{M}_{BD}$ is satisfied by B and D), the resulting mapping $\mathcal{M}_{AD}$ is also satisfied and, as a consequence, the function $\alpha(v_1)$ is an *inclusion*. If $\mathcal{M}_{AD}$ *is not* satisfied then $\alpha(v_1)$ is not an inclusion. Consequently, the function $\alpha(v_1)$ distinguishes when the mapping is satisfied or not, while the function $\alpha(q_{A,1})$ represents the computation of the query $\mathtt{Emp}(x_e) \wedge (x_e \doteq f_1(x_e))$ for the instance-database A (so that $\alpha(r_q) = \|\mathtt{Emp}(x_e) \wedge (x_e \doteq f_1(x_e))\|_A$ is the image of the function $\alpha(q_{A,1})$) that corresponds to the left-hand side of the implication of the operad's operation q_1.

The example above introduced the important properties of mapping-interpretations (R-algebras) and the way of recognizing when they are *models* of the schema mappings (that is, when they satisfy the schema mappings). Thus, we can formalize this property of mapping-interpretations by the following corollary:

Corollary 4 *Let $\mathcal{M}_{AB} : \mathcal{A} \rightarrow \mathcal{B}$ be a schema mapping. Then, for a given R-algebra α that is a mapping-interpretation, the function $\alpha(v_i)$ of each operad's operation $q_i = v_i \cdot q_{A,i} \in \mathbf{M}_{AB} = MakeOperads(\mathcal{M}_{AB})$ is an injection iff the mapping $\mathcal{M}_{AB}$ is satisfied by the instances $A = \alpha^*(\mathcal{A})$ and $B = \alpha^*(\mathcal{B})$ (i.e., when $(\alpha^*(\mathcal{A}), \alpha^*(\mathcal{B})) \in Inst(\mathcal{M}_{AB})$). That is, $\mathcal{M}_{AB}$ is satisfied iff for each $q_i = v_i \cdot q_{A,i} \in O(r_1, \ldots, r_k, r_B)$ in $\mathbf{M}_{AB}$, with $q_{A,i} \in O(r_1, \ldots, r_k, r_q)$, $v_i \in O(r_q, r_B)$, it holds that the image of the function $f = \alpha(q_{A,i}) : R_1 \times \cdots \times R_k \rightarrow \alpha(r_q)$ (where for each $1 \le i \le k$, $R_i = \mathcal{U}^{ar(r_i)} \backslash \alpha(r_i)$ if the place symbol $(\_)_i \in q_i$ is preceded by negation operator $\neg$; $\alpha(r_i)$ otherwise) is a subset of $\alpha(r_B)$, i.e., $im(f) \subseteq \alpha(r_B)$.*

Proof Let $\phi_{Ai}(\mathbf{x}) \Rightarrow r_B(\mathbf{t})$ be an implication χ in the normalized SOtgd of the mapping $\mathcal{M}_{AB}$ and $\mathbf{t}$ a tuple of terms with variables in $\mathbf{x} = \langle x_1, \ldots, x_m\rangle$. Then, based on Definition 11 for the mapping-interpretation α, we have:

If for a tuple $\mathbf{d} = \langle d_1, \ldots, d_m \rangle$ of values (in a given universe $\mathcal{U}$, which defines an assignment $g : \{x_1, \ldots, x_m\} \rightarrow \mathcal{U}$ such that $g(x_j) = d_j$, $1 \leq j \leq m$), the sentence $\phi_{Ai}(\mathbf{x})/g$ is true then $g^*(\mathbf{t})$ is a tuple in the relation $\alpha(r_B) \in B$ as it follows from Proposition 2.

Consequently, for the operad's operation $q_i \in O(r_1, \ldots, r_k, r_B)$ obtained from the implication χ, where $q_i = v_i \cdot q_{A,i}$ with $q_{A,i} \in O(r_1, \ldots, r_k, r_q)$ and $v_i \in O(r_q, r_B)$, the function $f = \alpha(q_{A,i}) : R_1 \times \cdots \times R_k \rightarrow \alpha(r_q)$ is well defined for each $\mathbf{d}_i \in R_i$ with $\mathbf{d} = Cmp(S, \langle \mathbf{d}_1, \ldots, \mathbf{d}_k \rangle)$ and, from Definition 11,

$$f\big(\langle \mathbf{d}_1, \ldots, \mathbf{d}_k \rangle\big) = g^*(\mathbf{t}) \in \alpha(r_q),$$

with $\alpha(v_i)(g^*(\mathbf{t})) = g^*(\mathbf{t}) \in \alpha(r_B)$. Otherwise, if $\phi_{Ai}(\mathbf{x})/g$ is false then $f(\langle \mathbf{d}_1, \ldots, \mathbf{d}_k \rangle) = \langle \rangle$, the empty tuple in the relation $\alpha(r_q)$, with $\alpha(v_i)(\langle \rangle) = \langle \rangle \in \alpha(r_B)$.

Consequently, the function $\alpha(v_i) : \alpha(r_q) \rightarrow \alpha(r_B)$ is an injection. □

Let us consider the case when an operad's operation is obtained from the integrity constraints of a schema:

Example 12 Let $(\phi_{Ai}(\mathbf{x}) \wedge (\mathbf{y} \neq \mathbf{z})) \Rightarrow r_\top(\overline{0}, \overline{1})$ be an implication χ in the SOtgd obtained from the algorithm *EgdsToSOtgd* of a given egd $(\phi_{Ai}(\mathbf{x}) \Rightarrow (\mathbf{y} \doteq \mathbf{z})) \in \Sigma_A$ of a schema $\mathcal{A} = (S_A, \Sigma_A)$, with $\{r_1, \ldots, r_k\} \subseteq S_A$ the set of all relational symbols of $\mathcal{A}$ that appear in the conjunctive formula $\phi_{Ai}(\mathbf{x})$ and $\mathbf{y} = \langle x_{j_1}, \ldots, x_{j_m} \rangle \subseteq \mathbf{x}$, $\mathbf{z} = \langle x_{l_1}, \ldots, x_{l_m} \rangle \subseteq \mathbf{x}$, with $j_i \neq l_i$ for $1 \leq i \leq m$.

We recall that the formula $(\mathbf{y} \neq \mathbf{z})$ is an abbreviation for the disjunctive formula $(x_{j_1} \neq x_{l_1}) \vee \cdots \vee (x_{j_m} \neq x_{l_m})$.

Then, by the algorithm *MakeOperads*, from this implication χ we obtain the operad's operation $q_{A,i} \in O(r_1, \ldots, r_k, r_{q_i})$, $v_i \in O(r_{q_i}, r_\top)$ such that $q_{A,i}$ is the expression obtained from the implication $(\phi_{Ai}(\mathbf{x}) \wedge (\mathbf{y} \neq \mathbf{z})) \Rightarrow r_{q_i}(\overline{0}, \overline{1})$, that is, the expression $(e \wedge (\mathbf{y} \neq \mathbf{z})) \Rightarrow (\_)(\overline{0}, \overline{1})$ where e is the expression on the left-hand side of the implication, obtained from the formula $\phi_{Ai}(\mathbf{x})$, where each relational symbol r_m is replaced by a place symbol $(\_)_m$, for $1 \leq m \leq k$. Thus:

1. If the integrity condition, given by the egd $\forall \mathbf{x}(\phi_{Ai}(\mathbf{x}) \Rightarrow (\mathbf{y} \doteq \mathbf{z}))$, is satisfied then $\phi_{Ai}(\mathbf{x}) \wedge (\mathbf{y} \neq \mathbf{z})$ is false for each assignment g to variables in $\mathbf{x}$. Thus, from Definition 11 for the mapping-interpretation α, for each $\langle \mathbf{d}_1, \ldots, \mathbf{d}_k \rangle \in \alpha(r_1) \times \cdots \times \alpha(r_k)$, $\alpha(q_{A,i})(\langle \mathbf{d}_1, \ldots, \mathbf{d}_k \rangle) = \langle \rangle$ so that $\alpha(q_{A,i})$ is a constant function and $\alpha(r_{q_i}) = \{\langle \rangle\}$ and hence $\alpha(v_i)(\langle \rangle) = \langle \rangle \in \alpha(r_\top) = R_=$. Consequently, the function $\alpha(v_i)$ is an injection.
2. If this integrity constraint *is not* satisfied then there exists a tuple $\mathbf{d}$ which defines an assignmentg for the variables in $\mathbf{x}$ with $\mathbf{d}_y = g^*(\mathbf{y}) = \langle d_{j_1}, \ldots, d_{j_m} \rangle$ and $\mathbf{d}_z = g^*(\mathbf{z}) = \langle d_{l_1}, \ldots, d_{l_m} \rangle$ such that $\phi_{Ai}(\mathbf{x})/g \wedge (\mathbf{d}_y \neq \mathbf{d}_z)$ is true. That is, there exist at least one index $1 \leq i \leq m$ such that $d_{j_i} \neq d_{l_i}$ and for the operad's operation $q_i = v_i \cdot q_{A,i}$ (the expression $e \Rightarrow (\_)(\overline{0}, \overline{1})$), from Definition 11, for $\langle \mathbf{d}_1, \ldots, \mathbf{d}_k \rangle \in \alpha(r_1) \times \cdots \times \alpha(r_k)$ such that $\mathbf{d} = Cmp(S, \langle \mathbf{d}_1, \ldots, \mathbf{d}_k \rangle)$ and $\bigwedge\{\pi_{j_h}(\mathbf{d}_j) = \pi_{n_h}(\mathbf{d}_n) \mid \{(j_h, j), (n_h, n)\} \in S\}$ is true, $\alpha(q_{A,i})(\langle \mathbf{d}_1, \ldots, \mathbf{d}_k \rangle) = g^*(\langle \overline{0}, \overline{1} \rangle) = \langle g(\overline{0}), g(\overline{1}) \rangle = \langle 0, 1 \rangle \in \alpha(r_{q_i})$ and hence $\alpha(v_i)(\langle 0, 1 \rangle) = \langle \rangle$ (because $\langle 0, 1 \rangle \notin \alpha(r_\top) = R_=$). Consequently, the function $\alpha(v_i)$ *is not* an injection.

Let for a schema $\mathcal{A} = (S_A, \Sigma_A)$, $(\phi_{Ai}(\mathbf{x}) \wedge \neg r(\mathbf{t})) \Rightarrow r_\top(\overline{0}, \overline{1})$ be an implication χ in the SOtgd obtained from the algorithm *EgdsToConSOtgd* of a given normalized implication $(\phi_{Ai}(\mathbf{x}) \Rightarrow r(\mathbf{t}))$ of a given tgd in Σ_A, with $\{r_1, \dots, r_k\} \subseteq S_A$ the set of all relational symbols of $\mathcal{A}$ that appear in the conjunctive formula $\phi_{Ai}(\mathbf{x})$.

Then, based on the algorithm *MakeOperads*, from this implication χ we obtain the operad's operations $q_{A,i} \in O(r_1, \dots, r_k, r, r_{q_i})$ and $v_i \in O(r_{q_i}, r_\top)$ such that $q_{A,i}$ is the expression obtained from the implication $(\phi_{Ai}(\mathbf{x}) \wedge \neg r(\mathbf{t})) \Rightarrow r_{q_i}(\overline{0}, \overline{1})$, that is, the expression $(e \wedge \neg(\_)_{k+1}(\mathbf{t})) \Rightarrow (\_)(\overline{0}, \overline{1})$ (the expression e on the left-hand side of the implication is obtained from the formula $\phi_{Ai}(\mathbf{x})$ where each relational symbol r_m is replaced by a place symbol $(\_)_m$, for $1 \leq m \leq k$). Thus:

1. If the integrity condition given by the tgd $\forall \mathbf{x}(\phi_{Ai}(\mathbf{x}) \Rightarrow r(\mathbf{t}))$ is satisfied then $\phi_{Ai}(\mathbf{x}) \wedge \neg r(\mathbf{t})$ is false for each assignment g to variables in $\mathbf{x}$ and hence, from Definition 11 for the mapping-interpretation α, for each $\langle \mathbf{d}_1, \dots, \mathbf{d}_{k+1} \rangle \in \alpha(r_1) \times \dots \times \alpha(r_k) \times (\mathcal{U}^{ar(r)} \backslash \alpha(r))$, $\alpha(q_{A,i})(\langle \mathbf{d}_1, \dots, \mathbf{d}_{k+1} \rangle) = \langle \rangle$, so that $\alpha(q_{A,i})$ is a constant functions and $\alpha(r_q) = \{\langle \rangle\}$ and hence $\alpha(v_i)(\langle \rangle) = \langle \rangle \in \alpha(r_\top) = R_=$. Consequently, the function $\alpha(v_i)$ is an injection.
2. If this integrity constraint *is not* satisfied then there exists a tuple $\mathbf{d}$ which defines an assignment g for the variables in $\mathbf{x}$ such that $\phi_{Ai}(\mathbf{x})/g \wedge \neg r(\mathbf{t})$ is true and, for the operad's operation $q_i = v_i \cdot q_{A,i}$ (i.e., the expression $(e \wedge \neg(\_)_{k+1}(\mathbf{t})) \Rightarrow (\_)(\overline{0}, \overline{1})$), from Definition 11, for $\langle \mathbf{d}_1, \dots, \mathbf{d}_{k+1} \rangle \in \alpha(r_1) \times \dots \times \alpha(r_k) \times (\mathcal{U}^{ar(r)} \backslash \alpha(r))$ such that $\mathbf{d} = Cmp(S, \langle \mathbf{d}_1, \dots, \mathbf{d}_{k+1} \rangle)$ and $\bigwedge\{\pi_{j_h}(\mathbf{d}_j) = \pi_{n_h}(\mathbf{d}_n) \mid \{(j_h, j), (n_h, n)\} \in S\}$ is true,
$$\alpha(q_{A,i})\big(\langle \mathbf{d}_1, \dots, \mathbf{d}_{k+1} \rangle\big) = g^*\big(\langle \overline{0}, \overline{1} \rangle\big) = \big\langle g(\overline{0}), g(\overline{1}) \big\rangle = \langle 0, 1 \rangle \in \alpha(r_q),$$
so that $\alpha(v_i)(\langle 0, 1 \rangle) = \langle \rangle$ (because $\langle 0, 1 \rangle \notin \alpha(r_\top) = R_=$). Consequently, the function $\alpha(v_i)$ *is not* an injection.

Consequently, Corollary 4 can be applied to a schema mapping that represents the integrity constraints $\Sigma_A = \Sigma_A^{\text{egd}} \cup \Sigma_A^{\text{tgd}}$ over a given schema $\mathcal{A}$, that is, to the mapping $\top_{AA_\top} = \{EgdsToSOtgd(\Sigma_A^{\text{egd}}) \wedge TgdsToConSOtgd(\Sigma_A^{\text{tgd}})\} : \mathcal{A} \to \mathcal{A}_\top$, from a schema $\mathcal{A}$ into the auxiliary schema $\mathcal{A}_\top = (\{r_\top\}, \emptyset)$.

Let us now consider the examples of Corollary 4 for the schema mappings, based on the SOtgds obtained from the set of tgds. The first one is a continuation of Example 8.

Example 13 For the operads defined in Example 8, let a mapping-interpretation (an R-algebra) α be an extension of Tarski's interpretation I_T of the source schema $\mathcal{A} = (S_A, \Sigma_A)$ that satisfies all constraints in Σ_A and defines its database instance $A = \alpha^*(S_A) = \{\alpha(r_i) \mid r_i \in S_A\}$ and, analogously, an interpretation of $\mathcal{C}$.

Let α satisfy the SOtgd of the mapping $\mathcal{M}_{AC}$ by the Tarski's interpretation for the functional symbols f_i, for $1 \leq i \leq 2$, in this SOtgd (denoted by $I_T(f_i)$).

Then we obtain the relations $\alpha(\texttt{EmpAcme})$, $\alpha(\texttt{EmpAjax})$, $\alpha(\texttt{Local})$, $\alpha(\texttt{Office})$, and $\alpha(\texttt{CanRetire})$. The interpretation of f_{Over65} is the characteristic function of the relation $\alpha(\texttt{Over65})$ in the instance $B = \alpha^*(S_B)$ of the database $\mathcal{B} = (S_B, \Sigma_B)$, so that $\overline{f}_{Over65}(a) = 1$ if $\langle a \rangle \in \alpha(\texttt{Over65})$.

Then this mapping interpretation α defines the following functions:

1. The function $\alpha(q_{A,1}) : \alpha(\texttt{EmpAcme}) \times \alpha(\texttt{Local}) \rightarrow \alpha(r_{q_1})$ such that for any tuple $\langle a \rangle \in \alpha(\texttt{EmpAcme})$ and $\langle b \rangle \in \alpha(\texttt{Local})$,
$$\alpha(q_{A,1})\big(\langle a \rangle, \langle b \rangle\big) = \big\langle a, I_T\big(f_1(a)\big)\big\rangle \quad \text{if } a = b; \qquad \langle\rangle \quad \text{otherwise.}$$
And for any $\langle a, b \rangle \in \alpha(r_{q_1})$, $\alpha(v_1)(\langle a, b \rangle) = \langle a, b \rangle$ if $\langle a, b \rangle \in \alpha(\texttt{Office})$; $\langle\rangle$ otherwise.
2. The function $\alpha(q_{A,2}) : \alpha(\texttt{EmpAjax}) \times \alpha(\texttt{Local}) \rightarrow \alpha(r_{q_2})$ such that for any tuple $\langle a \rangle \in \alpha(\texttt{EmpAjax})$ and $\langle b \rangle \in \alpha(\texttt{Local})$,
$$\alpha(q_{A,2})\big(\langle a \rangle, \langle b \rangle\big) = \big\langle a, I_T\big(f_2(a)\big)\big\rangle \quad \text{if } a = b; \qquad \langle\rangle \quad \text{otherwise.}$$
And for any $\langle a, b \rangle \in \alpha(r_{q_2})$, $\alpha(v_2)(\langle a, b \rangle) = \langle a, b \rangle$ if $\langle a, b \rangle \in \alpha(\texttt{Office})$; $\langle\rangle$ otherwise.
3. The function $\alpha(q_{A,3}) : \alpha(\texttt{EmpAcme}) \times \alpha(\texttt{Over65}) \rightarrow \alpha(r_{q_3})$ such that for any tuple $\langle a \rangle \in \alpha(\texttt{EmpAcme})$ and $\langle b \rangle \in \alpha(\texttt{Over65})$,
$$\alpha(q_{A,3})\big(\langle a \rangle, \langle b \rangle\big) = \langle a \rangle \quad \text{if } a = b; \qquad \langle\rangle \quad \text{otherwise.}$$
And for any $\langle a \rangle \in \alpha(r_{q_3})$, $\alpha(v_3)(\langle a \rangle) = \langle a \rangle$ if $\langle a \rangle \in \alpha(\texttt{CanRetire})$; $\langle\rangle$ otherwise.
4. The function $\alpha(q_{A,4}) : \alpha(\texttt{EmpAjax}) \times \alpha(\texttt{Over65}) \rightarrow \alpha(r_{q_4})$ such that for any tuple $\langle a \rangle \in \alpha(\texttt{EmpAjax})$ and $\langle b \rangle \in \alpha(\texttt{Over65})$
$$\alpha(q_{A,4})\big(\langle a \rangle, \langle b \rangle\big) = \langle a \rangle \quad \text{if } a = b; \qquad \langle\rangle \quad \text{otherwise.}$$
And for any $\langle a \rangle \in \alpha(r_{q_4})$, $\alpha(v_4)(\langle a \rangle) = \langle a \rangle$ if $\langle a \rangle \in \alpha(\texttt{CanRetire})$; $\langle\rangle$ otherwise.

From the fact that the mapping-interpretation satisfies the schema mappings, based on Corollary 4, all functions $\alpha(v_i)$, for $1 \leq i \leq 4$, are injections.

The second example is a continuation of Example 9:

Example 14 For the operads defined in Example 9, let a mapping-interpretation (an R-algebra) α be an extension of Tarski's interpretation I_T of the source schema $\mathcal{A} = (S_A, \Sigma_A)$ that satisfies all constraints in Σ_A and defines its database instance $A = \alpha^*(S_A) = \{\alpha(r_i) \mid r_i \in S_A\}$ and, analogously, an interpretation of $\mathcal{C}$.

Let α satisfy the SOtgd of the mapping $\mathcal{M}_{AC}$ by Tarski's interpretation for the functional symbol f_1 in this SOtgd (denoted by $I_T(f_1)$).

Then we obtain the relations $\alpha(\texttt{Local})$, $\alpha(\texttt{Office})$ and $\alpha(\texttt{CanRetire})$. The interpretation of f_{Over65} is the characteristic function of the relation $\alpha(\texttt{Over65})$ in the instance $B = \alpha^*(S_B)$ of the database $\mathcal{B} = (S_B, \Sigma_B)$ so that $\overline{f}_{Over65}(a) = 1$ if $\langle a \rangle \in \alpha(\texttt{Over65})$ and, analogously, the interpretation of f_{Emp} is the characteristic function of the relation $\alpha(\texttt{Emp})$ in the instance B with $\overline{f}_{Emp}(a) = 1$ if $\langle a \rangle \in \alpha(\texttt{Emp})$.

Then this mapping interpretation α defines the following functions:

1. The function $\alpha(q_{A,1}) : \alpha(\texttt{Local}) \times \alpha(\texttt{Emp}) \rightarrow \alpha(r_{q_1})$ such that for any tuple $\langle a\rangle \in \alpha(\texttt{Local})$ and $\langle b\rangle \in \alpha(\texttt{Emp})$,

$$\alpha(q_{A,1})(\langle a\rangle, \langle b\rangle) = \langle a, I_T(f_1(a))\rangle \quad \text{if } a = b; \qquad \langle\rangle \quad \text{otherwise.}$$

And for any $\langle a, b\rangle \in \alpha(r_{q_1})$, $\alpha(v_1)(\langle a, b\rangle) = \langle a, b\rangle$ if $\langle a, b\rangle \in \alpha(\texttt{Office})$; $\langle\rangle$ otherwise.

2. The function $\alpha(q_{A,2}) : \alpha(\texttt{Emp}) \times \alpha(\texttt{Over65}) \rightarrow \alpha(r_{q_2})$ such that for any tuple $\langle a\rangle \in \alpha(\texttt{Emp})$ and $\langle b\rangle \in \alpha(\texttt{Over65})$

$$\alpha(q_{A,2})(\langle a\rangle, \langle b\rangle) = \langle a\rangle \quad \text{if } a = b; \qquad \langle\rangle \quad \text{otherwise.}$$

And for any $\langle a\rangle \in \alpha(r_{q_2})$, $\alpha(v_2)(\langle a\rangle) = \langle a\rangle$ if $\langle a\rangle \in \alpha(\texttt{CanRetire})$; $\langle\rangle$ otherwise.

From the fact that mapping-interpretation satisfies the schema mappings, based on Corollary 4, all functions $\alpha(v_i)$, for $1 \leq i \leq 2$, are injections.

2.4.2 Query-Answering Abstract Data-Object Types and Operads

We consider the views as a universal property for the databases: they are the possible observations of the information contained in an instance-database, and we can use them in order to establish an equivalence relation between databases.

In the theory of *algebraic specifications*, an Abstract Data Type (ADT) is specified by a set of operations (constructors) that determine how the values of the carrier set are built up and by a set of formulae (in the simplest case, the equations) stating which values should be identified. In the standard initial algebra semantics, the defining equations impose a congruence on the initial algebra. Dually, a *coagebraic specification* of a class of systems, i.e., Abstract Object Types (AOT), is characterized by a set of operations (destructors) that specify what can be *observed* out of a system-*state* (i.e., an element of the carrier) and how a state can be transformed to a successor-state.

We start by introducing the class of coalgebras for database query-answering systems for a given instance-database (a set of relations) A. They are presented in an algebraic style by providing a co-signature. In particular, the sorts include one single "hidden sort" corresponding to the carrier of the coalgebra and other "visible" sorts, for the inputs and outputs, that have a given fixed interpretation. Visible sorts will be interpreted as sets without any algebraic structure defined on them. For us, the coalgebraic terms built over destructors are interpreted as the basic *observations* that one can make on the states of a coalgebra. Input sorts are considered as a set $\mathcal{L}_A$ of the finite unions of conjunctive finite-length queries $q(\mathbf{x})$ for a given instance-database A, as specified in Sect. 1.4.1.

Based on the theory of database observations and its power-view operator T, defined in Sect. 1.4.1, the output sort of this database AOT is the set TA of all resulting views (i.e., resulting n-ary relations) obtained by computation of queries $q(\mathbf{x}) \in \mathcal{L}_A$. It is considered as the carrier of a coalgebra as well.

Definition 12 AOT for a database query-answering system, for a given instance-database A, is a pair (S, Σ_{AOT}) such that:

1. The carrier set $S = (X_A, \mathcal{L}_A, \underline{\Upsilon})$ of the sorts where X_A is a hidden sort (a set of states of this database system), $\mathcal{L}_A$ is an input sort (a set of the unions of conjunctive queries over A), and $\underline{\Upsilon}$ is the set of all finitary relations for a given universe $\mathcal{U}$.
2. The signature $\Sigma_{AOT} = \{Next, Out\}$ is a set of operations:
   2.1. A `method` $Next : X_A \times \mathcal{L}_A \to X_A$ that corresponds to an execution of a next query $q(\mathbf{x}) \in \mathcal{L}_A$ in a current state $s \in X_A$ of a database A such that a database A passes to the next state; and
   2.2. An `attribute` $Out : X_A \times \mathcal{L}_A \to \underline{\Upsilon}$ such that for each $s \in X_A$, $q(\mathbf{x}) \in \mathcal{L}_A$, $Out(s, q(\mathbf{x}))$ is a relation computed by a query $q(\mathbf{x})$.

   The Data Object Type for a query-answering system is given by a coalgebra:
3. $\langle \lambda Next, \lambda Out\rangle : X_A \to X_A^{\mathcal{L}_A} \times \underline{\Upsilon}^{\mathcal{L}_A}$ of the polynomial endofunctor $(\_)^{\mathcal{L}_A} \times \underline{\Upsilon}^{\mathcal{L}_A} : \mathbf{Set} \to \mathbf{Set}$, where λ is the currying operator for functions.

In an object-oriented terminology, the coalgebras are expressive enough in order to specify the parametric methods and the attributes for a database (conjunctive) query answering systems. In a transition system terminology, such coalgebras can model a deterministic, non-terminating, transition system with inputs and outputs. In [4], a complete equational calculus for such coalgebras of restricted class of polynomial functors has been defined.

Here we will consider only the database query-answering systems without the side effects. That is, the obtained results (views) *will not be materialized* as a new relation of this database A but only visualized. Thus, when a database answers to a query, it remains in the same initial state. Thus, the set X_A is a singleton $\{A\}$ for a given database A and, consequently, it is isomorphic to the terminal object 1 in the **Set** category. As a consequence, from $1^{\mathcal{L}_A} \simeq 1$, we obtain that a method $Next$ is just an identity function $id : 1 \to 1$. Thus, the only interesting part of this AOT is the attribute part $Out : X_A \times \mathcal{L}_A \to \underline{\Upsilon}$, with the fact that $X_A \times \mathcal{L}_A = \{A\} \times \mathcal{L}_A \simeq \mathcal{L}_A$.

Consequently, we obtain an attribute mapping $Out : \mathcal{L}_A \to \underline{\Upsilon}$, whose graph is equal to the query-evaluation surjective mapping $ev_A : \mathcal{L}_A \twoheadrightarrow TA$, introduced in Sect. 1.4.1.

This mapping $ev_A : \mathcal{L}_A \twoheadrightarrow TA$ will be used as a semantic foundation for the database mappings.

Corollary 5 *A canonical method for the construction of the power-view database TA can be obtained by an Abstract Data-Object Type (S, Σ_{AOT}) for a query-answering system without side-effects as follows*:

$$TA \triangleq \{Out(q_i(\mathbf{x})) \mid q_i(\mathbf{x}) \in \mathcal{L}_A\}.$$

Proof In fact, from the reduction of Out to ev_A (for an AOT without side-effects), for a given database instance A, $\{Out(q_i(\mathbf{x})) \mid q_i(\mathbf{x}) \in \mathcal{L}_A\} = \{ev_A(q_i(\mathbf{x})) \mid q_i(\mathbf{x}) \in \mathcal{L}_A\} = TA$, from the fact that (in Sect. 1.4.1) ev_A is a surjective function. □

This corollary is a direct proof that the power-view database operator T, introduced in Sect. 1.4.1, represents the *observational* point of view for the instance-databases.

In fact, a conjunctive query is an implication from the body of a query $q_i(\mathbf{x}) \in \mathcal{L}_A$ with relational symbols in $\{r_{i1}, \ldots, r_{ik}\}$ (a subset of relational symbols in a schema $\mathcal{A}$) into the head $r_q(\mathbf{x})$, that is, a formula $q_i(\mathbf{x}) \Rightarrow r_q(\mathbf{x})$ that can be represented equivalently by an operad's operation $q_{A,i} = O(r_{i1}, \ldots, r_{ik}, r_q)$ of the mapping $\mathcal{M} : \mathcal{A} \to T\mathcal{A}$ with an SOtgd $\forall \mathbf{x}(q_i(\mathbf{x}) \Rightarrow r_q(\mathbf{x}))$.

Thus, for a given R-algebra α that is a mapping-interpretation of operads (Definitions 10 and 11) such that the instance $A = \alpha^*(\mathcal{A})$ of the schema $\mathcal{A}$ satisfies the SOtgd of the mapping $\mathcal{M} = \{\forall \mathbf{x}(q_i(\mathbf{x}) \Rightarrow r_q(\mathbf{x}))\} : \mathcal{A} \to T\mathcal{A}$ above, $Out(q_i(\mathbf{x})) = \|q_i(\mathbf{x})\|_A$ is the image of the function $\alpha(q_{A,i}) : \alpha(r_{i1}) \times \cdots \times \alpha(r_{ik}) \to \alpha(r_q)$ obtained from the operad's operation $q_{A,i} = O(r_{i1}, \ldots, r_{ik}, r_q)$, that is,

$$Out(q_i(\mathbf{x})) = \{\alpha(q_{A,i})(\mathbf{d}_1, \ldots, \mathbf{d}_m) \mid \mathbf{d}_j \in \alpha(r_{ij}), \text{for } 1 \le j \le k\} = \alpha(r_q).$$

Consequently, the AOT without side-effects of the instance database $A = \alpha^*(\mathcal{A})$ is the observational point of view for the basic database mappings and closely interrelated with mapping-operads.

Note that a single view-mapping at instance-database level can be defined as a T-coalgebra $f = \alpha^*(MakeOperads(\mathcal{M})) = \{\alpha(q_{A,i}), q_\bot\} : A \to TA$ that, obviously, *is not* a function but a set of functions. These arguments will be analyzed in detail after the definition of the database category **DB**, and hence it will be demonstrated that f is a morphism in such a database category.

2.4.3 Strict Semantics of Schema Mappings: Information Fluxes

As it was explained previously, the SOtgds represent the logical language syntax for the mapping compositions and use the existentially quantified functional symbols whose interpretation introduces the data values not contained in the active data domain of the source schema.

In fact, these functional symbols, introduced in the new algorithm (defined in Sect. 2.3) for composition of mappings, are used in order to replace the data contained in 'intermediate' databases (the hidden databases between the source and target database schemas) in order to guarantee the logical equivalence of *composed* SOtgd and *the set* of SOtgds used in such a composition for the determination of the target database. This fact in the data-exchange setting guarantees that each query over a target database will give the same resulting view regardless of whether we are using the set of intermediate mappings from the source into this target database, or the single SOtgd of the resulting composition obtained from the algorithm in Sect. 2.3.

From this logical point of view, the composed SOtgd represents the *complete* information of all 'intermediate' databases used in this composition and includes the strict subset of information of the source database that is mapped into the target database.

In the categorial semantics of mappings and their composition, we are only interested in this strict subset of information that is mapped (only) *from the source* database.

Differently from the data-exchange settings, we are not interested in the 'minimal' instance B of the target schema $\mathcal{B}$ such that, for a given instance (model) A of the source schema $\mathcal{A}$, (A, B) is an instance of the SOtgd of the schema mapping $\mathcal{M}_{AB}$. In our more general framework, we do not intend to determine 'exactly', 'canonically' or 'completely' the instance of the target schema $\mathcal{B}$ (for a fixed instance A of the source schema $\mathcal{A}$). Such a setting is more general and the target database is *only partially* determined by the source database $\mathcal{A}$: the other part of this target database can be, for example, determined by another database $\mathcal{C}$ (that is not any of the 'intermediate' databases between $\mathcal{A}$ and $\mathcal{B}$), or by the software programs which update the information contained in this target database $\mathcal{B}$.

In other words, the Data-exchange (and Data integration) settings are only special particular cases of this *general framework* of database mappings, where each database can be mapped from other sources, to map a part of its own information into other targets, and to be locally updated as well.

This last feature of the local update of a given database, in the database-mapping systems, will be considered in full detail in Chap. 7 dedicated to Operational semantics for database mappings. The process of an update of a database-mapping system begins with one local update of a given database and, after that, this update has to be propagated through the whole network of inter-mapped databases in order to guarantee that every (atomic) schema mapping in this network has to be satisfied at the end of this update-processing.

In our case, two equal compositions are not necessarily logically equivalent formulae, and their equality is considered only at the instance-database level.

Let us suppose that $\mathcal{M}_{AB_1} : \mathcal{A} \to \mathcal{B}_1$ is an atomic mapping, based on the set of tgds that 'transfer', for a given instance A of $\mathcal{A}$, the set S of views from $\mathcal{A}$ to $\mathcal{B}_1$ (based on the left-hand side of implications in the tgds), and let $\mathcal{M}_{B_1B_n} : \mathcal{B}_1 \to \mathcal{B}_n$ be a (possibly non-atomic) mapping composed of a set of atomic mappings $\mathcal{M}_{B_iB_{i+1}} : \mathcal{B}_i \to \mathcal{B}_{i+1}$, for $1 \leq i \leq n$, $n \geq 2$. Let S_i be the set of views of the ith atomic mapping for a given instance B_i of the schema $\mathcal{B}_i$ (based on the left-hand side of implications in the tgds of this atomic mapping). This second (possibly composed) mapping $\mathcal{M}_{B_1B_n}$ filters the information contained in the set of all possible views TS that can be obtained from the set of relations in S. A view $v \in TS$ will be propagated into the target database B_n if $v \in TS_i$ for all intermediate atomic mappings, i.e., for $1 \leq i \leq n$.

Consequently, the *strict semantics* of a composed mapping $\mathcal{M}_{B_1B_n} \circ \mathcal{M}_{AB_1} : \mathcal{A} \to \mathcal{B}_n$ represents the subset of all views in TS that are transferred to the target database $\mathcal{B}_n$, that is, it is the intersection $TS \cap TS_1 \cap \cdots \cap TS_n$ and hence equal to $TS \cap TS_R$ where $TS_R = TS_1 \cap \cdots \cap TS_n$ is the strict semantics of the composed mapping $\mathcal{M}_{B_1B_n}$.

Based on these considerations, we are now able to define an abstraction of this transferred information from a given source to a target schema, called *information flux*.

Moreover, each R-algebra α of a given set of mapping-operads between a source schema $\mathcal{A}$ and a target schema $\mathcal{B}$ determines a particular information flux from the source into the target schema.

Definition 13 (INFORMATION FLUX) Let α be a mapping-interpretation (an R-algebra in Definition 11) of a given set $\mathbf{M}_{AB} = \{q_1, \dots, q_n, 1_{r_\emptyset}\} = MakeOperads(\mathcal{M}_{AB})$ of mapping-operads, obtained from an *atomic* mapping $\mathcal{M}_{AB} : \mathcal{A} \rightarrow \mathcal{B}$, and $A = \alpha^*(S_A)$ be an instance of the schema $\mathcal{A} = (S_A, \Sigma_A)$ that satisfies all constraints in Σ_A.

For each operation $q_i \in \mathbf{M}_{AB}$, $q_i = (e \Rightarrow (\_)(\mathbf{t}_i)) \in O(r_{i,1}, \dots, r_{i,k}, r'_i)$, let $\mathbf{x}_i$ be its tuple of variables which appear at least one time free (not as an argument of a function) in $\mathbf{t}_i$ and appear as variables in the atoms of relational symbol of the schema $\mathcal{A}$ in the formula $e[(\_)_j/r_{i,j}]_{1\leq j\leq k}$. Then, we define

(i) $Var(\mathbf{M}_{AB}) = \bigcup_{1\leq i\leq n}\{\{x\} \mid x \in \mathbf{x}_i\}$.

We define the kernel of the information flux of $\mathbf{M}_{AB}$, for a given mapping-interpretation α, by (we denote the image of a function f by '$im(f)$')

(ii) $\Delta(\alpha, \mathbf{M}_{AB}) = \{\pi_{\mathbf{x}_i}(im(\alpha(q_i))) \mid q_i \in \mathbf{M}_{AB}$, and $\mathbf{x}_i$ is not empty$\} \cup \bot^0$, if $Var(\mathbf{M}_{AB}) \neq \emptyset$; $\bot^0$ otherwise.

We define the information flux by its kernel by

(iii) $Flux(\alpha, \mathbf{M}_{AB}) = T(\Delta(\alpha, \mathbf{M}_{AB}))$.

The flux of composition of $\mathbf{M}_{AB}$ and $\mathbf{M}_{BC}$ is defined by

(iv) $Flux(\alpha, \mathbf{M}_{BC} \circ \mathbf{M}_{AB}) = Flux(\alpha, \mathbf{M}_{AB}) \cap Flux(\alpha, \mathbf{M}_{BC})$.

We say that an information flux is *empty* if it is equal to $\bot^0 = \{\bot\}$ (and hence it is not the empty set), analogously as for an empty instance-database.

We recall that the *kernels* of the information fluxes, defined in point (ii), will be used as the actions for the Labeled Transition Systems (LTS) in the operational semantics for the database mappings (in Sect. 7.3).

The information flux of the SOtgd of the mapping $\mathcal{M}_{AB}$ for the instance-level mapping $f = \alpha^*(\mathbf{M}_{AB}) : A \rightarrow \alpha^*(\mathcal{B})$ composed of the set of functions $f = \alpha^*(\mathbf{M}_{AB}) = \{\alpha(q_1), \dots, \alpha(q_n), q_\bot\}$ is denoted by $\widetilde{f}$. Notice that $\bot \in \widetilde{f}$, and hence the information flux $\widetilde{f}$ is an instance-database as well.

From this definition, each instance-mapping is a set of functions whose information flux is the intersection of the information fluxes of all atomic instance-mappings that compose this composed instance-mapping. These basic properties of the instance-mappings will be used in order to define the database **DB** category where the instance-mappings will be the morphisms (i.e., the arrows) of this category, while the instance-databases (each instance-database is a set of relations of a schema also with the empty relation $\bot$) will be its objects.

In the case of an atomic mapping, obtained from a set of egds of a given schema $\mathcal{A}$, we have the following particular property:

Corollary 6 *Let Φ be the SOtgd equal to EgdsToSOtgd(Σ_A^{egd}) and Ψ be the SOtgd equal to TgdsToConSOtgd(Σ_A^{tgd}) for a given schema $\mathcal{A} = (S_A, \Sigma_A^{\text{egd}} \cup \Sigma_A^{\text{tgd}})$. Then, for every R-algebra α, the information flux Flux(α, MakeOperads($\{\Phi \wedge \Psi\}$)) $= \bot^0$ is empty.*

Proof From Definition 7, the integrity constraints are representable by an atomic mapping $\top_{AA_\top} = \{\Phi \wedge \Psi\} : \mathcal{A} \to \mathcal{A}_\top$.

From the fact that each abstract operad-operation q_i in *MakeOperads*($\{\Phi\}$) and in *MakeOperads*($\{\Psi\}$) is of the form $e \Rightarrow r_\top(\overline{0}, \overline{1})$, we have no free variables on the right-hand side of these implications, so that $\mathbf{x}_i$ in Definition 13 for the information flux is empty and $Var(\mathbf{T}_{AA_\top}) = \emptyset$, so that, from Definition 13 of its kernel, $\Delta(\alpha, \mathbf{T}_{AA_\top}) = \bot^0$ and hence $Flux(\alpha, \mathbf{T}_{AA_\top}) = T\,\bot^0 = \bot^0$ is empty. □

Note that this corollary confirms that, for any database schema $\mathcal{A} = (S_A, \Sigma_A)$ where $\Sigma_A = \Sigma_A^{\text{egd}} \cup \Sigma_A^{\text{tgd}}$, we can define the integrity-constraints mapping, as it was demonstrated by Example 12, by a schema mapping $\top_{AA_\top} = \{\Phi \wedge \Psi\} : \mathcal{A} \to \mathcal{A}_\top$ (where Φ is the SOtgd equal to *EgdsToSOtgd*(Σ_A^{egd}) and Ψ is the SOtgd equal to *TgdsToConSOtgd*(Σ_A^{tgd}) and $\mathcal{A}_\top = (\{r_\top\}, \emptyset))$ and, consequently, by equivalent operads-mapping $\mathbf{T}_{AA_\top} = MakeOperads(\{\Phi\}) \cup MakeOperads(\{\Psi\}) : \mathcal{A} \to \mathcal{A}_\top$ with the *empty* information flux.

We have no mapping from $\mathcal{A}_\top$ into other schema mappings, so that this integrity-constraint mapping does not participate in any significant composition with other mappings in a given database mapping system. Moreover, from the fact that the information flux of composed mappings is equal to the intersection of the information fluxes of all atomic arrows which compose this mapping, such a composed mapping which contains an atomic integrity-constraint mapping will always have an empty flux. Consequently, the role of the integrity-constraint mappings is only "logical", used to express the integrity constraints for schemas, and to verify if for a given mapping-interpretation α in Definition 11 they are satisfied (as it was specified by Corollary 4). The extension of the database mapping systems with the integrity-constraints mappings will not modify its original semantic structure, but in this extended framework not only the inter-schema mappings but also the schema integrity constraints will be the "first objects" of big data integration theory and will be presented in a uniform elegant manner.

Let us consider the following example, in the case when the schema mappings are satisfied by a given mapping-interpretation α:

Example 15 Let us consider Example 2 for composition of the atomic mappings $\mathcal{M}_{AB} : \mathcal{A} \to \mathcal{B}$, $\mathcal{M}_{BD} : \mathcal{B} \to \mathcal{D}$, $\mathcal{M}_{AC} : \mathcal{A} \to \mathcal{C}$ and $\mathcal{M}_{CD} : \mathcal{C} \to \mathcal{D}$, where

$$\begin{aligned}\mathcal{M}_{AB} = \big\{&\forall x_n \forall x_c \big(\texttt{Takes}(x_n, x_c) \Rightarrow \texttt{Takes1}(x_n, x_c)\big),\\ &\forall x_n \forall x_c \big(\texttt{Takes}(x_n, x_c) \Rightarrow \texttt{Student}\big(x_n, f_1(x_n)\big)\big)\big\},\\ \mathcal{M}_{BD} = \big\{&\forall x_n \forall x_c \forall y \big(\big(\texttt{Takes1}(x_n, x_c)\end{aligned}$$

$$\wedge\, \texttt{Student}(x_n, y)) \Rightarrow \texttt{Enrolment}(y, x_c))\big\}$$

$$\mathcal{M}_{AC} = \big\{\forall x_n \forall x_c\big(\texttt{Takes}(x_n, x_c) \Rightarrow \texttt{Learning}\big(x_n, x_c, f_2(x_n, x_c)\big)\big)\big\},$$

$$\mathcal{M}_{CD} = \big\{\forall x_n \forall x_c \forall x_p\big(\texttt{Learning}(x_n, x_c, x_p) \Rightarrow \texttt{Teaching}(x_p, x_c)\big)\big\}.$$

Hence, their composition is equal to the following mappings:

$$\begin{aligned}\mathcal{M}_{AD} &= \mathcal{M}_{BD} \circ \mathcal{M}_{AB}\\ &= \big\{\exists f_1\big(\forall x_1 \forall x_2\big(\texttt{Takes}(x_1, x_2) \Rightarrow \texttt{Enrolment}\big(f_1(x_1), x_2\big)\big)\big)\big\},\\ \mathcal{M}'_{AD} &= \mathcal{M}_{CD} \circ \mathcal{M}_{AC}\\ &= \big\{\exists f_2\big(\forall x_1 \forall x_2\big(\texttt{Takes}(x_1, x_2) \Rightarrow \texttt{Teaching}\big(f_2(x_1, x_2), x_2\big)\big)\big)\big\},\end{aligned}$$

which are not logically equivalent.

However, we obtain from Definition 13, for $A = \alpha^*(\mathcal{A})$, $B = \alpha^*(\mathcal{B})$ and $C = \alpha^*(\mathcal{C})$:

$$\begin{aligned}&\mathit{Flux}\big(\alpha, \big(\mathit{MakeOperads}(\mathcal{M}_{BD}) \circ \mathit{MakeOperads}(\mathcal{M}_{AB})\big)\big)\\ &\quad= \mathit{Flux}\big(\alpha, \mathit{MakeOperads}(\mathcal{M}_{AB})\big) \cap \mathit{Flux}\big(\alpha, \mathit{MakeOperads}(\mathcal{M}_{BD})\big)\\ &\quad= T\big(\big\|\texttt{Takes}(x_n, x_c)\big\|_A, \pi_{x_n}\big\|\texttt{Takes}(x_n, x_c)\big\|_A\big)\\ &\qquad \cap T\big(\big\{\pi_{x_c,y}\big\|\texttt{Takes1}(x_n, x_c) \wedge \texttt{Student}(x_n, y)\big\|_B\big\}\big),\end{aligned}$$

and

$$\begin{aligned}&\mathit{Flux}\big(\alpha, \big(\mathit{MakeOperads}(\mathcal{M}_{CD}) \circ \mathit{MakeOperads}(\mathcal{M}_{AC})\big)\big)\\ &\quad= \mathit{Flux}\big(\alpha, \mathit{MakeOperads}(\mathcal{M}_{AC})\big) \cap \mathit{Flux}\big(\alpha, \mathit{MakeOperads}(\mathcal{M}_{CD})\big)\\ &\quad= T\big(\big\|\texttt{Takes}(x_n, x_c)\big\|_A\big) \cap T\big(\big\{\pi_{x_p,x_c}\big\|\texttt{Learning}(x_n, x_c, x_p)\big\|_C\big\}\big).\end{aligned}$$

In the case when in the given universe $\mathcal{U} = \mathbf{dom}$, the domain $dom(y)$ for the attribute y of the atom $\texttt{Student}(x_n, y)$ is disjoint from $dom(x_n)$ and from $dom(x_c)$ of the atom $\texttt{Takes}(x_n, x_c)$, and if $dom(x_p)$ for the attribute x_p of the atom $\texttt{Learning}(x_n, x_c, x_p)$ is disjoint from $dom(x_n)$ and from $dom(x_c)$ of the atom $\texttt{Takes}(x_n, x_c)$, then it is easy to verify that

$$\begin{aligned}&\mathit{Flux}\big(\alpha, \big(\mathit{MakeOperads}(\mathcal{M}_{BD}) \circ \mathit{MakeOperads}(\mathcal{M}_{AB})\big)\big)\\ &\quad= \mathit{Flux}\big(\alpha, \big(\mathit{MakeOperads}(\mathcal{M}_{CD}) \circ \mathit{MakeOperads}(\mathcal{M}_{AC})\big)\big)\\ &\quad= T\big(\big\|\texttt{Takes}(x_n, x_c)\big\|_A\big).\end{aligned}$$

Consequently, in this case, the two composed mappings $\mathcal{M}_{AD}$ and $\mathcal{M}'_{AD}$ have the same information fluxes from $\mathcal{A}$ into the target schema $\mathcal{D}$, so that, from the strict semantics point of view, they are equal instance-level mappings.

We have shown in Corollary 2 that the schema-level mappings have the categorial morphism's properties (an associative composition, with an identity mapping for each schema) and now we will show that the information fluxes of the schema mappings have the categorial morphism's properties at the instance-mapping level as well:

Corollary 7 *The information fluxes of the schema mappings and the composition (i.e., the set intersection) of information fluxes satisfy categorial properties of the morphisms between instance-databases as objects of such a category.*

Proof For a given atomic schema mapping $\mathcal{M}_{AB} : \mathcal{A} \to \mathcal{B}$ and an instance $A = \alpha^*(\mathcal{A})$, the information flux $\widetilde{f} = \mathit{Flux}(\alpha, \mathit{MakeOperads}(\mathcal{M}_{AB})) \subseteq TA$ may represent the instance-mapping from the instance A into an instance $B = \alpha^*(\mathcal{B})$.

Note that if $(A, B) \in \mathit{Inst}(\mathcal{M}_{AB})$ (that is, when the instances A and B satisfy the schema mapping $\mathcal{M}_{AB}$) then *all* information contained in this information flux $\widetilde{f}$ is transferred from A into B.

It is easy to verify that for an identity schema mapping (which is always satisfied) $\mathit{Id}_A : \mathcal{A} \to \mathcal{A}$, defined in Lemma 5, $\mathit{Flux}(\alpha, \mathit{MakeOperads}(\mathit{Id}_A)) = TA$, so that

$$\begin{aligned} &\mathit{Flux}\big(\alpha, \mathit{MakeOperads}(\mathit{Id}_A) \circ \mathit{MakeOperads}(\mathcal{M}_{AB})\big) \\ &\quad = \mathit{Flux}\big(\alpha, \mathit{MakeOperads}(\mathit{Id}_A)\big) \cap \mathit{Flux}\big(\alpha, \mathit{MakeOperads}(\mathcal{M}_{AB})\big) \\ &\quad = TA \cap \mathit{Flux}\big(\alpha, \mathit{MakeOperads}(\mathcal{M}_{AB})\big) \\ &\quad = \mathit{Flux}\big(\alpha, \mathit{MakeOperads}(\mathcal{M}_{AB})\big), \end{aligned}$$

and hence the property of the categorial composition with identity morphisms (represented here by the information flux of an identity schema mapping) is satisfied.

The set intersection $\cap$ is an associative operation so that, together with the identity property above, it satisfies the categorial properties for composition of morphisms. □

Based on Corollary 2 and Corollary 7, it is possible to define the categorial semantics for the schema mappings, by defining a functor from the sketch category of a given schema-mapping graph into an instance-level category where the objects are the database-instances and the morphisms are characterized by the information fluxes of the schema-mappings.

The formalization of the schema mappings by means of operads, as it will be demonstrated in next two chapters, is useful in order to be able to extend each R-algebra α to a functor from the category sketch (obtained from a graph of (inter)schema mappings) into the base **DB** category, which represents the denotational semantics of this schema database mapping graph.

That is, each schema mapping $\mathcal{M}_{AB}$ transformed into its mapping-operad $\mathbf{M}_{AB} = \mathit{MakeOperads}(\mathcal{M}_{AB}) = \{q_1, \ldots, q_n, 1_{r_\emptyset}\} : \mathcal{A} \to \mathcal{B}$ may be seen as a morphism of a sketch category, and hence it will be mapped by a functor (R-algebra) α, which satisfies Definition 11 (i.e., such that it is a *mapping-interpretation*), into the

DB category morphism $\alpha^*(\mathbf{M}_{AB}) = \{f_1, \ldots, f_n, q_\bot\} : A \to B$, where $A = \alpha(S_A)$ is an instance-database of the schema $\mathcal{A}$, $B = \alpha(S_B)$ is an instance-database of the schema $\mathcal{B}$, and $f_i = \alpha(q_i)$ are the functions obtained from the operads, with the domain equal to the Cartesian product of a subset of relations in the instance A and the codomain is a relation in the instance B.

The functors such that, for every operad's operation $q_i = v_i \cdot q_{A,i}$ in each sketch's mapping, the function $\alpha(v_i)$ is an inclusion (injection) will define a *model* of the schema database mapping system expressed by this sketch.

This is the principal idea for the *categorical semantics* of the schema database mapping systems and will be discussed with more details in next two chapters.

2.5 Algorithm for Decomposition of SOtgds

In this section, we will present an algorithm for decomposition of a given SOtgd into two SOtgds. This decomposition will be used for the demonstration of the extended symmetry properties of the database mappings at the instance-level (Sect. 3.2.3 in Chap. 3). Let us define this algorithm that transforms an SOtgd Φ of a given schema mapping into two SOtgds Φ_E and Φ_M such that the composition of them is equivalent to the original SOtgd Φ:

Decomposition algorithm *DeCompose*($\mathcal{M}_{AB}$)
Input. A schema mapping $\mathcal{M}_{AB} : \mathcal{A} \to \mathcal{B}$ given by a SOtgd Φ.

$$\exists \mathbf{f}\big(\big(\forall \mathbf{x}_1(\phi_{A,1} \Rightarrow \psi_{B,1})\big) \wedge \cdots \wedge \big(\forall \mathbf{x}_n(\phi_{A,n} \Rightarrow \psi_{B,n})\big)\big).$$

Output. The pair (Φ_E, Φ_M) of SOtgds.

1. (*Normalize the SOtgd*)
   Initialize S, S_E and S_M to be the empty sets ($\emptyset$). Let $\mathcal{M}_{AB}$ be the singleton set
   $$\big\{\exists \mathbf{f}\big(\big(\forall \mathbf{x}_1(\phi_{A,1} \Rightarrow \psi_{B,1})\big) \wedge \cdots \wedge \big(\forall \mathbf{x}_n(\phi_{A,n} \Rightarrow \psi_{B,n})\big)\big)\big\}.$$
   Put each of n implications $\phi_{A,i} \Rightarrow \psi_{B,i}$, for $1 \le i \le n$, into S.
   Each implication χ in S has the form $\phi_{A,i}(\mathbf{x}_i) \Rightarrow \bigwedge_{1 \le j \le k} r_j(\mathbf{t}_j)$ where every variable in $\mathbf{x}_i$ is universally quantified, and each $\mathbf{t}_j$, for $1 \le j \le k$, is a sequence (tuple) of terms with variables in $\mathbf{x}_i$. We then replace each such implication χ in S with k implications: $\phi_{A,i}(\mathbf{x}_i) \Rightarrow r_1(\mathbf{t}_1), \ldots, \phi_{A,i}(\mathbf{x}_i) \Rightarrow r_k(\mathbf{t}_k)$.
2. (*Transformation into two new SOtgds*)
   Let $S = \{\chi_1, \ldots, \chi_m\}$ be the set of implications obtained in the previous step. Then for each implication χ_i, equal to the formula $q_{Ai}(\mathbf{x}_i) \Rightarrow r_i(\mathbf{t}_i)$, do as follows:
   2.1. Let $\mathbf{y}_i$ be the subset of variables $x_l \in \mathbf{x}_i$ that appear in some atom in q_{Ai} (i.e., of a relational symbol in $\mathcal{A}$) and in the atom $r_i(\mathbf{t}_i)$ as a variable $x_l \in \mathbf{t}_i$ (i.e., when $r_i(\ldots, x_l, \ldots)$). If $\mathbf{y}_i$ is not empty then we add the implication $q_{Ai}(\mathbf{x}_i) \Rightarrow r_{q_i}(\mathbf{y}_i)$ in S_E, by introducing a new fresh symbol r_{q_i}; otherwise we add the implication $q_{Ai}(\mathbf{x}_i) \Rightarrow r_\emptyset$ if $\mathbf{y}_i$ in S_E.

2.2. We add the implication $(r_{q_i}(\mathbf{y}_i) \wedge \psi_i'(\mathbf{x}_i)) \Rightarrow r_i(\mathbf{t}_i)$ (or $(r_\emptyset \wedge \psi_i'(\mathbf{x}_i)) \Rightarrow r_i(\mathbf{t}_i)$ if $\mathbf{y}_i$ is empty) in S_M, where $\psi'(\mathbf{x}_i)$ is the formula obtained from $q_{Ai}(\mathbf{x}_i)$ by substituting each atom $r(\mathbf{t})$ in it (where $\mathbf{t}$ is a tuple of terms with variables in $\mathbf{x}_i$) by a logically equivalent atom $(f_r(\mathbf{t}) \doteq \overline{1})$, where $f_r \notin \mathbf{f}$ is a new fresh introduced symbol for the *characteristic function* of r (hence, $\psi'(\mathbf{x}_i))$ and $q_{Ai}(\mathbf{x}_i)$ are logically equivalent).

3. (*Construct* SOtgd Φ_E and Φ_M)

3.1. If $S_M = \{\chi_1, \ldots, \chi_m\}$ and $\chi_1, \ldots, \chi_m$ are the implications from the previous step then Φ_M is a singleton set composed of an SOtgd

$$\exists \mathbf{h}(\forall \mathbf{x}_1 \chi_1 \wedge \cdots \wedge \forall \mathbf{x}_m \chi_m)$$

where $\mathbf{h} \supseteq \mathbf{f}$ is the tuple of all functional symbols that appear in any of the implications in S_M and the variables in $\mathbf{x}_i$ are all the variables found in the formula χ_i, for $1 \leq i \leq m$.

3.2. If $S_E = \{\overline{\chi}_1, \ldots, \overline{\chi}_m\}$ and $\overline{\chi}_1, \ldots, \overline{\chi}_m$ are the implications from the previous step then Φ_E is a singleton set composed of an SOtgd

$$\exists \mathbf{g}(\forall \mathbf{z}_1 \chi_1 \wedge \cdots \wedge \forall \mathbf{z}_m \chi_m)$$

where $\mathbf{g} \subseteq \mathbf{f}$ is the tuple of all functional symbols that appear in any of the implications in S_E and the variables in $\mathbf{z}_i$ are all the variables found in the formula $\overline{\chi}_i$, for $1 \leq i \leq m$.

Return the pair of SOtgds (Φ_M, Φ_E).

It is easy to verify that for a mapping $\mathcal{M}_{AB} = \{\Phi\} : \mathcal{A} \to \mathcal{B}$, with this decomposition $(\Phi_M, \Phi_E) = DeCompose(\mathcal{M}_{AB})$, we obtain two mappings, $\mathcal{M}_{AC} = \{\Phi_E\} : \mathcal{A} \to \mathcal{C}$ and $\mathcal{M}_{CB} = \{\Phi_M\} : \mathcal{C} \to \mathcal{B}$, where $\mathcal{C} = (S_C, \emptyset)$ and S_C is the set of all new introduced relational symbols (that appear on the right-hand side of implications in Φ_E and on the left-hand side of implications in Φ_M), such that $\mathcal{M}_{AB} = \mathcal{M}_{CB} \circ \mathcal{M}_{AC} = Compose(\mathcal{M}_{CB}, \mathcal{M}_{AC})$ (from the fact that for each par of implications χ_i and $\overline{\chi_i}$, equal to $q_{Ai}(\mathbf{x}_i) \Rightarrow r_{q_i}(\mathbf{y}_i)$ and $(r_{q_i}(\mathbf{y}_i) \wedge \psi_i'(\mathbf{x}_i)) \Rightarrow r_i(\mathbf{t}_i)$, respectively, $\psi_i'(\mathbf{x}_i)$ is logically equivalent to $q_{Ai}(\mathbf{x}_i)$).

Example 16 Let us consider the following three cases:

1. The mapping $\mathcal{M}'_{AD} = \{\Phi\} : \mathcal{A} \to \mathcal{D}$ in Example 5 with Φ equal to SOtgd

$$\exists f_1 \big(\forall x_1 \forall x_2 \big(\texttt{Takes}(x_1, x_2) \Rightarrow \texttt{Enrolment}\big(f_1(x_1), x_2\big)\big)\big).$$

Then Φ_E is equal to $\forall x_1 \forall x_2(\texttt{Takes}(x_1, x_2) \Rightarrow r_{q_1}(x_2))$, and Φ_M is equal to

$$\exists f_1 \exists f_{Takes} \big(\forall x_1 \forall x_2 \big(\big(r_{q_1}(x_2) \wedge \big(f_{Takes}(x_1, x_2) \doteq \overline{1}\big)\big) \Rightarrow \texttt{Enrolment}\big(f_1(x_1), x_2\big)\big)\big),$$

where f_{Takes} is a new functional symbol introduced in step 2 of the algorithm *DeCompose* because the variable x_1 on the right-hand side of the implication is not contained in the atom $r_{q_1}(x_2)$ and, consequently, we replaced the original atom $\mathtt{Takes}(x_1, x_2)$ with the equivalent to it equation $(f_{Takes}(x_1, x_2) \doteq \overline{1})$, because the relational symbol Takes is of schema $\mathcal{A}$ and not of the new schema $\mathcal{C} = (\{r_{q_1}\}, \emptyset)$ with a unary relational symbol r_{q_1}. Let us define these two mappings, $\mathcal{M}_{AC} = \{\Phi_E\} : \mathcal{A} \to \mathcal{C}$ and $\mathcal{M}_{CD} = \{\Phi_M\} : \mathcal{C} \to \mathcal{D}$.

For any mapping-interpretation α such that $A = \alpha^*(\mathcal{A})$ and both schema mappings $\mathcal{M}_{AC}$ and $\mathcal{M}'_{AD}$ are satisfied, we have that

$$Flux(\alpha, MakeOperads(\mathcal{M}_{AC})) = Flux(\alpha, MakeOperads(\mathcal{M}'_{AD})).$$

If for such an interpretation (R-algebra) α,

$$\alpha(r_{q_1}) = \pi_{x_2}(\|\mathtt{Takes}(x_1, x_2)\|_A) = \pi_{x_2}(\alpha(\mathtt{Takes}))$$

then also

$$Flux(\alpha, MakeOperads(\mathcal{M}_{CD})) = Flux(\alpha, MakeOperads(\mathcal{M}'_{AD})).$$

2. The mapping $\mathcal{M}'_{AD} = \{\Phi\} : \mathcal{A} \to \mathcal{D}$ in Example 5 with Φ equal to SOtgd

$$\exists f_{Student}(\forall x_1 \forall x_2 \forall y_3((\mathtt{Takes}(x_1, x_2) \\ \wedge (f_{Student}(x_1, y_3) \doteq \overline{1})) \Rightarrow \mathtt{Enrolment}(y_3, x_2))).$$

Then Φ_E is equal to

$$\exists f_{Student}(\forall x_1 \forall x_2 \forall y_3((\mathtt{Takes}(x_1, x_2) \\ \wedge (f_{Student}(x_1, y_3) \doteq \overline{1})) \Rightarrow r_{q_2}(y_3, x_2)))$$

and Φ_M is equal to

$$\exists f_{Student} \exists f_{Takes}(\forall x_1 \forall x_2 \forall y_3((r_{q_2}(y_3, x_2)) \wedge (f_{Takes}(x_1, y_2) \doteq \overline{1}) \\ \wedge (f_{Student}(x_1, y_3) \doteq \overline{1})) \Rightarrow \mathtt{Enrolment}(y_3, x_2)),$$

with a new schema $\mathcal{C} = (\{r_{q_2}\}, \emptyset)$ composed of a binary relational symbol r_{q_2}.

Let us define these two mappings, $\mathcal{M}_{AC} = \{\Phi_E\} : \mathcal{A} \to \mathcal{C}$ and $\mathcal{M}_{CD} = \{\Phi_M\} : \mathcal{C} \to \mathcal{D}$.

For any mapping-interpretation α such that $A = \alpha^*(\mathcal{A})$ and both schema mappings $\mathcal{M}_{AC}$ and $\mathcal{M}'_{AD}$ are satisfied, we have that

$$Flux(\alpha, MakeOperads(\mathcal{M}_{AC})) = Flux(\alpha, MakeOperads(\mathcal{M}'_{AD})).$$

If for such an interpretation (R-algebra) α,

$$\alpha(r_{q_2}) = \pi_{x_2, y_3}(\|\mathtt{Takes}(x_1, x_2) \wedge (f_{Student}(x_1, y_3) \doteq \overline{1})\|_A)$$

then also

$$Flux(\alpha, MakeOperads(\mathcal{M}_{CD})) = Flux(\alpha, MakeOperads(\mathcal{M}'_{AD})).$$

3. The mapping $\mathcal{M}_{AC} = \{\Phi\} : \mathcal{A} \to \mathcal{C}$ in Example 9 with Φ equal to SOtgd

$$\exists f_1 \exists f_{Emp} \exists f_{Over65} ($$
$$\forall x_e (((f_{Emp}(x_e) \doteq \overline{1}) \wedge \texttt{Local}(x_e)) \Rightarrow \texttt{Office}(x_e, f_1(x_e))$$
$$\wedge \forall x_e (((f_{Emp}(x_e) \doteq \overline{1}) \wedge (f_{Over65}(x_e) \doteq \overline{1})) \Rightarrow \texttt{CanRetire}(x_e)))).$$

Then Φ_E is equal to

$$\exists f_{Emp} \exists f_{Over65} ($$
$$\forall x_e (((f_{Emp}(x_e) \doteq \overline{1}) \wedge \texttt{Local}(x_e)) \Rightarrow r_{q3}(x_e))$$
$$\wedge \forall x_e (((f_{Emp}(x_e) \doteq \overline{1}) \wedge (f_{Over65}(x_e) \doteq \overline{1})) \Rightarrow r_{q4}(x_e)))$$

and Φ_M is equal to

$$\exists f_1 \exists f_{Emp} \exists f_{Over65} \exists f_{Local} ($$
$$\forall x_e ((r_{q3}(x_e) \wedge (f_{Emp}(x_e) \doteq \overline{1}) \wedge (f_{Local}(x_e) \doteq \overline{1})) \Rightarrow \texttt{Office}(x_e, f_1(x_e))$$
$$\wedge \forall x_e ((r_{q4}(x_e) \wedge (f_{Emp}(x_e) \doteq \overline{1}) \wedge (f_{Over65}(x_e) \doteq \overline{1})) \Rightarrow \texttt{CanRetire}(x_e)))),$$

with a new schema $\mathcal{C}' = (\{r_{q3}, r_{q4}\}, \emptyset)$ composed of two unary relational symbols r_{q3} and r_{q4}.

Lets us define these two mappings, $\mathcal{M}_{AC'} = \{\Phi_E\} : \mathcal{A} \to \mathcal{C}'$ and $\mathcal{M}_{C'C} = \{\Phi_M\} : \mathcal{C}' \to \mathcal{C}$.

For any mapping-interpretation α such that $A = \alpha^*(\mathcal{A})$ and both schema mappings $\mathcal{M}_{AC}$ and $\mathcal{M}_{AC'}$ are satisfied, we have that

$$\mathit{Flux}(\alpha, \mathit{MakeOperads}(\mathcal{M}_{AC'})) = \mathit{Flux}(\alpha, \mathit{MakeOperads}(\mathcal{M}_{AC})).$$

If for such an interpretation (R-algebra) α,

$$\alpha(r_{q3}) = \pi_{x_e}(\|(f_{Emp}(x_e) \doteq \overline{1}) \wedge \texttt{Local}(x_e)\|_A) = \alpha(\texttt{Emp}) \cap \alpha(\texttt{Local})$$

and

$$\alpha(r_{q4}) = \pi_{x_e}(\|(f_{Emp}(x_e) \doteq \overline{1}) \wedge (f_{Over65}(x_e) \doteq \overline{1})\|_A)$$
$$= \alpha(\texttt{Emp}) \cap \alpha(\texttt{Over65})$$

then also

$$\mathit{Flux}(\alpha, \mathit{MakeOperads}(\mathcal{M}_{C'C})) = \mathit{Flux}(\alpha, \mathit{MakeOperads}(\mathcal{M}_{AC})).$$

Based on this schema-mapping decomposition, we can obtain the instance-mapping decomposition as well:

Proposition 3 *For each schema mapping $\mathcal{M}_{AB} : \mathcal{A} \to \mathcal{B}$ with*

$$(\Phi_E, \Phi_M) = DeCompose(\mathcal{M}_{AB}),$$

let us define the mapping-operads

$$\mathbf{M}_E = MakeOperads(\{\Phi_E\}) \quad and \quad \mathbf{M}_M = MakeOperads(\{\Phi_M\}).$$

Let $S = \{\chi_1, \ldots, \chi_m\}$ be the set of implications of the normalized SOtgd of $\mathcal{M}_{AB}$. If for a mapping-interpretation α (with $A = \alpha^*(\mathcal{A})$ and $B = \alpha^*(\mathcal{B})$) such that for each implication $\chi_i \in S$, $q_{Ai}(\mathbf{x}_i) \Rightarrow r_i(\mathbf{t}_i)$, and its derived corresponding implications $q_{Ai}(\mathbf{x}_i) \Rightarrow r_{q_i}(\mathbf{y}_i)$ in S_E and $(r_{q_i}(\mathbf{y}_i) \wedge \psi_i(\mathbf{x}_i)) \Rightarrow r_i(\mathbf{t}_i)$ in S_M (in the algorithm Decompose), we have $\alpha(r_{q_i}) = \pi_{\mathbf{y}_i} \| q_{Ai}(\mathbf{x}_i) \|_A$, then:

$$Flux(\alpha, MakeOperads(\mathcal{M}_{AB})) = Flux(\alpha, \mathbf{M}_E) \cap Flux(\alpha, \mathbf{M}_M).$$

For such a mapping-interpretation α with the instance-mapping

$$f = \alpha^*(MakeOperads(\mathcal{M}_{AB})) : A \to B,$$

we denote the decomposition of f by two instance-mappings $ep(f) = \alpha^*(\mathbf{M}_E) : A \to TC$ and $in(f) = \alpha^*(\mathbf{M}_M) : TC \to B$, with schema $\mathcal{C} = (S_C, \emptyset)$ and $S_C = \bigcup_{q'_i \in \mathbf{M}_E} \partial_1(q'_i) = \{r_{q1}, \ldots, r_{qm}\}$ and $C = \alpha^*(S_C)$.

Proof Let $\mathbf{M}_{AB} = MakeOperads(\mathcal{M}_{AB}) = \{q_1, \ldots, q_m, 1_{r_\emptyset}\}$ with $q_i = v_i \cdot q_{A,i} \in O(r_{i,1}, \ldots, r_{i,k}, r_i^B)$, for $1 \le i \le m$, so that $MakeOperads(\Phi_E) = \{q'_1, \ldots, q'_m, 1_{r_\emptyset}\}$, with $q'_i = v'_i \cdot q'_{A,i} \in O(r_{i,1}, \ldots, r_{i,k}, r_{q_i})$, for $1 \le i \le m$, and $MakeOperads(\Phi_M) = \{q''_1, \ldots, q''_m, 1_{r_\emptyset}\}$ with $q''_i \in O(r_{q_i}, r_i^B)$.

First of all, from the fact that $\alpha(r_{q_i}) = \pi_{\mathbf{y}_i} \| q_{Ai}(\mathbf{x}_i) \|_A$, for $1 \le i \le m$, it follows, from Definition 11 for a mapping-interpretation, that (a.1.) the functions $\alpha(v_i)$ and $\alpha(v''_i)$ are equal. Then, from Definition 13,

(i) $Flux(\alpha, \mathbf{M}_E) = Flux(\alpha, \mathbf{M}_{AB}) = T\{\pi_{\mathbf{y}_i} \| e[(\_)_j / r_{i,j}]_{1 \le j \le k} \|_A \mid q'_i = v'_i \cdot q'_{A,i} = (e \Rightarrow (\_)(\mathbf{y}_i)) \in \mathbf{M}_E, q'_i \in O(r_{i_1,1}, \ldots, r_{i_k,k}, r_{q_i})$ and $\mathbf{y}_i \neq \emptyset\} = T\{\pi_{\mathbf{y}_i} \| q_{A,i}(\mathbf{x}_i) \|_A \mid 1 \le i \le m$ and $\mathbf{y}_i \neq \emptyset\} = T\{\alpha(r_{q_i}) \mid 1 \le i \le m$ and $\mathbf{y}_i \neq \emptyset\} = TC$ (because $\alpha(v'_i)$ is an identity function and hence an *injection* as well, for $1 \le i \le m$ such that $\mathbf{y}_i \neq \emptyset$).

(ii) $Flux(\alpha, \mathbf{M}_M) = T\{\pi_{\mathbf{y}_i} \| e[(\_)_1 / r_{q_i}] \|_C \mid q''_i = v''_i \cdot q''_{C,i} = (e \Rightarrow (\_)(\mathbf{t}_i)) \in \mathbf{M}_M, q''_i \in O(r_{q_i}, r_i^B)$ and $\mathbf{y}_i \neq \emptyset\} = T\{\pi_{\mathbf{y}_i} \| r_{q_i}(\mathbf{y}_i) \wedge \psi_i(\mathbf{x}_i) \|_C \mid 1 \le i \le m$ and $\mathbf{y}_i \neq \emptyset\} = T\{\pi_{\mathbf{y}_i} \| r_{q_i}(\mathbf{y}_i) \|_C \mid 1 \le i \le m$ and $\mathbf{y}_i \neq \emptyset\} =$
(since $\psi_i(\mathbf{d})$ is true when $q_{A,i}(\mathbf{d})$ is true, and $q_{A,i}(\mathbf{x}_i) \Rightarrow r_{q_i}(\mathbf{y}_i)$ is satisfied by α)
$= T\{\alpha(r_{q_i}) \mid 1 \le i \le m$ and $\mathbf{y}_i \neq \emptyset\}$ if $\alpha(v''_i)$ is an injection for all $1 \le i \le m$ such that $\mathbf{y}_i \neq \emptyset$; $\bot^0$ otherwise,
$=$ (from (a.1.)) $= T\{\alpha(r_{q_i}) \mid 1 \le i \le m$ and $\mathbf{y}_i \neq \emptyset\}$ if $\alpha(v_i)$ (equal to $\alpha(v''_i)$) is an injection for all $1 \le i \le m$ such that $\mathbf{y}_i \neq \emptyset$; $\bot^0$ otherwise,
$= T\{\pi_{\mathbf{y}_i} \| q_{Ai}(\mathbf{x}_i) \|_A \mid 1 \le i \le m$ and $\mathbf{y}_i \neq \emptyset\}$ if $\alpha(v_i)$ is an injection for all

$1 \leq i \leq m$ such that $\mathbf{y}_i \neq \emptyset$; $\bot^0$ otherwise,
$= T\{\pi_{\mathbf{x}_i} \| e[(\_)_j / r_{i,j}]_{1 \leq j \leq k} \|_A \mid q_i = v_i \cdot q_{A,i} = (e \Rightarrow (\_)(\mathbf{t}_i)) \in \mathbf{M}_{AB}, q_i \in O(r_{i_1,1}, \ldots, r_{i_k,k}, r_i^B)$ and $\mathbf{y}_i \neq \emptyset\}$, if $\alpha(v_i)$ is an injection for all $1 \leq i \leq m$ such that $\mathbf{y}_i \neq \emptyset$; $\bot^0$ otherwise,
$= \mathit{Flux}(\alpha, \mathbf{M}_{AB})$.
Thus, from (i) and (ii), $\mathit{Flux}(\alpha, \mathbf{M}_E) \cap \mathit{Flux}(\alpha, \mathbf{M}_M) = \mathit{Flux}(\alpha, \mathbf{M}_{AB})$.
The instance mappings $in(f)$ and $ep(f)$ are well defined because $C \subseteq TC$ (i.e., all relations in C are the relations in TC as well). □

In Example 16, we considered the decomposition of a mapping between two database schemas. Let us now consider the case of a decomposition of the integrity-constraint mapping for a given schema:

Example 17 Let as consider a schema $\mathcal{A} = (S_A, \Sigma_A)$ with the integrity constraints in $\Sigma_A = \Sigma_A^{\text{egd}} \cup \Sigma_A^{\text{tgd}}$, which can be represented (see Example 12) by a schema mapping $\mathcal{T}_{AA_\top} = \{\Phi \wedge \Psi\} : \mathcal{A} \to \mathcal{A}_\top$, where Φ is the SOtgd equal to $\mathit{EgdsToSOtgd}(\Sigma_A^{\text{egd}})$ and Ψ is the SOtgd equal to $\mathit{TgdsToConSOtgd}(\Sigma_A^{\text{tgd}})$ and $\mathcal{A}_\top = (\{r_\top\}, \emptyset)$ is an auxiliary schema, and, consequently, by equivalent operads-mapping

$$\mathbf{T}_{AA_\top} = \mathit{MakeOperads}\big(\{\Phi \wedge \Psi\}\big) : \mathcal{A} \to \mathcal{A}_\top$$

with the *empty* information flux.

That is, from Corollary 6, for each mapping-interpretation α, $\mathit{Flux}(\alpha, \mathbf{T}_{AA_\top}) = \bot^0$, because each operad's operation $q_i \in \mathbf{T}_{AA_\top}$ has a form $e \Rightarrow (\_)(\overline{0}, \overline{1})$ without free variables on the right-hand side of this implication and hence, from Definition 13, $\mathit{Var}(\mathbf{T}_{AA_\top}) = \emptyset$.

Thus, from the fact that $\mathit{Var}(\mathbf{T}_{AA_\top}) = \emptyset$ and the Decomposition algorithm, we obtain that $(\Phi_E, \Phi_M) = \mathit{DeCompose}(\mathcal{T}_{AA_\top})$ where Φ_E is composed of implications $q_{Ai}(\mathbf{x}_i) \Rightarrow r_\emptyset$ and Φ_M is composed of implications $(r_\emptyset \wedge \psi_{Ai}(\mathbf{x}_i)) \Rightarrow r_\top(\overline{0}, \overline{1})$.

Thus, from the fact that $r_\emptyset$ is a tautology (truth propositional letter, Definitions 1 and 8), each implication in Φ_E, $q_{Ai}(\mathbf{x}_i) \Rightarrow r_\emptyset$ is a tautology and hence Φ_E is a tautology as well, so that we can substitute it by a trivial tautology $r_\emptyset \Rightarrow r_\emptyset$.

From the fact that $r_\emptyset$ is a tautology, each implication $(r_\emptyset \wedge \psi_{Ai}(\mathbf{x}_i)) \Rightarrow r_\top(\overline{0}, \overline{1})$ is equivalent to the implication $\psi_{Ai}(\mathbf{x}_i) \Rightarrow r_\top(\overline{0}, \overline{1})$, that is, to the implication $q_{Ai}(\mathbf{x}_i) \Rightarrow r_\top(\overline{0}, \overline{1})$ (from the fact that $q_{Ai}(\mathbf{x}_i)$ is logically equivalent to $\psi_{Ai}(\mathbf{x}_i)$). Consequently, Φ_M is equal to the normalized Φ, so that SOtgd Φ_M and the original SOtgd Φ are logically equivalent.

Consequently, $\Phi_E \wedge \Phi_M$ is logically equivalent to the original SOtgd Φ.

Thus, we obtain a trivial mapping (see Example 7 for the trivial mappings with the empty database schema $\mathcal{A}_\emptyset$)

$$\mathcal{M}_{AA_\emptyset} = \{\Phi_E\} = \{r_\emptyset \Rightarrow r_\emptyset\} : \mathcal{A} \to \mathcal{A}_\emptyset \quad \text{and} \quad \mathcal{M}_{A_\emptyset A_\top} = \{\Phi_M\} : \mathcal{A}_\emptyset \to \mathcal{A}_\top.$$

Consequently, the obtained mapping-operads are $\mathbf{M}_{AA_\emptyset} = \mathit{MakeOperads}(\{\Phi_E\}) = \{1_{r_\emptyset}\} : \mathcal{A} \to \mathcal{A}_\emptyset$, and $\mathbf{M}_{A_\emptyset A_\top} = \mathit{MakeOperads}(\{\Phi_M\}) : \mathcal{A}_\emptyset \to \mathcal{A}_\top$. This decomposition can be represented by the following commutative diagram

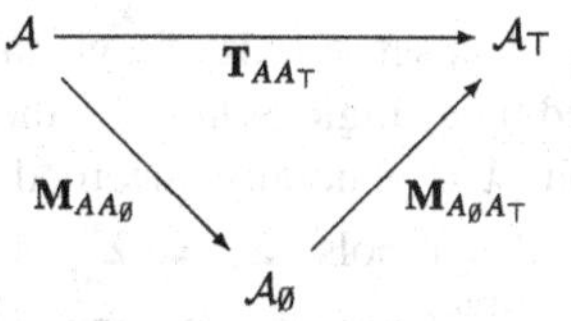

The commutativity $\mathbf{T}_{AA_T} = \mathbf{M}_{A_\emptyset A_T} \circ \mathbf{M}_{AA_\emptyset}$ comes from the fact that $\mathsf{T}_{AA_T} = \mathcal{M}_{A_\emptyset A_T} \circ \mathcal{M}_{AA_\emptyset} = Compose(\mathcal{M}_{A_\emptyset A_T}, \mathcal{M}_{AA_\emptyset})$ and, in this case, from the fact that Φ is logically equivalent to Φ_M and to $\Phi_E \wedge \Phi_M$.

Note that for both obtained mappings $Var(\mathbf{M}_{AA_\emptyset}) = Var(\mathbf{M}_{A_\emptyset A_T}) = \emptyset$, and hence they have also empty information fluxes for each mapping-interpretation α. That is, $Flux(\alpha, \mathbf{M}_{AA_\emptyset}) = Flux(\alpha, \mathbf{M}_{A_\emptyset A_T}) = \bot^0$.

The decomposition of any trivial mapping (in Example 7) $\mathcal{M} = \{r_\emptyset \Rightarrow r_\emptyset\}$ will produce the pair of two trivial mappings as well, that is, we will not produce new mappings by their decomposition. If we apply the decomposition to the graph in the previous example, by using the original mappings instead of derived mapping-operads, we will obtain the following graph:

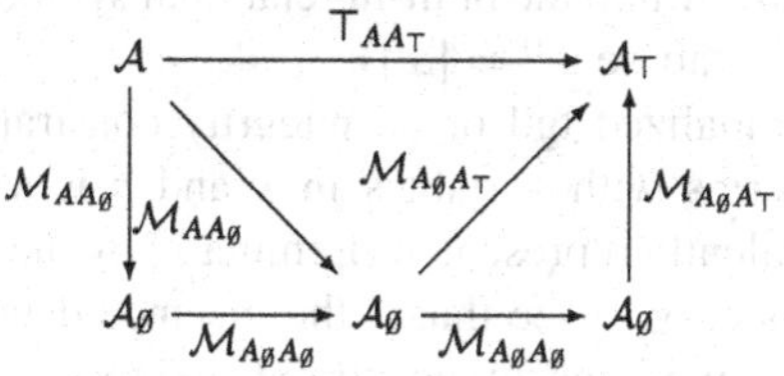

where $\mathcal{M}_{A_\emptyset A_\emptyset} = Id_{A_\emptyset} = \{r_\emptyset \Rightarrow r_\emptyset\} : \mathcal{A}_\emptyset \to \mathcal{A}_\emptyset$ is the identity mapping for the empty schema $\mathcal{A}_\emptyset$ (see Example 7).

2.6 Database Schema Mapping Graphs

In this section, we will summarize all the results of this chapter by a formal translation of the logic theory, of a given database mapping system, into a graph's formalism which represents a categorification provided in the next two chapters, that is, a translation of the logic schema-mapping theory into a small sketch category with a functorial semantics for its models.

We recall that any tuple-generating *mapping* from a schema $\mathcal{A}$ into a schema $\mathcal{B}$ is represented by a set of tgds $\forall \mathbf{x}(\phi_A(\mathbf{x}) \Rightarrow \exists \mathbf{z}(\psi_B(\mathbf{y}, \mathbf{z})))$ where $\mathbf{y} \subseteq \mathbf{x}$, by taking out the universal quantification $\forall \mathbf{x}$ from the head of this tgd logical sentence, that is, by the *view mapping* $q_A(\mathbf{y}) \Rightarrow q_B(\mathbf{y})$ where $q_A(\mathbf{y})$ (equivalent to $\exists \mathbf{y}_1 \phi_A(\mathbf{y}, \mathbf{y}_1)$ with $\mathbf{x}$ equal to the tuple of all variables in $\mathbf{y}$ and $\mathbf{y}_1$) is a conjunctive query over the schema $\mathcal{A}$ and $q_B(\mathbf{y})$ (equivalent to $\exists \mathbf{z}\, \psi_B(\mathbf{y}, \mathbf{z})$) is a conjunctive query over the schema $\mathcal{B}$. These view-based mappings are interpreted as follows:

- When a sentence $q_A(\mathbf{d})$ is true for a tuple of values $\mathbf{d}$ then $q_B(\mathbf{d})$ has to be a true sentence as well, so that the information $\mathbf{d}$ of an instance-database A (of the schema $\mathcal{A}$) is "transferred" by this mapping into the instance-database B (of the schema $\mathcal{B}$).

We have demonstrated that a set of tgds of a given inter-schema mapping can be equivalently represented by a single SOtgd (by the algorithm *TgdsToSOtgd*). A relational database schema $\mathcal{A}$ is generally specified by a pair (S_A, Σ_A) where S_A is a set of n-ary relational symbols, $\Sigma_A = \Sigma_A^{\text{tgd}} \cup \Sigma_A^{\text{egd}}$ with the set of the database integrity constraints Σ_A^{egd} expressed by *equality-generating dependencies* (egds) and the set of the *tuple-generating dependencies (tgds)* Σ_A^{tgd} in Definition 2.

Any integrity constraint (egd) $\forall \mathbf{x}(\phi_A(\mathbf{x}) \Rightarrow (\mathbf{y} \doteq \mathbf{z}))$ of a given schema database $\mathcal{A}$, with $\mathbf{y} = \langle y_1, \ldots, y_k\rangle \subseteq \mathbf{x}$ and $\mathbf{z} = \langle z_1, \ldots, z_k\rangle \subseteq \mathbf{x}$, will be represented (Lemma 3) by the new mapping $\forall \mathbf{x}((\phi_A(\mathbf{x}) \wedge (\mathbf{y} \neq \mathbf{z})) \Rightarrow r_\top(\overline{0}, \overline{1}))$ (from Lemma 3), where $r_\top$ is the built-in binary relational symbol for equality of FOL. The interpretation of this mapping is the same as for the standard inter-schema mappings:

When $\phi_A(\mathbf{x}) \wedge (\mathbf{y} \neq \mathbf{z})$ is true for a tuple of values $\mathbf{d}$, from the fact that the ground atom $r_\top(\overline{0}, \overline{1})$ is a false, it holds that the mapping $(\phi_A(\mathbf{x}) \wedge (\mathbf{y} \neq \mathbf{z})) \Rightarrow r_\top(\overline{0}, \overline{1})$ is not satisfied. It is easy to see that if the instance-database A is a *model* of the schema $\mathcal{A}$ (i.e., when the integrity constraints expressed by the corresponding egds are satisfied) then $q_A(\mathbf{x}) \wedge (\mathbf{y} \neq \mathbf{z})$ cannot be satisfied. The "transferred" information flux by this mapping (from A into the built-in relational symbol $r_\top$) is always empty (i.e., equal to the empty database $\bot^0 = \{\bot\}$).

Analogously, any normalized tgd of an integrity constraint $\forall \mathbf{x}(\phi_A(\mathbf{x}) \Rightarrow r(\mathbf{t}))$ where $\mathbf{t}$ is a tuple of terms with variables in $\mathbf{x}$ and r is a relational symbol of schema $\mathcal{A}$, can be equivalently represented (Lemma 2) by the formula $\forall \mathbf{x}((\phi_A(\mathbf{x}) \wedge \neg r(\mathbf{t})) \Rightarrow r_\top(\overline{0}, \overline{1}))$. It is easy to see that if the instance-database A is a *model* of the schema $\mathcal{A}$ (i.e., when the integrity constraints expressed by the corresponding tgds are satisfied) then $q_A(\mathbf{x}) \wedge r(\mathbf{t})$ cannot be satisfied. The "transferred" information flux by this mapping (from A into the built-in relational symbol $r_\top$) is always empty (i.e., equal to the empty database $\bot^0 = \{\bot\}$).

It is consistent with the representation of the integrity constraints (both egds and tgds) by the inter-schema mappings because the sentences as integrity-constraints do not transfer any data from source to target database and hence their information flux has to be empty. Note that in the case of ordinary query mappings, the minimal information flux is $\{\bot\} = \bot^0$ as well. We have demonstrated that a set of tgds in Σ_A^{tgd} can be equivalently represented by a single SOtgd (by the algorithm *TgdsToConSOtgd* in Sect. 2.2.1) and that a set of egds in Σ_A^{egd} can be equivalently represented by a single SOtgd (by the algorithm *EgdsToSOtgd* in Sect. 2.2.2).

Moreover, in order to translate this particular second-order logic (based on SOtgds sentences) into the categorial setting, we explained how we can translate the SOtgds into the set of abstract operad's operations that specify a mapping between database schemas and hence to use the functorial semantics for the database mappings based on R-algebras for operads (provided in the previous Sect. 2.4). Based on this translation into operads, we have seen that each operad's operation $q_i \in O(r_1, \ldots, r_m, r)$ obtained from an *atomic* mapping (Definition 7) is an algebraic specification for an implication conjunct in a normalized SOtgd.

Based on these considerations and Example 12 for the integrity constraints, we will formally define a graph used to specify a database schema-mapping system:

Definition 14 For a given set S of non-empty database schemas, we define the graph $G = (V_G, E_G)$ with $S \subseteq V_G$ as follows:

1. For each database schema $\mathcal{A} = (S_A, \Sigma_A)$ where $\Sigma_A = \Sigma_A^{\text{tgd}} \cup \Sigma_A^{\text{egd}}$ is not empty, we define the mapping edge $\mathsf{T}_{A A_\top} = \{\Phi\} : \mathcal{A} \to \mathcal{A}_\top$ in E_G where the database schema $\mathcal{A}_\top = (\{r_\top\}, \emptyset)$ is defined as a new distinct vertex in V_G and Φ is defined as follows:
   - If Σ_A^{egd} is an empty set of egds then we define Φ to be
   $$TgdsToConSOtgd\big(\Sigma_A^{\text{tgd}}\big);$$
   - If Σ_A^{tgd} is an empty set of tgds then we define Φ to be $EgdsToSOtgd(\Sigma_A^{\text{egd}})$;
   - Otherwise we define Φ to be
   $$EgdsToSOtgd\big(\Sigma_A^{\text{egd}}\big) \wedge TgdsToConSOtgd\big(\Sigma_A^{\text{tgd}}\big).$$
2. For each inter-schema mapping defined by a non-empty set S of tgds, between two database schemas $\mathcal{A} = (S_A, \Sigma_A)$ and $\mathcal{B} = (S_B, \Sigma_B)$, we define the edge $\mathcal{M}_{AB} = \{\Phi\} : \mathcal{A} \to \mathcal{B}$ in E_G where Φ is the SOtgd obtained from this set of tgds by *TgdsToSOtgd(S)*.

All defined edges in E_G of this graph G will be called *atomic* mappings.

Note that a mapping graph is a necessary step in order to define a small sketch category, as it follows from Corollary 2, and to make the full embedding of the database mapping system in the Categorical logic, based on functors (i.e., R-algebras) from such a sketch category into the denotational (instance database's) category. In fact, every directed graph can be transformed into a category as follows: the objects are the vertices of the graph and the arrows are paths in the graph.

Thus, the first next step is to examine if for this denotational database category we can use the standard **Set** category or, instead, we need a more appropriate one.

2.7 Review Questions

1. Can you explain the semantic difference between the ordinary tgds and *full* tgds? Why do we need a subset of the Second Order Logic? Can we deal with incomplete knowledge with the 2-valued logics and without the Second Order Logic?
2. Can we also deal with the inconsistent information by this kind of SOtgds? In which way are we able to resolve the problems with the mutually-inconsistent data in the standard GLAV semantics of an integration of a number of the source databases into a unique global schema?
3. The many-valued logics are used for the paraconsistent logics as well. What is gained by using the many-valued logics in order to deal with incomplete and inconsistent information, and what are the drawbacks? Which is a minimal many-valued logic able to also deal with incomplete and inconsistent data? It

is well known that the logic negation as an antitonic truth operator, and hence cannot be used in order to obtain the solutions for the schema mappings by using standard fix-point semantics. How we can overcome such a problem by the many-valued logics where we also use incomplete (unknown) information?
4. Are the provided algorithms for transformation of logic formulae (tgds) into the algebraic (operad's based) expressions able to deal with the logic negation operators as well? Is there any example presented in this chapter where we need negation logic operators, and, if so, why?
5. Why it is important to express the integrity constraints over the database schemas as a kind of the schema mappings? What is main difference between such integrity-mappings and standard inter-schema mappings based on the tgds? Are the integrity-constraint mappings important for the composition of the mappings? If not then explain why and which "trick" was used in order to make an effective separation of them and of inter-schema mappings. Do we have other technical possibilities to obtain an analogous result and compare it with the method proposed here?
6. Why are we using the typed operads as an effective algebraic language in order to provide the "algebraization" of the SOtgds logics used for the schema mappings? What is the difference of such an operad's algebra and the standard algebraic semantics presented in the introduction? What are the main operations in the operad's algebra? Can you reformulate the operad's algebra by the ordinary semantics of algebras and their operators, in the way as it was done with the category theory in Sect. 1.5.1?
7. What is the main reason for a definition of the semantics of the schema mappings in Definition 11, based on the Tarski's FOL semantics? Does this definition extend the FOL semantics into the Second Order Semantics for the tgds? Why did we introduce the negation operators in this method, if we did not use the logic negation in the tgds? Can we use this semantics of the mappings for the tgds with the queries that are using the logic negation operator as well? Try to elaborate such an extension and show some simple example.
8. An Abstract Data-Object is a basic concept for the observational point of view where the resulting relation (a view), obtained for a given query, is seen as an observation of the information contained in a given database instance. What is the relationship between the power-view database TA obtained as the (also infinite) set of observations on a given instance database A, with the syntax of the used query-language? If we are using the SQL for the relational databases then we obtain $A \subseteq TA$. Is this power-view operator T a monotonic operator, and, if it is, what is the reason for it? What do we obtain if we apply the operator T to the instance database TA?
9. What is intuitively the strict semantics of the schema mappings? Do we need it in a special case of the Data exchange setting, and if not, is it the main generalization of the Data exchange setting? Is the information flux a database as well, and can it contain the data which are not provided by the source database of a composed mapping with a number of the intermediate databases between the source and target database? If an atomic-data in an RDB database A is any

simple value contained in one of its relations, does it belong to the database TA as a single-attribute single-tuple relation?

10. Why do we provide the definition of equality of two database mappings at the instance level only, and how does this fact generalize the Data Exchange framework and the equivalence of the logic formulae used in SOtgds? Why can the category semantics presented in the introduction be satisfied by the information flux for a given instance mapping between two instance databases? Is the empty information flux between two database instances (of a source and target schema) equivalent to the fact that there is no mapping between them? How can it be explained in the case of an integrity-constraint mapping for the database schemas?
11. What is the relationship between the categorial symmetry applied to the database mapping systems and the algorithm for decomposition of SOtgds? Is the decomposition of the mappings with the empty information flux meaningful? Is the decomposition of the integrity-constraints mappings meaningful? Why is it useful to represent the database mapping systems by the graphs, from the user point of view and from the semantic (denotational) point of view? Why is the compositional property of the schema mappings so important, and which relationship does it have with the compositional properties of the information fluxes at the instance level of the database mappings? Is it a necessary condition in order to obtain a denotational semantics of the database mappings based on categories?

References

1. J.F. Adams, *Infinite Loop Spaces* (Princeton U. Press, Princeton, 1978)
2. J.C. Baez, J. Dofan, Categorification, in *Workshop on Higher Category Theory and Physics*, March 28–30, ed. by E. Getzler, M. Kapranov (Northwestern University, Evanston, 1997)
3. J.M. Boardman, R.M. Vogt, *Homotopy Invariant Structures on Topological Spaces*. Lecture Notes in Mathematics, vol. 347 (Springer, Berlin, 1973)
4. A. Corradini, A complete calculus for equational deduction in coalgebraic specification. Report SEN-R9723, National Research Institute for Mathematics and Computer Science, Amsterdam (1997)
5. R. Fagin, P.G. Kolaitis, L. Popa, W. Tan, Composing schema mappings: second-order dependencies to the rescue. ACM Trans. Database Syst. **30**(4), 994–1055 (2005)
6. L. Libkin, C. Sirangelo, Data exchange and schema mappings in open and closed worlds, in *Proc. of PODS'08*, Vancuver, Canada (2008)
7. J.P. May, *Simplicial Objects in Algebraic Topology* (Van Nostrand, Princeton, 1968)

Definition of DB Category

3

3.1 Why Do We Need a New Base Database Category?

For the composition of *complex* database mapping graphs, it is important to distinguish the following two basic compositions of database schemas $\mathcal{A}$ and $\mathcal{B}$ with respect to DBMSs:

- In the composed schema, in order to make it impossible to write a query over the composition with relations of both databases, the two database schemas should be mutually *separated* by two independent DBMSs: it is common case when two databases are separated, and this binary separation-composition at the schema level is denoted by $\mathcal{A} \dagger \mathcal{B}$ (disjoint union where all elements are indexed by schema symbols (denoted by a label $\ddot{\mathcal{A}}$, for a schema $\mathcal{A} = (S_A, \Sigma_A)$), equal to the complex schema $(S_{A\dagger B}, \Sigma_{A\dagger B})$ where $S_{A\dagger B} = (\{\ddot{\mathcal{A}}\} \times S_A) \cup (\{\ddot{\mathcal{B}}\} \times S_B)$ and $\Sigma_{A\dagger B} = (\{\ddot{\mathcal{A}}\} \times \Sigma_A) \cup (\{\ddot{\mathcal{B}}\} \times \Sigma_B)$, and $\times$ is the Cartesian product of sets (if $\mathcal{B} = \mathcal{A}$, then we set $\ddot{\mathcal{A}} = 1$ and $\ddot{\mathcal{B}} = 2$).
- In the composed schema, when the two database schemas are *connected* into the same DBMS (without any change of the two original database schemas): in this case, we are able to use the queries over this composed schema with relations *of both* databases for inter database mappings. This binary federation-composition, denoted by $\mathcal{A} \oplus \mathcal{B}$, is the (simple) *union* but with *renaming* (if we have the same relational symbol $r \in S_A \cap S_B$, then, before the union, we rename this symbol in S_B by a new fresh relational symbol of the same type), so that $S_{\mathcal{A}\oplus\mathcal{B}} = S_A \frac{\cup}{rn} S_B$, where $\frac{\cup}{rn}$ denotes the union with renaming.

The equality '=' for database schemas is naturally defined as follows: for any two schemas $\mathcal{A}, \mathcal{B} \in \mathbb{S}$, $\mathcal{A} = \mathcal{B}$ if $\pi_i(\mathcal{A}) = \pi_i(\mathcal{B})$, $i = 1, 2$. Let us consider the mappings $\mathcal{M} : \mathcal{A} \dagger \mathcal{B} \to \mathcal{C}$ and $\mathcal{M} : \mathcal{A} \oplus \mathcal{B} \to \mathcal{C}$. In the first case, in any query mapping $q(\mathbf{x}) \Rightarrow q_C(\mathbf{x})$ in the SOtgd of $\mathcal{M}$, all relational symbols in the query $q(\mathbf{x})$ must be of the database $\mathcal{A}$ or (mutually exclusive) of the database $\mathcal{B}$, that is,

$$\mathcal{M} = \big\{\exists \mathbf{f}\big(\forall \mathbf{x}_1\big(q_{A,1}(\mathbf{x}_1) \Rightarrow q_{C,1}(\mathbf{t}_1)\big) \wedge \cdots \wedge \forall \mathbf{x}_k\big(q_{A,k}(\mathbf{x}_k)$$
$$\Rightarrow q_{C,k}(\mathbf{t}_k)\big) \wedge \forall \mathbf{y}_1\big(q_{B,1}(\mathbf{y}_1)$$

Z. Majkić, *Big Data Integration Theory*, Texts in Computer Science,
DOI 10.1007/978-3-319-04156-8_3,

$$\Rightarrow q_{C,k+1}\big(\mathbf{t}'_1\big)\big) \wedge \cdots \wedge \forall \mathbf{y}_m\big(q_{B,m}(\mathbf{y}_m)$$

$$\Rightarrow q_{C,k+m}\big(\mathbf{t}'_m\big)\big)\big)\big\}, \qquad (3.1)$$

with $\mathbf{t}_i$ the tuple of terms with variables in $\mathbf{x}_i$, for $1 \le i \le k$, and $\mathbf{t}'_i$ the tuple of terms with variables in $\mathbf{y}_i$, for $1 \le i \le m$. Consequently, this mapping is the graph:

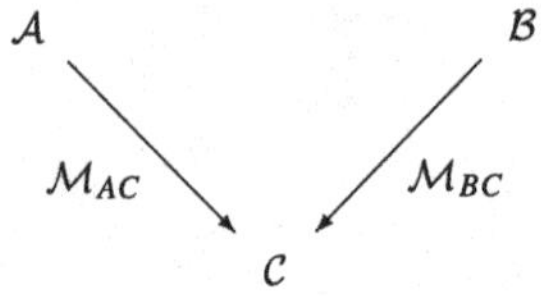

with

$$\mathcal{M}_{AC} = \big\{\exists \mathbf{f}_A\big(\forall \mathbf{x}_1\big(q_{A,1}(\mathbf{x}_1) \Rightarrow q_{C,1}(\mathbf{t}_1)\big) \wedge \cdots \wedge \forall \mathbf{x}_k\big(q_{A,k}(\mathbf{x}_k) \Rightarrow q_{C,k}(\mathbf{t}_k)\big)\big)\big\}$$

and

$$\mathcal{M}_{BC} = \big\{\exists \mathbf{f}_B\big(\forall \mathbf{y}_1\big(q_{B,1}(\mathbf{y}_1) \Rightarrow q_{C,k+1}\big(\mathbf{t}'_1\big)\big) \wedge \cdots \wedge \forall \mathbf{y}_m\big(q_{B,m}(\mathbf{y}_m)$$
$$\Rightarrow q_{C,k+m}\big(\mathbf{t}'_m\big)\big)\big)\big\},$$

and $\mathbf{f}_A, \mathbf{f}_B \subseteq \mathbf{f}$, while in the case of mapping $\mathcal{M} : \mathcal{A} \oplus \mathcal{B} \to \mathcal{C}$ such a decomposition is not possible because we can have a query mapping $q(\mathbf{x}) \Rightarrow q_C(\mathbf{x})$ in the SOtgd of $\mathcal{M}$ with relational symbols in $q(\mathbf{x})$ from both databases $\mathcal{A}$ and $\mathcal{B}$.

If we introduce the mappings $\mathcal{M}_1 = \{\bigwedge\{\forall \mathbf{x}_i(r_{Ai}(\mathbf{x}_i) \Rightarrow r_{Ai}(\mathbf{x}_i)) | r_{Ai} \in \mathcal{A}\}\}$ and $\mathcal{M}_2 = \{\bigwedge\{\forall \mathbf{y}_i(r_{Bi}(\mathbf{y}_i) \Rightarrow r_{Bi}(\mathbf{y}_i)) | r_{Bi} \in \mathcal{B}\}\}$ then we obtain the mapping graph

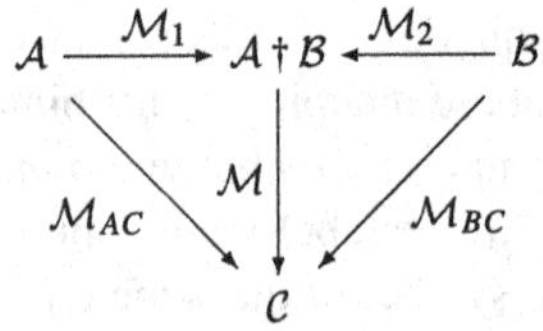

that can be seen as a cocone diagram for schema database mappings.

Let us consider another dual example, a mapping $\mathcal{M} : \mathcal{C} \to \mathcal{A} \dagger \mathcal{B}$. In this case, in any query mapping $q_C(\mathbf{x}) \Rightarrow q(\mathbf{x}) \in \mathcal{M}$, all relational symbols in the query $q(\mathbf{x})$ must be of database $\mathcal{A}$ or (mutually exclusive) of database $\mathcal{B}$. That is,

$$\mathcal{M} = \big\{\exists \mathbf{f}\big(\forall \mathbf{x}_1\big(q_{C,1}(\mathbf{x}_1) \Rightarrow q_{A,1}(\mathbf{t}_1)\big) \wedge \cdots \wedge \forall \mathbf{x}_k\big(q_{C,k}(\mathbf{x}_k)$$

$$\Rightarrow q_{A,k}(\mathbf{t}_k)\big) \wedge \forall \mathbf{y}_1\big(q_{C,k+1}(\mathbf{y}_1)$$

$$\Rightarrow q_{B,1}(\mathbf{t}'_1)) \wedge \cdots \wedge \forall \mathbf{y}_m (q_{C,k+m}(\mathbf{y}_m)$$
$$\Rightarrow q_{B,m}(\mathbf{t}'_m)))\}, \tag{3.2}$$

Consequently, this mapping can be equivalently represented by the graph:

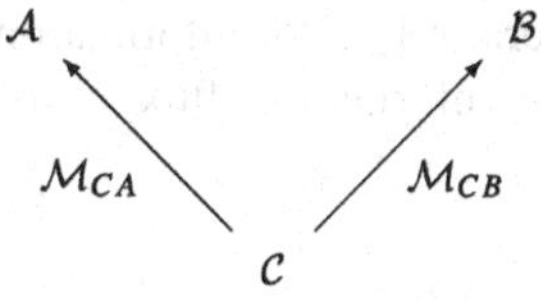

where

$$\mathcal{M}_{CA} = \{\exists \mathbf{f}_A (\forall \mathbf{x}_1 (q_{C,1}(\mathbf{x}_1) \Rightarrow q_{A,1}(\mathbf{t}_1)) \wedge \cdots \wedge \forall \mathbf{x}_k (q_{C,k}(\mathbf{x}_k) \Rightarrow q_{A,k}(\mathbf{t}_k)))\}$$

and

$$\mathcal{M}_{BC} = \{\exists \mathbf{f}_B (\forall \mathbf{y}_1 (q_{B,1}(\mathbf{y}_1) \Rightarrow q_{C,k+1}(\mathbf{t}'_1)) \wedge \cdots \wedge \forall \mathbf{y}_m (q_{B,m}(\mathbf{y}_m)$$
$$\Rightarrow q_{C,k+m}(\mathbf{t}'_m)))\}.$$

If we again introduce the mappings $\mathcal{M}_1 = \{\wedge\{\forall \mathbf{x}_i (r_{Ai}(\mathbf{x}_i) \Rightarrow r_{Ai}(\mathbf{x}_i)) | r_{Ai} \in \mathcal{A}\}\}$ and $\mathcal{M}_2 = \{\wedge\{\forall \mathbf{y}_i (r_{Bi}(\mathbf{y}_i) \Rightarrow r_{Bi}(\mathbf{y}_i)) | r_{Bi} \in \mathcal{B}\}\}$ then

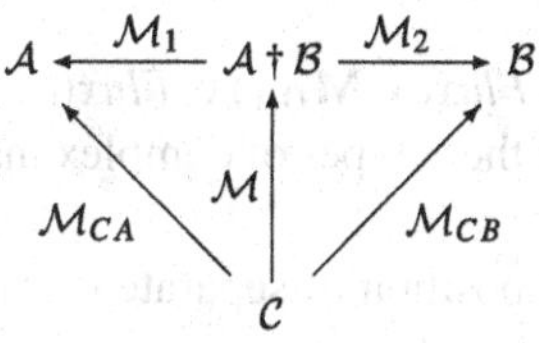

is a dual graph that can be seen as a cone diagram for the schema database mappings.

It is important to notice that differently from the mapping graphs, where the schema mappings are expressed by SOtgds, the small sketch categories derived from such a graph will have the equivalent schema mappings but represented algebraically by the sets of operad's operations, in order to be able to represent each R-algebra α as a *functor* from this sketch category (at the schema-level) into the denotational (instance-level) **DB** category. For the mappings that involve the composed source schema (by separation operator $\dagger$) $\mathcal{M} = \{\Phi\} : \mathcal{A} \dagger \mathcal{B} \to \mathcal{C}$, we cannot directly use the algorithm *MakeOperads*($\mathcal{M}$) because we would obtain operad's operations possibly with relations of both schemas. Consequently, we have to divide this SOtgd Φ into a conjunction $\Phi_A \wedge \Phi_B$, where Φ_A is an SOtgd whose left-hand sides of implications contain only relational symbols in $\mathcal{A}$ and Φ_B only relational symbols in

$\mathcal{B}$ (as in the mapping (3.1)). Consequently, we define the complex mapping-operad $\mathbf{M}_{(A\dagger B)C}$ by:

$$\mathbf{M}_{(A\dagger B)C} \triangleq [\mathbf{M}_{AC}, \mathbf{M}_{BC}] : \mathcal{A} \dagger \mathcal{B} \to \mathcal{C} \tag{3.3}$$

where $\mathbf{M}_{AC} \triangleq \mathit{MakeOperads}(\{\Phi_A\}) = (q_1, \dots, q_k, 1_{r\emptyset}\} : \mathcal{A} \to \mathcal{C}$ and $\mathbf{M}_{BC} \triangleq \mathit{MakeOperads}(\{\Phi_B\} = (q'_1, \dots, q'_m, 1_{r\emptyset}\}) : \mathcal{B} \to \mathcal{C}$ are two simple mapping-operads by introducing a structural-operator [_ , _]. Information flux of complex mapping-operads is different from the information flux of ordinary (simple) mapping-operads:

$$\begin{aligned} &\mathit{Flux}\big(\alpha, [\mathbf{M}_{AC}, \mathbf{M}_{BC}]\big) \\ &\quad \triangleq \mathit{Flux}(\alpha, \mathbf{M}_{AC}) \uplus \mathit{Flux}(\alpha, \mathbf{M}_{BC}) \\ &\quad = \big\{(1, a) \mid a \in \mathit{Flux}(\alpha, \mathbf{M}_{AC})\big\} \cup \big\{(2, b) \mid b \in \mathit{Flux}(\alpha, \mathbf{M}_{BC})\big\}. \end{aligned} \tag{3.4}$$

Analogously, each complex mapping-operad $\mathbf{M}_{D(A\dagger B)}$ from a simple (non-separated) schema $\mathcal{D}$ into separated schema $\mathcal{A} \dagger \mathcal{B}$ is denoted as $\langle \mathbf{M}_{DA}, \mathbf{M}_{DB} \rangle : \mathcal{D} \to \mathcal{A} \dagger \mathcal{B}$ where $\mathbf{M}_{DA} : \mathcal{D} \to \mathcal{A}$ and $\mathbf{M}_{DB} : \mathcal{D} \to \mathcal{B}$ are ordinary (simple) mapping-operads (equal to $\{1_{r\emptyset}\}$ if they are not defined), with

$$\mathit{Flux}\big(\alpha, \langle \mathbf{M}_{DA}, \mathbf{M}_{DB} \rangle\big) \triangleq \mathit{Flux}(\alpha, \mathbf{M}_{DA}) \uplus \mathit{Flux}(\alpha, \mathbf{M}_{DB}). \tag{3.5}$$

Any pair of simple mapping-operads $\mathbf{M}_{AC}, \mathbf{M}_{BD}$ can be expressed by single complex mapping-operad:

$$\mathbf{M}_{(A\dagger B)(C\dagger D)} = \mathbf{M}_{AC} \dagger \mathbf{M}_{BD} \triangleq \mathbf{M}_{AC} \uplus \mathbf{M}_{BD} : \mathcal{A} \dagger \mathcal{B} \to \mathcal{C} \dagger \mathcal{D}, \tag{3.6}$$

with $\mathit{Flux}(\alpha, \mathbf{M}_{AC} \uplus \mathbf{M}_{BD}) \triangleq \mathit{Flux}(\alpha, \mathbf{M}_{DA}) \uplus \mathit{Flux}(\alpha, \mathbf{M}_{DB})$.

Now we introduce formally these types of complex mapping-operads, as follows:

Definition 15 The n-ary composition of separated simple schemas define the following four types of complex morphisms:

1. A mapping-operad $\biguplus(\mathbf{M}_{A_1C_1}, \dots, \mathbf{M}_{A_nC_n}) : \mathcal{A}_1 \dagger \cdots \dagger \mathcal{A}_n \to \mathcal{C}_1 \dagger \cdots \dagger \mathcal{C}_n$ with

$$\begin{aligned} &\mathit{Flux}\big(\alpha, \biguplus(\mathbf{M}_{A_1C_1}, \dots, \mathbf{M}_{A_nC_n})\big) \\ &\quad = \biguplus\big(\mathit{Flux}(\alpha, \mathbf{M}_{A_1C_1}), \dots, \mathit{Flux}(\alpha, \mathbf{M}_{A_nC_n})\big); \end{aligned}$$

2. A mapping-operad $[\mathbf{M}_{A_1C}, \dots, \mathbf{M}_{A_nC}] : \mathcal{A}_1 \dagger \cdots \dagger \mathcal{A}_n \to \mathcal{C}$ with

$$\mathit{Flux}\big(\alpha, [\mathbf{M}_{A_1C}, \dots, \mathbf{M}_{A_nC}]\big) = \biguplus\big(\mathit{Flux}(\alpha, \mathbf{M}_{A_1C}), \dots, \mathit{Flux}(\alpha, \mathbf{M}_{A_nC})\big);$$

3. A mapping-operad $\langle \mathbf{M}_{DA_1}, \dots, \mathbf{M}_{DA_n} \rangle : D \to \mathcal{A}_1 \dagger \cdots \dagger \mathcal{A}_n$ with

$$\mathit{Flux}\big(\alpha, \langle \mathbf{M}_{DA_1}, \dots, \mathbf{M}_{DA_n} \rangle\big) = \biguplus\big(\mathit{Flux}(\alpha, \mathbf{M}_{DA_1}), \dots, \mathit{Flux}(\alpha, \mathbf{M}_{DA_n})\big).$$

4. The composition of last two types of complex morphisms is a mapping-operad from $\mathcal{D}$ into $\mathcal{C}$:

$$\begin{aligned}\mathbf{M}_{DC} &= [\mathbf{M}_{A_1C}, \ldots, \mathbf{M}_{A_nC}] \circ \langle \mathbf{M}_{DA_1}, \ldots, \mathbf{M}_{DA_n} \rangle \\ &= \lceil \mathbf{M}_{A_1C} \circ \mathbf{M}_{DA_1}, \ldots, \mathbf{M}_{A_nC} \circ \mathbf{M}_{DA_n} \rfloor : \mathcal{D} \to \mathcal{A}_1 \dagger \cdots \dagger \mathcal{A}_n \to \mathcal{C},\end{aligned}$$

with a set of non-empty mappings $\mathbf{M}_{A_iC} \circ \mathbf{M}_{DA_i} : \mathcal{D} \to \mathcal{C}, i = 1, \ldots, n$, enclosed by structural-operator $\lceil\_, \ldots, \_\rfloor$, all with the same source and target schemas, so that the resulting information flux is a flux of the union of all these mappings:

$$Flux(\alpha, \mathbf{M}_{DC}) = T\Big(\bigcup \big\{Flux(\alpha, \mathbf{M}_{A_iC} \circ \mathbf{M}_{DA_i}) \mid 1 \le i \le n\big\}\Big).$$

5. Each schema mapping above, with structural operators $\uplus$, $[,]$, $\langle,\rangle$ and $\lceil,\rfloor$, is denominated as a *complex* $\uplus$-atomic sketch's schema mapping.

For a given mapping-interpretation α, we obtain the instance-database complex morphisms, for example, $\alpha^*([\mathbf{M}_{A_1C}, \ldots, \mathbf{M}_{A_nC}]) = [\alpha^*(\mathbf{M}_{A_1C}), \ldots, \alpha(\mathbf{M}_{A_nC})]$. It will be analyzed in details in Sect. 3.3. We also introduce another kind of binary symmetric operations for a composition of the schemas and schema mappings, denoted by $\underset{\alpha}{\overset{\cap}{-}}$, where α is an R-algebra introduced in Sect. 2.4.1, Definition 10:

Definition 16 For each R-algebra α, we define an α-intersection symmetric binary operator $\underset{\alpha}{\overset{\cap}{-}}$ for the schema mappings, as follows:
1. For any two simple schemas $\mathcal{A}$ and $\mathcal{B}$, their α-intersection is defined by:

$$\mathcal{A} \underset{\alpha}{\overset{\cap}{-}} \mathcal{B} = (S, \emptyset), \quad \text{where } S = \big\{r \in \mathbb{R} \mid \alpha(r) \in T\big(\alpha^*(\mathcal{A})\big) \cap T\big(\alpha^*(\mathcal{B})\big)\big\}.$$

2. For any two simple schema mappings $\mathcal{M}_{AC} : \mathcal{A} \to \mathcal{C}$ and $\mathcal{M}_{BD} : \mathcal{B} \to \mathcal{D}$, the α-intersection of these two mappings is defined by:

$$\mathcal{M}_{AC} \underset{\alpha}{\overset{\cap}{-}} \mathcal{M}_{BD} = \{\Phi\} : \mathcal{A} \underset{\alpha}{\overset{\cap}{-}} \mathcal{B} \to \mathcal{C} \underset{\alpha}{\overset{\cap}{-}} \mathcal{D},$$

where Φ is a tgd

$$\bigwedge \Big\{\forall \mathbf{x}\big(r(\mathbf{x}) \Rightarrow r(\mathbf{x})\big) \mid r \in \mathcal{A} \underset{\alpha}{\overset{\cap}{-}} \mathcal{B} \text{ and } \alpha(r) \in Flux\big(\alpha, MakeOperads(\mathcal{M}_{AC})\big) \cap Flux\big(\alpha, MakeOperads(\mathcal{M}_{BD})\big)\Big\}.$$

It is easy to show that for a simple schema $\mathcal{A}$ and R-algebra α, with the instance-database $A = \alpha^*(\mathcal{A})$, the schema of the instance-database TA is equal to $\mathcal{A} \underset{\alpha}{\overset{\cap}{-}} \mathcal{A}$, that is, $TA = \alpha^*(\mathcal{A} \underset{\alpha}{\overset{\cap}{-}} \mathcal{A})$.

More over, for $A = \alpha^*(\mathcal{A})$ and $B = \alpha^*(\mathcal{B})$, $TA \cap TB = \alpha^*(\mathcal{A} \underset{\alpha}{\overset{\cap}{-}} \mathcal{B})$.

Analogously, for the α-intersection of schema mappings, it holds that

$$Flux\left(\alpha, \mathcal{M}_{AC} \underset{\alpha}{\overset{\cap}{-}} \mathcal{M}_{BD}\right) = Flux(\alpha, MakeOperads(\mathcal{M}_{AC}))$$

$$\cap\, Flux(\alpha, MakeOperads(\mathcal{M}_{BD})).$$

The schema separation, connection and α-intersection are necessary in order to manage complex database mappings as it will be demonstrated in what follows. However, their introduction is a reason for the introduction of a new database category as well. This fact will be explained in the next section.

3.1.1 Introduction to Sketch Data Models

The category-theoretic data model that we use has come to be known as the *sketch data model.*

As the sketch data model has been applied more widely, and particularly with the development of new approaches to problems like view update problem, theoreticians have increasingly been seeking for a way of translating between sketch data models and other data models. Potentially, sketch data model techniques and results might be translated into other models more familiar to practitioners.

Much of the theory of relational databases has been based around normalization, and it has been shown that the relational states corresponding to sketch data models (instances of sketch data model schema) are in at least third normal form.

The detailed presentation of sketches for the database mappings and their functorial semantics will be given in Sect. 4.1.

Sketches are developed by Ehresmann's school, especially by R. Guitartand and C. Lair [3, 6, 9]. Sketch is a category together with a distinguished class of cones and cocones. A *model* of the sketch is a set-valued functor turning all distinguished cones into limit cones, and all distinguished cocones into colimit cocones, in the category **Set** of sets.

There is an elementary and basic connection between sketches and logic [15]. Given any sketch, we can consider the underlying graph of the sketch as a (many-sorted) language, and we can write down axioms in the $\mathcal{L}_{\infty,\infty}$-logic (the infinitary FOL with finite quantifiers) over this language, so that the models of the axioms become exactly the models of the sketch.

The category of models of a given sketch has models as objects and the arrows that represent all natural transformations between the models as functors. A category is sketchable (esquissable) or accessible iff it is equivalent to the category of set-valued models of a small sketch.

Recall that a graph $G = (V_G, E_G)$ consists of a set of vertices denoted by V_G and a set of arrows (edges) denoted by E_G together with the operators $dom, cod : E_G \to V_G$ which assigns to each arrow its source and target. (Co)cones and diagrams are defined for graphs in exactly the same way as they are for categories, but commutative co(cones) and diagrams, of course, make no sense for graphs.

By a sketch me mean a 4-tuple (G, u, D, C) where G is a graph, $u : V_G \to E_G$ is a function which takes each vertex (node) $\mathcal{A}$ in V_G to an arrow from $\mathcal{A}$ to $\mathcal{A}$, D is a class of diagrams in G, and C is a class of (co)cones in G. Each (c)cone in G goes (to)from some vertex (from)to some diagram; that diagram need not be in D and, in fact, it is necessary to allow diagrams which are not in D as bases of (co)cones.

Notice that, differently from the work dedicated to categorical semantics of Entity-Relationship internal relational database models where nodes of sketches are single relations, here, at a higher level of abstraction, the nodes are whole databases. Consequently, in such a framework, we do not use commutative database mapping systems and hence D is an empty set. In fact, in a database mapping system, the (co)cone diagrams above will never be used in practical representations of database mapping systems. Instead of that, they will be alternatively used only for their self-consistent parts, as a first diagram above, or, equivalently, as a single arrow $\mathcal{M} : \mathcal{A} \dagger \mathcal{B} \to \mathcal{C}$.

However, for the introduced schema composition operator $\dagger$, the above cone and cocone diagrams have to be presented in C for our sketches.

Consequently, we obtain the following fundamental lemma for the categorial modeling of database mappings:

Lemma 6 *The* **Set** *cannot be used as the base category for the models of database-mapping sketches.*

Proof Let $\mathbf{E} = (G, u, D, C)$ be a database sketch where C is a set of (co)cones of the two diagrams introduced for the database schema composition operator $\dagger$, and a model of this sketch be a functor $F : \mathbf{E} \to \mathbb{B}$ where $\mathbb{B}$ is a base category. Then all cones in C have to be functorially translated into limit commutative diagrams in $\mathbb{B}$ and all cocones in C have to be functorially translated into colimit commutative diagrams in $\mathbb{B}$, i.e., the cocone in the figure in Sect. 3.1 has to be translated into a coproduct diagram and the cone has to be translated into a product diagram in $\mathbb{B}$. Consequently, the object $F(\mathcal{A} \dagger \mathcal{B})$ has to be both the product $A \times B$ and coproduct $A + B$, where $A = F(\mathcal{A})$ and $B = F(\mathcal{B})$ are two objects in $\mathbb{B}$; $\times$ and $+$ are the product and coproduct operators in $\mathbb{B}$. However, this cannot be done in **Set**. In fact, the product $A \times B$ in **Set** is the Cartesian product of these two sets A and B, while the coproduct $A + B$ is the disjoint union. Hence, $A \times B \simeq A + B$ is not an isomorphism in **Set**. Another reason is that any morphism in **Set** is a single function, while in our case it is a set of functions (or a *binary relation of tuples* obtained by union of graphs of this set of functions). □

Remark The fundamental consequence of this lemma is that we need to define a new base category for the categorial semantics of database mappings.

In fact, this is the first task that has to be done and we will define this new base category $\mathbb{B}$, denoted by **DB** (DataBase) category, and will show that it satisfies the duality property where the product and coproduct diagrams are dual diagrams and hence for any two objects (instance-databases) in **DB**, the objects $A \times B$ and $A + B$ are equal (up to isomorphism).

Given a database mapping system represented by a logic-graph $G = (V_G, E_G)$, in Definition 14, with a set of vertices V_G that are the database schemas and a set of edges E_G such that each schema mapping $\mathcal{M} \in E_G$ is a singleton set with SOtgd as the unique element (or alternatively, the empty set), we derive its sketch-graph G' (by substituting each edge $\mathcal{M}_{AB} = \{\Phi\} : \mathcal{A} \to \mathcal{B}$ of G by the algebraic-edge represented by a mapping-operad $\mathbf{M}_{AB} = MakeOperads(\mathcal{M}_{AB}) : \mathcal{A} \to \mathcal{B}$, which is a set of operad's operations) and we denote the sketch category obtained from this graph G by $\mathbf{Sch}(G)$.

This is a fundamental choice in order to have a coherent functorial semantics for the database mappings in the Categorical logic in which the models of a given database-mapping sketch-graph G are the R-algebras (mapping-interpretations in Definition 11) for the set of all operad's operations used in the morphisms of the sketch category $\mathbf{Sch}(G)$.

More details of the transformation of a database-mapping graph G (where the edges are SOtgds) into its sketch category $\mathbf{Sch}(G)$ (where each morphism is the set of operad's operations obtained from the corresponding mapping SOtgd) will be presented in Chap. 4 dedicated to the functorial semantics for database schema mappings, with a number of applications as well.

3.1.2 Atomic Sketch's Database Mappings

Based on considerations in Sect. 2.6 for schema mappings and in Sect. 3.1.1 for the sketches, we can formally introduce a theory for the *basic* (or atomic) mappings of sketches. In what follows, for each schema $\mathcal{A} = (S_A, \Sigma_A)$ we introduce its power-schema $\widehat{\mathcal{A}} = (S_{\widehat{A}}, \emptyset)$ such that $S_{\widehat{A}} = \{r \mid r = rsym(q(\mathbf{x})), q(\mathbf{x}) \in \mathcal{L}_A\}$ is the set of relational symbols obtained from a countable set $\mathcal{L}_A$ of views (each view is a union of conjunctive queries) $q(\mathbf{x})$ for a given schema $\mathcal{A}$ (introduced in Sect. 1.4.1). The function $rsym : \mathcal{L}_A \to \mathbb{R}$ assigns a particular relational k-ary symbol to each such a view so that for each $r \in S_A$, $rsym(r(\mathbf{x})) = r$. The interpretation for a power-schema relations is fixed by $\alpha(rsym(q(\mathbf{x}))) = \|q(\mathbf{x})\|_{\alpha^*(\mathcal{A})}$. Thus, for a given interpretation α such that $A = \alpha^*(\mathcal{A})$, $\alpha^*(\widehat{\mathcal{A}}) = \alpha^*(\{rsym(q(\mathbf{x})) \mid q(\mathbf{x}) \in \mathcal{L}_A\}) = \{\|q(\mathbf{x})\|_A \mid q(\mathbf{x}) \in \mathcal{L}_A\} = TA = \alpha^*(\mathcal{A} \underset{\alpha}{\widehat{\sqcap}} \mathcal{A})$.

Definition 17 We define the following two types of simple *atomic sketch's* arrows (or schema mappings) from a source schema $\mathcal{A} = (S_A, \Sigma_A)$:

1. VIEW-MAPPING—used for the queries and inter-schema database mappings:
   - For any single conjunctive *query* $q_i(\mathbf{x}_i)$ over a schema $\mathcal{A}$ with relational symbols in $\{r_{i1}, \ldots, r_{ik}\} \subseteq S_A$, we can define a schema mapping-operad's operation $q_i = v_i \cdot q_{A,i}$ with $q_{A,i} \in O(r_{i1}, \ldots, r_{ik}, r_q)$.

     If this operad's operation q_i is only a query over $\mathcal{A}$ then $v_i = 1_{r_q} \in O(r_q, r_q)$ is the identity operation, and this query can be represented by an atomic mapping $\mathbf{M}_{A\widehat{A}} = \{q_i, 1_{r_\emptyset}\} = MakeOperads(\{\forall \mathbf{x}_i (q_i(\mathbf{x}_i) \Rightarrow r_q(\mathbf{x}_i))\})$:

$\mathcal{A} \to \widehat{\mathcal{A}}$ where the $r_q = rsym(q_i(\mathbf{x}_i)) \in S_{\widehat{\mathcal{A}}}$ is a new introduced relational symbol for this query with $ar(r_q) = |\mathbf{x}_i|$ (the number of variables in the tuple $\mathbf{x}_i$). For a given mapping interpretation α with $A = \alpha^*(\mathcal{A})$, $Flux(\alpha, \mathbf{M}_{\mathcal{A}\widehat{\mathcal{A}}}) = T(\|q_i(\mathbf{x}_i)\|_A)$.

The union of conjunctive queries $q_1(\mathbf{x}_i), \dots, q_k(\mathbf{x}_i)$ with the same head is represented by an atomic mapping $\mathbf{M}_{\mathcal{A}\widehat{\mathcal{A}}} = \{q_1, \dots, q_k, 1_{r_\emptyset}\} : \mathcal{A} \to \widehat{\mathcal{A}}$, where each q_i, $1 \le i \le k$, is defined above and $Flux(\alpha, \mathbf{M}_{\mathcal{A}\widehat{\mathcal{A}}}) = T(\bigcup_{1 \le i \le k} \|q_i(\mathbf{x}_i)\|_A)$.

- For more complex atomic mappings that use a set of tgds S from $\mathcal{A}$ into $\mathcal{B}$ (each one is an implication with left-hand side equal to a conjunctive query $q_m(\mathbf{x}_m)$ over a schema $\mathcal{A}$), the atomic mapping is equal to the set of operad's operations in

$$\mathbf{M}_{AB} = MakeOperads\big(TgdsToSOtgd(S)\big) : \mathcal{A} \to \mathcal{B}.$$

2. Integrity-constraints mapping—used for specification of schema integrity constraints (in Definition 2, Lemma 2 and Lemma 3):

   2.1. By SOtgd $\Psi = TgdsToConSOtgd(\Sigma_A^{\mathrm{tgd}})$ for tgds, with implications $(\phi_A(\mathbf{x}) \wedge \neg r(\mathbf{t})) \Rightarrow r_\top(\overline{0}, \overline{1})$.

   2.2. By SOtgd $\Phi = EgdsToSOtgd(\Sigma_A^{\mathrm{egd}})$ for egds, with implications $(\phi_A(\mathbf{x}) \wedge (\mathbf{y} \neq \mathbf{z})) \Rightarrow r_\top(\overline{0}, \overline{1})$.

   Consequently, we obtain an atomic integrity-constraint schema mapping

$$\mathbf{T}_{AA_\top} = \big(\{\Phi \wedge \Psi\}\big) = \{q_1, \dots, q_k, 1_{r_\emptyset}\} : \mathcal{A} \to \mathcal{A}_\top,$$

   with operad's operations $q_i = v_i \cdot q_{A,i}$, such that $q_{A,i} \in O(r_1, \dots, r_n, r_q)$, $v_i \in O(r_q, r_\top)$, and $\{r_1, \dots, r_n\} \subseteq S_A$, for $1 \le i \le k$.

The semantics of the integrity-constraint mappings and the conditions for a mapping-interpretation α (i.e., R-algebra that satisfies the conditions in Definition 11) such that it corresponds to a given *model* (an instance that satisfies all integrity constraints) of a database schema $\mathcal{A}$ will be defined in Sect. 4.1. This integrity-constraints mapping over a given database schema is used only in order to specify the subset of interpretations α that are the models of this schema database. Consequently, the integrity-constraints mappings do not have any rule in the sequential compositions of database mappings (differently from the view-mappings). We recall that the integrity-constraints arrows (for any mapping-interpretation α) in the **DB** category have an empty information flux as it was demonstrated in Sect. 2.4.3.

Corollary 8 *Each simple atomic sketch's arrow* $\mathbf{M}_{AB} : \mathcal{A} \to \mathcal{B}$ *is obtained from an atomic schema mapping* $\mathcal{M}_{AB} : \mathcal{A} \to \mathcal{B}$. *Each complex* $\uplus$*-atomic sketch's arrow is obtained by structural operators in Definition* 15 *and a set of simple sketch's atomic arrows.*

Proof The single query sketch's mapping

$$\mathbf{M}_{\mathcal{A}\widehat{\mathcal{A}}} = \{q, 1_{r_\emptyset}\} = MakeOperads\big(\big\{\forall \mathbf{x}\big(q(\mathbf{x}) \Rightarrow r_q(\mathbf{x})\big)\big\}\big)$$

is obtained from the schema mapping $\mathcal{M}_{A\widehat{A}} = \{\forall \mathbf{x}(q(\mathbf{x}) \Rightarrow r_q(\mathbf{x}))\} : \mathcal{A} \to \widehat{\mathcal{A}}$ that is a simple atomic mapping as well. All other atomic sketch's mappings in Definition 17 are obtained from the atomic schema mappings. For complex $\uplus$-atomic sketch's schema mappings it holds from point 5 in Definition 15. □

Let us show the following important property for a schema database mapping system represented by a graph $G = (V_G, E_G)$ in Definition 14:

Proposition 4 *For a given non-empty schema database mapping system specified by a graph* $G = (V_G, E_G)$ *(Definition* 14), *for each graph edge* $\mathcal{M}_{AB} \in E_G$, *the morphism MakeOperads*($\mathcal{M}_{AB}$) *in the sketch category* **Sch**(G) *is an atomic mapping.*

Proof From Definition 14, each edge in a schema mapping graph G is obtained by the set of tgds or by the set of egds, so that by *MakeOperads* algorithm applied to such an edge we obtain an atomic view-mapping or an atomic integrity-constraint mapping in Definition 17. □

Notice that in the sketch categories we introduce the schema-separation compositions $\mathcal{A} \dagger \mathcal{B}$ and the $\uplus$-atomic sketch's schema mappings in Definition 15 between composed database schemas. As usual in the category theory (sketches are small categories), the composition of $\uplus$-*atomic* sketch's schema mappings can generate any kind of $\uplus$-complex sketch's schema mappings as explained in Definition 15 and will be presented by more details in the next section.

3.2 DB (Database) Category

Based on an observational point of view for relational databases and on the theory of R-algebras for mapping-operads, presented in Sect. 2.4, Definition 11, we are able to introduce a category **DB** [14] for instance-databases and view-based mappings between them, with the set of its objects Ob_{DB} and the set of its morphisms Mor_{DB}.

For the instance-database mappings, in Sect. 2.4.1 and Definition 10, we considered the R-algebras that transform the set of schema-mapping operad's operations $\mathbf{M}_{AB} = MakeOperads(\mathcal{M}_{AB}) = \{q_1, \ldots, q_k, 1_{r_\emptyset}\}$, of a given schema mapping $\mathcal{M}_{AB} : \mathcal{A} \to \mathcal{B}$, into the set of functions $\alpha^*(\mathbf{M}_{AB}) = \{\alpha(q_1), \ldots, \alpha(q_k), q_\perp\}$. Consequently, each morphism in the base denotational **DB** category has to be a set of functions from an instance-database A into an instance-database B (of schemas $\mathcal{A}$ and $\mathcal{B}$, respectively).

However, we need to characterize such an instance-mapping morphism also by its information flux (introduced in Definition 13) corresponding to this set of functions obtained from a mapping-interpretation α (in Definition 11) of operad's operations q_i above.

In order to facilitate the presentation of the properties of this category, we will begin with simple objects and simple arrows, and then extend all to complex objects and complex arrows. In this section, we will establish the principal properties for the objects and morphisms in **DB** and, after that, we will consider in dedicated sections the other particular properties.

Theorem 1 *Let us define, for a fixed enough big universe* $\mathcal{U} = \mathbf{dom} \cup SK$ *where* **dom** *is a finite set of values and* $SK = \{\omega_0, \omega_1, \dots\}$ *an infinite set of indexed Skolem constants, the sets of objects* Ob_{DB} *and morphisms* Mor_{DB} *of this* **DB** *category*:

1. *A simple object is an instance-database* A*, composed of a set of* n*-ary (finite* $n \geq 0$*) relations* $R_i = a_i \in A$, $i = 1, 2, \dots$ *also called the "elements of* A*" with unique nullary empty relation* $\bot \in A$ *such that* $A \cap \bot^0 = \bot^0 = \{\bot\}$.

   Complex objects are obtained by disjunctive union of simple objects.

   A closed object *is an instance-database* A *such that* $A = TA$.

   By Υ *we denote the top-closed object in* **DB** *such that* $\Upsilon = \bigcup_{A \in Ob_{DB}} A$.
2. *Let us define the set*

$$S_{DB} = \bigcup_{A,B \in Ob_{DB}} \big\{ \big(\alpha^*(\mathbf{M}_{AB}), Flux(\alpha, \mathbf{M}_{AB})\big) \mid \mathbf{M}_{AB} : \mathcal{A} \to \mathcal{B} \text{ is a simple atomic sketch's mapping or a complex } \uplus\text{-atomic sketch's mapping, and } \alpha \text{ is a mapping-interpretation such that } A = \alpha^*(\mathcal{A}), B = \alpha^*(\mathcal{B}) \big\}.$$

   The set of morphisms includes atomic morphisms in $\pi_1(S_{DB})$ *and all other morphisms that can be obtained by well-defined compositions of atomic morphisms.*
3. *We define recursively the mapping* $B_T : Mor_{DB} \to Ob_{DB}$ *such that for any simple morphism* $f : A \to B$:

$$B_T(f) \triangleq \widetilde{f} \quad \text{if } (f, \widetilde{f}) \in S_{DB}; \qquad B_T(g) \cap B_T(h) \quad \text{otherwise}$$
$$\big(\text{when } f = g \circ h, h, g \in Mor_{DB} \text{ with } cod(h) = dom(g)\big).$$

These sets of objects and morphisms define the principal part of the database category **DB**. *The extension of the mapping* B_T *to complex arrows (and their compositions) obtained by structural-operators* $\_\uplus\_$, $[\_,\_]$, $\langle\_,\_\rangle$ *and* $\lceil\_,\_\rfloor$ *is given in this section by Definition* 20.

Proof For each simple object $A = \alpha^*(\mathcal{A})$, we have its identity morphism $id_A : A \to A$ where (from Lemma 5)

$$\begin{aligned} id_A &= \alpha^*\Big(MakeOperads\Big(\Big\{\bigwedge\big\{\forall \mathbf{x}_i \big(r_i(\mathbf{x}_i) \Rightarrow r_i(\mathbf{x}_i)\big) \mid r_i \in \mathcal{A}\big\}\Big\}\Big)\Big) \\ &= \alpha^*\big(\big\{1_{r_i} \in O(r_i, r_i) \mid r_i \in \mathcal{A}\big\} \cup \{1_{r_\emptyset}\}\big) \\ &= \big\{id_{r_i} : \alpha(r_i) \to \alpha(r_i) \mid r_i \in \mathcal{A}\big\} \cup \{q_\bot\} \end{aligned}$$

and $\widetilde{id_A} = TA$. For a complex object $A = \alpha^*(\mathcal{A}_1 \dagger \cdots \dagger \mathcal{A}_k) = \biguplus_{1 \le i \le k} A_i$, where $A_i = \alpha^*(\mathcal{A}_i)$, its identity arrow is $id_A = \biguplus_{1 \le i \le k} id_{A_i}$ with simple atomic identity arrows $id_{A_i} : A_i \to A_i$. Moreover, $\widetilde{id_A} = TA = \biguplus_{1 \le i \le k} \widetilde{id_{A_i}} = \biguplus_{1 \le i \le k} TA_i$ which is a closed object as well.

Let $\mathbf{M}_{AB} = \mathit{MakeOperads}(\mathcal{M}_{AB}) = \{q_1^A, \ldots, q_m^A, 1_{r\emptyset}\}$ and

$$\mathbf{M}_{BC} = \mathit{MakeOperads}(\mathcal{M}_{BC}) = \{q_1^B, \ldots, q_n^B, 1_{r\emptyset}\}.$$

Then

$$\begin{aligned}
&\alpha^*(\mathbf{M}_{BC} \circ \mathbf{M}_{AB}) \\
&\quad = \alpha^*\big(\{q_i^B \cdot (q_{i_1}^A, \ldots, q_{i_{k_i}}^A) \mid 1 \le i \le n \text{ and } 1 \le i_j \le m \text{ for } 1 \le j \le k_i\} \\
&\qquad \cup \{1_{r\emptyset} \cdot 1_{r\emptyset}\}\big) \quad \text{(from definition of R-algebras in Definition 10)} \\
&\quad = \big\{\alpha(q_i^B)\big(\alpha(q_{i_1}^A) \times \cdots \times \alpha(q_{i_{k_i}}^A)\big) \mid 1 \le i \le n \text{ and } 1 \le i_j \le m \text{ for } 1 \le j \le k_i\big\} \\
&\qquad \cup \{q_\perp\} \quad (\text{from } \alpha(1_{r\emptyset}) = q_\perp) \\
&\quad = \big\{\alpha(q_1^B), \ldots, \alpha(q_n^B), q_\perp\big\} \circ \big\{\alpha(q_1^A), \ldots, \alpha(q_M^A), q_\perp\big\} \\
&\quad = \alpha^*\big(\{q_1^B, \ldots, q_n^B, 1_{r\emptyset}\}\big) \circ \alpha^*\big(\{q_1^A, \ldots, q_m^A, 1_{r\emptyset}\}\big) \\
&\quad = \alpha^*(\mathbf{M}_{BC}) \circ \alpha^*(\mathbf{M}_{AB}).
\end{aligned}$$

Thus, for given two morphisms $f = \alpha^*(\mathbf{M}_{AB}) : A \to B$ and $g = \alpha^*(\mathbf{M}_{BC}) : B \to C$ their composition is equal to:

$$g \circ f = \alpha^*(\mathbf{M}_{BC}) \circ \alpha^*(\mathbf{M}_{AB}) = \alpha^*(\mathbf{M}_{BC} \circ \mathbf{M}_{AB}) : A \to C.$$

Let as show the associativity of composition of morphisms in **DB**, with $h = \alpha^*(\mathbf{M}_{CD})$:

$$\begin{aligned}
&h \circ (g \circ f) \\
&\quad = \alpha^*(\mathbf{M}_{CD}) \circ \big(\alpha^*(\mathbf{M}_{BC}) \circ \alpha^*(\mathbf{M}_{AB})\big) = \alpha^*\big(\mathbf{M}_{CD} \circ (\mathbf{M}_{BC} \circ \mathbf{M}_{AB})\big) \\
&\qquad \text{(from the associativity of the operad-mapping composition in Corollary3)} \\
&\quad = \alpha^*\big((\mathbf{M}_{CD} \circ \mathbf{M}_{BC})\big) \circ \alpha^*(\mathbf{M}_{AB}) = \big(\alpha^*(\mathbf{M}_{CD}) \circ \alpha^*(\mathbf{M}_{BC})\big) \circ \alpha^*(\mathbf{M}_{AB}) \\
&\quad = (h \circ g) \circ f.
\end{aligned}$$

The left and right compositions of $\uplus$-atomic arrows in Definition 15 are based on component-to-component composition of atomic arrows and are specified in detail in the rest of this section. Thus, the properties above hold for the $\uplus$-atomic arrows as well. Consequently, **DB** is a well-defined category. □

The object composed of the empty relation only is denoted by $\perp^0$ and $T\perp^0 = \perp^0 = \{\perp\}$. We will show in Lemma 10 that the empty database (a database with only empty relations) is isomorphic to this bottom object $\perp^0$. Notice that this statement introduces, for each two simple objects A and B, the morphism $\bot^1 = \alpha^*(\mathbf{M}_{A,B})$ where $\mathbf{M}_{A,B} = \mathit{MakeOperads}(\{r_\emptyset \Rightarrow r_\emptyset\}) = \{1_{r_\emptyset}\}$ (see Example 7 in Sect. 2.4), with the empty information flux $\widetilde{\bot^1} = \bot^0$ (from Definition 13). The objects $\bot^0$ and Υ are the bottom and the top closed objects in **DB** (much more about them and the lattice of all databases (objects in **DB**) is presented in Sect. 8.1.5).

From definitions of morphisms, each morphism $f \in Mor_{DB}$ is a *set* of functions in $\alpha^*(\mathbf{M}_{AB})$ such that $\alpha(q_i) = \alpha(v_i)(\alpha(q_{A,i}))$ is a function that satisfies the mapping-interpretation conditions of Definition 11. Consequently, **DB** is different from the **Set** category (where each morphism is a single function) as we previously established in Sect. 3.1 and Lemma 6. As we have seen in the algorithm *MakeOperads*, each operad's operation q_i of a schema mapping between a schema $\mathcal{A}$ and a schema $\mathcal{B}$ is generally composed of two components: the first one corresponds to a conjunctive query $q_{A,i}$ over a source database $\mathcal{A}$ that defines this view-based mapping and the second component v_i defines which contribution of this mapping is transferred into the target relation, i.e., a kind of Global-or-Local-As-View (GLAV) mapping (sound or exact) [12].

Based on the atomic schema mappings introduced by Definition 17, we can introduce the *atomic morphisms* for instance-mappings in **DB** category as follows:

Definition 18 (ATOMIC MORPHISMS in **DB** category) An atomic morphism in **DB** is a set of functions $f = \alpha^*(\mathbf{M}_{AB})$ with the information flux $\widetilde{f} = B_T(f)$ such that $\mathbf{M}_{AB} = \{q_1, \ldots, q_k, 1_{r_\emptyset}\} : \mathcal{A} \to \mathcal{B}$ is an *atomic sketch's* schema mapping (Definition 17) and α is a mapping-interpretation (Definition 11) with $A = \alpha^*(\mathcal{A})$, $B = \alpha^*(\mathcal{B})$.

Let the schema mapping $\mathcal{M}_{AB} = \mathit{InverseOperads}(\mathbf{M}_{AB}) : \mathcal{A} \to \mathcal{B}$ be satisfied by α. Then for each operad's operation $q_i = v_i \cdot q_{A,i} \in \mathbf{M}_{AB}$, with $q_{A,i} \in O(r_{i1}, \ldots, r_{im}, r_q)$ and $v_i \in O(r_q, r_i), r_i \in \mathcal{B}$, the function $\alpha(v_i)$ is an injection (from Corollary 4) used to distinguish sound and exact assumption on the views as follows:

1. *Sound* case, i.e., when $\alpha(r_q) \subseteq \alpha(r_i)$;
2. *Exact* case, i.e., special inclusion case when $\alpha(r_q) = \alpha(r_i)$.

We extend the operators ∂_0 and ∂_1 to **DB** morphisms as well, so that

$$\partial_k(f) = \bigcup_{q_i \in \alpha^*(\mathbf{M}_{AB})} \partial_k(q_i), \quad \text{for } k = 0, 1.$$

A *complete morphism* (c-morphism) $f : A \to B$ satisfies the condition $\partial_0(f) \subseteq A$.

Every atomic morphism is a complete morphism. Thus, each view-map q_{A_i}, i.e., an atomic morphism $f = \{q_{A_i}, q_\bot\} : A \longrightarrow TA$, is a complete morphism (the case when $B = TA$ and $\alpha(v_i)$ belongs to the "exact case").

As we have seen from Definition 15 for the $\uplus$-atomic sketch's mappings,

$$\alpha^*\Big(\biguplus(\mathbf{M}_{DA_1}, \ldots, \mathbf{M}_{DA_n})\Big) = \biguplus\big(\alpha^*(\mathbf{M}_{DA_1}), \ldots, \alpha^*(\mathbf{M}_{DA_n})\big),$$

and hence they are compositions of atomic simple morphisms as well.

Remark In the rest of this book, we will use $\mathcal{A} = (S_A, \Sigma_A)$ also for the set of relational symbols in S_A and an instance-database A for $\alpha^*(\mathcal{A}) = \alpha^*(S_A)$: it will simply be called a "database" when it is clear from the context. The functions ∂_0 and ∂_1 are different from *dom* and *cod* functions used for the category arrows. For an *atomic* morphism or, more generally, for a *complete* morphism $f : A \to B$, the $\partial_0(f)$ specifies exactly the subset of relations in a database A used by f, while $\partial_1(f)$ defines the target relations in a database B for this mapping. Thus, for a complete mapping $\partial_0(f) \subseteq dom(f) = A$ and $\partial_1(f) \subseteq cod(f) = B$. In the case when f is a simple view-mapping, $\partial_1(f)$ is a singleton.

The fact that each atomic arrow $f : A \to B$ is a set of functions and has the information flux $\widetilde{f} = B_T(f)$ is important for the composition of the arrows in **DB**: any two simple arrows $f, g : A \to B$ are equal ($f \equiv g$) not only when the two sets of functions f and g are equal sets but also when they are different, but the information fluxes $\widetilde{f}$ and $\widetilde{g}$ are equal. More about this for complex arrows will be presented in Definition 23.

From the proof of Theorem 1, for the composition of any two morphisms in **DB**, $f = \alpha^*(\mathbf{M}_{AB}) : A \to B$ and $g = \alpha^*(\mathbf{M}_{BC}) : B \to C$,

$$\begin{aligned} h &= g \circ f = \alpha^*(\mathbf{M}_{BC}) \circ \alpha^*(\mathbf{M}_{AB}) = \alpha^*(\mathbf{M}_{BC} \circ \mathbf{M}_{AB}) \\ &= \{\alpha(q_i^B)(\alpha(q_{i_1}^A) \times \cdots \times \alpha(q_{i_{k_i}}^A)) \mid 1 \le i \le n, \text{ and } 1 \le i_j \le m \text{ for } 1 \le j \le k_i\} \\ &\quad \cup \{\alpha(1_{r\emptyset})\} \\ &= \{q_{B_i} \cdot (q_{Ai_1}, \ldots, q_{Ai_{k_i}}) \mid 1 \le i \le n, \text{ and } 1 \le i_j \le m \text{ for } 1 \le j \le k_i\} \cup \{q_\bot\}, \end{aligned}$$

where q_{B_i} denotes the function $\alpha(q_i^B)$, for $1 \le i \le n$, q_{A_j} denotes the function $\alpha(q_j^A)$, for $1 \le j \le m$, and $\cdot$ denotes the composition of functions. Graphically, such a composition of morphisms can be represented as a composition of trees (see the examples bellow where the 'graphical-tree' morphisms are denoted by $f_T, g_T, \ldots$).

Generally, a composed morphism $h : A \to C$ is not a complete morphism, that is, it can be graphically represented by a general tree such that not all its leaves are in A. Such an "*incomplete*" morphism is called a (partial) p-arrow. A p-arrow corresponds to a morphism obtained from a schema mapping $\mathcal{M}_{AB} = \{\Phi\} : \mathcal{A} \to \mathcal{B}$ where Φ is an SOtgd such that at least one left-hand side of its implications has the characteristic-functional symbols (for the relations that are not in the source schema $\mathcal{A}$).

By the graphical tree-composition of two trees in Fig. 3.1, f_T (incomplete) and g_T (complete), we obtain the tree h_T (of the p-arrow $h = g \circ f : A \longrightarrow C$):

$$\begin{aligned} h_T &= (g \circ f)_T \\ &= \bigcup_{q_{B_j} \in \alpha^*(\mathbf{M}_{BC})} \Big\{ q_{B_j} \cdot \Big(\bigcup_{q_{A_i} \in \alpha^*(\mathbf{M}_{AB}) \,\&\, \partial_1(q_{A_i}) \in \partial_0(q_{B_j}))} \{q_{A_i}(tree)\} \Big) \Big\} \\ &= \{q_{B_j} \cdot \{q_{A_i}(tree) \mid \partial_1(q_{A_i}) \in \partial_0(q_{B_j})\} \mid q_{B_j} \in \alpha^*(\mathbf{M}_{BC})\} \end{aligned}$$

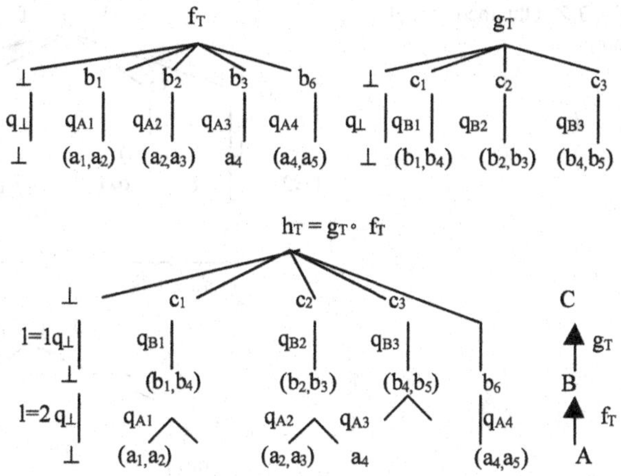

Fig. 3.1 Composed tree

$$= \{q_{B_j}(tree) \mid q_{B_j} \in \alpha^*(\mathbf{M}_{BC})\}$$

where $q_{A_i}(tree)$ is the tree bellow the branch q_{A_i}.

The difference between this graphical tree-representation of morphisms in Fig. 3.1 and of a standard operads-based representation is that in the graphical tree-representation, for the relations in $\partial_0(h)$ that are not relations in $\partial_0(f)$ (called *hidden relations*), we do not use the identity functions $\alpha(1_r)$ for these hidden relations (relations that are not in the source database A but in some intermediate database as 1_{Over65} in Example 8 or 1_{Over65} and 1_{Emp} in Example 9). Let us consider the following example in order to facilitate the understanding of the composition of morphisms:

Example 18 Let us consider the morphisms $f : A \longrightarrow B$ and $g : B \longrightarrow C$ such that

$$A = \{a_1, \dots, a_6\}, \qquad B = \{b_1, \dots, b_7\}, \qquad C = \{c_1, \dots, c_4\},$$

where $f = \{q_{A_1}, \dots, q_{A_4}, q_\perp\}$ with

$$\partial_0(q_{A_1}) = \{a_1, a_2\}, \qquad \partial_0(q_{A_2}) = \{a_2, a_3\},$$

$$\partial_0(q_{A_3}) = \{a_4\}, \qquad \partial_0(q_{A_4}) = \{a_4, a_5\},$$

$$\partial_1(q_{A_1}) = \{b_1\}, \qquad \partial_1(q_{A_2}) = \{b_2\},$$

$$\partial_1(q_{A_3}) = \{b_3\}, \qquad \partial_1(q_{A_4}) = \{b_6\}$$

and $g = \{q_{B_1}, \dots, q_{B_3}, q_\perp\}$ with

$$\partial_0(q_{B_1}) = \{b_1, b_4\}, \qquad \partial_0(q_{B_2}) = \{b_2, b_3\},$$

$$\partial_0(q_{B_3}) = \{b_4, b_5\}, \qquad \partial_1(q_{B_1}) = \{c_1\},$$

$$\partial_1(q_{B_2}) = \{c_2\}, \qquad \partial_1(q_{B_3}) = \{c_3\}.$$

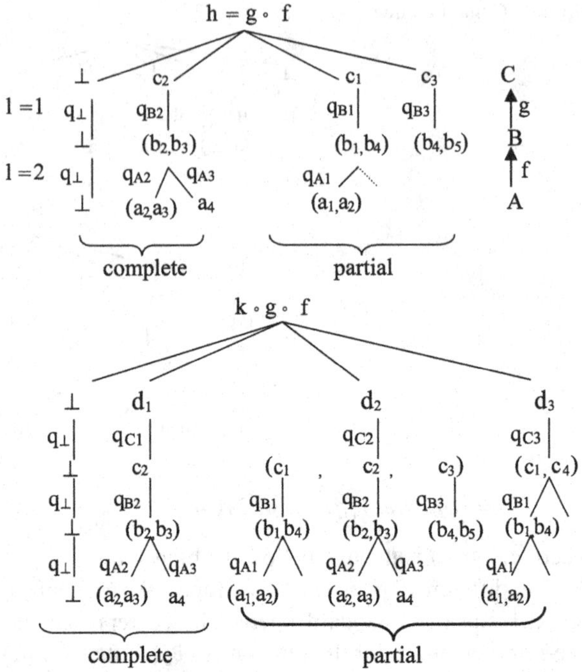

Fig. 3.2 Obtained partial morphism

These two morphisms are represented by the trees f_T and g_T and their sequential composition by h_T in Fig. 3.1.

From the point of view based on the information fluxes of these morphisms, the composition of morphisms $h = g \circ f : A \longrightarrow C$ can be graphically represented as *a part of the tree* h_T in Fig. 3.2. It only provides the strict information contribution from the object A (i.e., the source) into the object C (target of this composed morphism). Hence $\partial_0(f) = \{a_1, a_2, a_3, a_4, a_5\}$, $\partial_1(f) = \{b_1, b_2, b_3, b_6\}$, $\partial_0(g) = \{b_1, b_2, b_3, b_4, b_5\}$, $\partial_1(g) = \{c_1, c_2, c_3\}$, while $\partial_0(h) = \partial_0(g \circ f) = \{a_1, a_2, a_3, a_4\} \neq \partial_0(f)$, $\partial_1(h) = \partial_1(g \circ f) = \{c_1, c_2, c_3\} = \partial_1(g)$.

Let us consider, for example, the composition of the c-arrow $h : C \longrightarrow D$ with the composed arrow $g \circ f$ in the previous example, where $D = \{d_1, \ldots, d_4\}$, $h = \{q_{C_1}, q_{C_2}, q_{C_3}, q_\perp\}$,

$$\partial_0(q_{C_1}) = \{c_2\}, \qquad \partial_1(q_{C_1}) = \{d_1\},$$

$$\partial_0(q_{C_2}) = \{c_1, c_2, c_3\}, \qquad \partial_1(q_{C_2}) = \{d_2\},$$

$$\partial_0(q_{C_3}) = \{c_1, c_4\}, \qquad \partial_1(q_{C_3}) = \{d_3\},$$

with $q_{B_2}(tree) = q_{B_2} \cdot \{q_{A_2}, q_{A_3}\}$ a complete and $q_{B_1}(tree) = q_{B_1} \cdot \{q_{A_1}, -\}$ a partial (incomplete) component of this tree, as represented in Fig. 3.2. A composition of (complete) morphisms generally produces a partial (incomplete) morphism (only a part of the tree h_T represents a real contribution from A into C) with *hidden elements* (in the diagram of the composed morphism h, the elements b_4 and b_5 are a hidden relations).

In such a representation, for the consideration of the information fluxes, we "forgot" the parts of the tree $g_T \circ f_T$ that are not involved in a real information contribution of composed mappings from the source into the target object. In fact, the hidden relations do not appear on the left-hand side of SOtgd's implications of a given schema mapping, but only as the characteristic functions of these hidden relations. The hidden relations appear always in the operad's operations and as a part of the domain of the set of functions that compose a given morphism $f : A \to B$. Consequently, operad's based representation of morphisms are equivalent to the logical representation based on the SOtgds.

Based on Definition 13, the information flux for a composition of simple morphisms is given by the set-intersection of the information fluxes of all its simple morphisms.

Remark The information flux $\widetilde{f} = B_T(f)$ of a given morphism (instance-database mapping) $f : A \longrightarrow B$ is an instance-database as well, thus, an object in **DB**: the minimal information flux is equal to the bottom object $\bot^0$ so that, given any two database instances A and B in **DB**, there exists at least an arrow (morphism) between them $f : A \longrightarrow B$ such that $\widetilde{f} = \bot^0$ (for example, $\bot^1 : A \to B$).

Definition 19 A complex object in **DB** is expressed by a disjoint union $\uplus$ of simple instance-databases and hence each query over such a complex object can have the relational symbols belonging only to one of these simple databases. Let $A = A_1 \uplus \cdots \uplus A_m = \biguplus_{1\le j\le m} A_j = \{\{1\} \times A_1\} \cup \cdots \cup \{\{m\} \times A_m\}$ if $m \ge 2$; A_1 otherwise, and $B = \biguplus_{1\le i\le k} B_i = \{\{1\} \times B_1\} \cup \cdots \cup \{\{k\} \times B_k\}$ if $k \ge 2$; B_1 otherwise, where all A_j and B_i are simple (non-composed) databases. Consequently, from the fact that all simple databases are separated, $T(\biguplus_{1\le j\le m} A_j) = \biguplus_{1\le j\le m} TA_j$.

1. We say that A is *strictly complex* if $A_j \neq \bot^0$ for all $1 \le j \le m$.
2. We can define the new composition $\otimes$ by $\Upsilon \otimes A = A \otimes \Upsilon = TA$ and

$$A \otimes B \triangleq \begin{cases} TA \cap TB & \text{if } m = k = 1; \\ \biguplus_{1\le j\le m \ \&\ 1\le i\le k}(TA_j \cap TB_i) & \text{otherwise.} \end{cases}$$

3. We define an "observational" PO relation $\preceq$ such that $\Upsilon \preceq B$ iff $B = \Upsilon$ and for $A = \biguplus_{1\le j\le m} A_j \neq \Upsilon$, $A \preceq B = \biguplus_{1\le i\le k} B_i$ iff there exists a mapping $\sigma : \{1, \dots, m\} \to \{1, \dots, k\}$ such that for each $1 \le j \le m$, $TA_j \subseteq TB_{\sigma(j)}$.

   The equivalence in this ordering (i.e., $A \preceq B$ and $B \preceq A$) we denote by $A \approx B$.

Notice that $\preceq$ is a partial order (PO): $A \preceq A$ (for σ an identity function) and if $A \preceq B$ (with σ_1) and $B \preceq C$ (with σ_2) then $A \preceq C$ (with $\sigma = \sigma_2 \cdot \sigma_1$). For a given a tuple of objects (i.e., databases) in **DB** category $(S_1, \dots, S_n)$, $n \ge 2$, its disjoint union (from Definition 19) is $\biguplus_{1\le i\le n} S_i = \biguplus(S_1, \dots, S_n) = \bigcup_{1\le i\le n}\{(i, a) \mid a \in S_i\}$ (the union *indexed by positions* of the sets in a given tuple), denoted also by $S_1 \uplus S_2 \uplus \cdots \uplus S_n$. This indexing of simple databases that compose a complex object has to be maintained in the case when they are equal as well. For example, a complex object $C \uplus C$ has to be represented formally as an indexed complex object $A_1 \uplus A_2$ with $A_1 = A_2 = C$. This "*indexing by position*" is fundamental for the formalization of complex morphisms between such complex objects. It will be demonstrated that a disjoint union of two objects (databases) is isomorphic to a separation-composition of these objects and it will be considered in more details in Sect. 3.3.2.

For any two simple morphisms $f = \alpha^*(\mathbf{M}_{AC}) = \{\alpha(q_1), \ldots, \alpha(q_k), q_\perp\} : A \to C$ and $g = \alpha^*(\mathbf{M}_{BD}) = \{\alpha(q'_1), \ldots, \alpha(q'_m), q_\perp\} : B \to D$, we define a complex morphism

$$(f \uplus g) : A \uplus B \to C \uplus D$$

by the disjunctive union of them, that is, by the set

$$\{(1, \alpha(q_i)) \mid q_i \in \mathbf{M}_{AC}\} \cup \{(2, \alpha(q'_j)) \mid q'_j \in \mathbf{M}_{BD}\} \cup \{q_\perp\} = \alpha^*(\mathbf{M}_{AC} \uplus \mathbf{M}_{BD}).$$

Thus, from Definition 15, its flux is equal to

$$\begin{aligned} \widetilde{f \uplus g} &= \widetilde{\alpha^*(\mathbf{M}_{AC} \uplus \mathbf{M}_{BD})} \\ &= Flux(\alpha, \mathbf{M}_{AC}) \uplus Flux(\alpha, \mathbf{M}_{BD}) = \widetilde{f} \uplus \widetilde{g} \in Ob_{DB}. \end{aligned}$$

Thus, we can consider the following extension of morphisms in **DB**, due to Theorem 1, based on Definition 15:

Definition 20 Complex arrows in **DB** presented by Theorem 1 are obtained by the structural-operations $\_ \uplus \_$, $[\_ , \_]$, $\langle \_, \_ \rangle$ and $\lceil \_ , \_ \rfloor$ (all A, B, C and D are simple objects), as follows:

1. Morphism $f \uplus g : A \uplus B \to C \uplus D$ for any $f : A \to C$ and $g : B \to D$;
2. Morphism $[f, g] : A \uplus B \to C$ for any $f : A \to C$ and $g : B \to C$;
3. Morphism $\langle f, g \rangle : A \to C \uplus D$ for any $f : A \to C$ and $g : A \to D$;
4. Morphism $\lceil f \circ k, g \circ l \rfloor = [f, g] \circ \langle k, l \rangle : A \to B \uplus D \to C$ with simple arrows $f \circ k : A \to B \to C$ and $g \circ l : A \to D \to C$ such that $\widetilde{f \circ k} \neq \perp^0$ and $\widetilde{g \circ l} \neq \perp^0$ and with reductions: $\lceil k, g \rfloor = k$ if $\widetilde{k} \neq \perp^0$ and $\widetilde{g} = \perp^0$; $\lceil k, g \rfloor = \perp^1$ if $\widetilde{k} = \widetilde{g} = \perp^0$.

Hence, a *simple* arrow (or morphism) is an arrow between two simple (non-separation-composed) objects. Notice that a complex arrow $\lceil k, l \rfloor : A \to C$ exists only as a composition of two complex arrows defined in point 4. It means that both simple arrows $k, l : A \to C$ are composed arrows with *different* intermediate databases.

We recall that an arrow $\perp^1 = \{q_\perp : \perp \to \perp\} : A \to C$, which represents an empty function, means that, in fact, we have no mapping between A and B (the information flux of $\perp^1$ is empty, i.e., $\widetilde{\perp^1} = \perp^0 = \{\perp\}$). This fact explains why we are using only the arrows with nonempty fluxes in point 4 in a representation of a set of arrows between two fixed objects by $\lceil, \rfloor$. Based on Definitions 15 and 20 for binary compositions, we have the following cases for composition of complex morphisms (all $A_i, B_i, C_i, i = 1, 2$, and A, B, C are simple objects):

1. For $k: A_1 \rightarrow B_1, l: A_2 \rightarrow B_2, f: B_1 \rightarrow C_1$ and $g: B_2 \rightarrow C_2$,
$$(f \uplus g) \circ (k \uplus l) = (f \circ k) \uplus (g \circ l) : A_1 \uplus A_2 \rightarrow C_1 \uplus C_2;$$
2. For $k: A \rightarrow B_1, l: A \rightarrow B_2, f: B_1 \rightarrow C_1$ and $g: B_2 \rightarrow C_2$,
$$(f \uplus g) \circ \langle k, l\rangle = \langle f \circ k, g \circ l\rangle : A \rightarrow C_1 \uplus C_2;$$
3. For $k: A_1 \rightarrow B_1, l: A_2 \rightarrow B_2, f: B_1 \rightarrow C$ and $g: B_2 \rightarrow C$,
$$[f, g] \circ (k \uplus l) = [f \circ k, g \circ l] : A_1 \uplus A_2 \rightarrow C;$$
4. For $k: A \rightarrow B_1, l: A \rightarrow B_2, f: B_1 \rightarrow C$ and $g: B_2 \rightarrow C$, $[f, g] \circ \langle k, l\rangle = \lceil f \circ k, g \circ l\rfloor$ (if $f \circ k \neq \bot^1, g \circ l \neq \bot^1$); $g \circ l$ (if $f \circ k = \bot^1, g \circ l \neq \bot^1$); $f \circ k$ (if $f \circ k \neq \bot^1, g \circ l = \bot^1$); $\bot^1$ otherwise (an arrow from A to C);
5. For $k: A_1 \rightarrow B$, $f: B \rightarrow C_1$ and $g: B \rightarrow C_2$
$$\langle f, g\rangle \circ k = \langle f \circ k, g \circ k\rangle : A_1 \rightarrow C_1 \uplus C_2;$$
6. For $k: A_1 \rightarrow B, l: A_2 \rightarrow B$ and $f: B \rightarrow C$,
$$f \circ [k, l] = [f \circ k, g \circ k] : A_1 \uplus A_2 \rightarrow C;$$
7. For $k: A \rightarrow B, l: A \rightarrow B$, $f: B \rightarrow C_1$, and $g: B \rightarrow C_2$
$$\langle f, g\rangle \circ \lceil k, l\rfloor = \big\langle \lceil f \circ k, f \circ l\rfloor, \lceil g \circ k, g \circ l\rfloor \big\rangle : A \rightarrow C_1 \uplus C_1;$$
8. For $k: A_1 \rightarrow B, l: A_2 \rightarrow B$, $f: B \rightarrow C$ and $g: B \rightarrow C$,
$$\lceil f, g\rfloor \circ [k, l] = \big[\lceil f \circ k, g \circ k\rfloor, \lceil f \circ l, g \circ l\rfloor\big] : A_1 \uplus A_2 \rightarrow C;$$
9. For $k: A \rightarrow B, l: A \rightarrow B$, $f: B \rightarrow C$ and $g: B \rightarrow C$,
$$\lceil f, g\rfloor \circ \lceil k, l\rfloor = \lceil f \circ k, g \circ k, f \circ l, g \circ l\rfloor : A \rightarrow C.$$

Now we can extend these binary structural-operators in Definition 20 to n-ary operators:

Definition 21 We generalize the binary structure-operators in Definition 20 into n-ary operators, for any $n \geq 2$, by:
1. $\biguplus_{1 \leq i \leq n} f_i = \biguplus(f_1, \ldots, f_n) : \biguplus_{1 \leq i \leq n} A_i \rightarrow \biguplus_{1 \leq i \leq n} B_i$ where $f_i : A_i \rightarrow B_i$ for each $1 \leq i \leq n$;
2. $[f_1, \ldots, f_n] : \biguplus_{1 \leq i \leq n} A_i \rightarrow B$, where $f_i : A_i \rightarrow B$ for each $1 \leq i \leq n$;
3. $\langle f_1, \ldots, f_n\rangle : A \rightarrow \biguplus_{1 \leq i \leq n} B_i$, where $f_i : A \rightarrow B_i$ for each $1 \leq i \leq n$;
4. $\lceil f_1, \ldots, f_n\rfloor : A \rightarrow B$, where $\widetilde{f_i} \neq \bot^0$ for each $1 \leq i \leq n$.

Remark In order to simplify the presentation, we extend the structure-operators also to unary cases: for example, $\biguplus_{1\le i\le 1} A_i = \biguplus(A_1) = A_1$, $\biguplus_{1\le i\le 1} f_i = \biguplus(f_1) = f_1$.

We extend the application of $\biguplus$ also to (unordered) sets, so that for a given nonempty set (of sets) S, $\biguplus S$ represents an application of $\biguplus$ to *any* ordered list (tuple) of sets in S. It will be demonstrated that $\biguplus$ is a commutative and associative binary operator (up to isomorphism) so that, given any ordered list (a permutation) $(s_1, \dots, s_n)$ of all elements (which are the sets) in S, the following isomorphism is valid in **DB**:

$$\biguplus S \simeq \biguplus(s_1, \dots, s_n) = (s_1 \uplus \cdots \uplus s_n) = \bigcup_{1\le i\le n} \{(i,a) \mid a \in s_i\}.$$

Thus, for example, we have the following cases for a composition of complex morphisms obtained by using n-ary separation-compositions:

1. For $\biguplus_{1\le i\le n} f_i : \biguplus_{1\le i\le n} A_i \to \biguplus_{1\le i\le n} B_i$ and $\biguplus_{1\le i\le n} g_i : \biguplus_{1\le i\le n} B_i \to \biguplus_{1\le i\le n} C_i$,

$$\left(\biguplus_{1\le i\le n} g_i\right) \circ \left(\biguplus_{1\le i\le n} f_i\right) = \biguplus_{1\le i\le n} (g_i \circ f_i) : \biguplus_{1\le i\le n} A_i \to \biguplus_{1\le i\le n} C_i;$$

2. For $\langle f_1, \dots, f_n\rangle : A \to \biguplus_{1\le i\le n} B_i$ and $\biguplus_{1\le i\le n} g_i : \biguplus_{1\le i\le n} B_i \to \biguplus_{1\le i\le n} C_i$,

$$\left(\biguplus_{1\le i\le n} g_i\right) \circ \langle f_1, \dots, f_n\rangle = \langle g_1 \circ f_1, \dots, g_n \circ f_n\rangle : A \to \biguplus_{1\le i\le n} C_i;$$

3. For $\biguplus_{1\le i\le n} f_i : \biguplus_{1\le i\le n} A_i \to \biguplus_{1\le i\le n} B_i$ and $[g_1, \dots, g_n] : \biguplus_{1\le i\le n} B_i \to C$,

$$[g_1, \dots, g_n] \circ \left(\biguplus_{1\le i\le n} f_i\right) = [g_1 \circ f_1, \dots, g_n \circ f_n] : \biguplus_{1\le i\le n} A_i \to C;$$

4. For $\langle f_1, \dots, f_n\rangle : A \to \biguplus_{1\le i\le n} B_i$ and $[g_1, \dots, g_n] : \biguplus_{1\le i\le n} B_i \to C$,

$$[g_1, \dots, g_n] \circ \langle f_1, \dots, f_n\rangle = \lceil S \rfloor : A \to C \text{ if } |S| \ge 1; \quad \bot^1 : A \to C \quad \text{otherwise}$$

(where $S = \{g_i \circ f_i \mid \widetilde{g_i \circ f_i} = \widetilde{g_i} \cap \widetilde{f_i} \neq \bot^0, 1 \le i \le n\}$);

5. For $f : A \to B$ and $\langle g_1, \dots, g_n\rangle : B \to \biguplus_{1\le i\le n} C_i$,

$$\langle g_1, \dots, g_n\rangle \circ f = \langle g_1 \circ f, \dots, g_n \circ f\rangle : A \to \biguplus_{1\le i\le n} C_i;$$

6. For $[f_1, \dots, f_n] : \biguplus_{1\le i\le n} A_i \to B$ and $g : B \to C$,

$$g \circ [f_1, \dots, f_n] = [g \circ f_1, \dots, g \circ f_n] : \biguplus_{1\le i\le n} A_i \to C;$$

7. For $f_i : A \to B, 1 \le i \le n$ and $g : B \to C$,

$$g \circ \lceil f_1, \ldots, f_n \rfloor = \lceil S \rfloor : A \to C \quad \text{if } |S| \ge 1; \qquad \bot^1 : A \to C \quad \text{otherwise}$$

(where, from Definition 20, $S = \{g \circ f_i \mid \widetilde{g \circ f_i} \neq \bot^0, 1 \le i \le n\}$);

8. For $g_i : B \to C, 1 \le i \le n$ and $f : A \to B$,

$$\lceil g_1, \ldots, g_n \rfloor \circ f = \lceil S \rfloor : A \to C \quad \text{if } |S| \ge 1; \qquad \bot^1 : A \to C \quad \text{otherwise}$$

(where, from Definition 20, $S = \{g_i \circ f \mid \widetilde{g_i \circ f} \neq \bot^0, 1 \le i \le n\}$).

By these rules of composition of complex arrows, we obtain that the resulting arrow is a composition of structural-operators $\biguplus(\_, .., \_)$, $[\_, .., \_]$, $\langle \_, .., \_ \rangle$, and $\lceil \_, .., \_ \rfloor$. The 'point-to-point' (ptp) arrow is a *nonempty* (with flux different from $\bot^0$) simple arrow (a path of only simple atomic arrows) between two simple objects that compose the source and target database of this complex arrow: if we have more than one simple arrow between the same source and target simple objects then we fuse them (by union) into a single ptp arrow, as follows:

Definition 22 For any complex arrow $h : A \to B$ between 'indexed by position' complex objects $A = \biguplus_{1 \le j \le m} A_j$ and $B = \biguplus_{1 \le i \le k} B_i$, $m, k \ge 1$, we define its set of ptp morphisms by:

$$\frac{h}{\uplus} \triangleq \Big\{ h_{ji} = \bigcup \{ f_l : A_j \to B_i \mid f_l \text{ is a composition of simple arrows in } h$$
$$\text{such that } \widetilde{f_l} \neq \bot^0 \} \neq \emptyset \mid 1 \le j \le m, 1 \le i \le k \Big\}.$$

Then we extend the mapping B_T to complex arrows, too. $B_T(h) = \widetilde{h}$ is equal (up to isomorphism) to the object

$$\widetilde{\biguplus \frac{h}{\uplus}} \triangleq \begin{cases} \biguplus \{B_T(h_{ji}) \mid h_{ji} \in \frac{h}{\uplus}\} = \biguplus \{\widetilde{h_{ji}} \mid h_{ji} \in \frac{h}{\uplus}\} & \text{if } |\frac{h}{\uplus}| \ge 2; \\ B_T(h_{ji}) = \widetilde{h_{ji}} & \text{if } \frac{h}{\uplus} = \{h_{ji}\}; \\ \bot^0 & \text{otherwise.} \end{cases}$$

Notice that for a given complex arrow h each of its ptp arrows has a nonempty information flux (different from $\bot^0$) so that a ptp arrow is different from $\bot^1$ arrow. In the previous example for composition of complex arrows, the resulting arrow is composed of point-to-point arrows. For example, in rule 5, we obtained the complex arrow $\langle f \circ k, g \circ k \rangle : A_1 \to C_1 \uplus C_2$ where $f \circ k : A_1 \to C_1$ and $g \circ k : A_1 \to C_2$ are two point-to-point simple arrows. In rule 3, we obtained the complex arrow $[f \circ k, g \circ l] : A_1 \uplus A_2 \to C$ where $f \circ k : A_1 \to C$ and $g \circ l : A_2 \to C$ are two point-to-point simple arrows, etc.

This property extends by induction to any composition of complex arrows. The fact that the source and target complex objects A and B of a complex morphism $h : A \to B$ are 'indexed by position' is fundamental for the definition (indexing) of its ptp arrows. For example, the complex arrow $h = [id_C, id_C, id_C] : C \uplus C \uplus C \to C$ has to be represented formally by $h : A_1 \uplus A_2 \uplus A_3 \to B_1$, with $A_1 = A_2 = A_3 = B_1 = C$, so that $\frac{h}{\uplus} = \{h_{11} = id_C : A_1 \to B_1, h_{21} = id_C : A_2 \to B_1, h_{31} = id_C : A_3 \to B_1\}$ is the set of three ptp arrows and not the singleton $\{id_C : C \to C\}$. This requirement is fundamental for the definition of ptp arrows because it takes into consideration the mutually separated databases in the complex source object A and mutually independent morphisms from them into the target database B.

Based on Definitions 15, 20 and 21, for $\frac{h}{\uplus} \neq \emptyset$ we obtain $\widetilde{h} \simeq \biguplus\{\widetilde{f_i} \mid f_i \in \frac{h}{\uplus}\}$ (that is, $\widetilde{h}$ is not equal to $\widetilde{\biguplus \frac{h}{\uplus}}$ but only *isomorphic* to it). This isomorphism (instead of an equality) is based on the fact that $\biguplus(A, B) = A \uplus B \simeq B \uplus A = \biguplus(B, A)$ (formally demonstrated by point 2 of the following Lemma 9) and ptp arrows in h, as explained in Definition 21, are ordered while the set $\frac{h}{\uplus}$ is unordered: when $\biguplus$ is applied to a set then *any* ordering can be taken in this set.

Notice that the mapping B_T for complex morphisms continues to generate the closed objects. In fact, we can show it recursively. Let $\widetilde{f} = B_T(f), \widetilde{g} = B_T(g)$ be closed objects (i.e., from the fact that power-view operator T (Sect. 1.4.1) is idempotent, $T\widetilde{f} = \widetilde{f}, T\widetilde{g} = \widetilde{g}$). Then, $T(\widetilde{f} \uplus \widetilde{g}) = T\widetilde{f} \uplus T\widetilde{g} = \widetilde{f} \uplus \widetilde{g}$ and hence closed as well.

Remark As we have specified in Definition 22, the path-arrows with the same simple source and target objects are fused (by union) into a single morphism. For example, for given two simple databases A and B, the complex arrows $\lceil f, g \rfloor : A \to B$, where $f = \alpha(\mathbf{M}_{AB}^{(1)})$ and $g = \alpha(\mathbf{M}_{AB}^{(2)})$ (with sketch's arrows $\mathbf{M}_{AB}^{(n)} = \mathit{MakeOperads}(\mathcal{M}_{AB}^{(n)}) : \mathcal{A} \to \mathcal{B}$, $n = 1, 2$, and $\mathcal{M}_{AB}^{(1)} = \{\Phi\} : \mathcal{A} \to \mathcal{B}$, $\mathcal{M}_{AB}^{(2)} = \{\Psi\} : \mathcal{A} \to \mathcal{B}$), will be fused into a single arrow $k : A \to B$, where $k = f \cup g = \alpha(\mathbf{M}_{AB})$ is obtained by fusing the sketch's mappings $\mathbf{M}_{AB}^{(1)}$ and $\mathbf{M}_{AB}^{(1)}$ into $\mathbf{M}_{AB} = (\mathbf{M}_{AB}^{(1)} \cup \mathbf{M}_{AB}^{(1)}) : \mathcal{A} \to \mathcal{B}$. Thus, based on Definition 13, the flux of $f \cup g$ is equal to (as in point 3 of Definition 15):

$$
\begin{aligned}
\widetilde{\lceil f, g \rfloor} = \widetilde{f \cup g} &= T\Big(\Delta\big(\alpha, \mathcal{M}_{AB}^{(1)}\big) \cup \Delta\big(\alpha, \mathcal{M}_{AB}^{(2)}\big)\Big) \\
&= T\Big(T\big(\Delta\big(\alpha, \mathcal{M}_{AB}^{(1)}\big)\big) \cup T\big(\Delta\big(\alpha, \mathcal{M}_{AB}^{(2)}\big)\big)\Big) \\
&= T\Big(\mathit{Flux}\big(\alpha, \mathcal{M}_{AB}^{(1)}\big) \cup \mathit{Flux}\big(\alpha, \mathcal{M}_{AB}^{(2)}\big)\Big) \\
&= T(\widetilde{f} \cup \widetilde{g}) \neq \widetilde{f} \uplus \widetilde{g}.
\end{aligned}
$$

Generally, when $g = f$, the complex arrow $\lceil f, g \rfloor : A \to B$ may represent the two composed independent mapping paths f and g with different intermediate databases. Thus, $\widetilde{\lceil f, g \rfloor} = T(\widetilde{f} \cup \widetilde{f}) = T\widetilde{f} = \widetilde{f}$ (a closed object), that is, this pair of arrows $\lceil f, g \rfloor$ transfers the same information flux from the source to target database as a single mapping $f : A \to B$. However, they cannot be considered equal: the *redundant* transferring of the same information flux represents a more robust database mapping system in which if the first mapping is interrupted by a failure in one of its intermediate databases then the second (alternative) mapping may continue to work and to guarantee the general security mapping requirements.

This database mapping system requirement explains that, in order to establish the equality of two mappings $k = \lceil f, f \rfloor : A \to B$ and $f : A \to B$, their observational equality (when they have equal information fluxes, which are the set of all transferred views from A into B) is not a sufficient condition.

In fact, let us consider the mappings $\langle l_1, l_2 \rangle : A \to C \uplus C$ and $[m_1, m_2] : C \uplus C \to B$, so that their composition is $\lceil f_1, f_1 \rfloor = [m_1, m_2] \circ \langle l_1, l_2 \rangle = \lceil m_1 \circ l_1, m_2 \circ l_2 \rfloor : A \to B$, also when $m_1 = m_2 = m$ and $l_1 = l_2 = l$, is different from a single mapping $f = m \circ l : A \to B$, i.e., $\lceil f, f \rfloor \neq f$. However, if $m_2 = \bot^1$ or $l_2 = \bot^1$, then $[m_1, m_2] \circ \langle l_1, l_2 \rangle = \lceil f_1, \bot^1 \rfloor = f_1$ because the alternative mapping $\bot^1 : A \to B$ has empty information flux (it does not transfer anything from A into B) and can be eliminated from the mappings (as specified by point 2.4 in Definition 20).

Lemma 7 *For any complex arrow $h : A \to B$ with $A = \biguplus_{1 \le j \le m} A_j, m \ge 1$ and $B = \biguplus_{1 \le i \le k} B_i, k \ge 1$, any its substructure $\lceil h_{l_1}, \ldots, h_{l_n} \rfloor : A_j \to B_i, n \ge 2$, where each h_{l_i} is a simple arrow (with simple source and target objects) is fused into a single ptp arrow $(h_{ji} : A_j \to B_i) \in \frac{h}{\uplus}$. Thus, its information flux is $\widetilde{h_{ji}} = T(\bigcup\{\widetilde{h_{l_i}} \mid 1 \le i \le n\})$. Given any pair of simple source and target databases A_j and B_i, there is at most one ptp arrow $h_{ji} \in \frac{h}{\uplus}$ between them.*

Proof If each h_{l_i} is a simple arrow with simple source and target objects (databases) A_j and B_i, respectively, then $\lceil h_{l_1}, \ldots, h_{l_n} \rfloor : A_j \to B_i$ is fused into a single ptp, as it directly follows from definition of $\frac{h}{\uplus}$. The flux of these fused (by union) arrows into one ptp arrow has then the information flux as explained above and specified in the previous remark.

The fact that, for any pair of simple source and target databases A_j and B_i, there is at most one ptp arrow $h_{ji} \in \frac{h}{\uplus}$ between them is a direct consequence of the definition of $\frac{h}{\uplus}$. It means that h cannot have more than $k \cdot m$ ptp arrows. □

The logical equivalence of the SOtgds of two schema mappings is an insufficient requirement for the equality of the mappings at the instance-level (it was explained in Sect. 2.4.3 about the strict mapping semantics), and we need a more appropriate refinement of it based on the information fluxes. For any two logically equivalent schema mappings $\mathcal{M}_{AB} = \{\Phi\} : \mathcal{A} \to \mathcal{B}$ and $\mathcal{M}'_{AB} = \{\Phi'\} : \mathcal{A} \to \mathcal{B}$, where the SOtgds Φ and Φ' are logically equivalent, for any mapping-interpretation α we will have that the morphism (a set of functions) $f = \alpha^*(\mathit{MakeOperads}(\mathcal{M}_{AB}))$ is *equal* to the set of functions $g = \alpha^*(\mathit{MakeOperads}(\mathcal{M}'_A B))$. It holds from the fact that operad's based representation of the schema mappings is equivalent to the logical representation given by SOtgds. However, this kind of strong equality is not useful in our case, and we are considering a more week equality based on observational equivalence and information fluxes of instance-database mappings $f, g : A \to B$, where $A = \alpha^*(\mathcal{A})$ and $B = \alpha^*(\mathcal{B})$, as follows:

Definition 23 (EQUALITY OF ARROWS IN **DB**) In the observational-based semantics, the equality '$\equiv$' of two morphisms $f, g : A \to B$ (used for the commutativity of diagrams) in **DB** is defined as follows:
1. If both arrows are simple then $f \equiv g$ iff $\widetilde{f} = \widetilde{g}$;
2. If $f = \lceil f_1, \ldots, f_m \rfloor$ and $g = \lceil g_1, \ldots, g_k \rfloor$, with $m, k \geq 1$ and A and B simple databases, then $f \equiv g$ iff $m = k$ and there is a bijection $\sigma : \lceil f_1, \ldots, f_m \rfloor \to \lceil g_1, \ldots, g_k \rfloor$ such that $\widetilde{f_j} = \widetilde{\sigma(f_j)}$ for all $j = 1, \ldots, m$;
3. Otherwise, $f \equiv g$ iff $\frac{f}{\uplus} = \frac{g}{\uplus}$ (have the same set of ptp arrows).

Thus, if the morphisms f and g are equal in **DB**, they are not necessarily equal set-theoretically. By having in mind this difference between the common set-equality '$=$' and the **DB** arrows equality '$\equiv$', we will continue to use the same notation $f = g$ also for the arrows (instead of a more appropriate notation $f \equiv g$).

Consequently, the equality of arrows (morphisms) in **DB** is not the set-theoretical equality (by considering that each morphism is mathematically just a set of functions).

This observational equality of instance-database mappings means that at this level of DB abstraction, two instance-database mappings are equal iff the information transferred from the source instance-database into a target instance-database for both mappings is equivalent (or equal in special cases, because if $\widetilde{f} = \widetilde{g}$ then the identity morphism $id : \widetilde{f} \to \widetilde{g}$ is an isomorphism as well, so that $\widetilde{f} \simeq \widetilde{g}$) and in cases 2 above satisfy some additional requirements.

Remark In what follows, we will use the common symbol '$=$' for the *equality of morphisms* in the **DB** category by considering this observational meaning (and not the set-theoretic meaning). For all other cases, the equality symbol '$=$' has the set-theoretical meaning (for example, for the information fluxes $\widetilde{f} = B_T(f) = \widetilde{g} = B_T(g)$ or in definitions like of morphisms like $f = \alpha^*(\mathit{MakeOperads}(\mathcal{M}))$, the meaning for '$=$' is standard set-theoretic). All *commutativity diagrams* in **DB** will be based on this abstracted (non-set-based) equality for the morphisms.

Example 19 It is easy to verify the validity of composition rules for complex arrows (we denote by $=^n$ the equation obtained by application of the nth rule of compositions of arrows defined previously):
1. Let us show for the case $k : A_1 \to B, l : A_2 \to B, f : B \to C_1$ and $g : B \to C_2$ that $h = \langle f, g \rangle \circ [k, l] = \langle [f \circ k, f \circ l], [g \circ k, g \circ l] \rangle = [\langle f \circ k, g \circ k \rangle, \langle f \circ l, g \circ l \rangle] : A_1 \uplus A_2 \to C_1 \uplus C_2$, with the four obtained point-to-point arrows $f \circ k : A_1 \to C_1, g \circ k : A_1 \to C_2, f \circ l : A_2 \to C_1$ and $g \circ l : A_2 \to C_2$. That is, $\frac{h}{\uplus} = \{f \circ k, g \circ k, f \circ l, g \circ l\}$.

We have $\langle f,g\rangle \circ [k,l] =^5 \langle f\circ[k,l], g\circ[k,l]\rangle =^6 \langle [f\circ k, f\circ l],[g\circ k, g\circ l]\rangle$ and

$$\langle f,g\rangle \circ [k,l] =^6 \big[\langle f,g\rangle\circ k, \langle f,g\rangle \circ l\big] =^5 \big[\langle f\circ k, g\circ k\rangle, \langle f\circ l, g\circ l\rangle\big].$$

We can show that it is true by considering the ptp arrows. In fact,

$$\frac{h}{\uplus} = \frac{\langle [f\circ k, f\circ l],[g\circ k, g\circ l]\rangle}{\uplus} = \frac{[\langle f\circ k, g\circ k\rangle, \langle f\circ l, g\circ l\rangle]}{\uplus}$$
$$= \{f\circ k : A_1 \to C_1, g\circ k : A_1 \to C_2, f\circ l : A_2 \to C_1, g\circ l : A_2 \to C_2\}.$$

2. Let us show that $f = \langle id_C, id_C\rangle \circ [id_C, id_C] = \langle [id_C, id_C],[id_C, id_C]\rangle : C\uplus C \to C \to C\uplus C$ is different from $g = id_C \uplus id_C : C\uplus C \to C\uplus C$. In fact, in order to define the ptp arrows of these two morphisms, we have to use the 'indexing by position' for the source database $A = A_1 \uplus A_2$ and the target database $B = B_1 \uplus B_1$ where $A_1 = A_2 = B_1 = B_2 = C$. Thus,

$$\frac{\langle id_C, id_C\rangle \circ [id_C, id_C]}{\uplus}$$
$$= \{f_{11} = id_C : A_1 \to B_1, f_{12} = id_C : A_1 \to B_2,$$
$$f_{21} = id_C : A_2 \to B_1, f_{22} = id_C : A_2 \to B_2\}$$
$$\neq \{g_{11} = id_C : A_1 \to B_1, g_{22} = id_C : A_2 \to B_2\} = \frac{id_C \uplus id_C}{\uplus}.$$

Consequently, by point 3 of Definition 23, the identity arrow

$$id_C \uplus id_C \neq \langle id_C, id_C\rangle \circ [id_C, id_C].$$

In fact, $id_A = id_{A_1} \uplus id_{A_2}$ because for any $f : A \to X$ we have from Definition 22 that

$$\frac{f\circ(id_{A_1}\uplus id_{A_2})}{\uplus} = \left\{ f_{i_2}\circ id \mid f_{i_2} \in \frac{f}{\uplus}, id \in \{id_{A_1}, id_{A_2}\} \right.$$
$$\left. = \frac{id_{A_1}\uplus id_{A_2}}{\uplus}, dom(f_{i_2}) = cod(id) \right\}$$
$$= \left\{ f_{i_2} \mid f_{i_2} \in \frac{f}{\uplus} \right\} = \frac{f}{\uplus}$$

so that, by point 3 of Definition 23, $f = f\circ(id_{A_1}\uplus id_{A_2})$ and analogously $f = (id_{A_1}\uplus id_{A_2})\circ f$, so that $(id_{A_1}\uplus id_{A_2})$ is the identity arrow of $A = A_1 \uplus A_2$. That is, $id_A = id_{A_1\uplus A_2} = id_{A_1}\uplus id_{A_2}$.

Let us consider the associativity of composition of complex morphisms (that **DB** is well defined for complex arrows as well):

3. For $k : A_1 \to B, l : A_2 \to B, f : B \to C_1, g : B \to C_2, j : C_1 \to D$ and $m : C_2 \to D$, we consider the composition $[j,m]\circ\langle f,g\rangle\circ[k,l] : A_1 \uplus A_2 \to B \to C_1 \uplus C_2 \to D$. Let us consider the two different orders of composition:

3.1.

$$h_1 = [j,m] \circ (\langle f,g\rangle \circ [k,l]) =^5 [j,m] \circ \langle [f \circ k, f \circ l], [g \circ k, g \circ l] \rangle$$
$$=^4 \lceil [j \circ f \circ k, j \circ f \circ l], [m \circ g \circ k, m \circ g \circ l] \rfloor : A_1 \uplus A_2 \to D;$$

3.2.

$$h_2 = ([j,m] \circ \langle f,g\rangle) \circ [k,l] =^4 \lceil j \circ f, m \circ g \rfloor \circ [k,l]$$
$$=^7 \lceil [j \circ f \circ k, j \circ f \circ l], [m \circ g \circ k, m \circ g \circ l] \rfloor : A_1 \uplus A_2 \to D,$$

both of them with the *same set* of two fused point-to-point arrows

$$\frac{h_1}{\uplus} = \frac{h_2}{\uplus} = \{(j \circ f \circ k) \cup (m \circ g \circ k) : A_1 \to D,$$
$$(j \circ f \circ l) \cup (m \circ g \circ l) : A_2 \to D\},$$

so that $h_1 = h_2$, that is, this composition of complex arrows is associative as well.

4. For $k : A \to B_1, l : A \to B_2, f : B_1 \to C, g : B_2 \to C, j : C \to D_1$ and $m : C \to D_2$, we consider the compositions $\langle j,m\rangle \circ [f,g] \circ \langle k,l\rangle : A \to B_1 \uplus B_2 \to C \to D_1 \uplus D_2$. Let us consider the two different orders of composition:
4.1.

$$h_1 = \langle j,m\rangle \circ ([f,g] \circ \langle k,l\rangle) =^4 \langle j,m\rangle \circ \lceil f \circ k, g \circ l \rfloor$$
$$=^7 \lceil \langle j \circ f \circ k, j \circ g \circ l\rangle, \langle m \circ f \circ k, m \circ g \circ l\rangle \rfloor : A \to D_1 \uplus D_2,$$

4.2.

$$h_2 = (\langle j,m\rangle \circ [f,g]) \circ \langle k,l\rangle =^6 [\langle j \circ f, m \circ f\rangle, \langle j \circ g, m \circ g\rangle] \circ \langle k,l\rangle$$
$$=^4 \lceil \langle j \circ f \circ k, m \circ f \circ k\rangle, \langle j \circ g \circ l, m \circ g \circ l\rangle \rfloor : A \to D_1 \uplus D_2.$$

They are equal, $h_1 = h_2$, because

$$\frac{h_1}{\uplus} = \frac{h_2}{\uplus} = \{(j \circ f \circ k) \cup (j \circ g \circ l) : A \to D_1,$$
$$(m \circ f \circ k) \cup (m \circ g \circ l) : A \to D_2\}.$$

From the examples above, we have seen that there are different structural representations for the same arrow between two complex objects. That is, for a given set of ptp arrows between two complex objects, we can have a number of different structural representations that are mutually equal. Thus, it is natural to fix a canonical representation of such equivalent classes of arrows and to use them whenever possible. The fact is that each arrow in **DB** can be substituted by such canonical arrow, in an analogous way as any object can be substituted by an isomorphic object.

Lemma 8 *Let us define the function* $\circledast : Mor_{DB} \rightarrow Mor_{DB}$ *such that for any arrow* $f : A \rightarrow B$ *where* $A = \biguplus_{1\leq j\leq m} A_j$ *and* $B = \biguplus_{1\leq i\leq k} B_i$, $m, k \geq 1$,

$$\circledast(f) \triangleq \left[\langle h_{11}, \dots, h_{1k}\rangle, \dots, \langle h_{m1}, \dots, h_{mk}\rangle\right] : A \rightarrow B,$$

where $h_{ji} = f_{ji}$ *if* $(f_{ji} : A_j \rightarrow B_i) \in \frac{f}{\uplus}$; $\bot^1$ *otherwise.*

This mapping is an identity for the simple arrows. For any complex arrow it returns an equal arrow that contains the maximal number of simple arrows in its structural representation. For a given set $S = \frac{f}{\uplus}$ *of ptp arrows between* A *and* B, *by* $\widehat{\circledast}(A, B, S) \triangleq \circledast(f)$ *we denote a canonical arrow obtained from* S.

Proof There are the following cases:

1. The case of $m = k = 1$, so that $f : A_1 \rightarrow B_1$ is a simple arrow. Then $\frac{f}{\uplus} = \{f\}$ if $\widetilde{f} \neq \bot^0$; $\emptyset$ otherwise.

   Consequently, $\circledast(f) = f$ if $\frac{f}{\uplus} = \{f\}$, or $\circledast(f) = \bot^1$ if $\frac{f}{\uplus} = \emptyset$, but from the fact that in this case $\widetilde{f} = \bot^0 = \widetilde{\bot^1}$, from Definition 23, we obtain again that $\circledast(f) = f$.
2. The case of $m \cdot k \geq 2$ (a complex arrow) and $\frac{f}{\uplus} = \emptyset$. Let us consider the following cases:
   2.1. $m = 1$. Then, $\circledast(f) = [\langle h_{11}, \dots, h_{1k}\rangle] =$ (in [_] we have only one element) $= \langle h_{11}, \dots, h_{1k}\rangle$ where $h_{1i} = \bot^1$ for all $i = 1, \dots, k$, so that $\frac{\circledast(f)}{\uplus} = \emptyset = \frac{f}{\uplus}$ and hence $\circledast(f) = f$ is maximal with $1 \cdot k = k \geq 2$ simple arrows.
   2.2. $k = 1$. Then, $\circledast(f) = [\langle h_{11}\rangle, \dots, \langle h_{m1}\rangle] = [h_{11}, \dots, h_{m1}]$ where $h_{j1} = \bot^1$ for all $j = 1, \dots, m$, so that $\frac{\circledast(f)}{\uplus} = \emptyset = \frac{f}{\uplus}$ and hence $\circledast(f) = f$ is maximal with $m \cdot 1 = m \geq 2$ simple arrows.
   2.3. Both $m, k \geq 2$. Then, $\circledast(f) = [\langle h_{11}, \dots, h_{1k}\rangle, \dots, \langle h_{m1}, \dots, h_{mk}\rangle]$ where $h_{ji} = \bot^1$ for all $j = 1, \dots, m$ and $i = 1, \dots, k$, so that $\frac{\circledast(f)}{\uplus} = \emptyset = \frac{f}{\uplus}$ and hence $\circledast(f) = f$ is maximal $k \cdot m \geq 2$ simple arrows.
3. The case of $m \cdot k \geq 2$ (a complex arrow) and $\frac{f}{\uplus} \neq \emptyset$. Then, as in case 2, we have tree possible cases, and in each of them we obtain that $\frac{\circledast(f)}{\uplus} = \frac{f}{\uplus}$ and hence $\circledast(f) = f$ is maximal $k \cdot m \geq 2$ simple arrows. □

It is easy to see that for any f, $\circledast$ inserts only the new empty arrows $\bot^1$ in this canonical structural representation where, for each simple object A_j in a A, we have a full cone of simple arrows $\langle h_{j1}, \dots, h_{jk}\rangle : A_j \rightarrow B$ from A_j into each simple object in B_i for $i = 1, \dots, k$. For example, consider $\circledast(k)$ for $k = k_{11} \uplus [k_{2,2}, k_{3,2}] : A_1 \uplus A_2 \uplus A_3 \rightarrow B_1 \uplus B_2$.

Moreover, we are able to construct a well-defined complex arrow between any two complex objects by specifying the set S of ptp arrows. This is useful also when the set S of ptp arrows is empty. For example, $\widehat{\circledast}(A_1 \uplus A_2, B_1, \{f_{21}\}) = [\bot^1, f_{21}] : A_1 \uplus A_2 \rightarrow B_1$, or $\widehat{\circledast}(A_1, B_1 \uplus B_2, \emptyset) = \langle \bot^1, \bot^1\rangle : A_1 \rightarrow B_1 \uplus B_2$.

We can show the following important isomorphisms for (also complex) instance-databases in **DB**:

Lemma 9 *For any* $A, B, C \in Ob_{DB}$, *we have the following isomorphisms:*

1. *For any object A, we have the isomorphism* $is_A : A \to TA$ *with its inverse* is_A^{-1} : $TA \to A$ *such that* $\widetilde{is_A} = \widetilde{is_A^{-1}} = \widetilde{id_A} = TA$. *Thus,* $A \simeq TA$.
2. (*Commutativity*) $A \uplus B \simeq B \uplus A$.
3. (*Associativity*) $A \uplus (B \uplus C) \simeq A \uplus B \uplus C \simeq (A \uplus B) \uplus C$.
4. (*Composition with zero object*) $C \uplus \bot^0 \simeq C \simeq \bot^0 \uplus C$.

Proof Claim 1. Let us consider a simple object (instance-database) A. The SOtgd Φ_A of the schema mappings, corresponding to the morphisms $id_A : A \to A$, $is_A : A \to TA$ and $is_A^{-1} : TA \to A$, is equal to the *InverseOperads*($\{1_{r_i} \in O(r_i, r_i) \mid r_i \in S_A\} \cup \{1_{r_\emptyset}\}$), i.e., to the logical formula $\bigwedge\{\forall \mathbf{x}_i (r_i(\mathbf{x}_i) \Rightarrow r_i(\mathbf{x}_i)) \mid r_i \in S_A\}$ where $A = \alpha^*(\mathcal{A})$ of a schema $\mathcal{A}$ and $\mathcal{A} \subseteq \widehat{\mathcal{A}}$ (thus, $A = \alpha^*(\mathcal{A}) \subseteq \alpha^*(\widehat{\mathcal{A}}) = TA$). Thus, $id_A = \alpha^*(MakeOperads(\{\Phi_A\})) : A \to A$, $is_A = \alpha^*(MakeOperads(\{\Phi_A\})) : A \to TA$, and $is_A^{-1} = \alpha^*(MakeOperads(\{\Phi\}_A)) : TA \to A$, are *atomic* morphisms (Definition 18). That is, these three morphisms are based on the same SOtgd Φ_A and, consequently, for each object A in **DB** ($\widetilde{id_A} = TA$ is demonstrated by Theorem 1) that $\widetilde{is_A} = \widetilde{is_A^{-1}} = \widetilde{id_A} = TA$. For a complex object $A = \biguplus_{1 \le j \le m} A_j$, $m \ge 2$, $is_A = \biguplus_{1 \le j \le m} is_{A_j}$, $is_A^{-1} = \biguplus_{1 \le j \le m} is_{A_j}^{-1}$: In fact, $is_A^{-1} \circ is_A = \biguplus_{1 \le i \le k} is_{A_i}^{-1} \circ is_{A_i} = \biguplus_{1 \le i \le k} id_{A_i} = id_A$ (i.e., $\frac{id_A}{\uplus} = \frac{\biguplus_{1 \le i \le k} id_{A_i}}{\uplus} = \{id_{Ai} : A_i \to A_i \mid 1 \le i \le k\}$). Analogously, $is_{A_i} \circ is_{A_i}^{-1} = \biguplus_{1 \le i \le k} id_{TA_i} = id_{TA}$. Hence $\widetilde{is_A} = \widetilde{is_A^{-1}} = \widetilde{id_A} = \biguplus_{1 \le j \le m} TA_j = TA$.

Claim 2. (Commutativity) For any two objects in **DB** there is a commutativity isomorphism $A \uplus B \simeq B \uplus A$.

Let us show that this isomorphism is represented by the complex morphism $i_1 = \langle [\bot^1, id_B], [id_A, \bot^1] \rangle : A \uplus B \to B \uplus A$ (where $[\bot^1, id_B] : A \uplus B \to B$ and $[id_A, \bot^1] : A \uplus B \to A$) and its inverse by $i_2 = i_1^{OP} = [[\bot^1, id_B]^{OP}, [id_A, \bot^1]^{OP}] = [\langle \bot^1, id_B \rangle, \langle id_A, \bot^1 \rangle] : B \uplus A \to A \uplus B$.

In fact,

$$\begin{aligned} i_1 \circ i_2 &= \langle [\bot^1, id_B], [id_A, \bot^1] \rangle \circ [\langle \bot^1, id_B \rangle, \langle id_A, \bot^1 \rangle] \\ &= \langle [[\bot^1, id_B] \circ \langle \bot^1, id_B \rangle, [\bot^1, id_B] \circ \langle id_A, \bot^1 \rangle], \\ &\qquad [[id_A, \bot^1] \circ \langle \bot^1, id_B \rangle, [id_A, \bot^1] \circ \langle id_A, \bot^1 \rangle]\rangle \\ &= \langle [\lceil id_B \rceil, \lceil \bot^1 \rfloor], [\lceil \bot^1 \rfloor, \lceil id_A \rfloor] \rangle = \langle [id_B, \bot^1], [\bot^1, id_A] \rangle. \end{aligned}$$

Thus, $\frac{i_1 \circ i_2}{\uplus} = \{id_A, id_B\} = \frac{id_B \uplus id_A}{\uplus}$, and hence, by Definition 23,

$$i_1 \circ i_2 = id_B \uplus id_A = (\text{point 2 of Example 19}) = id_{B \uplus A} : B \uplus A \to B \uplus A.$$

Analogously,

$$\begin{aligned} i_2 \circ i_1 &= [\langle \bot^1, id_B \rangle, \langle id_A, \bot^1 \rangle] \circ \langle [\bot^1, id_B], [id_A, \bot^1] \rangle \\ &= \lceil \langle \bot^1, id_B \rangle \circ [\bot^1, id_B], \langle id_A, \bot^1 \rangle \circ [id_A, \bot^1] \rfloor \end{aligned}$$

$$= \lceil \langle \bot^1 \circ [\bot^1, id_B], id_B \circ [\bot^1, id_B] \rangle, \langle id_A, \circ [id_A, \bot^1], \bot^1 \circ [id_A, \bot^1] \rangle \rfloor$$
$$= \lceil \langle \bot^1, [\bot^1, id_B] \rangle, \langle [id_A, \bot^1], \bot^1 \rangle \rfloor : A \uplus B \to A \uplus B.$$

Thus, $\frac{i_2 \circ i_1}{\uplus} = \{id_A, id_B\} = \frac{id_A \uplus id_B}{\uplus}$, and hence we obtain an identity

$$i_2 \circ i_1 = id_A \uplus id_B = id_{A \uplus B} : A \uplus B \to A \uplus B.$$

Consequently, $i_1 = \langle [\bot^1, id_B], [id_A, \bot^1] \rangle : A \uplus B \to B \uplus A$ is an isomorphism, that is, $A \uplus B \simeq B \uplus A$.

Claim 3. (Associativity) It holds that $id_A \uplus \langle [id_B, \bot^1], [\bot^1, id_C] \rangle : A \uplus (B \uplus C) \to A \uplus B \uplus C$ is an isomorphism.

It is enough to show that $i_3 = \langle [id_B, \bot^1], [\bot^1, id_C] \rangle : B \uplus C \to B \uplus C$ is an isomorphism, as follows: $\frac{i_3}{\uplus} = \{id_B, id_C\} = \frac{id_{B \uplus C}}{\uplus}$ (from the fact that $id_{B \uplus C} = id_B \uplus id_C$), thus (by point 3 of Definition 23) $i_3 = id_{B \uplus C} : B \uplus C \to B \uplus C$ and from the fact that each identity morphism is an isomorphism, it holds that i_3 is an isomorphism.

Claim 4. $T(C \uplus \bot^0) = TC \uplus T\bot^0 = TC \uplus \bot^0 \neq TC$, and hence we cannot use Proposition 8 in order to demonstrate the isomorphism $C \uplus \bot^0 \simeq C$, but the standard categorial method for verification of the isomorphisms. For a given schema $\mathcal{C}$ with the interpretation α such that $C = \alpha^*(\mathcal{C})$, we have the identity morphism $id_C = \alpha^*(\mathbf{M}_{\mathcal{C}\mathcal{C}}) : C \to C$ with $\mathbf{M}_{\mathcal{C}\mathcal{C}} = \alpha^*(\{1_r \mid r \in \mathcal{C}\} \cup \{1_{r_\emptyset}\})$.

Let us define the morphisms $i_1 = \langle \alpha^*(\mathbf{M}_{\mathcal{C}\mathcal{C}}), \alpha^*(\{1_{r_\emptyset}\}) \rangle : C \to C \uplus \bot^0$ and $i_2 = [\alpha^*(\mathbf{M}_{\mathcal{C}\mathcal{C}}), \alpha^*(\{1_{r_\emptyset}\})] : C \uplus \bot^0 \to C$. Thus,

$$\begin{aligned}
i_2 \circ i_1 &= [\alpha^*(\mathbf{M}_{\mathcal{C}\mathcal{C}}), \alpha^*(\{1_{r_\emptyset}\})] \circ \langle \alpha^*(\mathbf{M}_{\mathcal{C}\mathcal{C}}), \alpha^*(\{1_{r_\emptyset}\}) \rangle \\
&= [id_C, \bot^1] \circ \langle id_C, \bot^1 \rangle \\
&= \lceil id_C \circ id_C, \bot^1 \circ \bot^1 \rfloor = \lceil id_C, \bot^1 \rfloor \\
&\quad \text{(from point 2.4 of Definition 20)} \\
&= id_C : C \to C,
\end{aligned}$$

i.e., the identity morphism for C. While,

$$\begin{aligned}
i_1 \circ i_2 &= \langle \alpha^*(\mathbf{M}_{\mathcal{C}\mathcal{C}}), \alpha^*(\{1_{r_\emptyset}\}) \rangle \circ [\alpha^*(\mathbf{M}_{\mathcal{C}\mathcal{C}}), \alpha^*(\{1_{r_\emptyset}\})] \\
&= \langle id_C, \bot^1 \rangle \circ [id_C, \bot^1] \\
&= \langle [id_c, \bot^1], [\bot^1, \bot^1] \rangle.
\end{aligned}$$

Thus, $\frac{i_1 \circ i_2}{\uplus} = \frac{id_{C \uplus \bot^0}}{\uplus} = \{id_C\}$ (from $id_{C \uplus \bot^0} = id_C \uplus \bot^1$) and hence, from point 3 of Definition 23, $i_1 \circ i_2 = id_{C \uplus \bot^0}$. Consequently, this morphism is the identity morphism of the object $C \uplus \bot^0$. Therefore, both morphisms $i_1 \circ i_2$ and $i_2 \circ i_1$ are the identity morphisms and so $i_1 = \langle id_C, \bot^1 \rangle : C \to C \uplus \bot^0$ is an isomorphism and hence, (a) $C \simeq C \uplus \bot^0$ and, analogously, by commutativity, $\bot^0 \uplus C \simeq C \uplus \bot^0 \simeq C$.

Let us show that i_1 is monic and epic as well. For any two morphisms $g_1, g_2 : D \rightarrow C$ such that $i_1 \circ g_1 = i_1 \circ g_2$, that is, $i_1 \circ g_1 = \langle id_C, \bot^1 \rangle \circ g_1 = \langle g_1, \bot^1 \rangle = i_1 \circ g_2 = \langle g_2, \bot^1 \rangle$, we obtain $g_1 = g_2$ (from $\{g_1\} = \frac{i_1 \circ g_1}{\uplus} = \frac{i_1 \circ g_2}{\uplus} = \{g_2\}$ and by point 3 of Definition 23) and hence i_1 is a monomorphism. For any two $g_1 = [k_1, \bot^1]$ and $g_2 = [k_2, \bot^1] : C + \bot^0 \rightarrow D$ such that $g_1 \circ i_1 = g_2 \circ g_1$, that is, $g_1 \circ i_1 = [k_1, \bot^1] \circ \langle id_C, \bot^1 \rangle = \lceil k_1, \bot^1 \rfloor = k_1$ (component $\bot^1$ is eliminated according to point 4 in Definition 20) $= g_2 \circ i_1 = [k_2, \bot^1] \circ \langle id_C, \bot^1 \rangle = \lceil k_2, \bot^1 \rfloor = k_2$, we obtain $k_1 = k_2$, thus $g_1 = g_2$ and hence i_1 is an epimorphism. □

Remark From the fact that $C \uplus \bot^0$ is isomorphic to C and that two isomorphic objects in a category can be substituted, we can eliminate for complex objects all their components which are equal to the empty database $\bot^0$ and, generally, consider only the *strictly-complex* objects. Now we obtain the first fundamental property of **DB** category which confirms that it is fundamentally different from **Set**:

Proposition 5 *The morphisms $[id_C, id_C] : C \uplus C \rightarrow C$ are epic and monic but not isomorphic. Consequently,* **DB** *is not a topos.*

Proof Let for any two morphisms $g_1 = \langle h_1, k_1 \rangle : C \rightarrow C \uplus C$ and $g_2 = \langle h_2, k_2 \rangle : C \rightarrow C \uplus C$, $[id_C, id_C] \circ g_1 = [id_C, id_C] \circ g_2 : C \rightarrow C$. Let us show that $g_1 = g_2$. In fact, $[id_C, id_C] \circ g_1 = [id_C, id_C] \circ \langle h_1, k_1 \rangle = \lceil id_C \circ h_1, id_C \circ k_1 \rfloor = \lceil h_1, k_1 \rfloor : C \rightarrow C$ and $[id_C, id_C] \circ g_2 = \lceil h_2, k_2 \rfloor : C \rightarrow C$. Consequently, $\lceil h_1, k_1 \rfloor = \lceil h_2, k_2 \rfloor$ and, from Definition 23, there exists a bijection σ such that $\sigma(h_1) = h_2$, $\sigma(k_1) = k_2$ with $\widetilde{h_1} = \widetilde{\sigma(h_1)} = \widetilde{h_2}$ and $\widetilde{k_1} = \widetilde{\sigma(k_1)} = \widetilde{k_2}$, so that $h_1 = h_2$ and $k_1 = k_2$ and hence $g_1 = \langle h_1, k_1 \rangle = \langle h_2, k_2 \rangle = g_2$. Consequently, $[id_C, id_C] : C \uplus C \rightarrow C$ is a monomorphism.

Let for any two morphisms $g_1 : C \rightarrow C$ and $g_2 : C \rightarrow C$, $g_1 \circ [id_C, id_C] = g_2 \circ [id_C, id_C]$ (the case when $g_1 = g_2 = \bot^1$ is trivial and hence we consider the cases when g_1 and g_2 are different from $\bot^1$). Let us show that $g_1 = g_2$. In fact, $g_1 \circ [id_C, id_C] = [g_1, g_1] : C \uplus C \rightarrow C$ and $[g_2 \circ id_C, id_C] = [g_2, g_2] : C \rightarrow C$. Consequently, $[g_1, g_1] = [g_2, g_2]$ and, from point 3 of Definition 23, $g_1 = g_2$. Therefore, $[id_C, id_C] : C \uplus C \rightarrow C$ is an epimorphism. Let us show that $f = [id_C, id_C] : C \uplus C \rightarrow C$ is not an isomorphism: If we suppose that it is an isomorphism then $f \circ f^{-1} : C \rightarrow C$ has to be the identity arrow id_C. From the fact that $f^{-1} = [id_C, id_C]^{OP} = \langle id_C^{OP}, id_C^{OP} \rangle = \langle id_C, id_C \rangle : C \rightarrow C \uplus C$, so that $f \circ f^{-1} = [id_C, id_C] \circ \langle id_C, id_C \rangle = \lceil id_C \circ id_C, id_C \circ id_C \rfloor = \lceil id_C, id_C \rfloor \neq id_C$ (by point 2 of Definition 23), i.e., the properties of an isomorphism are not satisfied.

From the fact that in any topos category if a morphism is both epic and monic then it must also be an isomorphism, we conclude that **DB** is not a topos. More about this will be considered in Sect. 9.1 dedicated to topological properties of **DB**. □

Thus, **DB** is not a topos. However, there is a subset of morphisms for which the topological properties hold so that if a morphism is both epic and monic then it must also be an isomorphism: it is the set of all simple arrows, as we will show in the next corollary, so that the first step is to analyze the topological properties of the simple morphisms in **DB** and only after that the general properties of (also complex) morphisms.

Corollary 9 *For each simple arrow* $f : A \to B$ *the following are valid*:
1. f *is monic iff* $\widetilde{f} = TA$.
2. f *is epic iff* $\widetilde{f} = TB$.
3. f *is an isomorphism iff it is monic and epic (i.e., when* $\widetilde{f} = TA = TB$).

Proof Claim 1. (From right to left) Let us show that if $\widetilde{f} = TA$ then $f : A \to B$ is monic. A simple arrow $f : A \longrightarrow B$ is monic iff for any (possibly complex) two arrows $h, g : \biguplus_{1 \le i \le n} X_i \longrightarrow A$, $f \circ h = f \circ g$ implies $h = g$. Let us show that it is satisfied when $\widetilde{f} = TA$. We have (from Proposition 6) that for each point-to-point simple arrow $(g_i : X_i \to A) \in \frac{g}{\uplus}$, $\widetilde{g_i} \subseteq TA$ (analogously for each $(h_i : X_i \to A) \in \frac{h}{\uplus}$, $\widetilde{h_i} \subseteq TA$). Thus, if $f \circ h = f \circ g$ then, from Definition 22, $\frac{f \circ g}{\uplus} = \frac{f \circ h}{\uplus}$, i.e., they have the same ptp arrows, that is, $f \circ g_i = f \circ h_i : X_i \to B$, so that they have the same fluxes, i.e., $\widetilde{f \circ g_i} = \widetilde{f \circ h_i}$. But we have that $\widetilde{f \circ g_i} = \widetilde{f} \cap \widetilde{g_i} = TA \cap \widetilde{g_i} = \widetilde{g_i}$ and $\widetilde{f \circ h_i} = \widetilde{f} \cap \widetilde{h_i} = TA \cap \widetilde{h_i} = \widetilde{h_i}$, so that $\widetilde{h_i} = \widetilde{g_i}$, i.e., they are equal ptp arrows $h_i = g_i : X_i \to A$. It holds for all ptp arrows in g and h, so that $g = h$, and, consequently, $f : A \longrightarrow B$ is a monomorphism.

(From left to right) If a simple arrow $f : A \hookrightarrow B$ is monic then for any (possibly complex) two arrows $h, g : \biguplus_{1 \le i \le n} X_i \longrightarrow A$, $f \circ h = [f \circ h_1, \dots, f \circ h_n] = f \circ g = [f \circ g_1, \dots, f \circ g_n] : \biguplus_{1 \le i \le n} X_i \longrightarrow B$, implies $h = g$, that is, $\forall 1 \le l \le n(f \circ g_l = f \circ h_l : X_l \to B$ implies $g_l = h_l : X_l \to A)$. Let us suppose that $\widetilde{f} \subset TA$, so that we can have, for example, $\widetilde{g_l} = \widetilde{f} \neq TA = \widetilde{h_l}$, that is, $g_l \neq h_l$ but with $\widetilde{f \circ g_l} = \widetilde{f} = \widetilde{f \circ h_l}$, i.e., $f \circ g_l = f \circ h_l$, which is a contradiction. Thus, it must hold that $\widetilde{f} = TA$.

Claim 2. The proof is analogous to that of Claim 1.

Claim 3. (From left to right) It holds for every category.

(From right to left) If $f : A \to B$ is both monic and epic then, from the first two points, $\widetilde{f} = TA = TB$. Let us show that $f = is_B^{-1} \circ is_A$. In fact, $\widetilde{is_B^{-1} \circ is_A} = \widetilde{is_B^{-1}} \cap \widetilde{is_A} =$ (from Lemma 9) $= TB \cap TA = TA = TB$, so that, from Definition 23, $f = is_B^{-1} \circ is_A : A \to B$, that is, f is a composition of two isomorphisms, and hence it is an isomorphism as well. □

Now we will also analyze the properties of complex arrows.

Notice that the complex morphism $f = [id_C, id_C] : C \uplus C \to C$ has to be formally represented by 'indexing by position' formalism, that is, $f = [id_C, id_C] : A_1 \uplus A_2 \to B_1$ where $A_1 = A_2 = B_1 = C$, so that $\frac{f}{\uplus} = \{f_{11} = id_C : A_1 \to B_1, f_{21} = id_C : A_2 \to B_1\}$, that is, by *two* mutually independent mapping-morphisms $id_C : C \to C$. Thus, the morphisms $g = [id_C, \bot^1] : C \uplus C \to C$ and $h = [\bot^1, id_C] : C \uplus C \to C$ have $\frac{g}{\uplus} = \{g_{11} = id_C : A_1 \to B_1\} = \{id_C : C \to C\} \neq \frac{f}{\uplus}$, $\frac{h}{\uplus} = \{h_{21} = id_C : A_2 \to B_1\} = \{id_C : C \to C\} \neq \frac{f}{\uplus}$. But $\frac{g}{\uplus} = \{id_C : C \to C\} = \frac{h}{\uplus}$ so that, from Definition 23, $g = h$. Thus, $[id_C, \bot^1] = g = h = [\bot^1, id_C]$.

Proposition 6 *We have the following fundamental properties for the morphisms in* **DB**

1. *If a morphism* $f : A \longrightarrow B$ *has a property* $P \in \{isomorphic, monic, epic\}$ *then each of its ptp arrows* $f_l \in \frac{f}{\uplus}$ *has the same property* P;
2. *If* $f : A \hookrightarrow B$ *is monic and* $A = \biguplus_{1 \leq j \leq m} A_j$, $m \geq 1$, *is strictly complex then for each* A_j *there must exist a ptp arrow* $f_n \in \frac{f}{\uplus}$ *with* $dom(f_n) = A_j$;
3. *If* $f : A \twoheadrightarrow B$ *is epic and* $B = \biguplus_{1 \leq i \leq k} B_i$, $k \geq 1$, $k \geq 1$, *is strictly complex then for each* B_i *there must exist a ptp arrow* $f_n \in \frac{f}{\uplus}$ *with* $cod(f_n) = B_i$.
4. *If* $C = \biguplus_{1 \leq j \leq m} C_j$ *and* $B = \biguplus_{1 \leq i \leq k} B_i$ *are strictly-complex objects and there is an isomorphism* $is : C \simeq B$, *then* $m = k$.

Proof Claim 1. Its inverted arrow is $\frac{f^{OP}}{\uplus} = \{f_{ji}^{OP} : B_i \rightarrow A_j \mid f_{ji} : A_j \rightarrow B_i \text{ in } \frac{f}{\uplus} \text{ and } \widetilde{f_{ji}^{OP}} = \widetilde{f_{ji}}\}$. Let $f : \biguplus_{1 \leq j \leq m} A_j \rightarrow \biguplus_{1 \leq i \leq k} B_i$ be an isomorphism, so that $f \circ f^{-1} = id_B$, thus $\frac{f \circ f^{-1}}{\uplus} = \frac{f \circ f^{OP}}{\uplus} = \{f_{ji} \circ f_{ji}^{OP} = id_i : B_i \rightarrow B_i \mid f_{ji} \in \frac{f}{\uplus}, f_{ji}^{OP} \in \frac{f^{OP}}{\uplus}$ with $cod(f_{ji}^{OP}) = dom(f_{ji}) = A_j$ and $cod(f_{ji}) = B_i\} = \frac{\biguplus_{1 \leq i \leq k} id_{B_i}}{\uplus} = \frac{id_B}{\uplus}$. Let us suppose that $f_{ji} \in \frac{f}{\uplus}$ is not an isomorphism. Thus, $f_{ji} \circ f_{ji}^{OP} \neq id_i : B_i \rightarrow B_i$ (with $cod(f_{ji}^{OP}) = dom(f_{ji})$ and $cod(f_{ji}) = B_i$), so that $f \circ f^{-1} \neq id_B$ and f would not be an isomorphism, which is a contradiction. Consequently, each $f_{ji} \in \frac{f}{\uplus}$ must be an isomorphism. The converse does not hold: let us consider the morphism $f = [id_C, id_C] : A_1 \uplus A_2 \rightarrow B_1$, with $A_1 = A_2 = B_1 = C$ where $\frac{f}{\uplus} = \{f_{11} = id_C : A_1 \rightarrow B_1, f_{21} = id_C : A_2 \rightarrow B_1\}$ where the point-to-point arrows id_C are the identity arrows (thus isomorphisms as well) but f is not an isomorphism (from Proposition 5). Let $f : A \hookrightarrow B$ be a monomorphism. Then, for every two arrows $g, h : X \rightarrow A$ if $f \circ g = f \circ h$ then we obtain $g = h$. Let us suppose that there exists $(f_{ji} : A_j \rightarrow B_i) \in \frac{f}{\uplus}$ which is not monic, that is, there exist g_{lj}, h_{lj} such that $f_{ji} \circ g_{lj} = f_{ji} \circ h_{lj}$ with $g_{lj} \neq h_{lj} : X_l \rightarrow A_j$. However, from the fact that f is a monomorphism it must hold that $g = h$ and, by Definition 23, $\frac{g}{\uplus} = \frac{h}{\uplus}$ with $g_{lj} = h_{lj} : X_l \rightarrow A_j$, which is a contradiction. Thus, each point-to-point arrow f_{ji} of a monic arrow f must be monic as well.

The proof for an epimorphism $f : A \twoheadrightarrow B$ is similar.

Claim 2. Let A be a strictly-complex object (with all $A_j \neq \bot^0$, $j = 1, \dots, m$) and let us suppose that there exists A_j such that $A_j \notin \{dom(f_n) \mid f_n \in \frac{f}{\uplus}\}$. Let $X = X_1$ be a simple object and $g, h : X_1 \rightarrow A$ be two arrows such that $f \circ g = f \circ h$ so that from the fact that f is a monomorphism, $g = h$ with two ptp arrows $(g_{1j} : X_1 \rightarrow A_j) \in \frac{g}{\uplus}$ and $(h_{1j} : X_1 \rightarrow A_j) \in \frac{h}{\uplus}$ with $g_{1j} = h_{1j}$. However, we can now take any ptp arrow $g'_{1j} : X_1 \rightarrow A_j$ such that $g'_{1j} \neq h_{1j}$ and change g into g' by substituting old g_{1j} by g'_{1j}, so that still $f \circ g' = f \circ h$ is valid, that is, $f \circ g' = f \circ g$ but not $g' = g$ and hence, we obtain that f is not monomorphic, which is a contradiction. Consequently, for each A_j there must exist at least one ptp arrow $f_n \in \frac{f}{\uplus}$ with $dom(f_n) = A_j$ if f is a monomorphism (note that $f_n \neq \bot^1$).

Claim 3. The proof when f is epic is similar.

Claim 4. For any two strictly-complex objects $C = \biguplus_{1\le j\le m} C_j$ and $B = \biguplus_{1\le i\le k} B_i$, let $is : C \to B$ be an isomorphism. Thus, $\{dom(is_i) \mid is_i \in \frac{is}{\uplus}\} = \{C_1, \dots, C_m\}$ and $\{cod(is_i) \mid is_i \in \frac{is}{\uplus}\} = \{B_1, \dots, B_k\}$ (from the fact that $is \circ is_1 = id_B = \biguplus_{1\le i\le k} id_{B_i}$ and $is^{-1} \circ is = id_A = \biguplus_{1\le j\le m} id_{A_j}$. Then, from point 1 of this proposition, each point-to-point arrow $is_i \in \frac{is}{\uplus}$ is an isomorphism with $|\frac{is}{\uplus}| \ge \max(m, k)$. Let us show that $m = k$. Let us suppose that $m > k$, so that we have two different isomorphisms $is_i, is_j \in \frac{is}{\uplus}$ with $dom(is_i) \ne dom(is_j)$ and $cod(is_i) = cod(is_j) = B_l$. Thus, $\langle is_i, is_j\rangle \circ [is_i^{OP}, is_j^{OP}] = \lceil is_i \circ is_i^{OP}, is_j \circ is_j^{OP} \rfloor = \lceil id_{B_l}, id_{B_l} \rfloor \ne id_{B_l}$. Hence is cannot be an isomorphism, which is a contradiction. Analogously, if we suppose that $m < k$, we again obtain a contradiction. Consequently, it must be that $m = k$. □

Remark Claim 2 of this proposition explains the fact that when $C \ne \bot^0$, the morphism $k = [id_C, id_C] : C \uplus C \to C$ where both point-to-point arrows are identities (thus, isomorphisms) is not an isomorphism (but only a monic and epic arrow, as established by Proposition 5). While, for non-strictly-complex objects, we have the isomorphisms when $m \ne k$ as well: let us consider the isomorphism $C \uplus \bot^0 \simeq C$, when $m = 2$ ($A_1 = C$, $A_2 = \bot^0$) and $k = 1$ with $B_1 = C$.

We obtain the following important results for the *monomorphisms* in **DB**:

Proposition 7 *The following monomorphisms in* **DB** *category are derivable from the morphisms and from the inclusions of the objects*:

1. *The conceptualized object* $\widetilde{f} = B_T(f)$, *obtained from a morphism* $f : A \longrightarrow B$, *is a closed object in* **DB** *such that* $\widetilde{f} = T\widetilde{f} \preceq A \otimes B$. *This PO relation may be represented internally in* **DB** *by the monomorphism* $in : \widetilde{f} \hookrightarrow \widetilde{id_A} \otimes \widetilde{id_B}$ *and its derivations* $in_A : \widetilde{f} \hookrightarrow \widetilde{id_A}$ *and* $in_B : \widetilde{f} \hookrightarrow \widetilde{id_B}$ *where* $\widetilde{id_A} = TA = \biguplus_{1\le j\le m} TA_j$, $m \ge 1$, *and* $\widetilde{id_B} = TB = \biguplus_{1\le i\le k} TB_i$, $k \ge 1$, *with ptp arrows*:
$$\left\{ in_{ji} = \{id_R \mid R \in \widetilde{f_{ji}}\} : \widetilde{f_{ji}} \hookrightarrow S_{ji} \mid (f_{ji} : A_j \to B_i) \in \frac{f}{\uplus} \right\},$$
where $S_{ji} = TA_j \cap TB_i$ *for* $\frac{in}{\uplus}$, $S_{ji} = TA_j$ *for* $\frac{in_A}{\uplus}$, *and* $S_{ji} = TB_i$ *for* $\frac{in_B}{\uplus}$.
2. *For any two objects* $TA \subseteq TB$ *there exists a monomorphism* $in_{TA} : TA \hookrightarrow TB$.

Proof Claim 1. Each conceptualized object (information flux) of an atomic arrow is a closed object. In fact, from Definition 13, for a simple atomic morphism $f = \alpha^*(\mathbf{M}_{AB}) : A \to B$, both databases A and B are simple objects (sets of relations) and so are TA and TB. Thus, if $\widetilde{f} = \bot^0$ (empty flux) then $T\bot^0 = \bot^0$. Otherwise, if $\widetilde{f} \ne \bot^0$ then

$$T\widetilde{f} = T\big(T\{\pi_{\mathbf{x}_i} \| e[(\_)_j/r_{i,j}]_{1\le j\le k} \|_A \mid$$
$$q_i \in O(r_{i,1}, \dots, r_{i,k}, r_i') \text{ is equal to } (e \Rightarrow (\_)(\mathbf{t}_i)) \in \mathbf{M}_{AB}\}\big)$$

$$= T\{\pi_{\mathbf{x}_i} \| e[(\_)_j/r_{i,j}]_{1\le j\le k} \|_A \mid$$
$$q_i \in O(r_{i,1}, \dots, r_{i,k}, r_i') \text{ is equal to } (e \Rightarrow (\_)(\mathbf{t}_i)) \in \mathbf{M}_{AB}\}$$
$$= \widetilde{f} \subseteq TA.$$

From the fact that this information flux is transferred into B, $\widetilde{f} \subseteq TB$, so that $\widetilde{f} \subseteq TA \cap TB$, i.e., by Definition 19, $\widetilde{f} \preceq A \otimes B$. Each arrow is a composition of a number of atomic arrows and the intersection of closed objects (their information fluxes) is always a closed object.

If $f : \biguplus_{1\le j\le m} A_j \to \biguplus_{1\le i\le k} B_i$ is a complex morphism with a set of more than one point-to-point arrow in $\frac{f}{\uplus}$ then, for each $f_{ji} : A_j \to B_i \in \frac{f}{\uplus}$, $\widetilde{f_{ji}} \subseteq TA_j \cap TB_i$. Hence $\widetilde{\biguplus \frac{f}{\uplus}} \preceq A \otimes B$ and $\widetilde{f} \preceq \widetilde{\biguplus \frac{f}{\uplus}}$ (from Definition 22, $\widetilde{f} \simeq \widetilde{\biguplus \frac{f}{\uplus}}$), so that $\widetilde{f} \preceq A \otimes B$. From Lemma 9, $\widetilde{id_A} = TA$ and, from a Definition 19 and $T(TA) = TA$, $A \otimes B = TA \otimes TB = \widetilde{id_A} \otimes \widetilde{id_B}$. Thus, we obtain the PO relation $\widetilde{f} \preceq \widetilde{id_A} \otimes \widetilde{id_B}$ and we will show that it can be internally represented in **DB** by a monomorphism $in : \widetilde{f} \hookrightarrow \widetilde{id_A} \otimes \widetilde{id_B}$. The fact that in, in_A and in_B are the morphisms defined by Theorem 1 and hence they can be obtained from particular schema mappings and a given R-algebra α will be shown in Theorem 4.

Let us consider a morphism $f : \biguplus_{1\le j\le m} A_j \to \biguplus_{1\le i\le k} B_i$, $k, m \ge 1$, with point-to-point (simple) arrows $(f_{ji} : A_j \to B_i) \in \frac{f}{\uplus}$. Let us define the morphism $in : \widetilde{f} \to TA \otimes TB$ with $\frac{in}{\uplus} = \{in_{ji} = \{id_R : R \to R \mid R \in \widetilde{f_{ji}}\} : \widetilde{f_{ji}} \hookrightarrow TA_j \cap TB_i \mid (f_{ji} : A_j \to B_i) \in \frac{f}{\uplus}\}$. Thus, there is a bijection $\sigma : \frac{f}{\uplus} \to \frac{in}{\uplus}$ with $in_{ji} = \sigma(f_{ji}) : \widetilde{f_{ji}} \to TA_j \cap TB_i$ and $\widetilde{in_{jk}} = \widetilde{f_{jk}}$ for any ptp arrow, and hence $\widetilde{in} \simeq \widetilde{\biguplus \frac{in}{\uplus}} \simeq \widetilde{\biguplus \frac{f}{\uplus}} \simeq \widetilde{f}$.

For each pair $g, h : \biguplus_{1\le l\le n} X_l \to \widetilde{f}$ of morphisms (with $(g_{l,ji} : X_l \to \widetilde{f_{ji}}) \in \frac{g}{\uplus} \neq \emptyset$ and $(h_{l,ji} : X_l \to \widetilde{f_{ji}}) \in \frac{h}{\uplus} \neq \emptyset$ so that, from Proposition 6, $\widetilde{g_{l,ji}} \subseteq T\widetilde{f_{ji}} = \widetilde{f_{ji}}$ and $\widetilde{h_{l,ji}} \subseteq T\widetilde{f_{ji}} = \widetilde{f_{ji}}$ and hence $\sigma(f_{ji}) \circ g_{l,ji} \in \frac{in \circ g}{\uplus}$ and $\sigma(f_{ji}) \circ h_{l,ji} \in \frac{in \circ h}{\uplus}$) such that $in \circ g = in \circ h : X \to B$, i.e., from point 3 of Definition 23, $\frac{in \circ g}{\uplus} = \frac{in \circ h}{\uplus}$. Thus, they have the same ptp arrows, that is, $\sigma(f_{ji}) \circ g_{l,ji} = \sigma(f_{ji}) \circ h_{l,ji} : X_l \to TA_j \cap TB_i$ so that they have the same fluxes, i.e., $\widetilde{\sigma(f_{ji}) \circ g_{l,ji}} = \widetilde{\sigma(f_{ji}) \circ h_{l,ji}}$. However, $\widetilde{\sigma(f_{ji}) \circ g_{l,ji}} = \widetilde{\sigma(f_{ji})} \cap \widetilde{g_{l,ji}} = \widetilde{f_{ji}} \cap \widetilde{g_{l,ji}} = \widetilde{g_{l,ji}}$ and, analogously, $\widetilde{\sigma(f_{ji})} \cap \widetilde{h_{l,ji}} = \widetilde{f_{ji}} \cap \widetilde{h_{l,ji}} = \widetilde{h_{l,ji}}$ so that $\widetilde{h_{l,ji}} = \widetilde{g_{l,ji}}$, i.e., ptp arrows $h_{l,ji} = g_{l,ji} : X_l \to \widetilde{f_{ji}}$ are equal. This holds for all ptp arrows in g and h so that $g = h$. That is, for each g, h such that $in \circ g = in \circ h : X \to B$, we obtained that $g = h$, so that

(i) $in : \widetilde{f} \hookrightarrow \widetilde{id_A} \cap \widetilde{id_B}$ is a monomorphism and, analogously,
(ii) $in_A : \widetilde{f} \hookrightarrow \widetilde{id_A} = TA$ and $in_B : \widetilde{f} \hookrightarrow \widetilde{id_B} = TB$.

Claim 2. Let us show that for any two simple objects A and B, $TA \subseteq TB$ implies the existence of a monomorphism $in_{TA} : TA \hookrightarrow TB$.

If $TA \subseteq TB$ then we have a morphism $f = \{id_R : R \to R \mid R \in TA\} : TA \hookrightarrow TB$ with $\widetilde{f} = TA$. Then, from (ii), we have a monomorphism $in_{TA} : \widetilde{f} \hookrightarrow \widetilde{id_{TB}} = T(TB) = TB$ so, by substituting $\widetilde{f}$ by TA, we obtain $in_{TA} : TA \hookrightarrow TB$.

Notice that for any two complex objects A and B such that $TA \subseteq TB$, these complex objects must have the *same* number $k \geq 2$ of simple databases, i.e., it must be true that $A = \biguplus_{1 \leq i \leq k} A_i$ and $B = \biguplus_{1 \leq i \leq k} B_i$ with $A_i \subseteq B_i$, for $1 \leq i \leq k$, and $TA = \biguplus_{1 \leq i \leq k} TA_i \subseteq \biguplus_{1 \leq i \leq k} TB_i = TB$, and hence we obtain the monomorphism $in_{TA} = \biguplus_{1 \leq i \leq k}(in_{TA_i} : TA_i \hookrightarrow TB_i)$. □

Thus, we are able to define all kinds of morphisms in the **DB** category as follows:

Proposition 8 *The following properties for morphisms are valid*:
1. *Each arrow* $f : A \to B$ *such that* $\widetilde{f} \simeq TA$ *(and in special case when* $\widetilde{f} = TA$*) is a monomorphism (monic arrow).*
2. *Each arrow* $f : A \to B$ *such that* $\widetilde{f} \simeq TB$ *(and in special case when* $\widetilde{f} = TB$*) is an epimorphism (epic arrow).*
3. *Each arrow* $f : A \to B$ *such that* $\widetilde{f} \simeq TA \simeq TB$ *(and in special case when* $\widetilde{f} = TA = TB$*) is an isomorphism (iso arrow).*

Proof Claim 1. For a simple arrow it holds from Corollary 9, so consider a strictly-complex source object A. For a given complex arrow $f : \biguplus_{1 \leq j \leq m} A_j \to \biguplus_{1 \leq i \leq k} B_i$, let there be an isomorphism $TA \simeq \widetilde{f}$. Then, from this isomorphism and the fact that $\widetilde{f}$ is strictly complex, also $TA = \biguplus_{1 \leq j \leq m} A_j$ is strictly complex, so we obtain $m = |\frac{f}{\uplus}|$ and a bijection $\sigma' : \frac{f}{\uplus} \to \{1, \ldots, m\}$ such that for each $f_{ji} : A_j \to B_i$, $\widetilde{f_{ji}} = TA_{\sigma'(f_{ji})} = TA_j$. Thus, from above, this ptp (simple) arrow f_{ji} is monic.

For each pair of morphisms $g, h : \biguplus_{1 \leq i \leq k} X_i \to A$ (with $(g_{lj} : X_l \to A_j) \in \frac{g}{\uplus} \neq \emptyset$ and $(h_{lj} : X_l \to A_j) \in \frac{h}{\uplus} \neq \emptyset$, it holds from above that $\widetilde{g_{lj}} \subseteq TA_j$ and $\widetilde{h_{lj}} \subseteq TA_j$) if $f \circ g = f \circ h : X \to B$ then, from point 3 of Definition 23, $\frac{f \circ g}{\uplus} = \frac{f \circ h}{\uplus}$, i.e., they have the same ptp arrows, that is, $f_{ji} \circ g_{lj} = f_{ji} \circ h_{lj} : X_l \to B_i$ and hence they have the same fluxes, i.e., $\widetilde{f_{ji} \circ g_{lj}} = \widetilde{f_{ji} \circ h_{lj}}$. However, $\widetilde{f_{ji} \circ g_{lj}} = \widetilde{f_{ji}} \cap \widetilde{g_{lj}} = TA_j \cap \widetilde{g_{lj}} = \widetilde{g_{lj}}$ and $\widetilde{f_{ji} \circ h_{lj}} = \widetilde{f_{ji}} \cap \widetilde{h_{lj}} = TA_j \cap \widetilde{h_{lj}} = \widetilde{h_{lj}}$, so that $\widetilde{h_{lj}} = \widetilde{g_{lj}}$, i.e., $h_{lj} = g_{lj} : X_l \to A_j$. It holds for all ptp arrows in g and h so that $g = h$ and, consequently, f is an monomorphism. If A is not strictly complex, then we have an isomorphism (which is monic as well) $is : A \to A'$ (thus, $TA \simeq TA'$) where A' is strictly complex, so that $f = f' \circ is : A \to B$ (where $f' : A' \to B$ is a monomorphism above with $\widetilde{f'} \simeq TA' \simeq TA$) is a composition of two monomorphism, with $\widetilde{f} \simeq \widetilde{f'}$, i.e., $\widetilde{f} \simeq TA$ and hence it is monic as well.

Claim 2. For a simple arrow it holds from Corollary 9 and so consider a strictly complex target object B. For a given complex arrow $f : \biguplus_{1 \leq j \leq m} A_j \to \biguplus_{1 \leq i \leq k} B_i$, let there be an isomorphism $TB \simeq \widetilde{f}$. Then, from this isomorphism and the fact that $\widetilde{f}$ and also $TB = \biguplus_{1 \leq j \leq k} B_j$ are strictly complex, we obtain $k = |\frac{f}{\uplus}|$ and a bijection $\sigma' : \frac{f}{\uplus} \to \{1, \ldots, k\}$ such that for each $f_{ji} : A_j \to B_i$, $\widetilde{f_{ji}} = TB_{\sigma'(f_{ji})} = TB_i$. Thus, from above, this ptp (simple) arrow f_{ji} is epic. For each pair of morphisms $g, h : B \to \biguplus_{1 \leq i \leq k} X_i$ (with $(g_{il} : B_i \to X_l) \in \frac{g}{\uplus} \neq \emptyset$ and $(h_{il} : B_i \to X_l) \in \frac{h}{\uplus} \neq \emptyset$, it holds from above that $\widetilde{g_{il}} \subseteq TB_i$ and $\widetilde{h_{il}} \subseteq TB_i$) if $g \circ f = h \circ f : A \to X$ then, from point 3 of Definition 23, $\frac{g \circ f}{\uplus} = \frac{h \circ f}{\uplus}$, i.e., they have the same ptp arrows, that is,

$g_{il} \circ f_{ji} = h_{il} \circ f_{ji} : A_j \to X_l$ and hence they have the same fluxes, i.e., $\widetilde{g_{il} \circ f_{ji}} = \widetilde{h_{il} \circ f_{ji}}$. However, $\widetilde{g_{il} \circ f_{ji}} = \widetilde{g}_{il} \cap \widetilde{f}_{ji} = \widetilde{g}_{il} \cap TB_i = \widetilde{g}_{il}$ and $\widetilde{h_{il} \circ f_{ji}} = \widetilde{h}_{il} \cap \widetilde{f}_{ji} = \widetilde{h}_{il} \cap TB_i = \widetilde{h}_{il}$, so that $\widetilde{h}_{il} = \widetilde{g}_{il}$, i.e., are equal ptp arrows $h_{il} = g_{il} : B_i \to X_l$. It holds for all ptp arrows in g and h so that $g = h$ and, consequently, f is an epimorphism. If B is not strictly complex, then we have an isomorphism (which is epic as well) $is : B' \to B$ (thus, $TB' \simeq TB$) where B' is strictly complex, so that $f = is \circ f' : A \to B$ (where $f' : B' \to B$ is an epimorphism above with $\widetilde{f'} \simeq TB \simeq TB'$) is a composition of two epimorphism, with $\widetilde{f} \simeq \widetilde{f'}$, i.e., $\widetilde{f} \simeq TB$ and hence it is epic as well.

Claim 3. For a simple arrow it holds from Corollary 9. So let us consider strictly complex objects $A = \biguplus_{1 \le j \le m} A_j$ and $B = \biguplus_{1 \le i \le k} B_i$ so that $m = k \ge 2$ and, from the isomorphism $TA = \biguplus_{1 \le j \le m} A_j \simeq TB = \biguplus_{1 \le i \le k} B_i$, we have a bijection (permutation) $\sigma : \{1, \dots, m\} \to \{1, \dots, k\}$ such that $TA_j = TB_{\sigma(j)}$, for $j = 1, \dots, m$. Let us consider the most general case when $TA_j \neq TA_i$ for all $i \neq j$, $i, j \in \{1, \dots, m\}$ (i.e., all TA_j are distinct). From $\widetilde{f} \simeq TA$, the number of ptp arrows in f has to be equal to m and hence $\frac{f}{\uplus} = \{f_1, \dots, f_m\}$ with $\widetilde{f} = \biguplus_{1 \le i \le m} \widetilde{f}_i \simeq TA$, thus, with a permutation $\sigma_1 : \{1, \dots, m\} \to \{1, \dots, k\}$ such that $\widetilde{f}_i = TA_{\sigma_1(i)} = TB_{\sigma(\sigma_1(i))}$, for $i = 1, \dots, m$. Thus, from the fact that all TA_j are distinct values, each ptp arrow is (a) $f_i : A_{\sigma_1(i)} \to B_{\sigma(\sigma_1(i))}$.

Let us define a morphism $g : TA \to TB$ such that $\frac{g}{\uplus} = \{g_{j\sigma(j)} : TA_j \to TB_{\sigma(j)} \mid 1 \le j \le m\}$ where each ptp arrow $g_{j\sigma(j)}$ is the identity arrow id_{TA_j} (from the fact that $TA_j = TB_{\sigma(j)}$). From $\frac{g^{-1}}{\uplus} = \{g^{-1}_{j\sigma(j)} : TB_{\sigma(j)} \to TA_j \mid 1 \le j \le m\}$, we obtain that $g \circ g^{-1} = id_{TB}$ and $g^{-1} \circ g = id_{TA}$ and hence g is an isomorphism.

It is easy to show that $\frac{is_B^{-1} \circ g \circ is_A}{\uplus} = \frac{f}{\uplus}$ where $is_A = \biguplus_{1 \le j \le m} is_{A_j} : A_j \to TA_j$ and $is_B^{-1} = \biguplus_{1 \le i \le m} is_{B_i}^{-1} : TB_i \to B_i$, so that for each $1 \le j \le m$, (b) $f_{\sigma_1^{-1}(j)} = is^{-1}_{B_{\sigma(j)}} \circ g_{j\sigma(j)} \circ is_{A_j} : A_j \to B_{\sigma(j)}$ (it is equal to (a) for $i = \sigma_1^{-1}(j)$, that is, $j = \sigma_1(i)$).

Consequently, from point 3 of Definition 23, $f = is_B^{-1} \circ g \circ is_A$, that is, f is a composition of three isomorphisms, and hence it is an isomorphism $f : A \simeq B$ as well.

Let us show that this is valid also when A and B are not strictly complex. Let A' and B' be the strictly complex objects obtained from A and B by elimination of all $\bot^0$. Then, from point 4 of Lemma 9 and commutativity (point 2 of Lemma 9), $f_1 : A \simeq A'$ is an isomorphism with $TA \simeq TA'$ and $f_2 : B' \simeq B$ is an isomorphism with $TB \simeq TB'$. Thus, we obtain $TA' \simeq TA \simeq TB \simeq TB'$, i.e., $TA' \simeq TB'$, so that $f' : A' \to B'$ for $\widetilde{f'} \simeq TA' \simeq TB'$ is an isomorphism and $f = f_2 \circ f' \circ f_1$ is a composition of the isomorphisms and hence an isomorphism as well. □

Notice that in this proposition we provided the *sufficient* conditions for monic, epic and isomorphic arrows in **DB** and not the necessary conditions: we have seen other cases of monic and epic arrows in the case of complex arrows, for example, $[id_A, id_A] : A \uplus A \to A$, which *are not* isomorphic. We can have the cases when A and B are isomorphic objects also when $TA \neq TB$ (as in the case when C is isomorphic to the separation-composed object $C \uplus \bot^0$ in Corollary 9. In what follows, we will show that any two isomorphic simple objects (simple databases) in **DB** are observationally equivalent.

Remark For each two objects A and B, $A \simeq B$ iff $TA \simeq TB$.

In fact, if $is : A \to B$ is an isomorphism then $is_B \circ is \circ is_A^{-1} : TA \to TB$ is a composition of three isomorphism and hence it is an isomorphism as well. Vice versa, if $is : TA \to TB$ is an isomorphism then $is_B^{-1} \circ is \circ is_A : A \to B$ is a composition of three isomorphism and hence it is an isomorphism as well.

Example 20 Consider the schema $\mathcal{A} = (S_A, \Sigma_A)$ with $S_A = \{\mathtt{StudentRome}(x_1, x_2), \mathtt{StudentMilan}(x_1, x_2), \mathtt{StudentFlorence}(x_1, x_2), \mathtt{StudentRMF}(x_1, x_2)\}$ and with binary relational symbols for students in the cities Rome, Milan and Florence (that associate each student name with a student id) and for all students in these three cities, and the schema $\mathcal{B} = (S_B, \emptyset)$ such that $S_B = \{\mathtt{Student}(x_1, x_2)\}$ intended to unify all students from these three cities. Σ_A is a set of tuple-generating constraints

$$\big\{\forall x_1 \forall x_2 \big(\mathtt{StudentRome}(x_1, x_2) \wedge \neg\mathtt{StudentRMF}(x_1, x_2) \Rightarrow r_\top(\overline{0}, \overline{1})\big),$$
$$\forall x_1 \forall x_2 \big(\mathtt{StudentMilan}(x_1, x_2) \wedge \neg\mathtt{StudentRMF}(x_1, x_2) \Rightarrow r_\top(\overline{0}, \overline{1})\big),$$
$$\forall x_1 \forall x_2 \big(\mathtt{StudentFlorence}(x_1, x_2) \wedge \neg\mathtt{StudentRMF}(x_1, x_2) \Rightarrow r_\top(\overline{0}, \overline{1})\big)\big\}.$$

Consequently, we will consider only mapping interpretations α such that

$$\begin{aligned}\alpha(\mathtt{Student}) &= \alpha(\mathtt{StudentRMF}) \\ &= \alpha(\mathtt{StudentRome}) \\ &\quad \cup \alpha(\mathtt{StudentMilan}) \cup \alpha(\mathtt{StudentFlorence}).\end{aligned}$$

Then,

$$\begin{aligned}TA &= T\big(\alpha(\mathtt{StudentRome}) \cup \alpha(\mathtt{StudentMilan}) \\ &\quad \cup \alpha(\mathtt{StudentFlorence}) \cup \alpha(\mathtt{StudentRFM})\big) \\ &= T\big(\alpha(\mathtt{Student})\big) = TB\end{aligned}$$

and, consequently, the database-instances A and B are isomorphic, that is, $A \simeq B$.

Let us consider how to define the isomorphic arrow $f : A \to B$ in the category **DB**.

First of all, while in the construction of TA we can use the *union* of conjunctive queries, i.e., the *disjunctive* logical formula

$$\begin{aligned}&\mathtt{StudentRome}(x_1, x_2) \vee \mathtt{StudentMilan}(x_1, x_2) \\ &\quad \vee \mathtt{StudentFlorence}(x_1, x_2),\end{aligned}$$

as explained in Sect. 1.4.1 in the Introduction, we cannot use the schema mapping

$$\forall x_1 \forall x_2 \big((\mathtt{StudentRome}(x_1, x_2) \vee \mathtt{StudentMilan}(x_1, x_2) \vee \mathtt{StudentFlorence}(x_1, x_2)) \Rightarrow \mathtt{Student}(x_1, x_2) \big)$$

because the left-hand sides of implications in tgds (and SOtgds) are only conjunctive formulae.

Consequently, we have to make a decomposition of such a disjunction into the three conjunctive components. Hence, the SOtgd for this schema mapping is a formula Φ equal to:

$$\forall x_1 \forall x_2 \big(\mathtt{StudentRome}(x_1, x_2) \Rightarrow \mathtt{Student}(x_1, x_2) \big)$$
$$\wedge \forall x_1 \forall x_2 \big(\mathtt{StudentMilan}(x_1, x_2) \Rightarrow \mathtt{Student}(x_1, x_2) \big)$$
$$\wedge \forall x_1 \forall x_2 \big(\mathtt{StudentFlorence}(x_1, x_2) \Rightarrow \mathtt{Student}(x_1, x_2) \big),$$

and, consequently, we define the mapping $\mathcal{M}_{AB} = \{\Phi\} : \mathcal{A} \to \mathcal{B}$.

Thus, we obtain the sketch's mapping

$$\mathbf{M}_{AB} = \mathit{MakeOperads}(\mathcal{M}_{AB}) = \{q_1, q_2, q_3, 1_{r_\emptyset}\} : \mathcal{A} \to \mathcal{B}$$

with $q_1 \in O(\mathtt{StudentRome}, \mathtt{Student})$, $q_2 \in O(\mathtt{StudentMilan}, \mathtt{Student})$, $q_3 \in O(\mathtt{StudentFlorence}, \mathtt{Student})$.

Consequently, we obtain the isomorphic arrow $f = \alpha^*(\{q_1, q_2, q_3, 1_{r_\emptyset}\}) : A \to B$ where $f = \{\alpha(q_1), \alpha(q_2), \alpha(q_3), q_\bot\}$ is the set of injective functions:

$$in_1 = \alpha(q_1) : \alpha(\mathtt{StudentRome}) \to \alpha(\mathtt{Student}),$$
$$in_2 = \alpha(q_2) : \alpha(\mathtt{StudentMilan}) \to \alpha(\mathtt{Student}),$$
$$in_3 = \alpha(q_3) : \alpha(\mathtt{StudentFlorence}) \to \alpha(\mathtt{Student}),$$

with the information flux (here $\mathbf{x}$ denotes the tuple (x_1, x_2)) equal to:

$$\begin{aligned}
\widetilde{f} &= B_T(f) \\
&= T\big\{\pi_{\mathbf{x}} \|\mathtt{StudentRome}\|_A, \pi_{\mathbf{x}} \|\mathtt{StudentMilan}\|_A, \\
&\qquad \pi_{\mathbf{x}} \|\mathtt{StudentFlorence}\|_A\big\} \\
&= T\big\{\alpha(\mathtt{StudentRome}), \alpha(\mathtt{StudentMilan}), \alpha(\mathtt{StudentFlorence})\big\} \\
&= TA,
\end{aligned}$$

i.e., it is monic and, from the fact that $TA = TB$ and hence $\widetilde{f} = TB$, it is also epic. Consequently, from Proposition 8, $f : A \to B$ is an isomorphism.

If we enlarge the schema $\mathcal{B}$ with a new relation $\mathtt{Takes1}(x_1, x_3)$ that associates student names with courses (as in Example 2) such that $\alpha(\mathtt{Takes1})$ is not the empty relation then $\widetilde{f} = TA \subset TB$, and hence in this case the morphism above $f : A \to B$ is a monic arrow only.

Remark It is important to take clearly in mind that the equality of arrows in **DB** and hence the commutativity of the diagrams in **DB** is based on the observational equivalence in Definition 23 and not on the set-theoretic equality. The (observational) equality of morphisms is more general then the set-theoretic equality:

Corollary 10 *For each two set-theoretically equal simple morphisms $f, g : A \to B$, they are equal in category* **DB** *(in the sense of observational equivalence in Definition* 23*). But not vice versa.*

Proof If f and g are set-theoretically equal then they have the same operads mapping $\mathbf{M}_{AB}$ and hence they have the same information flux $\mathit{Flux}(\alpha, \mathbf{M}_{AB})$, i.e., $\widetilde{f}$ is set-theoretically equal to $\widetilde{g}$ and, from Definition 23, $f = g$ in **DB** as well. Let us consider the SOtgd for this schema mapping

$$\mathcal{M}'_{AB} = \{\forall x_1 \forall x_2 (\mathtt{StudentRMF}(x_1, x_2) \Rightarrow \mathtt{Student}(x_1, x_2))\} : \mathcal{A} \to \mathcal{B},$$

in Example 20 above, with

$$\mathbf{M}'_{AB} = \mathit{MakeOperads}(\mathcal{M}'_{AB}) = \{q_4, 1_{r_\emptyset}\} : \mathcal{A} \to \mathcal{B}$$

and $q_4 \in O(\mathtt{StudentRMF}, \mathtt{Student})$. Then we obtain the isomorphism $f_1 = \alpha^*(\{q_4, 1_{r_\emptyset}\}) : A \to B$ where $f_1 = \{\alpha(q_4), q_\perp\}$ with the injective function

$$in_4 = \alpha(q_1) : \alpha(\mathtt{StudentRMF}) \to \alpha(\mathtt{Student}),$$

so that $\widetilde{f_1} = \widetilde{f}$, i.e., $f_1 = f$ in **DB**, while they are not set-theoretically equal. □

3.2.1 Power-View Endofunctor and Monad T

Let us extend the power-view operator T for objects into an endofunctor of **DB** category:

Theorem 2 *There exists the endofunctor $T = (T^0, T^1) : \mathbf{DB} \longrightarrow \mathbf{DB}$ such that:*

1. *For any object A, the object component T^0 is equal to the type power-view operator T, i.e., $T^0(A) \triangleq TA$;*
2. *For any morphism $f : A \longrightarrow B$, the arrow component T^1 is defined by*

$$Tf = T^1(f) \triangleq is_B \circ f \circ is_A^{-1} : TA \to TB$$

with $\frac{Tf}{\uplus} = \{Tf_i \mid f_i \in \frac{f}{\uplus}\}$, so that $\widetilde{Tf} \simeq \widetilde{f}$ and $T(Tf) = Tf$;

3. *Functor T preserves the properties of arrows, i.e., if a morphism f has a property $P \in \{monic, epic, isomorphic\}$ then also $T(f)$ has the same property P.*

Proof From the fact that all arrows are composed of atomic arrows, it is enough to consider only them. We have for composition of morphisms (derived from a composition of schema mappings and their information fluxes in Definition 13) that for a simple arrow f, $\widetilde{Tf} = \widetilde{is_B \circ f \circ is_A^{-1}} = \widetilde{is_B} \cap \widetilde{f} \cap \widetilde{is_A^{-1}} = TB \cap \widetilde{f} \cap TA =$ (from Proposition 7 if f is simple arrow then $\widetilde{f} \subseteq TA \cap TB) = \widetilde{f}$. If f is a complex arrow with $A = \biguplus_{1\leq j\leq m} A_j$ and $B = \biguplus_{1\leq i\leq k} B_i$ then $is_A^{-1} = \biguplus_{1\leq j\leq m} is_{A_j}^{-1} : TA_j \to A_j$ with $is_{A_j}^{-1} = \{id_R : R \to R \mid R \in A_j\}$, so that $\frac{is_A^{-1}}{\uplus} = \frac{id_A}{\uplus}$ and $is_B = \biguplus_{1\leq i\leq k} is_{B_i}$: $B_i \to TB_i$ with $is_{B_i} = \{id_R : R \to R \mid R \in B_i\}$ and hence $\frac{is_B}{\uplus} = \frac{id_B}{\uplus}$. Consequently, for each $f'_{ji} = Tf_{ji} = is_{B_i} \circ f_{ji} \circ is_{A_j}^{-1} \in \frac{Tf}{\uplus}$ with $f_{ji} : A_j \to B_i \in \frac{f}{\uplus}$, $\widetilde{f'_{ji}} = \widetilde{f_{ji}}$ so that, from Definition 22, $\widetilde{Tf} \simeq \widetilde{\biguplus \frac{Tf}{\uplus}} \simeq \widetilde{\biguplus \frac{f}{\uplus}} \simeq \widetilde{f}$.

For each identity arrow $id_A : A \to A$ with $\widetilde{id_A} = TA$, we obtain the identity arrow $id_{TA} = Tid_A : TA \to TA$ with $\widetilde{id_{TA}} = \widetilde{Tid_A} = \widetilde{id_A} = TA$. For the composition of any two arrows $f : A \to B$ and $g : B \to C$, $T(g \circ f) = is_C \circ (g \circ f) \circ is_A^{-1} = is_C \circ g \circ id_B \circ f \circ is_A^{-1} = is_C \circ g \circ (is_B^{-1} \circ is_B) \circ f \circ is_A^{-1} = Tg \circ Tf$. Thus, from Definition 23, $T(g \circ f) = Tg \circ Tf$. Consequently, the power-view operator T is a **DB** endofunctor.

Thus, $T(Tf) = T(is_B \circ f \circ is_A^{-1}) = Tis_B \circ Tf \circ Tis_A^{-1} =$ (from Example 21 bellow) $= id_{TB} \circ Tf \circ id_{TA} = Tf$. It is easy to verify that T is a 2-endofunctor and that preserves the properties of arrows: if f is an isomorphism then Tf is a composition of three isomorphism and hence an isomorphism as well. If f is monic then $Tf = is_B \circ f \circ is_A^{-1}$, that is, a composition of three monic arrows (each isomorphic arrow is monic and epic as well) and hence is a monomorphism as well. The proof when f is epic is similar. The functorial properties for different cases of complex arrows are presented with details in point 2 of Corollary 12 in Sect. 3.2.4. □

Let us show that the morphism $Tf : TA \to TB$, for a given morphisms $f = \alpha^*(MakeOperads(\mathcal{M}_{AB}))$ with $\mathcal{M}_{AB} = \{\Psi\}$, $\mathbf{M}_{AB} = MakeOperads(\mathcal{M}_{AB}) = \{q_1, \dots, q_k, 1_{r_\emptyset}\}$ and $S = \{\partial_1(q_i) \mid 1 \leq i \leq k\} \subseteq S_B$, corresponds to the definition of arrows in **DB** category, specified by Theorem 1, that is, there exists an SOtgd Φ of a schema mapping such that $Tf = \alpha^*(MakeOperads(\{\Phi\}))$.

We will show that Φ is just the SOtgd of the schema mapping $\mathcal{M}_{AB} : \mathcal{A} \to \mathcal{B}$, both with a mapping-interpretation α, used in definition of the morphism f.

From Theorem 2 above we obtain

$$\begin{aligned}
Tf &= is_B \circ f \circ is_A^{-1} \\
&= \alpha^*\big(MakeOperads(\{\Phi_B\})\big) \circ \alpha^*(\mathbf{M}_{AB}) \circ \alpha^*\big(MakeOperads(\Phi_A)\big) \\
&= \alpha^*\big(MakeOperads(\{\Phi_B\}) \circ \mathbf{M}_{AB} \circ MakeOperads(\Phi_A)\big) \\
&= \alpha^*\big((\{1_r \mid r \in S_B\} \cup \{1_{r_\emptyset}\}) \circ \mathbf{M}_{AB} \circ (\{1_r \mid r \in S_A\} \cup \{1_{r_\emptyset}\})\big) \\
&= \alpha^*\big((\{1_r \mid r \in S_B\} \cup \{1_{r_\emptyset}\}) \circ \mathbf{M}_{AB}\big) \\
&= \alpha^*\big((\{1_r \mid r \in S_B\} \cup \{1_{r_\emptyset}\}) \circ \{q_1, \dots, q_k, 1_{r_\emptyset}\}\big) \\
&= \alpha^*\big((\{1_r \mid r \notin S\} \cup \{1_r \mid r \in S\} \cup \{1_{r_\emptyset}\}) \circ \{q_1, \dots, q_k, 1_{r_\emptyset}\}\big) \\
&= \alpha^*\big((\{1_r \mid r \in S_B\} \cup \{1_{r_\emptyset}\}) \circ \{q_1, \dots, q_k, 1_{r_\emptyset}\}\big) \\
&= \alpha^*\big(\{1_r \mid r \notin S\} \cup \{q_1, \dots, q_k, 1_{r_\emptyset}\}\big) = \alpha^*\big(MakeOperads(\{\Phi\})\big),
\end{aligned}$$

where Φ is equal to SOtgd $\Psi \bigwedge\{\exists f_{r_i}\forall \mathbf{x}_i((f_{r_i}(\mathbf{x}_i)=1)\Rightarrow r_i(\mathbf{x}_i))\,|\,r_i\notin S\}$ and f_{r_i} is the characteristic function of the relation $\alpha(r_i)$. Consequently, from the fact that the information flux of the component $\bigwedge\{\exists f_{r_i}\forall \mathbf{x}_i((f_{r_i}(\mathbf{x}_i)=1)\Rightarrow r_i(\mathbf{x}_i))\,|\,r_i\notin S\}$ is empty, the information fluxes of f and Tf are equal, that is, $\widetilde{Tf}=\widetilde{f}$.

Thus, $Tf = f$ in the observational sense of Definition 23, while in the set-theoretic sense we have that $f \subseteq Tf$.

Example 21 Let us consider the definition of this endofunctor for simple isomorphic arrows $is_A : A \to TA$, $is_A^{-1} : TA \to A$ where, from the proof of Proposition 8, $is_A^{-1} = is_A = \mathbb{S}_A = \{id_R : R \to R \,|\, R \in A\}$.

Let us show that $T^1(is_A^{-1}) = T^1(is_A) = is_{TA} = id_{TA} : TA \to TA$.

In fact, from the definition of the isomorphic arrow $is_{TA} : TA \to T(TA)$ and from the idempotence $TTA = TA$, we obtain $is_{TA} = \mathbb{S}_{TA} = \{id_R : R \to R \,|\, R \in TA\} : TA \to TA$, that is, $is_{TA} = id_{TA} : TA \to TA$. Analogously, $is_{TA}^{-1} = \mathbb{S}_{TA} = \{id_R : R \to R \,|\, R \in TA\} : T(TA) \to TA$, that is,
(a) $is_{TA}^{-1} = is_{TA} = id_{TA} : TA \to TA$.
From Theorem 2 above (here $B = TA$) we have
(b)

$$\begin{aligned} T^1(is_A) &= is_B \circ is_A \circ is_A^{-1} \\ &\quad \bigl(\text{from } is_A \circ is_A^{-1} = id_{TA}\bigr) \\ &= is_B \circ id_{TA} = is_{TA} \circ id_{TA} = is_{TA}, \end{aligned}$$

and (here we replace A by TA and B by A)
(c)

$$\begin{aligned} T^1\bigl(is_A^{-1}\bigr) &= is_A \circ is_A^{-1} \circ is_{TA}^{-1} \\ &\quad \bigl(\text{from } is_A \circ is_A^{-1} = id_{TA}\bigr) \\ &= id_{TA} \circ is_{TA}^{-1} = is_{TA}^{-1}. \end{aligned}$$

Thus, from (a), (b), and (c) we obtain
(d) $T^1(is_A^{-1}) = T^1(is_A) = is_{TA}^{-1} = is_{TA} = id_{TA} : TA \to TA$, where $id_{TA} = T^1(id_A)$.

The endofunctor T is a right and left adjoint to the identity functor I_{DB}, i.e., $T \simeq I_{DB}$. Thus, we have the equivalence adjunction $\langle T, I_{DB}, \eta^C, \eta\rangle$ where the unit $\eta^C : T \simeq I_{DB}$ (such that for any object A the arrow $\eta_A^C \triangleq \eta^C(A) = is_A^{-1} : TA \longrightarrow A$) and the counit $\eta : I_{DB} \simeq T$ (such that for any A the arrow $\eta_A \triangleq \eta(A) = is_A : A \longrightarrow TA$) are isomorphic arrows in **DB** (by duality Theorem 3 it holds that $\eta^C = \eta^{OP}$).

We have already explained that the views of a database may be seen as its *observable computations*: what we need, in order to obtain an expressive power of computations in the category **DB**, are the categorial computational properties which are, as it is well known [17], based on monads:

Proposition 9 *Power-view endofunctor $T = (T^0, T^1) : \mathbf{DB} \longrightarrow \mathbf{DB}$ defines the monad (T, η, μ) and the comonad (T, η^C, μ^C) in* **DB** *such that $\eta : I_{DB} \simeq T$ and $\eta^C : T \simeq I_{DB}$ are two natural isomorphisms, while $\mu : TT \stackrel{\cdot}{\longrightarrow} T$ and $\mu^C : T \stackrel{\cdot}{\longrightarrow} TT$ are equal to the natural identity transformation $id_T : T \stackrel{\cdot}{\longrightarrow} T$ (from the idempotence, $T = TT$).*

Proof It is easy to verify that all commutative diagrams of the monad ($\mu_A \circ \mu_{TA} = \mu_A \circ T\mu_A$, $\mu_A \circ \eta_{TA} = id_{TA} = \mu_A \circ T\eta_A$) and of the comonad are diagrams composed of the identity arrows. Notice that by duality (in Sect. 3.2.2) $\eta_{TA} = T\eta_A = \mu_A^{OP}$. □

The notion of a monad is one of the most general mathematical notions. For instance, *every* algebraic theory, that is, every set of operations satisfying equational laws, can be seen as a monad (which is also a *monoid* in a category of endofunctors of a given category: the "operation" μ being the associative multiplication of this monoid and η its unit). Thus monoid laws of the monad do subsume all possible algebraic laws.

We will use monads [8, 10, 11] to provide *denotational semantics to database mappings*, and more specifically as a way of modeling computational/collection types [4, 16–18]. In order to interpret a database mappings (morphisms) in the category **DB**, we distinguish the object A (a database instance of type $\mathcal{A}$) from the object TA of observations (computations of type $\mathcal{A}$ without side-effects) and take as a denotation of (view) mappings the elements of TA (which are the views of type $\mathcal{A}$).

In particular, we identify A with the object of values (of type $\mathcal{A}$) and we obtain the object of observations by applying the unary type-constructor T (power-view operator) to A.

It is well known that each endofunctor defines the algebras and coalgebras, represented by the left and right commutative diagrams:

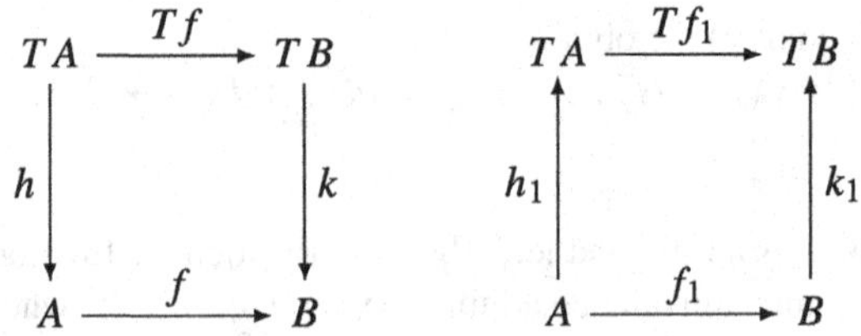

We will use the following well-known definitions in the category theory (we recall that the set of all arrows in a category **C** from A to B is denoted by $\mathbf{C}(A, B)$):

Definition 24 The categories CT_{alg} of T-algebras and CT_{coalg} of T-coalgebras, derived from the endofunctor T, are defined [1] as follows:

1. The objects of CT_{alg} are the pairs (A, h) with $A \in Ob_{DB}$ and $h \in \mathbf{DB}(TA, A)$; the arrows between objects (A, h) and (B,k)d are all arrows $f \in \mathbf{DB}(A, B)$ such that $k \circ Tf = f \circ h : TA \longrightarrow B$.
2. The objects of CT_{coalg} are the pairs (A, h_1) with $A \in Ob_{DB}$ and $h_1 \in \mathbf{DB}(A, TA)$; the arrows between objects (A, h_1) and (B, k_1) are all arrows $f_1 \in \mathbf{DB}(A, B)$ such that $Tf_1 \circ h_1 = k_1 \circ f_1 : A \longrightarrow TB$.

The (co)algebras of an endofunctor such that it is a monad as well satisfy more structural properties and are denominated as *monadic* (co)algebras(for (co)monads introduced in Sect. 1.5):

Definition 25 The *monadic* algebras/coalgebras are defined [1, 11] as follows:

- Each T-algebra $(A, h : TA \longrightarrow A)$ derived from a monad (T, η, μ), where h is a "structure map" such that $h \circ \mu_A = h \circ Th$ and $h \circ \eta_A = id_A$, is a monadic T-algebra. The Eilenberg–Moore category of all monadic algebras T_{alg} is a full subcategory of CT_{alg}.
- Each T-coalgebra $(A, k : A \longrightarrow TA)$ derived from a comonad (T, η^C, μ^C) such that $Tk \circ k = \mu_A^C \circ k$ and $\eta_A^C \circ k = id_A$ is a monadic T-coalgebra. The category of all monadic coalgebras T_{coalg} is a full subcategory of CT_{coalg}.

Thus, the monadic algebras satisfy the following commutative diagram:

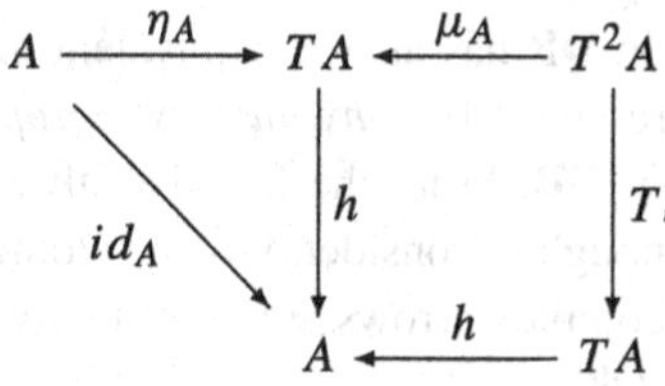

In the **DB** category, $\widetilde{\eta_A} = \widetilde{\mu_A} = \widetilde{id_A} = TA$, while for a simple arrow h, $\widetilde{Th} = \widetilde{h} \subseteq TA \otimes A = TA$ (from Proposition 7), so that the intersection of the fluxes of the arrows in this diagram commute.

Corollary 11 *For each object A in* **DB** *there is exactly one T-algebra (and its dual T-coalgebra)* $h = TA \rightarrow A$ *which is equal to the isomorphism* is_A^{-1}.

Proof If $A = \bot^0$ then $TA = \bot^0$, so that all arrows in the diagram above are the identity arrows $id_{\bot^0} = \bot^1$ (which is isomorphism as well) with $h = \bot^1$.

Let $A = \biguplus_{1 \leq j \leq m} A_j$. Thus, $TA = \biguplus_{1 \leq j \leq m} TA_j$ with $\mu_A = Tid_A = id_{TA} = \biguplus_{1 \leq j \leq m} id_{TA_j} : TA_j \rightarrow TA_j$ and the isomorphism

$$\eta_A = \biguplus_{1 \leq j \leq m} \eta_{A_j} = \biguplus_{1 \leq j \leq m} is_{A_j} : A_j \rightarrow TA_j.$$

From the commutativity of the triangle above, $id_A = h \circ \eta_A$, it must hold, from Definition 23, that $\frac{h \circ \eta_A}{\uplus} = \frac{id_A}{\uplus}$ and hence for each ptp arrow $id_{A_j} : A_j \to A_J$ in $\frac{id_A}{\uplus}$, we must have the ptp arrow $h_{jj} : TA_j \to A_J$ in $\frac{h}{\uplus}$. Consequently, $id_{A_j} = h_{jj} \circ \eta_{A_j}$, that is, $\widetilde{id_{A_j}} = TA_j = \widetilde{h_{jj} \circ \eta_{A_j}} = \widetilde{h_{jj}} \cap \widetilde{\eta_{A_j}} = \widetilde{h_{jj}} \cap TA_j = \widetilde{h_{jj}}$ (from the fact that, from Proposition 7, $\widetilde{h_{jj}} \subseteq T(TA_j) \cap TA_j = TA_j$). Thus, from Proposition 8, $h_{jj} : TA_j \to A_J$ is an isomorphism and $\frac{h}{\uplus} \supseteq \{h_{jj} : TA_j \to A_j \mid 1 \leq j \leq m\}$. However, from the fact that the number if ptp arrows in h cannot be greater than the number of ptp arrows in id_A (otherwise, the composed arrow $h \circ \eta_A = h \circ is_A$ would have more ptp arrows than is_A, i.e., more ptp arrows than in id_A, and we would have that $\frac{id_A}{\uplus} \neq \frac{h \circ \eta_A}{\uplus}$, i.e., $id_A \neq h \circ \eta_A$), we obtain this unique monadic T-algebra $h = is_A^{-1} : TA \to A$. □

3.2.2 Duality

The following duality theorem shows that for any commutative diagram in **DB** there is an equal commutative diagram, composed of equal objects and inverted equivalent arrows. This duality property of **DB** is a consequence of the fact that a composition of simple arrows (i.e., morphisms) is semantically based on the set-intersection commutativity of information fluxes of these arrows, and the information flux of inverted arrow is equal to the information flux of the original arrow.

Thus, *any limit diagram* in **DB** has also its equivalent *inverted colimit diagram* with equal objects and, more generally, *any universal property* also has its *equivalent couniversal property* in **DB**. From the fact that all arrows are composed of simple atomic arrows, it is enough to consider only the atomic arrows. Duality properties for different cases of complex arrows are presented with details also in point 3 of Corollary 12 in Sect. 3.2.4.

Theorem 3 *There exists a contravariant functor* $\underline{S} = (\underline{S}^0, \underline{S}^1) : \mathbf{DB} \longrightarrow \mathbf{DB}$ *such that*:

1. $\underline{S}^0$ *is an identity function on objects.*
2. *For any simple arrow* $f : A \longrightarrow B$ *with* $(Tf)^{OP} \triangleq \{id_R \mid R \in \widetilde{f}\} : TB \to TA$, *we define* $\underline{S}^1(f) \triangleq f^{OP} : B \to A$, *where* $f^{OP} = is_A^{-1} \circ (Tf)^{OP} \circ is_B$ *is an inverted morphism equivalent to* f *(i.e., with* $\widetilde{f^{OP}} = \widetilde{f}$*).*

   The inverse of an inverted morphism f^{OP} *is equal to the original morphism, that is,* $(f^{OP})^{OP} = f$. *For a complex arrow, we have* $\frac{f^{OP}}{\uplus} = \{f_{ji}^{OP} \mid f_{ji} \in \frac{f}{\uplus}\}$.
3. *The category* **DB** *is isomorphic to its dual category* $\mathbf{DB}^{OP}$.

Proof $\widetilde{(Tf)^{OP}} = \widetilde{f}$ and $\widetilde{f^{OP}} = \widetilde{is_A^{-1} \circ (Tf)^{OP} \circ is_B} = \widetilde{is_A^{-1}} \cap \widetilde{(Tf)^{OP}} \cap \widetilde{is_B} = TA \cap \widetilde{(Tf)^{OP}} \cap TB = TA \cap \widetilde{f} \cap TB = \widetilde{f}$.

Inverted arrow of any identity arrow is equal to the identity arrow. For the composition of f with $g : B \to C$ we have

$$\begin{aligned} S^1(g \circ f) &= is_A^{-1} \circ \big(T(g \circ f)\big)^{OP} \circ is_C \\ &= is_A^{-1} \circ (Tg \circ Tf)^{OP} \circ is_C \\ &= is_A^{-1} \circ (Tf)^{OP} \circ (Tg)^{OP} \circ is_C \\ &= is_A^{-1} \circ (Tf)^{OP} \circ is_B \circ is_B^{-1} \circ (Tg)^{OP} \circ is_C \\ &= S^1(f) \circ S^1(g). \end{aligned}$$

Thus, this contravariant functor is well defined.

It is convenient to represent this contravariant functor as a covariant functor $S : \mathbf{DB}^{OP} \longrightarrow \mathbf{DB}$ or a covariant functor $S^{OP} : \mathbf{DB} \longrightarrow \mathbf{DB}^{OP}$. It is easy to verify that for compositions of these covariant functors $SS^{OP} = I_{DB}$ and $S^{OP}S = I_{DB^{OP}}$ w.r.t. the adjunction $\langle S, S^{OP}, \phi\rangle : \mathbf{DB}^{OP} \longrightarrow \mathbf{DB}$, where ϕ is a bijection: for each pair of objects A, B in $\mathbf{DB}$ we have the bijection of hom-sets, $\phi_{A,B} : \mathbf{DB}(A, S(B)) \simeq \mathbf{DB}^{OP}(S^{OP}(A), B)$, i.e., $\phi_{A,B} : \mathbf{DB}(A, B) \simeq \mathbf{DB}(B, A)$, such that for any arrow $f \in \mathbf{DB}(A, B)$, $\phi_{A,B}(f) = S^1(f) = f^{OP}$. The unit and counit of this adjunction are the identity natural transformations, $\eta_{OP} : I_{DB} \longrightarrow SS^{OP}$ and $\epsilon_{OP} : S^{OP}S \longrightarrow I_{DB^{OP}}$, respectively, such that for any object A they return by its identity arrow. Thus, from this adjunction, we obtain that $\mathbf{DB}$ is isomorphic to $\mathbf{DB}^{OP}$. □

Let us show that the morphism $(Tf)^{OP} : TB \to TA$, for a given morphism $f = \alpha^*(\mathbf{M}_{AB}) : A \to B$ with $\mathbf{M}_{AB} = \mathit{MakeOperads}(\mathcal{M}_{AB}) : \mathcal{A} \to \mathcal{B}$, is well defined due to Theorem 1, that is, there exists an SOtgd Φ of a schema mapping such that $(Tf)^{OP} = \alpha^*(\mathit{MakeOperads}(\{\Phi\}))$.

Let us define the set of relational symbols $S_{TB} = \{r_i \mid r_i \in \mathbb{R}$ and $\alpha(r_i) \in \Delta(\alpha, \mathbf{M}_{AB}) \subseteq \widetilde{f}\}$ and the α-intersection schemas $\mathcal{A}\underset{\alpha}{\cap}\mathcal{A}$ and $\mathcal{B}\underset{\alpha}{\cap}\mathcal{B}$ with $TA = \alpha^*(\mathcal{A}\underset{\alpha}{\cap}\mathcal{A})$ and $TB = \alpha^*(\mathcal{B}\underset{\alpha}{\cap}\mathcal{B})$ (in Definition 16). Then we can define the SOtgd Φ by a conjunctive formula $\bigwedge\{\forall \mathbf{x}_i(r_i(\mathbf{x}_i) \Rightarrow r_i(\mathbf{x}_i)) \mid r_i \in S_{TB}\}$.

In fact, $\mathit{MakeOperads}(\{\Phi\}) = \{1_{r_i} \in O(r_i, r_i) \mid r_i \in S_{TB}\} \cup \{1_{r_\emptyset}\}$ is a sketch's mappings between schemas $\widehat{\mathcal{B}}$ and $\widehat{\mathcal{A}}$. Note that $r_\emptyset \in S_{TB}$ because $\bot \in \widetilde{f}$. Consequently,

$$\begin{aligned} (Tf)^{OP} &= \alpha^*\big(\mathit{MakeOperads}(\{\Phi\})\big) = \alpha^*\big(\big\{1_{r_i} \in O(r_i, r_i) \mid r_i \in S_{TB}\big\} \cup \{1_{r_\emptyset}\}\big) \\ &= \big(\big\{id_R \mid R \in \Delta(\alpha, \mathbf{M}_{AB}) \subseteq \widetilde{f}\big\} \cup \{q_\bot\}\big) : TB \to TA. \end{aligned}$$

Thus, both inverted arrows $(Tf)^{OP}$ and $f^{OP} = is_A^{-1} \circ (Tf)^{OP} \circ is_B$ (the demonstration is equal to the demonstration for arrow Tf after the proof of Theorem 2) are morphisms in the $\mathbf{DB}$ category as specified in Theorem 1.

Example 22 Let us consider the inverted arrows of the isomorphic arrows $is_A : A \to TA$ and $is_A^{-1} : TA \to A$.

For is_A (here $B = TA$) we have from Theorem 3 the following:

(i)

$$
\begin{aligned}
is_A^{OP} &= is_A^{-1} \circ (T is_A)^{OP} \circ is_{TA} \\
&\quad \big(\text{from (d) in Example 21}\big) \\
&= is_A^{-1} \circ (id_{TA})^{OP} \circ id_{TA} \\
&= is_A^{-1} \circ (id_{TA})^{OP} \\
&\quad (\text{inverted arrow of an identity arrow is the same identity arrow}) \\
&= is_A^{-1} \circ id_{TA} = is_A^{-1}.
\end{aligned}
$$

Analogously (here B is replaced by A and A by TA),
(ii)

$$
\begin{aligned}
\big(is_A^{-1}\big)^{OP} &= is_{TA}^{-1} \circ \big(T is_A^{-1}\big)^{OP} \circ is_A \\
&\quad \big(\text{from (d) in Example 21}\big) \\
&= id_{TA} \circ (id_{TA})^{OP} \circ is_A = (id_{TA})^{OP} \circ is_A \\
&\quad (\text{inverted arrow of an identity arrow is the same identity arrow}) \\
&= id_{TA} \circ is_A = is_A = \big(is_A^{-1}\big)^{-1}.
\end{aligned}
$$

Thus, from (i) and (ii), the label '^{OP}' for these isomorphic arrows is equivalent to inverse-arrow label '$^{-1}$'.

Let us show the validity of the definition of the functor $\underline{S}$ and $T^1(f^{OP}) = (Tf)^{OP}$, based on Theorem 2, for $f : A \rightarrow B$:
(iii)

$$
\begin{aligned}
(Tf)^{OP} &= \big(is_B \circ f \circ is_A^{-1}\big)^{OP} = \big(is_A^{-1}\big)^{OP} \circ f^{OP} \circ is_B^{OP} \\
&\quad \big(\text{from (i) and (ii)}\big) \\
&= is_A \circ f^{OP} \circ is_B^{-1} : TB \rightarrow TA.
\end{aligned}
$$

Thus, by composing on the left with is_A^{-1}, we obtain $is_A^{-1} \circ (Tf)^{OP} = is_A^{-1} \circ is_A \circ f^{OP} \circ is_B^{-1} = id_A \circ f^{OP} \circ is_B^{-1} = f^{OP} \circ is_B^{-1}$. By composing on the right with is_B, we obtain $is_A^{-1} \circ (Tf)^{OP} \circ is_B = f^{OP} \circ is_B^{-1} \circ is_B = f^{OP}$, that is, the definition $\underline{S}^1(f) \triangleq f^{OP} : B \rightarrow A$ in Theorem 3 is valid.

From Theorem 2 (because of $f^{OP} : B \rightarrow A$, A and B are exchanged):

$$
T^1\big(f^{OP}\big) = is_A \circ f^{OP} \circ is_B^{-1} \overset{\text{(from (iii))}}{=} (Tf)^{OP} : TB \rightarrow TA.
$$

Remark The object $\bot^0 = \{\bot\}$ (i.e., empty database instance) is the *zero object* (both terminal and initial) in **DB**. From any real object A in **DB** there is a unique arrow (which is an epimorphism) $f = \alpha^*(\mathbf{M}_{AA_\emptyset}) = \alpha^*(\{1_{r\emptyset}\}) = \{q_\bot\} = \bot^1 : A \twoheadrightarrow \bot^0$

(obtained from the sketch's arrow $\mathbf{M}_{AA_\emptyset} : A \to A_\emptyset$ in Example 17) and its dual arrow (which is a monomorphism) $f^{OP} = \bot^1 : \bot^0 \hookrightarrow A$, both with empty information flux. It is easy to verify that each *empty database* (with all empty relations) is isomorphic to the zero object $\bot^0$:

Lemma 10 *The bottom object $\bot^0 = \{\bot\}$ is the zero (terminal and initial) closed object in* **DB**. *Each database $A = \{R_1, \ldots, R_n, \bot\}$ with all empty relations $R_i = \{\langle\rangle\}, i = 1, \ldots, n$, is isomorphic to the zero object $\bot^0$. That is, $A \simeq \bot^0$.*

Proof For each simple object A and $f : A \to \bot^0$, we have from Proposition 7 that $\widetilde{f} \subseteq TA \cap T\bot^0 = TA \cap \bot^0 = \bot^0$ (From point 1 of Theorem 1), thus $f = \bot^1$ is the unique arrow from A into $\bot^0$, so that $\bot^0$ is a terminal object (analogously, $\bot^1 : \bot^0 \hookrightarrow A$ is the unique arrow, so that $\bot^0$ is an initial object as well).

Thus, for a complex object $A \uplus B$, the unique arrow into zero object is $[\bot^1, \bot^1] : A \uplus B \to \bot^0$ while its opposite is $\langle \bot^1, \bot^1 \rangle : \bot^0 \to A \uplus B$.

For each database $A = \{R_1, \ldots, R_n, \bot\}$ where every $R_i = \{\langle\rangle\}$, $k = ar(R_i) \geq 1$, is a k-ary empty relation, for $i = 1, \ldots, n$, we have the unique arrow $\bot^1 : A \to \bot^0$ (into terminal object) with $\bot^1 = \{q_\bot : \bot \to \bot\}$ and its unique inverted arrow $(\bot^1)^{OP} : \bot^0 \to A$ (from initial object) with $(\bot^1)^{OP} = \{q_\bot : \bot \to \bot\}$, hence

$$(\bot^1)^{OP} \circ \bot^1 = \bot^1 : A \to A.$$

The identity arrow for A is $id_A = \{id_{R_1} : R_i \to R_i \mid R_i \in A\} \cup \{q_\bot\}$ with $\widetilde{id_A} = \bot^0 = \widetilde{\bot^1}$ so that, from Definition 23, $(\bot^1)^{OP} \circ \bot^1 = id_A$ and, analogously, $\bot^1 \circ (\bot^1)^{OP} = id_{\bot^0}$. Thus, $\bot^1 : A \to \bot^0$ is an isomorphism, that is, $A \simeq \bot^0$. □

3.2.3 Symmetry

The mapping $B_T : Mor_{DB} \longrightarrow Ob_{DB}$, specified in the definition of the **DB** category in Theorem 1, is a fundamental concept for the categorial symmetry [13] introduced in Definition 4 (Sect. 1.5.1). Let us demonstrate the following equality for any morphism in **DB**, which will be used in the proof of the next theorem.

Lemma 11 *For any morphism $f : A \to B$, $f = f \circ f^{OP} \circ f$. Let $in_{\widetilde{f}} = in_B : \widetilde{f} \hookrightarrow TB$ and $in_{\widetilde{f^{OP}}} = in_A : \widetilde{f} \hookrightarrow TA$ be two monomorphisms defined by Proposition 7. Then, $in_{\widetilde{f^{OP}}} \circ in_{\widetilde{f}}^{OP} = Tf^{OP} : TB \to TA$.*

Proof If f is a simple arrow then $\widetilde{f} = \widetilde{f^{OP}}$ (by duality), so that $\widetilde{f \circ f^{OP} \circ f} = \widetilde{f} \cap \widetilde{f^{OP}} \cap \widetilde{f} = \widetilde{f}$ and hence, from Definition 23:
(i) $f = f \circ f^{OP} \circ f$.

For any two given objects $A = \biguplus_{1 \le j \le m} A_j, m \geq 1$ and $B = \biguplus_{1 \le i \le k} B_i, k \geq 1$ and a complex arrow f between them, $\frac{f^{OP}}{\uplus} = \{f_{ji}^{OP} \mid f_{ji} \in \frac{f}{\uplus}\}$, so that

$$\frac{f^{OP} \circ f}{\uplus} = \left\{ f_{ji}^{OP} \circ f_{ji} : A_j \to A_j \mid f_{ji} : A_j \to B_i \in \frac{f}{\uplus} \right\}.$$

Thus,

$$\begin{aligned}\frac{f \circ f^{OP} \circ f}{\uplus} &= \left\{ f_{ji} \circ h : A_j \to B_i \mid f_{ji} : A_j \to B_i \in \frac{f}{\uplus}, \right.\\ &\qquad \left. dom(f_{ji}) = A_j = cod(h), h \in \frac{f^{OP} \circ f}{\uplus} \right\}\\ &= \left\{ f_{ji} \circ \left(f_{ji}^{OP} \circ f_{ji} \right) : A_j \to B_i \mid f_{ji} : A_j \to B_i \in \frac{f}{\uplus} \right\}\\ &= \left\{ f_{ji} : A_j \to B_i \mid f_{ji} \in \frac{f}{\uplus} \right\}\\ &\qquad \text{(from (i) and fact that } f_{ji} \text{ are simple arrows)}\\ &= \frac{f}{\uplus}.\end{aligned}$$

Consequently, from Definition 23, $f = f \circ f^{OP} \circ f$ is valid for complex morphisms as well.

From point 1 of Proposition 7, we have for any morphism $f : A \to B$ the monomorphisms $in_{\widetilde{f}} = in_B : \widetilde{f} \hookrightarrow TB$ and $in_{\widetilde{f^{OP}}} = in_A : \widetilde{f} \hookrightarrow TA$ with

$$\frac{in_A}{\uplus} = \left\{ in_{ji} = \{id_R \mid R \in \widetilde{f_{ji}}\} : \widetilde{f_{ji}} \hookrightarrow TA_j \mid (f_{ji} : A_j \to B_i) \in \frac{f}{\uplus} \right\},$$

$$\frac{in_B}{\uplus} = \left\{ in_{ji} = \{id_R \mid R \in \widetilde{f_{ji}}\} : \widetilde{f_{ji}} \hookrightarrow TB_i \mid (f_{ji} : A_j \to B_i) \in \frac{f}{\uplus} \right\}.$$

Thus,

$$\frac{in_{\widetilde{f}}^{OP}}{\uplus} = \left\{ in_{ji}^{OP} = \{id_R \mid R \in \widetilde{f_{ji}}\} : TB_i \twoheadrightarrow \widetilde{f_{ji}} \mid (f_{ji} : A_j \to B_i) \in \frac{f}{\uplus} \right\}$$

and

$$\begin{aligned}\frac{in_{\widetilde{f^{OP}}} \circ in_{\widetilde{f}}^{OP}}{\uplus} &= \left\{ in_{ji}^{OP} \circ in_{ji} = \{id_R \circ id_R \mid R \in \widetilde{f_{ji}}\} : TB_i \twoheadrightarrow f_{ji} \hookrightarrow TA_j \mid \right.\\ &\qquad \left. (f_{ji} : A_j \to B_i) \in \frac{f}{\uplus} \right\}\end{aligned}$$

$$= \left\{ \{id_R \mid R \in \widetilde{f_{ji}}\} : TB_i \to TA_j \mid (f_{ji} : A_j \to B_i) \in \frac{f}{\uplus} \right\}$$

(from Theorem 3)

$$= \left\{ Tf_{ji}^{OP} : TB_i \to TA_j \mid (f_{ji} : A_j \to B_i) \in \frac{f}{\uplus} \right\} = \frac{Tf^{OP}}{\uplus}.$$

Consequently, from Definition 23, we obtain $in_{\widetilde{f^{OP}}} \circ in_{\widetilde{f}}^{OP} = Tf^{OP} : TB \to TA.$ □

Now we are ready to demonstrate that the **DB** category is a conceptually closed extended symmetric category, w.r.t. Definition 5 in Sect. 1.5.1, with introductions of the encapsulation operator J such that for each morphisms $f : A \to B$ in **DB**, $J(f) = \langle A, B, f \rangle$ is an object in the arrow (or comma) category **DB** $\downarrow$ **DB** with its inverse operator ψ such that $\psi(J(f)) = f$.

Theorem 4 *The category* **DB** *is an extended symmetric category, closed by the functor* $T_e = (T_e^0, T_e^1) : (\mathbf{DB} \downarrow \mathbf{DB}) \longrightarrow \mathbf{DB}$, *where* $T_e^0 = B_T\psi$ *is the object component of this functor such that for any arrow* $f : A \to B$ *in* **DB**, $T_e^0(J(f)) = \widetilde{f}$, *while its arrow component* T_e^1 *is defined as follows:*

For any arrow $(h_1; h_2) : J(f) \longrightarrow J(g)$ *in* **DB** $\downarrow$ **DB** (*so that* $g \circ h_1 = h_2 \circ f$ *in* **DB**), $T_e^1(h_1; h_2) \triangleq in_{\widetilde{g}}^{OP} \circ T(h_2 \circ f) \circ in_{\widetilde{f^{OP}}}$, *where* $in_{\widetilde{f^{OP}}} : \widetilde{f} \hookrightarrow TA$ *and* $in_{\widetilde{g}} : \widetilde{g} \hookrightarrow TD$ *are two monomorphisms (Proposition 7). This is represented by*

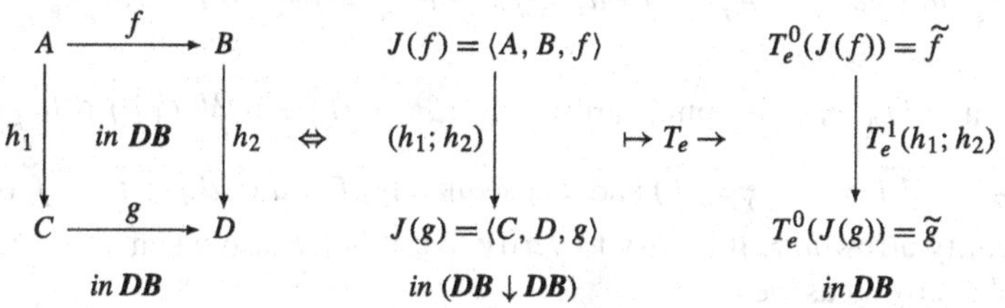

The associative composition operator for objects $*$, *defined for any fitted pair* $g \circ f$ *of arrows, satisfies* $B_T(g) * B_T(f) \triangleq \widetilde{g \circ f} = B_T(g \circ f)$.

Proof Let us consider a simple morphism $f = \alpha^*(\mathbf{M}_{AB}) : A \to B$. In accordance with Theorem 1, for $\mathbf{M}_{AB} = \mathit{MakeOperads}(\mathcal{M}_{AB})$, $A = \alpha^*(\mathcal{A})$, $B = \alpha^*(\mathcal{B})$ with $(\Phi_E, \Phi_M) = \mathit{DeCompose}(\mathcal{M}_{AB})$, and let us define the mapping-operads $\mathbf{M}_E = \mathit{MakeOperads}(\{\Phi_E\})$ and $\mathbf{M}_M = \mathit{MakeOperads}(\{\Phi_M\})$.

Let us show that the monomorphism $in_{\widetilde{f^{OP}}} = in_A : \widetilde{f} \hookrightarrow TA$ (from Lemma 11), for this morphisms $f = \alpha^*(\mathit{MakeOperads}(\mathcal{M}_{AB})) : A \to B$, corresponds to the definition of arrows in **DB** category, specified by Theorem 1, that is, there exists an SOtgd Φ of a schema mapping such that $in_{\widetilde{f^{OP}}} = \alpha^*(\mathit{MakeOperads}(\{\Phi\}))$.

Let us define the set of relational symbols

$$S=\{r_i \mid r_i \in \mathbb{R} \text{ and } \alpha(r_i) \in \Delta(\alpha, \mathbf{M}_{AB}) \subseteq \widetilde{f}\} \subseteq S_{\widetilde{f}} = \{r_i \mid r_i \in \mathbb{R} \text{ and } \alpha(r_i) \in \widetilde{f}\}$$

with $\widetilde{f} = TS$ and its schema $\mathcal{C} = (S_{\widetilde{f}}, \emptyset)$, so that $\widetilde{f} = \alpha^*(\mathcal{C})$ (analogously, from Definition 16, for the instance database TA we define the schema $(S_{TA}, \emptyset) = \mathcal{A} \underset{\alpha}{\sqcap} \mathcal{A}$). Consequently, we define the SOtgd Φ by a conjunctive formula $\bigwedge\{\forall \mathbf{x}_i (r_i(\mathbf{x}_i) \Rightarrow r_i(\mathbf{x}_i)) \mid r_i \in S\}$.

In fact, $MakeOperads(\{\Phi\}) = \{1_{r_i} \in O(r_i, r_i) \mid r_i \in S\} \cup \{1_{r_\emptyset}\}$ is a sketch's mapping between schemas $\mathcal{C}$ and $T\mathcal{A}$. Consequently,

$$\begin{aligned} in_{\widetilde{fOP}} &= \alpha^*\big(MakeOperads(\{\Phi\})\big) = \alpha^*\big(\{1_{r_i} \in O(r_i, r_i) \mid r_i \in S\} \cup \{1_{r_\emptyset}\}\big) \\ &= \big(\{id_R = \alpha(1_{r_i}) : R \to R \mid R = \alpha(r_i) \in S \subseteq \widetilde{f}\} \cup \{q_\perp\}\big) : \widetilde{f} \hookrightarrow TA, \end{aligned}$$

with $\widetilde{in_{\widetilde{fOP}}} = \widetilde{f}$.

Thus, from the duality theorem, this property holds for the epimorphism $in^{OP}_{\widetilde{f}} : TB \twoheadrightarrow \widetilde{f}$ (which is a dual (inverted arrow) of the monomorphism $in_{\widetilde{f}} = in_B : \widetilde{f} \hookrightarrow TB$ in Lemma 11) as well.

Consequently, the morphism $T^1_e(h_1; h_2) = in^{OP}_{\widetilde{g}} \circ T(h_2 \circ f) \circ in_{\widetilde{fOP}}$ is well defined in **DB** due to Theorem 1. It is easy to verify that also T_e is a well defined functor. In fact, for any identity arrow $(id_A; id_B) : J(f) \longrightarrow J(f)$,

$$T^1_e(id_A; id_B) = in^{OP}_{\widetilde{f}} \circ T(id_B \circ f) \circ in_{\widetilde{fOP}} = in^{OP}_{\widetilde{f}} \circ Tf \circ in_{\widetilde{fOP}}.$$

Thus, when f, id_A, id_B are simple arrows, $\widetilde{T^1_e(id_A; id_B)} = \widetilde{in^{OP}_{\widetilde{f}}} \cap \widetilde{Tf} \cap \widetilde{in_{\widetilde{fOP}}} = \widetilde{f}$ (from $\widetilde{in^{OP}_{\widetilde{f}}} = \widetilde{Tf} = \widetilde{in_{\widetilde{fOP}}} = \widetilde{f}$) and, consequently, $T^1_e(id_A; id_B) : \widetilde{f} \to \widetilde{f}$ is equal to the identity arrow $id_{\widetilde{f}}$. It is easy to verify that it holds also when f, id_A and id_B are complex arrows as well.

For any two arrows $(h_1; h_2) : J(f) \longrightarrow J(g)$ and $(l_1; l_2) : J(g) \longrightarrow J(k)$ if all arrows f, g, k, h_1, h_2, l_1 and l_2 are simple arrows then

$$\begin{aligned} \widetilde{T^1_e(l_1; l_2) \circ T^1_e(h_1; h_2)} &= \widetilde{T^1_e(l_1; l_2)} \cap \widetilde{T^1_e(h_1; h_2)} \\ &= (\widetilde{k} \cap \widetilde{l_2} \cap \widetilde{g}) \cap (\widetilde{g} \cap \widetilde{h_2} \cap \widetilde{f}) \\ &= \widetilde{k} \cap \widetilde{l_2} \cap \widetilde{g} \cap \widetilde{h_2} \cap \widetilde{f} \\ &\quad (\text{since } h_2 \circ f = g \circ h_1) \\ &= \widetilde{k} \cap \widetilde{l_2} \cap \widetilde{g} \cap \widetilde{h_1} \cap \widetilde{h_2} \cap \widetilde{f} \\ &\quad (\text{since } h_2 \circ f = g \circ h_1) \\ &= \widetilde{k} \cap \widetilde{l_2} \cap \widetilde{h_2} \cap \widetilde{f} \end{aligned}$$

$$= \widetilde{k} \cap \widetilde{l_2 \circ h_2} \cap \widetilde{f}$$

$$= \overline{in_{\widetilde{k}}^{OP} \circ T\big((l_2 \circ h_2) \circ f\big) \circ in_{\widetilde{f^{OP}}}}$$

$$= T_e^1(\widetilde{l_1 \circ h_1; l_2 \circ h_2})$$

$$= T_e^1\big(\widetilde{(l_1; l_2) \circ (h_1; h_2)}\big).$$

Thus, $T_e^1((l_1; l_2) \circ (h_1; h_2)) = T_e^1(l_1; l_2) \circ T_e^1(h_1; h_2)$.

Let us show this functorial composition property also when we have the complex arrows. We have (from Lemma 11):
(i) $in_{\widetilde{g^{OP}}} \circ in_{\widetilde{g}}^{OP} = Tg^{OP}$.
Thus,

$$\begin{aligned}
T_e^1(l_1; l_2) \circ T_e^1(h_1; h_2) &= in_{\widetilde{k}}^{OP} \circ T(l_2 \circ g) \circ in_{\widetilde{g^{OP}}} \circ in_{\widetilde{g}}^{OP} \circ T(h_2 \circ f) \circ in_{\widetilde{f^{OP}}} \\
&= in_{\widetilde{k}}^{OP} \circ T(l_2 \circ g) \circ Tg^{OP} \circ T(h_2 \circ f) \circ in_{\widetilde{f^{OP}}} \\
&\quad \text{(from (i))} \\
&= in_{\widetilde{k}}^{OP} \circ T\big(l_2 \circ g \circ g^{OP} \circ h_2 \circ f\big) \circ in_{\widetilde{f^{OP}}} \\
&= in_{\widetilde{k}}^{OP} \circ T\big(l_2 \circ \big(g \circ g^{OP} \circ g\big) \circ h_1\big) \circ in_{\widetilde{f^{OP}}} \\
&\quad \text{(since } h_2 \circ f = g \circ h_1) \\
&= in_{\widetilde{k}}^{OP} \circ T(l_2 \circ g \circ h_1) \circ in_{\widetilde{f^{OP}}} \\
&\quad \text{(from Lemma 11)} \\
&= in_{\widetilde{k}}^{OP} \circ T\big((l_2 \circ h_2) \circ f\big) \circ in_{\widetilde{f^{OP}}} \\
&\quad \text{(since } h_2 \circ f = g \circ h_1) \\
&= T_e^1(l_1 \circ h_1; l_2 \circ h_2) = T_e^1\big((l_1; l_2) \circ (h_1; h_2)\big).
\end{aligned}$$

For any identity arrow id_A, $T_e^0 J(id_A) = \widetilde{id_A} = TA \simeq A$. Thus, as required by Definition 5, the natural isomorphism $\varphi : T_e \circ \blacktriangle \simeq I_{DB}$ is valid.

Let us show that **DB** is an *extended symmetric* category, i.e., $\tau^{-1} \bullet \tau = \psi : F_{st} \stackrel{\cdot}{\longrightarrow} S_{nd}$ for vertical composition of natural transformations $\tau : F_{st} \stackrel{\cdot}{\longrightarrow} T_e$ and $\tau^{-1} : T_e \stackrel{\cdot}{\longrightarrow} S_{nd}$, so that for each object $J(f) = \langle A, B, f\rangle$ in **DB** $\downarrow$ **DB**, the following diagram commutes (we recall that $\psi_{J(f)} = \psi(J(f)) = f$),

where

$$T_e\big(J(f)\big) = \widetilde{f} \simeq \biguplus \widetilde{\frac{f}{\uplus}} = \biguplus_{f_{ji} \in \frac{f}{\uplus}} \widetilde{f_{ji}},$$

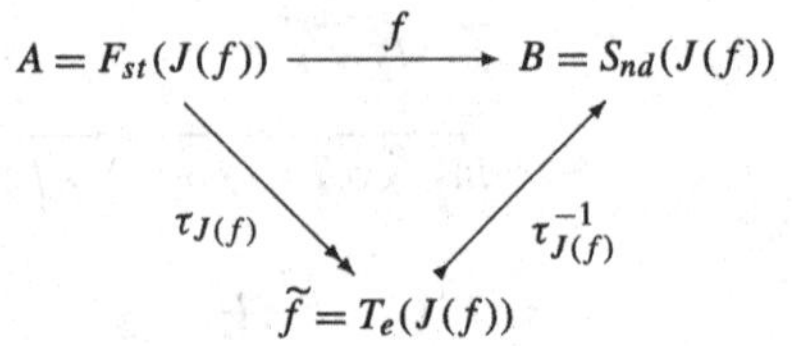

$$\frac{\tau_{J(f)}}{\uplus} = \left\{\tau_{J(f_{ji})} : A_j \twoheadrightarrow \widetilde{f_{ji}} \mid (f_{ji} : A_j \to B_i) \in \frac{f}{\uplus}\right\}, \quad \text{and}$$

$$\frac{\tau^{-1}_{J(f)}}{\uplus} = \left\{\tau^{-1}_{J(f_{ji})} : \widetilde{f_{ji}} \hookrightarrow B_i \mid (f_{ji} : A_j \to B_i) \in \frac{f}{\uplus}\right\}.$$

In fact, from Proposition 3 in Sect. 2.5, for a simple morphism

$$f = \alpha^*\big(MakeOperads(\mathcal{M}_{AB})\big) : A \to B,$$

we have two instance-mappings $ep(f) = \alpha^*(\mathbf{M}_E) : A \to TC$ and $in(f) = \alpha^*(\mathbf{M}_M) : TC \to B$ with $C = (S_C, \emptyset)$, $S_C = \bigcup_{q'_i \in \mathbf{M}_E} \partial_1(q'_i) = \{r_{q1}, \ldots, r_{qm}\}$ and $C = \alpha^*(S_C)$. From the fact that $TC = T\alpha^*(S_C) \subseteq \widetilde{f}$, we can define the equivalent morphisms to these two morphisms, by substituting the instance-database TC by $\widetilde{f}$ and hence defining the epimorphism $\tau_{J(f)} \triangleq \alpha^*(\mathbf{M}_E) : A \twoheadrightarrow \widetilde{f}$ (from Proposition 3, $\widetilde{\tau_{J(f)}} = \widetilde{ep(f)} = \widetilde{f}$) and the monomorphism $\tau^{-1}_{J(f)} \triangleq \alpha^*(\mathbf{M}_M) : \widetilde{f} \hookrightarrow B$ (from Proposition 3, $\widetilde{\tau^{-1}_{J(f)}} = \widetilde{in(f)} = \widetilde{f}$).

Notice that the existence of this monomorphism is also predicted by Proposition 7 with monic "inclusion" $\widetilde{f} \preceq TB$ (i.e., $in_B : \widetilde{f} \hookrightarrow TB$ and isomorphism $is_B^{-1} : TB \simeq B$ which is monic as well and hence their composition is a monomorphism equal to $\tau^{-1}_{J(f)} = is_B^{-1} \circ in_B : \widetilde{f} \hookrightarrow B$. In the same way, from Proposition 3, from the epimorphism $in_A^{OP} : TA \twoheadrightarrow \widetilde{f}$ (the dual (inverted) of the monomorphism $in_A : \widetilde{f} \hookrightarrow TA$) and isomorphism $is_A : A \simeq TA$ which is epic as well, we obtain that their composition is epic as well. That is, $\tau_{J(f)} = in_A^{OP} \circ is_A : A \twoheadrightarrow \widetilde{f}$ is an epimorphism. □

Remark For any given morphism $f : A \longrightarrow B$ in **DB**, the arrow $f_{ep} = \tau(J(f)) : A \twoheadrightarrow \widetilde{f}$ is an epimorphism, and the arrow $f_{in} = \tau^{-1}(J(f)) : \widetilde{f} \hookrightarrow B$ is a monomorphism, so that *any* morphism f in **DB** is a composition of *an epimorphism and monomorphism* $f = f_{in} \circ f_{ep}$ (as in **Set**, in Example 1), with the intermediate object equal to its information flux $\widetilde{f}$.

The particular cases with $\uplus$-composed complex objects and complex arrows between them will be presented in more detail also in point 4 of Corollary 12 in Sect. 3.3.2.

3.2.4 (Co)products

We recall that for a tuple of objects (i.e., instance-databases) in **DB** category $(S_1, \ldots, S_n)$, $n \geq 2$, we define its disjoint union (see Definition 19) as

$$\biguplus_{1 \leq i \leq n} S_i = \biguplus(S_1, \ldots, S_n) = \bigcup_{1 \leq i \leq n} \{(i, a) \mid a \in S_i\},$$

the union *indexed by positions* of the sets in a given tuple, denoted also as $S_1 \uplus S_2 \uplus \cdots \uplus S_n$. It will be demonstrated that the disjoint union of two objects is isomorphic to separation-composition of these objects.

Theorem 5 *There exists an idempotent coproduct bifunctor* $+ : \mathbf{DB} \times \mathbf{DB} \longrightarrow \mathbf{DB}$ *which is a disjoint union* $\biguplus$ *for objects and arrows in* **DB**.

The category **DB** *is cocartesian with initial (and terminal) object* $\perp^0$ *and for every pair of objects A and B the object* $A + B$ *is a coproduct with monomorphisms (injections)* $in_A = \langle id_A, \perp^1\rangle : A \hookrightarrow A + B$ *and* $in_B = \langle \perp^1, id_B\rangle : B \hookrightarrow A + B$.

Based on duality property, **DB** *is also a Cartesian category with a terminal object* $\perp^0$. *For each pair of objects A and B there exists a categorial product* $A \times B$ *with epimorphisms (projections)* $p_A = in_A^{OP} : A \times B \twoheadrightarrow A$ *and* $p_B = in_B^{OP} : A \times B \twoheadrightarrow B$. *Thus, the product bifunctor is equal to the coproduct bifunctor, i.e.,* $\times \equiv +$.

Proof Note that the coproduct $+$ is just the disjoint union $\biguplus$, so that:

1. For any identity arrow (id_A, id_B) in $\mathbf{DB} \times \mathbf{DB}$ where id_A and id_B are the identity arrows of A and B, respectively, $\frac{id_A + id_B}{\biguplus} = \frac{id_{A+B}}{\biguplus} = \{id_A, id_B\}$. Thus, $+^1(id_A, id_B) = id_A + id_B = id_{A+B}$ is an identity arrow of the object $A + B$.
2. For any $k : A \longrightarrow A_1$, $k_1 : A_1 \longrightarrow A_2$, $l : B \longrightarrow B_1$, $l_1 : B_1 \longrightarrow B_2$, it holds

$$\frac{+^1(k_1, l_1) \circ +^1(k, l)}{\uplus} = \{k_1 \circ k : A \rightarrow A_2, l_1 \circ l : B \rightarrow B_2\} = \frac{+^1(k_1 \circ k, l_1 \circ l)}{\uplus}.$$

   Thus, from point 3 of Definition 23, the functorial property for composition of morphisms $+^1(k_1 \circ k, l_1 \circ l) = +^1(k_1, l_1) \circ +^1(k, l)$ is valid.
3. Let us demonstrate the coproduct property of this bifunctor. For any two simple arrows $f : A \longrightarrow C$ and $g : B \longrightarrow C$, there exists a unique arrow $k = [f, g] : A + B \longrightarrow C$ such that $f = k \circ in_A$ and $g = k \circ in_B$ where $in_A = \langle id_A, \perp^1\rangle : A \hookrightarrow A + B$ and $in_B = \langle \perp^1, id_B\rangle : B \hookrightarrow A + B$ are the monomorphisms (injections). In fact, for any two arrows $h, k : D \rightarrow A$, if $in_A \circ h = in_A \circ k$ then $h = k$ (indeed $in_A \circ h = \langle id_A, \perp^1\rangle \circ h = \langle id_A \circ h, \perp^1 \circ h\rangle = \langle h, \perp^1\rangle = in_A \circ k = \langle h, \perp^1\rangle$, so that $h = k$). Analogously, it holds also for in_B. That is, the following coproduct diagram in **DB** commutes

   Assume that, for given schema $\mathcal{A}$, $\mathcal{B}$ and $\mathcal{C}$, we have the sketch's mappings $\mathbf{M}_{AC} : \mathcal{A} \rightarrow \mathcal{C}$ and $\mathbf{M}_{BC} : \mathcal{B} \rightarrow \mathcal{C}$ and an interpretation α such that $A = \alpha^*(\mathcal{A})$, $B = \alpha^*(\mathcal{B})$, $C = \alpha^*(\mathcal{C})$, $f = \alpha^*(\mathbf{M}_{AC})$ and $g = \alpha^*(\mathbf{M}_{BC})$. Then we define the morphisms $in_A = \alpha^*(\langle \mathbf{M}_{AA}, \{1_{r_\emptyset}\}\rangle) = \langle id_A, \perp^1\rangle$ and $in_B = \alpha^*(\langle \{1_{r_\emptyset}\}, \mathbf{M}_{BB}\rangle) = \langle \perp^1, id_B\rangle$ where $\mathbf{M}_{AA} = \{1_r \mid r \in \mathcal{A}\} \cup \{1_{r_\emptyset}\}$ and $\mathbf{M}_{BB} = \{1_r \mid r \in \mathcal{B}\} \cup \{1_{r_\emptyset}\}$.

$+ \equiv \uplus,$

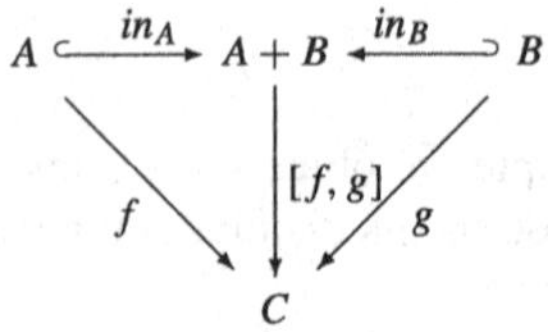

Thus, $[f,g] \circ in_A = [f,g] \circ \langle id_A, \bot^1 \rangle = \lceil f \circ id_A, g \circ \bot^1 \rfloor = \lceil f, \bot^1 \rfloor = \lceil f \rfloor = f$ (from Definition 20) and hence this triangle is commutative. Analogously, we show that $g = [f,g] \circ in_B$.

Let us suppose that there is another arrow $k' = [k_1, k_2] : A + B \to C$ such that the diagram above commutes, that is, $k' \circ in_A = f$ and $k' \circ in_B = g$. However, $k' \circ in_A = \lceil k_1, \bot^1 \rfloor = k_1$ and $k' \circ in_B = \lceil \bot^1, k_2 \rfloor = k_2$, so that $k_1 = f$ and $k_2 = g$, that is, $k' = k = [f,g]$ is a unique arrow that satisfies this commutativity. □

The product diagram is its dual with $k = [f,g]^{OP} = \langle f^{OP}, g^{OP} \rangle$ (from Corollary 12):

$\times \equiv \uplus,$

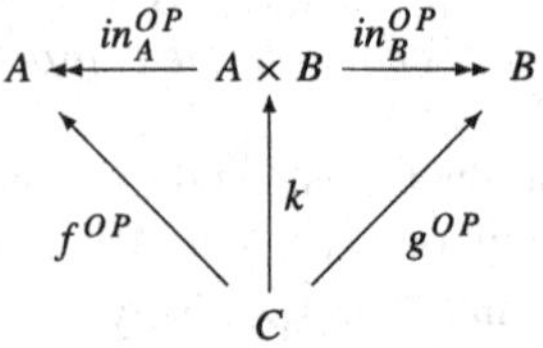

Let us show that the complex morphisms with coproducts satisfy the powerview functor T, the duality and the symmetry property in **DB**:

Corollary 12 *The complex morphisms satisfy the following* **DB** *properties*:

1. (*Power-view functorial properties*) $T(f+g) = Tf + Tg$, $T([f,g]) = [Tf, Tg]$, $T(\langle f,g \rangle) = \langle Tf, Tg \rangle$ *and* $T(\lceil f,g \rfloor) = \lceil Tf, Tg \rfloor$.
2. (*The duality properties*) $(f+g)^{OP} = f^{OP} + g^{OP}$, $[f,g]^{OP} = \langle f^{OP}, g^{OP} \rangle$, $\langle f,g \rangle^{OP} = [f^{OP}, g^{OP}]$ *and* $\lceil f,g \rfloor^{OP} = \lceil f^{OP}, g^{OP} \rfloor$.
3. (*The symmetry property of Theorem* 4 *with* $\tau^{-1} \bullet \tau = \psi$)

$$\tau(J(f+g)) = \tau_{J(f)} + \tau_{J(g)}, \qquad \tau^{-1}(J(f+g)) = \tau^{-1}_{J(f)} + \tau^{-1}_{J(g)},$$

$$\tau(J([f,g])) = \tau_{J(f)} + \tau_{J(g)}, \qquad \tau^{-1}(J([f,g])) = [\tau^{-1}_{J(f)}, \tau^{-1}_{J(g)}],$$

$$\tau(J(\langle f,g \rangle)) = \langle \tau_{J(f)}, \tau_{J(g)} \rangle, \qquad \tau^{-1}(J(\langle f,g \rangle)) = \tau^{-1}_{J(f)} + \tau^{-1}_{J(g)},$$

$$\tau(J(\lceil f,g \rfloor)) = \tau_{J(f)} \cup \tau_{J(g)}, \qquad \tau^{-1}(J(\lceil f,g \rfloor)) = \tau^{-1}_{J(f)} \cup \tau^{-1}_{J(g)}.$$

Proof Claim 1. (Power-view properties)

1.1. For $f+g : A+B \rightarrow C+D$, from Theorem 2, $T(f+g) = is_{C+D}^{-1} \circ (f+g) \circ is_{A+B} = (is_C^{-1} + is_D^{-1}) \circ (f+g) \circ (is_A + is_B) = (is_C^{-1} \circ f \circ is_A) + (is_D^{-1} \circ g \circ is_B) = Tf + Tg$.

1.2. For $[f,g] : A+B \rightarrow C$, from Theorem 2, $T([f,g]) = is_C \circ [f,g] \circ is_{A+B}^{-1} = is_C \circ [f,g] \circ (is_A^{-1} + is_B^{-1}) = [is_C \circ f \circ is_A^{-1}, is_C \circ g \circ is_B^{-1}] = [Tf, Tg]$.

The other two cases can be proved analogously.

Claim 2. (Duality) It holds from the fact that $\widetilde{f^{OP}} = \widetilde{f}$. Let us show that Theorem 3 is applicable. For example, consider $[f,g] : A+B \rightarrow C$:

$$\begin{aligned} [f,g]^{OP} &\overset{\text{(from Th. 2)}}{=} \left[is_C^{-1} \circ Tf \circ is_A, is_C^{-1} \circ Tg \circ is_B\right]^{OP} \\ &= \left\langle \left(is_C^{-1} \circ Tf \circ is_A\right)^{OP}, \left(is_C^{-1} \circ Tg \circ is_B\right)^{OP} \right\rangle \\ &= \left\langle is_A^{-1} \circ (Tf)^{OP} \circ is_C, is_B^{-1} \circ (Tg)^{OP} \circ is_C \right\rangle \\ &\overset{\text{(from Th. 3)}}{=} \left\langle f^{OP}, g^{OP} \right\rangle. \end{aligned}$$

Claim 3. The extended symmetry $\tau^{-1} \bullet \tau = \psi$ can be shown by the following four commutative diagrams;

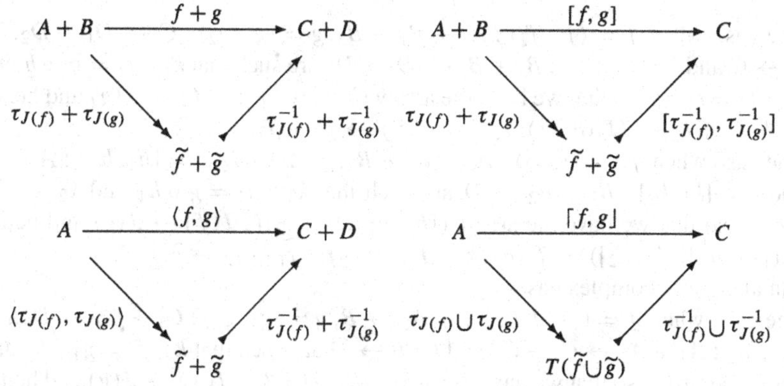

based on the results in point 1 and the facts of Theorem 4, i.e., $\widetilde{\tau_{J(f)}} = \widetilde{\tau_{J(f)}^{OP}} = \widetilde{f}$.

For example, in the second case we obtain

$$\begin{aligned} \left[\tau_{J(f)}^{-1}, \tau_{J(g)}^{-1}\right] \circ (\tau_{J(f)} + \tau_{J(g)}) &= \left[\tau_{J(f)}^{-1}, \tau_{J(g)}^{-1}\right] \circ (\tau_{J(f)} \uplus \tau_{J(g)}) \\ &\quad \text{(by rule 3 for composition of complex arrows)} \\ &= \left[\tau_{J(f)}^{-1} \circ \tau_{J(f)}, \tau_{J(g)}^{-1} \circ \tau_{J(g)}\right] = [f,g]. \end{aligned}$$ □

Let us show some examples for the extended symmetry functor T_e of Theorem 4, applied to complex morphisms and (co)products:

Example 23 Let us consider a number of "conceptualizations" of complex morphisms in Theorem 4, expressed by the functor $T_e = (T_e^0, T_e^1) : (\mathbf{DB} \downarrow \mathbf{DB}) \longrightarrow \mathbf{DB}$:

1. The case when $f = f_1 + f_2 : A_1 + A_2 \to B_1 + B_2$, $g = g_1 + g_2 : C_1 + C_2 \to D_1 + D_2$, $h = h_1 + h_2 : A_1 + A_2 \to C_1 + C_2$ and $k = k_1 + k_2 : B_1 + B_2 \to D_1 + D_2$, are such that $k_1 \circ f_1 = g_1 \circ h_1$ and $k_2 \circ f_2 = g_2 \circ h_2$, so that we have the arrow $(h_1 + h_2; k_1 + k_2) : J(f) \to J(g)$ and hence
$$T_e(h_1 + h_2; k_1 + k_2) = T_e(h_1, k_1) + T_e(h_2, k_2) : \widetilde{f_1} + \widetilde{f_2} \to \widetilde{g_1} + \widetilde{g_2}.$$
2. The case when $f = [f_1, f_2] : A_1 + A_2 \to B$, $g : C \to D$, $h = [h_1, h_2] : A_1 + A_2 \to C$ and $k : B \to D$ are such that $k \circ f_1 = g \circ h_1$ and $k \circ f_2 = g \circ h_2$, so that we have the arrow $([h_1, h_2]; k) : J(f) \to J(g)$ and hence
$$T_e\big([h_1, h_2]; k\big) = \big[T_e(h_1, k), T_e(h_2, k)\big] : \widetilde{f_1} + \widetilde{f_2} \to \widetilde{g}.$$
3. The case when $f : A \to B$, $g = [g_1, g_2] : C_1 + C_2 \to D$, $h = \langle h_1, h_2 \rangle : A \to C_1 + C_2$ and $k : B \to D$ are such that $k \circ f = g_1 \circ h_1$ and $k \circ f = g_2 \circ h_2$, so that we have the arrow $(\langle h_1, h_2 \rangle; k) : J(f) \to J(g)$ and hence
$$T_e\big(\langle h_1, h_2 \rangle; k\big) = \big\langle T_e(h_1, k), T_e(h_2, k)\big\rangle : \widetilde{f} \to \widetilde{g_1} + \widetilde{g_2}.$$
4. The case when $f = \langle f_1, f_2 \rangle : A \to B_1 + B_2$, $g = \langle g_1, g_2 \rangle : C \to D_1 + D_2$, $h : A \to C$ and $k = k_1 + k_2 : B_1 + B_2 \to D_1 + D_2$ are such that $k_1 \circ f_1 = g_1 \circ h$ and $k_2 \circ f_2 = g_2 \circ h$, so that we have the arrow $(h; k_1 + k_2) : J(f) \to J(g)$ and hence $T_e(h; k_1 + k_2) = \langle T_e(h, k_1), T_e(h, k_2)\rangle : \widetilde{f} \to \widetilde{g_1} + \widetilde{g_2}$.
5. The case when $f = \langle f_1, f_2 \rangle : A \to B_1 + B_2$, $g : C \to D$, $h = \lceil h_1, h_2 \rfloor : A \to C$ and $k = [k_1, k_2] : B_1 + B_2 \to D$ are such that $k_1 \circ f_1 = g \circ h_1$ and $k_2 \circ f_2 = g \circ h_2$, so that we have the arrow $(\lceil h_1, h_2 \rfloor; [k_1, k_2]) : J(f) \to J(g)$ and hence $T_e(\lceil h_1, h_2 \rfloor; [k_1, k_2]) = [T_e(h_1, k_1), T_e(h_2, k_2)] : \widetilde{f_1} + \widetilde{f_2} \to \widetilde{g}$.
But also more complex cases:
6. The case when $f = [f_1, f_2] : A_1 + A_2 \to B$, $g = [g_1, g_2] : C_1 + C_2 \to D$, $h = h_1 + h_2 : A_1 + A_2 \to C_1 + C_2$ and $k : B \to D$ are such that $k \circ f_1 = g_1 \circ h_1$ and $k \circ f_2 = g_2 \circ h_2$, so that we have the arrow $(h_1 + h_2; k) : J(f) \to J(g)$ and hence
$$\begin{aligned} T_e(h_1 + h_2; k) &= \langle in_{g_1}, in_{g_2} \rangle \circ T\big(k \circ [f_1, f_2]\big) \circ (in_{f_1} + in_{f_2}) \\ &= \big[\big\langle T_e(h_1, k), in_{g_2} \circ T(k \circ f_1) \circ in_{f_1}\big\rangle, \\ &\quad \big\langle in_{g_1} \circ T(k \circ f_2) \circ in_{f_2}, T_e(h_2, k)\big\rangle\big] : \widetilde{f_1} + \widetilde{f_2} \to \widetilde{g_1} + \widetilde{g_2}. \end{aligned}$$
7. The case when $f = \langle f_1, f_2 \rangle : A_1 + A_2 \to B$, $g = [g_1, g_2] : C_1 + C_2 \to D$, $h = \langle h_1, h_2 \rangle : A_1 + A_2 \to C_1 + C_2$ and $k = [k_1, k_2] : B_1 + B_2 \to D_1 + D_2$, so that we obtain
$$\begin{aligned} &T_e\big(h_1 + h_2; [k_1, k_2]\big) \\ &\quad = \big\lceil\big\langle\big[T_e(h_1, k_1), in_{g_1} \circ T(k_1 \circ f_1) \circ in_{f_2}\big], \end{aligned}$$

$$\big[in_{g_2} \circ T(k_1 \circ f_1) \circ in_{f_1}, in_{g_2} \circ T(k_1 \circ f_1) \circ in_{f_2}\big]\big\rangle,$$
$$\big\langle\big[in_{g_1} \circ T(k_2 \circ f_2) \circ in_{f_1}, in_{g_1} \circ T(k_2 \circ f_2) \circ in_{f_2}\big],$$
$$\big[in_{g_2} \circ T(k_2 \circ f_2) \circ in_{f_1}, T_e(h_2, k_2)\big]\big\rangle\big] : \widetilde{f}_1 + \widetilde{f}_2 \to \widetilde{g}_1 + \widetilde{g}_2.$$

3.2.5 Partial Ordering for Databases: Top and Bottom Objects

Let us consider the "observational" Partial Order (PO) relation introduced in Definition 19, its extension to morphisms and a 2-category property of **DB**:

Theorem 6 *PO subcategory* $\mathbf{DB}_I \subseteq \mathbf{DB}$ *is defined by* $Ob_{DB_I} = Ob_{DB}$ *and* Mor_{DB_I} *is a union of all isomorphisms and initial arrows* $\langle\overbrace{\bot^1, \ldots, \bot^1}^{m}\rangle : \bot^0 \hookrightarrow \biguplus_{1\le j\le m} A_j$ *and* $S_I = \{in : A \hookrightarrow B \,|\, $*for two strictly-complex objects* $A = \biguplus_{1\le j\le m} A_j \preceq B = \biguplus_{1\le i\le k} B_i$ *with a mapping (from Definition* 19) $\sigma : \{1, \ldots, m\} \to \{1, \ldots, k\}$ *and* $\frac{in}{\uplus} = \{in_{j\sigma(j)} : A_j \hookrightarrow B_{\sigma(j)} \,|\, j = 1, \ldots, m\}\}$.

This ordering can be extended by categorial symmetry property to morphisms as well:

$$f \preceq g \quad \textit{iff} \quad \widetilde{f} \preceq \widetilde{g}.$$

The power-view endofunctor $T : \mathbf{DB} \longrightarrow \mathbf{DB}$ *is a 2-endofunctor and a closure operator.*

DB *is a 2-category, and 1-cells are its ordinary morphisms, while 2-cells (denoted by* $\sqrt{\_}$*) are the arrows between ordinary morphisms: for any two morphisms* $f, g : A \longrightarrow B$ *such that* $f \preceq g$*, a 2-cell arrow is the "inclusion"* $\sqrt{\beta} : f \xrightarrow{\preceq} g$*. Such a 2-cell arrow is represented by an ordinary monomorphism in* **DB**, $\beta : \widetilde{f} \hookrightarrow \widetilde{g}$, *with* $\beta = T_e^1(f_{ep}^{OP} \circ f_{ep}; f_{in} \circ f_{in}^{OP})$, *where* $f_{ep} = \tau(J(f)) : A \twoheadrightarrow \widetilde{f}$ *is an epimorphism, and the arrow* $f_{in} = \tau^{-1}(J(f)) : \widetilde{f} \hookrightarrow B$ *is a monomorphism.*

Proof $\mathbf{DB}_I$ is well defined: for any object A we have that its identity arrow $id_A : A \to A$ is a monomorphism (we have $A \preceq A$ with σ identity function, and $in = id_A$), thus an arrow in $\mathbf{DB}_I$. Each isomorphism is a monomorphism as well, and any initial arrow is monic as well, so that all arrows and their compositions in $\mathbf{DB}_I$ are monic arrows. For each isomorphism $A \simeq B$ represented by arrow $is : A \to B$ in $\mathbf{DB}_I$, we define $\sigma : \{1, \ldots, m\} \to \{1, \ldots, k\}$ such that $\sigma(j) = i$ if $(is_{ji} : A_j \to B_i) \in \frac{is}{\uplus} = \bot^0$; 1 otherwise (when $A_j = \bot^0$). Thus for each $j = 1, \ldots, m$, $TA_j \subseteq TB_{\sigma(j)}$, hence $A \preceq B$. For each initial arrow $\langle\overbrace{\bot^1, \ldots, \bot^1}^{m}\rangle : \bot^0 \hookrightarrow \biguplus_{1\le j\le m} A_j$ and any $\sigma : \{1\} \to \{1, \ldots, m\}$, $T\bot^0 = \bot^0 \subseteq TA_{\sigma(1)}$, thus it represents the PO relation $\bot^0 \preceq A$. Any monomorphism in S_I represents a PO relation $A \preceq B$ between two strictly-complex objects. Each composed arrow by the monic arrows above is a composition of PO relations of each of its components and hence a PO relation as well. Consequently, each arrow f in $\mathbf{DB}_I$ represents the PO relation $dom(f) \preceq cod(f)$. Let us show that each PO relation $A \preceq B$ is represented by a monomorphism in $\mathbf{DB}_I$ as well: if $A \simeq \bot^0$ then it is represented by a composition of the isomorphism (monic as well) $is : A \to \bot^0$ and initial arrow from $\bot^0$ into B which is monic as well. Otherwise $A \preceq B$ is generally representable as $A \simeq A_S \preceq B_S \simeq B$ where A_S is a strictly-complex object obtained by eliminating all objects $\bot^0$ from A (analogously for B_S). Thus $A \preceq B$ is represented in $\mathbf{DB}_I$ as a composition of these two isomorphisms (that are also monic) and a monomorphism in S_I between two strictly-complex objects (the composition of monic arrows is monic as well). Consequently, between any two given objects in $\mathbf{DB}_I$ there can exist at most one arrow, so this is a PO category. The power-view operator T satisfies the following:

(i) $TA = T(TA)$, as explained in the introduction, Sect. 1.4.1.

(ii) $A \subseteq B$ implies $TA \subseteq TB$: T is a monotonic operator w.r.t. $\subseteq$ (the free SPRJU algebra of terms $\mathcal{L}_A$ is monotonic with respect to its carrier set A; see Sect. 1.4.1).

(iii) $A \subseteq TA$, each element of A is also a view of A.

Thus, T is a closure operator and an object A such that $A = TA$ is a closed object.

For any two simple morphisms $f, g : A \to B$ such that $\widetilde{f} \subseteq \widetilde{g}$, it is easy to show that there are monomorphisms $\beta = T_e^1(f_{ep}^{OP} \circ f_{ep}; f_{in} \circ f_{in}^{OP}) : \widetilde{f} \hookrightarrow \widetilde{g}$. In fact, for $h_1 = f_{ep}^{OP} \circ f_{ep} : A \to A$ with $\widetilde{h_1} = \widetilde{f_{ep}^{OP}} \cap \widetilde{f_{ep}} = \widetilde{f} \cap \widetilde{f} = \widetilde{f}$ and $h_2 = f_{in} \circ f_{in}^{OP}$: $B \to B$ with $\widetilde{h_2} = \widetilde{f_{in}} \cap \widetilde{f_{in}^{OP}} \widetilde{f} \cap \widetilde{f} = \widetilde{f}$, we obtain that $\widetilde{h_2 \circ f} = \widetilde{f} \cap \widetilde{f} = \widetilde{f}$ and $\widetilde{g \circ h_1} = \widetilde{g} \cap \widetilde{f} = \widetilde{f}$ (for $\widetilde{f} \subseteq \widetilde{g}$), so that $h_2 \circ f = g \circ h_1$, and, consequently, the morphism $T_e^1(h_1; h_2) = T_e^1(f_{ep}^{OP} \circ f_{ep}; f_{in} \circ f_{in}^{OP}) : \widetilde{f} \to \widetilde{g}$ is well defined. Thus, $\beta = T_e^1(f_{ep}^{OP} \circ f_{ep}; f_{in} \circ f_{in}^{OP}) = in_{\widetilde{g}}^{OP} \circ T(f_{in} \circ f_{in}^{OP} \circ f) \circ in_{\widetilde{f}}$, so that (from $\widetilde{T f_{in}^{OP}} = T\widetilde{f_{in}} = T\widetilde{f} = \widetilde{f}$), $\widetilde{\beta} = \widetilde{in_{\widetilde{g}}^{OP}} \cap T\widetilde{f_{in}} \cap T\widetilde{f_{in}^{OP}} \cap \widetilde{Tf} \cap \widetilde{in_{\widetilde{f}}} = \widetilde{g} \cap \widetilde{f} = \widetilde{f}$, so that $\beta = T_e^1(f_{ep}^{OP} \circ f_{ep}; f_{in} \circ f_{in}^{OP})$ is a monomorphism (from Corollary 9).

It is easy to verify that **DB** is a 2-category with 0-cells (its objects), 1-cells (its ordinary morphisms) and 2-cells ("inclusions" arrows between mappings). The horizontal and vertical composition of 2-cells is just the composition of PO relations. Given $f, g, h : A \longrightarrow B$ with 2-cells $\sqrt{\beta} : f \stackrel{\preceq}{\longrightarrow} g$, $\sqrt{\delta} : g \stackrel{\preceq}{\longrightarrow} h$, their vertical composition is $\sqrt{\gamma} = \sqrt{\delta} \circ \sqrt{\beta} : f \stackrel{\preceq}{\longrightarrow} h$. Given $f, g : A \longrightarrow B$ and $h, l : B \longrightarrow C$ with 2-cells $\sqrt{\beta} : f \stackrel{\preceq}{\longrightarrow} g$, $\sqrt{\delta} : h \stackrel{\preceq}{\longrightarrow} l$, for a given composition functor $\bullet : \mathbf{DB}(A, B) \times \mathbf{DB}(B, C) \longrightarrow \mathbf{DB}(A, C)$, their horizontal composition is $\sqrt{\gamma} = \sqrt{\delta} \bullet \sqrt{\beta} : h \circ f \stackrel{\preceq}{\longrightarrow} l \circ g$. □

For example, the equivalence $q_{A_i} \approx q_{B_j}$ (i.e., $q_{A_i} \preceq q_{B_j}$ and $q_{B_j} \preceq q_{A_i}$) of two view mappings $q_{A_i} : A \longrightarrow TA$ and $q_{B_j} : B \longrightarrow TB$, for the simple databases A and B, is obtained when they produce the same view.

The categorial symmetry operator $T_{\mathfrak{e}}^0 J : Mor_{DB} \longrightarrow Ob_{DB}$ for any morphism f in **DB** produces its information flux $\widetilde{f}$ (i.e., the "conceptualized" database of this mapping). Consequently, we can define a "mapping between mappings" (which are

2-cells "inclusions") and also all higher n-cells [2], with their direct representation by 1-cell morphisms.

Example 24 Let us consider the two ordinary simple morphisms (1-cells) in **DB**, $f : A \longrightarrow B$ and $g : C \longrightarrow D$ such that $\widetilde{f} \subseteq \widetilde{g}$ (thus, $f \preceq g$) and $\widetilde{g} \subseteq TC \cap TD$. We want to show that the 1-cell corresponding monomorphism $\beta : \widetilde{f} \hookrightarrow \widetilde{g}$ is a result of the symmetric closure functor T_e. We define two arrows $h_1 = is_C^{-1} \circ inc_C \circ \tau(J(f))$ and $h_2 = is_D^{-1} \circ in_D \circ (\tau_{J(f)}^{-1})^{OP}$ (where $inc_C : \widetilde{f} \hookrightarrow TC$ is a monomorphism, well defined in Proposition 7 because $\widetilde{f} \subseteq \widetilde{g} \subseteq TC$, $is_C^{-1} : TC \longrightarrow C$ is an isomorphism, $in_D : \widetilde{f} \hookrightarrow TD$ is a monomorphism from Proposition 7 and $\widetilde{f} \subseteq \widetilde{g} \subseteq TD$, and $is_D^{-1} : TD \longrightarrow D$ is an isomorphism). Hence, $\widetilde{h_1} = \widetilde{is_C^{-1}} \cap \widetilde{inc_C} \cap \widetilde{\tau(J(f))} = TC \cap T\widetilde{f} \cap \widetilde{f} = \widetilde{f}$ (because $TC \supseteq \widetilde{g} \supseteq \widetilde{f}$ and $T\widetilde{f} = \widetilde{f}$) and, analogously, $\widetilde{h_2} = \widetilde{is_D^{-1}} \cap \widetilde{in_D} \cap \widetilde{(\tau_{J(f)}^{-1})^{OP}} = TD \cap T\widetilde{f} \cap \widetilde{f} = \widetilde{f}$ (because $TD \supseteq \widetilde{g} \supseteq \widetilde{f}$ and $T\widetilde{f} = \widetilde{f}$). Thus, $\widetilde{g \circ h_1} = \widetilde{h_2 \circ f} = \widetilde{f}$ and hence $g \circ h_1 = h_2 \circ f$.

Thus, there exists the arrow $\beta = T_e(h_1; h_2) : J(f) \longrightarrow J(g)$ in **DB** $\downarrow$ **DB**, and the following commutative diagram in **DB**

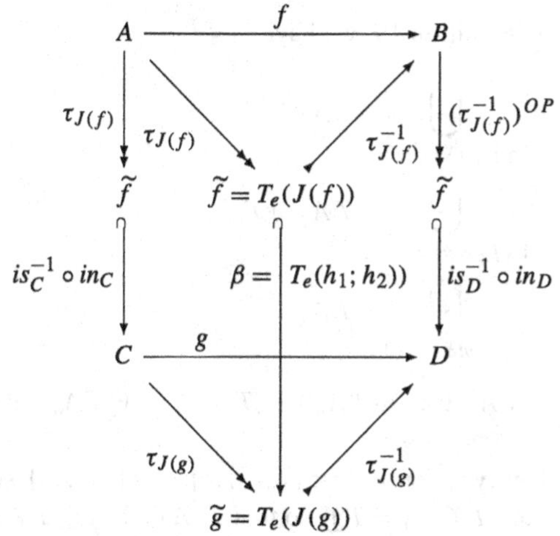

Let us show that also $T_e(h_1; h_2)$ is a monomorphism with $\beta = T_e(h_1; h_2) : \widetilde{f} \hookrightarrow \widetilde{g}$. In fact, $\widetilde{h_2} = \widetilde{f}$ and, by definition in Theorem 4, $T_e(h_1; h_2) \triangleq in_{\widetilde{g}}^{OP} \circ T(h_2 \circ f) \circ in_{\widetilde{f}}$. Thus, $\widetilde{T_e(h_1; h_2)} = \widetilde{in_{\widetilde{g}}^{OP}} \cap \widetilde{Th_2} \cap \widetilde{Tf} \cap in_{\widetilde{f}} = \widetilde{g} \cap \widetilde{f} = \widetilde{f}$ and, consequently, $T_e(h_1; h_2)$ is a monomorphism (from Corollary 9). In the particular case of Theorem 6 when $A = C$ and $B = D$ and $is_C^{-1} \circ inc_C = \tau_{J(f)}^{OP} = f_{ep}^{OP}$ and $is_D^{-1} \circ in_D = \tau_{J(f)}^{-1} = f_{in}$, we obtain that the 2-cell's arrow $\sqrt{\beta} : f \xrightarrow{\preceq} g$ is represented by the 1-cell monomorphism $\beta = T_e^1(f_{ep}^{OP} \circ f_{ep}; f_{in} \circ f_{in}^{OP}) : \widetilde{f} \hookrightarrow \widetilde{g}$.

Let us define the full subcategory of **DB** composed of only simple objects:

Definition 26 The full subcategory of **DB** composed of only simple objects will be denoted as $\underline{\mathbf{DB}}$, with the power-view endofunctor equal to the restriction of T to simple objects and arrows only.

The bottom object in $\underline{\mathbf{DB}}$ is that of **DB**, equal to $\bot^0$ (zero object), while the top object is equal to $\underline{\Upsilon} = \bigcup_{A \in Ob_{\underline{\mathbf{DB}}}} A \subset \Upsilon = \bigcup_{A \in Ob_{\mathbf{DB}}} A$.

The exact relationship between $\underline{\Upsilon}$ and the top complex object Υ is given in Proposition 55, Sect. 8.1.5, dedicated to algebraic database lattice, $\Upsilon = \underline{\Upsilon} \cup (\biguplus_{\omega} \underline{\Upsilon}) = \underline{\Upsilon} \cup (\overbrace{\underline{\Upsilon} \uplus \underline{\Upsilon} \uplus \cdots}^{\omega})$. Let us show that the objects Υ and $\bot^0$ are the top and bottom objects in **DB** w.r.t. the PO relation $\preceq$.

Proposition 10 *For any object $A \in Ob_{DB}$ (for $A = \Upsilon$ as well) the following are valid*:

1. $A \otimes \Upsilon \preceq \Upsilon$ *and* $A \uplus \Upsilon \preceq \Upsilon$.
2. Υ *and* $\bot^0$ *are the top and bottom objects, respectively, (up to isomorphism) in* **DB**.

Proof Claim 1. From Definition 19, we have

$$\begin{aligned} A \otimes \Upsilon &= \biguplus_{1 \le j \le m \& i=1,2,\ldots} TA_j \cap \Upsilon_i \\ &= \biguplus_{1 \le j \le m \& i=1,2,\ldots} TA_j \cap \underline{\Upsilon} \\ &= \biguplus_{1 \le j \le m \& i=1,2,\ldots} TA_j \\ &= (TA_1 \uplus \cdots \uplus TA_m) \uplus (TA_1 \uplus \cdots \uplus TA_m) \uplus \cdots. \end{aligned}$$

Thus, we have an identity mapping $\sigma : \{1, 2, \ldots\} \to \{1, 2, \ldots\}$ such that $T(A \otimes \Upsilon)_j \in \{A_1, \ldots, A_m\}$ and $T\Upsilon_{\sigma(j)} = T\underline{\Upsilon}$ so that $T(A \otimes \Upsilon)_j \subseteq T\Upsilon_{\sigma(j)}$ for all $j \ge 1$ and hence $A \otimes \Upsilon \preceq \Upsilon$.

Hence $A \uplus \Upsilon = A_1 \uplus \cdots \uplus A_m \uplus \Upsilon$ and for the identity mapping $\sigma : \{1, 2, \ldots\} \to \{1, 2, \ldots\}$, for $1 \le j \le m$, $T(A \uplus \Upsilon)_j = TA_j \subseteq T\Upsilon\sigma(j) = T\underline{\Upsilon}$, and for $j > m$, $T(A \uplus \Upsilon)_j = T\underline{\Upsilon} \subseteq T\Upsilon\sigma(j) = T\underline{\Upsilon}$. Thus, $A \uplus \Upsilon \preceq \Upsilon$.

Claim 2. For any object A in **DB**, $TA \subseteq \bigcup_{B \in Ob_{DB}} B = \Upsilon = T\Upsilon$. Thus, $A \preceq \Upsilon$ and $TA \preceq \Upsilon$, i.e., the closed object Υ is a top object. Notice that from point 2 of Proposition 7 there exist a monomorphism $in_{TA} : TA \hookrightarrow \Upsilon$, and $in_{TA} \circ is_A : A \hookrightarrow \Upsilon$ is a composition of two monomorphisms ($is_A : A \to TA$ is an isomorphism, thus monic arrow as well) and hence a monomorphism as well.

From the fact that $\bot \in A$ for any object A in **DB** and the fact that $\bot^0 = \{\bot\}$, $\bot^0 \subseteq A$, so that $\bot^0 \preceq A$, we obtain that $\bot^0$ is a bottom object. □

3.3 Basic Operations for Objects in DB

In this section, we will introduce the set of basic operations for instance-databases used during the definition of database mappings between two (or more) databases. They are corresponding operators in this base **DB** category for two schema basic operations, separation $\dagger$ and connection $\oplus$, introduced in Sect. 3.1, used in order to define the schema data mappings.

Consequently, from the fact that the schema separations and schema connections will be present in the sketch's categories obtained from the schema mapping graphs, we need, in order to satisfy the functorial semantics for database mapping, their corresponding operations at the instance-level (i.e., **DB** category) as well.

As we will see, these operations correspond to important database integration concepts used in practice.

3.3.1 Data Federation Operator in DB

The basic database operation is the DBMS's *Data federation* of two database instances A and B. A federated database system is a type of meta-database management system (DBMS) which transparently integrates multiple autonomous database systems into a *single federated database*. The constituent databases are interconnected via a computer network, and may be geographically decentralized. Since the constituent database systems remain autonomous, a federated database system is a contrastable alternative to the (sometimes daunting) task of merging together several disparate databases. A federated database, or virtual database, is the fully-integrated, logical composite of all constituent databases in a federated database system. In this way, we are able to compute the queries with the relations of *both* databases. In fact, Data Federation technology is just used for such an integration of two previously separated databases.

Consequently, given any two databases (objects in **DB**) A and B, the federation of them (under the common DBMS) corresponds to a *union*, denoted by $\underline{\cup}_{rn}$ (i.e., the union $\cup$ with the renaming (rn) of the relational symbols in $\mathcal{A}$ and $\mathcal{B}$ that have originally the same names) of them under the same DBMS, thus, equals to the database $A\underline{\cup}_{rn}B$.

That is, for any two schemas $\mathcal{A}$ and $\mathcal{B}$ and an interpretation α, such that $A = \alpha^*(\mathcal{A})$ and $B = \alpha^*(\mathcal{B})$, we have that $A\underline{\cup}_{rn}B = \alpha^*(\mathcal{A} \oplus \mathcal{B})$, and:

Proposition 11 *For any two simple databases A and B, their federation $A\underline{\cup}_{rn}B$ is isomorphic to their simple union $A \cup B$.*

Proof Notice that for any relation in $A \cup B$ we have possibly a set of copies of the same relation, but only with different name: thus, $T(A\underline{\cup}_{rn}B) = T(A \cup B)$ and hence we have the isomorphism $is^{-1} \circ is_{rn} : A\underline{\cup}_{rn}B \to A \cup B$ obtained from the isomorphisms $is_{rn} : A\underline{\cup}_{rn}B \to T(A\underline{\cup}_{rn}B)$ and $is : A \cup B \to T(A \cup B)$ (from Lemma 9). □

Thus, the data federation of two simple databases is isomorphic to the *merging* of these two databases. The merging of databases is formally defined in Sect. 8.1.2.

3.3.2 Data Separation Operator in DB

We have defined the separation-composition of schemas $\mathcal{A} \dagger \mathcal{B} = (S_{\mathcal{A}\dagger\mathcal{B}}, \Sigma_{\mathcal{A}\dagger\mathcal{B}})$ in Sect. 3.1 as a *disjunctive union* with elements indexed by schema labels ($\ddot{\mathcal{A}}$ and $\ddot{\mathcal{B}}$, respectively), i.e., with $S_{\mathcal{A}\dagger\mathcal{B}} = \{(\ddot{\mathcal{A}}, r_i) \mid r_i \in S_{\mathcal{A}}\} \cup \{(\ddot{\mathcal{B}}, r_j) \mid r_j \in S_{\mathcal{B}}\}$, $\Sigma_{\mathcal{A}\dagger\mathcal{B}} = \{(\ddot{\mathcal{A}}, \phi_i) \mid \phi_i \in \Sigma_{\mathcal{A}}\} \cup \{(\ddot{\mathcal{B}}, \psi_j) \mid \psi_j \in \Sigma_{\mathcal{B}}\}$ (if $\mathcal{B} = \mathcal{A}$ then $\ddot{\mathcal{A}} = 1$ and $\ddot{\mathcal{B}} = 2$).

In what follows, we will express a finite separation-composition by using the binary operator $\dagger$ and the parenthesis (for example, $(\mathcal{A} \dagger (\mathcal{B} \dagger \mathcal{C})) \dagger \mathcal{D}$), so that in **DB** we will use such a binary version $\ddagger$ of this operator at instance-database level:

Definition 27 The separation-composition of any two instance-databases $A = \alpha^*(\mathcal{A})$ and $B = \alpha^*(\mathcal{B})$, obtained from the schemas $\mathcal{A}$ and $\mathcal{B}$, is denoted by $A \ddagger B$:

$$A \ddagger B \triangleq \alpha^*(S_{\mathcal{A}\dagger\mathcal{B}}) = \{(\ddot{\mathcal{A}}, \alpha(r_i)) \mid r_i \in S_{\mathcal{A}}\} \cup \{(\ddot{\mathcal{B}}, \alpha(r_j)) \mid r_j \in S_{\mathcal{B}}\}$$
$$= \{(\ddot{\mathcal{A}}, R_i) \mid R_i \in A\} \cup \{(\ddot{\mathcal{B}}, R_j) \mid R_j \in B\}.$$

It corresponds to the instance-database of two separated schemas, i.e., to two mutually isolated instance-databases with separated database management systems, so that it is impossible to compute the queries with the relations from both schemas.

For any two morphisms $f : A \to C$ and $g : B \to D$, we define the morphism $(f \uplus g) : A \ddagger B \to C \ddagger D$ as for coproducts.

The separation property for morphisms in **DB** is represented by the facts that

$$\partial_0(f \uplus g) \triangleq \partial_0(f) \uplus \partial_0(g), \qquad \partial_1(f \uplus g) \triangleq \partial_1(f) \uplus \partial_1(g).$$

Remark For any database A, the *replication* of this database (over different DB servers) can be denoted by the separation-composed object $A \ddagger A$ in this category **DB**. From definition above, for any two databases A and B, $A \ddagger B = B \ddagger A$, that is, as we intended, the *separation-composition* for the databases *is a commutative* operation. We are able, analogously to Definition 15, to define any finite separation-composition

$$\ddagger^n(A_1, \dots, A_n) = A_1 \ddagger (A_2 \ddagger (\cdots \ddagger A_n) \dots) \triangleq \alpha^*(\mathcal{A}_1 \dagger \cdots \dagger \mathcal{A}_n).$$

From this definition, we have for any two instance databases A and B, $A \ddagger B \neq A \uplus B$ (if $\mathcal{A} \neq \mathcal{B}$) because the disjoint union introduced above is strictly noncommutative, and in the **Set** category the objects $A \uplus B$ and $B \uplus A$ are not equal but only isomorphic. However, it can be shown that they are isomorphic objects in **DB** so that the definition of **DB** category in Theorem 1 is still valid after this introduction of the separation-composition:

Lemma 12 *For any two simple instance-databases A and B,*

$$A \ddagger B = B \ddagger A \quad and \quad A \ddagger B \simeq A \uplus B.$$

Proof If $A = B$, the claim is clear because all relationships above are identities. From Definition 27,

(i) $A \ddagger B = \{(\ddot{\mathcal{A}}, \phi_i) \mid \phi_i \in \Sigma_A\} \cup \{(\ddot{\mathcal{B}}, \psi_j) \mid \psi_j \in \Sigma_B\} = B \ddagger A.$

Let us show that $id_A \uplus id_B$ (where $id_A : A \to A, id_B : B \to B$ are the identity morphisms) is the identity morphism $id_{A \ddagger B}$ for $A \ddagger B$. In fact, for any morphisms $f \uplus g : A \ddagger B \to C \ddagger D$ (with $f : A \to C, g : B \to D$), $(f \uplus g) \circ (id_A \uplus id_B) = (f \circ id_A) \uplus (g \circ id_B) = f \uplus g$, and for any morphism $h \uplus k : C \ddagger D \to A \ddagger B$ (with $h : C \to A, k : D \to B$), $(id_A \uplus id_B) \circ (k \uplus h) = (id_A \circ k) \uplus (id_B \circ k) = h \uplus k$. Thus, $id = (id_A \uplus id_B) : A \ddagger B \to A \ddagger B$ is equal to the identity arrow of $A \ddagger B$, i.e.,

(ii) $id_A \uplus id_B = id_{A \ddagger B}$.

In point 2 of Example 19, it was demonstrated that $(id_A \uplus id_B) : A \uplus B \to A \uplus B$ is the identity arrow $id_{A \uplus B}$ of the object $A \uplus B$ as well. That is,

(iii) $id_A \uplus id_B = id_{A \uplus B}$.

Let us show now that the morphism $(id_A \uplus id_B) : A \ddagger B \to A \uplus B$ is an isomorphism, with its inverse morphism

$$(id_A \uplus id_B)^{OP} = id_A^{OP} \uplus id_B^{OP} = id_A \uplus id_B : A \uplus B \to A \ddagger B.$$

In fact,

$$\begin{aligned}(id_A \uplus id_B) \circ (id_A \uplus id_B)^{OP} &= (id_A \uplus id_B) \circ (id_A \uplus id_B) \\ &= (id_A \circ id_A) \uplus (id_B \circ id_B) = id_A \uplus id_B \\ &\quad \text{(from (iii))} \\ &= id_{A \uplus B} : A \uplus B \to A \uplus B \quad \text{(the identity)},\end{aligned}$$

and analogously,

$$\begin{aligned}(id_A \uplus id_B)^{OP} \circ (id_A \uplus id_B) &= id_A \uplus id_B \\ &\quad \text{(from (ii))} \\ &= id_{A \ddagger B} : A \ddagger B \to A \ddagger B \quad \text{(the identity)}.\end{aligned}$$

Consequently, $(id_A \uplus id_B) : A \ddagger B \to A \uplus B$ is an isomorphism, i.e.,

(iv) $A \ddagger B \simeq A \uplus B$.

Thus, we have from (iv) and (i) that $A \uplus B \simeq A \ddagger B = B \ddagger A \simeq B \uplus A$ and hence $A \uplus B \simeq B \uplus A$ as it was demonstrated in Lemma 9. □

Notice that $T(A \ddagger B) = \{(\ddot{\mathcal{A}}, v_i) \mid v_i \in TA\} \cup \{(\ddot{\mathcal{B}}, v_j) \mid v_j \in TB\} = TA \ddagger TB \neq T(A \uplus B) = \{(1, v_i) \mid v_i \in TA\} \cup \{(2, v_j) \mid v_j \in TB\} = \widetilde{id_A} \uplus \widetilde{id_B} = \widetilde{id_A \uplus id_B} = \widetilde{id_{A \uplus B}}$. However, from (iv), $\widetilde{id_{A \ddagger B}} = T(A \ddagger B) \simeq T(A \uplus B) = \widetilde{id_{A \uplus B}}$, so that for the identity morphism $id : A \ddagger B \to A \ddagger B$ with $\widetilde{id} = T(A \ddagger B) \simeq \widetilde{id_{A \uplus B}}$, we obtain that $id = id_{A \ddagger B}$ is well defined identity morphism.

Remark From the fact that each object $A \uplus B$, obtained by disjoint union, is isomorphic to the object $A \ddagger B$, so that $A \uplus B$ can be equivalently substituted by the separation-composed object $A \ddagger B$, we will use only this second version of separation-composition in **DB** category (the disjoint union of objects in **DB** will be explicitly used prevalently for the demonstration purposes).

Corollary 13 *The following three isomorphisms for the separation-composition of objects in* **DB** *are valid:*

- *(Composition with zero (initial and final) object)* $C \ddagger \bot^0 \simeq C \simeq \bot^0 \ddagger C$.
- *(Commutativity)* $A \ddagger B \simeq B \ddagger A$.
- *(Associativity)* $A \ddagger (B \ddagger C) \simeq A \ddagger B \ddagger C \simeq (A \ddagger B) \ddagger C$.

Proof Claim 1. We have

$$C \ddagger \bot^0 \simeq C \uplus \bot^0 \text{ (from Lemma 12)} \simeq C \quad \text{(from Lemma 9) and,}$$

$$\bot^0 \ddagger C \simeq \bot^0 \uplus C \text{ (from Lemma 12)} \simeq C \quad \text{(from Lemma 9).}$$

For a given schema $\mathcal{C}$ with the interpretation α such that $C = \alpha^*(\mathcal{C})$, we have the identity morphism $id_C = \alpha^*(\mathbf{M}_{CC}) : C \to C$ with $\mathbf{M}_{CC} = \alpha^*(\{1_r \mid r \in \mathcal{C}\} \cup \{1_{r_\emptyset}\})$.

Let us define, as in the proof of point 4 of Lemma 9, the morphisms

$$i_1 = \langle \alpha^*(\mathbf{M}_{CC}), \alpha^*(\{1_{r_\emptyset}\}) \rangle = \langle id_C, \bot^1 \rangle : C \to C \uplus \bot^0 \quad \text{and}$$

$$i_2 = [\alpha^*(\mathbf{M}_{CC}), \alpha^*(\{1_{r_\emptyset}\})] = [id_C, \bot^1] : C \uplus \bot^0 \to C.$$

Then, by using the isomorphism $id_C \uplus \bot^1 : C \ddagger \bot^0 \to C \uplus \bot^0$, we obtain the composition

$$i_2 \circ (id_C \uplus \bot^1) = [id_C, \bot^1] \circ (id_C \uplus \bot^1) = [id_C \circ id_C, \bot^1 \circ \bot^1]$$
$$= [id_C, \bot^1] = i_2 : C \ddagger \bot^0 \to C$$

which is an isomorphism (it has been shown in the proof of point 4 of Lemma 9) and so is its inverse $(id_C \uplus \bot^1) \circ i_1 = i_1 : C \to C \ddagger \bot^0$.

Claim 2. (Commutativity)

$$A \ddagger B \simeq A \uplus B \quad \text{(from Lemma 12)}$$
$$\simeq B \uplus A \quad \text{(from point 4 of Lemma 9)}$$
$$\simeq B \ddagger A. \quad \text{(from Lemma 12)}$$

Let us show that this commutativity is represented by the isomorphism

$$i_1 = \langle [\bot^1, id_B], [id_A, \bot^1] \rangle : A \ddagger B \to B \ddagger A,$$

where $[\bot^1, id_B] : A \ddagger B \to B, [id_A, \bot^1] : A \ddagger B \to A$. In fact, in the proof of Lemma 9, we have shown that $i_1 : A \uplus B \to B \uplus A$ is an isomorphism. Let us show that $[\langle id_A, \bot^1\rangle, \langle \bot^1, id_B\rangle] : A \ddagger B \to A \uplus B$ is equal to the isomorphism $id_A \uplus id_B : A \ddagger B \to A \uplus B$. In fact, $\frac{[\langle id_A, \bot^1\rangle, \langle \bot^1, id_B\rangle]}{\uplus} = \{id_A, id_B\} = \frac{id_A \uplus id_B}{\uplus}$ and hence, from Definition 23, $[\langle id_A, \bot^1\rangle, \langle \bot^1, id_B\rangle] = id_A \uplus id_B$. Consequently, we can define the composition

$$(id_B \uplus id_A) \circ i_1 \circ \bigl[\langle id_A, \bot^1\rangle, \langle \bot^1, id_B\rangle\bigr] : A \ddagger B \to A \uplus B \to B \uplus A \to B \ddagger A,$$

and to show that it is equal to i_1 as follows:

$$\begin{aligned}
&(id_B \uplus id_A) \circ i_1 \circ \bigl[\langle id_A, \bot^1\rangle, \langle \bot^1, id_B\rangle\bigr] \\
&\quad = (id_B \uplus id_A) \circ \bigl\langle [\bot^1, id_B], [id_A, \bot^1]\bigr\rangle \circ \bigl[\langle id_A, \bot^1\rangle, \langle \bot^1, id_B\rangle\bigr] \\
&\quad = \bigl\langle id_B \circ [\bot^1, id_B], id_A \circ [id_A, \bot^1]\bigr\rangle \circ \bigl[\langle id_A, \bot^1\rangle, \langle \bot^1, id_B\rangle\bigr] \\
&\quad = \bigl\langle [\bot^1, id_B], [id_A, \bot^1]\bigr\rangle \circ \bigl[\langle id_A, \bot^1\rangle, \langle \bot^1, id_B\rangle\bigr] \\
&\qquad \text{(from case 1 in Example 19)} \\
&\quad = \bigl\langle \bigl[[\bot^1, id_B] \circ \langle id_A, \bot^1\rangle, [\bot^1, id_B] \circ \langle \bot^1, id_B\rangle\bigr], \\
&\qquad \bigl[[id_A, \bot^1] \circ \langle id_A, \bot^1\rangle, [id_A, \bot^1] \circ \langle \bot^1, id_B\rangle\bigr]\bigr\rangle \\
&\quad = \bigl\langle \bigl[\ulcorner \bot^1 \lrcorner, \ulcorner id_B \lrcorner\bigr], \bigl[\ulcorner id_A \lrcorner, \ulcorner \bot^1 \urcorner\bigr]\bigr\rangle \\
&\quad = \bigl\langle [\bot^1, id_B], [id_A, \bot^1]\bigr\rangle = i_1.
\end{aligned}$$

Claim 3. (Associativity)

$$\begin{aligned}
A \ddagger (B \ddagger C) &\simeq A \uplus (B \ddagger C) \simeq A \uplus (B \uplus C) \quad \text{(from Lemma 12)} \\
&\simeq A \uplus B \uplus C = \uplus(A, B, C) \quad \text{(from Lemma 9)} \\
&\simeq \ddagger(A, B, C) = A \ddagger B \ddagger C \quad \text{(from Lemma 12)}
\end{aligned}$$

Similarly to the proof above, $id_A \uplus \langle [id_B, \bot^1], [\bot^1, id_C]\rangle : A \ddagger (B \ddagger C) \to A \ddagger B \ddagger C$ represents this isomorphism. □

3.4 Equivalence Relations in DB Category

We can introduce a number of different equivalence relations for instance-databases:

- *Identity* relation—two instance-databases (sets of relations) A and B are identical when the set identity $A = B$ holds.
- *Isomorphism* relation '$\simeq$' in **DB**.
- *Behavioral equivalence* relation '$\approx$' in Definition 19—two instance-databases A and B are behaviorally equivalent when each view obtained from a database A can also be obtained from a database B, and viceversa.

- *Weak observational equivalence* relation '$\approx_w$'—two instance-databases A and B are weakly equivalent when each "certain" view (without Skolem constants) obtained from a database A can be also obtained from a database B, and viceversa.

We will show the following inclusion relationships between them:

$$`=' \subseteq `\simeq' \subseteq `\approx' \subseteq `\approx_w'.$$

It is also possible to define other kinds of equivalences for databases. In the rest of this chapter, we will consider only the last two equivalences defined above.

3.4.1 The (Strong) Behavioral Equivalence for Databases

Let us consider now the problem of defining the equivalent objects (instance-databases) from a *behavioral point of view based on observations*: as we have seen, each arrow (morphism) is composed of a number of "queries" (view-maps) and each query may be considered as an *observation* over some database instance (i.e., an object of **DB**). Thus, we can characterize each object in **DB** (i.e., an instance-database) by its behavior according to a given set of observations.

Indeed, if one object A is considered as a black-box, the object TA is the set of all observations on A. Thus, given two objects A and B, we are able to define the relation of equivalence between them based on the notion of the bisimulation relation. If the observations (i.e., resulting views of queries) of the instance-databases A and B are always equal, independently of their particular internal structure, then they look equivalent to an observer. In fact, any database can be seen as an Abstract Object Type (AOT), presented in Sect. 2.4.2 by Definition 12, with a number of internal states that can be observed by using query operators (i.e., programs without side-effects). Consequently, two simple databases A and B are equivalent (or bisimilar) if they provide the same set of observations, i.e., when the set of views TA is equal to TB.

For the complex databases with separation operator ‡, represented equivalently (up to isomorphisms) by non-symmetric disjoint union $\uplus$, the behavioral equivalence is more general relation than the isomorphism relation in **DB** category:

Definition 28 The relation of (strong) behavioral equivalence '$\approx$' between objects (databases) in **DB** is defined by the PO relation '$\preceq$' (in Definition 19) on the set Ob_{DB},

$$A \approx B \quad \text{iff} \quad A \preceq B \text{ and } B \preceq A.$$

This equivalence relation for morphisms is defined by, $f \approx g$ iff $\widetilde{f} \approx \widetilde{g}$.

Notice that for any arrow f in **DB**, $f \uplus \bot^1 \approx \bot^1 \uplus f \approx f$, where $\bot^1 = \alpha^*(\{1_{r\emptyset}\}) = \{\alpha^*(1_{r\emptyset})\} = \{q_\bot : \bot \rightarrow \bot\}$ is an empty morphism between objects (i.e., for any two objects A and B, we have an empty morphism $\bot^1 : A \rightarrow B$ with

$\widetilde{\bot^1} = \{\bot\} = \bot^0$, obtained from empty set of SOtgds between the schemas $\mathcal{A}$ and $\mathcal{B}$).

Moreover, any two simple arrows $f, g : A \to B$ are equal ($f \equiv g$, in Definition 23) iff they are observationally equivalent, that is, if $f \approx g$.

This relation of behavioral equivalence between the objects does not strictly correspond to the notion of isomorphism in the category **DB** (see Proposition 8), in fact, we have:

Corollary 14 *Any two isomorphic objects A and B in* **DB** *are behaviorally equivalent, that is, if* $A \simeq B$ *then* $A \approx B$*, but not vice versa.*

Only for the simple objects (databases) the behavioral equivalence corresponds to the isomorphism in **DB**.

Proof If there is an isomorphism $A \simeq B$, with an iso arrow $is : A \to B$ then, from the proof of Theorem 6, we obtain $A \preceq B$. Moreover, from duality, the arrow $is^{OP} : B \to A$ is an isomorphism and hence $B \preceq A$.

The converse does not hold: for example, $A \uplus A \approx A$, but *not* $A \uplus A \simeq A$. For two simple objects A and B, $A \preceq B$ means $TA \subseteq TB$. Thus $A \approx B$ means $TA = TB$, so that $A \simeq TA = TB \simeq B$, i.e., $A \simeq B$. □

Notice that a database $A \ddagger A$, which is composed of two copies of a database A, each one separated from another (each one has a different and independent DBMS), is behaviorally equivalent to a database A but not isomorphic to A.

The behavioral equivalence is derived from the PO relation '$\preceq$' so that it will have an important role in the generation of the database complete lattice in Sect. 8.1.5 based on this ordering. It is well known that any two isomorphic objects in a given category can be mutually substituted and, from the fact that they are also behaviorally equivalent, it means that in this database lattice two isomorphic object can be substituted as well by preserving PO ordering in this lattice.

3.4.2 Weak Observational Equivalence for Databases

A database instance can also have some relations with tuples containing *Skolem constants*. For example, the minimal Herbrand models of a Global (virtual) schema of a Data integration system with incomplete information [5, 7, 12] will have the Skolem constants for such a missed information. This example will be provided and discussed in Sect. 4.2.4. We consider (see Definition 1 for FOL and Sect. 1.4 in the introduction) a recursively enumerable set of all Skolem constants as marked (labeled) nulls $SK = \{\omega_0, \omega_1, \dots\} \subseteq \mathcal{U}$, disjoint from a database domain set $\mathbf{dom} \subseteq \mathcal{U}$ of all values for databases. Moreover, we introduce a unary predicate *Val*(_) such that $Val(d_i)$ is true iff $d_i \in \mathbf{dom}$ and hence $Val(\omega_i)$ is false for any $\omega_i \in SK$. Thus, we can define a new weak power-view operator for databases as follows:

Definition 29 Weak power-view operator $T_w : Ob_{DB} \longrightarrow Ob_{DB}$ is defined as follows: for any simple database A in **DB** category,

$$T_w(A) \triangleq \{R \mid R \in T(A) \text{ and } \forall_{1\le k\le|R|}\forall(d_i \in \pi_k(R))Val(d_i)\}$$

where $|R|$ is the number of attributes of the view (relation) R, and π_k is a kth projection operator on relations. For $A = \biguplus_{1\le j\le m} A_j, m \ge 1$, $T_w(A) = \biguplus_{1\le j\le m} T_w(A_j)$. We define a partial order relation '$\preceq_w$' for databases and a weak observational equivalence relation '$\approx_w$' (by $A \approx_w B$ iff $A \preceq_w B$ and $B \preceq_w A$) as in Definition 19 by substituting operator T by T_w.

The following properties are valid for the weak partial order $\preceq_w$ w.r.t. the partial order $\preceq$ (we denote '$A \prec B$' iff $A \preceq B$ and not $A \approx B$):

Proposition 12 *Let $A = \biguplus_{1\le j\le m} A_j, m \ge 1$ and $B = \biguplus_{1\le i\le k} B_i, k \ge 1$ be any two databases (objects in* **DB** *category), then:*
1. *$T_w(A) \simeq A$, if A is a database without Skolem constants; $T_w(A) \prec A$, otherwise;*
2. *$A \preceq B$ implies $A \preceq_w B$ (thus, $A \approx B$ implies $A \approx_w B$);*
3. *$T_w(T_w(A)) = T(T_w(A)) = T_w(TA) = T_w(A) \subseteq TA$, thus, each object $D = T_w(A)$ is a closed object (i.e., $D = TD$) such that $D \approx_w A$.*

Proof Claim 1. From $T_w(A) \subseteq TA$ ($T_w(A) = TA$ only if A is without Skolem constants).

Claim 2. If $A \prec B$ then $TA_j \subset TB_{\sigma(j)}$ with $T_w(TA_j) \subseteq T_w(TB_{\sigma(j)})$, for all $1 \le j \le m$, i.e., $A \preceq_w B$.

Claim 3. It holds from the definition of the operator T and T_w: $T_w(T_w(A)) = T(T_w(A))$ because $T_w(A)$ is the set of views of A without Skolem constants and due to Claim 1. $T_w(TA) = \{R \mid R \in TT(A) \text{ and } \forall_{1\le k\le|R|}\forall(d_i \in \pi_k(R))Val(d_i)\} = \{R \mid R \in TA \text{ and } \forall_{1\le k\le|R|}\forall(d_i \in \pi_k(R))Val(d_i)\} = T_w(A)$, from $T = TT$. Let us show that $T_w(T_w(A)) = T_w(A)$. For every view $v \in T_w(T_w(A))$, from $T_w(T_w(A)) = T(T_w(A)) \subseteq TA$, it holds that $v \in TA$ and hence, from the fact that v is without Skolem constants, $v \in T_w(A)$. The converse is obvious. □

Notice that from point 2, the partial ordering '$\preceq$' is a stronger discriminator for databases than the weak partial order '$\preceq_w$', i.e., we can have two objects $A \prec B$ that are weakly equivalent, $A \approx_w B$ (for example, when $A = T_w(B)$ and B is a database with Skolem constants). Let us extend the notion of the type operator T_w into the notion of the endofunctor of **DB** category:

Theorem 7 *There exists a weak power-view endofunctor $T_w = (T_w^0, T_w^1)$:* **DB** $\longrightarrow$ **DB** *such that the object component T_w^0 is equal to the type operator T_w and:*
1. *For any simple morphism $f : A \longrightarrow B$, the arrow component T_w^1 is defined by*

$$T_w(f) = T_w^1(f) \triangleq \{id_R : R \to R \mid R \in \widetilde{f} \text{ and } \psi\}$$

where ψ is a statement $\forall_{1\leq k\leq|R|}\forall(d_i \in \pi_k(R))Val(d_i)$, $|R|$ is the number of attributes of the relation (view) R, and π_k is a kth projection. We define $T_w(f)$ for a complex arrow f by the set of ptp arrows $\frac{T_w(f)}{\uplus} = \{T_w(f_{ji}) \mid f_{ji} \in \frac{f}{\uplus}\}$.

2. *There exist the natural transformations $\xi : T_w \stackrel{\bullet}{\longrightarrow} T$ (natural monomorphism) and $\xi^{-1} : T \stackrel{\bullet}{\longrightarrow} T_w$ (natural epimorphism) such that for any simple object A, $\xi(A) = inc_A$ is a monomorphism and $\xi^{-1}(A) = inc_A^{OP}$ is an epimorphism, which are equivalent, i.e., $\xi(A) \approx \xi^{-1}(A)$.*

Proof Let us show that the morphism $T_w(f) : T_w(A) \to T_w(B)$, for a given morphism $f = \alpha^*(\mathbf{M}_{AB}) : A \to B$, corresponds to the definition of arrows in **DB** category, specified by Theorem 1, that is, there is an SOtgd Φ of a schema mapping such that $T_w(f) = \alpha^*(MakeOperads(\{\Phi\}))$. Let us define the set of relational symbols in the instance-database $T_w(A)$ by $S_{T_wA} = \{r_i \mid r_i \in \mathbb{R}$ and $\alpha(r_i) \in T_w(A)\}$, and its schema $\mathcal{C} = (S_{T_wA}, \emptyset)$, so that $T_w(A) = \alpha^*(\mathcal{C})$ (for the instance-database $T_w(B)$, we define analogously the schema $\mathcal{D} = (S_{T_wB}, \emptyset)$). Let us define the set of relational symbols $S_{T_wf} = \{r_i \mid r_i \in \mathbb{R}$ and $R = \alpha(r_i) \in \widetilde{f}$ and $\forall_{1\leq k\leq|R|}\forall(d_i \in \pi_k(R))Val(d_i)\}$. Then we define the SOtgd Φ by a formula $\bigwedge\{\forall \mathbf{x}_i(r_i(\mathbf{x}_i) \Rightarrow r_i(\mathbf{x}_i)) \mid r_i \in S_{T_wf}\}$.

In fact, $MakeOperads(\{\Phi\}) = \{1_{r_i} \in O(r_i, r_i) \mid r_i \in S_{T_wf}\} \cup \{1_{r_\emptyset}\}$ is a sketch's mapping from the schema $\mathcal{C}$ into $\mathcal{D}$ and

$$\begin{aligned} T_w(f) &= \alpha^*\bigl(MakeOperads\bigl(\{\Phi\}\bigr)\bigr) \\ &= \alpha^*\bigl(\bigl\{1_{r_i} \in O(r_i, r_i) \mid r_i \in S_{T_wf}\bigr\} \cup \{1_{r_\emptyset}\}\bigr) \\ &= \bigl\{id_{r_i} : \alpha(r_i) \to \alpha(r_i) \in O(r_i, r_i) \mid r_i \in S_{T_wf}\bigr\} \cup \{q_\perp\}. \end{aligned}$$

For any identity arrow $id_A : A \to A$,

$$\begin{aligned} T_w(id_A) &= \bigl\{id_R : R \to R \mid R \in \widetilde{id_A} \text{ and } \forall_{1\leq k\leq|R|}\forall\bigl(d_i \in \pi_k(R)\bigr)Val(d_i)\bigr\} \\ &= \bigl\{id_R : R \to R \mid R \in TA \text{ and } \forall_{1\leq k\leq|R|}\forall\bigl(d_i \in \pi_k(R)\bigr)Val(d_i)\bigr\}, \end{aligned}$$

with $\widetilde{T_w(id_A)} = \{R \mid R \in TA$ and $\forall_{1\leq k\leq|R|}\forall(d_i \in \pi_k(R))Val(d_i)\} = T_w(A)$, and hence we obtain the identity arrow $T_w(id_A) = id_{T_wA} : T_w(A) \to T_w(A)$. It is extended to identity arrows of complex objects analogously (the disjoint union of simple identity arrows).

It is easy to verify that for any two simple arrows $f : A \longrightarrow B$ and $g : B \longrightarrow C$,

$$\begin{aligned} T_w(g \circ f) &= \bigl\{id_R : R \to R \mid R \in \widetilde{g \circ f} \text{ and } \forall_{1\leq k\leq|R|}\forall\bigl(d_i \in \pi_k(R)\bigr)Val(d_i)\bigr\} \\ &= \bigl\{id_R : R \to R \mid R \in \widetilde{g} \cap \widetilde{f} \text{ and } \forall_{1\leq k\leq|R|}\forall\bigl(d_i \in \pi_k(R)\bigr)Val(d_i)\bigr\}. \end{aligned}$$

Thus,

$$\begin{aligned} \widetilde{T_w(g \circ f)} &= \bigl\{R \mid R \in \widetilde{g} \cap \widetilde{f} \text{ and } \forall_{1\leq k\leq|R|}\forall\bigl(d_i \in \pi_k(R)\bigr)Val(d_i)\bigr\} \\ &= \bigl\{R \mid R \in \widetilde{g} \text{ and } \forall_{1\leq k\leq|R|}\forall\bigl(d_i \in \pi_k(R)\bigr)Val(d_i)\bigr\} \end{aligned}$$

$$\cap \{R \mid R \in \widetilde{f} \text{ and } \forall_{1\le k\le |R|}\forall(d_i \in \pi_k(R))Val(d_i)\}$$
$$= \widetilde{T_w(g)} \cap \widetilde{T_w(f)}.$$

Consequently, $T_w(g \circ f) = T_w(g) \circ T_w(f)$. It holds for complex arrows as well (by considering the composition of their ptp arrows in $\frac{f}{\uplus}$ and $\frac{g}{\uplus}$).

Let us show that the monomorphism $inc_A : T_w(A) \hookrightarrow TA$, for a given morphism $f = \alpha^*(\mathbf{M}_{AB}) : A \to B$, corresponds to the definition of arrows in **DB** category, specified by Theorem 1, that is, there is an SOtgd Ψ of a schema mapping such that $inc_A = \alpha^*(MakeOperads(\{\Psi\}))$.

Let us define the set of relational symbols in the instance-database $T_w(A)$ by $S_{T_wA} = \{r_i \mid r_i \in \mathbb{R} \text{ and } \alpha(r_i) \in T_w(A)\}$, and its schema $\mathcal{C} = (S_{T_wA}, \emptyset)$, so that $T_w(A) = \alpha^*(\mathcal{C})$ (we define analogously the schema $T\mathcal{A} = (S_{TA}, \emptyset)$ for the instance database TA). Then we define the SOtgd Φ by a conjunctive formula

$$\bigwedge\{\forall \mathbf{x}_i\big(r_i(\mathbf{x}_i) \Rightarrow r_i(\mathbf{x}_i)\big) \mid r_i \in S_{T_wA}\}.$$

In fact, $MakeOperads(\{\Psi\}) = \{1_{r_i} \in O(r_i, r_i) \mid r_i \in S_{T_wA}\} \cup \{1_{r_\emptyset}\}$ is a sketch's mapping from the schema $\mathcal{C}$ into $T\mathcal{A}$ and

$$\begin{aligned} inc_A &= \alpha^*\big(MakeOperads(\{\Psi\})\big) \\ &= \alpha^*\big(\{1_{r_i} \in O(r_i, r_i) \mid r_i \in S_{T_wA}\} \cup \{1_{r_\emptyset}\}\big) \\ &= \{id_{r_i} : \alpha(r_i) \to \alpha(r_i) \in O(r_i, r_i) \mid r_i \in S_{T_wA}\} \cup \{q_\perp\}. \end{aligned}$$

Thus, $\widetilde{inc_A} = T\{\alpha(r_i) \in O(r_i, r_i) \mid r_i \in S_{T_wA}\} = T(T_w(A)) =$ (from point 4 of Proposition 12) $= T_w(A)$ and hence $inc_A : T_w(A) \hookrightarrow TA$ is a monomorphism (for the inclusion $T_w(A) \subseteq TA$) and, by duality, $inc_A^{OP} : TA \twoheadrightarrow T_w(A)$ is an epimorphism. It is easy to verify that the natural transformations ξ and ξ^{-1} are well defined. □

Analogously to the monad (T, η, μ) and comonad (T, η^C, μ^C) of the endofunctor T, we can define such structures for the weak endofunctor T_w as well:

Proposition 13 *The weak power-view endofunctor* $T_w = (T_w^0, T_w^1) : \mathbf{DB} \longrightarrow \mathbf{DB}$ *defines the monad* (T_w, η_w, μ_w) *and the comonad* (T_w, η_w^C, μ_w^C) *in* **DB** *such that* $\eta_w = \xi^{-1} \bullet \eta : I_{DB} \stackrel{\cdot}{\longrightarrow} T_w$ *is a natural epimorphism and* $\eta_w^C = \eta^C \bullet \xi : T_w \stackrel{\cdot}{\longrightarrow} I_{DB}$ *is a natural monomorphisms ('•' is a vertical composition for natural transformations), while* $\mu_w : T_wT_w \stackrel{\cdot}{\longrightarrow} T_w$ *and* $\mu_w^C : T_w \stackrel{\cdot}{\longrightarrow} T_wT_w$ *are equal to the natural identity transformation* $id_{T_w} : T_w \stackrel{\cdot}{\longrightarrow} T_w$ *(because* $T_w = T_wT_w$*).*

Proof It is easy to verify that all commutative diagrams of the monad and the comonad are composed of identity arrows. □

3.5 Review Questions

1. From the DBMS and from the query-answering points of view, what is the meaning of the separation $\mathcal{A} \dagger \mathcal{B}$ and the connection $\mathcal{A} \oplus \mathcal{B}$ of two database schemas $\mathcal{A}$ and $\mathcal{B}$? Why is it necessary to introduce the complex mapping-operads with the structural-operators in our setting, for a fixed source and target databases in the case when we have a set of intermediate databases between them? Why does the information flux of the complex mapping-operads have to be disjoint union of the information fluxes of its simple mapping components in Definition 15 for all structural operators different from $\lceil\lfloor, \ldots, \_\rfloor$, instead of a simple union?
2. What is the fundamental difference between the simple atomic sketch's arrows: View-mappings and the Integrity-constraints mappings? What is the reason that they can be used for the database sketches and category theory? It holds from Theorem 1 that each category **DB** is determined by a given universe $\mathcal{U}$ where the domain **dom** can be big enough for any database mapping system but finite: why must the set of Skolem marked null values SK be infinite instead, with the result that we can have the instance database with finitary relations (a finite number of attributes) but with infinite number of tuples? Can you provide an example with a cyclic mapping between the databases with incomplete information which requires an introduction of the Skolem constants, such that this cyclic mapping produces the relations with an infinite number of tuples? Which kind of tgds needs the Skolem constants, and why does the incomplete information in a database mapping need the SOtgds? Does it mean that in all situations, when we use the SOtgds, we must obtain the databases with some relation composed by an infinite number of tuples?
3. How is it possible that the n-ary structural operators for the morphisms in Definitions 20 and 21 are enough to express any possible complex arrow, based on the fact that the unique operator $\uplus$ for composition of complex objects (databases) is the disjoint union (use the fact that point 1 of Definition 21 defines the parallel composition of the arrows, points 2 and 3 define the ingoing and outgoing branching, and point 4 defines the union of different simple arrows). The definition of the ptp arrows is dual to the structural way of the representation of the complex arrows with the set of structural operators. Is there a possibility to define a canonical structural expression, for a complex arrow composed by a given set of ptp, different for that defined by Lemma 8? Can you provide an example?
4. The equality of the arrows in the **DB** category is not the set-theoretical equality, but an observational one, based on the relationships of the information fluxes of the simple arrows. Can you give an explanation why, for two given simple instance databases A and B, the simple mapping $f : A \to B$ is different from the mapping $\lceil f, f\rfloor : A \to B$?
5. In Proposition 5, we demonstrated that **DB** category is not a topos. Which kind of degeneration of the **DB** category would we obtain if we force the morphism $[id_C, id_C] : C \uplus C \to C$ to be an isomorphism, by considering the fact that $\uplus$ is the coproduct for the objects in **DB** category?

6. The power-view endofunctor T is a monad as well for the **DB** category. We have seen that object TA is composed by the observations on a given instance database A, obtained by the finite-length SPJRU queries. Based on this, which kind of the monad and initial algebra semantics is in strict relationship with this endofunctor? What is the relationship between the duality property of the **DB** category and the information fluxes of the composed atomic morphisms? Why can each morphism be represented as a composition of one epic and one monic arrows in the **DB** category?
7. Two basic operations for the objects in **DB** category are Data Federation (isomorphic to union) and Data Separation (isomorphic to disjoint union). Can we use any of them for the merging operation? If no then explain why and provide examples.
8. The strong behavioral equivalence for the databases is obtained from the basic observational PO relation $\preceq$ with the bottom object $\bot^0$ (an empty database) and the top object Υ. This equivalence for the simple objects (databases) corresponds to the isomorphism of the objects in **DB**. Two isomorphic complex objects are behaviorally equivalent as well. What would happen if we force that also the converse be valid (consider the relationship between the object C and the object, which is a coproduct in **DB** category, $C \uplus C$)?
9. $C \uplus C$ is a coproduct of two equal databases. What does it mean from the DBMS point of view for this complex database, and can we execute a query which uses the relations from both databases? The same query may be executed over only one of these two equal databases: why are $C \uplus C$ and C observationally equivalent? Moreover, why in this case are they not also isomorphic (so that one can be equivalently substituted by another). How can this difference from the isomorphism and observational equivalence express the *redundancy* in the database theory?
10. In Lemma 10, it was shown that the bottom object $\bot^0$ is a zero and closed object. Show formally that the total simple object $\underline{\Upsilon}$ in Definition 26 is a closed object, that is, $\underline{\Upsilon} = T\underline{\Upsilon}$. Then, based on the fact that the top complex object in **DB** category is $\Upsilon = \underline{\Upsilon} \cup (\biguplus_\omega \underline{\Upsilon}) = \underline{\Upsilon} \cup (\overbrace{\underline{\Upsilon} \uplus \underline{\Upsilon} \uplus \cdots}^{\omega})$, show that it is a closed object as well.
11. Two strongly equivalent databases are also weakly equivalent from the observational point of view. By applying the weak power-view operator T_w to an instance database A with the schema $\mathcal{A}$ and incomplete information, that is, with the relational tables which can have the Skolem constants, we obtain the subset of all views of A with only tuples without Skolem constants. In which way a SQL query to this database A has to be modified (extended) in order to obtain only the results without the Skolem constants? By considering the *certain answers* for a given query over a database A (the answers which are true in every model of this database), are the relations in $T_w A$ and, if it is so, why?

References

1. A. Asperti, G. Longo, *Categories, Types and Structures* (MIT Press, Cambridge, 1991)
2. J. Baez, J. Dolan, Higher-dimensional algebra III: n-categories and the algebra of opetopes. Adv. Math. **135**, 145–206 (1998)
3. M. Barr, C. Wells, The formal description of data types using sketches, in *Mathematical Foundations of Programming Language Semantics*. LNCS, vol. 298 (Springer, Berlin, 1988)
4. P. Buneman, S. Naqui, V. Tanen, L. Wong, Principles of programming with complex objects and collection types. Theor. Comput. Sci. **149**(1), 3–48 (1995)
5. A. Calì, D. Calvanese, G. Giacomo, M. Lenzerini, Data integration under integrity constraints, in *Proc. of the 14th Conf. on Advanced Information Systems Engineering (CAiSE 2002)* (2002), pp. 262–279
6. C. Ehresmann, Introduction to the theory of structured categories. Technical Report 10, University of Kansas (1966)
7. R. Fagin, P.G. Kolaitis, R.J. Miller, L. Popa, DATA exchange: semantics and query answering, in *Proc. of the 9th Int. Conf. on Database Theory (ICDT 2003)* (2003), pp. 207–224
8. G.M. Kelly, A.J. Power, Adjunctions whose counits are coequalizers, and presentations of finitary enriched monads. J. Pure Appl. Algebra **89**, 163–179 (1993)
9. C. Lair, Sur le genre d'esquissibilite des categories modelables (accessibles) possedant les produits de deux. Diagrammes **35**, 25–52 (1996)
10. J. Lambek, P. Scott, *Introduction to Higher Order Categorial Logic* (Cambridge University Press, Cambridge, 1986)
11. S.M. Lane, *Categories for the Working Mathematician* (Springer, Berlin, 1971)
12. M. Lenzerini, Data integration: a theoretical perspective, in *Proc. of the 21st ACM SIGACT SIGMOD SIGART Symp. on Principles of Database Systems (PODS 2002)* (2002), pp. 233–246
13. Z. Majkić, Categories: symmetry, n-dimensional levels and applications. PhD Thesis, University "La Sapienza", Roma, Italy (1998)
14. Z. Majkić, Abstract database category based on relational-query observations, in *International Conference on Theoretical and Mathematical Foundations of Computer Science (TMFCS-08)*, Orlando FL, USA, July 7–9 (2008)
15. M. Makkai, R. Pare, *Accessible Categories: The Foundations of Categorical Model Theory*. Contemporary Mathematics, vol. 104 (Am. Math. Soc., Providence, 1989)
16. E. Moggi, Computational lambda-calculus and monads, in *Proc. of the 4th IEEE Symp. on Logic in Computer Science (LICS'89)* (1989), pp. 14–23
17. E. Moggi, Notions of computation and monads. Inf. Comput. **93**(1), 55–92 (1991)
18. G.D. Plotkin, A.J. Power, Adequacy for algebraic effects, in *Proc. FOSSACS 2001*. LNCS, vol. 2030 (2001), pp. 1–24

References

Functorial Semantics for Database Schema Mappings

4

4.1 Theory: Categorial Semantics of Database Schema Mappings

A relational database schema $\mathcal{A}$ is generally specified by a pair (S_A, Σ_A), where S_A is a set of n-ary relation symbols, $\Sigma_A = \Sigma_A^{\text{tgd}} \cup \Sigma_A^{\text{egd}}$ with a set of the database integrity constraints expressed by *equality-generating dependencies* (egds) Σ_A^{egd} and a set of the *tuple-generating dependencies* (*tgds*) Σ_A^{tgd} introduced by Definition 2.

Note that any atomic inter-schema database *mapping* from a schema $\mathcal{A}$ into a schema $\mathcal{B}$ is represented by a *set* of general tgds $\forall \mathbf{x}\,(\phi_A(\mathbf{x})) \Rightarrow \exists \mathbf{z}\,(\psi_B(\mathbf{y}, \mathbf{z}))$ where $\mathbf{y} \subseteq \mathbf{x}$, with elimination of the universal quantification $\forall \mathbf{x}$ from the head of each tgd sentence, that is, by the open (with free variables) *view mapping* formulas $q_A(\mathbf{x}) \Rightarrow q_B(\mathbf{y})$ where $q_A(\mathbf{x})$ (equivalent to $\phi_A(\mathbf{x})$) is a conjunctive query over the schema $\mathcal{A}$ and $q_B(\mathbf{y})$ (equivalent to $(\exists \mathbf{z}\; \psi_B(\mathbf{y}, \mathbf{z}))$) is a conjunctive query over the schema $\mathcal{B}$.

In Sect. 2.6, we explained what the models of the schema mappings based on tgds and of the integrity-constraint schema mappings are. We have demonstrated that a set of tgds can be equivalently represented by a single SOtgd (by Skolemization of existential quantifiers on the right-hand side implications of tgds) and how the set of integrity constraints of a database schema can be equivalently represented by a single SOtgd (by the algorithms *TgdsToConSOtgd* and *EgdsToSotgd* in Sect. 2.2).

Moreover, in order to translate this particular second-order logic (based on SOtgds sentences) into the categorial setting, we explained how we can translate the SOtgds into the set of abstract operad's operations that specify a mapping between database schemas, and hence to use the functorial semantics for database mappings based on R-algebras for operads (in Sect. 2.4).

Based on this translation into operads, we have seen that each operad's operation $q_i \in O(r_1, \ldots, r_m, r)$, obtained from an *atomic* mapping (introduced in Definition 7), is an algebraic specification for an implication conjunct in a normalized SOtgd such that $\partial_0(q_i) = \{r_1, \ldots, r_m\} \subseteq S_A$ is a subset of relations of a source schema $\mathcal{A} = (S_A, \Sigma_A)$ and $\partial_1(q_i) = \{r\}$ is a singleton set with a relation $r \in S_B$ of a target schema $\mathcal{B} = (S_B, \Sigma_B)$. Consequently, the representation of a database

Z. Majkić, *Big Data Integration Theory*, Texts in Computer Science,
DOI 10.1007/978-3-319-04156-8_4,

schema $\mathcal{A} = (S_A, \Sigma_A)$ in the operad's setting is provided by its set of its relational symbols in S_A and the representation of its integrity-constraints Σ_A is provided by an inter-schema mapping from $\mathcal{A}$ into $\mathcal{A}_\top$.

In what follows, we will explain how the logical *model* theory for database schemas and their mappings based on views [13], which corresponds to particular mapping-interpretations α (R-algebras presented in Sect. 2.4.1 by Definition 11 and specified by Corollary 4, in order to satisfy the schema mappings as in Example 12), can be translated into the category theory by using the **DB** category defined in the previous chapter. However, the approach used here is slightly different from the method used in [13].

Based on this semantics for the logical formulae without free variables (i.e., the integrity constraints expressed by egds and tgds, and the query inter-schema mappings expressed by tgds), we are able to define the categorial interpretations for database schema mappings, based on functors from the sketch category of a given database mapping system (i.e., a sketch-graph introduced by Definition 14 in Sect. 2.6) into the instance-level **DB** category.

4.1.1 Categorial Semantics of Database Schemas

As we explained in Sect. 3.1, in order to define the database mapping systems, we use two fundamental operators for the database schemas: the data federation $\oplus$ and data separation $\dagger$, with two corresponding monoids, $((\mathbb{S}, =), \dagger, \mathcal{A}_\emptyset)$ and $((\mathbb{S}, =), \oplus, \mathcal{A}_\emptyset)$, where $\mathcal{A}_\emptyset = (\{r_\emptyset\}, \emptyset)$ is the empty schema such that for any α, from Definition 10, $\alpha^*(\mathcal{A}_\emptyset) = \{\perp\} = \perp^0$ and the distribution law $\mathcal{A} \oplus (\mathcal{B} \dagger \mathcal{C}) = (\mathcal{A} \oplus \mathcal{B}) \dagger (\mathcal{A} \oplus \mathcal{C})$ holds.

In fact, from Sect. 3.3.1, $\mathcal{A} \oplus \mathcal{A}_\emptyset = \mathcal{A} \underset{rn}{\overset{\cup}{-}} \mathcal{A}_\emptyset = \mathcal{A}$ for any schema $\mathcal{A}$ (because $r_\emptyset$ is a propositional constant, so that it does not change a database schema). Analogously, for the separation composition $\mathcal{A} \dagger \mathcal{A}_\emptyset = \mathcal{A}$. Consequently, each vertex in a graph G of a database mapping system is a term of the combined algebra of these two monoids, $\mathbb{S}_{\mathrm{Alg}} = ((\mathbb{S}, =), \oplus, \dagger, \mathcal{A}_\emptyset)$, i.e., any well formed *term* (i.e., an algebraic expression) of this algebra for schemas $\mathbb{S}_{\mathrm{Alg}}$ is a database schema. Hence, a database schema $\mathcal{A} \in \mathbb{S}_{\mathrm{Alg}}$ is an *atomic* schema or *composed* schema by a finite number of atomic schemas and two symmetric algebraic operators $\oplus$ and $\dagger$ of the algebra $\mathbb{S}_{\mathrm{Alg}}$.

For each atomic schema database and interpretation α, $A = \alpha^*(\mathcal{A})$ is an instance-database of this schema and hence an object in **DB** category. The interpretation of the (composite) schemas (i.e., the terms of the algebra $\mathbb{S}_{\mathrm{Alg}}$) in **DB** category is provided by the following proposition:

Proposition 14 *Let α be a given interpretation. Then there exists the following homomorphism from the schema-database level into the instance-database level*:

$$\alpha^* : \Big((\mathbb{S}, =), \oplus, \dagger, \mathcal{A}_\emptyset \Big) \to \Big((Ob_{DB}, \simeq), \underset{rn}{\overset{\cup}{-}}, \ddagger, \perp^0 \Big).$$

Proof The interpretation of a given schema $\mathcal{A}$ is an instance $A = \alpha^*(\mathcal{A})$ of this database, that is, an object in **DB**; for every interpretation $\alpha^*(\mathcal{A}_\emptyset) = \bot^0$.

From the monoidal property we have the equation $\mathcal{A} \oplus \mathcal{A}_\emptyset = \mathcal{A}$ in the algebra $\mathbb{S}_{\mathrm{Alg}}$. By the above homomorphism, $\alpha^*(=)$ is equal to the isomorphism $\simeq$ in **DB**, $\alpha^*(\oplus) = \underset{rn}{\uplus}$, so that $\alpha^*(\mathcal{A} \oplus \mathcal{A}_\emptyset) = \alpha^*(\mathcal{A}) \underset{rn}{\uplus} \alpha^*(\mathcal{A}_\emptyset) = A \underset{rn}{\uplus} \bot^0 \simeq A$. From the monoidal property we have the equation $\mathcal{A} \dagger \mathcal{A}_\emptyset = \mathcal{A}$ in the algebra $\mathbb{S}_{\mathrm{Alg}}$. By the above homomorphism, $\alpha^*(\dagger) = \ddagger$, so that, based on Definition 27, $\alpha^*(\mathcal{A} \dagger \mathcal{A}_\emptyset) = \alpha^*(\mathcal{A}) \ddagger \alpha^*(\mathcal{A}_\emptyset) = A \ddagger \bot^0$, and from Corollary 13 $A \ddagger \bot^0 \simeq A$ is an isomorphism in **DB**. □

Let $\mathcal{A} = (S_A, \Sigma_A)$ be the nonempty database schema where S_A is a set of relational symbols with a given list of attributes and $\Sigma_A = \Sigma_A^{\mathrm{tgd}} \cup \Sigma_A^{\mathrm{egd}}$ is the set of its integrity constraints which can be an empty set as well.

We can represent this schema database by a logic-graph composed by the integrity-constraints mapping $\mathsf{T}_{AA_\top} : \mathcal{A} \to \mathcal{A}_\top$ ("truth" arrow), where $\mathcal{A}_\top = (\{r_\top\}, \emptyset)$ is the "FOL-identity" schema (introduced in Sect. 2.2 and the sketch's graphs by Definition 14) such that for every α, $\alpha^*(\mathcal{A}_\top) = \{\alpha(r_\top)\} = \{I_T(r_\top)\} = R_=$ (identity relation from Sect. 1.3, Definition 11, proof of Lemma 2 and Example 12). For the mapping-interpretations α^* which are also the models of a given database schema (or database inter-schema mapping), the information flux of $\mathsf{T}_{AA_\top}$ is empty (from Corollary 6) as well, i.e., $\alpha^*(\mathit{MakeOperads}(\mathsf{T}_{AA_\top})) = \bot^0 = \{\bot\}$ ($\bot = \{\langle\rangle\}$ is an empty relation with the empty tuple $\langle\rangle$) and hence there is no information transferring which explains their pure "logical" semantics in the database mapping systems. We recall that when we pass from a "logic-mapping" graph G into the sketch-graph G' then we substitute "logical" edge $\mathcal{M} \in G$ by an algebraic edge $\mathit{MakeOperads}(\mathcal{M}) \in G'$.

Proposition 15 *Given a nonempty database schema $\mathcal{A} = (S_A, \Sigma_A)$ with a set of integrity constraints $\Sigma_A = \Sigma_A^{\mathrm{tgd}} \cup \Sigma_A^{\mathrm{egd}}$, we define the sketch-graph $G(\mathcal{A})$ derived from this schema, composed of the vertices $\mathcal{A}$ and $\mathcal{A}_\top = (\{r_\top\}, \emptyset)$ (more precisely, by S_A and $\{r_\top\}$, respectively), and the following edges:*

- *Identity mapping-operad for $\mathcal{A}$, $\mathbf{M}_{AA} = \{1_r \mid r \in S_A\} \cup \{1_{r_\emptyset}\} : \mathcal{A} \to \mathcal{A}$, and for the "FOL-identity" schema $\mathcal{A}_\top$, $\mathbf{M}_{A_\top A_\top} = \{1_{r_\top}, 1_{r_\emptyset}\} : \mathcal{A}_\top \to \mathcal{A}_\top$.*
- *Integrity-constraint mapping-operad:*
  - *If $\Sigma_A^{\mathrm{egd}} \neq \emptyset$ and $\Sigma_A^{\mathrm{tgd}} \neq \emptyset$, then Φ is equal to*

$$\mathit{EgdsToSOtgd}\big(\Sigma_A^{\mathrm{egd}}\big) \wedge \mathit{TgdsToConSOtgd}\big(\Sigma_A^{\mathrm{tgd}}\big);$$

  - *If $\Sigma_A^{\mathrm{egd}} \neq \emptyset$ and $\Sigma_A^{\mathrm{tgd}} = \emptyset$, then Φ is equal to $\mathit{EgdsToSOtgd}(\Sigma_A^{\mathrm{egd}})$;*
  - *If $\Sigma_A^{\mathrm{egd}} = \emptyset$ and $\Sigma_A^{\mathrm{tgd}} \neq \emptyset$, then Φ is equal to $\mathit{TgdsToConSOtgd}(\Sigma_A^{\mathrm{tgd}})$;*

Hence, we define the mapping-operad

$$\mathbf{T}_{AA_\top} = \mathit{MakeOperads}(\mathsf{T}_{AA_\top}) : \mathcal{A} \to \mathcal{A}_\top \quad \text{where } \mathsf{T}_{AA_\top} = \{\Phi\}.$$

We denote the obtained sketch category from this sketch-graph $G(\mathcal{A})$ by $\mathbf{Sch}(G(\mathcal{A}))$.

Each mapping-interpretation α for the schema-mapping graph $G(\mathcal{A})$ is a functor $\alpha^* : \mathbf{Sch}(G(\mathcal{A})) \longrightarrow \mathbf{DB}$ *as well, such that it generates an instance database* $A = \alpha^*(\mathcal{A})$, *the unique instance of "FOL-identity" schema* $A_T = \alpha^*(\mathcal{A}_\top) \triangleq \{\alpha(r_\top), \bot\}$, *where* $\alpha(r_\top) = R_=$, *and the following arrows in* $\mathbf{DB}$:

1. $\{\alpha(1_r) \mid r \in S_A\} \cup \{q_\bot\} \triangleq \alpha^*(\mathbf{M}_{AA}) : A \longrightarrow A$, *and* $\{\alpha(1_{r_\top}), q_\bot\} \triangleq \alpha^*(\mathbf{M}_{A_\top A_\top}) : A_T \to A_T$.
2. $f_{\Sigma_A} \triangleq \alpha^*(\mathbf{T}_{AA_\top}) : A \longrightarrow A_T$ *with empty flux* $\widetilde{f_{\Sigma_A}} = \widetilde{\alpha^*(\mathbf{T}_{AA_\top})} = \bot^0$.

Proof The sketch-graph $G(\mathcal{A}) \subseteq \mathbf{Sch}(G(\mathcal{A}))$ is a particular case of a database-mapping sketch-graph in Definition 14 with the mapping-operads edges defined above. We have to show that each mapping-interpretation α (satisfying Definition 11) is a functor $\alpha^* : \mathbf{Sch}(G(\mathcal{A})) \longrightarrow \mathbf{DB}$:

Claim 1. The arrows $\mathbf{M}_{AA} : \mathcal{A} \to \mathcal{A}$ and $\mathbf{M}_{A_\top A_\top} : \mathcal{A}_\top \to \mathcal{A}_\top$ are the identity mapping-operads (obtained from identity schema mappings as shown in the proof of Proposition 1) in Sect. 2.4. Hence, they are the identity arrows in the sketch category $\mathbf{Sch}(G(\mathcal{A}))$. Let us show that their functorial translation in $\mathbf{DB}$, that is, the morphisms in point 1 of this proposition, are the identity arrows as well.
Indeed, the morphism $\{\alpha(1_r) \mid r \in S_A\} \cup \{q_\bot\}$ is the set of identity functions $\alpha(1_r) : \alpha(r) \to \alpha(r)$ and hence this morphism is equal to its inverse and, consequently, it is the identity morphism $id_A : A \to A$ (as it has been shown in the proof of Theorem 1). Analogously, the morphism $\{\alpha(1_{r_\top}), q_\bot\}$ is the identity morphism $id_{A_\top} : A_\top \to A_\top$.

Claim 2. The functorial property of α^* for the composition of arrows is valid (see the proof of Theorem 1), i.e., $\alpha^*(\mathbf{M}_{BC} \circ \mathbf{M}_{AB}) = \alpha^*(\mathbf{M}_{BC}) \circ \alpha^*(\mathbf{M}_{AB})$. The property for the empty flux is satisfied, $\widetilde{f_{\Sigma_A}} = \bot^0 = \{\bot\}$ of the arrow $f_{\Sigma_A} : A \longrightarrow A_\top$ in the $\mathbf{DB}$ category, as it holds from Example 12, Definition 13 and Corollary 6. □

From the fact that the morphism $f_{\Sigma_A} : A \longrightarrow A_\top$ in the **DB** category has *always* the empty information flux, i.e., also when A is a model of a database schema $\mathcal{A}$ (when all integrity constraints of a database schema are satisfied), the sequential composition of these integrity-constraint arrows with other arrows is meaningless because such a composed arrow in the **DB** category has the empty information flux as well, while in the operational semantics we are interested only in the arrows in G that have a nonempty information flux. Thus, we will use only non-integrity-constraint arrows in a graph G for their meaningful mutual compositions.

Remark From Definition 13 and Corollary 6, the information flux of the integrity-constraint mapping operad $\mathbf{T}_{AA_\top}$ of a schema $\mathcal{A} = (S_A, \Sigma_A)$ is always empty (equal to $\bot^0$). Consequently, for every mapping interpretation α, $f_{\Sigma_A} = \alpha^*(\mathbf{T}_{AA_\top})$ has the empty flux, and hence f_{Σ_A} is equivalent to the empty morphism $\bot^1 : A \to A_\top$. Obviously, we will not use the empty morphism $\bot^1 = \{q_\bot : \bot \to \bot\}$ instead of the atomic morphism (in Definition 18), $f_{\Sigma_A} = \alpha^*(\mathbf{T}_{AA_\top}) = \alpha^*(\{q_i = v_i \cdot q_{A,i} \mid q_i \in$

$\mathbf{T}_{AA_\top}$ with $q_{A,i} \in O(S, r_q)$, $S \subseteq S_A$, $v_i \in O(r_q, r_\top)\}) \cup \{q_\perp\}$ because we can verify if α satisfies this integrity constraint whenever $\alpha(v_i)$ is an injection function or not (see Example 12 in Sect. 2.4.1). It is particularly important for the Operational semantics of database mapping presented in Chap. 7.

4.1.2 Categorial Semantics of a Database Mapping System

It is easy to verify that a graph G of a given database mapping system (introduced in Definition 14) can be extended into a sketch category $\mathbf{Sch}(G)$ analogously as in Proposition 15 for a single database schema, in the way that $\bigcup_{\mathcal{A}\in V_G} \mathbf{Sch}(G(\mathcal{A})) \subseteq \mathbf{Sch}(G)$.

The formalization of the embedding $\gamma : G \to \mathbf{Sch}(G)$ of a graph G into the sketch category $\mathbf{Sch}(G)$ can be provided by Proposition 15 for the vertices of G and for the inter-schema edges in G as follows:

Proposition 16 *Let us consider a graph $G = (V_G, E_G)$ of a database mapping system in Definition* 14. *Then we define the embedding $\gamma : G \to \mathbf{Sch}(G)$ of this graph G into the sketch $\mathbf{Sch}(G)$, as follows*:

1. $\bigcup_{\mathcal{A}\in V_G} \mathbf{Sch}(G(\mathcal{A})) \subseteq \mathbf{Sch}(G)$, *from Proposition* 15.
2. *For any schema mapping edge $\mathcal{M}_{AB} : \mathcal{A} \to \mathcal{B}$ in G, we introduce the mapping-operad arrow $\mathbf{M}_{AB} = MakeOperad(\mathcal{M}_{AB}) : \mathcal{A} \to \mathcal{B}$ in $\mathbf{Sch}(G)$.*
3. *All other arrows in this sketch category $\mathbf{Sch}(G)$ are obtained by composition of atomic arrows specified in the points above.*

Each mapping-interpretation α for the schema-mapping graph G is also a functor $\alpha^* : \mathbf{Sch}(G) \longrightarrow \mathbf{DB}$ such that it generates an instance database $A = \alpha^*(\mathcal{A})$ for each schema $\mathcal{A} \in \mathbf{Sch}(G)$, the unique instance of "truth" schema $A_T = \alpha^*(\mathcal{A}_\top) \triangleq \{\alpha(r_\top), \perp\}$, where $\alpha(r_\top) = R_=$, and the following arrows in $\mathbf{DB}$:

1. *For each schema $\mathcal{A} \in \mathbf{Sch}(G)$, the identity morphism $id_A : A \to A$ and, eventually, its integrity-constraint morphism $f_{\Sigma_A} : A \to A_\top$ as specified by Proposition* 15.
2. *For each mapping-operad arrow $\mathbf{M}_{AB} : \mathcal{A} \to \mathcal{B}$ in $\mathbf{Sch}(G)$, we define the morphism $f = \alpha^*(\mathbf{M}_{AB}) : A \to B$, where $A = \alpha^*(\mathcal{A})$ and $B = \alpha^*(\mathcal{B})$.*

Consequently, the set $Int(G)$ of all mapping-interpretations of a given mapping system graph G is a subset of all functors from $\mathbf{Sch}(G)$ into $\mathbf{DB}$, i.e., $Int(G) \subseteq \mathbf{DB}^{\mathbf{Sch}(G)}$.

Proof Same as the proof of Proposition 15. The fact that each mapping-interpretation α of a given graph G is a functor from $\mathbf{Sch}(G)$ into $\mathbf{DB}$ comes directly from Theorem 1, where each morphism in $\mathbf{DB}$ is a mapping-interpretation of some schema mapping. Thus, for any schema mapping $\mathcal{M} : \mathcal{A} \to \mathcal{B}$ in G, we have that

$$\alpha^*\big(MakeOperads(\mathcal{M})\big) : \alpha^*(\mathcal{A}) \to \alpha^*(\mathcal{B})$$

is a morphism in $\mathbf{DB}$; this holds for the composition of mappings in G, and hence this mapping-interpretation α is a functor. □

But we have the graphs G such that $Int(G) \subset \mathbf{DB}^{\mathbf{Sch}(G)}$, that is, we can have the functors from $\mathbf{Sch}(G)$ into $\mathbf{DB}$ such that they are not the mapping-interpretations (as provided by Definition 11), for example:

Example 25 Let us consider the graph G composed of two schemas $\mathcal{A} = (\{r_1, r_2\}, \emptyset)$ and $\mathcal{B} = (\{r_3\}, \emptyset)$, where $ar(r_i) = 2$, $i = 1, 2, 3$, with unique mapping $\mathcal{M} = \{\Phi\} : \mathcal{A} \to \mathcal{B}$ where Φ is the SOtgd $\forall x, y, z((r_1(x, y) \wedge (y, z)) \Rightarrow r_3(a, z))$. We can define another graph G_1 with the same schemas $\mathcal{A}$ and $\mathcal{B}$ but with the mapping $\mathcal{M}_1 = \{\Psi\} : \mathcal{A} \to \mathcal{B}$ where Ψ is the SOtgd $\forall x, y, z, w((r_1(x, y) \wedge (w, z)) \Rightarrow r_3(a, z))$. Let us consider an R-algebra α such that $A = \alpha^*(\mathcal{A}) = \{R_1, R_2, \bot\}$ and $B = \alpha^*(\mathcal{B}) = \{R_3, \bot\}$ with

$$R_1 = \alpha(r_1) = \{(a_1, b_1), (a_1, b_2)\},$$

$$R_2 = \alpha(r_2) = \{(b_1, c_1), (b_3, c_2)\},$$

$$R_3 = \alpha(r_3) = \{(a, c_1), (a, c_2)\},$$

and $g = \alpha^*(\mathcal{M}) = \{\alpha(q_1) \cdot \alpha(v_1), q_\bot\} : A \to B$ where $\alpha(v_1) : \alpha(r_q) \hookrightarrow R_3$ is an injection and $f = \alpha(q_1) : R_1 \times R_2 \to \alpha(r_q)$ is the function with the graph $\{((a_1, b_1), (a, c_1)), ((a_1, b_1), (a, c_2)), ((a_1, b_2), (a, c_1)), ((a_1, b_2), (a, c_2))\}$, and where $\alpha(r_q) = R_3$ (i.e., image of f).

It is easy to verify that α *is not* a mapping-interpretation of $\mathcal{M} \in G$. In fact, in order for it to be a mapping-interpretation, the graph of f has to be equal to $\{((a_1, b_1), (a, c_1))\}$. But α is a well defined functor from the fact that $f \in Mor_{DB}$, and other arrows in $\mathbf{Sch}(G)$ are only the identity arrows of $\mathcal{A}$ and $\mathcal{B}$. In fact, we have that α is a mapping-interpretation of the graph G_1, so that from Theorem 1, $f = \alpha^*(MakeOperads(\mathcal{M}_2)) : A \to B$, so that f is a well defined morphism in $\mathbf{DB}$.

It is easy to verify that there is at most *one* arrow between any given two nodes in the obtained sketch $\mathbf{Sch}(G)$. For any object $\mathcal{A}$ in $\mathbf{Sch}(G)$, we have its identity arrow and possibly an integrity-constraint arrow $\mathbf{T}_{AA_\top}$.

4.1.3 Models of a Database Mapping System

There is a fundamental functorial-*interpretation* relationship between the schema mappings and their models in the instance-level category $\mathbf{DB}$. It is based on the Lawvere categorial theories [1, 9], where he introduced a way of describing algebraic structures using categories for theories, functors into base category $\mathbf{Set}$ (which we substitute by a more adequate category $\mathbf{DB}$) and natural transformations for morphisms between models. For example, Lawvere's seminal observation is that the theory of groups is a category with a group object, that a group in $\mathbf{Set}$ is a product preserving functor, and that a morphism of groups is a natural transformation of functors. This observation was successively extended to define the categorial semantics for different algebraic and logic theories. This work is based on the theory of *sketches*, which are fundamentally small categories obtained from graphs enriched with concepts such as (co)cones mapped by functors in (co)limits of the base category **Set**. It was demonstrated that, for every sentence in basic logic, there is a sketch with the same category of models and vice versa [14]. Accordingly, sketches are called graph-based logic and provide very clear and intuitive specification of computational data and activities. For any small sketch **E**, the category of models *Mod*(**E**) is an accessible category by Lair's theorem and reflexive subcategory of $\mathbf{Set}^{\mathbf{E}}$ by Ehresmann–Kennison theorem. A generalization to base categories other than **Set** was proved by Freyd and Kelly in 1972 [5].

In what follows, we substitute the base category **Set** by this new database category **DB**. In fact, we translate each database mapping logic theory based on SOtgds into an algebraic theory expressed by a sketch-category **Sch**(G) where all arrows are R-algebra terms. Then we describe R-algebraic structures using these sketch-categories for theories and α functors into base category **DB** in order to obtain the models of the database-mapping theories.

For instance, the *sketch* category **Sch**(G) for the separation-composition mapping cocone diagram (graph G), introduced in Sect. 3.1, is presented on the left-hand side commutative diagram below. Notice that the mapping arrow $\mathcal{M}$ in a graph G is replaced by the morphism $\mathbf{M} = MakeOperads(\mathcal{M})$ in this sketch, while the nodes (objects) are eventually augmented by introducing another auxiliary schema $\mathcal{A}_{\top}$ as explained in Proposition 16. The functorial translation (with a R-algebra α^*) of this sketch into the **DB** category has to be a coproduct diagram in **DB** (see Definition 27 and Theorem 5):

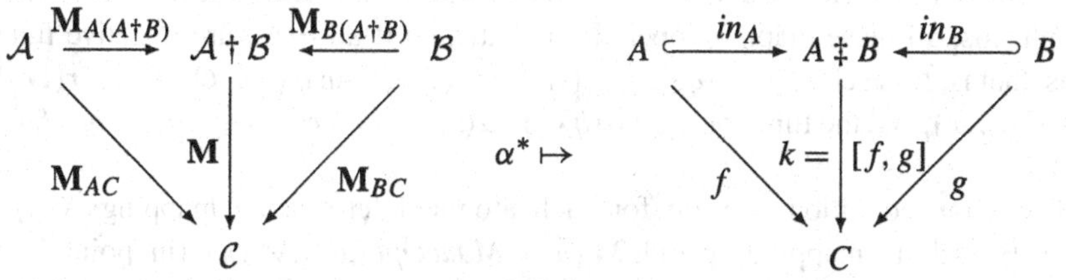

where $\mathbf{M} = [\mathbf{M}_{AC}, \mathbf{M}_{BC}]$, $k = \alpha^*(\mathbf{M}) = [\alpha^*(\mathbf{M}_{aC}), \alpha^*(\mathbf{M}_{BC})] = [f, g]$, with the monomorphisms $in_A = \alpha^*(\mathbf{M}_{A(A\dagger B)})$ and $in_B = \alpha^*(\mathbf{M}_{B(A\dagger B)})$, and the instance-databases $A = \alpha^*(\mathcal{A})$, $B = \alpha^*(\mathcal{B})$ and $C = \alpha^*(\mathcal{C})$.

As we explained in the introduction to sketch data models (Sect. 3.1.1), in a database mapping system expressed by a graph G, we never use the "commutative diagrams" as the left diagram above (but only the arrow $\mathbf{M} = [\mathbf{M}_{AC}, \mathbf{M}_{BC}] : \mathcal{A} \dagger \mathcal{B} \to \mathcal{C}$ or, more frequently, two simple arrows $\mathbf{M}_{AC} = \gamma(\mathcal{M}_{AC}) = MakeOperads(\mathcal{M}_{AC}) : \mathcal{A} \to \mathcal{C}$ and $\mathbf{M}_{BC} = \gamma(\mathcal{M}_{BC}) = MakeOperads(\mathcal{M}_{BC}) : \mathcal{B} \to \mathcal{C}$) and hence our sketch $\mathbf{E} = \mathbf{Sch}(G)$ is a simple small category, i.e., 4-tuple (G, u, D, C) where D and C are empty sets. Consequently, these database-mapping sketches are more simple than the sketches used for definition of Entity-Relationship models of single relational databases.

Proposition 17 *Let* $\mathbf{Sch}(G)$ *be a sketch category generated from a schema mapping graph G by applying the method in Proposition* 16 *and let* α^* *be a functor* $\alpha^* : \mathbf{Sch}(G) \longrightarrow \mathbf{DB}$ *(that is, a mapping-interpretation* $\alpha^* \in Int(G) \subseteq \mathbf{DB}^{\mathbf{Sch}(G)}$*).*

This functor is a model *of a database mapping system defined by the graph G if each atomic sketch's mapping-operad arrow* $\mathbf{M}_{AB} = \{v_1 \cdot q_{A,i}, \ldots, v_n \cdot q_{A,n}, 1_{r_\emptyset}\} : \mathcal{A} \to \mathcal{B}$ *in* $\mathbf{Sch}(G)$, *all* $\alpha(v_i)$, $1 \leq i \leq n$ *are the injective functions.*

Consequently, the category of models of a database schema mapping graph G is a subcategory of functors $\mathbf{DB}^{\mathbf{Sch}(G)}$ *and is denoted by* $Mod(\mathbf{Sch}(G))$.

For a given model (i.e., a functor) $\alpha^* \in Mod(\mathbf{Sch}(G))$ *of a graph G, its image in* **DB** *composed of only atomic sketch's arrows is called a DB-mapping system and is denoted by* $\mathcal{M}_G$.

Proof It is easy to verify based on general theory for sketches [1]. Each arrow in a sketch (obtained from a schema mapping graph G) is a mapping-operad that can be converted by R-algebra α of a given functor $\alpha^* \in Int(G)$ into a morphism in **DB**. The functorial property for the identity mappings follows from Proposition 15. For any two atomic mappings $\mathbf{M}_{AB} : \mathcal{A} \longrightarrow \mathcal{B}$ and $\mathbf{M}_{BC} : \mathcal{B} \longrightarrow \mathcal{C}$ and their atomic morphisms in **DB**, $f = \alpha^*(\mathbf{M}_{AB})$ and $g = \alpha^*(\mathbf{M}_{BC})$, $\alpha^*(\mathbf{M}_{BC} \circ \mathbf{M}_{AB}) = g \circ f$.

It remains to show that such a functor satisfying the conditions specified in this proposition is a *model* of a database mapping system given by a graph G. In fact, such a mapping interpretation $\alpha^* \in Int(\mathbf{Sch}(G))$ is a model of G if it is a model of each database schema (a vertex in G) and hence the schema integrity-constraints of all schemas in G are satisfied. i.e., all integrity-constraints mapping-operads in Proposition 16 are satisfied as well. From Corollary 4 and Definition 18 for atomic morphisms, all these mapping-operads are satisfied if all $\alpha(v_i)$ are injective functions, that is, for each $q_i = v_i \cdot q_{A,i} \in O(r_1, \ldots, r_k, r)$, with $q_{A,i} \in O(r_1, \ldots, r_k, r_q)$, $v_i \in O(r_q, r)$, and the function $f_i = \alpha(q_{A,i}) : \alpha(r_1) \times \cdots \times \alpha(r_k) \to \alpha(r_q)$, $im(f_i) \subseteq \alpha(r)$.

The same condition is valid for each atomic inter-schema mapping $\mathcal{M}_{AB} : \mathcal{A} \to \mathcal{B}$ and its mapping-operad $\mathbf{M}_{AB} = MakeOperads(\mathcal{M}_{AB})$ (in point 2 of Proposition 16) and hence they are satisfied as well. Consequently, this mapping-interpretation functor α^* is a model of a database mapping system given by a graph G. □

Based on these results, for each model of a database mapping system given by a graph G, we may omit the integrity-constraint arrows (that have the empty information flux) from the DB-mapping system $\mathcal{M}_G$, as well as identity arrows (which are always satisfied for each mapping-interpretation α), so that such a reduced instance-level DB-mapping system $\mathcal{M}_G$ will have *the same* user-meaningful structure as the specification schema-level mapping system of the graph G. More about arrows (natural transformations) and their compositions in $Mod(\mathbf{Sch}(G))$, in the case of schema-graphs is presented in Sect. 6.2.2 (Proposition 30) for the database transactions, and in the chapter dedicated to the operational semantics.

4.2 Application: Categorial Semantics for Data Integration/Exchange

Data *exchange* [4] is a problem of taking data structured under a source *schema* and creating an instance of a target *schema* that reflects the source data as accurately as possible.

Data *integration* [10] instead is a problem of combining data residing at different sources, and providing the user with a unified global *schema* of this data.

We describe the architecture of a tool for semantic data integration (depicted in Fig. 4.1), based on commercial systems for information integration, and explain its behavior. Such a system can be view as a plug-in that is able to cooperate with the commercial systems for data federation in order to achieve effective data integration. This architecture of a system is able to manage complex integration environments with a number of heterogeneous sources being integrated under a common view of it. The system provides the user with a classical database interface through which the sources are transparently integrated and queried. The independence and autonomy of data sources are also preserved by means of a fully virtual approach.

In order to deal with the information integration problem, they mainly act as *data federation* tool, enabling users to access remote data sources as if they were contained in the local database. However, they do not let the designer define an arbitrary description of the domain of interest. In particular, there are two main limitations affecting the effective employability of the typical data federation tools:

- External data source can be modeled as aliases that can be queried as if they were local tables. However, the correspondence between remote data sources and local alias is always one-to-one, instead of letting the designer define more expressive correspondence between a virtual concept of interest and, for example, a view or a query over multiple source data sets.
- Furthermore, if the sources are relational, the aliases' schema are identical to those of the modeled data source relations, which means that both have the same number, name and type of attributes. Even if sometimes the definition of aliases is a little more flexible for non-relational data sources, there are still several rules that the designer has to follow, which limit the expressiveness of the correspondence even inside a single source data set.

The above mentioned limits are typical of data federation tools. Indeed, in this kind of tools, the designer is simply provided with a view of data sources that is strongly source-dependent, since it basically reflects the source structure.

The commercial solutions for Data Integration can use the following features:

- Database Federation Tools allow for seeing multiple databases as a single resource;
- Heterogeneous Source Access capability to access simultaneously relational and non-relational-data;
- Grid Computing Tools overcoming the sources location problem.

Some of experimental tools, based on query rewriting, which extend the limited features of Database Federation Tools, resolving the problems of consistent query answering in the Data integration framework with key, exclusion and inclusion dependencies over global schema, can be found in [3].

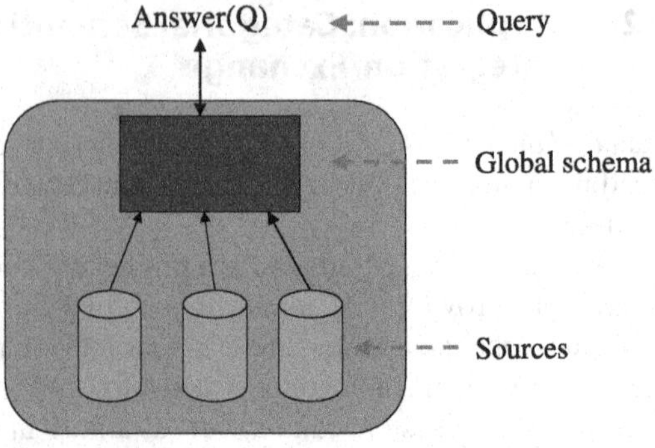

Fig. 4.1 Conceptual architecture

In what follows, we will consider the translation of the data exchange/integration systems into our semantic framework. In this framework, the concepts are defined in a more abstract way than in the instance database framework represented in the "computational" **DB** category. Consequently, we require an interpretation mapping from the scheme into the instance level, which will be given categorially by functors.

4.2.1 Data Integration/Exchange Framework

The task of a data integration system [10] is to provide the user with a unified view, called *global schema*, of a set of heterogeneous data sources. Once the user issues a query over the global schema, the system carries out the task of suitably accessing the different sources and assemble the retrieved data into the final answer to the query. In this context, a crucial issue is the specification of the relationship between the global schema and the sources, which is called *mapping* [7, 10]. In this section, we use a more complex mapping, called GLAV [4, 6], which consists in associating views over the global schema to views over the sources.

Since the global schema is a representation of the domain of interest of the system, it needs to be represented by means of a flexible an expressive formalism: to this aim, *integrity constraints* are expressed on it. The data at the sources may not satisfy the constraints on the global schema; in this case, a common assumption (which is the one adopted in what follows) is to consider the sources as *sound*, i.e., they provide a *subset* of the data that satisfy the global schema.

We formalize a *relational data integration system* $\mathcal{I}$ in terms of a triple $\langle \mathcal{G}, \mathcal{S}, \mathcal{M} \rangle$, where

- $\mathcal{G} = (S_G, \Sigma_G)$ is the *target global schema*, expanded by the *new unary predicate* *Val*(_) such that *Val*(c) is true if $c \in$ **dom**, expressed in a FOL.
- $\mathcal{S}$ is the *source schema*, expressed in a FOL. While the source integrity constraints may play an important role in deriving dependencies in $\mathcal{M}$, they do not play any direct role in the data integration/exchange framework and we may ignore them.
- $\mathcal{M}$ is the *mapping* between $\mathcal{G}$ and $\mathcal{S}$, constituted by a set of *assertions*

$$q_{\mathcal{S}} \rightsquigarrow q_{\mathcal{G}}, \qquad (4.1)$$

where $q_{\mathcal{S}}$ and $q_{\mathcal{G}}$ are two queries of the same arity, over the source schema $\mathcal{S}$ and over the target schema $\mathcal{G}$, respectively. Intuitively, an assertion $q_{\mathcal{S}} \leadsto q_{\mathcal{G}}$ specifies that the concept represented by the query $q_{\mathcal{S}}$ over the sources corresponds to the concept in the target schema represented by the query $q_{\mathcal{G}}$.

- Queries $q_{\mathcal{C}}(\mathbf{x})$, where $\mathbf{x} = (x_1, \ldots, x_k)$ is a nonempty list of variables, over the global schema are *conjunctive queries*. We will use, for every original query $q_{\mathcal{C}}(\mathbf{x})$, only a *lifted query* over the global schema, denoted by q, such that $q := q_{\mathcal{C}}(\mathbf{x}) \wedge Val(x_1) \wedge \cdots \wedge Val(x_k)$.

In order to define the semantics of a data integration system, we start from the data at the sources, and specify which are the data that satisfy the global schema. A *source instance database* D for $\mathcal{I} = \langle \mathcal{G}, \mathcal{S}, \mathcal{M} \rangle$ is constituted by one relation $\|r\|_D$ for each source relational symbol r in $\mathcal{S}$ (sources that are not relational may be suitably presented in the relational form by wrapper's programs). We call a *global database* for $\mathcal{I}$, or simply a *database* for $\mathcal{I}$, any (instance) database for $\mathcal{G}$. A database G for $\mathcal{I}$ is said to be *legal* with respect to D if:

- G satisfies the integrity constraints of $\mathcal{G}$;
- G satisfies $\mathcal{M}$ with respect to D.
- We restrict our attention to sound views only, which are typically considered the most natural ones in a data integration setting [7, 10].

In order to obtain an answer to a lifted query q from a data integration system, a tuple of constants is considered as an answer to this query only if it is a *certain* answer, i.e., it satisfies the query in *every legal* global database.

We may try to infer all the legal databases for $\mathcal{I}$ and compute the tuples that satisfy the lifted query q in all such legal databases. However, the difficulty here is that, in general, there is an infinite number of legal databases. Fortunately, we can define another *universal* (*canonical*) database $can(\mathcal{I}, D)$ that has the interesting property of faithfully representing all legal databases. The construction of the canonical database is similar to the construction of the *restricted chase* of a database described in [8].

4.2.2 GLAV Categorial Semantics

Let us consider the following Global-and-Local-As-View (GLAV) case when each dependency in $\mathcal{M}$ is a *tuple-generating dependency* (*tgd*), introduced in Sect. 1.4.2,

$$\forall \mathbf{x} \left(\exists \mathbf{y}\, q_{\mathcal{S}}(\mathbf{x}, \mathbf{y}) \Longrightarrow \exists \mathbf{z}\, \mathbf{q}_{\mathcal{G}}(\mathbf{x}, \mathbf{z}) \right) \tag{4.2}$$

where the formula $q_{\mathcal{S}}(\mathbf{x})$ is a conjunction of atomic formulas over $\mathcal{S}$ and $q_{\mathcal{G}}(\mathbf{x}, \mathbf{z})$ is a conjunction of atomic formulas over $\mathcal{G}$.

In the Global-As-View (GAV) approach, for each relation r in the global (mediated) schema), we write a query over the source relations specifying how to obtain r's tuples from the sources. However, adding sources to the data integration system is not trivial. In particular, given a new source, we need to figure out all the ways in which it can be used to obtain tuples for each of the relations in the global schema.

Therefore, we need to consider the possible interactions of the new source with each of the existing sources, and this limits the ability of the GAV approach to scale to a large collection of sources. In the Local-As-View (LAV) approach, the descriptions of the sources are given in the opposite direction. That is, the contents of a data source are described as a query over the global schema relation.

Notice that GLAV includes as special cases both LAV (when each assertion is of the form $q_{\mathcal{S}}(\mathbf{x}) = r_s(\mathbf{x})$, for some relation r_s in $\mathcal{S}$) and GAV (when each assertion is of the form $q_{\mathcal{G}}(\mathbf{x}, \mathbf{z}) = r_g(\mathbf{x}, \mathbf{z})$, for some relation r_g in $\mathcal{G}$) data integration mapping in which the views are sound. The GLAV system in data integration corresponds to the data exchange setting. Both data exchange and data integration use the certain answers as the standard semantics of queries over the target (global) schema. However, it should be noted that there are important differences: in data integration, we do not materialize a target (global) schema and hence the source instances are used to compute the certain answers by using the query-rewriting algorithms. In contrast, in our generalization of the data exchange setting, it may not be feasible to couple applications as in data integration because a global schema is contemporarily a source schema for other schemas. This may occur, for instance, in peer-to-peer applications that must share data, yet maintain a high degree of autonomy. Hence, queries over a given schema may have to be answered using its materialized instance alone.

Moreover, each integrity constraint for the global schema $\mathcal{G}$ in Σ_T will be one of the possible cases:

1. A *tuple-generating dependency* (*tgd*) of the form

$$\forall \mathbf{x} \, (\exists \mathbf{y} \, \phi_{\mathcal{G}}(\mathbf{x}, \mathbf{y}) \Longrightarrow \exists \mathbf{z} \, (\psi_{\mathcal{G}}(\mathbf{x}, \mathbf{z})). \tag{4.3}$$

   In the Data Integration setting, we will consider only the class of weakly-full tgd for which query answering is decidable, i.e., when the right-hand side of the implication has no existentially quantified variables, and if each $y_i \in \mathbf{y}$ appears at most once on the left side), for example, in the inclusion constraints (IC) and its more specific case of the foreign-key (FK) constraints.
2. An *equality-generating dependency* (*egd*)

$$\forall \mathbf{x} \, (\phi_{\mathcal{G}}(\mathbf{x}) \Longrightarrow (x_1 \doteq x_2)) \tag{4.4}$$

   where the formulae $\phi_{\mathcal{G}}(\mathbf{x})$ and $\psi_{\mathcal{G}}(\mathbf{x}, \mathbf{y})$ are conjunctions of atomic formulae over $\mathcal{G}$, and $\mathbf{x}_1, \mathbf{x}_2$ are among the variables in $\mathbf{x}$, used, for example, for the key-constraints (PK) and the exclusion constraints (EC).

   Let us consider the most general case of GLAV mapping:

Definition 30 For a general GLAV data integration/exchange system $\mathcal{I} = \langle \mathcal{G}, \mathcal{S}, \mathcal{M} \rangle$ where $\mathcal{G} = (S_T, \Sigma_T)$ is the global and $\mathcal{S} = (S_S, \emptyset)$ is the source database schema and $\mathcal{M}$ a set of tgds which map a view of source database into a view of global database, we define the following graph $G_{\mathcal{I}}$:

For the set of tgds mappings in $\mathcal{M}$, the construction of this graph $G_{\mathcal{I}}$ is the same as that presented in Definition 14, with the mapping edge

$$\mathcal{M}_{SG} = \{TgdsToSOtgd(\mathcal{M})\} : \mathcal{S} \to \mathcal{G}.$$

The integrity-constraint mapping

$$\mathsf{T}_{GA_{\mathsf{T}}} = \{EgdsToSOtgd(\Sigma_G^{\text{egd}}) \wedge TgdsToConSOtgd(\Sigma_G^{\text{tgd}})\} : \mathcal{G} \to \mathcal{A}_{\mathsf{T}},$$

if $\Sigma_G = \Sigma_G^{\text{egd}} \cup \Sigma_G^{\text{tgd}}$ is a nonempty set of integrity constraints over the global schema $\mathcal{G}$, is not explicitly defined as an edge in this graph $G_{\mathcal{I}}$, but it will be introduced in the form of a mapping-operad arrow in the sketch category of this system (in Proposition 15). Let $\mathbf{Sch}(\mathcal{I})$ be the category generated by this graph $G_{\mathcal{I}}$ (in Proposition 16). We can now define a mapping functor from the schema-level sketch category into the instance-level category **DB**:

Proposition 18 *Let* $\mathbf{Sch}(\mathcal{I})$ *be the sketch category of a given data integration system* $\mathcal{I} = \langle \mathcal{G}, \mathcal{S}, \mathcal{M} \rangle$*, generated from a graph* $G_{\mathcal{I}}$ *in Definition* 30 *and Proposition* 16*, with the mapping arrow obtained from edge* $\mathcal{M}_{SG} = \{TgdsToSOtgd(\mathcal{M})\} \in G_{\mathcal{I}}$,

$$\mathbf{M}_{SG} = MakeOperads(\mathcal{M}_{SG}) : \mathcal{S} \to \mathcal{G},$$

with the identity arrows for $\mathcal{S}, \mathcal{G}$ *and* $\mathcal{A}_{\mathsf{T}}$ *and, if* $\Sigma_T = \Sigma_G^{\text{egd}} \cup \Sigma_G^{\text{tgd}}$ *is a nonempty set of integrity constraints over the global schema* $\mathcal{G}$*, with the integrity-constraint arrow* $\mathbf{T}_{GG_{\mathsf{T}}} = MakeOperads(\{\Phi\}) : \mathcal{G} \to \mathcal{A}_{\mathsf{T}}$*, where*:

1. *If* Σ_G^{egd} *and* Σ_G^{tgd} *are both nonempty then* Φ *is equal to*

$$EgdsToSOtgd(\Sigma_G^{\text{egd}}) \wedge TgdsToConSOtgd(\Sigma_G^{\text{tgd}});$$

2. *If* Σ_G^{egd} *is empty and* Σ_G^{tgd} *is nonempty then* Φ *is* $TgdsToConSOtgd(\Sigma_G^{\text{tgd}})$;
3. *If* Σ_G^{egd} *is nonempty and* Σ_G^{tgd} *is empty then* Φ *is equal to* $EgdsToSOtgd(\Sigma_G^{\text{egd}})$.

If for this data integration system $\mathcal{I} = \langle \mathcal{G}, \mathcal{S}, \mathcal{M} \rangle$*, for a given instance* D *of the data source schema* $\mathcal{S}$*, there exists the universal (canonical) instance* $G = can(\mathcal{I}, D)$ *of the global schema* $\mathcal{G}$ *legal w.r.t.* D*, then there exists the mapping-interpretation R-algebra* α *and the functor (categorial Lawvere's model)* $\alpha^* : \mathbf{Sch}(\mathcal{I}) \longrightarrow \mathbf{DB}$ *such that* $\alpha^*(\mathcal{S}) = D$ *and* $\alpha^*(\mathcal{G}) = G = can(\mathcal{I}, D)$.

Proof Directly from Propositions 16, 17, Definition 30 and the properties of **DB**, each morphism in **DB** represents a denotational semantics for a well defined exchange problem between two database instances, so we can define a functor for such an exchange problem. In fact, if there is a canonical solution $G = can(\mathcal{I}, D)$ of the global schema $\mathcal{G}$, that is legal w.r.t. the source data instance $D = \alpha^*(\mathcal{S})$ then, by assuming that $\alpha^*(\mathcal{G}) = can(\mathcal{I}, D)$, all integrity constraints in Σ_G are satisfied and the set of mapping tgds in $\mathcal{M}$ are satisfied (i.e., the SOtgd *TgdsToSOtgd*($\mathcal{M}$) is a true sentence) and hence from Proposition 17 $\alpha^* \in Mod(G_{\mathcal{I}})$ is a functorial model of the data integration system $\mathcal{I}$. Such a functor $\alpha^* : \mathbf{Sch}(\mathcal{I}) \longrightarrow \mathbf{DB}$, between the schema integration level (theory) and the instance-level (which is a model of this theory) is just a particular R-algebra for operads. □

Remark A solution for a data integration/exchange system does not always exist (if there exists a failing finite chase, see [2, 4] for more information). However, if it exists then it is a *canonical universal solution* and in that case there exists also a mapping model-functor of the theorem above. But there always are the functors (which are mapping-interpretations) α^* such that $D = \alpha^*(\mathcal{S})$ which do not satisfy the mapping $\mathbf{M}_{SG}$. Moreover, in Sect. 4.2.4, we will consider one of them for which the global schema will give the same certain answers to the queries as an infinite canonical model. The results for Data Exchange systems in [4] are restricted to a class of tgds (the *weakly-acyclic* tgds) that ensures that the models of the target schema are always finite (otherwise the target schema could not be materialized). So, the problem of dealing with infinite models is not addressed.

In this book, we do not make such a restriction so that we are able to consider the infinite models when **dom** is finite as well: these infinite models are obtained by using the tgds with existentially quantified right side, by introducing an infinite sequence $SK = \{\omega_0, \omega_1, \dots\}$ of Skolem constants in the universe $\mathcal{U} = \mathbf{dom} \cup SK$. Consequently, if a model is infinite it means that we are using all Skolem constants, by introducing them in the fixed ordering $\omega_0, \omega_1, \omega_2, \dots$, in a given set of *cyclic* tgds.

So, this theorem can be abbreviated by:
"given a data mapping graph $\mathcal{I} = \langle \mathcal{G}, \mathcal{S}, \mathcal{M} \rangle$, there is a model-functor α^* : $\mathbf{Sch}(\mathcal{I}) \longrightarrow \mathbf{DB}$ if there exists a universal (canonical) solution for a corresponding data integration/exchange problem".

Obviously, we can have other model-functors $\alpha^* : \mathbf{Sch}(\mathcal{I}) \longrightarrow \mathbf{DB}$ that are different from the canonical one. In what follows, for the GAV system with key constraints, we will consider in more details its canonical (universal) model.

Notice that in the complex integration/exchange systems composed of a number of integrations (sub)systems $\mathcal{I} = \{\langle \mathcal{G}_k, \mathcal{S}_k, \mathcal{M}_k \rangle | k \in N\}$ we can use their representation by a database mapping system graphs in Definition 14.

In the rest of this chapter, we will consider Data Integration applications by using GAV mappings with the (PK) and (IC) integrity constraints over the global schema. Unfortunately, PKs and ICs interact reciprocally so that the (decision) problem of query answering in this setting becomes undecidable [2]. In fact, we have the following table for the complexity of query answering (for the union of conjunctive queries) in GAV systems with sound semantics and with PK and IC constraints:

PKs	ICs	Complexity
no	GEN	PTIME/PSPACE
yes	no	PTIME/NP
yes	FK	PTIME/PSPACE
yes	NKC	PTIME/PSPACE
yes	GEN	undecidable

where GEN stands for general ICs, and NKC for non-key-conflicting ICs (in [2]). The FK is in the class of NKC, so that its complexity is PTIME/PSPACE (PTIME in data complexity, and PSPACE w.r.t. the size of the query), and consequently, in the next section we will use this case of IC.

4.2.3 Query Rewriting in GAV with (Foreign) Key Constraints

This system provided in [2] has the following characteristics of the components of a data integration system $\mathcal{I} = \langle \mathcal{G}, \mathcal{S}, \mathcal{M} \rangle$:

- The *global schema* is expressed in the relational model with the key and foreign key integrity constraints $\Sigma_G = \Sigma_G^{\text{egd}} \cup \Sigma_G^{\text{tgd}}$. We assume that in such a global schema $\mathcal{G} = (S_G, \Sigma_G)$ there is exactly one key constraint for each relation.
  1. (*Key constraints*) Given a relational symbol r in the schema with arity n, a key constraint over r is expressed in the form $key(r) = \{i_1, \dots, i_k\} \subset \{1, 2, \dots, n\}$, the subset of key-attribute indexes. We denote the ordered sequence of indexes in $key(r)$ by $\mathbf{K}$ and define its complement set

     $$\{j_1, \dots, j_{n-k}\} = \{1, 2, \dots, n\} \backslash key(r)$$

     so that $\langle x_{i_1}, \dots, x_{i_k} \rangle$ is a tuple of a subset of key-attributes of r. Such a constraint is satisfied in an instance-database A if for each $\mathbf{t}_1, \mathbf{t}_2 \in \|r\|_A$, $\mathbf{t}_1 \neq \mathbf{t}_2$ iff $\pi_{\mathbf{K}}(\{\mathbf{t}_1\}) \neq \pi_{\mathbf{K}}(\{\mathbf{t}_2\})$, where $\pi_{\mathbf{K}}$ is the projection over attributes with indexes in $\mathbf{K}$.

     Such a key constraint can be represented by the egd

     $$(PK) \quad \forall \mathbf{y} \forall \mathbf{z} \big(r(\mathbf{y}) \wedge r(\mathbf{z}) \wedge \big((y_{1_1} = z_{1_1}) \wedge \dots \wedge (y_{1_k} = z_{1_k}) \big) \\ \Rightarrow \big((y_{j_1} = z_{j_1}) \wedge \dots \wedge (y_{j_{n-k}} = z_{j_{n-k}}) \big) \big) \in \Sigma_G^{\text{egd}},$$

     where $\mathbf{y} = \langle y_1, \dots, y_n \rangle$ and $\mathbf{z} = \langle z_1, \dots, z_n \rangle$.
  2. (*Foreign key constraints*) Given two relations r_1 and r_2, a foreign key constraint is represented by the constraint $\pi_{\mathbf{K}'}(r_1) \subseteq \pi_{\mathbf{K}}(r_2)$ where $\mathbf{K}$ is an ordered sequence of indexes in $key(r_2) = \{i_1, \dots, i_h\} \subseteq \{1, \dots, m\}$, $m = ar(r_2)$ and $h \leq n = ar(r_1)$, with the sequence $\mathbf{x} = \langle x_{i_1}, \dots, x_{i_h} \rangle$ constituting the key of r_2, and $\mathbf{K}'$ is the correspondent sequence of the same attributes in r_1. Such

a constraint is satisfied in an instance-database A if for each tuple $\mathbf{t}_1 \in \|r_1\|_A$ there exists a tuple $\mathbf{t}_2 \in \|r_2\|_A$ such that $\pi_{\mathbf{K}'}(\{\mathbf{t}_1\}) = \pi_{\mathbf{K}}(\{\mathbf{t}_2\})$.

It can be provided by a tgd $\forall \mathbf{x}(\exists \mathbf{y} r_1(\mathbf{x}, \mathbf{y}) \Rightarrow \exists \mathbf{z} r_2(\mathbf{x}, \mathbf{z})) \in \Sigma_G^{\text{tgd}}$.

Let $\{j_1, \ldots, j_{m-h}\} = \{1, \ldots, m\} \backslash \{i_1, \ldots, i_h\}$ be the set of attribute indexes of r_2 that are not key-attributes and let us define the set of terms $S = \{t_{j_k} \mid t_{j_k} = f_{r_2, j_k}(x_{i_1}, \ldots, x_{i_h})$ for $1 \le k \le m - h\}$ and the tuple $\mathbf{t}_1$ of terms (the variables are terms as well) in $\{x_{i_1}, \ldots, x_{i_h}\} \cup S$ ordered by their indexes from 1 to m. Hence, this foreign key constraint can be given by the SOtgd

$$\exists f_{r_2, j_1}, \ldots, f_{r_2, j_{m-h}} \big(\forall x_1, \ldots, x_n \big(r_1(x_1, \ldots, x_n) \Rightarrow r_2(\mathbf{t}_1) \big) \big),$$

i.e., by

$$(FK) \quad \exists \mathbf{f}_1 \big(\forall \mathbf{x}_1 \big(r_1(\mathbf{x}_1) \Rightarrow r_2(\mathbf{t}_1) \big) \big),$$

where $\mathbf{f}_1$ is the tuple of Skolem function symbols that appear in the tuple of terms $\mathbf{t}_1$ and $\mathbf{x}_1$ is the tuple of variables (i.e., of attributes) of r_1.

- The *mapping* $\mathcal{M}$ is defined following the GAV approach: to each relation r of the global schema $\mathcal{G}$ we associate a query $\rho(r)$ over the source schema $\mathcal{S}$. We assume that this query preserves the key constraint of r.
  Thus, a map $\rho(r) \rightsquigarrow r$ is a tgd $\forall \mathbf{x}(\rho(r)(\mathbf{x}) \Rightarrow r(\mathbf{x}))$, i.e., $\mathcal{M}$ is a set of tgds.
- For each relational symbol r of the global schema $\mathcal{G}$, we may compute the relation $\|r\|_G$ by evaluating the query $\rho(r)$ over the source database D. The various relations so obtained define what we call the *retrieved global database ret*($\mathcal{I}$, D). Notice that, since we assume that $\rho(r)$ has been designed so as to resolve all key conflicts regarding r, the retrieved global database satisfies all the key constraints in $\mathcal{G}$.

In our case, with the *integrity constraints* and with *sound mapping*, the semantics of a data integration system $\mathcal{I}$ is specified in terms of a set of *legal* global instance-databases, namely, those databases (they exits iff $\mathcal{I}$ *is consistent* w.r.t. D, i.e., iff *ret*($\mathcal{I}$, D) does not violate any key constraint in $\mathcal{G}$) that are supersets of the *retrieved* global database *ret*($\mathcal{I}$, D).

For a given retrieved global database *ret*($\mathcal{I}$, D), we may inductively construct the canonical database *can*($\mathcal{I}$, D) by starting from *ret*($\mathcal{I}$, D) and repeatedly applying the following rule (FR) [2]:

> if $\langle d_{i_1}, \ldots, d_{i_h}\rangle \in \pi_{\mathbf{K}'}(\|r_1\|_{can(\mathcal{I},D)})$ and $\langle d_{i_1}, \ldots, d_{i_h}\rangle \notin \pi_{\mathbf{K}}(\|r_2\|_{can(\mathcal{I},D)})$, where $\mathbf{K} = \langle i_1, \ldots, i_h\rangle$ is an ordered sequence of indexes in *key*(r_2), $\mathbf{K}'$ is the correspondent sequence of the same attributes in r_1, and the foreign key constraint $\pi_{\mathbf{K}'}(r_1) \subseteq \pi_{\mathbf{K}}(r_2)$ is in $\Sigma_T \subseteq \mathcal{G}$, then insert in $\|r_2\|_{can(\mathcal{I},D)}$ the tuple $\mathbf{t} = \langle d_1, \ldots, d_{ar(r_2)}\rangle$ such that
> - $\pi_{\mathbf{K}}(\{\mathbf{t}\}) = \langle d_{i_1}, \ldots, d_{i_h}\rangle$, and
> - for each i such that $1 \le i \le ar(r_2)$ and i not in $\mathbf{K}$, $\pi_i(\{\mathbf{t}\}) = f_{r_2,i}(d_{i_1}, \ldots, d_{i_h})$, where π_i is the ith projection.

Notice that the this rule (FR) does enforce the satisfaction of the foreign key constraint $\pi_{\mathbf{K}'}(r_1) \subseteq \pi_{\mathbf{K}}(r_2)$ by adding a suitable tuple in r_2: the key of the new tuple is determined by the values in $\pi_{\mathbf{K}'}(r_1)$, and the values of the non-key attributes are provided by means of the Skolem function symbols $f_{r_2,i}$.

Based on the results in [2], $can(\mathcal{I}, D)$ is an appropriate database for answering queries in a data integration system. Notice that the terms involving Skolem functions are never part of *certain answers*. If we consider the case when the output of Skolem functions are marked null-values (disjoint from **dom**), then we can consider the lifted queries q which use the *Val*(_) predicate in order to eliminate the tuples with a Skolem values in $can(\mathcal{I}, D)$. The built-in *new unary predicate Val*(_) satisfies that $Val(c)$ is true iff $c \in \mathbf{dom}$.

Consequently, the database mapping graph at the logical sketch's level of this GAV data integration system can be represented by the graph composed of two mapping-operad arrows:

(a) $\mathbf{M}_{S\mathcal{G}_T} = MakeOperads(TgdsToSOtgd(\mathcal{M})) : \mathcal{S} \longrightarrow \mathcal{G}_T$, where $\mathcal{G}_T = (S_G, \emptyset)$ will be used for the retrieved (from the source database $\mathcal{S}$ only) data and, consequently, has the same global database schema as $\mathcal{G}$ but with an empty set of integrity constraints, and the new inter-schema mapping-operad;

(b) $\mathbf{M}_{\mathcal{G}_T G} = \{q_i \mid q_i = ((\_)_1(\mathbf{x}_i) \Rightarrow (\_)(\mathbf{x}_i)) \in O(r_i, r_i')$ for each $r_i \in \mathcal{G}_T$ and the same relation $r_i' \in \mathcal{G}\} \cup \{1_{r_\emptyset}\} : \mathcal{G}_T \longrightarrow \mathcal{G}$ that represents the inclusion between retrieved global database $ret(\mathcal{I}, D)$ and the complete canonical global database $can(\mathcal{I}, D)$ (obtained by satisfaction of the integrity constraints Σ_G, as explained above for foreign key constraints).

The global schema integrity-constraints mapping operad

$$\mathbf{T}_{GA_\top} = MakeOperads\big(\big\{EgdsToSOtgd\big(\Sigma_G^{\mathrm{egd}}\big)$$
$$\wedge\, TgdsToCanSOtgd\big(\Sigma_G^{\mathrm{tgd}}\big)\big\}\big) : \mathcal{G} \to \mathcal{A}_\top$$

is satisfied because we assumed that all primary key-integrity constraints for the global schema are satisfied by the retrieved database $ret(\mathcal{I}, D)$ and the foreign key constraints are satisfied by insertion of new tuples in the relations (as explained above).

Based on Example 12 in Sect. 2.4.1, for each egd (PK) (in point 1 above) for a relation r in $\mathcal{G}$, we will obtain the $n-k$ operators $\{q_1, \ldots, q_{n-k}\} \subseteq \mathbf{T}_{GA_\top}$ such that for any $1 \le i \le n-k$, $q_i \in O(r, r, r_\top)$, with

$$q_i = \big((\_)_1(\mathbf{y}) \wedge (\_)_2(\mathbf{z}) \wedge \big((y_{1_1} = z_{1_1}) \wedge \cdots \wedge (y_{1_k} = z_{1_k})\big)$$
$$\wedge \big((y_{j_1} \neq z_{l_1}) \vee \cdots \vee (y_{j_{n-k}} \neq z_{l_{n-k}})\big)\big) \Rightarrow (\_)(\overline{0}, \overline{1}).$$

For a given set of k foreign key constraints, represented by tgds (FK)

$$\exists \mathbf{f}_i \big(\forall \mathbf{x}_i \big(r_{i,1}(\mathbf{x}_i) \Rightarrow r_{i,2}(\mathbf{t}_i)\big)\big),$$

$i = 1, \ldots, k$, where $\mathbf{t}_i$ is a tuple of terms with variables in $\mathbf{x}_i$ and functional symbols in $\mathbf{f}_i$, we obtain $q_i \in O(r_{i,1}, r_{i,2}, r_\top)$ with $q_i = ((\_)_1(\mathbf{x}_i) \wedge \neg(\_)_2(\mathbf{t}_i)) \Rightarrow (\_)(\overline{0}, \overline{1})$.

The sketch category derived by this data integration system is denoted by $\mathbf{Sch}(\mathcal{I})$ in the left part of the following diagram:

$$
\begin{array}{ccccc}
\mathcal{G} & \xrightarrow{\mathbf{T}_{GA_{\top}}} & \mathcal{A}_{\top} & \quad & \alpha^*(\mathcal{G}) = can(\mathcal{I}, D) \xrightarrow{f_{\Sigma G}} A_{\top} \\
\uparrow \mathbf{M}_{G_T G} & & & & \uparrow \alpha^*(\mathbf{M}_{G_T G}) = f_{in} \\
\mathcal{G}_T & & \alpha^* \mapsto & & \alpha^*(\mathcal{G}_T) = ret(\mathcal{I}, D) \\
\uparrow \mathbf{M}_{SG_T} & & & & \uparrow \alpha^*(\mathbf{M}_{SG_T}) = f_M \\
\mathcal{S} & & & & \alpha^*(\mathcal{S}) = D \\
 & & & & \\
\text{in } \mathbf{Sch}(\mathcal{I}) & & & & \text{in } \mathbf{DB}
\end{array}
$$

where $A_{\top} = \alpha^*(\mathcal{A}_{\top}) = \{\alpha(r_{\top})\}$ is the "equality" database with the built-in relation $\alpha(r_{\top}) = R_{=}$.

Corollary 15 *If there is the canonical model of a GAV data integration system* $\mathcal{I}$ *with FK constraints, described in this section and expressed by the sketch category* $\mathbf{Sch}(\mathcal{I})$, *then its functorial model* $\alpha^* : \mathbf{Sch}(\mathcal{I}) \longrightarrow \mathbf{DB}$ *(where the mapping-interpretation* α *in Definition* 11 *is an extension of Tarski's interpretation* I_T*) is provided by the corresponding arrows*:

$$
\begin{aligned}
f_M &= \alpha^*(\mathbf{M}_{SG_T}) : D \longrightarrow ret(\mathcal{I}, D), \\
f_{in} &= \alpha^*(\mathbf{M}_{G_T G}) : ret(\mathcal{I}, \mathcal{D}) \longrightarrow can(\mathcal{I}, \mathcal{D}), \\
f_{\Sigma G} &= \alpha^*(\mathbf{T}_{GA_{\top}}) : can(\mathcal{I}, \mathcal{D}) \longrightarrow A_{\top},
\end{aligned}
$$

where $\alpha^*(\mathcal{S}) = D$ *is the instance of the source database* $\mathcal{S}$, $G_T = \alpha^*(\mathcal{G}_T) = ret(\mathcal{I}, D)$ *is the retrieved global database,* $G = \alpha^*(\mathcal{G}) = can(\mathcal{I}, D)$ *is the universal (canonical) instance of the global schema with the integrity constraints, and*

- $f_M = \{q_{\bot}\} \cup \{f : \alpha(r_1) \times \cdots \times \alpha(r_l) \to \alpha(r) \mid where\ (r_1, \ldots, r_l)$ *is the set of all relational symbols in the query* $\rho(r)$, $(\rho(r) \rightsquigarrow r) \in \mathcal{M}\}$, *where* $\alpha(r_i) = \|r_i\|_D$, $1 \le i \le l$, *and* $\alpha(r) = \|r\|_{ret(\mathcal{I},D)}$.

  For each map $(\rho(r) \rightsquigarrow r) \in \mathcal{M}$ *(i.e., a tgd* $\forall \mathbf{x}(\rho(r)(\mathbf{x}) \Rightarrow r(\mathbf{x}))$*) and its operad* $q = (e \Rightarrow (\_)(\mathbf{x})) \in O(r_1, \ldots, r_l, r)$, *where e is an operad's expression obtained from the query formula* $\rho(r)(\mathbf{x})$, *we obtain the function*

$$
f = \alpha(q) : \alpha(r_1) \times \cdots \times \alpha(r_l) \to \alpha(r).
$$

- $f_{in} = \{in_r : \|r\|_{ret(\mathcal{I},D)} \to \|r\|_{can(\mathcal{I},D)} \mid r \in S_T\}$, *where each function* in_r *is an injection. Thus, we obtain the monomorphism* $f_{in} : ret(\mathcal{I}, D) \hookrightarrow can(\mathcal{I}, D)$.
- $f_{\Sigma G} = \{g_1, \ldots, g_n, q_{\bot}\}$ *where each* g_i *is:*

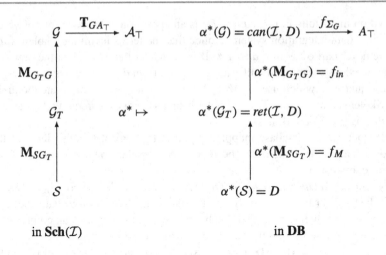

1. *(PK) function for* $1 \leq i \leq n-k$, $g_i = \alpha(q_i) : \alpha(r_i) \times \alpha(r_i) \to \alpha(r_\top)$ *is a constant function (where* $q_i = ((\_)_1(\mathbf{y}) \wedge (\_)_2(\mathbf{z}) \wedge ((y_{1_1} = z_{1_1}) \wedge \cdots \wedge (y_{1_k} = z_{1_k})) \wedge ((y_{j_1} \neq z_{l_1}) \vee \cdots \vee (y_{j_{n-k}} \neq z_{l_{n-k}}))) \Rightarrow (\_)(\overline{0}, \overline{1}) \in O(r_i, r_i))$, *that is, for every tuple* $\langle \mathbf{d}_1, \mathbf{d}_2 \rangle \in \alpha(r_i) \times \alpha(r_i) = \|r_i\|_{can(\mathcal{I},D)} \times \|r_i\|_{can(\mathcal{I},D)}$, $g_i(\langle \mathbf{d}_1, \mathbf{d}_2 \rangle) = \langle \rangle$.
2. *(FK) function when* $n-k+1 \leq i \leq n$, $g_i = \alpha(q_i) : \alpha(r_{i,1}) \times \alpha(r_{i,2}) \to \alpha(r_\top)$ *(where* $q_i = ((\_)_1(\mathbf{x}_i) \wedge (\_)_2(\mathbf{t}_i)) \Rightarrow (\_)(\overline{0}, \overline{1}) \in O(r_{i,1}, r_{i,2}))$ *is a constant function, that is, for every tuple* $\langle \mathbf{d}_1, \mathbf{d}_2 \rangle \in \alpha(r_{i,1}) \times \alpha(r_{i,2}) = \|r_{i,1}\|_{can(\mathcal{I},D)} \times \|r_{i,2}\|_{can(\mathcal{I},D)}$, $g_i(\langle \mathbf{d}_1, \mathbf{d}_2 \rangle) = \langle \rangle$.

Proof Directly from definitions presented in this section and by Propositions 16 and 17. From the definition of f_{in}, we obtain that for each relational symbol $r \in S_G$ (which is a relation both in $\mathcal{G}$ and $\mathcal{G}_T$) we obtain that $\|r\|_{ret(\mathcal{I},D)} \subseteq \|r\|_{can(\mathcal{I},D)}$, that is, the $ret(\mathcal{I}, D)$ is a subdatabase of the $can(\mathcal{I}, D)$, that is, we obtain the monomorphism $f_{in} : ret(\mathcal{I}, D) \hookrightarrow can(\mathcal{I}, D)$ and hence $\widetilde{f_{in}} = T\,ret(\mathcal{I}, D)$. □

Remark If we were to interpret the foreign key integrity constraints in Σ^{tgd} as the standard inter-schema mapping based on tgds, that is, by the mapping operad $\mathbf{M}_{GG} = MakeOperads(TgdsToSOtgd(\Sigma_G^{\mathrm{tgd}})) : \mathcal{G} \to \mathcal{G}$ then we would obtain the morphisms in **DB**,

$$f_{\Sigma_G^{\mathrm{tgd}}} = \{g_1, \ldots, g_k, q_\perp\} \cup \{id_r : \alpha(r) \to \alpha(r) \mid r \in S_T\} : can(\mathcal{I}, D) \Rightarrow can(\mathcal{I}, D),$$

where each function $g_i = \alpha((\_)_1(\mathbf{x}_i)) \Rightarrow (\_)(\mathbf{t}_i) : \alpha(r_{i,1}) \to \alpha(r_{i,2})$, for $1 \leq i \leq k$, is defined by: for every tuple $\langle d_1, \ldots, d_n \rangle \in \alpha(r_{i,1}) = \|r_{i,1}\|_{ret(\mathcal{I},D)}$, $g_i(\langle d_1, \ldots, d_n \rangle) = \mathbf{t}$ such that $\pi_{\mathbf{K}}(\{\mathbf{t}\}) = \langle d_{i_1}, \ldots, d_{i_h} \rangle$ and for each $i \notin \mathbf{K} = \langle i_1, \ldots, i_h \rangle$ (ordered sequence of indexes in $key(r_{i,2})$) such that $1 \leq i \leq ar(r_{i,2})$, $\pi_i(\{\mathbf{t}\}) = I_T(f_{r_2,i})(v_{i_1}, \ldots, v_{i_h})$.

That is, the endomorphism $f_{\Sigma_G^{\mathrm{tgd}}} : \mathcal{G} \to \mathcal{G}$ would implement the process defined previously by the rules in (FR). But we have explained that we avoid representing the tgd's integrity constraint as an endomorphism (i.e., a morphism with the same source and target object, $\mathcal{G}$ in this case) with a nonempty information flux, in order to avoid the undesirable side-effects produced by their compositions with the real (desiderated) inter-schema mappings. We avoided it by representing the integrity constraints by "logic" mappings with target schema $\mathcal{A}_\top$, which have the empty information flux (as can be seen for f_{Σ_G} in this Corollary 15).

4.2.3.1 Query Rewriting Coalgebra Semantics

The computation of a query q over a global schema $\mathcal{G}$ requires the building of a canonical database $can(\mathcal{I}, D)$, which is generally infinite. In order to overcome this problem, a *query rewriting algorithm* [2] consists of two separate phases:

1. Instead of referring explicitly to the canonical database for query answering, this algorithm transforms the original lifted query q into a new query $exp_{\mathcal{G}}(q)$ over a global schema, called the *expansion of* q *w.r.t.* Σ_G^{tgd}, such that the answer to

Fig. 4.2 Query answering process

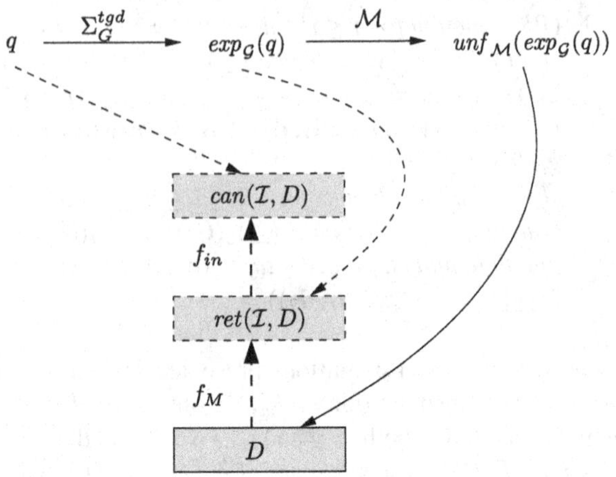

$exp_{\mathcal{G}}(q)$ over the retrieved global database is equal to the answer to q over the canonical database.

2. In order to avoid the building of the retrieved global database, the query does not evaluate $exp_{\mathcal{G}}(q)$ over the retrieved global database. Instead, this algorithm unfolds $exp_{\mathcal{G}}(q)$ to a new query, called $unf_{\mathcal{M}}(exp_{\mathcal{G}}(q))$, over the source relations on the basis of $\mathcal{M}$, and then uses the unfolded query $unf_{\mathcal{M}}(exp_{\mathcal{G}}(q))$ to access the sources.

Figure 4.2 shows the basic idea of this approach (taken from [2]). In order to obtain the *certain answers* $\|q\|_{\mathcal{I}}$, the user lifted query q could in principle be evaluated (dashed arrow) over the (possibly infinite) canonical database $can(\mathcal{I}, D)$, which is generated from the retrieved global database $ret(\mathcal{I}, D)$. In turn, $ret(\mathcal{I}, D)$ can be obtained from the source database D by evaluating the queries of the mapping. This query answering process instead expands the query according to the constraints Σ_G^{tgd} in $\mathcal{G}$, than unfolds it according to $\mathcal{M}$, and then evaluates it on the source database.

The architecture of such a query-rewriting GAV system [3], based on the Database Federation tool for DB2 (IBM), is presented in Fig. 4.3.

Let us show how the symbolic diagram in Fig. 4.2 can be effectively represented by the commutative diagrams in **DB**, corresponding to the homomorphisms between T-coalgebras that represent the equivalent queries over these three instance-databases: each query in the **DB** category is represented by an arrow and *can be composed* with the arrows that semantically denote mappings and integrity constraints.

Theorem 8 *Let $\mathcal{I} = \langle \mathcal{G}, \mathcal{S}, \mathcal{M} \rangle$ be a GAV data integration system with a canonical model provided by an R-algebra functor $\alpha^* : \mathbf{Sch}(\mathcal{I}) \longrightarrow \mathbf{DB}$ (Corollary 15), with the source database $\alpha^*(\mathcal{S}) = D$, the retrieved global database $\alpha(\mathcal{G}_T) = ret(\mathcal{I}, D)$ for $\mathcal{I}$ w.r.t. D, and the universal (canonical) database $\alpha(\mathcal{G}) = can(\mathcal{I}, D)$ for $\mathcal{I}$ w.r.t. D.*

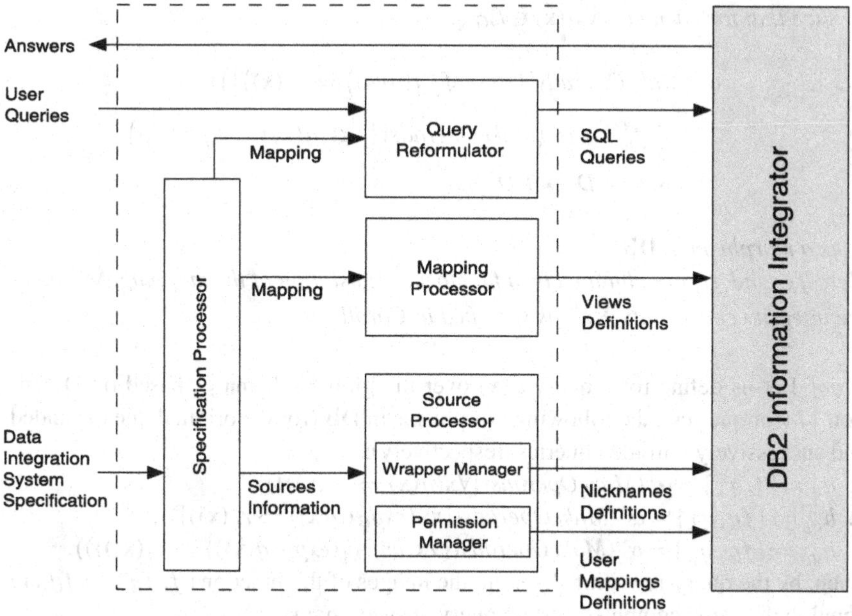

Fig. 4.3 Process architecture

Then, the denotational semantics for the query rewriting algorithms $exp_{\mathcal{G}}(q)$ and $unf_{\mathcal{M}}(q)$, for a query expansion and query unfolding, respectively, are given by two functions on the set $\mathcal{L}_{\mathcal{G},\mathbf{x}}$ of queries with variables in **x** *over the global schema $\mathcal{G}$:*

1.

$$\alpha^*\big(MakeOperads\big(\{\forall \mathbf{x}\big(exp_{\mathcal{G}}(\_) \Rightarrow r_q(\mathbf{x})\big)\}\big)\big) = Tf_{in}^{OP} \circ \alpha^*\big(MakeOperads\big(\{\forall \mathbf{x}\big((\_) \Rightarrow r_q(\mathbf{x})\big)\}\big)\big) \circ f_{in} : \mathcal{L}_{\mathcal{G},\mathbf{x}} \to \mathbf{DB}\big(ret(\mathcal{I}, D), Tret(\mathcal{I}, D)\big),$$

such that for each query $q(\mathbf{x}) \in \mathcal{L}_{\mathcal{G},\mathbf{x}}$,

$$\alpha^*\big(MakeOperads\big(\{\forall \mathbf{x}\big(exp_{\mathcal{G}}(q(\mathbf{x})) \Rightarrow r_q(\mathbf{x})\big)\}\big)\big) = Tf_{in}^{OP} \circ \alpha^*\big(MakeOperads\big(\{\forall \mathbf{x}\big((q(\mathbf{x})) \Rightarrow r_q(\mathbf{x})\big)\}\big)\big) \circ f_{in} : ret(\mathcal{I}, D) \to Tret(\mathcal{I}, D)$$

is a morphism in **DB**.

2.

$$\alpha^*\big(MakeOperads\big(\{\forall \mathbf{x}\big(unf_{\mathcal{M}}(\_) \Rightarrow r_q(\mathbf{x})\big)\}\big)\big) = Tf_{M}^{OP} \circ \alpha^*\big(MakeOperads\big(\{\forall \mathbf{x}\big((\_) \Rightarrow r_q(\mathbf{x})\big)\}\big)\big) \circ f_M : \mathcal{L}_{\mathcal{G},\mathbf{x}} \to \mathbf{DB}(D, TD),$$

such that for each query $q(\mathbf{x}) \in \mathcal{L}_{\mathcal{G},\mathbf{x}}$,

$$\alpha^*\big(MakeOperads(\{\forall\mathbf{x}(unf_{\mathcal{M}}(q(\mathbf{x})) \Rightarrow r_q(\mathbf{x}))\})\big)$$
$$= Tf_{in}^{OP} \circ \alpha^*\big(MakeOperads(\{\forall\mathbf{x}((q(\mathbf{x})) \Rightarrow r_q(\mathbf{x}))\})\big)$$
$$\circ f_{in} : D \to TD$$

is a morphism in **DB**.
Here f_M *and* f_{in} *are obtained by a functorial translation of the mapping* $\mathcal{M}$ *and of the integrity constraints* Σ_G^{tgd} *as specified in Corollary* 15.

Proof Let us define for a query $q(\mathbf{x})$ over the global schema $\mathcal{G}$, based on Definition 17 (for queries), the following morphisms in **DB** (for the original, the expanded and successively unfolded queries, respectively):
1. $h_q = \{f, q_\perp\} = \alpha^*(MakeOperads(\{\forall\mathbf{x}(q(\mathbf{x}) \Rightarrow r_q(\mathbf{x}))\}))$,
2. $h_{qE} = \{f_E, q_\perp\} = \alpha^*(MakeOperads(\{\forall\mathbf{x}(exp_{\mathcal{G}}(q(\mathbf{x})) \Rightarrow r_q(\mathbf{x}))\}))$,
3. $h_{qU} = \{f_U, q_\perp\} = \alpha^*(MakeOperads(\{\forall\mathbf{x}(unf_{\mathcal{M}}(exp_{\mathcal{G}}(q(\mathbf{x}))) \Rightarrow r_q(\mathbf{x}))\}))$.

Then, by the query-rewriting theorem, the images of the functions f, f_E and f_U are equal to the same computed certain query answer, that is,

$$im(f) = im(f_E) = im(f_U) = \|r_q(\mathbf{x})\|.$$

Hence, the following diagrams in **DB** (on the right) based on the composition of T-coalgebra homomorphisms

$$f_M : (D, h_{qU}) \longrightarrow \big(ret(\mathcal{I}, D), h_{qE}\big) \quad \text{and}$$
$$f_{in} : \big(ret(\mathcal{I}, D), h_{qE}\big) \hookrightarrow \big(can(I, \mathcal{D}), h_q\big)$$

commute:

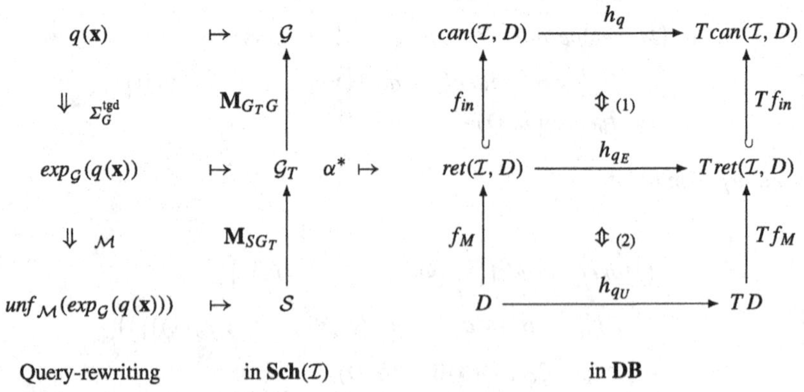

It is easy to verify the first two facts, from the equivalent query-rewriting,

$$\widetilde{h_q} = \widetilde{h_{q_E}} = \widetilde{h_{q_U}} = T\{\|r_q(\mathbf{x})\|, \bot\} \subseteq \widetilde{f_{in}} = Tcan(\mathcal{I}, D),$$

that is, $h_q \approx h_{q_E}$ (corresponding to the equivalence $\Updownarrow_{(1)}$ in diagram (1) above) and $h_{q_E} \approx h_{q_U}$ (corresponding to the equivalence $\Updownarrow_{(2)}$ in diagram (2) above).

Consequently, from the commutative diagrams above, $h_{q_E} = Tf_{in}^{OP} \circ h_q \circ f_{in}$, corresponding to point 1 of this proposition, and $h_{q_U} = Tf_M^{OP} \circ h_{q_E} \circ f_M$, corresponding to point 2 of this proposition. □

Note that for every $r \in \mathcal{G}$, $\|r\|_{ret(\mathcal{I},D)} \subseteq \|r\|_{can(\mathcal{I},D)}$. However, the only new information transmitted by the monomorphism f_{in} is that obtained by Skolem's functions, which is not considered by certain-answers to queries, and this fact explains why for each query $q(\mathbf{x})$ over $can(\mathcal{I}, D)$ we have the equivalent to it query $exp_{\mathcal{G}}(q(\mathbf{x}))$ with the *same* certain answer $\|r_q\|$.

From the definition of the weak observational equivalence and its weak power-view endofunctor T_w in Definition 29 and Proposition 12, we obtain $Tret(\mathcal{I}, D) = T_w(can(\mathcal{I}, D)) \preceq can(\mathcal{I}, D)$ so that

$$\widetilde{h_q} = T\{\|r_q(\mathbf{x})\|_{can(\mathcal{I},D)}, \bot\} \subseteq T_w\big(can(\mathcal{I}, D)\big) = Tret(\mathcal{I}, D)$$

(from the fact that a query $q(\mathbf{x})$ is lifted and hence $\|r_q(\mathbf{x})\|_{can(\mathcal{I},D)}$ has no Skolem constants) and $\widetilde{h_{q_E}} \subseteq Tret(\mathcal{I}, D)$: it explains why it is possible to have $\widetilde{h_{q_E}} = \widetilde{h_q}$.

Let us show that the relationships between an original and its rewritten queries over retrieved and source databases can be represented by the morphisms in **DB** category:

Example 26 Let us consider the two ordinary morphisms (1-cells) in **DB**, $f : A \longrightarrow B$, $g : C \longrightarrow D$ in Example 24, in the case of the commutative diagram (1) provided in the proof of Theorem 8, that is, when $A = can(\mathcal{I}, D)$, $B = Tcan(\mathcal{I}, D)$, $f = h_q$ and $g = h_{q_E}$ with C replaced by D and D replaced by TD.

From the fact that $h_q \approx h_{q_E}$ in the proof of Theorem 8, it is equivalently represented by $h_q \preceq h_{q_E}$ and $h_{q_E} \preceq h_q$ (from Theorem 6). However, we have shown in Theorem 6 that $h_q \preceq h_{q_E}$ (and similarly for other case) in **DB** can be represented by an ordinary 1-cell (morphisms) $\beta_1 : \widetilde{h_q} \hookrightarrow \widetilde{h_{q_E}}$ and hence (from Example 24) the following diagrams commute:

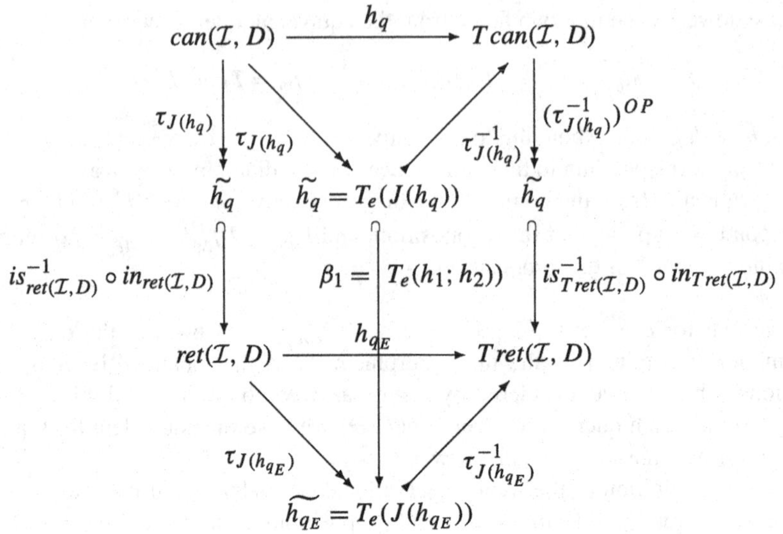

where $\widetilde{h_q} = \widetilde{h_{qE}} = T\{\|r_q(\mathbf{x})\|, \perp\}$, $h_1 = is^{-1}_{ret(\mathcal{I},D)} \circ in_{ret(\mathcal{I},D)} \circ \tau_{J(h_q)}$, $h_2 = is^{-1}_{Tret(\mathcal{I},D)} \circ in_{Tret(\mathcal{I},D)} \circ (\tau^{-1}_{J(h_q)})^{OP}$ and $\beta_1 = T_e(h_1; h_2) : \widetilde{h_q} \hookrightarrow \widetilde{h_{qE}}$, that is, $\beta_1 = \{id_{\|r_q(\mathbf{x})\|}, q_\perp\} : \{\|r_q(\mathbf{x})\|, \perp\} \hookrightarrow \{\|r_q(\mathbf{x})\|, \perp\}$.

Analogously, we obtain $\beta_2 = \{id_{\|r_q(\mathbf{x})\|}, q_\perp\} : \widetilde{h_{qE}} \hookrightarrow \widetilde{h_q}$, so that $\beta_2 = \beta_1$ are the identity morphism for the instance database composed of the single computed query-relation $\|r_q(\mathbf{x})\|$.

Let us show that the morphism $\beta_1 : \widetilde{h_q} \to \widetilde{h_{qE}}$ in the example above can be directly obtained from the commutative diagram (1) in **DB**, presented in the proof of Theorem 8:

Corollary 16 *It holds that* $\beta_1 = (T_e(f_{in}; Tf_{in}))^{OP} : \widetilde{h_q} \to \widetilde{h_{qE}}$.

Proof From the commutative diagram (1) in **DB** in the proof of Theorem 8, and from Theorem 4, we have that $T_e(f_{in}; Tf_{in}) = in^{OP}_{\widetilde{h_q}} \circ T(Tf_{in} \circ h_{qE}) \circ in_{\widetilde{h_{qE}}} : \widetilde{h_{qE}} \to \widetilde{h_q}$. Thus, its dual is equal to

$$
\begin{aligned}
T_e(f_{in}; Tf_{in})^{OP} &= \left(in^{OP}_{\widetilde{h_q}} \circ T(Tf_{in} \circ h_{qE}) \circ in_{\widetilde{h_{qE}}}\right)^{OP} \\
&= in^{OP}_{\widetilde{h_{qE}}} \circ T(Tf_{in} \circ h_{qE})^{OP} \circ in_{\widetilde{h_q}} \\
&= in^{OP}_{\widetilde{h_{qE}}} \circ T\left(h^{OP}_{qE} \circ Tf^{OP}_{in}\right) \circ in_{\widetilde{h_q}} \\
&= in^{OP}_{\widetilde{h_{qE}}} \circ Th^{OP}_{qE} \circ Tf^{OP}_{in} \circ in_{\widetilde{h_q}} : \widetilde{h_q} \to \widetilde{h_{qE}}.
\end{aligned}
$$

Consequently, its information flux is $T_e(\widetilde{f_{in}; Tf_{in}})^{OP} = \widetilde{in^{OP}_{\widetilde{h_{qE}}}} \cap \widetilde{Th^{OP}_{qE}} \cap \widetilde{Tf^{OP}_{in}} \cap \widetilde{in_{\widetilde{h_q}}}$ = (from the fact that $\widetilde{in_{\widetilde{h_q}}} = \widetilde{in^{OP}_{\widetilde{h_{qE}}}} = \widetilde{Th^{OP}_{qE}} = \widetilde{h_{qE}}$ and from Corollary 15 $\widetilde{h_{qE}} \subseteq \widetilde{f_{in}} = \widetilde{Tf^{OP}_{in}}) = \widetilde{h_{qE}}$ = (from Example 26) $= \widetilde{\beta_1}$ and hence $T_e(f_{in}; Tf_{in})^{OP} = \beta_1$. □

Let us now show how the query-rewriting *algorithms* can be represented in **DB** category. First define the set of all queries over canonical model of the global schema, which is represented in **DB** by the set of query-morphisms $S_{can(\mathcal{I},D)} = \{h_q \mid h_q = \{f, q_\perp\} \in \mathbf{DB}(can(\mathcal{I}, D), Tcan(\mathcal{I}, D))$ such that $\widetilde{h_q} \in T_w(can(\mathcal{I}, D))\}$. Then the query-expansion algorithm can be represented by the function $Exp_Q : S_{can(\mathcal{I},D)} \to \mathbf{DB}(ret(\mathcal{I}, D), Tret(\mathcal{I}, D))$ such that for each query-morphism over global database $h_q = \{f, q_\perp\} \in S_{can(\mathcal{I},D)}$ we obtain an equivalent query-morphism over retrieved database $Exp_Q(h_q) = \{f_E, q_\perp\} \in \mathbf{DB}(ret(\mathcal{I}, D), Tret(\mathcal{I}, D))$, with $im(f_e) = im(f)$ and hence $\widetilde{h_q} = \widetilde{Exp_Q(h_q)} = \widetilde{h_{qE}}$.

4.2.4 Fixpoint Operator for Finite Canonical Solution

The database instance $can(\mathcal{I}, \mathcal{D})$ which is a model of the global schema $\mathcal{G} = (S_G, \Sigma_G)$ can be an infinite database (see Example 27 bellow) and hence impossible to materialize for the real applications. Thus, in this subsection, we introduce a new approach to the canonical instance-database, closer to the data exchange approach [4]. It is not restricted to the existence of query-rewriting algorithms and hence can be used in order to define a Coherent Closed World Assumption for data integration systems also in the absence of query-rewriting algorithms [12].

The construction of the finite canonical instance-database for $\mathcal{G}$ that does not satisfy all the integrity constraints of the logical theory for a data integration system $\mathcal{I} = \langle \mathcal{G}, \mathcal{S}, \mathcal{M} \rangle$ is similar to the construction of the canonical model $can(\mathcal{I}, \mathcal{D})$ described in [2]. The *difference* lies in the fact that in the construction of this revisited canonical database for a global schema (which is not a model of the global schema), denoted by $can_F(\mathcal{I}, \mathcal{D})$, the fresh *marked null values* (from the set $SK = \{\omega_0, \omega_1, \dots\}$ of Skolem constants) are used instead of terms involving Skolem's functions in the SOtgds obtained from the foreign key constraints in Σ_G^{tgd}, following the idea of construction of the restricted chase of a database described in [8]. Thus, for a given universe $\mathcal{U} = \mathbf{dom} \cup SK$, we permit to use these Skolem constants for primary keys as well, differently from standard relational databases where we have only one NULL value that cannot be used for primary key attributes of a relation. Here it is possible just because we have the marked null values ω_i, so that we can use them also for the primary-key attributes because $\omega_i \neq \omega_j$ for all $i \neq j$.

Analogously to the method applied in Sect. 4.2.3, we will introduce a new auxiliary schema $\mathcal{G}_F = (S_G, \emptyset)$ that has the same relations of the global schema $\mathcal{G}$ but without integrity constraints, in order that the instance-database $can_F(\mathcal{I}, D)$ (which is not a model of $\mathcal{G}$ because does not satisfy all constraints in Σ_G) is a model of $\mathcal{G}_F$.

Another motivation for concentrating on canonical solutions is a view [15] that many logic programs are appropriately thought of as having two components, an *intensional* database (IDB) that represents the reasoning component, and the *extensional* database (EDB) that represents a collection of facts. Over the course of time, we can "apply" the same IDB to many quite different EDBs. In this context, it makes sense to think of the IDB as implicitly defining a transformation from an EDB to a set of derived facts: we would like the set of derived facts to be the canonical model.

Now we will construct inductively the revisited canonical database model $can_F(\mathcal{I}, \mathcal{D})$ over the domain $\mathcal{U} = \mathbf{dom} \cup SK$ by starting from $ret(\mathcal{I}, \mathcal{D})$ and repeatedly applying the following rule (analog to the rule (FR) in Sect. 4.2.3):

(FRF) if $\langle d_{i_1}, \ldots, d_{i_h}\rangle \in \pi_{\mathbf{K}'}(\|r_1\|_{can_F(\mathcal{I},D)})$, with $d_{i_m} \in \mathbf{dom}$ for each $1 \leq m \leq h$, and $\langle d_{i_1}, \ldots, d_{i_h}\rangle \notin \pi_{\mathbf{K}}(\|r_2\|_{can_F(\mathcal{I},D)})$, where $\mathbf{K} = \langle i_1, \ldots, i_h\rangle$ is an ordered sequence of indexes in $key(r_2)$, and the foreign key constraint $\pi_{\mathbf{K}'}(r_1) \subseteq \pi_{\mathbf{K}}(r_2)$ is in $\Sigma_T \subseteq \mathcal{G}$, then insert in $\|r_2\|_{can_F(\mathcal{I},D)}$ the tuple $\mathbf{t} = \langle d_1, \ldots, d_{ar(r_2)}\rangle$ such that

- $\pi_{\mathbf{K}}(\{\mathbf{t}\}) = \langle d_{i_1}, \ldots, d_{i_h}\rangle$, and
- for each i such that $1 \leq i \leq ar(r_2)$, and i not in $\mathbf{K}$, $\pi_i(\{\mathbf{t}\}) = \omega_k$, where π_i is the ith projection and ω_k is a fresh marked null value.

Note that the above rule does enforce the satisfaction of the foreign key constraint $\pi_{\mathbf{K}'}(r_1) \subseteq \pi_{\mathbf{K}}(r_2)$, by adding a suitable tuple in r_2: the key of the new tuple is determined by the values in $\pi_{\mathbf{K}'}(r_1)$, and the values of the non-key attributes are formed by means of the fresh marked values ω_k by ordering $k = 0, 1, \ldots$ during the application of the rule above.

Let us denote by Σ_{FRF} the set of all rules (FRF) above, obtained from foreign key constraints in Σ_G^{tgd} of the global schema $\mathcal{G}$. This set of rules Σ_{FRF} defines the "immediate consequence" monotonic operator T_B defined by:

$$T_B(I) = I \cup \big\{ A \mid A \in B_{\mathcal{G}}, A \leftarrow A_1 \wedge \cdots \wedge A_n \text{ is a ground instance}$$
$$\text{of a rule in } \Sigma_{\mathrm{FRF}} \text{ and } \{A_1, \ldots, A_n\} \in I \big\},$$

where at the beginning I is the set of ground atoms $r(\mathbf{d})$ for each tuple $\mathbf{d} \in \|r\|_{ret(\mathcal{I},\mathcal{D})}$, and $B_{\mathcal{G}}$ is a Herbrand base for the global schema $\mathcal{G}_F$. Thus, $can_F(\mathcal{I}, \mathcal{D})$ is a least fixpoint of this immediate consequence operator.

Hence, we define the model α of this data integration system $\mathcal{I}$, with finite source database $\alpha^*(\mathcal{S}) = D$, finite canonical instance-database $\alpha^*(\mathcal{G}_F) = can_F(\mathcal{I}, D)$ and (possibly infinite) $\alpha^*(\mathcal{G}) = can(\mathcal{I}, D)$, with the schema mappings similar to those defined in Sect. 4.2.3:

1. $\mathbf{M}_{\mathcal{S}\mathcal{G}_F} = MakeOperads(TgdsToSOtgd(\mathcal{M})) : \mathcal{S} \longrightarrow \mathcal{G}_F$,
2. $\mathbf{M}_{\mathcal{G}_F\mathcal{G}} = \{q_i \mid q_i = ((\_)_1(\mathbf{x}_i) \Rightarrow (\_)(\mathbf{x}_i)) \in O(r_i, r_i') \text{ for each } r_i \in \mathcal{G}_F \text{ and the same relation } r_i' \in \mathcal{G}\} \cup \{1_{r_\emptyset}\} : \mathcal{G}_F \longrightarrow \mathcal{G}$ that represents the inclusion between finite instance-database $can_F(\mathcal{I}, D)$ and the complete canonical global database $can(\mathcal{I}, D)$ which satisfies all constraints in Σ_G,

3. The integrity-constraints mapping operad,

$$\begin{aligned}\mathbf{T}_{GA_\top} = {} & \mathit{MakeOperads}\big(\mathit{EgdsToSOtgd}\big(\Sigma_G^{\mathrm{egd}}\big)\\ & \wedge \mathit{TgdsToCanSOtgd}\big(\Sigma_G^{\mathrm{tgd}}\big)\big) : \mathcal{G} \to \mathcal{A}_\top.\end{aligned}$$

The sketch category derived by this data integration system is denoted by $\mathbf{Sch}(\mathcal{I})$ and represented in the left side of the following diagram:

$$\begin{array}{ccccc}
\mathcal{G} \xrightarrow{\mathbf{T}_{GA_\top}} \mathcal{A}_\top & & \alpha^*(\mathcal{G}) = can(\mathcal{I}, D) \xrightarrow{f_{\Sigma_G}} \mathcal{A}_\top \\
\mathbf{M}_{G_F G} \uparrow & & \uparrow \alpha^*(\mathbf{M}_{G_F G}) = f_{in} \\
\mathcal{G}_F & \alpha^* \mapsto & \alpha^*(\mathcal{G}_F) = can_F(\mathcal{I}, D) \\
\mathbf{M}_{S G_F} \uparrow & & \uparrow \alpha^*(\mathbf{M}_{S G_F}) = f_M \\
\mathcal{S} & & \alpha^*(\mathcal{S}) = D \\
\text{in } \mathbf{Sch}(\mathcal{I}) & & \text{in } \mathbf{DB}
\end{array}$$

Example 27 Suppose that we have two relations r and s in $\mathcal{G}$, both of arity 2 and having as key the first attribute, with the following foreign key constraints in $\mathcal{G}$:

1. $\pi_2(r) \subseteq \pi_1(s)$, of the tgd $\forall x \forall y (r(x, y) \Rightarrow \exists z s(y, z))$, with $\mathbf{K} = \{1\}$ key indexes of s, and $\mathbf{K}' = \{2\}$.
2. $\pi_1(s) \subseteq \pi_1(r)$, of the tgd $\forall x \forall z (s(x, z) \Rightarrow \exists y r(x, y))$, with $\mathbf{K} = \{1\}$ key indexes of r, and $\mathbf{K}' = \{1\}$.

That is, $\mathcal{G} = (S_G, \Sigma_G)$ with $S_G = \{r, s\}$, $ar(r) = ar(s) = 2$ and

$$\Sigma_G^{\mathrm{tgd}} = \big\{\forall x \forall y \big(r(x, y) \Rightarrow \exists z s(y, z)\big), \forall x \forall z \big(s(x, z) \Rightarrow \exists y r(x, y)\big)\big\}.$$

Each *certain answer* of the original user query $q(\mathbf{x})$, $\mathbf{x} = (x_1, \ldots, x_k)$ over a global schema is equal to the answer $\|q_L(\mathbf{x})\|_{can(\mathcal{I}, \mathcal{D})}$ of the *lifted* query $q_L(\mathbf{x})$, equivalent to the formula $q(\mathbf{x}) \wedge Val(x_1) \wedge \cdots \wedge Val(x_k)$, over this canonical model (the unary predicate $Val(\_)$ was introduced in Sect. 3.4.2 for the weak equivalence of databases with Skolem marked null values in SK).

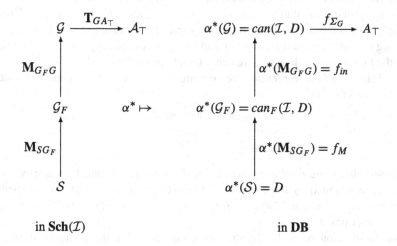

Thus, if it is possible to materialize such a canonical model, the certain answers could be obtained over such a database without query rewriting. Often it is not possible (as in this example) if this canonical model is *infinite*. In such a case, we can use the revisited fixpoint semantics described in [11], based on the fact that, after some point, the new tuples added into a canonical model insert only new Skolem constants which are not useful in order to obtain *certain* answers true in all models of a database (note that this criterion was introduced in the (FRF) rules for the foreign key constraints, defined previously in this section). In fact, the Skolem constants are not part of any certain answer to conjunctive query. Consequently, we are able to obtain a *finite subset* of a canonical *database*, which is large enough to obtain all certain answers as follows:

Suppose that the retrieved global database stores a single tuple $\langle a, b\rangle$ in r and with empty relation s, that is, $\|r\|_{ret(\mathcal{I},D)} = \{\langle a, b\rangle\}$ and $\|s\|_{ret(\mathcal{I},D)} = \{\langle\rangle\}$. Then, by applying the above rule 1, we insert the tuple (b, ω_0) in s and, successively, by applying the rule 2, we add (b, ω_1) in r and hence we can not apply more these rules from the fact that $\omega_1 \notin \mathbf{dom}$ and we cannot apply more the rule 1.

The following table represents the computation of the finite canonical instance database $can_F(\mathcal{I}, D)$:

$\|r\|_{can_F(\mathcal{I},\mathcal{D})}$	$\|s\|_{can_F(\mathcal{I},\mathcal{D})}$
a, b	b, ω_0
b, ω_1	

Obviously, this finite database $can_F(\mathcal{I}, \mathcal{D})$ does not satisfy the foreign key constraint 1, represented by the SOtgds $\exists f_1(\forall x\forall y(r(x, y) \Rightarrow s(y, f_1(x, y))))$ (while the constraint 2 by the SOtgd $\exists f_2(\forall x\forall z(s(x, z) \Rightarrow r(x, f_2(x, z)))))$ with two Skolem functions f_1 and f_2.

The satisfaction of both FK integrity constraints, for the retrieved database $ret(\mathcal{I}, D)$ defined above (by $\|r\|_{ret(\mathcal{I},D)} = \{\langle a, b\rangle\}$, $\|s\|_{ret(\mathcal{I},D)} = \{\langle\rangle\}$), will produce the following *infinite* canonical model $can(\mathcal{I}, D)$ of the global schema, for the Tarski interpretation of Skolem functions $I_T(f_1), I_T(f_2) : \mathcal{U}^2 \rightarrow SK$ such that $I_T(f_1)(a, b) = \omega_0$ and $I_T(f_2)(b, I_T(f_1)(a, b)) = \omega_1$ for the two fresh Skolem constants $\omega_0, \omega_1 \in SK$, in order to satisfy the monomorphism $\alpha^*(\mathbf{M}_{G_TG})$: $can_F(\mathcal{I}, D) \hookrightarrow can(\mathcal{I}, D)$ of the diagram above, as follows:

$\|r\|_{can(\mathcal{I},\mathcal{D})}$	$\|s\|_{can(\mathcal{I},\mathcal{D})}$
a, b	b, ω_0
b, ω_1	ω_1, ω_2
ω_1, ω_3	ω_3, ω_4
...	...

where $\omega_0 = f_1(a, b)$, $\omega_1 = f_2(b, \omega_0)$, $\omega_2 = f_1(b, \omega_1)$, $\omega_3 = f_2(\omega_1, \omega_2)$, $\omega_4 = f_1(\omega_1, \omega_3)$, etc.

Consequently, instead of the infinite database $can(\mathcal{I}, D)$ we can materialize the finite instance-database $can_F(\mathcal{I}, D)$, as a finite least fixpoint of the immediate-consequence monotonic operator T_B applied to the retrieved database $ret(\mathcal{I}, D)$ and use it in order to obtain the certain answers to lifted queries without query-rewriting. This technique was used also in the case for the queries with negation [12], when it is not possible to use the query-rewriting method described in the previous section.

Based on Definition 17, for any single *query* $q_i(\mathbf{x}_i)$ over the global schema $\mathcal{G}$ with relational symbols in $\{r_{i1}, \ldots, r_{ik}\} \subseteq S_G$, where $\mathbf{x}_i = \langle x_1, \ldots, x_k \rangle$ a tuple of variables, we can define the lifted query $q_L(\mathbf{x}_i)$ equivalent to the conjunctive formula $q_i(\mathbf{x}_i) \wedge Val(x_1) \wedge \cdots \wedge Val(x_k)$ and hence with the schema mapping operad's operation $q_i = v_i \cdot q_{G,i}$ where $q_{G,i} \in O(r_{i1}, \ldots, r_{ik}, r_q)$ and $v_i = 1_{r_q} \in O(r_q, r_q)$ is the identity operation. This lifted query can be represented by an atomic mapping

$$\mathbf{M}_{\mathcal{G}\widehat{\mathcal{G}}} = \{q_i, 1_{r_\emptyset}\} = MakeOperads\big(\big\{\forall \mathbf{x}_i\big(\big(q_i(\mathbf{x}_i) \wedge \big(h_V(x_1) \doteq \overline{1}\big) \wedge \cdots \\ \wedge \big(h_V(x_k) \doteq \overline{1}\big)\big) \Rightarrow r_q(\mathbf{x}_i)\big)\big\}\big) : \mathcal{G} \to \widehat{\mathcal{G}}$$

where $r_q \in S_{\widehat{\mathcal{G}}}$ is a new introduced relational symbol for this query with $k = ar(r_q) = |\mathbf{x}_i|$ (the number of variables in the tuple $\mathbf{x}_i$) and h_V is the functional symbol of the characteristic function of the built-in unary predicate *Val*, so that for every Tarski's interpretation I_T we obtain the fixed function $\overline{h}_V = I_T(h_V) : \mathcal{U} \to \mathbf{2}$ such that for each $d \in \mathcal{U} = \mathbf{dom} \cup SK$, $\overline{h}_V(d) = 1$ iff $d \in \mathbf{dom}$; 0 otherwise.

Analogously, this lifted query over $\mathcal{G}_F$ can be represented by an atomic mapping

$$\mathbf{M}_{\mathcal{G}_F\widehat{\mathcal{G}_F}} = \big\{q_i', 1_{r_\emptyset}\big\} = MakeOperads\big(\big\{\forall \mathbf{x}_i\big(\big(q_i(\mathbf{x}_i) \wedge \big(h_V(x_1) \doteq \overline{1}\big) \wedge \cdots \\ \wedge \big(h_V(x_k) \doteq \overline{1}\big)\big) \Rightarrow r_q'(\mathbf{x}_i)\big)\big\}\big) : \mathcal{G}_F \to \widehat{\mathcal{G}_F}$$

where $r_q' \in S_{\widehat{\mathcal{G}_F}}$ is a new introduced relational symbol for this query.

For a given mapping interpretation α with $\alpha^*(\mathcal{G}) = can(\mathcal{I}, D)$ and hence $\alpha^*(\widehat{\mathcal{G}}) = T can(\mathcal{I}, D)$,

$$\alpha\big(r_q'\big) = \alpha(r_q) = \big\|q_i(\mathbf{x}_i) \wedge Val(x_1) \wedge \cdots \wedge Val(x_k)\big\|_{can(\mathcal{I},D)} \\ = \big\|q_i(\mathbf{x}_i) \wedge Val(x_1) \wedge \cdots \wedge Val(x_k)\big\|_{can_F(\mathcal{I},D)}.$$

Hence, for this R-algebra α we obtain the morphism

$$h_q = \{f, q_\perp\} = \alpha^*(\mathbf{M}_{\mathcal{G}\widehat{\mathcal{G}}}) : can(\mathcal{I}, D) \to T can(\mathcal{I}, D)$$

and $h_{qF} = \{f_F, q_\perp\} = \alpha^*(\mathbf{M}_{\mathcal{G}_F\widehat{\mathcal{G}_F}}) : can_F(\mathcal{I}, D) \to T can_F(\mathcal{I}, D)$, with the functions:

$$f : \|r_{i1}\|_{can(\mathcal{I},D)} \times \cdots \times \|r_{ik}\|_{can(\mathcal{I},D)} \to \alpha(r_q),$$

$$f_F : \|r_{i1}\|_{can_F(\mathcal{I},D)} \times \cdots \times \|r_{ik}\|_{can_F(\mathcal{I},D)} \to \alpha\big(r_q'\big).$$

Consequently, we obtain that the images of the functions f and f_F are equal to the same computed certain query answer, that is, $im(f) = im(f_E) = \alpha(r_q') = \alpha(r_q)$, so that the following diagram in **DB** (on the right), based on the T-coalgebra homomorphism $f_{in} : (can_F(\mathcal{I}, D), h_{qF}) \hookrightarrow (can(I, \mathcal{D}), h_q)$, commutes.

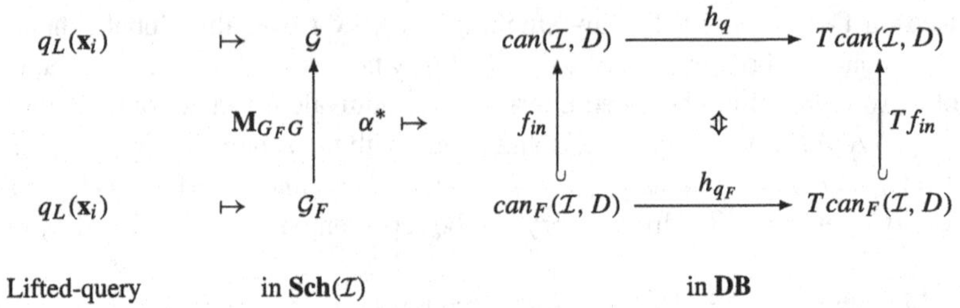
$q_L(\mathbf{x}_i)$
$\mapsto$
$\mathcal{G}$
$\mathbf{M}_{\mathcal{G}_F\mathcal{G}}$
$\alpha^* \mapsto$
$q_L(\mathbf{x}_i)$
$\mapsto$
$\mathcal{G}_F$
$can(\mathcal{I}, D)$
h_q
$T can(\mathcal{I}, D)$
fin
$\Updownarrow$
$T fin$
$can_F(\mathcal{I}, D)$
h_{q_F}
$T can_F(\mathcal{I}, D)$
Lifted-query
in **Sch**($\mathcal{I}$)
in **DB**

In fact, from the fact that $\widetilde{h_q} = \widetilde{h_{q_F}} = Flux(\alpha, \mathbf{M}_{\mathcal{G}\widehat{\mathcal{G}}}) \subseteq T can_F(\mathcal{I}, D) = \widetilde{fin} = \widetilde{T fin} = T\,\widetilde{fin}$, we obtain $\widetilde{T fin} \cap \widetilde{h_{q_F}} = \widetilde{h_q} \cap \widetilde{fin}$ and hence $T fin \circ h_{q_F} = h_q \circ fin$.

In fact, we introduced marked null values (instead of Skolem functions) in order to define and materialize such a *finite* database $can_F(\mathcal{I}, D)$. It *is not* a model of the data integration system (which is infinite); however, it has all necessary query-answering properties: it is able to provide all certain answers to conjunctive queries over a global schema. Consequently, it can be materialized and used for query answering, instead of methods based on the query-rewriting algorithms.

The procedure for computation of a finite canonical database for the global schema, based on "immediate consequence" monotonic operator $T_{\mathcal{B}}$ defined in precedence, can be intuitively described as follows: it starts with an instance $\langle I, G_\emptyset \rangle$ which consists of I, instance of the source schema, and of the empty instance $G_\emptyset$ for the target (global schema) such that $R = \{\langle\rangle\}$ (i.e., empty relation with only an empty tuple $\langle\rangle$) for each relation in $G_\emptyset$ (we recall that $G_\emptyset \simeq \bot^0$ in **DB**). Then we chase $\langle I, G_\emptyset \rangle$ by applying all the dependencies in $\mathcal{M}$ (a finite set of source-to-target dependencies) and Σ_G (a finite set of target (global schema) integrity dependencies) as long as they are applicable. This process may fail (if an attempt to identify two domain constants is made in order to define a homomorphism between two consecutive target instances) or it may never terminate. Let us denote two consecutive target instances of this process (with $J_0 = G_\emptyset$) by J_i and J_{i+1}. Then we introduce this "chase" function $C_h : \Theta \longrightarrow \Theta$ where Θ is the set of all pairs $\langle I, J \rangle$, I is a source instance and J one of target instances generated by I such that

$$\langle I, J_{i+1} \rangle = C_h\big(\langle I, J_i \rangle\big).$$

Let us define the partial ordering $\sqsubseteq$ over the set $\Theta_w = \{J_i \mid \langle I, J_i \rangle \in \Theta\}$ such that $J_i \sqsubseteq J_k$ iff $\forall r \in J_i (T_w\{r, \bot\} \subseteq T_w\{\sigma(r), \bot\})$, where $\sigma : J_i \rightarrow J_k$ is a bijection such that for each $r \in J_i$, $r \subseteq \sigma(r)$.

Proposition 19 *There exists the monotonic function $\Psi : \Theta_w \longrightarrow \Theta_w$ such that its least fixpoint is obtained from $J_0 = G_\emptyset$ in a finite number $n \geq 1$ of steps and is equal to $can_F(\mathcal{I}, D) = \Psi^n(J_0)$.*

Proof We can define the function $\Psi : \Theta_w \longrightarrow \Theta_w$ such that for each $J_i \in \Theta_w$: $\Psi(J_i) = J_{i+1}$ if $J_{i+1} \sqsubseteq J_i$ is not true; J_i otherwise.

This function is monotonic w.r.t. $\sqsubseteq$ from the fact that $J_i \sqsubseteq J_k$ iff $i \leq k$ and thus $\Psi(J_i) \sqsubseteq \Psi(J_k)$, with least fixed point $\Psi(J_n) = J_n = can_F(\mathcal{I}, \mathcal{D})$.

This chain is represented in **DB** by the following composition of monomorphisms:

$$G_{\emptyset} = J_0 \hookrightarrow J_1 \hookrightarrow J_2 \hookrightarrow \cdots \hookrightarrow J_n = can_F(\mathcal{I}, \mathcal{D}),$$

so that $G_{\emptyset} = J_0 \preceq J_1 \preceq J_2 \preceq \cdots \preceq J_n = can_F(\mathcal{I}, \mathcal{D})$. □

Notice that each infinite canonical database of a global database schema $\mathcal{G}$ is weakly equivalent to the finite instance-database $can_F(\mathcal{I}, \mathcal{D})$ that *is not a model* of $\mathcal{G}$ but is obtained as the least fixpoint of the operator Ψ.

Thus, $can(\mathcal{I}, \mathcal{D}) \approx_w can_F(\mathcal{I}, \mathcal{D})$, where $can(\mathcal{I}, \mathcal{D})$ is an infinite model of $\mathcal{G}$, and $can_F(\mathcal{I}, \mathcal{D})$ is a finite weakly equivalent object to it in the **DB** category.

4.3 Review Questions

1. What is the relationship between the functorial semantics for database schemas and Tarski's semantics? What is the relationship between the functorial semantics for the database mappings and Tarski semantics and Second-order Logic semantics? What is the meaning of the homomorphism α^* in Proposition 14? Can we define a similar homomorphism between schema database mappings and the morphisms in **DB** category and, if so, formalize such a homomorphism and explain what is its meaning in the database theory.
2. What is the fundamental difference between the graph of a given database mapping system and the sketch category obtained from it? Why are we using the FOL with identity? Is the identity binary predicate of the FOL used in this database logic theory different for any database schema and the sketch obtained from this schema in Proposition 15? If not, in which FOL theory is it a binary predicate, and what is exactly its interpretation? Can we have two different interpretations of this predicate for a given **DB** category?
3. For the homomorphism α^* in Proposition 14, which represents a functorial interpretation of the database schemas, is it possible to determine if such an interpretation is a model of a schema $\mathcal{A}$ without the definition of the sketch $\mathbf{Sch}(G(\mathcal{A}))$ in Proposition 15? For which kind of the schemas is every homomorphism α^* in Proposition 14 also a model of such schemas? How can we use the morphism f_{Σ_A} in Proposition 15 in order to differentiate the models from non-models?
4. What are the main differences between the Data Exchange and Data integration? How is the Data Exchange generalized in this new framework? What is the relationship between DATALOG and Data Integration? Explain the possible solutions for the data-inconsistency problems of information integrated from the different sources? In which way is the data-inconsistency, obtained from a database mapping system, checked by using the functorial semantics? Why is the **DB** category able to support also the data-inconsistency without the problems of the explosive inconsistency that such situation makes to the formalism based on the FOL?

5. The translation of the Data integration framework into the functorial sketch's based semantics, provided in example in Sects. 4.2.2 and 4.2.3, is not an equivalent representation because in Data Integration framework we do not provide the materialization of its global schema, while in the functorial semantics each schema is mapped into an instance data base of such a schema. Is it a unique semantic difference? How is the problem with incomplete information in Data Integration (with a class of possible models) resolved in the functorial semantics? The certain answers to the queries over a global schema in Data Integration framework are true in all models of the global schema. How can the certain answers be provided in the functorial semantics with SOtgds (with the introduction of the Skolem constants)?
6. Provide an example by modification of Example 27 of Data Integration in Sects. 4.2.3 and 4.2.4 by dividing the global schema into two separate databases, one containing only the binary relation r and the second only the binary relation s, and by substituting the FK constraints by two corresponding mappings between these two databases. The operational semantics for these two data mapping representations will be equal (by considering that in Example 27 we obtained an acyclic mapping graph, while in this new setting we obtain a cyclic graph) and, if not, why?
7. Consider the modified version of Example 27. Why can't the fixpoint operator for a finite solution be used in a standard Data Integration framework, while, instead, it can be realized in the functorial semantics with the operational semantics based on the fixpoint operator for such a cyclic database mapping system?

   Clearly, the model of such a mapping system will be composed by two databases above, with relations containing an infinite number of tuples. Is this fixpoint semantics, which do not produce a finite model of the database mapping system (but is equivalent for the certain-answer semantics for the query-answering system, where the Skolem constants are eliminated from resulting views), the way to resolve the cyclic database mapping systems containing incomplete information (and hence with the possibly infinite models) in practice? Is there another equivalently valid solution?
8. Can we obtain in this framework the infinite databases without using infinite number of Skolem constants? Why is it necessary to theoretically consider the infinite databases also for the cyclic database mapping systems, if we have the fixpoint semantics provided above in order to obtain only the finite databases with the same power to provide the certain-answers to the database queries? (Consider the whole process of the algebraization of the SOtgds and the logic semantics of the mappings.)

References

1. M. Barr, C. Wells, *Toposes, Triples and Theories*. Grundelehren der math. Wissenschaften, vol. 278 (Springer, Berlin, 1985)

2. A. Calì, D. Calvanese, G. Giacomo, M. Lenzerini, Data integration under integrity constraints, in *Proc. of the 14th Conf. on Advanced Information Systems Engineering (CAiSE 2002)* (2002), pp. 262–279
3. A. Calì, D. Lembo, R. Rosati, M. Ruzi, Experimenting data integration with DID@DIS, in *Proc. CAiSE* (2004), pp. 51–66
4. R. Fagin, P.G. Kolaitis, R.J. Miller, L. Popa, DATA exchange: semantics and query answering, in *Proc. of the 9th Int. Conf. on Database Theory (ICDT 2003)* (2003), pp. 207–224
5. P.J. Freyd, G.M. Kelly, Categories of continuous functors. I. J. Pure Appl. Algebra **2**, 169–191 (1972)
6. M. Friedman, A. Levy, T. Millstein, Navigational plans for data integration, in *Proc. of the 16th Nat. Conf. on Artificial Intelligence (AAAI'99)* (AAAI Press/The MIT Press, Cambridge, 1999), pp. 67–73
7. A.Y. Halevy, Answering queries using views: a survey. VLDB J. **10**(4), 270–294 (2001)
8. D.S. Johnson, A.C. Klug, Testing containment of conjunctive queries under functional and inclusion dependencies. J. Comput. Syst. Sci. **28**, 167–189 (1984)
9. F.W. Lawvere, Functorial semantics of algebraic theories. Proc. Natl. Acad. Sci. **50**, 869–872 (1963)
10. M. Lenzerini, Data integration: a theoretical perspective, in *Proc. of the 21st ACM SIGACT SIGMOD SIGART Symp. on Principles of Database Systems (PODS 2002)* (2002), pp. 233–246
11. Z. Majkić, Fixpoint semantics for query answering in data integration systems, in *AGP03—8. th Joint Conference on Declarative Programming*, Reggio Calabria (2003), pp. 135–146
12. Z. Majkić, Querying with negation in data integration systems, in *9th International Database Engineering and Application Symposium (IDEAS)*, Montreal, Canada, July 25–27, 2005 (IEEE Computer Society, Washington, 2005), pp. 58–70
13. Z. Majkić, Data base mappings and theory of sketches. arXiv:1104.4899v1 (2011), 26 April, pp. 1–21
14. M. Makkai, R. Pare, *Accessible Categories: The Foundations of Categorical Model Theory*. Contemporary Mathematics, vol. 104 (Am. Math. Soc., Providence, 1989)
15. R. Reiter, On closed world databases, in *Logic and Databases*, ed. by H. Gallaire, J. Minker (Plenum Press, New York, 1978), pp. 55–76

Extensions of Relational Codd's Algebra and DB Category

5

5.1 Introduction to Codd's Relational Algebra and Its Extensions

The standard terminology for a Relational Data Base (RDB) is slightly different from the terminology used for basic database concepts presented in Sect. 1.4 of the introduction, for instance, in RDB we use the term "table" as a special case of the more general concept "relation", and attributes instead of predicate variables, etc.

In this section, we will shortly introduce some additional RDB concepts that are used in E.F. Codd's relational algebra for RDBs [1].

A base table in a relational system consists of a row of column headings, together with zero or more rows of data values. The column heading row specifies one or more columns (giving, among other things, a data type for each). Each data row contains exactly one scalar value for each of the columns specified in the column heading row. Five primitive operators of Codd's algebra are: the selection, the projection, the Cartesian product (also called the cross-product or cross-join), the set union, and the set difference. Another operator, rename, was not noted by Codd, but the need for it is shown by the inventors of Information Systems Base Language (ISBL) for one of the earliest database management systems which implemented Codd's relational model of data. These six operators are fundamental in the sense that if we omit any one of them, we will lose expressive power. Many other operators have been defined in terms of these six. Among the most important are the set intersection, division, and the natural join. In fact, ISBL made a compelling case for replacing the Cartesian product with the natural join, of which the Cartesian product is a degenerate case.

Recall that two relations r_1 and r_2 are union-compatible iff $\{atr(r_1)\} = \{atr(r_2)\}$, where for a given list (or tuple) of the attributes $\mathbf{a} = atr(r) = \langle a_1, \ldots, a_k\rangle = \langle atr_r(1), \ldots, atr_r(k)\rangle$, we denote the *set* $\{a_1, \ldots, a_k\}$ by $\{atr(r)\}$, $k = ar(r)$, with the injective function $nr_r : \{1, \ldots, k\} \to SN$ which assigns distinct names to each column of this relation. If a relation r_2 is obtained from a given relation r_1 by permutating its columns, then we say that they are not equal (in the set-theoretic sense) but that they are equivalent. Notice that in the RDB theory the two equivalent re-

Z. Majkić, *Big Data Integration Theory*, Texts in Computer Science,
DOI 10.1007/978-3-319-04156-8_5,

lations are considered equal as well. In what follows, given any two lists (tuples), $\mathbf{d} = \langle d_1, \dots, d_k \rangle$ and $\mathbf{b} = \langle b_1, \dots, b_m \rangle$ their concatenation $\langle d_1, \dots, d_k, b_1, \dots, b_m \rangle$ is denoted by **d&b**, where '&' is the symbol for concatenation of the lists. By $\|r\|$ we denote the extension of a given relation (relational symbol) r; it is extended to any term t_R of Codd's algebra, so that $\|t_R\|$ is the relation obtained by computation of this term. Let us briefly define these basic operators, and their correspondence with the formulae of FOL:

1. Rename is a unary operation written as _ RENAME $name_1$ AS $name_2$ where the result is identical to input argument (relation) r except that the column i with name $nr_r(i) = name_1$ in all tuples is renamed to $nr_r(i) = name_2$.

   This operation is neutral w.r.t. the logic, where we are using the variables for the columns of relational tables and not their names.
2. Cartesian product is a binary operation _ TIMES _, written also as $\_ \otimes \_$, such that for the relations r_1 and r_2, first we do the rename normalization of r_2 (w.r.t. r_1), denoted by r_2^ρ, such that:

   For each kth copy of the attribute a_i (or, equivalently, $a_i(0)$) of the mth column of r_2 (with $1 \le m \le ar(r_2)$), denoted by $a_i(k) = atr_{r_2}(m) \in atr(r_2)$, such that the maximum index of the same attribute a_i in r_1 is $a_i(n)$, we change r_2 by:
   (a) $a_i(k) \mapsto a_i(k+n)$;
   (b) If $name_1 = nr_{r_2}(m)$ is a name that exists in the set of the column names in r_1, then we change the naming function $nr_{r_2} : \{1, \dots, ar(r_2)\} \to SN$, by $nr_{r_2}(m) = name_2$, where $name_2 \in SN$ is a new name distinct from all other used names, and we define the renaming normalization ρ by mapping $name_1 \mapsto name_2$.

   The relation obtained from r_2, after this renaming normalization, will be denoted by r_2^ρ. Then we define the new relation r (when both $\|r_1\| \neq \{\langle\rangle\}$ and $\|r_2\| \neq \{\langle\rangle\}$, i.e., when are not empty relations) by $r_1 \otimes r_2^\rho$, with $\|r\| \triangleq \{\mathbf{d}_1 \& \mathbf{d}_2 \mid \mathbf{d}_1 \in \|r_1\|, \mathbf{d}_2 \in \|r_2\|\}$, and with the naming function $nr_r : \{1, \dots, ar(r_1) + ar(r_2)\} \to SN$ such that $nr_r(i) = nr_{r_1}(i)$ for $1 \le i \le ar(r_1)$ and $nr_r(i) = nr_{r_2}(i)$ for $1 + ar(r_1) \le i \le ar(r_1) + ar(r_2)$, and the $atr_r : \{1, \dots, ar(r_1) + ar(r_2)\} \to \mathbf{att}$ function defined by $atr_r(i) = atr_{r_1}(i)$ for $1 \le i \le ar(r_1)$ and $atr_r(i) = atr_{r_2}(i)$ for $1 + ar(r_1) \le i \le ar(r_1) + ar(r_2)$.

   This Cartesian product is given by the following logical equivalence, by considering the relational symbols as predicates,

   $$r(x_1, \dots, x_{ar(r_1)}, y_1, \dots, y_{ar(r_2)}) \Leftrightarrow \big(r_1(x_1, \dots, x_{ar(r_1)}) \wedge r_2(y_1, \dots, y_{ar(r_2)})\big),$$

   so that $\|r\| = \|r_1(x_1, \dots, x_{ar(r_1)}) \wedge r_2(y_1, \dots, y_{ar(r_2)})\|$.

   (if $\|r_1\|$ is empty then $r_1 \otimes r_2^\rho = r_2$; if $\|r_2\|$ is empty then $r_1 \otimes r_2^\rho = r_1$).
3. Projection is a unary operation written as _ $[S]$, where S is a tuple of column names such that for a relation r_1 and $S = \langle nr_{r_1}(i_1), \dots, nr_{r_1}(i_k) \rangle$, with $k \ge 1$ and $1 \le i_m \le ar(r_1)$ for $1 \le m \le k$, and $i_m \neq i_j$ if $m \neq j$, we define the relation r by: $r_1[S]$, with $\|r\| = \|r_1\|$ if $\exists name \in S.name \notin nr(r_1)$; otherwise $\|r\| = \pi_{\langle i_1, \dots, i_k \rangle}(\|r_1\|)$, where $nr_r(m) = nr_{r_1}(i_m)$, $atr_r(m) = atr_{r_1}(i_m)$, for $1 \le m \le k$.

This projection is given by the following logical equivalence

$$r(x_{i_1},\ldots,x_{i_k}) \Leftrightarrow \exists x_{j_1}\ldots x_{j_n} r_1(x_1,\ldots,x_{ar(r_1)}),$$

where $n = ar(r_1) - k$ and for all $1 \leq m \leq n$, $j_m \notin \{i_1,\ldots,i_k\}$, so that $\|r\| = \|\exists x_{j_1}\ldots x_{j_n} r_1(x_1,\ldots,x_{ar(r_1)})\|$.

4. Selection is a unary operation written as _ WHERE C, where a condition C is a finite-length logical formula that consists of atoms '($name_i$ θ $name_j$)' or '($name_i$ θ $\overline{d}$)', with built-in predicates $\theta \in \Sigma_\theta \supseteq \{\doteq, >, <\}$, a constant $\overline{d}$, and the logical operators $\wedge$ (AND), $\vee$ (OR) and $\neg$ (NOT), such that for a relation r_1 and $name_i$, $name_j$ the names of its columns, we define the relation r by

$$r_1 \text{ WHERE } C,$$

as the relation with $atr(r) = atr(r_1)$ and the function nr_r equal to nr_{r_1}, where $\|r\|$ is composed by the tuples in $\|r_1\|$ for which C is satisfied.

This selection is given by the following logical equivalence:

$$r(x_{i_1},\ldots,x_{i_k}) \Leftrightarrow \big(r_1(x_1,\ldots,x_{ar(r_1)}) \wedge C(\mathbf{x})\big),$$

where $C(\mathbf{x})$ is obtained by substitution of each $name_i = nr_{r_1}(j)$ (of the jth column of r_1) in the formula C by the variable x_j, so that

$$\|r\| = \big\|r_1(x_1,\ldots,x_{ar(r_1)}) \wedge C(\mathbf{x})\big\|.$$

4.1. We assume as an *identity unary operation*, the operation _ WHERE C when C is the atomic condition $\overline{1} \doteq \overline{1}$ (i.e., a tautology).

5. Union is a binary operation written as _ UNION _ such that for two union-compatible relations r_1 and r_2, we define the relation r by: r_1 UNION r_2, where $\|r\| \triangleq \|r_1\| \cup \|r_2\|$, with $atr(r) = atr(r_1)$, and the functions $atr_r = atr_{r_1}$, and $nr_r = nr_{r_1}$. This union is given by the following logical equivalence:

$$r(x_1,\ldots,x_n) \Leftrightarrow \big(r_1(x_1,\ldots,x_n) \vee r_2(x_1,\ldots,x_n)\big),$$

where $n = ar(r) = ar(r_1) = ar(r_2)$, so that

$$\|r\| = \big\|r_1(x_1,\ldots,x_n) \vee r_2(x_1,\ldots,x_n)\big\|.$$

6. Set difference is a binary operation written as _ MINUS_ such that for two union-compatible relations r_1 and r_2, we define the relation r by r_1 MINUS r_2, where $\|r\| \triangleq \{\mathbf{t} \mid \mathbf{t} \in \|r_1\|$ such that $\mathbf{t} \notin \|r_2\|\}$, with $atr(r) = atr(r_1)$, and the functions $atr_r = atr_{r_1}$, and $nr_r = nr_{r_1}$.

Let r_1 and r_2 be the predicates (relational symbols) for these two relations. Then their difference is given by the following logical equivalence:

$$r(x_1,\ldots,x_n) \Leftrightarrow \big(r_1(x_1,\ldots,x_n) \wedge \neg r_2(x_1,\ldots,x_n)\big),$$

where $n = ar(r) = ar(r_1) = ar(r_2)$, and hence

$$\|r\| = \big\|r_1(x_1,\ldots,x_n) \wedge \neg r_2(x_1,\ldots,x_n)\big\|.$$

Natural join $\bowtie$ is a binary operator, written as $(r_1 \bowtie r_2)$, where r_1 and r_2 are relations. The result of the natural join is the set of all combinations of tuples in r_1 and r_2 that are equal on their common attribute names. In fact, $(r_1 \bowtie r_2)$ can be obtained by creating the Cartesian product $r_1 \otimes r_2$ and then by execution of the Selection with the condition C defined as a conjunction of atomic formulae $nr_{r_1}(i) = nr_{r_2}(j)$ (where i and j are the columns of the same attribute in r_1 and r_2, respectively, i.e., satisfying $atr_{r_1}(i) = atr_{r_2}(j)$) that represents the equality of the common attribute names of r_1 and r_2. The natural join is arguably one of the most important operators since it is the relational counterpart of logical AND. Note carefully that if the same variable appears in each of two predicates that are linked by AND, then that variable stands for the same thing and both appearances must always be substituted by the same value. In particular, the natural join allows the combination of relations that are associated by a foreign key. It can also be used to define a composition of binary relations. In category theory, the join is precisely the fiber product. Altogether, the operators of a relational algebra have identical expressive power to that of domain relational calculus or tuple relational calculus. However, relational algebra is less expressive than first-order predicate calculus without function symbols. Relational algebra corresponds to *a subset* of FOL (denominated *relational calculus*), namely Horn clauses without recursion and negation (or union of conjunctive queries introduced in Sect. 1.4). Consequently, relational algebra is essentially equivalent in expressive power to *relational calculus* (and thus FOL and queries defined in Sect. 1.4 in the introduction); this result is known as Codd's theorem. However, the negation, applied to a formula of the calculus, constructs a formula that may be true on an infinite set of possible tuples. To overcome this difficulty, Codd restricted the operands of relational algebra to finite relations only and also proposed restricted support for negation $\neg$ (NOT) and disjunction $\vee$ (OR). Codd defined the term "relational completeness" to refer to a language that is complete with respect to first-order predicate calculus apart from the restrictions he proposed. In practice, the restrictions have no adverse effect on the applicability of his relational algebra for database purposes. In our theoretical setting, we permit the relations with infinite number of tuples, so we do not use such restrictions.

Several papers have proposed new operators of an algebraic nature as candidates for addition to the original set. We choose the additional unary operator 'EXTEND_ ADD $a, name$ AS e' denoted shortly as $\_\ \langle a, name, e\rangle$, where a is a new added attribute as the new column (at the end of relation) with a new fresh name *name* and e is an expression (in the most simple cases it can be the value NULL or a constant $\overline{d}$, or the ith column name $nr(i)$ of the argument (i.e., relation) of this operation), for *update* relational algebra operators, in order to cover all of the basic features of data manipulation (DML) aspects of a relation models of data, so that

- We define a unary operator $\_\ \langle a, name, e\rangle$, for an attribute $a \in \textbf{att}$, its name, and expression e, as a function with a set of column names, such that for a relation r_1 and expression e composed of the names of the columns of r_1 with $n = ar(r_1)$, we obtain the $(ar(r_1) + 1)$-ary relation r by $\langle a, name, e\rangle(r_1)$, with the naming function $nr_r : \{ar(r_1) + 1\} \rightarrow SN$ such that $nr_r(i) = nr_{r_1}(i)$ if $i \leq ar(r_1)$ and

$nr_r(ar(r_1)+1) = name$ otherwise being a fresh new name for this column; with the attribute function $atr_r : \{ar(r_1)+1\} \rightarrow$ **att** such that $atr_r(i) = atr_{r_1}(i)$ if $i \leq ar(r_1)$ and $atr_r(ar(r_1)+1) = a$ otherwise, and

$$\|r\| = \{\langle\rangle\} \cup \{\mathbf{d}\&e(\mathbf{d}) \mid \mathbf{d} \in \|r_1\|\},$$

where $e(\mathbf{d}) \in dom(a)$ is a constant or the value obtained from the function e where each name $nr_r(i)$ is substituted by the value d_i of the tuple $\mathbf{d} = \langle d_1, \ldots, d_n\rangle \in \|r_1\|$; in the special cases, we can use nullary functions (constants) for the expression e (for example, for the NULL value).

(note that r is empty if e is an expression and r_1 empty as well).

Then, for a nonempty relation r_1, the EXTENDr_1 ADD $a, name$ AS e (i.e., $r_1\langle a, name, e\rangle$) can be represented by the following logical equivalence:

$$r(x_1, \ldots, x_{n+1}) \Leftrightarrow \big(r_1(x_1, \ldots, x_n) \wedge \big(x_{n+1} = e(\mathbf{x})\big)\big),$$

where $e(\mathbf{x})$ is obtained by substituting each $name_i = nr_{r_1}(j)$ (of the jth column of r_1) in the expression e by the variable x_j.

We are able to define a new relation with a single tuple $\langle \overline{d}_1, \ldots, \overline{d}_k\rangle, k \geq 1$ with the given list of attributes $\langle a_1, \ldots, a_k\rangle$, by the following finite length expression,

$$\text{EXTEND}\,\big(\ldots(\text{EXTEND}\; r_\emptyset \;\text{ADD}\; a_1, name_1 \;\text{AS}\; \overline{d}_1)\ldots\big)\;\text{ADD}\; a_k, name_k \text{AS}\; \overline{d}_k,$$

or equivalently, by $r_\emptyset\langle a_1, name_1, \overline{d}_1\rangle \otimes \cdots \otimes r_\emptyset\langle a_k, name_k, \overline{d}_k\rangle$, where $r_\emptyset$ is the empty type relation with $\|r_\emptyset\| = \bot, ar(r_\emptyset) = 0$ introduced in Definition 8, and empty functions $atr_{r_\emptyset}$ and $nr_{r_\emptyset}$. Such single tuple relations can be used for an insertion in a given relation (with the same list of attributes) in what follows.

Update Operators The three update operators, 'UPDATE', 'DELETE' and 'INSERT' of the Relational algebra, are derived operators from these previously defined operators in the following way:

1. Each algebraic formulae 'DELETE FROM r WHERE C' is equivalent to the formulae 'r WHERE $\neg C$', that is, to 'r MINUS (r WHERE C)'.
2. Each algebraic expression (a term) 'INSERT INTO $r[S]$ VALUES (list of values)', 'INSERT INTO $r[S]$ AS SELECT...', is equivalent to 'r UNION r_1' where the union compatible relation r_1 is a one-tuple relation (defined by list) in the first, or a relation defined by 'SELECT...' in the second case.

   In the case of a single tuple insertion (version with 'VALUES') into a given relation r, we can define a single tuple relation r_1 by using 'EXTEND...' operations.
3. Each algebraic expression 'UPDATE r SET $[nr_r(i_1) = e_{i_1}, \ldots, nr_r(i_k) = e_{i_k}]$ WHERE C', for $n = ar(r)$, where $e_{i_m}, 1 \leq i_m \leq n$ for $1 \leq m \leq k$ are the expressions and C is a condition, is equal to the formula '(r WHERE $\neg C$) UNION r_1', where r_1 is a relation expressed by

$$(\text{EXTEND } (\ldots (\text{EXTEND } (r \text{ WHERE } C) \text{ ADD } att_r(1), name_1 \text{ AS } e_1) \ldots) \\ \text{ADD } att_r(n), name_n \text{ AS } e_n)[S],$$

such that for each $1 \leq m \leq n$, if $m \notin \{i_1, \ldots, i_k\}$ then $e_m = nr_r(m)$, and $S = \langle name_1, \ldots, name_n \rangle$.
Consequently, all update operators of the relational algebra can be obtained by addition of these 'EXTEND _ ADD $a, name$ AS e' operations.

Definition 31 We denote the algebra of the set of operations, introduced previously in this section (points from 1 to 6 and EXTEND _ ADD $a, name$ AS e) with additional nullary operator (empty-relation constant) $\perp$, by Σ_{RE}. Its subalgebra without _ MINUS _ operator is denoted by Σ_R^+, and without $\perp$ and unary operators EXTEND _ ADD $a, name$ AS e is denoted by Σ_R (it is the "select–project–join–rename+union" (SPJRU) subalgebra). We define the set of terms $\mathcal{T}_P X$ with variables in X of this Σ_R-algebra (and analogously for the terms $\mathcal{T}_P^+ X$ of Σ_R^+-algebra), inductively as follows:

1. Each relational symbol (a variable) $r \in X \subseteq \mathbb{R}$ and a constant (i.e., a nullary operation) is a term in $\mathcal{T}_P X$;
2. Given any term $t_R \in \mathcal{T}_P X$ and an unary operation $o_i \in \Sigma_R$, $o_i(t_R) \in \mathcal{T}_P X$;
3. Given any two terms $t_R, t'_R \in \mathcal{T}_P X$ and a binary operation $o_i \in \Sigma_R$, $o_i(t_R, t'_R) \in \mathcal{T}_P X$.

We define the evaluation of terms in $\mathcal{T}_P X$, for $X = \mathbb{R}$, by extending the assignment (i.e., R-algebra) $\|\_\| : \mathbb{R} \to \underline{\Upsilon}$ which assigns a relation to each relational symbol (a variable) to all terms by the function $\|\_\|_\# : \mathcal{T}_P \mathbb{R} \to \underline{\Upsilon}$ (with $\|r\|_\# = \|r\|$), where $\underline{\Upsilon}$ is the universal database instance (set of all relations in Definition 26). For a given term t_R with relational symbols $r_1, \ldots, r_k \in \mathbb{R}$, $\|t_R\|_\#$ is the relational table obtained from this expression for the given set of relations $\|r_1\|, \ldots, \|r_k\| \in \underline{\Upsilon}$, with the constraint that $\|t_R \text{ UNION } t'_R\|_\# = \|t_R\|_\# \cup \|t'_R\|_\#$ if the relations $\|t_R\|_\#$ and $\|t'_R\|_\#$ are union compatible; $\perp$ otherwise.

Each R-algebra $\alpha : X \to \underline{\Upsilon}$ is a restriction of an assignment $\|\_\|$ to $X \subseteq \mathbb{R}$.

We say that two terms $t_R, t'_R \in \mathcal{T}_P X$ are equivalent (or equal), denoted by $t_R \approx t'_R$, if for all assignments $\|t_R\|_\# = \|t'_R\|_\#$.

Let us consider an example for terms of the Σ_{RE} algebra of this Definition 31:

Example 28 Consider the term t_R of the Σ_{RE}-algebra, equal to the algebraic expression

$$(((((r_1[S_1])\text{WHERE } C_1)\text{UNION}((r_2\text{WHERE } C_2)[S_2]))\text{WHERE } C_5)[S_5]) \\ \text{MINUS}((\text{EXTEND}((r_3[S_3])\text{UNION}((r_4\text{WHERE } C_4)[S_4])) \\ \text{ADD } a, name \text{ AS } e)[S_6]),$$

$$
\begin{array}{c}
\_\text{MINUS}\_ \\
\diagup \qquad\qquad \diagdown \\
\_[S_5] \qquad\qquad\qquad \_[S_6] \\
| \qquad\qquad\qquad\qquad | \\
\_\text{WHERE } C_5 \qquad \text{EXTEND}\_\text{ADD } a,\text{name AS } e \\
| \qquad\qquad\qquad\qquad | \\
\_\text{UNION}\_ \qquad\qquad\quad \_\text{UNION}\_ \\
\diagup \qquad \diagdown \qquad\qquad \diagup \qquad \diagdown \\
\_\text{WHERE } C_1 \quad \_[S_2] \qquad \_[S_3] \qquad \_[S_4] \\
| \qquad\qquad | \qquad\qquad | \qquad\qquad | \\
\_[S_1] \qquad \_\text{WHERE } C_2 \qquad r_3 \qquad \_\text{WHERE } C_4 \\
| \qquad\qquad | \qquad\qquad\qquad\qquad | \\
r_1 \qquad\qquad r_2 \qquad\qquad\qquad\qquad r_4
\end{array}
$$

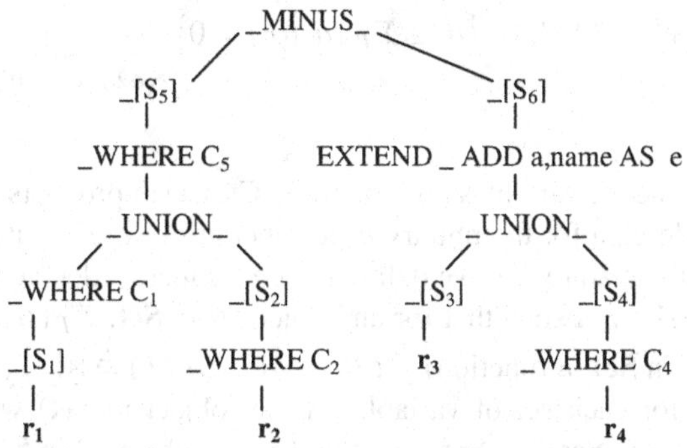

which is represented by the tree above, with the four different paths that end with leafs which are equal to relational tables.

In what follows, we will show that it is possible to transform each such a term-tree into the *single path* (composed of only unary operations) which ends with a single relation, obtained as a Cartesian product of the original relational tables updated (eventually) by the operations "EXTEND _ AD ...".

This term can be represented (in the figure) as the term-tree where the leafs are the relational symbols (the variables in $X \subseteq \mathbb{R}$), $r_1, r_2, r_3, r_4 \in X$.

In Sect. 5.2, we will show how such a complex tree-terms can be reduced to the simple path-term with only one leaf, obtained as a Cartesian product of (eventually updated) relational tables. In order to show that each assignment $\| \_ \| : X \to \Upsilon$ which assigns a relation to each relational symbol (a variable) can be *uniquely* extended to all terms, by a function $\| \_ \|_\# : \mathcal{T}_P X \to \Upsilon$, we need to introduce the theory of initial algebras and initial semantics in the next section.

5.1.1 Initial Algebras and Syntax Monads: Power-View Operator

The *syntax* of a programming language, given by a signature Σ_P (as, for example, Σ_{RE} or its subset Σ_R in Definition 31), can be identified with a monad, the *syntactical monad* $\mathcal{T}_P$ freely generated by the program constructors (operations) $o_i \in \Sigma_P$ (with arity $ar(o_i) \geq 0$ where the nullary operations are the constants, as $\bot$, for example).

We illustrate the link between a single-sorted Σ_P algebra signature and the $\mathcal{T}_P$-algebras of the endofunctor $\mathcal{T}_P$. The assumption that the signature Σ_P is finite is not essential for the correspondence between models of Σ_P and algebras of $\mathcal{T}_P$. If Σ_P is infinite one can define $\mathcal{T}_P$ via an infinite coproduct (in **Set** which corresponds to the disjoint union $\uplus$), commonly written for a set of program variables in X (in **Set** the '$\times$' denotes the Cartesian product)

$$\Sigma_P(X) = \biguplus_{o_i \in \Sigma_P} X^{ar(o_i)} \triangleq \{o_i \in \Sigma_P \mid ar(o_i) = 0\} \bigcup_{o_i \in \Sigma_P \ \& \ ar(o_i) \geq 1} \{X^{ar(o_i)} \times \{i\}\}$$

where X is the set of variables and the finite Cartesian products $X^n = \overbrace{X \times \cdots \times X}^{n}$ are used as a domain for the finitary n-ary operator $o_i \in \Sigma_P$ with $n = ar(o_i)$. More formally, for a signature Σ_P we define the endofunctor (denoted by the same symbol) $\Sigma_P : \mathbf{Set} \longrightarrow \mathbf{Set}$ such that for any object B in **Set**, $\Sigma_P(B) \triangleq \biguplus_{o_i \in \Sigma_P} B^{ar(o_i)}$, and any arrow in **Set** (a function) $f : B \longrightarrow C$, $\Sigma_P(f) \triangleq \biguplus_{o_i \in \Sigma_P} f^{ar(o_i)}$.

Moreover, for each set of variables X (an object in **Set**) we define the endofunctor $(X \uplus \Sigma_P) : \mathbf{Set} \longrightarrow \mathbf{Set}$ such that for any object B in **Set**, $(X \uplus \Sigma_P)(B) \triangleq X \uplus_{o_i \in \Sigma_P} B^{ar(o_i)}$, and any arrow in **Set** (a function) $f : B \longrightarrow C$, $(X \uplus \Sigma_P)(f) \triangleq \{id_X\} \uplus_{o_i \in \Sigma_P} f^{ar(o_i)}$, where id_X is the identity function for the set X.

Consequently, the *free* Σ_P-algebra $\mathcal{T}_P X$ on a set of variables X (X is called a set of *free generators*, and $\mathcal{T}_P X$ is said to be *freely generated* by X) may be computed as the union of the chain (so that the initial $X \uplus \Sigma_P$-algebra defined below is the *least fixed point* (lfp) of the functor $X \uplus \Sigma_P : \mathbf{Set} \to \mathbf{Set}$, with the isomorphism $\mathcal{T}_P X \simeq_I X \uplus \Sigma_P(\mathcal{T}_P X)$):

$$\emptyset \subseteq X \uplus \Sigma_P(\emptyset) \subseteq X \uplus \Sigma_P\big(X \uplus \Sigma_P(\emptyset)\big)$$
$$\subseteq X \uplus \Sigma_P\big(X \uplus \Sigma_P\big(X \uplus \Sigma_P(\emptyset)\big)\big) \subseteq \cdots$$

obtained by iterating the functor $X \uplus \Sigma_P$ on the empty set $\emptyset$, with $\Sigma_P(\emptyset)$ equal to the set of nullary operators (i.e., constants) in Σ_P. Hence, in **Set** we have the inclusion of variables in terms, $inl_X : X \hookrightarrow \mathcal{T}_P X$, and the injective mapping $inr_X : \Sigma_P(\mathcal{T}_P X) \to \mathcal{T}_P X$ which represents the composition of more complex terms from the more simple terms in $\mathcal{T}_P X$, that is, by using the operations $o_i \in \Sigma_P$ with $n = ar(o_i) \geq 1$ and hence for $t_1, \ldots, t_n \in \mathcal{T}_P X$, $inr_X((t_1, \ldots, t_n), i) = o_i(t_1, \ldots, t_n) \in \mathcal{T}_P X$ (if $ar(o_i) = 0$, $inr_X(o_i) = o_i \in \mathcal{T}_P X$).

The fact that standard mathematical constructions are inductive is mirrored by the common assumption the axioms of set theory include the 'axiom of foundation' which postulates that the set-membership relation '$\in$' is well-founded: for every set X, there is no infinitely descending chain $\cdots \in X_2 \in X_1 \in X_0 = X$. The axiom of foundation allows an inductive (idealized) construction of sets starting from the empty set $\emptyset$ (the base) and recursively applying the powerset operator mapping a set to all its subsets. For example, the inductive construction of natural (ordinal) numbers.

If instead of the ordinary category **Set** of sets we want to work with the classes (i.e., large sets) then we have to use the superlarge category **SET** with objects that can be classes as well and arrows the class-functions. We can form the *class* V of all sets in this case, namely the universe of sets. This class is a (strict) fixed point $\mathcal{P}_S(V) = V$ of the operator $\mathcal{P}_S$ mapping a class (i.e., possibly large set) to the class of all its (small) subsets. This operator can be extended to the endofunctor $\mathcal{P}_S : \mathbf{SET} \to \mathbf{SET}$ so that $\mathcal{P}_S(V) = V$ can be seen as an algebra structure $id : \mathcal{P}_S(V) \to V$ of this endofunctor. It is shown that the axiom of foundation is equivalent to postulating that "the universe $\mathcal{P}_S(V) = V$ is an initial $\mathcal{P}_S$-algebra." This provides the formal link between the initiality and a (generalized) induction on well-founded relations.

For the signature endofunctor Σ_P and every set of program variables X, the initial Σ_P-algebra (Lambek's Lemma) are isomorphisms (for the carrier set $\mathcal{T}_P X$),

$$(X \uplus \Sigma_P)(\mathcal{T}_P X) \simeq_I \mathcal{T}_P X$$

with the isomorphism (bijection in **Set**), $\simeq_I = [inl_X, inr_X] : X \uplus \Sigma_P(\mathcal{T}_P X) \longrightarrow \mathcal{T}_P X$, where $inl_X : X \hookrightarrow \mathcal{T}_P X$ and $inr_X : \Sigma_P(\mathcal{T}_P X) \longrightarrow \mathcal{T}_P X$ are injective arrows of the coproduct $X \uplus \Sigma_P(\mathcal{T}_P X)$. Consequently, the $(\mathcal{T}_P X, [inl_X, inr_X])$ is the initial $(X \uplus \Sigma_P)$-algebra and, consequently, for any other $(X \uplus \Sigma_P)$-algebra $(Z, [h, f])$ algebra (for each constant o_i, $ar(o_i) = 0$ and $h(o_i) = o_i \in Z$) there is a *unique* homomorphism $f_\# : (\mathcal{T}_P X, [inl_X, inr_X]) \to (Z, [h, f])$, as represented in the following commutative diagrams:

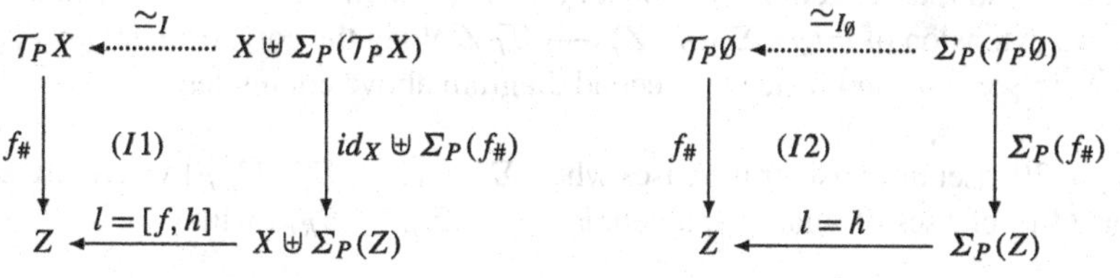

The left commutative diagram (I1) in **Set** corresponds to the unique homomorphism $f_\# : (\mathcal{T}_P X, \simeq_I) \to (Z, l)$, from the initial "syntax" $(X \uplus \Sigma_P)$-algebra to another $(X \uplus \Sigma_P)$-algebra)). If the set of variables X is the empty set $\emptyset$ (and hence f is the empty function, with empty graph) with $\emptyset \uplus S \simeq S$ in **Set**, we obtain the simpler commutative diagram above (I2) on the right (note that in **Set** the object $\emptyset \uplus S$ is isomorphic to S for any set S), where the unique homomorphism $f_\# : \mathcal{T}_P \emptyset \to Z$ represents the *meaning* of the ground terms in $\mathcal{T}_P \emptyset$. Note that the right commutative diagram (I2) represents the unique homomorphisms $f_\#$ from the initial Σ_P-algebra of *ground terms* $(\mathcal{T}_P \emptyset, \simeq_{I\emptyset})$ where $\simeq_{I\emptyset}$ is an isomorphism equal to into inr_X for $X = \emptyset$, into the Σ_P-algebra (Z, h).

This commutative diagram for the initial algebra with variables in X can be represented in the following way as well: by inductive extension (principle), for every Σ_P-algebra structure $h : \Sigma_P Z \longrightarrow Z$ and every mapping $f : X \longrightarrow Z$, there is the unique (inductive extension of h along the mapping f) arrow $f_\# : \mathcal{T}_P X \longrightarrow Z$ such that the following first diagram (in **Set**) commutes (the first diagram bellow is just recombination of the commutative diagram (I1). The second diagram is a particular case of the first diagram, where Z is replaced by $\mathcal{T}_P Z$ and the mapping $f : X \to Z$ by the mapping $inl_Z \circ f : X \to \mathcal{T}_P Z$):

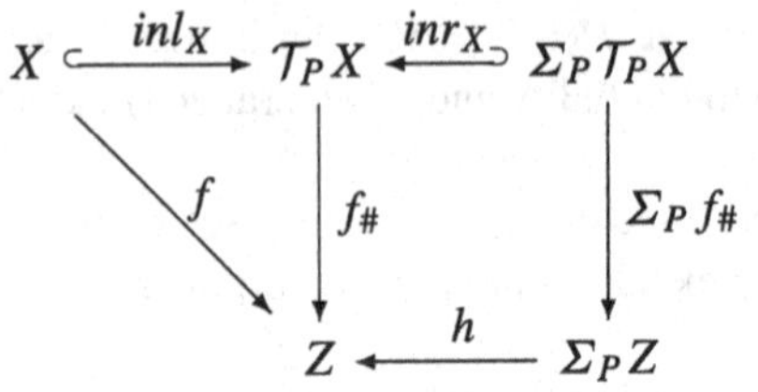

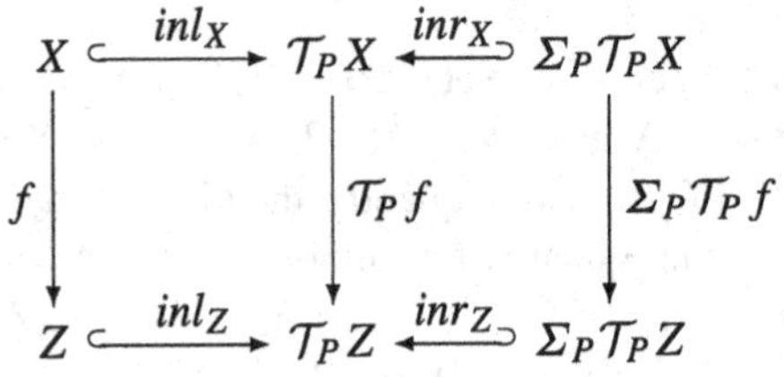

This inductive principle can be used to show that the operator $\mathcal{T}_P$ inductively extends to the *syntax* endofunctor $\mathcal{T}_P : \mathbf{Set} \to \mathbf{Set}$ (a monad freely generated by signature Σ_P), so that its action $\mathcal{T}_P f$ on an arrow (a function) $f : X \longrightarrow Z$ takes the inductive extension of $inr_Z : \Sigma_P(\mathcal{T}_P Z) \longrightarrow \mathcal{T}_P Z$ along the composite $inl_Z \circ f$, i.e., $\mathcal{T}_P f \triangleq (inl_Z \circ f)_\#$, and hence the second diagram above commutes.

Example 29 Let us consider the cases when $\Sigma_P \in \{\Sigma_R, \Sigma_R^+, \Sigma_{RE}\}$ where, by Definition 31 of classes of relational algebras, $\Sigma_R \subseteq \Sigma_R^+ \subseteq \Sigma_{RE}$, with

$$\Sigma_{RE}(X) = \biguplus_{o_k \in \Sigma_{RE}} X^{ar(o_k)} = \{\bot\} \bigcup_{o_i \in \Sigma_{RE}, ar(o_i)=1, i \geq 1} \{X \times \{i\}\} \bigcup_{i=-1,0,1} \{X^2 \times \{i\}\}$$

where $\bot \in \Sigma_{RE}$ (empty relation) is a constant (a unique nullary operator), unary operators are the family of projections _$[S_i]$, selections _ WHERE C_i, and EXTEND _ ADD a, *name* AS e operators, and the binary operators are Cartesian product _ TIMES _, _ UNION _ and _ MINUS _ operators.

Note that the nullary operator (a constant) $\bot$, together with the unary operators '*EXTEND...*' and the binary operation '*UNION*', is able to define any relation.

Then Z in the initial algebra semantics diagram above is an instance database (i.e., a simple object $Z \subseteq \underline{\Upsilon} \subset \Upsilon$ in **DB**, see Definition 26 in Sect. 3.2.5) and hence $X \subseteq \mathbb{R}$ and the mapping $f : X \to Z$ assigns the relations to variables (i.e., relational symbols) in X (with the fixed assignment $f(r_\emptyset) = \bot \in Z$ for the empty relation $r_\emptyset \in \mathbb{R}$). Thus, when f is a restriction of a general assignment $\|\_\| : X \to \underline{\Upsilon}$ introduced by Definition 31 for the class of relational algebras.

In all these cases, the unique $X \uplus \Sigma_P$-algebra homomorphisms $f_\# : \mathcal{T}_P X \to Z$ are the restrictions of the evaluation of terms mappings $\|\_\|_\# : \mathcal{T}_P \mathbb{R} \to \underline{\Upsilon}$.

Consequently, in the $X \uplus \Sigma_P$-algebra $[f, h] : X \uplus \Sigma_P(Z) \to Z$, the mapping $h : \Sigma_P(Z) \to Z$ has to satisfy $h(\bot) = \bot \in Z$ (for a given type (i.e., a signature) of algebras each carrier set Z must contain all constants, i.e., all nullary operators) and the following conditions (in order to make the above initial algebra semantics diagram (I1) commutative): for each unary (or binary) operator $o_i \in \Sigma_P$ and the relations $f_\#(t), f_\#(t_1), f_\#(t_2) \in Z$ (obtained from terms $t, t_1, t_2 \in \mathcal{T}_P X$)

1. $h(f_\#(t), i) = \|o_i(t)\|_\#$, where $o_i(t) \in \mathcal{T}_P X$ is a term, and $\|\_\ \|_\# : \mathcal{T}_P \mathbb{R} \rightarrow \underline{\Upsilon}$ in Definition 31 is the evaluation of terms (R-algebra f is a restriction of $\|\_\ \|$ to $X \subseteq \mathbb{R}$).
2. $h((f_\#(t_1), f_\#(t_2)), -1) = \|t_1 \quad \text{TIMES} \quad t_2\|_\#, h_A((f_\#(t_1), f_\#(t_2)), 0) = \|t_1 \text{ UNION } t_2\|_\#$ and $h_A((f_\#(t_1), f_\#(t_2)), 1) = \|t_1 \text{ MINUS } t_2\|_\#$ if $f_\#(t_1)$ and $f_\#(t_2)$ are union compatible; $\bot$ otherwise.

Let $\Sigma_P = \Sigma_R$ be SPRJU signature and $\mathcal{Z} = (S_Z, \Sigma_Z)$ be a schema of the instance database Z, with the set of relational symbols in S_Z. Let us show that Z has to be a closed object in **DB**. We have two possible cases:

1. When $S_Z \subseteq X$, the assignment $f : X \rightarrow Z$ must satisfy the condition for each $r \in X \subseteq \mathbb{R}$, $f(r) = \bot$ if $r \notin S_Z$. (In each object (instance database) in **DB** the empty relation $\bot$ is an element of it.)

   In this case, from the fact that the elements (i.e., relations) in Z are in the image of the mapping $f_\#$ (which is an evaluation of terms in $\mathcal{T}_P X$), this database instance has to be a closed object, that is, $Z = TD$ of an database instance $D \subseteq Z$.
2. When $S_Z \supset X$, let us define the database instance $C = \{f(r) \mid r \in X\}$. So, the image of $f_\#$ is the set of all views that can be obtained with SPRJU algebra from this instance database C and, consequently, $TC \subseteq Z$. However, Z is the image of the $h : \Sigma_R Z \rightarrow Z$ which computes all views over Z as well, so that, again Z has to be a closed object in **DB**, $Z = TD$, with $TC \subseteq TD$. The maximal case for Z is when $Z = \underline{\Upsilon} = T\underline{\Upsilon}$.

   In what follows, we will consider a particular case when $X = S_Z$, as in the case which defines the power-view operator T for the instance databases, based on the initial $X \uplus \Sigma_R$-algebra semantics.

Consequently, we obtain the *syntax* monad $(\mathcal{T}_P, \eta, \mu)$ with $\eta_X = inl_X$, while $\mu_X = (id_{\mathcal{T}_P X})_\#$ is the unique inductive extension of $inr_X : \Sigma_P(\mathcal{T}_P X) \longrightarrow \mathcal{T}_P X$ along the identity on $\mathcal{T}_P X$, as represented in the following commutative diagram:

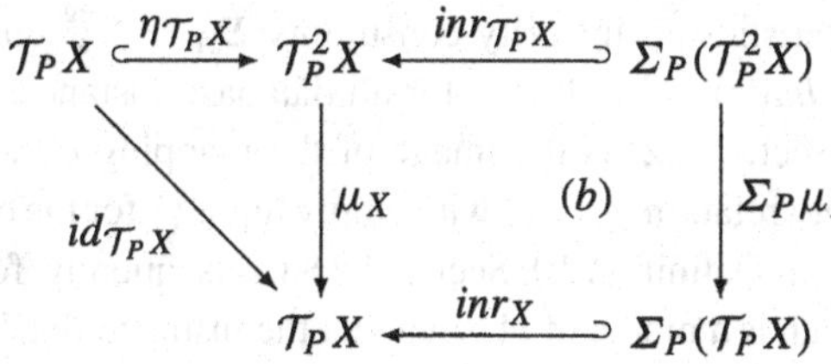

This syntax monad framework permits us to form the terms from any operator derivable from the signature Σ_P:

Example 30 Let us consider the set of integers $\mathbb{I}$ with the set of binary operators $\{+, -, *\}$ for addition, substraction, and multiplication, respectively, and the unary operator $|\_|$ (finding absolute value of integers), so that $\Sigma_P = \{\mathbb{I}, +, -, *, |\_|\}$.

Then we can define the set of only unary operators $\Sigma = \{\mathbb{I}, \_ + n, \_ - n, \_ * n \mid \text{for all finite positive integers } n \geq 1\} \cup \{|\_|\}$.

For instance, let us consider the *derived* unary operator '$(\_) + n$'. Hence, given any term $t \in \mathcal{T}_P X$, that is, a term $(t) \in \mathcal{T}_P^2 X$, one can form the term $t + n \in \mathcal{T}_P X$ by first applying $(\_) + n$ to t and then $\mu_X, t + n = \mu_X((t) + n)$.

This derivation can be seen from the commutative diagram (b) above: for $((t), n) \in (\mathcal{T}_P^2 X \times \mathcal{T}_P^2 X)$, for operation '+') $\subseteq \Sigma_P(\mathcal{T}_P^2 X)$, we have $inr_{\mathcal{T}_P X}((t), n) = (t) + n \in \mathcal{T}_P^2 X$, $\Sigma_P \mu_X((t), n) = (t, n) \in (\mathcal{T}_P X \times \mathcal{T}_P X$, for operation '+') $\subseteq \Sigma_P(\mathcal{T}_P X)$, so that $inr_X(t, n) = t + n = \mu_X((t) + n) \in \mathcal{T}_P X$. Derived operators can be seen as context and then the operator μ_X is formally needed to remove the brackets after plugging terms in the holes of a context.

Let us show the definition of the power-view operator T for a database instance A (a set of relational tables), introduced in Sect. 1.4.1, based on the free algebra of ground terms with the signature $\Sigma_P = \Sigma_R$ (a particular case of Example 29) where Σ_R is the signature of the "select–project–join+union" relational subalgebra in Definition 31, such that for a given database schema $\mathcal{A} = (S_F, \Sigma_A)$ its set of relational symbols is considered as a set of variables, $X = S_A \cup \{r_\emptyset\} = \{r_1, \ldots, r_n, r_\emptyset\}$ (or, as usual, $X = \mathcal{A}$), with

$$\Sigma_R(X) = \biguplus_{o_k \in \Sigma_R} X^{ar(o_k)} = \bigcup_{o_i \in \Sigma_R, ar(o_i)=1, i \geq 1} \{X \times \{i\}\} \bigcup_{i=-1,0} \{X^2 \times \{i\}\}$$

where the unary operators are: the family of projections $\_[S_i]$, selections $\_$ WHERE C_i operators, and two binary operators are Cartesian product $\_$ TIMES $\_$ and $\_$ UNION $\_$ operators (Join is a combination of the Cartesian product and selection). Thus, for a given "assignment" $\alpha : X \to A \subseteq TA$, we have $A = \alpha^*(\mathcal{A}) = \{R_1, \ldots, R_n, \bot\}$ with the relational tables $R_i = \alpha(r_i), i = 1, \ldots, n$, where α is a R-algebra which satisfies all integrity constraints $\Sigma_A = \Sigma_A^{\mathrm{egd}} \cup \Sigma_A^{\mathrm{tgd}}$ of this schema $\mathcal{A}$. So that $A = Im(\alpha)$ (with $\bot \in A$ for all database-instances (i.e., objects) in **DB** in Theorem 1, Sect. 3.1.2) is the image of the mapping α, with fixed assignment $\alpha(r_\emptyset) = \bot$ (empty relation $\bot = \langle\rangle$ with empty tuple $\langle\rangle$ for the empty type (relational symbol) $r_\emptyset \in \mathbb{R}$ (in Definition 10, Sect. 2.4.1). Consequently, for a schema $\mathcal{A}$ and an R-algebra α which is a model of $\mathcal{A}$, such that the instance-database $A = \alpha^*(\mathcal{A})$ is an object in **DB**, the set of views (i.e., relations) TA is determined from the *unique epic* $(X \uplus \Sigma_P)$-homomorphism $\alpha_\#$, from the initial $(X \uplus \Sigma_P)$-algebra $(\mathcal{T}_P X, \simeq)$ into this $(X \uplus \Sigma_P)$-algebra $(TA, [\alpha, h_A])$. This is represented by the following commutative diagram in **Set** (X is the set of rel. symbols in $\mathcal{A}$),

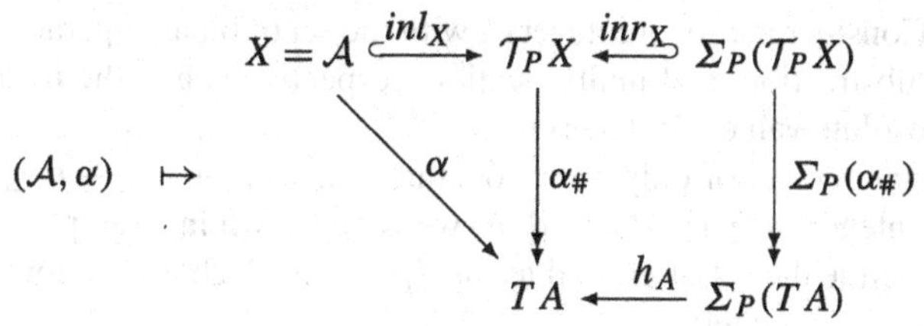

where the injective mapping inr_X represents the composition of terms with variables from the more simple terms, for each two terms $t, t' \in \mathcal{T}_P X$, as follows:

1. $inr_X(t, i) = o_i(t) \in \mathcal{T}_P X$, for each unary project/select operator $o_i \in \Sigma_R, i \geq 1$;
2. $inr_X((t, t'), -1) = t$ TIMES $t' \in \mathcal{T}_P X$, for Cartesian product;
3. $inr_X((t, t'), 0) = t$ UNION t'.

The mapping $h_A : \Sigma_P(TA) \to TA$ is defined in Example 29, for each unary (or binary) operator $o_i \in \Sigma_P = \Sigma_R$ and the relations $\alpha_\#(t), \alpha_\#(t_1), \alpha_\#(t_2) \in TA$, by:

1. $h_A(\alpha_\#(t), i) = \|o_i(t)\|_\#$, where $o_i(t) \in \mathcal{T}_P X$ is a term and evaluation of terms $\|\_\|_\# : \mathcal{T}_P \mathbb{R} \to \underline{\Upsilon}$ in Definition 31 (R-algebra α is a restriction of $\|\_\|$ to $X \Subset \mathbb{R}$).
2. $h_A((\alpha_\#(t_1), \alpha_\#(t_2)), -1) = \|t_1$ TIMES $t_2\|_\#$ and $h_A((\alpha_\#(t_1), \alpha_\#(t_2)), 0) = \|t_1$ UNION $t_2\|_\#$ if $\alpha_\#(t_1)$ and $\alpha_\#(t_2)$ are union compatible; $\bot$ otherwise.

Thus, $\alpha_\#$ is the unique epic inductive extension of h_A along the assignment $\alpha : X \to A \subseteq TA$ (for each A, $A \subseteq TA$) to all terms with variables in $\mathcal{T}_P X$ (the mapping $\alpha_\#$ is equal to α for the variables in $X \subseteq \mathcal{T}_P X$) with $\alpha_\#(r$ WHERE $\overline{1} \doteq \overline{1}) = \alpha(r) \in A \subseteq TA$ (because we have the identity unary operation "_ WHERE $\overline{1} \doteq \overline{1}$" (point 4.1 in Sect. 5.1)).

Notice that in the initial diagram above, in the case when $A = \underline{\Upsilon} \subset \Upsilon$ with $T\underline{\Upsilon} = \underline{\Upsilon}$ and $X = \mathbb{R}$, the R-algebra α corresponds to the evaluation $\|\_\|$ in Definition 31.

This initial semantics, which maps each pair $(\mathcal{A}, \alpha)$ with $A = \alpha^*(\mathcal{A})$ into the set of views TA, algebraically defines the power-view operator $T : Ob_{DB} \to Ob_{DB}$, which is then extended to the fundamental power-view endofunctor $T : \mathbf{DB} \to \mathbf{DB}$ in Sect. 3.2.1.

5.2 Action-Relational-Algebra RA Category

Action-based categories, for a given Σ-algebra, can be obtained if each arrow is a unary operation of this algebra and the target object of such an arrow is equal to the term of this algebra, obtained by applying this unary operation to the source object (which is another term of the same algebra) of this arrow. Hence, the set of objects in such a category is equal to the set of terms of this algebra, and for each arrow its target object is obtained from its source object by applying the action (defined by this arrow) to this source object. In this case, we have to transform the n-ary operators, $n \geq 2$, of this algebra into the set of unary operators, as follows in this simple example:

Example 31 Consider the set of integers $\mathbb{I}$ with the set of binary operators $\{+, -, *\}$ for addition, substraction, and multiplication, respectively, and the unary operator $|\_|$ (finding absolute value of integers).

The we define the set of only unary operators $\Sigma = \{\_ + n, \_ - n, \_ * n \mid \text{for all}$ finite positive integers $n \geq 1\} \cup \{\mathbb{I}, |\_|\}$, as we have shown in Example 30.

Hence, we define the set of ground terms $\mathcal{T}_I$ of this algebra as follows:

1. The constants in $\mathbb{I}$ are terms.
2. If t_I is a term and $\_ \odot \in \Sigma$ a unary operator then $t_I \odot$ is a term.

   So, for instance, when $\_ \odot$ is equal to the unary operation $\_ + 4$ and for a constant -1, the expression $-1 + 4$ is a term in $\mathcal{T}_I$.

Consequently, we can define an action-category for this Σ-algebra as follows:

1. The set of objects of this category is equal to the set of terms $\mathcal{T}_I$.
2. The set of arrows of this category is equal to the set of unary operators in Σ. The identity arrows are just the identity operators $(\_ + 0, \_ - 0, \_ * 1)$.

Consequently, for a source object $t_I \in \mathcal{T}_I$ and an operation $\_\odot \in \Sigma$, we have the arrow $\_\odot : t_I \rightarrow t_I \odot$ in this action-category as, for example, the arrow $\_+4 : (-1) \rightarrow (-1+4)$. Notice that in such an action-category the *nidification* of (unary) operators in a given term can be simply represented as a *composition of arrows*. For instance, the term $(((-1) + 4) - 3) * 6$, can be equivalently represented as the following composition of three arrows: $\_ * 6 \circ \_ - 3 \circ \_ + 4 : (-1) \rightarrow (((-1) + 4) - 3) * 6$.

For a given term $t_I \in \mathcal{T}_I$ we denote by $\|t_I\|$ the evaluated value of this term: for instance, let t_I be the term $4 - 5 + 7$, then $\|t_I\| = 6$. We say that two terms t_I, t'_I are equal if $\|t_I\| = \|t'_I\|$, and hence the terms $4 - 5 + 7$ and $9 - 3$ are equal.

As we may see from this example, in an action-category the nidification of (unary) operators of a given finite term of an algebra can be represented as a composition of finite number of arrows in the *same ordering* of the nidification of the atomic operators of this algebra. Or, equivalently, the composition of the algebraic operators is traduced into the composition of the arrows in such an action-category, with the property that each target object of a given arrow is equal to the application of the operator of this arrow to the source object of this arrow: the target object is exactly the result of application of this unary operator to the source object of this arrow.

Consequently, what we need in order to define an action-category **RA** for the Σ_{RE}-algebra introduced in Sect. 5.1 is to reduce all its binary operators to a unique binary operator and by introducing the new auxiliary unary operators to replace the eliminated binary operators. The natural choice for the unique binary operator is the Cartesian product _ TIMES _, with the additional requirement that the updates in any given term t_R are unfolded directly to the real relational tables.

The necessity to transform each term t_R of the Σ_{RE}-algebra into a composed arrow (by using unary operators only) in RA means that we must unfold the _ TIMES _ binary operations toward the source object of this composed object: it means that at the end of these transformations, the operators _ TIMES _ appear only at the most nidified level of each term, and consequently, the Cartesian products will be executed only over (eventually updated) real relational tables. Consequently, each original term-tree (with leafs equal to real relational tables, as in Example 28) has to be transformed into an equivalent *single-path-term*, where the source object is a Cartesian product of (eventually updated) real relational tables that appear in the original term.

The algorithm how to transform any term into such a single-path-term, by unfolding the updates (by 'EXTEND...') and Cartesian products to the most nidified level will be presented in what follows. Let us consider first the reduction of the '_ UNION _' and '_ MINUS _' binary operators in Σ_{RE}, by introducing two new unary operators '_ REDUCE S_2 TO S_1' and '_ DISJOINT S_2 FROM S_1'.

Definition 32 Let for a given r_1 let $S_1 = \langle nr_{r_1}(i), \dots, nr_{r_1}(i+k)\rangle$ and $S_2 = \langle nr_{r_1}(j), \dots, nr_{r_1}(j+k)\rangle$ with $i \geq 1$, $k \geq 0$, $j > i+k$ and $j+k \leq ar(r_1)$. Then we define the following two unary operators:

1. Reduction is a unary operation written as _ REDUCE S_2 TO S_1 such that for a relation r_1 and $S = \langle nr_{r_1}(1), \dots, nr_{r_1}(ar(r_1))\rangle$ we define the relation r by:

$$r_1 \text{ REDUCE } S_2 \text{ TO } S_1,$$

   where $\|r\| = \pi_{S \setminus S_2}(\|r_1\| \cup \{\mathbf{d}' \mid \text{for each } \mathbf{d} \in \|r_1\| \text{ the tuple } \mathbf{d}' \text{ is obtained by replacing } d_{i+m} \text{ by } d_{j+m} \text{ for } m = 0, \dots, k \text{ in } \mathbf{d}\})$.
2. Disjoint is a unary operation written as _ DISJOINT S_2 FROM S_1 such that for a relation r_1 and $S = \langle nr_{r_1}(1), \dots, nr_{r_1}(ar(r_1))\rangle$ we define the relation r by:

$$r_1 \text{ DISJOINT } S_2 \text{ FROM } S_1,$$

   where $\|r\| = \pi_{S \setminus S_2}(\{\mathbf{d} \mid \mathbf{d} \in \|r_1\| \text{ and } \nexists \mathbf{d}' \in \|r_1\| \text{ such that } (d_i = d'_j) \wedge \dots \wedge (d_{i+k} = d'_{j+k})\})$.

In both cases, $\pi_{S \setminus S_2}$ is the projection for all columns different from that in S_2 (or, equivalently, the elimination of all columns in S_2), so that $ar(r) = ar(r_1) - (k+1)$ and nr_r, atr_r are the reductions of nr_{r_1} and atr_{r_1}, respectively, obtained by this projection.

Proposition 20 *Let us define the Σ_{RA} algebra for the relations composed of the unique binary operator* _ TIMES _ *(denoted by* _ $\otimes$ _ *as well), the unary operators Renaming, Projection, Selection and Extend (i.e.,* _ $\langle a, name, e\rangle$*) in Sect.* 5.1 *and the unary operators Reduction and Disjoint in Definition* 32.

We denote the set of terms (with variables $X \subseteq \mathbb{R}$) of this Σ_{RA} algebra by $\mathcal{T}_{RA}X$, and the evaluation of terms by the function $\|\_\|_\# : \mathcal{T}_{RA}X \to \underline{\Upsilon}$ (which is a unique extension of a given assignment $\|\_\| : \mathbb{R} \to \underline{\Upsilon}$ as in Definition 31 *and Sect.* 5.1.1).

The strict subalgebra of Σ_{RA}, with only the Cartesian product binary operation _ $\otimes$ _ *and the unary operations* _ $\langle a, name, e\rangle$*, will be denoted by Σ_{RA}^-, and the set of its terms by $\mathcal{T}_{RA}^- X$ (i.e., $\mathcal{T}_{RA}^- X \subset \mathcal{T}_{RA}X$).*

There is a term-rewriting mapping $Trw : \mathcal{T}_{RE}X \to \mathcal{T}_{RA}X$ such that a term $t_R \in \mathcal{T}_{RE}X$ of the Σ_{RE} algebra in Definition 31 *can be transformed into an equivalent term in $Trw(t_R) \in \mathcal{T}_{RA}X$ with the operators of this new Σ_{RA} algebra.*

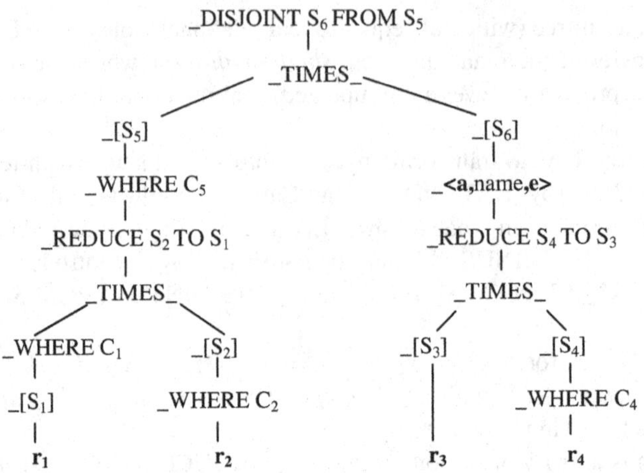

Proof We have to show that the binary operators 'UNION' and 'MINUS' can be equivalently expressed by operators in this new Σ_{RA} algebra. In fact, for any two terms t_R, t'_R of the Σ_{RE}-algebra with the union-compatible relations r_1 with $\|r_1\| = \|t_R\|_\#$, and r_2 with $\|r_2\| = \|t'_R\|_\#$, with $n = ar(r_1) = ar(r_2)$, we have that

't_R UNION t'_R' is equivalent to t_R if $\|t'_R\|_\#$ is empty, to t'_R if $\|t_R\|_\#$ is empty, and to '$(t_R \otimes t'_R)$ REDUCE S_2 TO S_1' otherwise;

't_R MINUS t'_R' is equivalent to t_R if $\|t'_R\|_\#$ is empty or $\|t_R\|_\#$ is empty, and to '$(t_R \otimes t'_R)$ DISJOINT S_2 FROM S_1' otherwise,

where $S_1 = \langle nr_r(1), \dots nr_r(n)\rangle$, $S_2 = \langle nr_r(n+1), \dots nr_r(2n)\rangle$, for the relation r equal to the Cartesian product $r_1 \otimes r_2^\rho$.

By application of these transformations to each term $t_R \in \mathcal{T}_{RE}X$ of the Σ_{RE}-algebra as far as possible, we obtain an equivalent to it term $t_{RA} \in \mathcal{T}_{RA}X$, and we define a mapping *Trw* by $Trw(t_R) = t_{RA}$, with $\|Trw(t_R)\|_\# = \|t_R\|_\#$. □

Example 32 Let us consider the term t_R given in Example 28 equal to the algebraic expression

$$(((((r_1[S_1])\text{WHERE } C_1)\text{UNION}((r_2\text{WHERE } C_2)[S_2]))\text{WHERE } C_5)[S_5])$$
$$\text{MINUS}((\text{EXTEND}((r_3[S_3])\text{UNION}((r_4\text{WHERE } C_4)[S_4]))$$
$$\text{ADD } a, \mathit{name} \text{ AS } e)[S_6]),$$

which is represented by the tree in Example 28, with the four different paths that end with leafs which are equal to relational tables.

We will use both 'TIMES' and $\otimes$ symbols for the Cartesian products.

By reduction of 'MINUS' and 'UNION' operations into the composition of operations in Σ_{RA}, we obtain the equivalent term $t_{RA} \in \mathcal{T}_{RA}$,

$$((((((((r_1[S_1])\text{WHERE } C_1))\text{TIMES}(((r_2[S_2])\text{WHERE } C_2))\text{REDUCE}S_2\text{TO}S_1)$$

$$\text{WHERE } C_5)[S_5])\text{TIMES}(((((((r_3[S_3])\text{TIMES}((r_4\text{WHERE } C_4)[S_4])))$$

$$\text{REDUCE}S_4\text{TO}S_3)\langle a, name, e\rangle))\text{WHERE } (name = e))[S_6]))$$

$$\text{DISJOINT}S_6\text{FROM}S_5$$

represented by the tree above (notice that with this reduction of the term-trees of the Σ_{RE}-algebra to the equivalent terms in $\mathcal{T}_{RA}X$, we replace the original binary nodes with the 'TIMES' nodes and we introduce the new unary operations, and hence the obtained tree is more complex than original tree).

Let us now define the category for this new action-relational algebra:

Proposition 21 *We define the action-relational category* **RA** *as follows*:
1. *The objects of this category are* (*here* '$\backslash$' *is the set difference*):
   1.1. $\mathcal{T}_{RA}^{-}X \subset Ob_{RA}$, *for* $X = \mathbb{R}$;
   1.2. *For each object* $t_{RA} \in Ob_{RA}$ *and each unary operator* $o_i \in \Sigma_{RA} \backslash \Sigma_{RA}^{-}$, *the term* $o_i(t_{RA})$ *is an object, i.e.*, $o_i(t_{RA}) \in Ob_{RA}$.
2. *The* atomic *arrows of this category are the set of all unary operators in* $\Sigma_{RA} \backslash \Sigma_{RA}^{-}$.

The category RA is skeletal.

Proof Notice that this is really an action-category, and hence for each unary operation o_i and term t_{RA} we have the arrow $o_i : t_{RA} \to o_i(t_{RA})$ because $o_i(t_{RA})$ is a also term in $\mathcal{T}_{RA}X$ and hence an object in **RA**.

For any object $t_{RA} \in Ob_{RA}$ (a term in $\mathcal{T}_{RA}X$) we have that its identity arrow $id_{t_{RA}}$ is equal to the identity unary operation $\_$ WHERE C when C is the atomic condition $\overline{1} \doteq \overline{1}$ (a tautology), so that $id_{t_{RA}} = (\_ \text{ WHERE } \overline{1} \doteq \overline{1}) : t_{RA} \to (t_{RA} \text{ WHERE } \overline{1} \doteq \overline{1})$, form the fact that the term (i.e., object) $(t_{RA} \text{ WHERE } \overline{1} \doteq \overline{1})$ is equal to t_{RA}.

Given any two arrows $o_i : t_{RA} \to o_i(t_{RA})$ and $o_j : o_i(t_{RA}) \to o_j(o_i(t_{RA}))$, their composition is still a unary operation $o_j o_i$ obtained by their composition, and hence the composition of these two arrows is an arrow equal to $o_j \circ o_i : t_{RA} \to o_j(o_i(t_{RA}))$.

From the fact that the composition of unary operations is associative, the composition of arrows in **RA** satisfies the associativity property.

Two objects (terms) t_{RA}, t'_{RA} are isomorphic, $t_{RA} \simeq t'_{RA}$, if they are equal (if for all assignments $\|\_\| : \mathbb{R} \to \Upsilon$, $\|t_{RA}\|_\# = \|t'_{RA}\|_\#$) and hence this category is skeletal. Thus, $t_{RA} \simeq t'_{RA}$ iff $t_{RA} \approx t'_{RA}$, and the iso arrows are $is = (\_\text{WHERE}\overline{1} \doteq \overline{1}) : t_{RA} \to t'_{RA}$ and $is^{-1} = (\_ \text{ WHERE } \overline{1} \doteq \overline{1}) : t'_{RA} \to t_{RA}$, so that $is^{-1} \circ is = (\_ \text{ WHERE } (\overline{1} \doteq \overline{1}) \wedge (\overline{1} \doteq \overline{1})) = (\_ \text{ WHERE } \overline{1} \doteq \overline{1}) = id : t_{RA} \to t_{RA}$ (analogously, $is \circ is^{-1} = id : t'_{RA} \to t'_{RA}$). Notice that is is well defined, that is,

$(t_{RA}$ WHERE $\overline{1} \doteq \overline{1}) \approx t_{RA} \approx$ (from the isomorphism) $\approx t'_{RA}$. Analogously, it holds for is^{-1} as well. □

The set of all arrows Mor_{RA} is obtained by the composition of its atomic arrows. Thus, each arrow in **RA** is just a composition of a number of unary operations in $\Sigma_{RA} \backslash \Sigma_{RA}^{-}$. From this proposition we can see that $Ob_{RA} \subset \mathcal{T}_{RA}X$, so that we have to show that this algebra is rich enough to represent *all* terms in $\mathcal{T}_{RA}X$ and, consequently, form Proposition 20, all terms of the original relational algebra Σ_{RE} in Definition 31.

5.2.1 Normalization of Terms: Completeness of RA

In this section, we will show that the **RA** category is complete w.r.t. the set of terms in $\mathcal{T}_{RA}X$, in the way that each term $t_{RA} \in \mathcal{T}_{RA}X$ is represented by a (composed) arrow $f : t'_{RA} \to f(t'_{RA})$ such that $t'_{RA} \in \mathcal{T}_{RA}^{-}X$ is a term represented as a particular Cartesian product of (eventually updated) relational tables that are the leafs in t_{RA} and $\| f(t'_{RA})\|_{\#} = \|t_{RA}\|_{\#}$. Intuitively, it means that each term-tree (as that in Example 32) with a number of leafs can be reduced to a single path-term with a unique leaf obtained as a Cartesian product of (eventually updated) relational tables.

Consequently, we have to show that each Cartesian product in a given term can be unfolded down to the unique leaf (represented by Cartesian product of the relational tables), and that update unary operations "EXTEND…", i.e., $\_\ \langle a, name, e\rangle$ (that are not atomic arrows in **RA**), can also be unfolded down to the unique leaf (represented by Cartesian product of the relational tables).

The first step to do it is to define the "Cartesian product" for the arrows in **RA**, as follows:

Proposition 22 *Given any two arrows in* **RA**, $f_1 : t_{RA,1} \to f_1(t_{RA,1})$ *and* $f_2 : t_{RA,2} \to f_2(t_{RA,2})$, *with* $t_{RA,1}, t_{RA,2} \in \mathcal{T}_{RA}^{-}X$ *and the relations* $R_1 = \|f_1(t_{RA,1})\|_{\#}$ *and* $R_2 = \|t_{RA,2}\|_{\#}$, *with the arrows,*

1. $f_1^+ : t_{RA,1} \otimes t^{\rho}_{RA,2} \to f_1(t_{RA,1}) \otimes t^{\rho}_{RA,2}$, *obtained from* f_1, *where each operation* $\_\ [S]$ *is substituted by the operation* $\_\ [S \& nr(R_2^{\rho})]$, *and*
2. $f_2^+ : f_1(t_{RA,1}) \otimes t^{\rho}_{RA,2} \to f_1(t_{RA,1}) \otimes f_2(t^{\rho}_{RA,2})$, *obtained from* f_2, *where each operation* $\_\ [S]$ *is substituted by the operation* $\_\ [nr(R_1) \& S]$.

Then we define the operation of "Cartesian product" $\otimes^1$ *for the arrows by:*

$$f_1 \otimes^1 f_2 \triangleq \left(f_2^+ \circ f_1^+\right) : t_{RA,1} \otimes t^{\rho}_{RA,2} \to f_1(t_{RA,1}) \otimes f_2\left(t^{\rho}_{RA,2}\right).$$

The Cartesian product for terms (objects) and for arrows defines the Cartesian tensor bifunctor $(\otimes, \otimes^1) : \mathbf{RA} \times \mathbf{RA} \to \mathbf{RA}$, *both with the isomorphisms, for any* $A, B, C \in Ob_{RA}$: $\alpha_{A,B,C} : (A \otimes B) \otimes C \simeq A \otimes (B \otimes C)$, $\beta_A : r_\emptyset \otimes A \simeq A$, *and* $\gamma_A : A \otimes r_\emptyset \simeq A$, *which are the identities. Hence,* **RA** *is a strict monoidal category with tensorial product* $\otimes$ *and the unit* $r_\emptyset \in \mathbb{R}$ *with the empty relation* $\|r_\emptyset\|_{\#} = \perp$.

Proof It is enough to demonstrate that $\|(f_1 \otimes^1 f_2)(t_{RA,1} \otimes t^{\rho}_{RA,2})\|_{\#} = \|f_1(t_{RA,1}) \otimes f_2(t^{\rho}_{RA,2})\|_{\#}$ and hence $f_1 \otimes^1 f_2 : t_{RA,1} \otimes t^{\rho}_{RA,2} \to (f_1 \otimes^1 f_2)((t_{RA,1} \otimes t^{\rho}_{RA,2}))$ is just an arrow in **RA**.

In fact, the first arrow f_1^+ executes all operations over the relation $\|t_{RA,1}\|_{\#}$ as the original arrow f_1, while all projection operations $\_[S]$ with a tuple of attributes in S in f_1 are transformed into $\_[S\&nr(R^{\rho}_2)]$ (concatenation of the tuple S and the tuple of names $nr(R^{\rho}_2)$, in the way that also all columns of the relation R^{ρ}_2 are propagated to the final result. Analogously, the second arrow f_2^+ executes all operations over the relation R^{ρ}_2 as the original arrow f_2, while all projection operations $\_[S]$ in f_2 are transformed into $\_[nr(R_1)\&S]$, in the way that all columns of the relation R_1 are propagated to the final result. Consequently, $\|(f_1 \otimes^1 f_2)(t_{RA,1} \otimes t^{\rho}_{RA,2})\|_{\#} = \|(f_2^+(f_1^+(t_{RA,1} \otimes t^{\rho}_{RA,2})))\|_{\#} = \|f_1(t_{RA,1}) \otimes f_2(t^{\rho}_{RA,2})\|_{\#}$.

From the fact that $f_1 \otimes^1 f_2 = f_2^+ \circ f_1^+$, for the cases when f_1 or f_2 (or both) are the identity arrows, we obtain that $f_2^+ = id_1 \otimes^1 f_2 = \otimes^1(id_1, f_2)$, $f_1^+ = f_1 \otimes^1 id_2 = \otimes^1(f_1, id_2)$ and hence $\otimes^1(id_1, id_2) = id$. Thus, $f_1 \otimes^1 f_2 = f_2^+ \circ f_1^+ = (id_1 \otimes^1 f_2) \circ (f_1 \otimes^1 id_2) = (id_1 \circ f_1) \otimes^1 (f_2 \circ id_2)$ (it is easy to show that this is a special case of the more general property $(g_2 \otimes^1 h_2) \circ (g_1 \otimes^1 h_1) = (g_2 \circ g_1) \otimes^1 (h_2 \circ h_1)$), and hence $\otimes^1$ satisfies the functorial properties.

Hence $\|(A \otimes B) \otimes C\|_{\#} = \|A \otimes (B \otimes C)\|_{\#}$, $\|\bot \otimes A\|_{\#} = \|A\|_{\#}$, and $\|A \otimes \bot\|_{\#} = \|A\|_{\#}$, so that the isomorphisms $\alpha_{A,B,C}$, λ_A and ϱ_A are the identity arrows in **RA**, and hence **RA** is a strict monoidal category. □

Based on the definition, we can reduce each term $f_1(t_{RA,1}) \otimes f_2(t^{\rho}_{RA,2})$ to a unique arrow in **RA** by an unfolding of $\otimes$, as follows:

$$
f_1(t_{RA,1}) \otimes f_2(t^{\rho}_{RA,2}) \quad \mapsto \quad
\begin{array}{c}
f_2^+(f_1^+(t_{RA,1} \otimes t^{\rho}_{RA,2})) \\
\Big\uparrow \; f_1 \otimes^1 f_2 = f_2^+ \circ f_1^+ \\
t_{RA,1} \otimes t^{\rho}_{RA,2}
\end{array}
$$

Definition 33 The unfolding of a simple binary tree-term in $\mathcal{T}_{RA}X$, with the node $\_$ TIMES $\_$ (or $\_ \otimes \_$) and the leafs $t_{RA,1} = dom(f_1)$ and $t_{RA,2} = dom(f_2)$ in $\mathcal{T}^-_{RA}X$ where the paths f_1 and f_2 are the compositions of the unary operations in $\Sigma_{RA} \backslash \Sigma^-_{RA}$, into a simple path tree with a unique leaf $t_{RA,1} \otimes t^{\rho}_{RA,2} \in \mathcal{T}^-_{RA}X$, is presented by:

tree-term $f_1(t_{RA,1}) \otimes f_2(t^{\rho}_{RA,2})$ *path-term* $f_2^+(f_1^+(t_{RA,1} \otimes t^{\rho}_{RA,2}))$

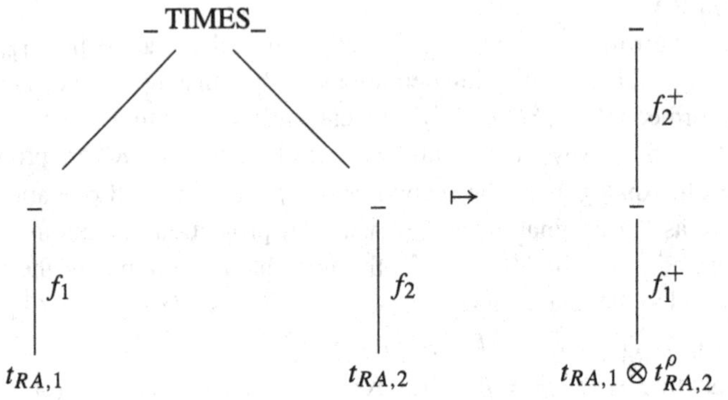

Let us consider now the unfolding of the 'update' operations _ $\langle a, name, e\rangle$ (i.e., 'EXTEND...') along the paths composed of unary operations in $\Sigma_{RA} \backslash \Sigma^-_{RA}$, as follows:

Definition 34 The unfolding of an operation _ $\langle a, name, e\rangle$ along a path composed of unary operations in $\Sigma_{RA} \backslash \Sigma^-_{RA}$, for a given term t_{RA} and corresponding relational symbol r, such that $\|r\| = \|t_{RA}\|$, $name \notin nr(r)$, is defined by the following cases:

1. $(t_{RA}$ RENAME $name_2$ AS $name_1)\langle a, name, e\rangle \mapsto (t_{RA})\langle a, name, e_{nr}\rangle$, where e_{nr} is obtained from e by substitution of $name_1$ with $name_2$.
2. $(t_{RA}[S])\langle a, name, e\rangle \mapsto ((t_{RA})\langle a, name, e\rangle)[S\&name]$.
3. $(t_{RA}$ WHERE $C)\langle a, name, e\rangle \mapsto ((t_{RA})\langle a, name, e\rangle)$ WHERE C.
4. Let $S_1 = \langle nr_r(i), \ldots, nr_r(i+k)\rangle$ and $S_2 = \langle nr_r(j), \ldots, nr_r(j+k)\rangle$ for $i \geq 1$, $k \geq 0$, $j > i+k$ and $j+k < ar(r)$, then
   4.1. $(t_{RA}$ REDUCE S_2 TO $S_1)\langle a, name, e\rangle \mapsto ((t_{RA}\langle a, name, e\rangle)\langle a(1), name_1, e_{rn}\rangle)$ REDUCE $[S_{2+}]$ TO $[S_{1+}]$, where $S_{2+} = \langle nr_r(i), \ldots, nr_r(i+k), name_1\rangle$, $S_{1+} = \langle nr_r(j), \ldots, nr_r(j+k), name\rangle$ and e_{nr} is obtained from e by replacing for every $1 \leq m \leq k$, $nr_r(i+m) \in S_1$ with $nr_r(j+m) \in S_2$ and $a(1) = a$, and $name_1$ a fresh new name not in $nr(r)$.
   4.2. $(t_{RA}$ DISJOINT S_2 FROM $S_1)\langle a, name, e\rangle \mapsto (t_{RA}\langle a, name, e\rangle)$ DISJOINT $[S_2]$ FROM $[S_1]$.

Let us show how the unfolding of the operations _ $\langle a, name, e\rangle$ (i.e., 'EXTEND...') along a path composed of unary operations in $\Sigma_{RA} \backslash \Sigma^-_{RA}$, given by Definition 34, and of the binary operation _ TIMES _ (or _ $\otimes$ _), can be used in order to transform any tree-term $t_{RA} \in \mathcal{T}_{RA}X$ into an equivalent single path-term with the unique leaf $t'_{RA} \in \mathcal{T}^-_{RA}X$ (a Cartesian product of (eventually updated) relational tables), by the following tree-term transformations, presented in the following example with a number of diagrams:

Example 33 Consider the term t_R given in Example 32, represented here in the first step of the tree-term transformations.

The next steps of the transformations in these diagrams are denoted by $\downarrow (n)$, where n is the index of the next transformation-step to be executed.

In fact, in Step 1, we have indicated by $\downarrow (2)$ and $\downarrow (3)$ the unfolding of binary nodes _ TIMES _ (or $\_ \otimes \_$), as specified by Definition 33. Notice that any two consecutive operations _ WHERE C_k and _ WHERE C_m, can be shortened by the equivalent operation _ WHERE $C_k \wedge C_m$.

After the execution of the Steps 2 and 3, we obtained the tree represented by the diagram Step 3. The next step of transformation $\downarrow (4)$ corresponds to the unfolding of the unary operation $\langle a, name, e\rangle\_$ along the path composed of unary operations in $\Sigma_{RA} \backslash \Sigma_{RA}^{-}$, equal to (_ REDUCE S_4 TO S_3) $\circ$ (_ $[S_3\&S_4]$) $\circ$ (_ WHERE C_4), as it is specified by Definition 34. The unfolding of $\langle a, name, e\rangle\_$ over (_ REDUCE S_4 TO S_3) is presented in Step 4, and corresponds to the transformation defined in point 4.1 of Definition 34 (note that the attribute $a(1)$ is a copy of the attribute a, while $nome_1$ is a new fresh name different from the original $name$).

By $\downarrow (5)$ we indicated the next steps for unfolding of $(\_ \langle a(1), name_1, e_{nr}\rangle) \circ (\_ \langle a, name, e\rangle)$ along the operations _ $[S_3\&S_4]$, and its result is presented in the diagram Step 5.

By $\downarrow (6)$ we indicated the next steps for unfolding of $(\_ \langle a(1), name_1, e_{nr}\rangle) \circ (\_ \langle a, name, e\rangle)$ along the operations _ WHERE C_4, with resulting diagram in Step 6.

In the last Step 7, we executed the unfolding of _ TIMES _ binary operators to leafs, as specified by tensorial ("Cartesian") product of the two paths in Proposition 22, and hence we obtained the final normalized one path-term with the unique leaf which is a term in $\mathcal{T}_{RA}^{-}X$ (a Cartesian product of relational tables with all updates).

Based on this example, we can generalize this process in the following proposition:

Proposition 23 (THE COMPLETENESS OF **RA**) *There is a normalization mapping $Nrm : \mathcal{T}_{RA}X \to Mor_{RA}$ with composed mapping $Rd_{RE} = Nrm \circ Trw : \mathcal{T}_{RE}X \to Mor_{RA}$ such that a term $t_R \in \mathcal{T}_{RE}X$ of the Σ_{RE}-algebra in Definition* 31 *is represented by an arrow $f : t'_{RA} \to f(t'_{RA})$, with:*

1. *The source object is a term $t'_{RA} \in \mathcal{T}_{RA}^{-}X$ composed as a Cartesian products of relational tables (that appear in the original term t_R) updated eventually by the 'EXTEND...' operations $\_ \langle a, name, e\rangle$.*
2. *The target object is a term $f(t'_{RA}) \in \mathcal{T}_{RA}X$, where f is a path composed of the unary operations (arrows) in $\Sigma_{RA} \backslash \Sigma_{RA}^{-}$, such that $f(t'_{RA})$ is equivalent to t_R.*

Proof The first step is to reduce a given term $t_R \in \mathcal{T}_{RE}X$ of the Σ_{RE}-algebra to the equivalent term t_{RA}, by using the term-rewriting algorithm $Trw : \mathcal{T}_{RE}X \to \mathcal{T}_{RA}X$ in Proposition 20. This binary tree-term t_{RA} with a given root operation can have as binary nodes only the Cartesian product operations _ *TIMES* _ (or $\_ \otimes \_$). The

Step 1

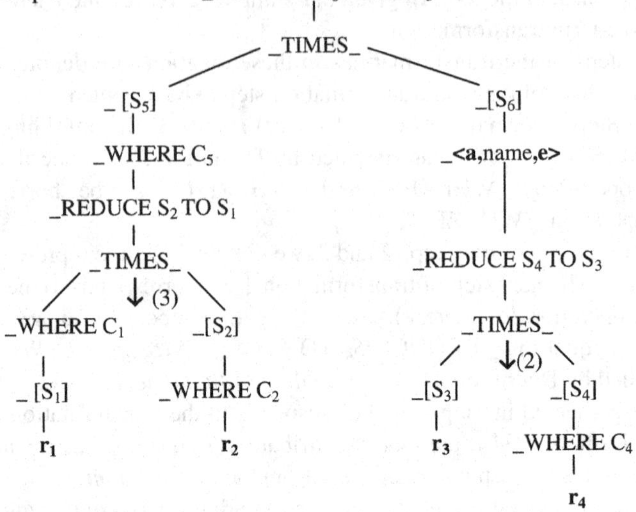

Step 3

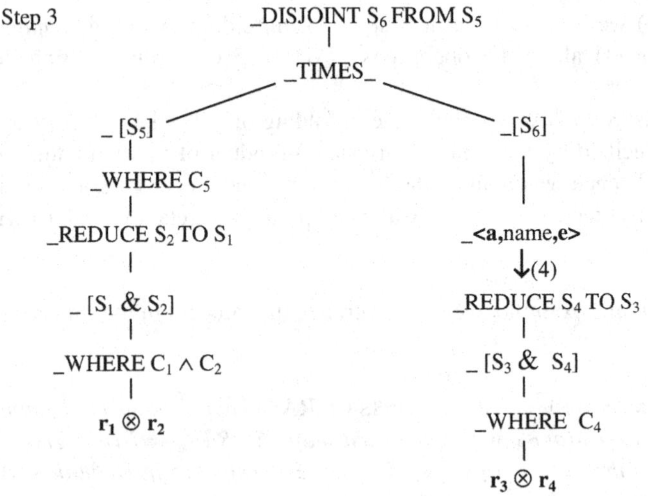

binary nodes such that both of its arguments are the terms *without* another binary nodes are denominated *atomic nodes* (in Example 33 the binary nodes denoted by $\downarrow n, n = 2, 3, 7$ are atomic).

Based on this consideration, we present the following algorithm for the reduction of a (tree) term $t_{RA} \in \mathcal{T}_{RA}X$ into an equivalent path-term with the unique leaf $t'_{RA} \in \mathcal{T}^{-}_{RA}X$:

1. If t_{RA} is a tree-term (with binary nodes), then go to Step 2. If this path-term contains the unary operations $\_\,\langle a, name, e\rangle$ go to Step 3; otherwise go to Step 4.

Step 4

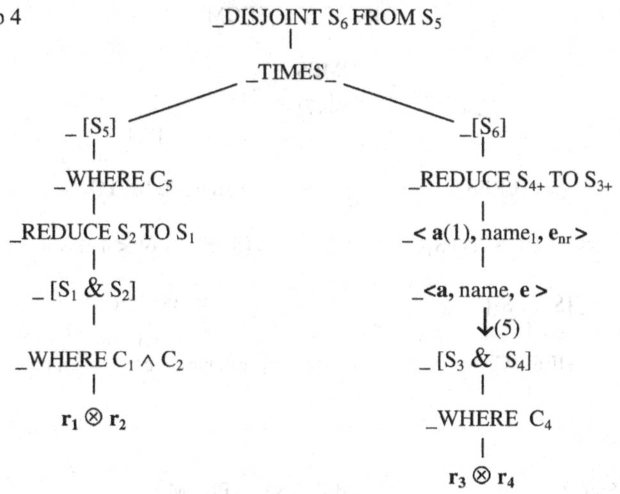

Step 5

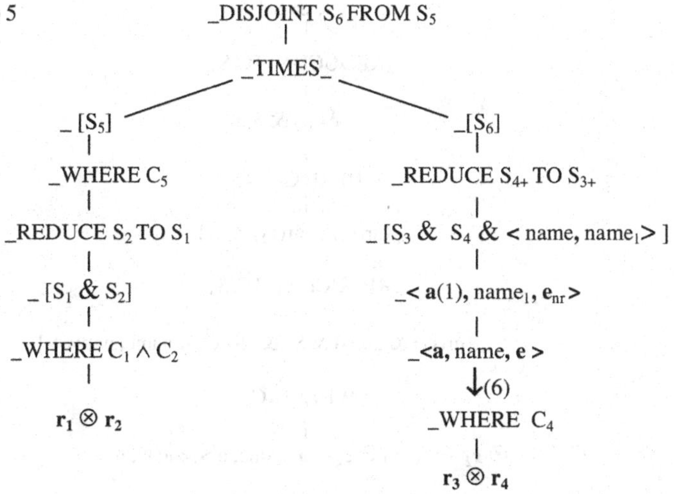

2. Find an atomic node in t_{RA}. If both paths from this node to their leafs are without unary operations _ $\langle a, name, e\rangle$, then eliminate this binary node and unify these two paths in a unique path, as specified by Definition 22, and hence go to Step 1.
   Otherwise, take in consideration the path with the unary operations _ $\langle a, name, e\rangle$ and continue with Step 3.
3. Here we consider the path of the current term t_{RA} such that it contains at least one operation _ $\langle a, name, e\rangle$. Take into consideration such an operation, which is closer to the leaf of this path, and unfold it along this path for only one step toward the leaf, as specified by one of the cases in Definition 34. Continue this process up to the situation when _ $\langle a, name, e\rangle$ reaches the leaf of this path. Then go to Step 1.

Step 6

_DISJOINT S_6 FROM S_5
|
TIMES
↓(7)

_ $[S_5]$
|
_WHERE C_5
|
_REDUCE S_2 TO S_1
|
_ $[S_1$ & $S_2]$
|
_WHERE $C_1 \wedge C_2$
|
$\mathbf{r}_1 \otimes \mathbf{r}_2$

_$[S_6]$
|
_REDUCE S_{4+} TO S_{3+}
|
_ $[S_3$ & S_4 & < name, name_1>]
|
_WHERE C_4
|
(($\mathbf{r}_3 \otimes \mathbf{r}_4$) **<a, name, e >**)< **a**(1), name_1, $\mathbf{e}_{nr}$ >

Step 7

_DISJOINT S_6 FROM S_5
|
_ $[S_5$ & $S_6]$
|
_REDUCE S_2 TO S_1
|
_ $[S_1$ & S_2 & $S_6]$
|
_WHERE $C_1 \wedge C_2$
|
_[nr($\mathbf{r}_1$) & nr($\mathbf{r}_2$) & S_6]
|
_REDUCE S_{4+} TO S_{3+}
|
_ [nr($\mathbf{r}_1$) & nr($\mathbf{r}_2$) & S_3 & S_4 & < name, name_1>]
|
_WHERE C_4
|
$\mathbf{r}_1 \otimes \mathbf{r}_2 \otimes$ ((($\mathbf{r}_3 \otimes \mathbf{r}_4$) **<a, name, e >**)< **a**(1), name_1, $\mathbf{e}_{nr}$ >)

4. At the end of this algorithm we obtain a path-term $f(t'_{RA})$, where f is a path of unary operations (different from $\_\ \langle a, name, e\rangle$), with the unique leaf $t'_{RA} \in \mathcal{T}^-_{RA}X$: this path-term is represented in **RA** by the arrow $f : t'_{RA} \to f(t'_{RA})$.

   This algorithm above defines the *normalization* function $Nrm : \mathcal{T}_{RA}X \to Mor_{RA}$. □

5.2.2 RA versus DB Category

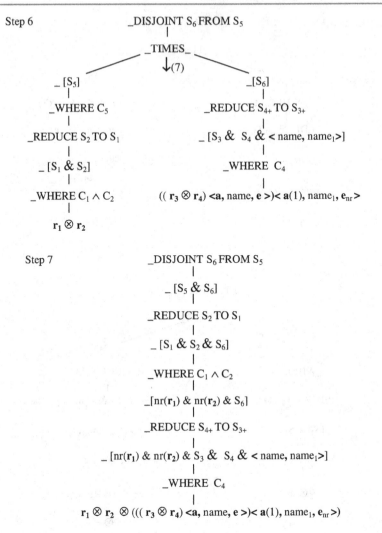

As we have demonstrated, all the terms of the "select–project–join+union" (SPJRU) Σ_R algebra which is used to define the views $q(\mathbf{x})$ (introduced in basic database concepts, Sect. 1.4) and basic SOtgds in the inter-schema database mappings (introduced in Sects. 1.4.2 and 2.1) are interpreted (instantiated) by the R-algebras in the denotational instance-level **DB** category. Thus, **DB** morphisms cover all SPJRU-algebra terms.

Now we will show that **DB** morphisms are able to support *all* Σ_{RE} algebra terms (from Definition 31, the SPJRU (Σ_R) is a strict subalgebra of the Σ_{RE} algebra) and hence the operations as _ MINUS _ and all update relational database operations. As we will see, in the *operational semantics* of the database mapping, we will use the update database operations as well. Consequently, the fact that such operations can be expressed in the **DB** category by the particular morphisms between instance databases is an important "full relational-algebra closure" property of the denotational database **DB** category.

Proposition 24 *There is an evaluation functor* $Eval : \mathbf{RA} \to \mathbf{DB}$ *which evaluates the terms in* **RA**, *as follows*:

1. *For each object* $t_{RA} \in Ob_{RA} = \mathcal{T}_{RA}X$ *we define*: $Eval(t_{RA}) \triangleq \{\|t_{RA}\|_\#, \perp\}$.
2. *For each atomic arrow* $o_i : t_{RA} \to o_i(t_{RA})$, *let* r_1 *and* r *be the relational symbols of the relational table* $R_1 = \|t_{RA}\|_\#$ *and* $R = \|o_i(t_{RA})\|_\#$, *respectively, with two auxiliary relational symbols* r_{q1} *and* r_{q2} *of the same type as* r. *Hence, we define the database schemas* $\mathcal{A} = (\{r_1\}, \emptyset)$ *and* $\mathcal{B} = (\{r_1\}, \emptyset)$ *and an R-algebra* α *such that* $\alpha(r_1) = R_1$ *and* $\alpha(r) = R$, *with the operations* $q_{A,1} \in O(r_1, r_{q1})$, $v_1 \in O(r_{q1}, r)$, $q_{A,2} \in O(r_1, r_{q2})$ *and* $v_2 \in O(r_{q2}, r)$.

   Then we define the component of Eval for the arrows, in the following two cases:

   2.1. *If* o_i *is a unary operation* _ REDUCESS_2 TO S_1 *where*

   $$S_1 = \langle nr_{r_1}(i), \ldots, nr_{r_1}(i+k)\rangle \quad \text{and} \quad S_2 = \langle nr_{r_1}(j), \ldots, nr_{r_1}(j+k)\rangle$$

   for $i \geq 1$, $k \geq 0$, $j > i+k$ *and* $j+k \leq ar(r_1)$, *then*

   $$Eval(o_i) \triangleq \{\alpha(v_1) \cdot \alpha(q_{A,1}), \alpha(v_2) \cdot \alpha(q_{A,2}), q_\perp\} : \alpha^*(\mathcal{A}) \to \alpha^*(\mathcal{B}),$$

   so that $\alpha(r_{q1}) = \|t_{RA}[S_1]\|_\#$ *and* $\alpha(r_{q2}) = \|t_{RA}[S_2]\|_\#$, *where* $\alpha(v_1) : \alpha(r_{q1}) \hookrightarrow R$ *and* $\alpha(v_2) : \alpha(r_{q2}) \hookrightarrow R$ *are two injective functions,* $R = \alpha(r_{q1}) \cup \alpha(r_{q2})$, $\alpha(q_{A,1}) = \pi_{\langle i,\ldots,i+k\rangle} : R_1 \to \alpha(r_{q1})$ *and* $\alpha(q_{A,2}) = \pi_{\langle j,\ldots,j+k\rangle} : R_1 \to \alpha(r_{q2})$.

   2.2. *Otherwise, when* o_i *is another unary operation, we define*:

   $$Eval(o_i) \triangleq \{\alpha(v_1) \cdot \alpha(q_{A,1}), q_\perp\} : \alpha^*(\mathcal{A}) \to \alpha^*(\mathcal{B}),$$

   so that $\alpha(r_{q1}) = R$ *and* $\alpha(v_1) = id_R : R \to R$ *is an identity function, while* $\alpha(q_{A,1})$ *is defined for the following remaining cases*:

   - o_i *is a unary operation* _ RENAME *name*$_2$ AS *name*$_1$. *Then* $r = r_1$ *with* $\alpha(r_{q1}) = R = R_1$ *and the identity function* $\alpha(q_{A,1}) = id_R : R \to R$.
   - o_i *is a unary operation* _ $[S]$ *with* $S = \langle nr_{r_1}(i_1), \ldots nr_{r_1}(i_k)\rangle$. *Then,* $\alpha(q_{A,1}) = \pi_{\langle i_1,\ldots,i_k\rangle} : R_1 \to R$.

- *o_i is a unary operation* _ WHERE C. *Then* $\alpha(q_{A,1}) : R_1 \rightarrow R$ *is a function such that for any tuple* $\mathbf{d} \in R_1$, $\alpha(q_{A,1})(\mathbf{d}) = \mathbf{d}$ *if the condition* C *holds for the values in* $\mathbf{d}$; $\langle\rangle$ *otherwise.*
- *o_i is a unary operation* _ DISJOINTS_2 FROM S_1, *where* $S_1 = \langle nr_{r_1}(i), \ldots, nr_{r_1}(i+k)\rangle$ *and* $S_2 = \langle nr_{r_1}(j), \ldots, nr_{r_1}(j+k)\rangle$ *for* $i \geq 1$, $k \geq 0$, $j > i+k$ *and* $j+k \leq ar(r_1)$. *Then,* $\alpha(q_{A,1}) : R_1 \rightarrow R$ *is a function such that for any tuple* $\mathbf{d} = \langle d_1, \ldots, d_{ar(r_1)}\rangle \in R_1$, $\alpha(q_{A,1})(\mathbf{d}) = \pi_{\langle i,\cdots+k\rangle}(\mathbf{d})$ *if* $\nexists \mathbf{d}' \in R_1$ *such that* $(d_i = d'_j) \wedge \cdots \wedge (d_{i+k} = d'_{j+k})$; $\langle\rangle$ *otherwise.*

By this functorial transformation, each atomic arrow o_i *in* **RA** *is mapped into an atomic morphism in* **DB** *(specified by Definition* 18).

Proof We have to show that for each atomic arrow $o_i : t_{RA} \rightarrow o_i(t_{RA})$ in **RA**, the morphism $Eval(o_i) : Eval(t_{RA}) \rightarrow Eval(o_i(t_{RA}))$ is an *atomic* morphism (Definition 18), that is, there is an atomic sketch's schema mapping $\mathbf{M}_{AB} : \mathcal{A} \rightarrow \mathcal{B}$ (Definition 17) such that $\alpha^*(\mathbf{M}_{AB}) = Eval(o_i)$.

In fact, for each case of an atomic arrow o_i we are able to make a simple sketch **Sch**(G), where a graph G is composed of two vertices (schemas) $\mathcal{A} = (\{r_1\}, \emptyset)$ and $\mathcal{B} = (\{r\}, \emptyset)$ introduced in this proposition for this arrow $o_i : t_{RA} \rightarrow o_i(t_{RA})$ in **RA**, with an atomic inter-schema mapping $\mathcal{M}_{AB} = \{\Phi\} : \mathcal{A} \rightarrow \mathcal{B}$, where (w.r.t. definitions in the proposition above):

1. For the case when o_i is a unary operation _ REDUCE S_2 TO S_1, with the tuple of variables $\mathbf{x} = \mathbf{x}_1\&(x_i, \ldots, x_{i+k})\&\mathbf{x}_2\&(x_j, \ldots, x_{j+k})\&\mathbf{x}_3$, the SOtgd Φ is equal to:
$$\begin{aligned}&\big(\forall\mathbf{x}\big(r_1(\mathbf{x}) \Rightarrow r\big(\mathbf{x}_1, (x_i, \ldots, x_{i+k}), \mathbf{x}_2, \mathbf{x}_2\big)\big)\big)\wedge\\ &\wedge\big(\forall\mathbf{x}\big(r_1(\mathbf{x}) \Rightarrow r\big(\mathbf{x}_1, (x_j, \ldots, x_{j+k}), \mathbf{x}_2, \mathbf{x}_2\big)\big)\big).\end{aligned}$$
2. For the case o_i is a unary operation _ RENAME $name_2$ AS $name_1$, the logical formula Φ is equal to the SOtgd (here r is equal to r_1) $\forall\mathbf{x}(r_1(\mathbf{x}) \Rightarrow r_1(\mathbf{x}))$.
3. For the case o_i is a unary operation _ $[S]$ with a tuple of indexes
$$S = \big\langle nr_{r_1}(i_1), \ldots, nr_{r_1}(i_k)\big\rangle,$$
the logical formula Φ is equal to the SOtgd $\forall\mathbf{x}(r_1(\mathbf{x}) \Rightarrow r(x_{i_1}, \ldots, x_{i_k}))$.
4. For the case o_i is a unary operation _ WHERE C, the logical formula Φ is equal to the SOtgd $\forall\mathbf{x}((r_1(\mathbf{x}) \wedge C(\mathbf{x})) \Rightarrow r(\mathbf{x}))$.
5. For the case when o_i is a unary operation _ DISJOINT S_2 FROM S_1 with $\mathbf{x} = \mathbf{x}_1\&(x_i, \ldots, x_{i+k})\&\mathbf{x}_2\&(x_j, \ldots, x_{j+k})\&\mathbf{x}_3$ and $\mathbf{y} = \mathbf{y}_1\&(y_i, \ldots, y_{i+k})\&\mathbf{y}_2\&(y_j, \ldots, y_{j+k})\&\mathbf{y}_3$, the logical formula Φ is equal to the SOtgd
$$\begin{aligned}&\forall\mathbf{x}\forall\mathbf{y}\big(\big(r_1(\mathbf{x}) \wedge r_1(\mathbf{y}) \wedge \big((x_i \neq y_j) \vee \cdots \vee (x_{i+k} \neq y_{j+k})\big)\big)\\ &\quad\Rightarrow r\big(\mathbf{x}_1, (x_i, \ldots, x_{i+k}), \mathbf{x}_2, \mathbf{x}_2\big)\big).\end{aligned}$$

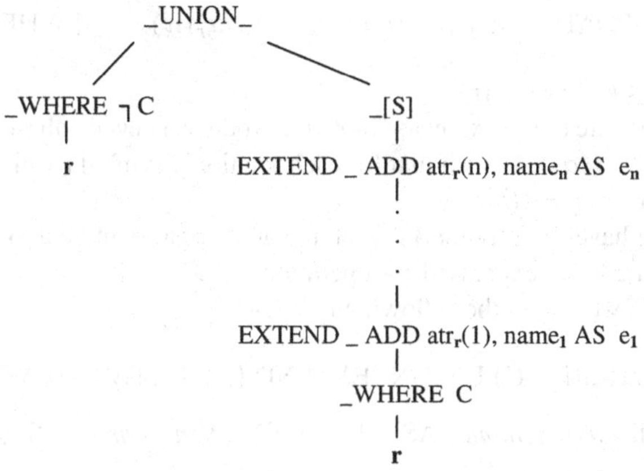

Consequently, the sketch **Sch**(G) will have an *atomic* arrow (in Definition 17), $\mathbf{M}_{AB} = \mathit{MakeOperads}(\{\Phi\}) : \mathcal{A} \rightarrow \mathcal{B}$, which for an R-algebra α will have the information flux $\mathit{Flux}(\alpha, \mathbf{M}_{AB})$, and hence $\alpha^*(\mathbf{M}_{AB}) : \alpha^*(\mathcal{A}) \rightarrow \alpha^*(\mathcal{B})$ is (from Theorem 1) an atomic morphism in **DB**.

It is easy to verify that for each of these cases $\alpha^*(\mathbf{M}_{AB}) = \mathit{Eval}(o_i)$. The identity arrow o_i is mapped into the identity morphism in **DB** and, from these sketch's interpretations, the functorial composition of the arrows is valid: For any composed arrow $o_k \circ o_i$, where $\mathbf{M}_{BC}$ is the sketch's arrow for o_k and $\mathbf{M}_{AB}$ is the sketch's arrow for o_i, for the atomic arrow composition $\mathbf{M}_{BC} \circ \mathbf{M}_{AB}$ in a sketch category,

$$\begin{aligned}\mathit{Eval}(o_k \circ o_i) &= \alpha^*(\mathbf{M}_{BC} \circ \mathbf{M}_{AB}) \\ &\quad \big(\text{from the functorial property of } \alpha^*\big) \\ &= \alpha^*(\mathbf{M}_{BC}) \circ \alpha^*(\mathbf{M}_{AB}) = \mathit{Eval}(o_k) \circ \mathit{Eval}(o_i).\end{aligned}$$

Thus, the functor *Eval* is valid for composed (non atomic) arrows as well. □

Example 34 In what follows, we will present the three basic atomic update operations, updating (i), inserting (ii) and deleting (iii) of tuples in a given instance database $\alpha(\mathcal{A})$ of a schema $\mathcal{A}$, and we will show that each of them is representable in **DB** by a morphism $f : \alpha(\mathcal{A}) \rightarrow \alpha_1(\mathcal{A})$ where $\alpha_1(\mathcal{A})$ is the resulting database instance obtained after such an update. This result is important because it demonstrates not only the denotational property of **DB**, but also its operational capability to represent the atomic database-updating transactions by the atomic morphisms of **DB** category:

(i) Let us consider the following representation in **DB** category of an update of a given relation $r \in S_A$ of a database instance $A = \alpha^*(\mathcal{A})$ of a schema $\mathcal{A} = (S_A, \emptyset)$, with $\alpha(r) = \|r\|_\#$ the extension of this relation before this update such that $\|r \text{ WHERE } C\|_\#$ *is a nonempty* set of tuples:

$$\text{UPDATE } r \text{ SET } \bigl[nr_r(i_1) = e_{i_1}, \ldots, nr_r(i_k) = e_{i_k}\bigr] \text{ WHERE } C,$$

where $1 \le k \le n = ar(r)$.

This update of the extension of r will produce a new database instance of $\mathcal{A}$, equal to $A_1 = \alpha_1^*(\mathcal{A})$ such that for each relational symbol r_A in $\mathcal{A}$ but different from r, $\alpha_1(r_A) = \alpha(r_A)$.

As we have seen (point 3 for the update operators in Sect. 5.1), this update operation can be expressed by operations in Σ_{RE} algebra (in Definition 31) and hence we obtain the following tree-term,

$$(r \text{ WHERE } \neg C) \text{ UNION } \bigl(\text{EXTEND } \bigl(\ldots \bigl(\text{EXTEND } (r \text{ WHERE } C)$$

$$\text{ADD } atr_r(1), name_1 \text{ AS } e_1\bigr) \ldots \text{ADD } atr_r(n), name_n \text{ AS } e_n\bigr)[S],$$

where for each $1 \le m \le n$, if $m \notin \{i_1, \ldots, i_k\}$ then $e_m = nr_r(m)$, and $S = \langle name_1, \ldots, name_n \rangle$. This term is represented in the diagram (tree) above.

Now we will transform this term into an equivalent Σ_{RA} term, by substituting the operators 'UNION' with the equivalent composition of operators in Σ_{RA}, as specified in the proof of Proposition 20.

So, the final part of the right-path of this term, EXTEND (r WHERE C) ADD $atr_r(1), name_1$ AS e_1 can be reduced (by the algorithm in the proof of Proposition 23) into the path-term $(r\langle atr_r(1), name_1, e_1\rangle)$ WHERE C.

By repeating this process to all 'EXTEND...' operations in the right-path of the tree-term in the figure above, we obtain an equivalent path-term of Σ_{RA} algebra (t_{RA} WHERE C)$[S]$, where the leaf term $t_{RA} = t^{(n)} \in \mathcal{T}_{RA}^{-}X$ is recursively defined by:

$$t^{(i)} = t^{(i-1)}\bigl\langle atr_r(i), name_i, e_i\bigr\rangle \quad \text{for } i = 1, 2, \ldots, n \quad \text{and} \quad t^{(0)} = r.$$

By substitution of the Σ_{RE} operation 'UNION' with the corresponding composition of operations in Σ_{RA}, the original tree-term in the figure above is reduced to the following Σ_{RA} tree-term:

$$\bigl((r \text{ WHERE } \neg C) \text{ TIMES } \bigl((t_{RA} \text{ WHERE } C)[S]\bigr)\bigr) \text{ REDUCE } S \text{ TO } S_1,$$

represented by the second diagram bellow (where $S_1 = nr(r) = \langle nr_r(1), \ldots, nr_r(n)\rangle$).

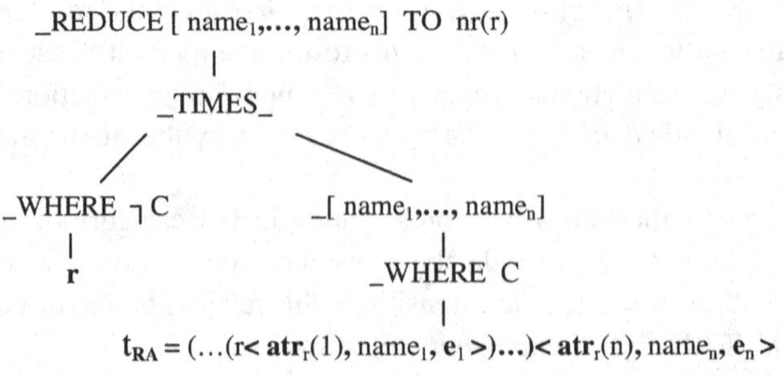

Thus, by applying the algorithm in the proof of Proposition 23, we transform this binary tree-term into the following path-term

$$((((r \otimes t_{RA}^{\rho}) \text{ WHERE } C^{\rho})[nr(r)\&S]) \text{ WHERE } \neg C) \text{ REDUCE } S \text{ TO } nr(r),$$

where $(r \otimes t_{RA}^{\rho}) \in \mathcal{T}_{RA}^{-}X$ is the unique leaf, C^{ρ} is the renamed version of C w.r.t. Cartesian product renaming ρ and $S = \langle name_1, \dots, name_n \rangle$.

Thus, this update of r is represented by the arrow $o_U : (r \otimes t_{RA}^{\rho}) \rightarrow o_U(r \otimes t_{RA}^{\rho})$ in **RA**, equal to the following composition of atomic arrows:

$$\begin{aligned}&(\_\text{ REDUCE } \langle name_1, \dots, name_n \rangle \text{ TO } nr(r)) \\ &\circ (\_\text{ WHERE } \neg C) \circ (\_[nr(r)\&\langle name_1, \dots, name_n \rangle]) \circ (\_\text{ WHERE } C^{\rho})\end{aligned}$$

such that the new extension of $r \in \mathcal{A}$, $\alpha_1(r) = \|o_U(r \otimes t_{RA}^{\rho})\|$, which is generally different from $\alpha(r)$.

Consequently, by the functor $Eval : \mathbf{RA} \rightarrow \mathbf{DB}$ of Proposition 24, we obtain the morphism $Eval(o_U) : B \rightarrow \alpha_1^*(\mathcal{A})$, where $B = \{\|r \otimes t_{RA}^{\rho}\|_{\#}, \bot\}$ is an instance database in **DB** of a schema $\mathcal{B} = (\{r'\}, \emptyset)$ and r' is a relational symbol of the leaf term $r \otimes t_{RA}^{\rho}$.

Let us show that there is a morphism $h_r : \alpha^*(\mathcal{A}) \rightarrow B$ such that $\|r \otimes t_{RA}^{\rho}\|$ is the result of an inter-schema mapping $\mathcal{M}_{AB} : \mathcal{A} \rightarrow \mathcal{B}$ and the R-algebra α is extended to $r' \in \mathcal{B}$ by $\alpha(r') = \|r \otimes t_{RA}^{\rho}\|_{\#}$.

In fact, based on the logical equivalence for the algebraic operation of Cartesian product in Sect. 5.1, and on the recursive definition of $t_{RA} = t^{(n)}$ (with $t^{(0)} = r$), we can define a mapping $\mathcal{M}_{AB} = \{\Psi\} : \mathcal{A} \rightarrow \mathcal{B}$ such that Ψ is the following SOtgd:

$$\forall \mathbf{x}_0 \forall \mathbf{x}_1 \big((r(\mathbf{x}_0) \wedge r(\mathbf{x}_1)) \Rightarrow r'(\mathbf{x}_0 \& \langle e_1(\mathbf{x}_1), \dots, e_n(\mathbf{x}_1)\rangle) \big)$$

where for $i = 0, 1$, $\mathbf{x}_i = \langle x_{i,1}, \dots, x_{i,n} \rangle$ are the tuples of variables and e_i are the functional symbols (also with arity zero for the constants), and $ar(r') = 2n$.

Consequently,

$$h_r = \alpha^*\big(MakeOperads(\mathcal{M}_{AB})\big) = \alpha^*\big(\{v_1 \cdot q_{A,1}, 1_{r_\emptyset}\}\big) : \alpha^*(\mathcal{A}) \rightarrow \alpha^*(\mathcal{B})$$

is a morphism in **DB**, where the R-algebra α is a mapping-interpretation (Definition 11) which satisfies this schema mapping $\mathcal{M}_{AB}$. That is, $q_{A,1} \in O(r, r, r_q)$ is an operation of the expression

$$\big((\_)_1(\mathbf{x}_0) \wedge (\_)_2(\mathbf{x}_1)\big) \Rightarrow (\_)\big(\mathbf{x}_0 \& \langle e_1(\mathbf{x}_1), \dots, e_n(\mathbf{x}_n)\rangle\big),$$

$v_1 \in O(r_q, r')$, and $\alpha(v_1)$ is the identity function for the relational table $\alpha(r') = \alpha(r_q) = \|r \otimes t_{RA}^{\rho}\|_{\#}$.

The information flux of this morphism in **DB** is equal to

$$\widetilde{h_r} = Flux\big(\alpha, MakeOperads(\mathcal{M}_{AB})\big).$$

Finally, the update of the relation $r \in \mathcal{A}$, which changes its extension from $\alpha(r) = \|r\|_\#$ into $\alpha_1(r) = \|o_U(r \otimes t^\rho_{RA})\|_\#$, is represented in the **DB** category as the composed morphism $f_U = Eval(o_U) \circ h_r : \alpha^*(\mathcal{A}) \to \alpha_1^*(\mathcal{A})$.

By generalization, the updates of $k \geq 1$ different relations $r_1, \ldots, r_k$ of a given schema $\mathcal{A}$ such that for each of them we have the morphisms $h_{r_i}, Eval(o_{U_i}), i = 1, \ldots, k$ can be represented in **DB** by a composed morphism

$$f_U = \Big(\bigcup_{1 \leq i \leq k} Eval(o_{U_i})\Big) \circ \Big(\bigcup_{1 \leq i \leq k} h_{r_i}\Big) : \alpha^*(\mathcal{A}) \to \alpha_1^*(\mathcal{A}), \qquad (5.1)$$

where $\alpha(r_i) = \|r_i\|_\#$, $\alpha_1(r_i) = \|o_{U_i}(r \otimes t^\rho_{RA_i})\|_\#$, for $i = 1, \ldots, k$, with the intermediate schema $\mathcal{B} = (\{r'_1, \ldots r'_k\}, \emptyset)$ and $\alpha(r'_i) = \|r_i \otimes t^\rho_{RA_i}\|_\#$, for $i = 1, \ldots, k$, and for each $r_A \in \mathcal{A}$ with $r_A \notin \{r_1, \ldots r_k\}$, $\alpha_1(r_A) = \alpha(r_A)$.

(ii) The operation of inserting the set $\{\langle d_{11}, \ldots, d_{1n}\rangle, \ldots, \langle d_{m1}, \ldots, d_{mn}\rangle\}$ of $m \geq 1$ tuples of a relation r_I into a given relation $r \in \mathcal{A}$ (used in the operation of update above) corresponds to the following term in Σ_{RE}

$$\text{INSERT INTO } r[S] \ldots,$$

equivalent (if $\|r\|$ in a nonempty relation) to the path-term, $(r \otimes r^\rho_I)$ REDUCE S_1 TO S, of the Σ_{RA} algebra, with the leaf r' equal to Cartesian product $(r \otimes r^\rho_I)$, where $S_1 = \langle nr_{r'}(n+1), \ldots, nr_{r'}(2n)\rangle$ and $S = nr(r) = \langle nr_{r'}(1), \ldots, nr_{r'}(n)\rangle$ (if $\|r\| = \{\langle\rangle\}$ is an empty relation, than this path-term is r_I WHERE $\overline{1} = \overline{1}$).

Hence, it corresponds to the arrow $o_I = \_$ REDUCE S_1 TO $S : (r \otimes r^\rho_I) \to o_I(r \otimes r^\rho_I)$ in **RA**. That is, from Proposition 24, for a schema $\mathcal{B} = (\{r'\}, \emptyset)$ with $\alpha(r') = \|r \otimes r^\rho_I\|_\#$, we obtain the **DB**-morphism $Eval(o_I) : \alpha^*(\mathcal{B}) \to \alpha_1^*(\mathcal{A})$, where α_1 is the R-algebra obtained from α after this insertion (with $\alpha_1(r) = \|o_I(r \otimes r^\rho_I)\|_\#$ and for each $r_A \in \mathcal{A}$ such that $r_A \neq r$, $\alpha_1(r_A) = \alpha(r_A)$).

Let us show that there is a morphism $h_r : \alpha^*(\mathcal{A}) \to \alpha^*(\mathcal{B})$ such that $\alpha(r')$ is the result of an inter-schema mapping $\mathcal{M}_{AB} : \mathcal{A} \to \mathcal{B}$. It is easy to show that $\mathcal{M}_{AB} = \{\Phi\}$, where Φ is equal to SOtgd:

$$\forall \mathbf{x}\big(r(\mathbf{x}) \Rightarrow r'\big(\mathbf{x}; \langle d_{11}, \ldots, d_{1n}\rangle\big)\big) \wedge \cdots \wedge \forall \mathbf{x}\big(r(\mathbf{x}) \Rightarrow r'\big(\mathbf{x}; \langle d_{m1}, \ldots, d_{mn}\rangle\big)\big),$$

so that we have the morphism $h_r = \alpha^*(\boldsymbol{M}_{AB}) : \alpha^*(\mathcal{A}) \to \alpha^*(\mathcal{B})$, where $\boldsymbol{M}_{AB} = MakeOperads(\{\Phi\}) = \{v_1 \cdot q_{A,1}, \ldots, v_m \cdot q_{A,m}, 1_{r_\emptyset}\}$ for which the R-algebra α is a mapping-interpretation, $q_{A,i} \in O(r, r_{q,i})$ is an operation of expression $(\_)_1(\mathbf{x}) \Rightarrow (\_)(\mathbf{x}; \langle d_{i1}, \ldots, d_{in}\rangle)$ and $v_i \in O(r_{q,i}, r')$ with $\alpha(v_i) : \alpha(r_{q,i}) \hookrightarrow \alpha(r')$ is an injective function for each $i = 1, \ldots, m$, so that $\bigcup_{1 \leq i \leq m} Im(\alpha(v_i)) = \bigcup_{1 \leq i \leq m} \alpha(r_{q,i}) = \alpha(r') = \|r \otimes r^\rho_I\|_\#$.

Consequently, this insertion is represented in the **DB** category by the composed morphism $f_I = Eval(o_I) \circ h_r : \alpha^*(\mathcal{A}) \to \alpha_1^*(\mathcal{A})$.

By generalization, the insertion of the tuples into the $k \geq 1$ different relations $r_1, \ldots, r_k$ of a given schema $\mathcal{A}$, by the set of INSERT INTO $r_i[S_i] \ldots$, for $i = 1, \ldots, k$ such that that for each of them we have the morphisms $h_{r_i}, Eval(o_{I_i}), i = 1, \ldots, k$, can be represented in **DB** by a composed morphism

$$f_I = \left(\bigcup_{1 \leq i \leq k} Eval(o_{I_i}) \right) \circ \left(\bigcup_{1 \leq i \leq k} h_{r_i} \right) : \alpha^*(\mathcal{A}) \to \alpha_1^*(\mathcal{A}), \qquad (5.2)$$

where $\alpha(r_i) = \|r_i\|_\#$, $\alpha_1(r_i) = \|o_{I_i}(r_i \otimes r^{\rho}_{I_i})\|_\#$, for $i = 1, \ldots, k$, with the intermediate schema $\mathcal{B} = (\{r'_1, \ldots r'_k\}, \emptyset)$ and $\alpha(r'_i) = \|r_i \otimes r^{\rho}_{I_i}\|_\#$, for $i = 1, \ldots, k$, and for each $r_A \in \mathcal{A}$ with $r_A \notin \{r_1, \ldots r_k\}$, $\alpha_1(r_A) = \alpha(r_A)$.

(iii) The operation of deleting a set of tuples of a relation r_I corresponds to the following term t_R of the Σ_{RE}-algebra:

DELETE FROM r WHERE C.

It is equivalent to the Σ_{RA} path-term r WHERE $\neg C$, that is, to the atomic arrow $o_D = \_$ WHERE $\neg C : r \to o_D(r)$ in **RA**, so that, from Proposition 24, its representation in **DB** is given by

$$Eval(o_D) = \alpha^*\big(MakeOperads(\{\Phi\})\big) = \alpha^*\big(\{v_1 \cdot q_{A,1}, 1_{r_\emptyset}\}\big) : \alpha^*(\mathcal{A}) \to \alpha_1^*(\mathcal{A}),$$

where $\alpha(v_1)$ is the identity function for $\alpha(r_q) = \|o_D(r)\|_\#$ and $\alpha(q_{A,1}) : \alpha(r) \to \alpha(r_q)$ is the function defined in Proposition 24 (by substituting C with $\neg C$), and $q_{A,1} \in O(r, r_q)$ is the expression $((\_)_1(\mathbf{x}) \wedge \neg C(\mathbf{x})) \Rightarrow (\_)(\mathbf{x})$, obtained from SOtgd Φ, $\forall \mathbf{x}((r(\mathbf{x}) \wedge \neg C(\mathbf{x}) \Rightarrow r'(\mathbf{x})))$.

However, analogously to the previous cases, we can express this deleting term t_R by an, equivalent to it, tree-term of Σ_{RA} algebra r MINUS (r WHERE C). Hence, it can be transformed into the path-term (($r \otimes r^{\rho}$) WHERE C^{ρ}) DISJOINT S FROM S_1, where C^{ρ} is obtained from C by renaming column names w.r.t. ρ of Cartesian product, and, equivalent to it, the following arrow in **RA**:

$$o_D = \_\ \text{DISJOINT } S \text{ FROM } S_1 \circ \_\ \text{WHERE } C^{\rho}) : \big(r \otimes r^{\rho}\big) \to o_D\big(r \otimes r^{\rho}\big),$$

where $S = nr(r^{\rho})$, $S_1 = nr(r)$. Consequently, by the functor $Eval : \mathbf{RA} \to \mathbf{DB}$ of Proposition 24, we obtain the morphism $Eval(o_D) : B \to \alpha_1^*(\mathcal{A})$, where $B = \{\|r \otimes r^{\rho}\|_\#, \perp\}$ is an instance database of a schema $\mathcal{B} = (\{r'\}, \emptyset)$ where r' is a relational symbol for the leaf term $r \otimes r^{\rho}$.

Let us show that there is a morphism $h_r : \alpha^*(\mathcal{A}) \to B$ such that $\|r \otimes r^{\rho}\|_\#$ is the result of an inter-schema mapping $\mathcal{M}_{AB} : \mathcal{A} \to \mathcal{B}$ and R-algebra α, extended to $r' \in \mathcal{B}$ by $\alpha(r') = \|r \otimes r^{\rho}\|_\#$.

In fact, we can define a mapping $\mathcal{M}_{AB} = \{\Psi\} : \mathcal{A} \to \mathcal{B}$ such that Ψ is equal to SOtgd $\forall \mathbf{x}_0 \forall \mathbf{x}_1((r(\mathbf{x}_0) \wedge r(\mathbf{x}_1)) \Rightarrow r'(\mathbf{x}_0 \& \mathbf{x}_1))$.

Consequently,

$$h_r = \alpha^*\big(MakeOperads(\mathcal{M}_{AB})\big) = \alpha^*\big(\{v_1 \cdot q_{A,1}, 1_{r_\emptyset}\}\big) : \alpha^*(\mathcal{A}) \to \alpha^*(\mathcal{B})$$

is a **DB** morphism, where R-algebra α is a mapping-interpretation (Definition 11) which satisfies this schema mapping $\mathcal{M}_{AB}$, that is, $q_{A,1} \in O(r, r, r_q)$ is operation of the expression $((\_)_1(\mathbf{x}_0) \wedge (\_)_2(\mathbf{x}_1)) \Rightarrow (\_)(\mathbf{x}_0\&\mathbf{x}_1)$, $v_1 \in O(r_q, r')$ and $\alpha(v_1)$ is the identity function for the relational table $\alpha(r') = \alpha(r_q) = \|r \otimes r^\rho\|_\#$.

Finally, the deleting of the tuples of $r \in \mathcal{A}$ which change the extension from $\alpha(r) = \|r\|_\#$ into $\alpha_1(r) = \|o_D(r \otimes r^\rho)\|_\#$ is represented in the **DB** category as a composed morphism $f_D = Eval(o_D) \circ h_r : \alpha^*(\mathcal{A}) \to \alpha_1^*(\mathcal{A})$.

By generalization, deleting of the tuples in $k \geq 1$ different relations $r_1, \ldots, r_k$ of a given schema $\mathcal{A}$ such that for each of them we have the morphisms $h_{r_i}, Eval(o_{D_i})$, $i = 1, \ldots, k$ can be represented in **DB** by a composed morphism

$$f_D = \left(\bigcup_{1 \leq i \leq k} Eval(o_{D_i}) \right) \circ \left(\bigcup_{1 \leq i \leq k} h_{r_i} \right) : \alpha^*(\mathcal{A}) \to \alpha_1^*(\mathcal{A}), \tag{5.3}$$

where $\alpha(r_i) = \|r_i\|$, $\alpha_1(r_i) = \|o_{D_i}(r \otimes r^\rho)\|_\#$, for $i = 1, \ldots, k$, with the intermediate schema $\mathcal{B} = (\{r_1', \ldots r_k'\}, \emptyset)$ and $\alpha(r_i') = \|r_i \otimes r_i^\rho\|_\#$, for $i = 1, \ldots, k$, and for each $r_A \in \mathcal{A}$ with $r_A \notin \{r_1, \ldots r_k\}$, $\alpha_1(r_A) = \alpha(r_A)$.

Let us show that the functor $Eval : \mathbf{RA} \to \mathbf{DB}$ is a functor between two extended symmetric categories:

Proposition 25 *The strict monoidal category* **RA** *with the tensor product* $\otimes$ *(i.e., Cartesian product _ TIMES _) is an extended symmetric category with* $B_T = S_{nd}^0$, *closed by the functor* $T_e = S_{nd} : (\mathbf{RA} \downarrow \mathbf{RA}) \longrightarrow \mathbf{RA}$ *where* $T_e^0 = B_T \psi$ *is the object component of this functor such that for any arrow* $f : t_{RA} \to f(t_{RA})$ *in* **RA**, $\widetilde{f} = T_e^0(J(f)) = S_{nd}^0 \psi(J(f)) = f(t_{RA})$, *and for any arrow* $(h_1; h_2) : J(f) \longrightarrow J(g)$ *in* **RA** $\downarrow$ **RA** *such that* $g \circ h_1 = h_2 \circ f$ *in* **RA**, $T_e^1(h_1; h_2) = S_{nd}^1((h_1; h_2)) = h_2$.

This is represented by

$$\begin{array}{ccccccc}
t_{RA} & \xrightarrow{f} & f(t_{RA}) & & J(f) = \langle t_{RA}, f(t_{RA}), f \rangle & & \widetilde{f} = T_e^0(J(f)) = f(t_{RA}) \\
h_1 \downarrow & \text{in } \mathbf{RA} & \downarrow h_2 & \Leftrightarrow & (h_1; h_2) \downarrow & \xrightarrow{T_e} & h_2 \downarrow = T_e^1(h_1; h_2) \\
h_1(t_{RA}) & \xrightarrow{g} & D & & J(g) = \langle h_1(t_{RA}), D, g \rangle & & \widetilde{g} = T_e^0(J(g)) = D \\
 & \text{in } \mathbf{RA} & & & \text{in } (\mathbf{RA} \downarrow \mathbf{RA}) & & \text{in } \mathbf{RA}
\end{array}$$

where $D = g(h_1(t_{RA})) = h_2(f(t_{RA}))$. *All arrows in* **RA** *are epic and monic. The associative composition operator for objects* $*$, *defined for any fitted pair* $g \circ h_1$ *of arrows, satisfies* $B_T(g) * B_T(h_1) = g \circ h_1 = B_T(g \circ h_1)$.

Proof From Proposition 22, the category **RA** is skeletal, that is, $t_{RA} \simeq t'_{RA}$ iff $t_{RA} = t'_{RA}$, that is, iff $\|t_{RA}\|_{\#} = \|t'_{RA}\|_{\#}$, and hence the iso arrows are the identity arrows. Moreover, from the skletal property, for any two objects we have at maximum one arrow between them. Consequently, each arrow is both monic and epic (however, not necessarily isomorphic): it may be an isomorphism only if it is an identity (only in this case for this pair of objects there is an inverted arrow as well).

The functor T_e is equal to the second comma functorial projection, introduced in Sect. 1.5.1. Thus, **RA** is an *extended symmetric* category, that is, $\tau^{-1} \bullet \tau = \psi : F_{st} \stackrel{\centerdot}{\longrightarrow} S_{nd}$ for the vertical composition of natural transformations $\tau = \psi : F_{st} \stackrel{\centerdot}{\longrightarrow} T_e$ and $\tau^{-1} = Id : T_e \stackrel{\centerdot}{\longrightarrow} S_{nd}$. Hence, for each object $J(f) = \langle t_{RA}, f(t_{RA}), f\rangle$ in **RA** $\downarrow$ **RA** there is the following functorial *Eval* translation (we recall that $\psi_{J(f)} = \psi(J(f)) = f$) from **RA** into **DB**:

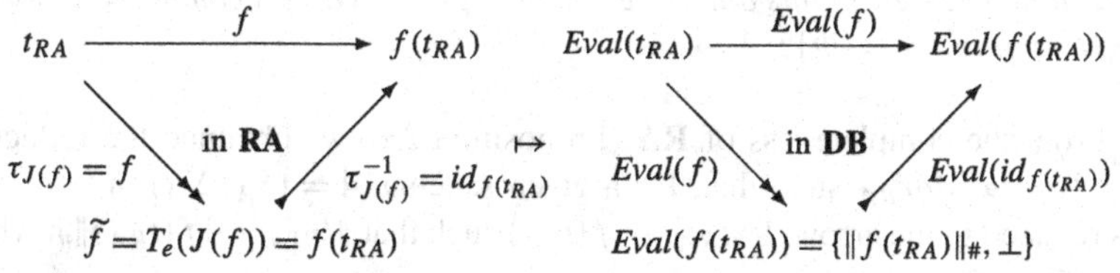

It is easy to see that the diagram on the right is just the analogous extended symmetric decomposition of the morphism $Eval(f)$, from the fact that

$$Eval\big(f(t_{RA})\big) = \big\{\big\|f(t_{RA})\big\|_{\#}, \perp\big\} \simeq T\big(\big\{\big\|f(t_{RA})\big\|_{\#}, \perp\big\}\big) = Eval\big(\widetilde{f(t_{RA})}\big). \quad \square$$

From Proposition 51 in Sect. 8.1.5, we can show that **DB** is a monoidal category $(\mathbf{DB}, \widehat{\oplus}, \perp^0, \alpha, \beta, \gamma)$, where $\perp^0 = \{\perp\}$ is the empty database, with the "merging" tensor product $\oplus$ (in Definition 57). The associative and idempotent properties of $\oplus$ is demonstrated by Proposition 51, where $\oplus$ is the join operation in the complete lattice $L_{DB} = (Ob_{DB}, \preceq, \otimes, \oplus)$ with the top and bottom objects Υ and $\perp^0$, respectively.

Thus, its full subcategory $\underline{\mathbf{DB}}$ composed of only *simple* databases with the top simple database $\underline{\Upsilon}$, introduced previously in Definition 26 in Sect. 3.2.5, is a monoidal category $(\underline{\mathbf{DB}}, \oplus, \perp^0, \alpha, \beta, \gamma)$, with a merging tensorial product $\oplus$ (equal to the join operation of the algebraic lattice $(Ob_{\underline{DB}}, \preceq, \otimes, \oplus)$ with the top and bottom objects $\underline{\Upsilon}$ and $\perp^0$, respectively; see Proposition 52 in Sect. 8.1.5) reduced to $A \oplus B = T(A \cup B)$ for any two simple databases A and B. It is easy to verify that for any two objects (terms) t_{RA} and t'_{RA} in **RA**, we have by functorial *Eval* translation:

$$\begin{aligned} Eval\big(t_{RA} \otimes t'_{RA}\big) &= \big\{\|t_{RA}\|_{\#} \otimes \big\|t'_{RA}\big\|_{\#}, \perp\big\} \\ &\simeq T\big(\big\{\|t_{RA}\|_{\#} \otimes \big\|t'_{RA}\big\|_{\#}, \perp\big\}\big) \\ &= T\big(\big\{\|t_{RA}\|_{\#}, \big\|t'_{RA}\big\|_{\#}, \perp\big\}\big) \end{aligned}$$

$$= T\big(Eval(t_{RA}) \cup Eval(t'_{RA})\big)$$
$$= Eval(t_{RA}) \oplus Eval(t'_{RA}).$$

5.3 Relational Algebra and Database Schema Mappings

In what follows, we will demonstrate that all relational algebra operations over the databases correspond to a specific subclass of inter-schema mappings, and hence every update of a given database schema can be equivalently represented as a composed inter-schema mapping.

Proposition 26 *Every term $t_R \in \mathcal{T}_{RE}$ with the relational symbols (i.e., the variables) of a given database schema $\mathcal{A}$ such that it is an update of a given relation or an extraction of a view of this database can be represented by a composed database schema mapping $\mathcal{M} = \{\Phi\} : \mathcal{A} \to \mathcal{A}$.*

Proof From the completeness of **RA** (Proposition 23), we obtained the reduction $Rd_{RE} : \mathcal{T}_{RE} \to Mor_{RA}$ such that, for a given schema $\mathcal{A} = (S_A, \Sigma_A)$, a term $t_R \in \mathcal{T}_{RE}$ is reduced to the arrow $f : t_{RA} \to f(t_{RA})$ such that $\|t_R\|_\# = \|f(t_{RA})\|_\#$, where $t_{RA} = t_1 \otimes \cdots \otimes t_m$ with

$$t_i \in \big\{r_i, \big(\ldots\big(r_i\langle a_{i1}, name_{i1}, e_{i1}\rangle\big)\ldots\big)\langle a_{ik}, name_{ik}, e_{ik}\rangle\big\}, \quad r_i \in S_A,$$

for $1 \le i \le m$. Each term $t_i = (\ldots(r_i\langle a_{i1}, name_{i1}, e_{i1}\rangle)\ldots)\langle a_{ik}, name_{ik}, e_{ik}\rangle$ can be represented by the SOtgd Ψ_i equal to

$$\forall \mathbf{x}_i \big(r_i(\mathbf{x}_i) \Rightarrow r_i^1\big(\mathbf{x}_i \& \big(e_{i1}(\mathbf{x}_i), \ldots, e_{ik}(\mathbf{x}_i)\big)\big)\big),$$

where each $e_{ij}(\mathbf{x}_i)$, for $1 \le j \le k$, is an expression (i.e., a function, usually nullary function (constant)) with the variables in $\mathbf{x}_i = (x_{i1}, \ldots, x_{il})$, and $ar(r_i) = l, ar(r_i^1) = l + k$.

Hence, we define the schema $\mathcal{A}_1 = (\{r_i \mid t_i = r_i \in t_{RA}\} \cup \{r_i^1 \mid ar(r_i^1) = ar(r_i) + k,$ for $t_i = (\ldots(r_i\langle a_{i1}, name_{i1}, e_{i1}\rangle)\ldots)\langle a_{ik}, name_{ik}, e_{ik}\rangle \in t_{RA}\}, \emptyset)$, and the schema $\mathcal{B}_0 = (\{r_{t_0}\}, \emptyset)$ with $ar(r_{t_0}) = \Sigma_{r \in \mathcal{A}_1} ar(r)$, so that the construction of t_{RA} is defined by the composition of schema mappings:

$$\mathcal{M}_{A_1 B_0} \circ \mathcal{M}_{AA_1} : \mathcal{A} \to \mathcal{B}_0, \tag{5.4}$$

where $\mathcal{M}_{AA_1} = \{(\bigwedge_{r_i^1 \in \mathcal{A}_1} \Psi_i) \wedge (\bigwedge_{r_i \in \mathcal{A}_i} \forall \mathbf{x}_i (r_i(\mathbf{x}_i) \Rightarrow r_i(\mathbf{x}_i)))\} : \mathcal{A} \to \mathcal{A}_1$ and $\mathcal{M}_{A_1 B_0} = \{\Phi\}$ where Φ is the SOtgd for the Cartesian product of the relations in $\mathcal{A}_1$ (see point 2 in Sect. 5.1).

Let $f = o_k \circ \cdots \circ o_2 \circ o_1$, with o_i, $1 \le i \le k$ the unary algebraic operators in Σ_{RA}. Then, from the proof of Proposition 24, each o_i can be expressed by an SOtgd Φ_i, thus by an inter-schema mapping $\mathcal{M}_{B_{i-1} B_i} : \mathcal{B}_{i-1} \to \mathcal{B}_i$, with $\mathcal{B}_i = (\{r_{t_i}\}, \emptyset)$ and $ar(r_{t_i}) \ge 1$ is the arity of the relation $\|o_i(r_{t_{i-1}})\|_\#$, for $i = 1, \ldots, k$. Consequently, the construction of $f(t_{RA})$ is defined by the following composition of schema mappings:

$$\mathcal{M}_{AB_k} = \mathcal{M}_{B_k B_{k-1}} \circ \cdots \circ \mathcal{M}_{B_1 B_0} \circ \mathcal{M}_{A_1 B_0} \circ \mathcal{M}_{AA_1} : \mathcal{A} \to \mathcal{B}_k. \tag{5.5}$$

Thus, we have the following two cases:

1. If this term $t_R \in \Sigma\mathcal{T}_{RE}$ is an updating of a relation $r \in \mathcal{A}$ then it can be equivalently represented by the schema mapping $\mathcal{M}_{B_k A} \circ \mathcal{M}_{AB_k} : \mathcal{A} \to \mathcal{A}$, where $\mathcal{M}_{B_k A} = \{\Psi\} : \mathcal{B}_k \to \mathcal{A}$ with Ψ equal to the SOtgd $\forall \mathbf{x}(r_{t_k}(\mathbf{x}) \Rightarrow r(\mathbf{x}))$.
2. Otherwise, t_R is an SPJRU term of a view over $\mathcal{A}$ and hence it can be represented by the schema mapping $\mathcal{M}_{A_2 A} \circ \mathcal{M}_{B_k A_2} \circ \mathcal{M}_{AB_k} : \mathcal{A} \to \mathcal{A}$, where $\mathcal{A}_2 = (S_A \cup \{r_{t_k}\}, \Sigma_A)$ (i.e., the extension of $\mathcal{A}$ by a relational symbol for this view), $\mathcal{M}_{B_k A_2} = \{\Phi_1\}$ with Φ_1 equal to SOtgd $\forall \mathbf{x}(r_{t_k}(\mathbf{x}) \Rightarrow r_{t_k}(\mathbf{x}))$, and $\mathcal{M}_{A_2 A} = \{\Phi_2\}$ where Φ_2 is equal to the SOtgd $\bigwedge_{r \in S_A} \forall \mathbf{x}(r(\mathbf{x}) \Rightarrow r(\mathbf{x}))$. □

Thus, each update of a given database, or the extraction of its views, may be specified as a particular schema mapping. The evaluation functor $Eval : \mathbf{RA} \to \mathbf{DB}$ formally expresses the semantics of the terms (i.e., their evaluation) of the Σ_{RA} algebra and, from the fact that all (also the update) terms of the Codd's algebra may be equivalently represented by the terms of the Σ_{RA} algebra, we conclude that the **DB** category is a powerful denotational semantics not only for the database schema mappings (based on SPJRU subalgebra of the Σ_R Codd's algebra) but also for the database updates.

Corollary 17 *For any given database schema $\mathcal{A}$, each update of its relations in a current instance A can be expressed in* **DB** *by a morphism $f : A \to A_1$, where A_1 is the resulting database of the schema $\mathcal{A}$ after these updates.*

Proof It is a consequence of the reduction to the inter-schema mappings in Proposition 26. We have shown in the previous Example 34 that for all three basic kinds of the data manipulation (of the relations of a given instance $A = \alpha^*(\mathcal{A})$), which are UPDATE, INSERT or DELETE, they are expressed in **DB** by a morphism $f : \alpha^*(\mathcal{A}) \to \alpha_1^*(\mathcal{A})$ ($f = f_U$ in (i), $f = f_I$ in (ii) and $f = f_D$ in (iii), in Example 34), where $A_1 = \alpha_1(\mathcal{A})$ is the database instance after such updates.

Consequently, *any* data manipulation of the relations of a given instance A of database schema $\mathcal{A}$ is a combination of these three basic kinds of data manipulations and hence a composition of these basic morphisms above, which is a morphism from A into the final obtained database instance of this schema $\mathcal{A}$ (after all updates). □

Based on this corollary, the whole collection of all the data manipulations of a given database schema $\mathcal{A}$ is represented as a composition of morphisms from the empty database instance A_0, with all empty relations (immediately after the creation of RDB schema of this database), into the final (current) instance of this database.

This 'completeness property' of the **DB** category, w.r.t. the full Codd's algebra, both with updating of the relational tables of the instance databases (with inserting and deleting, expressed in Example 34 by the morphisms in **DB** category), will be used in the next two chapters dedicated to the categorial RDB-machines and to the operational semantics for the database mappings.

5.4 DB Category and Relational Algebras

From the introduction to the symmetric categories and Corollary 1 (Sect. 1.5.1) and the results presented in Chap. 3, dedicated to the definition of the **DB** category, we obtain that its algebra $Alg_{\mathbf{DB}} = ((Ob_{DB}, Mor_{DB}), \Sigma_{\mathbf{DB}})$ has the signature $\Sigma_{\mathbf{DB}} = \{dom, cod, id, \circ\} \cup S_{DB}$, with the additional two operators $*$, $B_T \in S_{DB}$, where $* : Ob_{DB}^2 \rightarrow Ob_{DB}$ and the unary operator $B_T \in \Sigma_{\mathbf{DB}}$ is also a homomorphism between two algebras $B_T : (Mor_{DB}, \circ) \rightarrow (Ob_{DB}, *)$, and the following operators $\{\uplus, \frac{\cup}{rn}, \bigotimes, \bigoplus\} \subset S_{DB}$ for the composition of objects for any two simple database instances $A, B \in Ob_{DB}$:

1. Data-separation (i.e., (Co)product, disjoint union), $\times(A, B) = \biguplus(A, B)$;
2. Data-federation (union with renaming of the relations), $\frac{\cup}{rn}(A, B)$;
3. Power-view composition, $TA = T(A)$;
4. Matching (monoidal product), $\bigotimes(A, B) = \bigcap(T(A), T(B))$ (to be introduced formally in Sect. 8.1.1);
5. Merging, $\bigoplus(A, B) = \bigcup(T(A), T(B))$ (to be introduced formally in Sect. 8.1.2).

Another set of operators in S_{DB} was introduced for the composition (pairing) of the arrows in Chap. 3.

What we can discover, for the composition of the objects in **DB**, is the fact that all of them are the standard set-like operators (as the intersection, union, union with renaming and disjoint union). The only specifically relational DB operator is the unary power-view operator $T : Ob_{DB} \rightarrow Ob_{DB}$ and, as it was demonstrated at the end of Sect. 5.1.1, the object (i.e., instance-database) TA is obtained from the instance-database A by the set $\mathcal{T}_P X$ (where the set of variables $X = \mathcal{A}$ is the set of relational symbols of the database schema $\mathcal{A}$) of terms of Σ_R (SPJRU) relational algebra over the schema $\mathcal{A}$ (obtained by the unique surjective extension $\alpha_\# : \mathcal{T}_P X \rightarrow TA$ of the assignment (R-algebra) $\alpha : X \rightarrow TA$ which defines the instance-database $A = \{\alpha(r) \mid r \in X = \mathcal{A}\}$).

It means that all objects in **DB** are constructed by the standard set-like operators above and the Codd's relational (SPRJU) algebra Σ_R in Definition 31.

Similarly, all arrows in **DB** are constructed by simple arrows and by the set of composition (pairing) operations in $S_{DB} \subset \Sigma_{\mathbf{DB}}$, however, the open question is if the internal structure of the simple arrows based on the R-operads can be expressed equivalently by some extension of the Codd's relational algebra Σ_R. Here we will try to provide a proper answer to this question.

For each simple arrow $f = \{q_1, \ldots, q_n, q_\perp\} : A \rightarrow B$ in **DB** with $q_i \in O(r_{i_1}, \ldots, r_{i_k}, r_B)$, $r_{i_m} \in \mathcal{A}$, $1 \le m \le k$, for $1 \le i \le n$, and $r_B \in \mathcal{B}$, we can obtain its "conceptualized" object (the information flux), $\widetilde{f} = B_T(f) \subseteq TA \cap TB$, which is a closed object, that is, $\widetilde{f} = TC$, where $C \subset TA \cap TB$. Thus, the conceptualization of every simple arrow in **DB** is an instance database obtained by the set of Σ_R terms over a schema C with relations in C. Based on this conceptualization and duality between objects and arrows in the **DB** category, it is justified to ask if each simple arrow f is by itself composed in some way by the set of terms of some particular extension of the Codd's relational algebra Σ_R, that is, if each R-operad $q_i \in O(r_{i_1}, \ldots, r_{i_k}, r_B)$ is *computationally equivalent* to some relational-algebra term, and exactly to which one.

If there is such a relational algebra, then it has to include the Codd's (SPRJU) Σ_R algebra and hence in this case all elements (the objects and the arrows) in the **DB** category can be equivalently represented by the subsets of terms of such a relational algebra. Consequently, the category **DB**, also with different prime-objects (which for **DB** are the instance-databases and not the simple relations as for the relational algebras), would be equivalent (from the computational point of view) to such a relational algebra.

In what follows, we will make this investigation and will shortly resume the process of the "algebraization" of the schema database mappings presented in Chap. 2.

In Sect. 2.4, we discussed the process of the "algebraization" of the logic theory based on SOtgds into an algebraic theory based on R-operads $q \in O(r_1, \ldots, r_k, r)$, which are the abstract k-ary "operations" (functional symbols), but we did not consider the problem of what is a minimal subset of the elementary R-operads that composes the signature of this database-mapping algebra.

We mentioned that the left-hand sides of the SOtgds can be equivalently represented by SPRJU (Codd's) relational algebra (Σ_R in Definition 31 in Sect. 5.1) but augmented by the binary MINUS operator (for the predicates preceded by logical negation $\neg$, so that a logic formula $\neg r(x_1, \ldots, x_n)$ is transformed into the relational-algebra term $(\overbrace{r_\infty \otimes \cdots \otimes r_\infty}^{ar(r)}$ MINUS $r)$, where for every R-algebra α, from point 1 of Definition 10 in Sect. 2.4.1, $\overbrace{\alpha(r_\infty) \otimes \cdots \otimes \alpha(r_\infty)}^{ar(r)} = \overbrace{\mathcal{U} \times \cdots \times \mathcal{U}}^{ar(r)} = \mathcal{U}^{ar(r)}$ for a given universe $\mathcal{U}$, so that, $\overline{R} = \alpha_\#((\overbrace{r_\infty \otimes \cdots \otimes r_\infty}^{ar(r)}$ MINUS $r) = \mathcal{U}^{ar(r)} \backslash \alpha(r) = \mathcal{U}^{ar(R)} \backslash R$ (i.e., the complement of R). Notice that here we do not make the Codd's constraints about the necessity to have only the relations with a finite extensions (i.e., the finite number of tuples), so we permit the infinite universes as well (especially when $\mathcal{U} = \mathbf{dom} \cup SK$ where SK is an infinite set of marked null values, i.e., the Skolem constants) also for a finite domain **dom**. Hence, we do not make extensions of the Codd's original SPRJU algebra of finite relations with a set of new algebraic operations, but we permit to work in the full correspondence with the FOL for such relational algebra where we can have the infinite extensions of the predicates (i.e., relations) for a given infinite universe $\mathcal{U}$.

Then in Definition 11, Sect. 2.4.1, we considered the valid subset of R-algebra interpretations α for these R-operads (*mapping-interpretations*), such that $\alpha(q) : \alpha(r_1) \times \cdots \times \alpha(r_k) \to \alpha(r)$ is an n-ary operation (a function) whose arguments are the relations $\alpha(r_i) \in \underline{\Upsilon}$. Thus, in order to define the algebra (its signature and the semantics of its operators (R-operads) for the database mappings, we need to recognize a strict subfamily of functions $\alpha(q)$, from which, by a composition, we can obtain all other R-operads as composed terms: we can see that all of them are parts of the morphisms of the **DB** category and hence each morphism in **DB** would be a *set* of terms of this *database-mapping algebra*. However, such a database-mapping algebra would also have other operations which are introduced by existentially quantified functional symbols in the SOtgds and are used in the right-hand sides of the tgds implication as the terms with variables (i.e., the attributes of the relations in the left-hand side of the tgds material implication).

Proposition 27 *For any given schema mapping $\mathcal{M}_{AB}: \mathcal{A} \to \mathcal{B}$, each of its operads $q_i \in O(r_1, \ldots, r_k, r_B)$ in MakeOperads($\mathcal{M}_{AB}$) can be equivalently represented by a term t_R of the relational Σ_{RE} algebra in Definition* 31 *such that if the extension of r_B before this mapping was $R_B = \alpha(r_B) \in B = \alpha^*(\mathcal{B})$ then, after the execution of this mapping, its extension is $R'_B = \alpha_1(r_B) = R_B \cup \|t_R\|_\#$ (for the assignment $\|\_\|: \mathbb{R} \to \Upsilon$ in Definition* 31 *such that for each $r \in \mathcal{A}$, $\|r\| = \alpha(r)$).*

Proof Let $\phi_{Ai}(\mathbf{x}) \Rightarrow r_B(\mathbf{t})$ be an implication χ in the normalized SOtgd $\exists \mathbf{f}(\Psi)$ (where Ψ is an FOL formula, i.e., a conjunction of a number of such implications) of a mapping $\mathcal{M}_{AB}: \mathcal{A} \to \mathcal{B}$, and $\mathbf{t}$ be a tuple of terms with variables in $\mathbf{x} = \langle x_1, \ldots, x_m \rangle$ and functional symbols in $\mathbf{f}$ (at least one, as in Example 16, Sect. 2.5). Let $q_i \in \textit{MakeOperads}(\mathcal{M}_{AB})$ be an operad's operation obtained by algorithm *MakeOperads* from this implication and equal to the expression $((e \wedge C) \Rightarrow (\_\,)(\mathbf{t})) \in O(r_1, \ldots, r_k, r_B)$ where C is a condition (as in Example 16), where $q_i = v_i \cdot q_{A,i}$ with $q_{A,i} \in O(r_1, \ldots, r_k, r_q)$ and $v_i \in O(r_q, r_B)$ are such that for a new relational symbol r_q, $n = ar(r_q) = ar(r_B)$, so that the tuple of terms on the right-hand side of implication is $\mathbf{t} = (t_1, \ldots, t_n)$ and its sublist of terms with functional symbols $\mathbf{t}_1 = (t_{i_1}, \ldots t_{i_m}) \subseteq \mathbf{t}$, $m \leq n$.

From Definition 11 in Sect. 2.4.1, for an R-algebra α we obtain a mapping-interpretation $\alpha(v_i): \alpha(r_q) \to \alpha(r_B)$ which is an *injective* function (when a schema-mapping is satisfied), with the k-ary operation (a k-ary function):

$$\alpha(q_{A,i}): R_1 \times \cdots \times R_k \to \alpha(r_q),$$

where for each $1 \leq i \leq k$, $R_i = \mathcal{U}^{ar(r_i)} \backslash \alpha(r_i)$ (i.e., the complement of the relation $\alpha(r_i)$) if the place symbol $(\_\,)_i \in e$ is preceded by negation operator $\neg$; $\alpha(r_i)$ otherwise. Thus, the mapping $\alpha(q_i) = \alpha(v_i) \cdot \alpha(q_{A,i}): R_1 \times \cdots \times R_k \to R_B$, where $R_B = \alpha(r_B)$, is equivalent to the following Σ_{RE}-algebra term t_R:

$$[S]\big(\text{EXTEND}(\ldots(\text{EXTEND}(t'_R \text{WHERE } C \wedge C')\text{ADD } a_{N+1}, name_{N+1}, t_{i_1}) \ldots) \\ \text{ADD } a_{N+m}, name_{N+m}, t_{i_m}\big),$$

where
1. $N = ar(r_1) + \cdots + ar(r_k)$;
2. The Cartesian product of the relations $t'_R = t'_1 \otimes \cdots \otimes t'_k$, where for each $1 \leq i \leq k$, $t'_i = ((\overbrace{r_\infty \otimes \cdots \otimes r_\infty}^{ar(r_i)})$ MINUS $r_i)$, if the place symbol $(\_\,)_i \in e$ is preceded by negation operator $\neg$; r_i otherwise;

3. $[S] = [K_1, \ldots, K_n]$, where, for each $1 \leq l \leq n$, if $t_l \in \mathbf{t}$ is a variable then K_l is the position of this variable in the Cartesian product t'_R; otherwise, if $t_l = t_{i_j} \in \mathbf{t}_1$ then $K_l = N + j$;
4. C' is the condition $\alpha((q_{A,i})(\mathbf{x}_1, \ldots, \mathbf{x}_k)) \neq \langle\rangle$, where for $1 \leq i \leq k$ we have the tuples of variables $\mathbf{x}_i = (x_{i,1}, \ldots, x_{i,ar(r_i)})$ for each conjunct $(\_)_i(\mathbf{x}_i) \in e$. □

Based on this proposition, which demonstrated that each arrow in **DB** (a set of functions) is computationally equivalent to the corresponding set of terms of the Σ_{RE}-algebra, and on the fact that each object (a database instance) can be obtained by terms of $\Sigma_R \subset \Sigma_{RE}$ algebra, we obtain the following corollary:

Corollary 18 *The denotational database-mapping category* **DB** *is computationally equivalent to the relational* Σ_{RE} *algebra.*

Based on the fact that the extension Σ_{RE} of the Codd's relational algebra Σ_R is computationally strictly less expressive than the typed λ-calculus, it holds that **DB** cannot be a topos: in fact, in the last chapter, we will demonstrate that **DB** is less expressive and a kind of weak monoidal topos.

This corollary demonstrates that the evaluation functor $Eval : \mathbf{RA} \to \mathbf{DB}$ in Proposition 24 is a mapping between two computationally equivalent categories, and hence the database algebra $Alg_{\mathbf{DB}} = ((Ob_{DB}, Mor_{DB}), \Sigma_{\mathbf{DB}})$ is the strict generalization of the relational algebra Σ_{RE}, from the relations to the databases and their mappings.

Let us define the *inverse-inclusion* function $in^{-1} : R_2 \twoheadrightarrow R_1$ for an injective function $in : R_1 \hookrightarrow R_2$, for two relations $R_1 \subseteq R_2$, such that for each tuple $\mathbf{d} \in R_2$:

$$in^{-1}(\mathbf{d}) = \mathbf{d} \quad \text{if } \mathbf{d} \in R_1; \qquad \langle\rangle \text{ (empty tuple)} \quad \text{otherwise.}$$

We can see by Proposition 24 that the unary operators (i.e., atomic arrows in **RA**) are mapped by the functor *Eval* into the following family of elementary R-algebra functions: identities, projections, and inverse-inclusions.

The question is what the full family of the elementary R-algebra functions used in the **DB** category is, which compose the algebra signature the database-mappings R-algebra.

We recall that for a given set of formulae $S = \{\phi_1, \ldots, \phi_n\}$ by $\bigwedge S$ we denote the conjunction $\phi_1 \wedge \cdots \wedge \phi_n$.

Let us define an algorithm that transforms a mapping implication $\phi_{Ai}(\mathbf{x}) \Rightarrow r_B(\mathbf{t})$ in a n SOtgd $\exists \mathbf{f}(\Psi)$ (where Ψ is an FOL formula, i.e., a conjunction of a number of such implications of a mapping $\mathcal{M}_{AB} : \mathcal{A} \to \mathcal{B}$, and $\mathbf{t} = (t_1, \ldots, t_n)$ is a tuple of terms with variables in $\mathbf{x}$ and functional symbols in $\mathbf{f}$), into the logically equivalent implication, where all variables in each relational symbol on the left-hand side are unique (i.e., different from all other variables in all relational symbols) and the relation in the right-hand side has a tuple of the variables only:

Variable-normalization algorithm *VarTransform*$(\phi_{Ai}(\mathbf{x}) \Rightarrow r_B(\mathbf{t}))$

Input. A mapping implication $\phi_{Ai}(\mathbf{x}) \Rightarrow r_B(\mathbf{t})$, where $\mathbf{t} = (t_1, \ldots, t_n)$ is a tuple of terms with variables in $\mathbf{x}$ and $\mathbf{t}_1 = (t_{i_1}, \ldots, t_{i_m}) \subseteq \mathbf{t}$ is the (possibly empty) sublist of terms with at least one functional symbol in $\mathbf{f}$.
Output. A logically equivalent implication with extended set of variables.

1. (*Replacing of the characteristic functions with its relations*)
   If there are no equations in $\phi_{Ai}(\mathbf{x})$ with the characteristic functions for relations, go to Step 2. Otherwise, each condition in $\phi_{Ai}(\mathbf{x})$ with a characteristic function of some relation, $(f_{r_i}(\mathbf{x}_i) \doteq \overline{1})$ with $\mathbf{x}_i \subseteq \mathbf{x}$, is eliminated from $\phi_{Ai}(\mathbf{x})$ and replaced by $r_i(\mathbf{x}_i)$.
2. (*Introduction of the new variables for the join-conditions*)
   Here we have the following implication (logically equivalent to the implication in input), for $k \geq 1$,

   $$\Big(\bigwedge\{\sim_i r_i(\mathbf{x}_i) | 1 \leq i \leq k\} \wedge C(\mathbf{x})\Big) \Rightarrow r_B(\mathbf{t}),$$

   where $\sim_i$ is the identity logical operator or the negation logical operator $\neg$, the tuples of variables $\mathbf{x}_i \subseteq \mathbf{x}$, $1 \leq i \leq k$, and $C(\mathbf{x})$ is a condition with variables in $\mathbf{x}$ without characteristic functions of relations, or if there is no such a condition we take that it is a tautology $(\overline{1} \doteq \overline{1})$.
   Set S_J to be the empty set. For each variable x_{i_j} in a relation r_i, such that the same variable is present in another relations on the left-hand side of this implication, we introduce a fresh new variable y_{i_j}, then we substitute x_{i_j} in all other relations r_j, $1 \leq j \leq k$, $j \neq i$, by y_{i_j} and insert the equation $y_{i_j} = x_{i_j}$ in S_J. Continue this process as far as possible. At the end of this step (if S_J is empty then set $S_J = \{r_\top(\overline{0}, \overline{0})\}$), we obtain the resulting implication,

   $$\Big(\bigwedge\{\sim_i r_i(\mathbf{x}'_i) | 1 \leq i \leq k\} \wedge C(\mathbf{x}) \bigwedge S_J\Big) \Rightarrow r_B(\mathbf{t}),$$

   where the new tuples of variables $\mathbf{x}'_i$ are obtained from $\mathbf{x}_i$ by the substitutions of some variables by the new introduced variables above (by substitutions of old variables with new variables), so that for $1 \leq i, j \leq k, i \neq j, \mathbf{x}'_i \cap \mathbf{x}'_j = \emptyset$.
3. (*Elimination of the terms with functional symbols from the right-hand side of implication*)
   If $\mathbf{t}_1$ is empty then set $S_F = \{r_\top(\overline{1}, \overline{1})\}$ and go to Step 4.
   For each term $t_{i_j} \in \mathbf{t}_1$ we introduce a fresh new variable z_{i_j} and we insert the equation $z_{i_j} = t_{i_j}$ in S_F.
4. (*Construction of the new implication*)
   The new implication is a formula

   $$\Big(\bigwedge\{\sim_i r_i(\mathbf{x}'_i) | 1 \leq i \leq k\} \wedge C(\mathbf{x}) \bigwedge S_J \bigwedge S_F\Big) \Rightarrow r_B(\mathbf{t}_2),$$

   where $\mathbf{t}_2$ is obtained from $\mathbf{t}$ by substituting each term t_{i_j} by the variable z_{i_j}, for $1 \leq j \leq m$.
   Return the resulting implication.

Let us show this algorithm in the case of Example 16 in Sect. 2.5:

Example 35 Consider the following three cases:

1. The mapping $\mathcal{M}'_{AD} = \{\Phi\} : \mathcal{A} \to \mathcal{D}$ in Example 5, with Φ equal to SOtgd $\exists f_1(\forall x_1 \forall x_2(\texttt{Takes}(x_1, x_2) \Rightarrow \texttt{Enrolment}(f_1(x_1), x_2)))$. With the decomposition we obtained Φ_E equal to $\forall x_1 \forall x_2(\texttt{Takes}(x_1, x_2)) \Rightarrow r_{q_1}(x_2)$, and Φ_M equal to

$$\exists f_1 \exists f_{Takes}\big(\forall x_1 \forall x_2\big(\big(r_{q_1}(x_2) \wedge \big(f_{Takes}(x_1, x_2) \doteq \bar{1}\big)\big) \\ \Rightarrow \texttt{Enrolment}\big(f_1(x_1), x_2\big)\big)\big).$$

   By application of the algorithm *VarTransform* to the implications in Φ, Φ_E and Φ_M, we obtain the following respective implications:
   1.1. $(\texttt{Takes}(x_1, x_2) \wedge (\bar{1} \doteq \bar{1}) \wedge r_\top(\bar{0}, \bar{0}) \wedge (z_1 = f_1(x_1))) \Rightarrow \texttt{Enrolment}(z_1, x_2)$.
   1.2. $(\texttt{Takes}(x_1, x_2) \wedge (\bar{1} \doteq \bar{1}) \wedge r_\top(\bar{0}, \bar{0}) \wedge r_\top(\bar{1}, \bar{1})) \Rightarrow r_{q_1}(x_2)$.
   1.3. $(r_{q_1}(x_2) \wedge \texttt{Takes}(x_1, y_1) \wedge (\bar{1} \doteq \bar{1}) \wedge (y_1 = x_2) \wedge (z_1 = f_1(x_1))) \Rightarrow \texttt{Enrolment}(z_1, x_2)$.
2. The mapping $\mathcal{M}'_{AD} = \{\Phi\} : \mathcal{A} \to \mathcal{D}$ in Example 5, with Φ equal to SOtgd

$$\exists f_{Student}\big(\forall x_1 \forall x_2 \forall y_3\big(\big(\texttt{Takes}(x_1, x_2) \wedge \big(f_{Student}(x_1, y_3) \doteq \bar{1}\big)\big) \\ \Rightarrow \texttt{Enrolment}(y_3, x_2)\big)\big).$$

   We obtained Φ_E equal to

$$\exists f_{Student}\big(\forall x_1 \forall x_2 \forall y_3\big(\big(\texttt{Takes}(x_1, x_2) \wedge \big(f_{Student}(x_1, y_3) \doteq \bar{1}\big)\big) \Rightarrow r_{q_2}(y_3, x_2)\big)\big)$$

   and Φ_M equal to

$$\exists f_{Student}\big(\forall x_1 \forall x_2 \forall y_3\big(\big(r_{q_2}(y_3, x_2)\big) \wedge \big(f_{Student}(x_1, y_3) \doteq \bar{1}\big)\big) \\ \Rightarrow \texttt{Enrolment}(y_3, x_2)\big)\big).$$

   By application of the algorithm *VarTransform* to the implications in Φ, Φ_E and Φ_M, we obtain the following respective implications:
   2.1. $(\texttt{Takes}(x_1, x_2) \wedge \texttt{Student}(y_1, y_3) \wedge (\bar{1} \doteq \bar{1}) \wedge (y_1 = x_1) \wedge r_\top(\bar{1}, \bar{1})) \Rightarrow \texttt{Enrolment}(y_3, x_2)$.
   2.2. $(\texttt{Takes}(x_1, x_2) \wedge \texttt{Student}(y_1, y_3) \wedge (\bar{1} \doteq \bar{1}) \wedge (y_1 = x_1) \wedge r_\top(\bar{1}, \bar{1})) \Rightarrow r_{q_2}(y_3, x_2)$.
   2.3. $(r_{q_2}(y_3, x_2) \wedge \texttt{Student}(x_1, y_4) \wedge (\bar{1} \doteq \bar{1}) \wedge (y_4 = y_3) \wedge r_\top(\bar{1}, \bar{1})) \Rightarrow \texttt{Enrolment}(y_3, x_2)$.
3. The mapping $\mathcal{M}_{AC} = \{\Phi\} : \mathcal{A} \to \mathcal{D}$ in Example 9, with Φ equal to SOtgd with the two implications

$$\exists f_1 \exists f_{Emp} \exists f_{Over65} \big(\forall x_e \big(\big(\big(f_{Emp}(x_e) \doteq \overline{1}\big) \wedge \texttt{Local}(x_e)\big) \Rightarrow \texttt{Office}\big(x_e, f_1(x_e)\big)\big) \wedge \forall x_e \big(\big(\big(f_{Emp}(x_e) \doteq \overline{1}\big) \wedge \big(f_{Over65}(x_e) \doteq \overline{1}\big)\big) \Rightarrow \texttt{CanRetire}(x_e)\big)\big)\big).$$

By application of the algorithm *VarTransform* to these two implications in Φ, we obtain the following respective implications:

3.1. $(\texttt{Emp}(x_e) \wedge \texttt{Local}(y_1) \wedge (\overline{1} \doteq \overline{1}) \wedge (y_1 = x_e) \wedge (z_1 = f_1(x_e))) \Rightarrow \texttt{Office}(x_e, z_1)$.

3.2. $(\texttt{Emp}(x_e) \wedge \texttt{Over65}(y_1) \wedge (\overline{1} \doteq \overline{1}) \wedge (y_1 = x_e) \wedge r_\top(\overline{1}, \overline{1})) \Rightarrow \texttt{CanRetire}(x_e)$.

If we were to consider the case when an employee could retire when he reached the age of 65 but with only 80 % of the full pension, and only if he retired at the age of 70 or older he would receive the full pension, we could introduce the function $f_{\%} : \mathcal{N} \to [0, 1]$ such that $f_{\%}(x) = 0$ if $x \leq 64$, $f_{\%}(x) = 1$ if $x \geq 70$, and $f_{\%}(x) = 0.8 + 0.4 * (x - 65)$ for $65 \leq x \leq 70$.

Then we may replace the unary relation `Over65` by the binary relation `People` where the first attribute is equal to the attribute of `Over65` and the second is for the age when he asked for the retirement. Analogously, we add the second attribute to the relation `CanRetire` dedicated to the reduction index for its pension. Consequently, the second implication above will be substituted by the implication,

$$\forall x_e \forall x \big(\big(\big(f_{Emp}(x_e) \doteq \overline{1}\big) \wedge \big(f_{Person}(x_e, x) \doteq \overline{1}\big)\big) \Rightarrow \texttt{CanRetire}\big(x_e, f_{\%}(x)\big)\big).$$

By application of the algorithm *VarTransform* to this new implication, we obtain the following implication:

3.3. $(\texttt{Emp}(x_e) \wedge \texttt{Person}(y_1, x) \wedge (x \geq 65) \wedge (y_1 = x_e) \wedge (z_1 = f_{\%}(x))) \Rightarrow \texttt{CanRetire}(x_e, z_1)$,

and hence $C(x)$ is a condition $(x \geq 65)$, $S_j = \{(y_1 = x_e)\}$ and $S_F = \{(z_1 = f_{\%}(x))\}$.

Based on this algorithm, we can demonstrate the following theorem:

Theorem 9 *The signature of the R-algebra used to define the database mappings in the* **DB** *category is composed of the following family of operations*:

1. *The identity functions* $id_R : R \to R$, *for any relation* $R \in \Upsilon$;
2. *The injections (inclusion functions)* $in : R \hookrightarrow R_1$ *(with* $R \subseteq R_1$*) and the inverse-inclusions* $in^{-1} : R_1 \twoheadrightarrow R$ *(i.e., the surjective one-to-one functions)*;
3. *The projections* $\pi_S : R_1 \to R_2$ *such that the list* S *is a subset of the set of attributes of* R_1 *and the set attributes of* R_2 *is equal to this list* S;
4. *The n-ary* $(n \geq 2)$ *Cartesian functions* $f_\otimes^{(n)} : R_1 \times \cdots \times R_n \to R$ *such that* $ar(R) = \sum_{1 \leq i \leq n} ar(R_i)$, *and for any* $\mathbf{d}_i \in R_i$, $1 \leq i \leq n$,

$$f_\otimes^{(n)}(\mathbf{d}_1, \ldots, \mathbf{d}_n) = \mathbf{d}_1 \& \cdots \& \mathbf{d}_n;$$

5. *For each relation R with the tuple of attributes in* $\mathbf{x}$, *and a term* $t_i(\mathbf{x}_i)$ *with the functions and variables* $\mathbf{x}_i = \pi_S(\mathbf{x}) \subseteq \mathbf{x}$ *(determined by the projection on a sublist of the attributes S of R), the surjective function* $f_{R,t_i} : R \twoheadrightarrow R_1$ *such that for any tuple* $\mathbf{d} \in R$,

$$f_{R,t_i}\big(\pi_S(\mathbf{d})\big) = \mathbf{d}\&b, \quad \textit{where } b = t_i\big[\mathbf{x}_i, \pi_S(\mathbf{d})\big] \in \mathcal{U},$$

so that $ar(R_1) = ar(R) + 1$ *and* R_1 *is the image of this function.*

Proof Let us consider the following mapping implication $\phi_{Ai}(\mathbf{x}) \Rightarrow r_B(\mathbf{t})$, where $\mathbf{t} = (t_1, \ldots, t_n)$ is a tuple of terms with variables in $\mathbf{x}$ and $\mathbf{t}_1 = (t_{i_1}, \ldots, t_{i_m}) \subseteq \mathbf{t}$ is the (possibly empty) sublist of terms with at least one functional symbol in $\mathbf{f}$. Hence, we obtain as output of the algorithm *VarTransfom* the implication:

$$\Big(\bigwedge\{\sim_i r_i(\mathbf{x}'_i) | 1 \le i \le k\} \wedge C(\mathbf{x}) \bigwedge S_J \bigwedge S_F\Big) \Rightarrow r_B(\mathbf{t}_2),$$

where S_F is a set of equations $z_{i_j} = t_{i_j}$, $1 \le j \le m$, $N = ar(r_1) + \cdots + ar(r_k)$, and $\mathbf{x}' = \mathbf{x}'_1; \ldots; \mathbf{x}'_k$. We divide this implication into a number of "atomic" logical equivalences, by introducing a set of new auxiliary intermediate relations, as follows:

Implication/Equivalence	Σ_{RE} algebra term	R-algebra operation
$\bigwedge\{\sim_i r_i(\mathbf{x}'_i) \| 1 \le i \le k\}$ $\Leftrightarrow\ r_C(\mathbf{x}'_1\&\cdots\&\mathbf{x}'_k)$	$r_C = t_1 \otimes \cdots \otimes t_k$ (Cartesian product)	$f_\otimes^{(k)} : R_1 \times \cdots \times R_k \to R_C$ (Cartesian k-ary function)
$r_C(\mathbf{x}') \wedge C(\mathbf{x}) \bigwedge S_J$ $\Leftrightarrow\ r_{S_0}(\mathbf{x}')$	$r_{S_0} = r_C$ WHERE $C \bigwedge S_J$ (Select by $C(\mathbf{x})$ + Join by S_J)	$in^{-1} : R_C \twoheadrightarrow R_{S_0}$ (Inverse-inclusion function)
$r_{S_0}(\mathbf{x}') \Leftrightarrow r_{S_1}(\mathbf{x}'\&t_{i_1})$	$r_{S_1} =$ EXTEND r_{S_0} ADD $a_{N+1}, name_{N+1}, t_{i_1}$	$f_{R_{S_0}, t_{i_1}} : R_{S_0} \to R_{S_1}$, with $\pi_{[1,\ldots,N]}(R_{S_1}) = R_{S_0}$
... for $m \ge j \ge 2$ ...	...	...
$r_{S_{j-1}}(\mathbf{x}'\&\langle t_{i_1}, \ldots, t_{i_{j-1}}\rangle)$ $\Leftrightarrow\ r_{S_j}(\mathbf{x}'\&\langle t_{i_1}, \ldots, t_{i_j}\rangle)$	$r_{S_j} =$ EXTEND $r_{S_{j-1}}$ ADD $a_{N+j}, name_{N+j}, t_{i_j}$	$f_{R_{S_{j-1}}, t_{i_j}} : R_{S_{j-1}} \to R_{S_j}$
...	...	...
$r_{S_{m-1}}(\mathbf{x}'\&\langle t_{i_1}, \ldots, t_{i_{m-1}}\rangle)$ $\Leftrightarrow\ r_{S_m}(\mathbf{x}'\&\langle t_{i_1}, \ldots, t_{i_m}\rangle)$	$r_{S_m} =$ EXTEND $r_{S_{m-1}}$ ADD $a_{N+m}, name_{N+m}, t_{i_m}$	$f_{R_{S_{m-1}}, t_{i_m}} : R_{S_{m-1}} \to R_{S_m}$
$r_{S_m}(\mathbf{x}'\&\langle t_{i_1}, \ldots, t_{i_m}\rangle)$ $\Leftrightarrow\ r_q(\mathbf{t}_2)$	$r_q = [S]r_{S_m}$ (Project on S)	$\pi_S : R_{S_m} \to R_q$ (Projection function)
$r_q(\mathbf{t}_2) \Rightarrow r_B(\mathbf{t}_2)$	$(R_B$ UNION $R_q) \mapsto R_B$ (Union)	$in : R_q \hookrightarrow R_B$ (Injective function)

where in the Cartesian product of relations $t_1 \otimes \cdots \otimes t_k$, for each $1 \le i \le k$, $t_i = ((\overbrace{r_\infty \otimes \cdots \otimes r_\infty}^{ar(r_i)})$ MINUS $r_i)$, if the symbol $\sim_i$ is a negation operator $\neg$; r_i otherwise. So that $R_i = \|t_i\|_\#$, $1 \le i \le k$.

$[S] = [K_1, \dots, K_n]$, where, for each $1 \leq l \leq n$, if $t_l \in \mathbf{t}$ is a variable then K_l is the position of this variable in the Cartesian product r_C; otherwise if $t_l = t_{i_j} \in \mathbf{t}_1$ then $K_l = N + j$. In the third column of this table, we have the family of operations of the R-algebra used to define the database mappings in the **DB** category. The identity functions in this family can be considered as special cases of the injective functions. □

Based on this theorem, we are able to define a R-algebra α of operads in Definition 10, Sect. 2.4.1, in a standard way (as the relational algebras in Definition 31) as follows:

Corollary 19 *An R-algebra α with the variables in $\mathbb{R}$ (i.e., the relational symbols) has the signature Σ_α composed of k-ary operators $q_i \in O(r_{i1}, \dots, r_{im_i}, r_i)$ such that $\alpha(q_i)$ is one of the following functions (from Theorem 9):*

1. *If $m_i \geq 2$ then we have the Cartesian m_i-ary function*

$$\alpha(q_i) = f_\otimes^{(m_i)} : \alpha(r_{i1}) \times \dots \times \alpha(r_{im_i}) \to \alpha(r_i);$$

   otherwise ($m_i = 1$), we have the function $\alpha(q_i) : \alpha(r_{11}) \to \alpha(r_i)$ such that $\alpha(q_i) \in \{id, in, in^{-1}, f_{\alpha(r_{11}),t}\}$;
2. *Each relational symbol (a variable) $r \in X \subseteq \mathbb{R}$ is a term in $\mathcal{T}_\alpha X$;*
3. *Given the terms $t_i \in \mathcal{T}_\alpha X$, for $1 \leq i \leq m_i$, and the operator q_i,*

$$q_i(t_1, \dots, t_{m_i}) \in \mathcal{T}_\alpha X.$$

We define the evaluation of terms in $\mathcal{T}_\alpha X$, for $X = \mathbb{R}$, by extending the assignment $\|\_\| = \alpha : \mathbb{R} \to \underline{\Upsilon}$ to all terms by its unique extension $\|\_\|_\# : \mathcal{T}_\alpha X \to \underline{\Upsilon}$, such that:

4. *If $t_i = q_i(r_{i1}, \dots, r_{im_i}) \in \mathcal{T}_\alpha X$ then $\|t_i\|_\#$ is the image of the function*

$$\alpha(q_i) : \alpha(r_{i1}) \times \dots \times \alpha(r_{im_i}) \to \alpha(r_i).$$

5. *If $t_i = q_i(t_{i1}, \dots, t_{im_i}) \in \mathcal{T}_\alpha X$ then $\|t_i\|_\#$ is the image of the function*

$$\alpha(q_i)\big(\alpha(t_{i1}) \times \dots \times \alpha(t_{im_i})\big).$$

Proof Let us show that each function $f : R_1 \times \dots \times R_k \to R$, with $N = ar(R_1) + \dots + ar(R_k)$, $n = ar(R)$, is a composition of the operations of the signature Σ_α of the R-algebra α. In fact,

$$f = in \circ \pi_{[N+1,\dots,N+n]} \circ f_{R_{C_n},t_n} \circ \dots \circ f_{R_{C_1},t_1} \circ in^{-1} \circ f_\otimes^{(k)},$$

where

1. $f_\otimes^{(k)} : R_1 \times \dots \times R_k \to R_C$ is a Cartesian k-ary function;
2. $in^{-1} : R_C \twoheadrightarrow R_{C_0}$ is an inverse-inclusion function such that for any $\mathbf{d} = \mathbf{d}_1 \& \cdots \& \mathbf{d}_k \in R_C$, $\mathbf{d}_i \in R_i$ for $1 \leq i \leq k$, we define $in^{-1}(\mathbf{d}) = \langle\rangle$ if $f(\mathbf{d}_1, \dots, \mathbf{d}_k) = \langle\rangle$; $\mathbf{d}$ otherwise;

3. $f_{R_{C_j},t_j} : R_{C_{j-1}} \to R_{C_j}$, where $t_j = \pi_j(f(\mathbf{x}))$, $1 \leq j \leq n$;
4. $\pi_{[N+1,\ldots,N+n]} : R_{C_n} \to R_q$ is a projection function;
5. $in : R_q \hookrightarrow R$ is a inclusion (injective function).

Consequently, for any R-operad $q_i \in O(r_1, \ldots, r_k, r)$, the function

$$f = \alpha(q_i) : \alpha(r_1) \times \cdots \times \alpha(r_k) \to \alpha(r)$$

is composed of the operations of the Σ_α algebra. Thus, this signature Σ_α represents all compositions of R-operads. □

From this Corollary and from Theorem 9, we conclude that every R-algebra operad has an equivalent Σ_{RE} algebra term and hence Σ_α relational algebra is computationally equivalent to the Σ_{RE} (full extension of Codd's algebra). The significant difference is that the Σ_α-algebra terms are the functions *between* the relations, and can be used for a composition of the arrows (morphisms) in the **DB** category, while the evaluation of the Σ_{RE}-algebra terms are only the single relations. Consequently, the Σ_α relational algebra (of R-operads) is more appropriate for a specification of the database mappings.

From the fact that the **DB** category is computationally equivalent to the full extension of Codd's (SPRJU) algebra, exactly to Σ_{RE} algebra with all update statements for relational tables (as in SQL), based on this equivalence, in the next chapter we will develop a kind of the categorial RDB machines able to support not only all standard computations for the embedded SQL updates of RDBS but also all computations for the Big Data Integration systems based on the SOtgds database mappings.

5.5 Review Questions

1. In his SPJRU relational algebra, Codd restricted this algebra by eliminating the negation operator. In our case, the universe $\mathcal{U} = \mathbf{dom} \cup SK$ is the union of a finite (but large enough) set of values of a domain **dom** and an infinite enumerable set of Skolem constants (i.e., marked null values), so that each choice of **dom** defines a different version of the **DB** category. Consequently, the negation applied to a query formula may return with an infinite set of possible tuples. In our framework, where we deal with the infinite database with the relations that can have an infinite number of tuples, can we accept the negation operation as well? Have we already used the negation for the relations and where?
2. Prepare an example and explain how we are able to use the unary relational algebra operators 'EXTEND _ ADD a, *name* AS e' denoted shortly as $\_\langle a, name, e\rangle$ in order to represent the existentially quantified Skolem functions in the mapping SOtgds.
3. In order to be able do define any relation, or to extract any information (a view) from a given database instance by using only the *finite* length expressions of relational algebras, it is important to have only the finitary relations in our theoretical

framework, but we still permit that the relations can contain also an infinite number of tuples. Is it possible to have only a finite number of the relational algebra operators?

4. Why is the initiallity property for the syntax algebra semantics important? How are we using the inductive principle in such an initial semantics? In which way is the semantics of another algebra with the same set of algebraic operators in Σ_P but a different carrier set defined by the initial algebra semantics? (Consider Example 29 by providing a concrete carrier set Z.)
5. Why in the Σ_{RE} relational algebra is it enough to have only one nullary operator (a constant) equal to the empty relation $\perp$? Does it mean that we are able to construct any relation by using only the empty relation $\perp$ and the (infinite) set of unary operators in Σ_{RE} and its three binary operators? Can you show the inductive process of constructions of the relations from the relation $\perp$?
6. Describe and explain the inductive process of the construction of the power-object TA, for a given database schema $\mathcal{A}$ and a given interpretation α with the database instance $A = \alpha^*(\mathcal{A})$, based on the initial algebra semantics.
7. Why is it important to be able to transform any tree-term of a relational algebra (where the leafs are the simple relations) into a single path-term with only one complex leaf? Define the subclass of the single path-terms for the Σ_P equal to the Codd's SPJRU algebra.
8. In this chapter, we presented an example of transformation of the term-tree

$$\begin{aligned}&\Big(\Big(\Big(\big((r_1[S_1])\text{WHERE } C_1\big)\text{UNION}\big((r_2\text{WHERE } C_2)[S_2]\big)\Big)\text{WHERE } C_5\Big)[S_5]\Big)\\&\qquad\text{MINUS}\Big(\Big(\text{EXTEND}\Big((r_3[S_3])\text{UNION}\big((r_4\text{WHERE } C_4)[S_4]\big)\Big)\\&\qquad\text{ADD } a, \mathit{name} \text{ AS } e\Big)[S_6]\Big)\end{aligned}$$

into the single-path term in $\mathcal{T}_{RA}$ which is a morphism of the **RA** action-category. Choose other examples of the terms of Σ_{RE} algebra and show their transformation into a single path-term in $\mathcal{T}_{RA}$.
9. The evaluation functor $\mathit{Eval}: \mathbf{RA} \to \mathbf{DB}$ formally expresses the semantics of the terms (i.e., their evaluation) of the most expressive relational algebra Σ_{RA}, and hence the category **DB** is able to represent not only the query computations but all kinds of the RDB updates. Conversely, Proposition 27 demonstrates that each morphism in the **DB** category can be equivalently represented by a term of the relational algebra Σ_{RE}, and hence the computational power of the **DB** category is equivalent to the most powerful relational algebra with all kinds of update operations. Based on a part of the table in the proof of Theorem 9 given below, develop an algorithm for the inverse mapping from **DB** into **RA** (first, transform the morphism of **DB** into an equivalent term of Σ_{RE} algebra, by using the table above, and then transform this term into a single path-term and the corresponding morphism in **RA**).

Σ_{RE} algebra term	R-algebra operation
$r_C = t_1 \otimes \cdots \otimes t_k$ (Cartesian product)	$f_{\otimes}^{(k)} : R_1 \times \cdots \times R_k \rightarrow R_C$ (Cartesian k-ary function)
$r_{S_0} = r_C$ WHERE $C \bigwedge S_J$ (Select by $C(\mathbf{x})$ + Join by S_J)	$in^{-1} : R_C \twoheadrightarrow R_{S_0}$ (Inverse-inclusion function)
r_{S_1} = EXTEND r_{S_0} ADD $a_{N+1}, name_{N+1}, t_{i_1}$	$f_{R_{S_0}, t_{i_1}} : R_{S_0} \rightarrow R_{S_1}$, with $\pi_{[1,\ldots,N]}(R_{S_1}) = R_{S_0}$
...	...
r_{S_j} = EXTEND $r_{S_{j-1}}$ ADD $a_{N+j}, name_{N+j}, t_{i_j}$	$f_{R_{S_{j-1}}, t_{i_j}} : R_{S_{j-1}} \rightarrow R_{S_j}$ ...
... r_{S_m} = EXTEND $r_{S_{m-1}}$ ADD $a_{N+m}, name_{N+m}, t_{i_m}$	$f_{R_{S_{m-1}}, t_{i_m}} : R_{S_{m-1}} \rightarrow R_{S_m}$
$r_q = [S] r_{S_m}$ (Project on S)	$\pi_S : R_{S_m} \rightarrow R_q$ (Projection function)
$(R_B$ UNION $R_q) \mapsto R_B$ (Union)	$in : R_q \hookrightarrow R_B$ (Injective function)

Reference

1. E.F. Codd, Relational completeness of data base sublanguages, in *Data Base Systems*. Courant Computer Science Symposia Series, vol. 6 (Prentice Hall, Englewood Cliffs, 1972)

Categorial RDB Machines

6

6.1 Relational Algebra Programs and Computation Systems

The theory of computation has combined a number of elements drawn from quite different areas: mathematics, linguistics, biology, electrical engineering and, of course, computer science. Turing attempted to precisely formalize the first *abstract model of computation* by using the notion of an effective procedure, or algorithm. His approach was to identify the fundamental, primitive operations involved in the process to solve some problem, and to define an *abstract machine* capable of performing those operations according to a clearly specifiable *program*. It is not possible to prove that a particular mathematical definition is a correct formulation of an intuitive idea like that of computation; but it is possible to compare one mathematical definition with another (as it will be done in what follows). Turing machines are not the only abstract models of computation. A number of other researchers have attempted to capture in a mathematical definition the nature of computation. More general definitions [1] for these concepts are as follows:

A *computation machine* **M** (D. Scott, 1967) is a four-tuple

$$\mathbf{M} = (S_M, Oper, Test, In_M, Out_M),$$

where:

1. S_M is a set of memory-states;
2. *Oper* is a set of operators (functions) $f : S_M \to S_M$ (elementary operations of machine);
3. *Test* is a set of test-functions $f : S_M \to \{0, 1\}$;
4. In_M is an input function $In_M : INP_0 \to S_M$, where INP_0 is a set of all input values;
5. Out_M is an output function $Out_M : S_M \to OUT$, where OUT is a set of all output values.

A *program* P for a machine **M** is a finite set of instructions, where two different instructions have two different addresses. The syntax of each instruction is either:

1. A triple (i, f, j), where i is its address, j is the address of the next instruction and $f \in Oper$; or

Z. Majkić, *Big Data Integration Theory*, Texts in Computer Science,
DOI 10.1007/978-3-319-04156-8_6,

2. A quadruple (i, t, j, k) where i is its address, $t \in Test$, j is the address of the next instruction when t returns with 1, while k is the address of the next instruction when t returns with 0;
3. The initial instruction of a program has the address 0. Every next address, which appears in a program P, such that it is not an address of some instruction in P, is an address of the final state of P.

Intuitively, the semantics of a program P can be described in the following way: A machine begins to execute the first instruction of a program, with address equal to 0, and after that calls the next instruction. It is possible to have one of the two following cases:

- Called instruction is (i, f, j). The machine executes the operation f and passes from the current state $st \in S_M$ into the resulting state st'. After that it calls the next, jth instruction (if this instruction does not exist then the end of computation is reached).
- Called instruction is (i, t, j, k). The machine with a current state $st \in S_M$ executes the test t and after that calls the next instruction—j if t returns 1, k otherwise (if this instruction does not exist then the end of computation is reached).

The most powerful Σ_{RE} algebra, with its operations introduced by Definition 31 and the set of terms $\mathcal{T}_{RE}X$ with variables $r \in X$ (i.e., relational symbols), can be used as a database-sublanguage in general programming languages, to manipulate the data in relational form.

The same language $\mathcal{T}_{RE}X$ can be available at two different interfaces, namely an interactive interface and an application programming interface. Therefore, $\mathcal{T}_{RE}X$ is both an *interactive query language* and a *database programming language* (like SQL). That is, any $\mathcal{T}_{RE}X$ statement that can be entered at a terminal can alternatively be embedded in a program (in some general purpose programming language $\mathcal{L}_P$, where $\mathcal{T}_{RE}X \subset \mathcal{L}_P$).

The first point to be made is that there will normally be many users, of both kinds, all operating on the same data at the same time. Relational DataBase Management System (RDBMS) will automatically apply the necessary controls (basically locking) to ensure that those users are protected from one another, i.e., RDBMS will guarantee that one user's updates cannot cause another user's operations to produce an incorrect result. It is generally an objective of multi-user systems to allow individual users to behave as if the system were single-user instead. Note that tables, like users, also come in two kinds, namely base tables and views:

- A base table is a "real" table, i.e., a relational table (n-ary relation $R \in \underline{\Upsilon}$) that physically exists, in the sense that there exist physically stored records, and also physical indexes, that directly represent that table in storage as an extension $R = \|r\|$ of a given relational symbol r of a database schema.
- By contrast, a view is a "virtual" table, i.e., a table that does not directly exist in physical storage, but looks to the user as if it did. Formally, a view is a term $t \in \mathcal{T}_P X \subset \mathcal{T}_{RE}X$ of the SPJRU subalgebra $\Sigma_R \subset \Sigma_{RE}$ in Definition 31, different from a variable $r \in X$ (i.e., a relational symbol of a database schema), whose extension $R = \|t\|_{\#}$ is a "virtual" table. Views can be thought of as different ways of looking at the "real" tables: it is derived from one or more underlying base tables.

The major data definition functions (DDL—Data Definition Language) are:

- CREATE TABLE
- CREATE VIEW
- CREATE INDEX
- DROP TABLE
- DROP VIEW
- DROP INDEX

A general purpose programming language $\mathcal{L}_P$ can access the database by means of embedded statements of $\mathcal{T}_{RE}X$. This embedded $\mathcal{T}_{RE}X$ represents a "loose coupling" between $\mathcal{T}_{RE}X$ and the host language $\mathcal{L}_P$.

A statement in $\mathcal{T}_{RE}X$, like 'SELECT...' statements of SQL, requires a special treatment because it causes a table to be retrieved—a table that, in general, contains multiple records—and programming languages $\mathcal{L}_P$ usually are not equipped to handle more than one record at a time. It is therefore necessary to provide some kind of a bridge between the set-at-a-time (set of tuples (or "records") in $\|t\|_\#$ for a statement (a term) $t \in \mathcal{T}_{RE}X$) level of $\mathcal{T}_{RE}X$ and the record-at-a-time level of $\mathcal{L}_P$. Hence, the cursors provide such a bridge: a cursor consists essentially of a kind of pointer that can be used to run through a set of records, pointing to each of the records in the set in turn and thus providing addressability to those records at a time. When a program (a formula of $\mathcal{L}_P$) updates the database in some way, this update should initially be regarded as tentative only—tentative in the sense that, if something subsequently goes wrong, the update may be undone (either by the program itself or by the system). For example, if the program encounters an unexpected error, say an overflow condition, and terminates abnormally, then the system will automatically undo all such tentative updates on the program's behalf. Updates remain tentative until one of two things happens:

1. Either a COMMIT statement of $\mathcal{L}_P$ is executed, which makes all tentative updates firm ("committed"), or
2. A ROLLBACK statement of $\mathcal{L}_P$ is executed, which undoes all tentative updates.

Once committed, an update is guaranteed never to be undone. These problems of recovery in database system are tightly bonded with the notion of *transaction processing*.

Usually a definition of a general purpose programming language $\mathcal{L}_P$ is given as a set of all programs (well defined formulas of $\mathcal{L}_P$) expressed by this language. In the literature, a general definition is provided for a computation machine, programs and computation systems (execution of a program P on machine **M**) [1]. However, such definitions are limited for input features in the sense that the input of data is considered only at the initial state of a program execution. Let us now introduce more general definitions for these concepts, considering that input of data can be distributed in different parts of a program. We denote these "input-extended" machines that are able to compute such more complex programs by **MI**.

Definition 35 A *input-extended* computation machine **MI** is a 5-tuple

$$\mathbf{MI} = (S_M, Oper, Test, In_M, In_E, Out_M),$$

where:
1. S_M is a set of memory-states;
2. *Oper* is a set of operators (functions) $f : S_M \to S_M$ (i.e., the elementary operations of machine);
3. *Test* is a set of test-functions $f : S_M \to \{0, 1\}$;
4. In_M is an initial input-function $In_M : INP_0 \to S_M$, where INP_0 is a set of all initial input-values;
5. In_E is a (partial) input function $In_E : INP \times S_M \to S_M$, where *INP* is a set of all input values $z_i \in INP$, $z_i = \{z_{i1}, z_{i2}, \dots, z_{ik}\}$, $k \geq 1$;
6. Out_M is an output function $Out_M : S_M \to OUT$, where *OUT* is a set of all output values.

Notice that in our case $\Sigma_{RE} \subset Oper$ is the subset of operators of a relational algebra used as a database-sublanguage, to manipulate the data in relational form.

Definition 36 A program P for an input-extended machine **MI** is a finite set of instructions, where any two different instructions have two different addresses. Each instruction has one of the following syntax forms:
1. A triple (i, f, j) where i is its address, j is the address of the next instruction and $f \in Oper \cup \{In^*_E\}$, where $In^*_E : INP \times S_M \to S_M$ is a total function defined for any $(z, st) \in INP \times SM$ by:

$$In^*_E(z, st) = In_E(z, st) \quad \text{if } (z, st) \text{ is an element of the domain of } In_E;$$

$$st, \quad \text{otherwise.}$$

2. A quadruple (i, t, j, k) where i is its address, $t \in Test$, j is the address of the next instruction when t returns 1, while k is the address of the next instruction when t returns 0.
3. The initial instruction of a program has the address 0. Every next address, which appears in a program P, such that it is not an address of some instruction in P, is an address of the final state of P.

In order to define this operational effect of the execution of a program P on a machine **MI**, we use the following *Computation System*:

Definition 37 An *execution of a program* P on a machine **MI** is a Computation System $CS = (TrS, In_S, Out_S, Stop_S)$ as follows:
1. $TrS = (S_P, \to_P)$ is a deterministic transition system where

$$S_P = \{(i, st) \mid i \text{ is the address of some instruction of P},$$

$$st \in S_M \text{ is the state of } \mathbf{M} \text{ before execution of this } i\text{th instruction}\}$$

is a set of configurations, and $\to_P$ is a function defined for any (i, st) by

$$\big((i, st), \big(j, \lambda In^*_E(z)(st)\big)\big) \in \to_P \quad \text{if } \big(i, In^*_E, j\big) \in \to_P,$$

where $z \in INP$ is the input value taken by **MI** in this current state $st \in S_M$ and λ is the abstraction operator (i.e., currying) of the λ-calculus such that $\lambda In_E^*(z)(st) = I_E^*(z, st)$,

$$\big((i, st), (j, f(st))\big) \in \rightarrow_P \quad \text{if } (i, f, j) \in P,$$

$$\big((i, st), (j, st)\big) \in \rightarrow_P \quad \text{if } (i, t, j, k) \in P \text{ and } t(sk) = 1,$$

$$\big((i, st), (k, st)\big) \in \rightarrow_P \quad \text{if } (i, t, j, k) \in P \text{ and } t(sk) = 0;$$

2. $In_S(x) = (0, In_M(x))$;
3. $Out_S(i, st) = Out_M(st)$;
4. $Stop_S : S_P \rightarrow \{0, 1\}$ is a total stop-criteria function such that $Stop_S(i, st) = 0$ if i is not an address of any instruction of P.

Given any global program input value $z = \{z_k \mid (j, In_E^*(z_k)(st)) \in \rightarrow_P\} \cup \{z_0\}$, $z_0 \in INP_0$, the result of a computation is $Res_M(z) = Out_S \circ (\rightarrow_P)_{Stop_S} \circ In_S(z_0)$, where $(\rightarrow_P)_{Stop_S}$ denotes the iteration of $\rightarrow_P$ with respect to the $Stop_S$. The set of all **MI**-computable functions is defined by $F(\mathbf{MI}) = \{Res_M \mid P \text{ is a program of } \mathbf{MI}\}$.

It is desirable for a general-purpose programming language $\mathcal{L}_P$ to provide "computational completeness", i.e., it should be capable of computing all computable functions. The machine able to execute such programs has to be universal: the class of all computable functions on such machines has to be equal to F_μ (the class of all μ-recursive, partial recursive functions, defined inductively by Kleene (1936)). The Turing machine is the most well-known universal machine ($\mathbf{M}_U$). Some of other abstract models for universal machines resemble modern computers more closely in their details. The Register machine **RM** of Minsky (1961) is a good example for such a consideration, and hence will be generalized by an input-extension for our purposes:

Definition 38 A input-extended register machine **RMI** is a 6-tuple

$$\mathbf{RMI} = (S_U, Oper_U, Test_U, In_U, In_{EU}, Out_U),$$

where

- S_U is a set of memory-states of $n \gg 2$ registers $x = (x_1, x_2, \ldots, x_n)$;
- $Oper_U = \bigcup_{1 \le i \le n} \{a_i, s_i\}$, with $a_i(x_1, \ldots, x_i, \ldots, x_n) = (x_1, \ldots, x_i + 1, \ldots, x_n)$ and $s_i(x_1, \ldots, x_i, \ldots, x_n) = \begin{cases} (x_1, \ldots, x_i, \ldots, x_n) & \text{if } x_i = 0; \\ (x_1, \ldots, x_i - 1, \ldots, x_n) & \text{if } x_i > 0; \end{cases}$
- $Test_U = \{t_i \mid i = 1, 2, \ldots, n\}$ such that $t_i(x_1, \ldots, x_i, \ldots, x_n) = 1$ iff $x_i = 0$;
- In_U is an initial-input function $In_U : INP_0 \rightarrow S_U$, where INP_0 is a set of all initial-input values;
- In_{EU} is an (partial) input function $In_{EU} : INP \times S_U \rightarrow S_U$, where INP is a set of all input values $z \in INP$, $z = \{z_1, \ldots, z_k\}$, $1 \le k \le n$, and $In_{EU}(z, x) = y$, where $y_i = x_i$ or $y_i \in \{z_1, \ldots, z_k\}$, $1 \le i \le n$ (i.e., it changes a subset of registers by the input value z);

- Out_U is an output function $Out_U : S_U \to OUT$, where OUT is a set of all output values.

The idea is to use a universal machine as a general computer in order to support a general language $\mathcal{L}_P$, by passing the execution of database update statements (the terms of $\mathcal{T}_{RE}X$ directly to a kind of specialized RDB-machine $\mathbf{M}_R$ which will provide a DBMS for a stored data). The embedded RDB statements $t \in \mathcal{T}_{RE}X$ will not be executed by the universal machine, but only passed as a string to this specialized RDB machine $\mathbf{M}_R$.

Let us show that this extension of **RM** is a universal machine as well:

Proposition 28 *The input-extended register machine* **RMI** *is universal.*

Proof From Definition 37, $F(\mathbf{RM}) \subseteq F(\mathbf{RMI})$, by the fact that when the domain of In_{EU} is empty, then the **RMI** is equivalent to the ordinary **RM** which is well-known to be universal, i.e., $F(\mathbf{RM}) = F_\mu$, where F_μ is a class of μ-recursive (partially recursive) functions defined by:
1. $N(x) = x + 1$ (successor function);
2. $U_i^n(x_1, \dots, x_i, \dots, x_n) = x_i$ (ith projection);
3. $C_i^n(x_1, \dots, x_i, \dots, x_n) = i$ (constant function);
4. $f(x) = \mu y(g(x, y) = 0)$ (μ-operator (minimal natural number y such that $g(x, y) = 0$));
5. By substitution, $f(h_1(x_1), \dots, h_i(x_i), \dots, h_n(x_n)) \in F_\mu$, where $h_1, \dots, h_n \in F_\mu$;
6. By primitive recursion, $f(x, 0) = g(x)$, $f(x, y + 1) = h(x, y, f(x, y))$, where $g, h \in F_\mu$.

Thus, it is enough to also show that $F(\mathbf{RMI}) \subseteq F(\mathbf{RM})$, by induction on n-operators (n-ROI) of **RMI**, defined by [1]:
1. a_i, s_i and $In_E^*(z)$, for any input value $z \in INP$, are n-operators (n-ROI);
2. If M, M_1 and M_2 are n-ROI, also M_1M_2 ("concatenation") and $_i(M)_i$ ("iteration") of M unless the ith component (register) does not become the first time equal to 0) are n-ROI. That is, $M_1M_2(x) = M_2(M_1(x))$ and $_i(M)_i(x) = (M)_{non\ t_i}(x)$ with a predicate test $t_i(x) = 1$ iff $x_i = 0$.

   Let us introduce the Gödel's numbering of the formulas '$\langle\_\rangle$' as follows: for any list $\mathbf{x} = (x_1, \dots, x_i, \dots, x_n)$ let

$$x = \langle \mathbf{x} \rangle = \langle x_1, \dots, x_i, \dots, x_n \rangle = p_1^{x_1} p_2^{x_2} \cdots p_n^{x_n},$$

where $2 = p_1 < p_2 < \cdots < p_n$ are the first n prime numbers. Let us denote by $(x)_i$ the ith value of the memory at the ith address $(x)_i = x_i$, and let us introduce the length function $l \in F_{\text{prim}}$ (class of primitive recursive functions) such that $l(x) = n$ if $x = \langle x_1, \dots, x_i, \dots, x_n \rangle$ and $x_n \neq 0$.

Let us define $y * x = \langle \mathbf{y} \rangle * \langle \mathbf{x} \rangle \triangleq \langle \mathbf{y}\&\mathbf{x} \rangle$. Let us show that for any n-ROI M, $\langle M \rangle \in F_\mu$, where $\langle M \rangle(x) = y$ iff $M(\langle (x)_1, \ldots, (x)_n \rangle) = \langle (y)_1, \ldots, (y)_n \rangle$ and $l(y) \leq n$ (where $x = p_1^{x_1} p_2^{x_2} \ldots p_n^{x_n}$, $y = p_1^{y_1} p_2^{y_2} \ldots p_n^{y_n}$), and hence

$$Res_M(\mathbf{x}) = Res_M(x_1, \ldots, x_i, \ldots, x_n) = \big(\langle M \rangle(x)\big)_{m+1}$$
$$= \big(\langle M \rangle\big(\langle x_1, \ldots, x_i, \ldots, x_n \rangle\big)\big)_{m+1}.$$

The fact that $\langle M \rangle \in F_\mu$ holds from the following structural induction on M:

2.1. $\langle a_i \rangle(x) = \langle (x)_1, \ldots, (x)_i + 1, \ldots, (x)_n \rangle = p_1^{x_1} \ldots p_i^{x_i+1} \cdots p_n^{x_n}$;

2.2. $\langle s_i \rangle(x) = \begin{cases} \langle (x)_1, \ldots, (x)_i - 1, \ldots, (x)_n \rangle = p_1^{x_1} \ldots p_i^{x_i-1} \cdots p_n^{x_n} & \text{if } (x)_i \geq 1; \\ x, & \text{otherwise}; \end{cases}$

2.3. $\langle In_E^*(z) \rangle(x) = \langle In_E \rangle(z * x) = \langle (y)_1, \ldots, (y)_n \rangle$, such that for any $1 \leq i \leq n$, $(y)_i = (x)_i$ or $(y)_i = \{(z)_1, \ldots, (z)_m\}$, $m \leq n$;

2.4. $\langle M_1 M_2 \rangle(x) = \langle M_2 \rangle(\langle M_1 \rangle(x))$;

2.5. $\langle {}_i(M)_i \rangle(x) = Iter(\langle M \rangle)(x, y((Iter(\langle M \rangle)(x, y))_i = 0))$, where $Iter(g)(x, k) = g^k(x)$, i.e., $Iter(g)(x, 0) = x$, $Iter(g)(x, k+1) = g(Iter(g)(x, k))$.

Thus, $F(\mathbf{RMI}) = F(\mathbf{RM}) = F_\mu$, i.e., **RMI** is a universal machine (like the Turing machine). □

6.1.1 Major DBMS Components

The internal state of a DBMS is quite complex. From a high-level point of view, however, it can be regarded as having just three major components, each of which divides up into numerous subcomponents. The three major components are as follows:

1. The *system service* component, which supports system operation, operator communication, logging, and similar functions.
2. The *locking services* component, which provides the necessary controls for managing concurrent access to data.
3. The *database services* component, which supports the definition, retrieval, and update of user and system data.

Of these three components, only the last is directly relevant to the user. That component, in turn, divides into five principal subcomponents, which we refer to as the Precompiler, Bind, the Runtime Supervisor, the Stored Data Manager, and the Buffer Manager, whose functions are as follows (refer to Fig. 6.1):

- The *Precompiler* is a preprocessor for application programs that contain embedded database statements. It collects those statements into a DBRM (Data Base Request Module) replacing them in the original program by host language CALLs to the Runtime Supervisor, both with the list of values used in SQL statements INSERT and UPDATE (these initial values used in the original source program may be successively modified by the program during its execution). The host language is responsible for providing various non-database facilities such as local (temporary) variables, computational operations, if–then–else logic, and so

$$\mathbf{M_U} = (S_U, \text{Oper}_U, \text{Test}_U, \text{In}_U, \text{In}_{EU}, \text{Out}_U) \qquad \mathbf{M_R} = (S_R, \text{Oper}_R, \text{Test}_R, \text{In}_R, \text{In}_{ER}, \text{Out}_R)$$

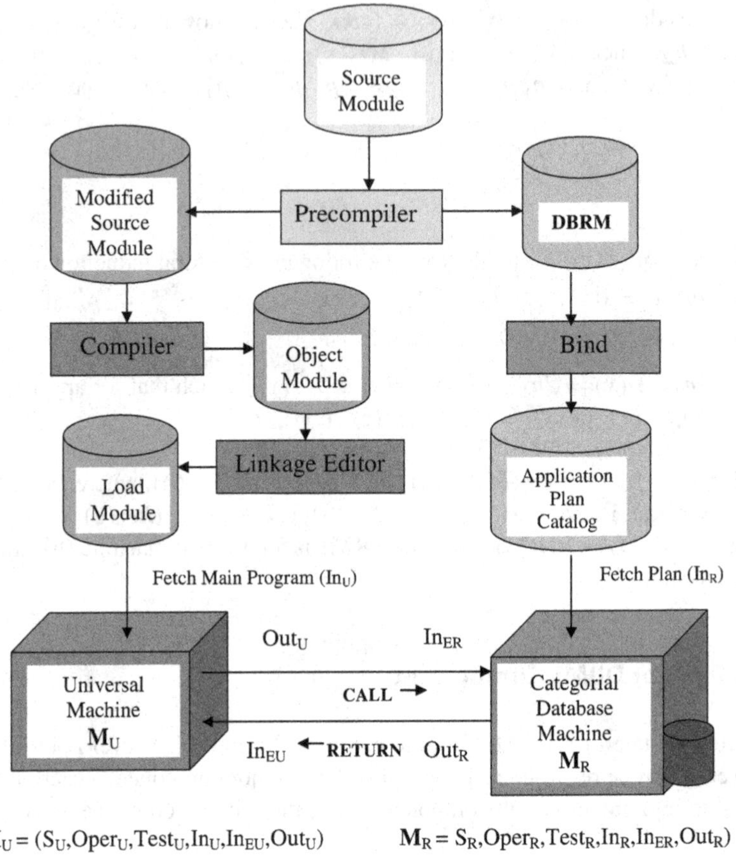

Fig. 6.1 Relational DBMS

on. In order to provide clear separation of database manipulation, the host language and embedded language are loosely coupled. The embedded formulas in the original general-purpose programming language are usually preceded by the statement 'EXEC...' like in SQL; the DBRM can be regarded as nothing more than the "pure" embedded formulas. It does not consist of executable code. We replace a standard SQL by $\mathcal{T}_{RE}$ terms (internal subalgebras of Σ_{RE} in Definition 31) in the following way (see Table 6.1). In the table,

1. t_S is a SPRJU (Σ_R) term;
2. $t_D = $ 'r MINUS (r WHERE C)';
3. $t_I = $ 'r UNION t_1', where t_1 is an SPRJU term $t_S = $ 'SELECT...' in case (ii); $t_1 = r'$ otherwise (with $ar(r') = ar(r)$, and r' is a relational symbol for a one-tuple relations obtained from the 'list of values' by substitution of each value in this list with a fresh new variable and by adding NULL values for the attributes of r which are not specified in the list S). The assignment of values to these variables will be done during the execution of the program P when

Table 6.1 SQL translations

Embedded SQL statements: 'EXEC t'	Embedded (internal) algebra statements
$t =$ 'SELECT...'	$t_S \in \mathcal{T}_P X$ in $\Sigma_R \subset \Sigma_{RE}$
$t =$ 'DELETE FROM r WHERE C'	$t_D \in \mathcal{T}_P^+ X$ in $\Sigma_R^+ \subset \Sigma_{RE}$
(i) $t =$ 'INSERT INTO $r[S]$ VALUES (list)'	$t_I \in \mathcal{T}_{RE} X$
(ii) $t =$ 'INSERT INTO $r[S]$ AS SELECT...'	$t_I \in \mathcal{T}_P X$
$t =$ 'UPDATE r SET $[nr_r(i_1) = e_{i_1}, \ldots, nr_r(i_k) = e_{i_k}]$ WHERE C'	$t_U \in \mathcal{T}_{RE} X$
$t =$ 'EXTEND r ADD $a, name$ AS e'	idem, $t_E = t \in \mathcal{T}_{RE} X$

will be executed the instruction 'CALL n' from host $\mathbf{M}_U$ by passing such a list of values to $\mathbf{M}_R$.

The $k = |S| \le ar(r)$ has to be equal to the number of values in the list in the case (1), or to the $ar(\|t_S\|_\#)$ in case (ii).

4. $t_U =$ '(r WHERE $\neg C$) UNION r_1', where r_1 is a relation expressed by

$$\big(\text{EXTEND}(\ldots(\text{EXTEND}(r \text{ WHERE } C)\text{ADD } att_r(1), name_1 \text{AS} e_1)\ldots) \\ \text{ADD } att_r(n), name_n \text{ AS } e_n\big)[S],$$

and for each $1 \le m \le n = ar(r)$ if $m \notin \{i_1, \ldots, i_k\}$ then $e_m = nr_r(m)$, and $S = \langle name_1, \ldots, name_n \rangle$.

Notice that in SQL each expression e_m, $m \in \{i_1, \ldots, i_k\}$, is a simple *value* in **dom**. However, after the *Bind* module, these values in the embedded (internal) algebra statements are treated as the variables: the assignment of values to them will be done during the execution of the program P when executing the instruction 'CALL n' from host $\mathbf{M}_U$ by passing such a list of values to $\mathbf{M}_R$.

Analogously, all constants used in 'SELECT...' or in the conditions 'C', will be substituted by new fresh variables, and their assignment will be done during the execution of a program P.

5. The *Bind* component compiles one or more related DBRMs to produce an application plan (a program for an RDB machine $\mathbf{M}_R$) composed principally of the *normal* forms $Rd_{RE}(t_R) = Nrm(Trw(t_R))$ (the arrows in the category **RA**), obtained from the internal algebra statements $t_R \in \mathcal{T}_{RE} X$ after precompilation. Each normal form formula is a composed n-operator of the database machine $\mathbf{M}_R$, i.e., exactly a composition of the atomic arrows in **RA** (which are unary operations in $\Sigma_{RA} \backslash \Sigma_{RA}^-$ in Proposition 20, Sect. 5.2), which is an executable code for this categorial database machine $\mathbf{M}_R$. The Bind component executes the following three steps for any statement $t_R \in \mathcal{T}_{RE} X$ in order to obtain a executable code for $\mathbf{M}_R$:
   5.1. Executes the term-rewriting algorithm $Trw : \mathcal{T}_{RE} X \to \mathcal{T}_{RA} X$ defined in Proposition 20, Sect. 5.2.

5.2. Executes the normalization (sequential translation) algorithm $Nrm : \mathcal{T}_{RA}X \to Mor_{RA}$ in Proposition 23, Sect. 5.2.1. Each source object $t_{RA} \in \mathcal{T}_{RA}^{-}X$, of the obtained arrow $f : t_{RA} \to f(t_{RA})$ in **RA**, is a composition of the unary operators $\_\langle a, name, e\rangle$ and $r \otimes \_$, where $r \in X$ is a relational symbol whose extension $\|r\|_{\#}$ is contained in data memory of $\mathbf{M}_R$, usually a table of a database; these operators are applied to the single (temporary) relation in the IO (Input/Output) memory of $\mathbf{M}_R$.

These two steps specify that the Bind module implements the mapping $Rd_{RE} = Nrm \circ Trw : \mathcal{T}_{RE}X \to Mor_{RA}$ in Proposition 23 (THE COMPLETENESS OF **RA**).

5.3. The last step is a transformation of the algebra operators contained in these **RA** arrows into the *variable-binding operators*: Let us consider an algebra operator '$\_$ WHERE $C(d_1, \ldots, d_n)$' $\in \Sigma_{RA}$, where the $d_i \in \mathbf{dom}$ are the values used in the condition C. Let us substitute these values in C by the variables, so that we obtain a condition expression $C(x_1, \ldots, x_n)$, and then we introduce a variable-binding operator '$\_$ WHERE $C(x_1, \ldots, x_n)_{(x_1,\ldots,x_n)\in S}$', so that in a new term 'r WHERE $C(x_1, \ldots, x_n)_{(x_1,\ldots,x_n)\in S}$' $\notin \mathcal{T}_{RA}$ only r is a free variable, while $x_i, 1 \le i \le n$ are the bounded variables, and the set of lists S is a parameter. The set of previously obtained arrows **RA** will be transformed in this way, by substituting the Σ_{RA} operators with these new variable-binding operators, and successively saved as an *Application plan* P_i (executable code for the database machine, called as 'n-operator') in the *Application Plan Catalog* in Fig. 6.1. The source object for each arrow is equal to the resulting table of the Source-object program (contained in IO memory) defined in step 2. During the execution of the host application program the triple $(n, 0, L)$, with the list of values $L = (d_1, \ldots, d_n)$, will be passed as input data to the machine $\mathbf{M}_R$ (into its IO memory) by the host language CALLs of the *Runtime Supervisor* of the universal machine $\mathbf{M}_U$ (see Fig. 6.2).

Thus in the case above, we will have that $S = \{L\}$, so that the variable-binding operator above '$\_$ WHERE $C(x_1, \ldots, x_n)_{(x_1,\ldots,x_n)\in S}$' becomes equal to the simple unary Σ_{RA} algebraic operator '$\_$ WHERE $C[x_1/d_1, \ldots, x_n/d_n]$' where x_i/d_i denotes the substitution of the bounded variable x_i by the value $d_i \in \mathbf{dom}$.

Figure 6.1 above is intended to show the major steps involved in the preparation and execution of an RDB application. We assume that the original source program P is written in a high-level general purpose language containing the embedded SQL formulas. The steps that program P must go through are as follows:

1. Before P can be compiled by the regular compiler of such high-level general purpose language, it must first be processed by the Precompiler. Its function is to analyze a source program in any one of those languages, stripping out the SQL statements it finds and replacing them by host language CALL statements. At execution time those CALLs will pass control to the Runtime Supervisor. From the embedded SQL statements it encounters the Precompiler constructs a DBRM

(Data Base Request Module) Σ_{RE}-statements, which subsequently becomes input to the Bind component.
2. Next, the modified source program is compiled and link-edited in the normal way. Output from this step we refer to as "Load Module P" (stored in $\mathbf{M}_U$).
3. Now we come to the Bind step. As already suggested, Bind is really a database compiler: It converts high-level database requests—in effect Σ_{RE}-statements—into executable (by $\mathbf{M}_R$) code.
4. Finally, we come to the execution time. The main program from a Load Module is fetched by $\mathbf{M}_U$ into its program memory by initial-input function In_U (we use a universal machine in order to be able to compute any computable function), the Application Plan (set of normal forms embedded programs P_i defined as n-operators) is fetched by Categorial DataBase machine $\mathbf{M}_R$ into its program memory by initial-input function In_R, and the current stored image of Database is also fetched by $\mathbf{M}_R$ into its data memory.

Since the original high-level program P has effectively been broken into two pieces (Load Module P_U (for $\mathbf{M}_U$) and Application Plan P_R (for $\mathbf{M}_R$)), those two pieces must somehow be brought back together again in the execution time. The load module P_U is loaded into program memory of $\mathbf{M}_U$ and it starts to execute in the usual way. Sooner or later it reaches the first of the CALLs inserted by the Precompiler. Control goes to the Runtime Supervisor which prepares a data (with eventually a list of values $L = (d_1, \dots, d_k)$ in IO memory for a list of variables $(e_{i_1}, \dots, e_{i_k})$; case of an simple 'INSERT INTO $r[S]$ VALUES $(e_{i_1}, \dots, e_{i_k})$', or 'UPDATE r SET $[nr_r(i_1) = e_{i_1}, \dots, nr_r(i_k) = e_{i_k}]$ WHERE C' statement) for a normal form n-operator of Application Plan (loaded into program memory of $\mathbf{M}_R$), derived by Bind from the original embedded SQL formulae, which has to be executed by $\mathbf{M}_R$. This data is exported by the function Out_U (see Fig. 6.1) and is imported into $\mathbf{M}_R$ by the function In_{ER}. Then $\mathbf{M}_U$ executes the statement 'CALL n' (n is a number of application plan of $\mathbf{M}_R$) and wait for returned data from $\mathbf{M}_R$. The $\mathbf{M}_R$ receives this CALL, fetch the input data (with list of values $L = (d_1, \dots, d_k)$ in the case of simple 'INSERT INTO $r[S]$ VALUES $(e_{i_1}, \dots, e_{i_k})$' or 'UPDATE r SET $[nr_r(i_1) = e_{i_1}, \dots, nr_r(i_k) = e_{i_k}]$ WHERE C' statement) into its IO memory, then transforms the nth application plan (a composed n-operator) into an arrow of **RA** (by reduction of the variable-binding operators of this application plan into simple Σ_{RA} algebra operators, obtained by substitution of bounded variables with the values in L) and perform its execution (by $Eval : \mathbf{RA} \to \mathbf{DB}$), and executes RETURN operation to restitute (by the function Out_R) the computed data saved in its IO memory (computed relation). If an embedded SQL statement is different from the case 'SELECT...' of Table 6.1, then this new computed table replaces the old version of this table in a database (data memory of $\mathbf{M}_R$; in that moment the consistency of a database is not guaranteed). After that it returns again into the wait state for the next CALL from $\mathbf{M}_U$. When $\mathbf{M}_U$ receives (by the function In_{EU}) the requested data from $\mathbf{M}_R$ into its IO memory, it continues to execute the next statement of its program and can use cursors to run through a set of received records in its IO memory.

In the next sections, we will consider in more details this Relational DBMS architecture represented in the two bottom boxes in Fig. 6.1.

6.2 The Categorial RBD Machine

The synchrony of the conversations between a universal and the categorial RDB machines can be schematically represented by Fig. 6.2:

- The *Runtime Supervisor* of $\mathbf{M}_U$ oversees SQL programs (embedded formulas) during execution. When such a program requests some database operation, control goes first to the Runtime Supervisor, thanks to the CALL inserted by the Precompiler. It transforms the values from $\mathbf{M}_U$ memory (constants and values of variables used in the main program) into a well-defined list L of values to be transferred to $\mathbf{M}_R$, where the tuples of relations will be generated to be inserted or updated. After that it executes the statement 'CALL n' as a request to the categorial RDB machine $\mathbf{M}_R$, to execute the program (Application Plan) P_n, and waits for the resulting relational table (after the execution of the statement RETURN by $\mathbf{M}_R$). Analogous operations are executed for the statements COMMIT and ROOLBACK used for a transaction processing and consistent database updating (Synchpoints), which are not embedded SQL formulas. Thus, Runtime Supervisor is a part of a program memory which starts with statements in COMMIT, ROOLBACK, CALL n and waits for return of data from $\mathbf{M}_R$.
- The *Stored Data Manager* of $\mathbf{M}_R$ manages the stored database, retrieving and updating records as requested by Application Plans. It is part of a program memory, and corresponds to the instructions (i, f, j) in the Load Module P of $\mathbf{M}_U$, where i is the current program address (i.e., value of Program Counter), j is the address of the next instruction, and $f = In^*_{EU}$ (see Definitions 38 and 36), and hence, by execution of this instruction, $\mathbf{M}_U$ takes back from $\mathbf{M}_R$ the relation with the tuples of one 'SELECT...' statement in the source program P.
- The *Buffer Manager* of $\mathbf{M}_R$ is the component responsible for a physical transferring of data between the disk and $\mathbf{M}_R$ system/data memory; it performs the actual I/O operations (for example, during COMMIT and ROLLBACK) toward a persistent Database. The abstract view of the Buffer Manager is limited to the specific definition for the input/output functions In_R, Out_R, of $\mathbf{M}_R$, and not as a codification of specific programs:
  1. In_R is the initial input for $\mathbf{M}_R$ program: the initial state of $\mathbf{M}_R$ is determined by $st'(0) = In_R(z)$, a database is a part of data in z and hence the physically stored (consistent) image of database is loaded into a data memory of $\mathbf{M}_R$ to be subsequently updated by the application programs.

     It is used also for each 'ROLLBACK' statement of the main program, so that after the execution of this statement, the current state of $\mathbf{M}_R$ is reimposed to be the last physically stored (consistent) image of a database.
  2. Out_{DB}—for each 'COMMIT' statement of the main program P the new *consistent* version of database is physically transferred to the disk memory.

Let us now define these two machines in order to support the previously described type of the synchrony conversations:

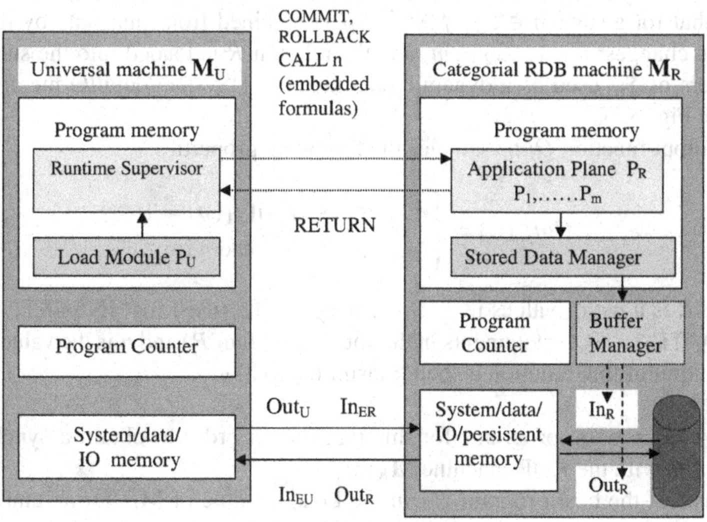

Fig. 6.2 Synchronization

Definition 39 A Universal general purpose programming language machine $\mathbf{M}_U$ is an

$$\mathbf{RMI} = (S_U, Oper_U, Test_U, In_U, In_{EU}, Out_U),$$

with the following features:

1. The operations COMMIT, ROLLBACK, CALL $n \in Oper_U$ are used to execute the Application Plans in $\mathbf{M}_R$. NOP $\in Oper_U$ is the operation which does not change the state of the machine.
2. x_1 is the register for synchronization. Its default value (imposed by the initial state st_0) is equal to 0. This value is changed in the following cases:
   - Imposed to 1 after an execution of the statement COMMIT;
   - Imposed to 2 after an execution of the statement ROLLBACK;
   - Imposed to $2+k$ after an execution of the statement CALL k (for kth embedded, compiled by Bind SQL statement, program P_k in the Application Plane of $\mathbf{M}_R$).
3. x_2 is the register for the error condition received from $\mathbf{M}_R$. Its default value (imposed by the initial state st_0) is equal to 0. This value is changed by the function In^*_{EU} (see the next point) and is also imposed to 0 after an execution of one of the operators described in the point 1 above.
4. The extended input function In^*_{EU} satisfies the following properties:

$$\lambda In^*_{EU}(z) = \begin{cases} \text{a function } f : S_U \to S_U & \text{if } z = (n, m, R) \in INP; \\ \text{identity function, } id : S_U \to S_U & \text{otherwise,} \end{cases}$$

such that for a state $st \in S_U$, $f(st) \in S_U$ is obtained from this state by the following changes: $x_1 = n, x_2 = m$, and the relation R is loaded into the subset of registers of S_U used as a dynamic data memory ('System/data/IO memory' of $\mathbf{M}_U$ in Fig. 6.2).

5. The output function Out_U satisfies the following properties:

$$Out_U(st) = \begin{cases} \emptyset \text{ (empty set)} & \text{if } t_1(st) = 1; \\ (n, 0, L), & \text{otherwise,} \end{cases}$$

where L is a list of values in I/O memory of $\mathbf{M}_U$ (used for 'INSERT...' and 'UPDATE...' SQL statements in the source program P) and n is the value of x_1. The output of this function is then transmitted to $\mathbf{M}_R$.

Now we can write programs for this machine in order to obtain a synchronic conversation with the RDB machine $\mathbf{M}_R$:

The part of the host program P_U in the Load Module of $\mathbf{M}_U$ for an embedded SQL statement $g \in \{\text{COMMIT, ROLLBACK}, \{\text{CALL } n\}_{n \geq 1}\}$ is as follows:

...	prepared triple $(n, 0, L)$ is in its I/O memory
$(i-1, f_1^n, i)$	f_1^n sets x_1 to value $n \geq 1$ (program number)
$(i, g, i+1)$	$g \in \{\text{COMMIT, ROLLBACK}, \{\text{CALL}n\}_{n \geq 1}\}$
$(i+1, Out_U, i+2)$	communicates (CALL's) the triple $(n, 0, L)$ to $\mathbf{M}_R$ to be executed with the program $n \geq 1$
$(i+2, f_3^N, i+3)$	f_3^N sets the register x_3 to the value $N \gg 1$ (time-counter)
$(i+3, f_3^-, i+4)$	f_3^- decrements the time-counter in x_3
$(i+4, t_3, i+k, i+5)$	tests the out-of-time response of $\mathbf{M}_R$
$(i+5, In_{EU}^*, i+6)$	takes the resulting data z from $\mathbf{M}_R$ given by $Out_R(st'(m))$, where $st'(m)$ is the current state of $\mathbf{M}_R$
$(i+6, t_1, i+7, i+3)$	tests if $\mathbf{M}_R$ has responded with the results ($z \neq \emptyset$)
$(i+7, t_2, i+8, i+m)$	tests the error condition of $\mathbf{M}_R$ (consistency)
$(i+8, \ldots, \ldots)$	continues with execution without errors
...	
$(i+k, \ldots, \ldots)$	continues with out-of-time error condition
...	
$(i+m, \ldots, \ldots)$	continues for error database-condition
...	

Program 1

Let us define a *categorial* RDB machine $\mathbf{M}_R$ which has to execute the SQL statements of the original source program P, translated by Precompiler and Bind modules into the Application Plans (n-operators, composed of the variable-binding operators obtained from the operators of the Σ_{RA} algebra in Proposition 20, Sect. 5.2).

The I/O memory (a subset of registers) will be used alternatively in order to receive the data from the host universal machine, and to send the obtained results (for example, obtained relation from a 'SELECT...' SQL statements in the source program P), while the data-memory is the subset of registers used to load the whole database from the permanent (disk) memory, and for their dynamic updates (by Buffer Manager) during the execution of a program P.

The System memory is the subset of registers used for the synchronization, Application Plane (set of embedded programs, i.e., the arrows in **RA** category), and all internal computations. Such a categorial machine has to be synchronized with the host universal machine $\mathbf{M}_R$ by additional set of operators in

$$\{\text{NOP}, \text{RETURN}, \text{ERR}, \text{VCON}, \text{COM}, \text{ACOM}, \text{DEC}, \text{RBACK}\},$$

and we will use the first four registers x_i and test functions t_i for such features, as follows:

Definition 40 A Categorial RDB machine $\mathbf{M}_R$ is defined by

$$\mathbf{RMI} = (S_R, Oper_R, Test_R, In_R, In_{ER}, Out_R)$$

with the following features (we assume that during the execution of a program all states in S_R are progressively enumerated in the way that $st'(k) \neq st'(k+1)$):

1. • x_1 is the memory register used to save the number n of the program P_n in Application Plane, requested to be executed from $\mathbf{M}_U$ by CALL n;
   • x_2 is the memory register for the error-condition. Its default value (imposed in $st'(0)$) is equal to 0. After the commitment or execution of 'SELECT...' for an inconsistent database it is imposed to 1;
   • x_3 is the memory register used to sign the synchpoints (of transactions) for a database with $x_3 = 1$;
   • x_4 is the memory register used to determine the address of a program to execute.
2. $\{t_1, t_2, t_3, t_4\} \subseteq Test_R$, where for a state $st'(k) \in S_R$ and $i = 1, 2, 3, 4$, $t_i(st'(k)) = 1$ if x_i is equal to 0; 1, otherwise.
3. $\Sigma_{RA} \cup \{\text{NOP}, \text{RETURN}, \text{ERR}, \text{VCON}, \text{COM}, \text{ACOM}, \text{DEC}, \text{RBACK}\} \subseteq Oper_R$, where Σ_{RA} are the algebra operators used in the category **RA**, and
   • NOP does not change the state of the machine;
   • RETURN is a function which assigns 0 to x_1;
   • ERR is function which assigns 1 to x_2;
   • VCON is function which assigns 0 to x_2 if the database in the dynamic memory of Buffer Manager is consistent in the current machine's state; 1 otherwise;
   • COM is function which assigns 1 to x_3, then changes the state by rewriting a database from the working memory (Buffer Manager) into the physical storage memory (which is also a part of the state of $\mathbf{M}_R$);

- ACOM is function which assigns 0 to x_3;
- DEC decrements the value of the register x_4;
- RBACK assigns 1 to x_3 and a physically stored consistent image of database is loaded into the dynamic data memory (registers of $\mathbf{M}_R$ controlled by Buffer Manager).

4. The initial input function In_R generates the initial state $st'(0)$ by assigning 0 to x_1 and x_2 and 1 to x_3, and a physically stored consistent image of database is loaded into the dynamic data memory (registers of $\mathbf{M}_R$ controlled by Buffer Manager so that $Out_{DB}(st'(0))$ is this loaded and consistent database).
5. The extended input function In^*_{ER} satisfies the following properties:

$$\lambda In^*_{ER}(z) = \begin{cases} f : S_R \to S_R & \text{if } z \text{ is } (n, m, L) \text{ or a database;} \\ id : S_R \to S_R \text{ (identity function)} & \text{otherwise,} \end{cases}$$

such that for a state $st'(k) \in S_R$, $st'(k+1) = f(st'(k)) \in S_R$ is obtained from $st'(k)$ by the following setting: $x_1 = x_4 = n$, $x_2 = m$ and the list of values in L is loaded in the subset of registers of S_R used for the I/O memory if $z = (n, m, L)$. Otherwise, if z is a database then it is loaded into the dynamic data-memory.
6. The output function Out_R satisfies the following properties:

$$Out_R\big(st'(k)\big) = \begin{cases} \emptyset \text{ (empty set)} & \text{if } t_1(st'(k)) = 0; \\ (0, n, R) & \text{otherwise,} \end{cases}$$

where R is a relation in I/O memory of $\mathbf{M}_R$ (obtained by last execution of a program P_k in the Application Plane) and n is the value of its register x_2, with $R = \bot$ if $n > 0$ or P_k is not obtained from a SQL statement 'SELECT…' in the source program P. If the output is a triple $(0, n, R)$ then it is transmitted to $\mathbf{M}_U$.
7. The DB output function $Out_{DB} : S_R \to Ob_{DB}$ such that, for each state $st'(k) \in S_R$, $Out_{DB}(st'(k))$ is the current database in the memory of Buffer Manager.

Notice that if $t_1(st'(k)) = 1$ and a database in data-memory is consistent then VCON in this state $st'(k)$ imposes the error condition register to $x_2 = 0$.

A database in the memory of Buffer Manager obtained by RBACK is a just the last version of database in the physical storage.

The name 'categorial machine' is based on the following theorem:

Theorem 10 (CATEGORIAL MACHINE) *Every Application Plan P_k in $\mathbf{M}_R$, $k \geq 3$, is obtained from an arrow $f_k : t_{RA} \to f_k(t_{RA})$ of the 'syntax' category* **RA** *by substitution of the Σ_{RA} operators with the corresponding variable-binding operators. An execution of P_k in $\mathbf{M}_R$ corresponds to the transformation of such an Application Plan into an arrow in* **RA** *(by binding the bounded variables in the variable-binding* *operators with the values in the list L received from* $\mathbf{M}_U$*) and, successively, to the evaluation functor* $Eval : \mathbf{RA} \to \mathbf{DB}$ *(in Proposition* 24*).*

Thus, any execution of such an Application Plan in $\mathbf{M}_R$ *is an arrow* $f_{DB} : A \to A_1$ *in* $\mathbf{DB}$*, where A is the instance-database of a schema* $\mathcal{A}$ *in the data-memory of* $\mathbf{M}_R$ *before the execution of this plan while* A_1 *is the instance-database of the same schema* $\mathcal{A}$ *in the data-memory of* $\mathbf{M}_R$ *after the execution of this plan.*

Proof Using the functor $Eval : \mathbf{RA} \to \mathbf{DB}$ (in Proposition 24), we have demonstrated that each Σ_{RE}-algebra tree-term t_R (a full relational algebra term) may be represented by the equivalent arrow $f_k : t_{RA} \to f_k(t_{RA})$ in $\mathbf{RA}$ where t_{RA} is a term with variables (i.e., relational names of a given database schema $\mathcal{A}$) and binary operator $\otimes$ (Cartesian product 'TIMES') and the unary operators $\_\langle a, name, e\rangle$ (i.e., 'EXTEND...'), while f_k is a composition of the unary operators in Σ_{RA}. Consequently, we obtained from Proposition 24 that *Eval* corresponds to the evaluation of such terms in the database category $\mathbf{DB}$, where each relational name r_i (a variable) in t_R is substituted by its current extension (a relational table) $R_i = \|r_i\|_\#$, and to each variable 'e' in a unary operator $\_\langle a, name, e\rangle$ a corresponding value is assigned in the received list L (in the tuple $(n, 0, L)$ in its I/O memory) from $\mathbf{M}_U$. After that the relation $R = Eval(t_R)$ is evaluated and, finally, the function (i.e., a composed operator) f_R is applied to it. In fact, in $\mathbf{M}_R$ this computation is done just in this ordering. The obtained relation $R = Eval(t_R)$ will be temporarily saved in the system memory of $\mathbf{M}_R$ and after that the relation $R' = f_R(R)$ is computed:

- If f_R is a part of the Application Plan of an 'SELECT...' statement (in the original source program P then the triple $(0, err, R')$ is saved in I/O memory, to be transmitted to the host $\mathbf{M}_U$ machine.
- Otherwise (the update SQL statements), the old relation (used in this SQL statement) of the database in the data memory of $\mathbf{M}_R$ will be replaced by its updated version R'.

Let us show that any execution of an arrow $f_k : t_{RA} \to f_k(t_{RA})$ (obtained from an Application Plan by binding the bounded variables in its variable-binding operators with the values in the list L received from $\mathbf{M}_U$) in $\mathbf{RA}$ is just an arrow $f_{DB} : A \to A_1$ in $\mathbf{DB}$, where A is the instance-database of a given schema $\mathcal{A}$ in the data-memory of $\mathbf{M}_R$ before the execution of this plan, and A_1 is the instance-database of the same schema $\mathcal{A}$ in the data-memory of $\mathbf{M}_R$ after the execution of this plan:

1. If the Application Plan is obtained from an update SQL statement, the proof for this is given in Corollary 17, Sect. 5.3.
2. If the Application Plan is obtained from a 'SELECT...' statement of SQL then we obtain the following: The term t_R is then only the Cartesian product of relational symbols in the schema $\mathcal{A}$ of this database, that is, $t_R = r_{i_1} \otimes \cdots \otimes r_{i_n}$, with a list $L_R = \langle r_{i_1}, \ldots, r_{i_n}\rangle$, such that we can have the same relational symbol of $\mathcal{A}$ repeated at different positions in this list. Let the current instance-database in data memory of $\mathbf{M}_R$ be $A = \alpha^*(\mathcal{A})$, and hence the extension of each r_{i_k} is

$R_{i_k} = \alpha(r_{i_k}) = \|r_{i_k}\|_\# \in A$, $k = 1, \ldots, n$, so that $R = \|t_R\|_\# = R_{i_1} \times \cdots \times R_{i_n}$ (here $\times$ is the Cartesian product (a function) of the relations, while $\otimes$ is its operational symbol used for relational symbols). Let us introduce a relational symbol r such that $ar(r) = \sum_{1 \le k \le n} ar(r_{i_1})$, with the schema $\mathcal{B} = (\{r\}, \emptyset)$ and $B = \alpha^*(\mathcal{B}) = Eval(t_R) = \{R, \bot\}$, so that SOtgd Ψ equal to

$$\forall \mathbf{x}_{i_1} \ldots \forall \mathbf{x}_{i_n} \big(\big(r_{i_1}(\mathbf{x}_{i_1}) \wedge \cdots \wedge r_{i_n}(\mathbf{x}_{i_n}) \big) \Rightarrow r(\mathbf{x}_{i_1}, \ldots, \mathbf{x}_{i_n}) \big) \tag{6.1}$$

corresponds to the Cartesian product in t_R.

Hence, we have the mapping-operad

$$\mathbf{M} = MakeOperads\big(\{\Psi\}\big) = \{q_1, 1_{r_\emptyset}\} : \mathcal{A} \to \mathcal{B},$$

with $q_1 \in O(r_{i_1}, \ldots, r_{i_n}, r)$ and, consequently, the arrow

$$f_q = \alpha^*(\mathbf{M}) = \big\{\alpha(q_1), q_\bot\big\} : A \to B,$$

so that we obtain the surjective mapping function

$$\alpha(q_i) : R_{i_1} \times \cdots \times R_{i_n} \twoheadrightarrow \alpha(r) = R = \|t_R\|_\#,$$

that is, we have the composition $Eval(f_R) \circ f_q : A \to Eval(f_R(t_R)) \subseteq TA$. Consequently, we obtain

$$f_{DB} = is_A^{-1} \circ in \circ Eval(f_R) \circ f_q : A \to A_1, \quad \text{with } A_1 = A, \tag{6.2}$$

where $is_A^{-1} : TA \simeq A$ is an isomorphism in **DB** and a monomorphism $in : Eval(f_R(t_R)) \hookrightarrow TA$ (the inclusion $Eval(f_R(t_R)) = \{R', \bot\} \subseteq TA$, with $R' = \|f_R(t_R)\|_\#$), with $in = \{id_{R'}, q_\bot\}$. □

Let us consider the following program for the categorial RDB machine $\mathbf{M}_R$, where the Application Plane is composed of $P_n, n \ge 3$ embedded compiled SQL statements in the host program, where P_1 is the plan for COMMIT, P_2 is the plan for ROLLBACK, and the rest of the plans are the embedded statements (n-operators, composed of the variable-binding operators, obtained from the arrows of category

RA) in a given ordering determined by their compilation from the original host source program P:

$(0, In_R, 1)$	Initial input, loads the physically stored database into data
(1, ACOM, 4)	memory and assigns $x_1 = x_2 = x_4 = 0$ and $x_3 = 1$
$(2, Out_R, 3)$	transfers a triple $(0, n, R)$ in I/O memory to $\mathbf{M}_U$, n is the value of x_2(error), $R \neq \perp$ only for execution of 'SELECT...'
(3, RETURN, 4)	End of every program
(4, NOP,5)	
$(5, In^*_{ER}, 6)$	Waits for CALLs from $\mathbf{M}_U$ (given by $Out_R(st(j))$, where $st(j)$ is the current state of M_U)
$(6, t_1, 4, 10)$	
...	$(n, 0, L)$ is received from $\mathbf{M}_U$ and $x_2 = 0$, $x_1 \geq 1$ is imposed
(10, DEC, 11)	
$(11, t_4, 12, 20)$	
(12, VCON, 13)	COMMIT; checks the consistency of a database in memory
$(13, t_2, 14, 2)$	
(14, COM, 1)	extracts a database from working memory (Buffer) and save a database in the physical storage
...	
(20, DEC, 21)	
$(21, t_4, 22, 30)$	
(22, RBACK, 1)	ROLLBACK
...	
(30, DEC, 31)	
$(31, t_4, 32, 40)$	
$(32, P_1, 2)$	Plan P_3
...	
...	
$(N-1,\ \text{DEC}, N)$	where $N = 1 + 10 * n$
$(N, t_4, N+1, N+2)$	
$(N+1, P_n, 2)$	last Plan P_n, $n \geq 4$ in the Application Plan
$(N+2,\ \text{ERR},\ 2)$	
...	

Program 2

Let us denote the parts of the Application Plan by $Ap_k = P_k \circ (t_4 \circ \text{DEC})^k$, $k = 1, 2, \ldots, n$, where:

- $P_1 = \text{ACOM} \circ \text{COM} \circ t_2 \circ \text{VCON}$ is the kernel for COMMIT;
- $P_2 = \text{ACOM} \circ \text{RBACK}$ is the kernel for ROLLBACK;
- $P_k, k = 3, \ldots, n$ are the kernels for embedded SQL formulas (P_k are the arrows in **RA**, composed of unary operators in Σ_{RA}).

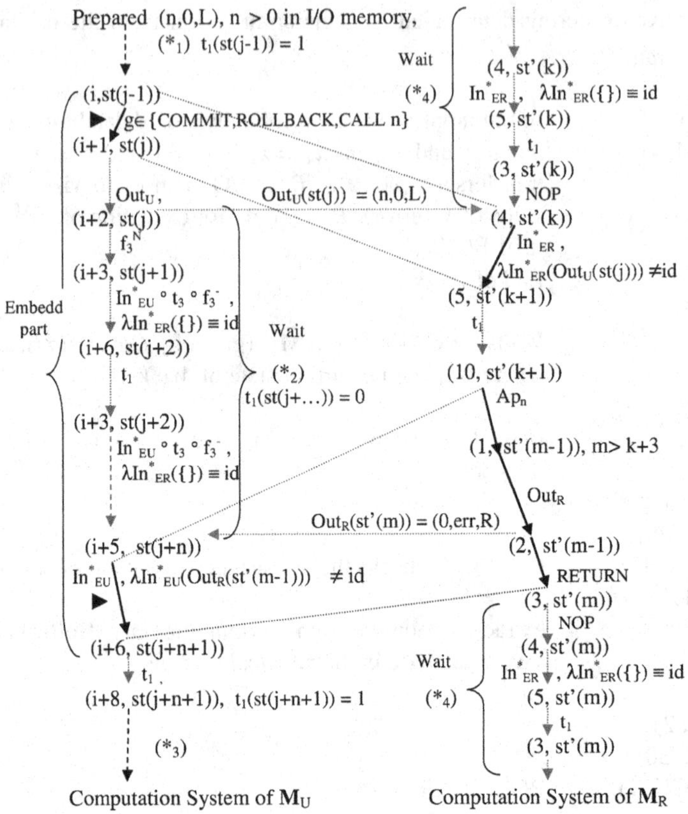

Fig. 6.3 Synchronization process

Thus, the synchrony of the conversations between $\mathbf{M}_U$ and $\mathbf{M}_R$ can be represented by the transition diagram in Fig. 6.3. The states of the machines are progressively enumerated from their initial states, w.r.t. a deterministic transition system $TrS = (S_P, \rightarrow_P)$ (in Definition 37), where

$$S_P = \big\{(i, st(j)) \mid i \text{ is the address of an instruction of } P,\; st(j) \in S_M \text{ is the state of a machine before the execution of this } i\text{th instruction}\big\}$$

is the set of the configurations and $\rightarrow_P$ is a function defined for any $(i, st(k)) \in S_P$ by $((i, st(j)), (k, f(st(j+1)))) \in \rightarrow_P$ if $(i, f, k) \in P$ and f is not the identity function, in the way that for any j, $st(j+1) \neq st(j)$, i.e., the index of the next state is incremented only if that state is different from the current state.

The $(*_1)$ and $(*_3)$ are the pre/post embedded statements of the host program (Program 1), $(*_2)$ is a repetition of the composition $t_1 \circ \lambda In^*_{EU}(\{\}) \circ t_3 \circ f_3^-$, where $\{\} = \emptyset$, and $(*_4)$ is a repetition of composition $t_1 \circ \lambda In^*_{ER}(\{\}) \circ \mathrm{NOP}$, which is the identity function.

Notice that in the transition system of $\mathbf{M}_U$ of the host program (Program 1) we consider the case when $\mathbf{M}_U$ receives the answer from $\mathbf{M}_R$ just in time and without error.

The significant transitions in the computation system of $\mathbf{M}_U$, for the synchronization of $\mathbf{M}_U$, are denoted by the symbol ▶ in Fig. 6.3, while the two horizontal red lines are the time-synchronization instances.

6.2.1 The Categorial Approach to SQL Embedding

The synchrony conversations, for a given source host program P, between $\mathbf{M}_U$ and $\mathbf{M}_R$, described by the transition diagram in Fig. 6.3, represent the execution of the embedding (of the compiled SQL statements). The states of the machines are progressively enumerated from their initial states $st(0) \in S_U, st'(0) \in S_R$, w.r.t. a deterministic transition system $TrS = (S_P, \rightarrow_P)$ (in Definition 37). We can abstract this deterministic transition system by forgetting the instruction addresses, that is, by substituting the transitions $((j, st(i)), (k, f(st(i)))) \in \rightarrow_P$ by the simple state-transitions $(st(i), f(st(i))) \in \rightarrow_P$.

Let $ATrS_U = (\pi_2(S_{PU}), \rightarrow_P)$ be an abstracted deterministic transition system obtained from an execution of the program P on a host machine $\mathbf{M}_U$, with $\pi_2(S_{PU}) = \{st(i) | 0 \leq i \leq N\} \subseteq S_U$ the set of its internal states progressively enumerated from 0 (initial state) to the last Nth state (end of this program) and $ATrS_R = (\pi_2(S_{PR}), \rightarrow_P)$ be the deterministic abstracted transitions system obtained from the execution of the program P on an RDB machine $\mathbf{M}_R$, represented in Fig. 6.3, Sect. 6.2. The most significant parts of the TrS (in what follows, we will use this simplified abstracted transition system) of the $\mathbf{M}_R$, underlined in Fig. 6.3 by delimited intervals from diagram on the left (signed by ▶) and right sides) are the following:

1. **Variable binding.** It transforms an Application Plan into an arrow in **RA**.

   The transition $st'(k) \xrightarrow{\lambda In^*_{ER}(Out_U(st(j)))} st'(k+1) \in ATrS_R$ that corresponds to the binding of bounded variables in the variable-binding operators of the called Application Plan.

   The Bind module during the compilation of the original source host program P generates the Application Plans for each SQL statement in P, which are the "variable-binding" **RA** arrows: it means that instead of the concrete operators of Σ_{RA} algebra, the constants (i.e., values) used for such operators are here substituted by the bounded variables. Only when the $\mathbf{M}_U$ calls an Application Plan, by passing the triple $(n, 0, L)$ with the list L of values for this nth plan, then the bounded variables of these value-binding operators are substituted (variable binding) with the values in L, and hence these operators become the operators of Σ_{RA} algebra, and the Application Plan becomes exactly an arrow in **RA**.
2. **Execution.** By an application of the evaluation functor $Eval : \mathbf{RA} \rightarrow \mathbf{DB}$.

   The transition $st'(k+1) \xrightarrow{f_n} st'(m) \in ATrS_R, m \geq k+3, n \geq 1$, which corresponds to the composition of transactions $f_n = \text{RETURN} \circ Out_R \circ Ap_n$, where $Ap_n = P_n \circ (t_4 \circ \text{DEC})^n$, and

- $P_1 = \text{ACOM} \circ \text{COM} \circ t_2 \circ \text{VCON}$ is the kernel for COMMIT;
- $P_2 = \text{ACOM} \circ \text{RBACK}$ is the kernel for ROLLBACK;
- $P_n, n \geq 3$ are the kernels for embedded SQL formulas (P_n are the arrows in **RA**, composed of unary operators in Σ_{RA}).

Based on this abstracted transition system, where the nodes are simply the states of the machines, we are able to define an algorithm of the resulting synchronization of the states between the host and RDB machines:

Synchronization mapping algorithm *Smap*($ATrS_U, ATrS_R$)
Input. The abstracted transition systems $ATrS_U = (\pi_2(S_{PU}), \rightarrow_P)$ and $ATrS_R = (\pi_2(S_{PR}), \rightarrow_P)$ for a given source program P.
Output. Mapping $F_P^0 : \pi_2(S_{PU}) \rightarrow \pi_2(S_{PR})$

1. Let $st'(k)$ be the idle state of $\mathbf{M}_R$ after the initialization process, when $\mathbf{M}_R$ is waiting for the calls from $\mathbf{M}_U$. Set $i = j = 0$.
2. $i = i + 1$. If $i \leq \max\{l \mid st(l) \in \pi_2(S_{PU})\}$ go to Step 3.
   Otherwise, $(st(m), st'(k)) \in F_P^0$, for $j \leq m \leq i - 1$, go to Step 8.
3. If $st(i-1) \neq \partial_0(g)$, for $g \in \{\text{COMMIT}, \text{ROLLBACL}, \text{CALL } n\}$, go to Step 2.
4. $(st(m), st'(k)) \in F_P^0$, for $j \leq m \leq i-1$, and $(st(i), st'(k+1)) \in F_P^0$. $j = i + 1$.
5. $i = i + 1$. If $i \leq \max\{l \mid st(l) \in \pi_2(S_{PU})\}$ go to Step 6.
   Otherwise, $(st(m), st'(k+1)) \in F_P^0$, for $j \leq m \leq i - 1$, go to Step 8.
6. $(st(i), st'(k+1)) \in F_P^0$. If
$$st(i) \neq \partial_0\big(\lambda In^*_{EU}\big(Out_R\big(st'(m)\big)\big)\big) \neq \partial_1\big(\lambda In^*_{EU}\big(Out_R\big(st'(m)\big)\big)\big)$$
   then go to Step 5.
7. $(st(i+1), st'(m)) \in F_P^0$, where $st'(m) = \partial_1(f_n)$ for $f_n = \text{RETURN} \circ Out_R \circ Ap_n$, with $Ap_n = P_n \circ (t_4 \circ \text{DEC})^n$, and $\partial_0(f_n) = st'(k+1)$.
   Set $k = m, i = i + 1, j = i + 1$. Go to Step 2.
8. **Return** the graph of the function $F_P^0 : \pi_2(S_{PU}) \rightarrow \pi_2(S_{PR})$.

Based on this synchronization mapping of the algorithm *Smap*($ATrS_U, ATrS_R$), we are able to define the following functorial semantics for the synchronization process in Fig. 6.3, Sect. 6.2:

Proposition 29 *For a given source program P with the abstract transitions systems $ATrS_U = (\pi_2(S_{PU}), \rightarrow_P)$ and $ATrS_R = (\pi_2(S_{PR}), \rightarrow_P)$, we define the following two categories:*

1. *The category $\mathbf{U}_P$ which is equal to the $ATrS_U$ where the objects are the machine states in $ATrS_U$, and the atomic arrows are the atomic transitions in $ATrS_U$, but where the transitions Out_U and the test transitions are substituted by the* NOP *transition.*
2. *The category $\mathbf{R}_P$ where the objects are the machine states in the image of the mapping $F_P^0 : \pi_2(S_{PU}) \rightarrow \pi_2(S_{PR})$ (obtained by the algorithm Smap($ATrS_U$, $ATrS_R$)), and the atomic arrows are the Application Plans P_n, $n \geq 3$ (i.e., the*

arrows in **RA**) *and the rest of the transitions in* $ATrS_R$ *between the states in the image of the mapping* F^0, *but where the transitions* Out_R *and the test transitions are substituted by the* NOP *transition.*

3. *Let us define the arrow-function* $F_P^1 : Mor_{\mathbf{U}_P} \to Mor_{\mathbf{R}_P}$ *such that for each atomic arrow* $g : \partial_0(g) \to \partial_1(g)$ *in* $Mor_{\mathbf{U}_P}$:

$$F_P^1(g) = \begin{cases} \lambda In^*_{ER}(Out_U(\partial_0(g))) & \textit{if } g \in \{\text{COMMIT}, \text{ROLLBACK}, \text{CALL } n\}; \\ f_n & \textit{if } g = \lambda In^*_{EU}(Out_R(F_P^0(\partial_1(g)))) \neq id; \\ \text{NOP} & \textit{otherwise}, \end{cases}$$

where f_n *is the composition of atomic transitions in* $ATrS_R$ *between the states* $F_P^0(\partial_0(g))$ *and* $F_P^0(\partial_1(g))$, *but where the test transitions are substituted by the* NOP *transition, as follows*:

- $f_1 = \text{RETURN} \circ \text{ACOM} \circ \text{COM} \circ \text{VCOM} \circ (\text{DEC})^1$;
- $f_2 = \text{RETURN} \circ \text{ACOM} \circ \text{RBACK} \circ (\text{DEC})^2$;
- $f_n = \text{RETURN} \circ P_n \circ (\text{DEC})^n$, *for* $n \geq 3$ (P_n *is an atomic arrow, as in* **RA**),

and for the composition of atomic arrows $g \circ f$, $F_P^1(g \circ f) = F_P^1(g) \circ F_P^1(f)$.

Then there exists the functor $F_P = (F_P^0, F_P^1) : \mathbf{U}_P \to \mathbf{R}_P$ *where the object-function* F_P^0 *is the mapping obtained from the algorithm Smap*($ATrS_U$, $ATrS_R$).

Proof It is easy to verify that $\mathbf{U}_P$ (and also $\mathbf{R}_P$) is a well defined category: its identity arrows are $\text{NOP} : S_U \to S_U$ (which are the identity functions and hence do not change a state), while all other arrows are the (unary) functions from S_U into S_U. Notice that also $In^*_{EU}(\{\}) = In^*_{EU}(\emptyset) = \text{NOP} : S_U \to S_U$ is an identity arrow as well (analogously, $In^*_{ER}(\{\}) = In^*_{ER}(\emptyset) = \text{NOP} : S_U \to S_U$ is an identity arrow in $\mathbf{R}_P$). Thus, the arrows of these two categories are the unary operators (the unary functions) of these two machines, so that their composition is another unary function, that is, a well defined arrow (for the composition of arrows the associativity holds, because they are the composition of functions).

Let us show that for the definition of F_P^1 for the atomic arrows, the condition $F_P^1(g \circ f) = F_P^1(g) \circ F_P^1(f)$ for the two atomic arrows f and g is satisfied. Let us consider the following possible cases for $f : st(l) \to st(l_1)$ and $g : st(l_1) \to st(l_2)$ in Fig. 6.3 (for one call from $\mathbf{M}_U$ to $\mathbf{M}_R$):

1. If $l_2 < j$ then, from definition of F_P^1, $F_P^1(g \circ f) = F_P^1(g) = F_P^1(f) = \text{NOP} : st'(k) \to st'(k)$, so that $F_P^1(g \circ f) = F_P^1(g) \circ F_P^1(f)$.
2. If $l_2 = j$ and $g \in \{\text{COMMIT}, \text{ROLLBACK}, \text{CALL}, n\}$ then we define

$$F_P^1(g \circ f) \triangleq F_P^1(g) = In^*_{ER}\big(Out_U\big(\partial_0(g)\big)\big) : st'(k) \to st'(k+1).$$

Thus, from definition of F_P^1, $F_P^1(f) = \text{NOP} : st'(k) \to st'(k)$, and we obtain $F_P^1(g \circ f) = F_P^1(g) = F_P^1(g) \circ \text{NOP} = F_P^1(g) \circ F_P^1(f)$.

3. If $l_2 = j$ and $f \in \{\text{COMMIT}, \text{ROLLBACK}, \text{CALL}, n\}$ then we define

$$F_P^1(g \circ f) \triangleq F_P^1(f) = In^*_{ER}\big(Out_U\big(\partial_0(g)\big)\big) : st'(k) \to st'(k+1).$$

Thus, from definition of F_P^1, $F_P^1(g) = \text{NOP} : st'(k+1) \rightarrow st'(k+1)$, and we obtain $F_P^1(g \circ f) = F_P^1(f) = \text{NOP} \circ F_P^1(f) = F_P^1(g) \circ F_P^1(f)$.

4. If $j \leq l$ and $l_2 \leq j+n$, then, from definition of F_P^1,

$$F_P^1(g \circ f) = F_P^1(g) = F_P^1(f) = \text{NOP} : st'(k) \rightarrow st'(k)$$

and hence $F_P^1(g \circ f) = F_P^1(g) \circ F_P^1(f)$.

5. If $l_2 = j+n+1$ and $l_1 = j+n$ then we define

$$F_P^1(g \circ f) \triangleq F_P^1(g) = \text{RETURN} \circ Ap_n : st'(k+1) \rightarrow st'(m).$$

Thus, from definition of F_P^1, $F_P^1(f) = \text{NOP} : st'(k+1) \rightarrow st'(k+1)$ and hence $F_P^1(g \circ f) = F_P^1(g) = F_P^1(g) \circ \text{NOP} = F_P^1(g) \circ F_P^1(f)$.

6. If $l = j+n$ and $l_1 = j+n+1$ then we define

$$F_P^1(g \circ f) \triangleq F_P^1(f) = \text{RETURN} \circ Ap_n : st'(k+1) \rightarrow st'(m).$$

Thus, from definition of F_P^1, $F_P^1(g) = \text{NOP} : st'(m) \rightarrow st'(m)$ and hence $F_P^1(g \circ f) = F_P^1(f) = \text{NOP} \circ F_P^1(f) = F_P^1(g) \circ F_P^1(f)$.

7. (Repetition of point 1, for the new idle (wait) state $st'(m)$ of $\mathbf{M}_R$) If $l > j+n$ and $l_2 < j'$ where $\partial(g) = st(j')$ and $g \in \{\text{COMMIT}, \text{ROLLBACK}, \text{CALL}, n\}$ then, from definition of F_P^1,

$$F_P^1(g \circ f) = F_P^1(g) = F_P^1(f) = \text{NOP} : st'(m) \rightarrow st'(m)$$

and hence $F_P^1(g \circ f) = F_P^1(g) \circ F_P^1(f)$.

The functor F_P is well defined: each identity arrow in $\mathbf{U}_P$ is mapped into the identity arrow (NOP) in $\mathbf{U}_P$. From the compositional property for the atomic arrows, for the composition of *all* arrows $F_P^1(g \circ f) = F_P^1(g) \circ F_P^1(f)$ is valid. □

Notice that the functor F_P, represented in Fig. 6.4, "forgets" all transitions of the host program of $\mathbf{M}_U$ which are not a part of the synchronized conversations between the universal host machine and RDB categorial machine $\mathbf{M}_R$.

Let $Inst_U : Ob_{U_P} \rightarrow Time$ and $Inst_R : Ob_{R_P} \rightarrow Time$ be two mapping that define the time-instance when a given state is achieved during the execution of the source program P of the machines $\mathbf{M}_U$ and $\mathbf{M}_R$, respectively. From the fact that the states are progressively enumerated during the execution of this program,

$$Inst_U\big(st(j)\big) \leq Inst_U\big(st(m)\big) \quad \text{iff} \quad j \leq m, \tag{6.3}$$

and, analogously, for the states $st'(k)$ of M_R.

Let $WS \subset Ob_{R_P}$ be the subset of idle wait-states of $\mathbf{M}_U$. We have that before the beginning of execution of the program P, the $\mathbf{M}_R$ is in one of this idle wait-states $st'(k_0) \in WS$. Consequently, for each state $st(j)$, $j \geq 0$, in $ATrS_U$, $Inst_U(st(j)) \geq Inst_R(st'(k_0))$. For example, in Fig. 6.3, Sect. 6.2, $st'(k), st'(m) \in WS$, with $Inst_U(st(i)) \geq Inst_R(st'(k))$ for $i \leq j+n$, and $Inst_U(st(i)) \geq Inst_R(st'(m))$ for $i \geq j+n+1$.

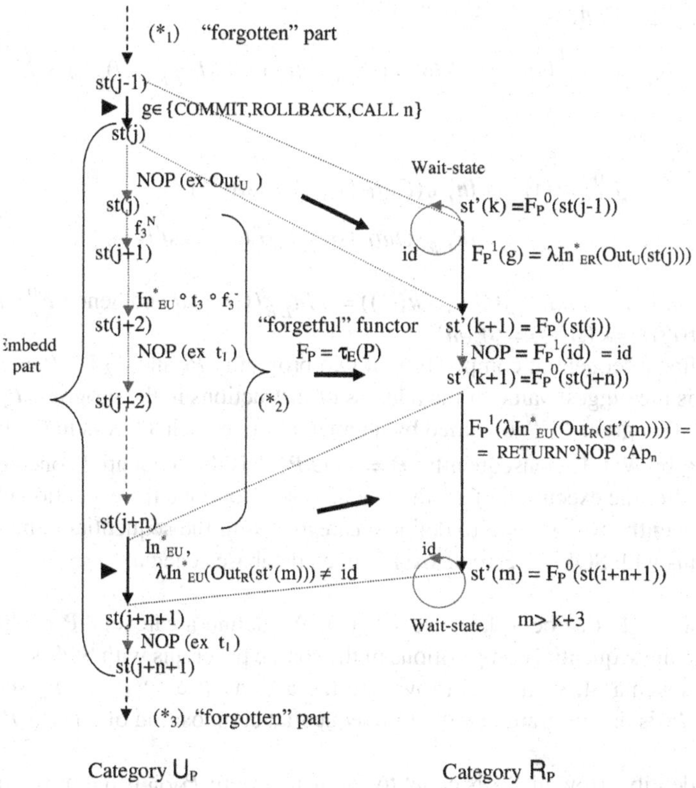

Fig. 6.4 Embedding: functorial semantics

Consequently, we are able to define a function

$$\kappa : Ob_{R_U} \to Ob_{R_P} \tag{6.4}$$

such that for any $st(j) \in Ob_{R_U}$, $\kappa(st(j)) = st'(k)$ with

$$k = \max\{i \mid st'(i) \in WS \text{ and } Inst_R(st'(i)) \le Inst_U(st(j))\}.$$

It is easy to verify that for each state $st(i) \in ATrS_U$

$$F_P^0(st(i)) = In_{ER}^*(Out_U(st(i), \kappa(st(i)))) = \lambda In_{ER}^*(Out_U(st(i))(\kappa(st(i))). \tag{6.5}$$

Let us consider the one synchronized conversation cycle in Fig. 6.3, Sect. 6.2, where:

1. For $i \le j-1$, $\lambda In_{ER}^*(Out_U(st(i))) = \lambda In_{ER}^*(\{\}) = id$ and hence

$$F_P^0(st(i)) = id(\kappa(st(i))) = \kappa(st(i)) = st'(k).$$

2. For $j \leq i \leq j+n$,

$$\lambda In^*_{ER}\big(Out_U\big(st(i)\big) = \lambda In^*_{ER}\big(Out_U\big(st(j)\big) = \lambda In^*_{ER}(n, 0, L) \neq id,$$

so that

$$\begin{aligned} F^0_P\big(st(i)\big) &= \lambda In^*_{ER}\big(Out_U\big(st(i)\big)\big(\kappa\big(st(i)\big)\big)\big) \\ &= \lambda In^*_{ER}\big(Out_U\big(st(j)\big)\big(st'(k)\big) = st'(k+1). \end{aligned}$$

3. For $j+n+1 \leq i$, $\lambda In^*_{ER}(Out_U(st(i))) = \lambda In^*_{ER}(\{\}) = id$ and hence $F^0_P(st(i)) = id(\kappa(st(i))) = \kappa(st(i)) = st'(m)$.

Let us define a sequential composition of two programs P_1 and P_2 by $P = P_2 \circ P_1$ where N is the biggest value of the address of instructions in the program P_1. Then we define a program P_2^{+N} obtained by augmentation of each address in the instructions of P_2 by $N+1$. Consequently, $P = P_1 \cup P_2^{+N}$ is the sequential concatenation such that after the execution of P_1 the machine will continue the execution of P_2.

Consequently, we are able to define a category with the sequential composition of programs with SQL statements over a given database schema:

Definition 41 Let $\mathcal{A}$ be a database schema. We define a category $\mathbf{P}_\mathcal{A}$ where the objects are the sequential compositions of the source programs with SQL statements over this schema $\mathcal{A}$, while the arrows are the actions $P \circ \_ : P_1 \rightarrow P_2$ such that $P_2 = P \circ P_1$ is the program obtained by sequential composition of P_1 and P.

Each identity arrow in $\mathbf{P}_\mathcal{A}$ is equal to $P_{NOP} \circ \_$ where P_{NOP} is the program composed of only one instruction NOP. The associativity of composition of the arrows is guaranteed by the associativity of the sequential composition of the programs, thus $\mathbf{P}_\mathcal{A}$ is well defined.

Corollary 20 *For every database schema $\mathcal{A}$ there are the following functors*:

1. $U_M : \mathbf{P}_\mathcal{A} \rightarrow \mathbf{Cat}$ *such that for each program $P \in Ob_{\mathbf{P}_\mathcal{A}}$, $U^0_M(P) = \mathbf{U}_P$ is the category obtained from $ATrS_U$ (by Proposition* 29), *and for an arrow $P \circ \_ :$ $P_1 \rightarrow P_2$ we obtain an inclusion functor $in = U^1_M(P \circ \_) : \mathbf{U}_{P_1} \hookrightarrow \mathbf{U}_{P_2}$.*
2. $R_M : \mathbf{P}_\mathcal{A} \rightarrow \mathbf{Cat}$ *such that for each program $P \in Ob_{\mathbf{P}_\mathcal{A}}$, $R^0_M(P) = \mathbf{R}_P$ is the category obtained from $ATrS_R$ (by Proposition* 29), *and for an arrow $P \circ \_ :$ $P_1 \rightarrow P_2$ we obtain an inclusion functor $in = R^1_M(P \circ \_) : \mathbf{R}_{P_1} \hookrightarrow \mathbf{R}_{P_2}$.*

Consequently, there is an "embedding" natural transformation τ_E : $U_M \stackrel{\bullet}{\longrightarrow} R_M$,

$$\mathbf{P}_\mathcal{A} \; \underset{R_M}{\overset{U_M}{\rightrightarrows}} \; \mathbf{Cat} \qquad \big(\downarrow \tau_E\big)$$

such that for any program P, we obtain the functor (an arrow in **Cat**) $\tau_E(P) = F_P : \mathbf{U}_P \rightarrow \mathbf{R}_P$ *(defined in point* 3 *of Proposition* 29).

Proof In fact, for any arrow $P \circ \_ : P_1 \rightarrow P_2$ in $\mathbf{P}_\mathcal{A}$ with $P_2 = P \circ P_1$, the following diagram of functors in **Cat** commutes:

$$\begin{array}{ccc}
U_M(P_1) = \mathbf{U}_{P_1} & \xrightarrow{\;in = U_M(P \circ _)\;} & U_M(P_2) = \mathbf{U}_{P_2} \\
\Big\downarrow \tau_E(P_1) = F_{P_1} & & \Big\downarrow \tau_E(P_2) = F_{P_2} \\
R_M(P_1) = \mathbf{R}_{P_1} & \xrightarrow{\;in = R_M(P \circ _)\;} & R_M(P_2) = \mathbf{R}_{P_2}
\end{array}$$

□

The vertical arrows in the commutative diagram in **Cat** above are the "forgetful" functors, whose example is represented in Fig. 6.4, while the horizontal arrows are the inclusion functors (the category $\mathbf{U}_{P_1}$ is a subcategory of the category $\mathbf{U}_{P_2}$, and $\mathbf{R}_{P_1}$ is a subcategory of the category $\mathbf{U}_{R_2}$).

Remark Each atomic arrow P_n, $n \geq 3$ in a category $\mathbf{R}_P = R_M(P)$ obtained for a given source program P, which is an arrow in **RA** as well, can be considered as a program composed of a set of instruction of $\mathbf{M}_R$ (see the table of Program 2).

Consequently, we have the following equations for the sequential composition with the 'identity program' P_{NOP}:

$$\text{(IDC)} \quad P_{\text{NOP}} \circ P_n = P_n \circ P_{\text{NOP}} = P_n, \quad \text{for } n \geq 3.$$

6.2.2 The Categorial Approach to the Transaction Recovery

The problems of recovery and concurrency in a database system are heavily bound up with the notion of transaction processing. In this section therefore, we first explain what a transaction is, what the term transaction processing (or transaction management) from a categorial point of view means, and how a database conserves the integrity constraints during execution of the application programs.

We can define a *transaction* as a logical unit of work. During an execution of programs, the database is not even consistent between two updates; thus a logical unit of work (i.e., transaction) is not necessarily just a single database operation; rather, it is a *sequence* of several such operations (each operation in our working framework is an arrow P_n of **RA**), which in general transforms a consistent state of the database (data memory of a categorial RDB machine $\mathbf{M}_R$) into another consistent state, without necessarily preserving consistency at all intermediate states.

We define the graph $G(\mathcal{A})$ for a schema $\mathcal{A}$ (by Proposition 15 in Sect. 4.1.1) and its sketch category $\mathbf{Sch}(G(\mathcal{A}))$. Let the current instance-database in the dynamic data memory of $\mathbf{M}_R$ be $\alpha^*(\mathcal{A}) = \{\alpha(r_i) \mid r_i \in S_A\} \cup \{\bot\}$, and after the execution of a given Application plan P_n (derived by compilation of the source program P), which is an arrow in **RA**, we obtain the new instance-database $A_1 = \alpha_1^*(\mathcal{A})$ in the data memory of $\mathbf{M}_R$: this changing of the extension of the schema $\mathcal{A}$ we will denote by a transition $\alpha^* \overset{\mathcal{M}_{P_n}}{\Longrightarrow} \alpha_1^*$.

Notice that $\alpha_1^*(\mathcal{A})$ is always consistent for the reduced schema $\underline{\mathcal{A}} = (S_A, \emptyset)$ (and hence the R-algebra α_1 is a model of $\underline{\mathcal{A}}$). However, it may be inconsistent for the original database schema $\mathcal{A} = (S_A, \Sigma_A)$ (thus, α_1 is not necessarily a model of $\mathcal{A}$).

In order to be able to consider the sequences of such transitions for a given schema $\mathcal{A}$, we introduce a category of such transitions, as follows:

Proposition 30 *For a given sketch* $\mathbf{Sch}(G(\mathcal{A}))$ *of a schema* $\mathcal{A} = (S_A, \Sigma_A)$, *the category of interpretations of* $\mathcal{A}$ *is the subcategory of functors* $Int(G(\mathcal{A})) \subseteq \mathbf{DB}^{\mathbf{Sch}(G(\mathcal{A}))}$, *from Proposition* 16 *(and Proposition* 17 *in Sect.* 4.1.3*), where*:

1. *Each object is an R-algebra (a mapping interpretation)* $\alpha^* : \mathbf{Sch}(G(\mathcal{A})) \to \mathbf{DB}$, *which is functor such that for the identity schema mapping*

$$\mathbf{M}_{AA} = \{1_{r_i} \in O(r_i, r_i) \mid r_i \in S_A\} \cup \{1_{r_\emptyset}\} : \mathcal{A} \to \mathcal{A},$$

   we obtain the identity morphism $\alpha^*(\mathbf{M}_{AA}) : A \to A$ *for* $A = \alpha^*(\mathcal{A})$.

   If $\Sigma_A \neq \emptyset$, *then for* $\mathbf{M}_{\mathcal{A}_\top \mathcal{A}_\top} = \{1_{r_\top}, 1_{r_\emptyset}\} : \mathcal{A}_\top \to \mathcal{A}_\top$, *we obtain the identity morphism* $\alpha^*(\mathbf{M}_{\mathcal{A}_\top \mathcal{A}_\top}) : A_T \to A_T$ *for* $A_T = \alpha^*(\mathcal{A}_\top) = \{R_=, \bot\}$, *and for the sketch arrow* $\mathbf{T}_{A\mathcal{A}_\top} : \mathcal{A} \to \mathcal{A}_\top$, *we obtain* $f_{\Sigma_A} = \alpha^*(\mathbf{T}_{A\mathcal{A}_\top}) : A \to A_T$ *with* $\widetilde{f_{\Sigma_A}} = \bot^0$ *(by Proposition* 15 *in Sect.* 4.1.1*).*
2. *Each atomic arrow is a natural transformation* $\eta_{P_n} : \alpha^* \xrightarrow{\bullet} \alpha_1^*$, *denoted by a transition* $\alpha^* \overset{\mathcal{M}_{P_n}}{\Longrightarrow} \alpha_1^*$, *where* $\mathcal{M}_{P_n} : \mathcal{A} \to \mathcal{A}$ *is the schema mapping (by Proposition* 26, *Sect.* 5.3*) such that* $\alpha_1^*(\mathcal{A})$ *is an instance-database obtained from the instance-database* $\alpha^*(\mathcal{A})$ *by execution of an arrow* P_n *in* **RA**, *such that*:
   2.1. $\eta_{P_n}(\mathcal{A}) : \alpha^*(\mathcal{A}) \to \alpha_1^*(\mathcal{A})$ *is defined by Corollary* 17 *as an arrow in* $\{f_U, f_I, f_D\}$, *specified in Example* 34 *for the cases when* P_n *is obtained from* 'UPDATE...', 'INSERT...' *or* 'DELETE...', *SQL statements, respectively.*
   2.2. $\eta_{P_n}(\mathcal{A}_\top) = id_{A_T} : A_T \to A_T$.

The compositions of the arrows in $Int(G(\mathcal{A})) \subseteq \mathbf{DB}^{\mathbf{Sch}(G(\mathcal{A}))}$ *satisfy the following property*:

$$\left(\alpha_1^* \overset{\mathcal{M}_{P_k}}{\Longrightarrow} \alpha_2^*\right) \circ \left(\alpha^* \overset{\mathcal{M}_{P_n}}{\Longrightarrow} \alpha_1^*\right) = \left(\alpha^* \overset{\mathcal{M}_{P_k \circ P_n}}{\Longrightarrow} \alpha_2^*\right).$$

Proof Let us show that this category of functors, for a given schema $\mathcal{A}$, is well defined:

1. The fact that each arrow in $Int(G(\mathcal{A})) \subseteq \mathbf{DB}^{\mathbf{Sch}(G(\mathcal{A}))}$ is a natural transformation $\eta_{P_n} : \alpha^* \xrightarrow{\bullet} \alpha_1^*$ holds from the following commutative diagram on the right:

$$
\begin{array}{ccccc}
\mathcal{A} & & A=\alpha^*(\mathcal{A}) & \xrightarrow{\eta_{P_n}(\mathcal{A})} & A_1=\alpha_1^*(\mathcal{A}) \\
\Big\downarrow id_{\mathcal{A}}=\mathbf{M}_{AA} & & \Big\downarrow id_A=\alpha^*(\mathbf{M}_{AA}) & & \Big\downarrow id_{A_1}=\alpha_1^*(\mathbf{M}_{AA}) \\
\mathcal{A} & & A=\alpha^*(\mathcal{A}) & \xrightarrow{\eta_{P_n}(\mathcal{A})} & A_1=\alpha_1^*(\mathcal{A}) \\
\Big\downarrow \mathbf{T}_{AA_{\top}} & & \Big\downarrow f_{\Sigma_A}=\alpha^*(\mathbf{T}_{AA_{\top}}) & & \Big\downarrow f_{\Sigma_A}=\alpha_1^*(\mathbf{T}_{AA_{\top}}) \\
\mathcal{A}_{\top} & & A_T=\alpha^*(\mathcal{A}_{\top}) & \xrightarrow{\eta_{P_n}(\mathcal{A}_{\top})} & A_T=\alpha_1^*(\mathcal{A}_{\top})
\end{array}
$$

Category $\mathbf{Sch}(G(\mathcal{A}))$ Category **DB**

Notice that the bottom square commutes (it exists only if the set of integrity constraints Σ_A is not empty) because the information flux of f_{Σ_A} is always empty (independently if α satisfies the integrity-constraint mapping $\mathbf{T}_{AA_{\top}}$ or *not*, from Corollary 6, Sect. 2.4.3), i.e., $\widetilde{f_{\Sigma_A}}=\bot^0$, and hence the composition with another simple arrows will have empty flux as well.

2. For each object (a functor) α^* in $Int(G(\mathcal{A}))\subseteq \mathbf{DB}^{\mathbf{Sch}(G(\mathcal{A}))}$, its identity arrow is a natural transformation, $id_{\alpha^*}=\eta_{P_{\mathrm{NOP}}}:\alpha^*\overset{\bullet}{\longrightarrow}\alpha^*$, that is, a transition $\alpha^*\overset{\mathcal{M}_{P_{\mathrm{NOP}}}}{\Longrightarrow}\alpha^*$.
3. The associativity of the compositions of the arrows (i.e., natural transformations) in $Int(G(\mathcal{A}))\subseteq \mathbf{DB}^{\mathbf{Sch}(G(\mathcal{A}))}$ is based on the property of composition of the arrows, and the fact that for the given three arrows, $\alpha^*\overset{\mathcal{M}_{P_n}}{\Longrightarrow}\alpha_1^*$, $\alpha_1^*\overset{\mathcal{M}_{P_k}}{\Longrightarrow}\alpha_2^*$ and $\alpha_2^*\overset{\mathcal{M}_{P_m}}{\Longrightarrow}\alpha_3^*$, their composition is equal to the arrow $\alpha^*\overset{\mathcal{M}_{P_m\circ P_k\circ P_n}}{\Longrightarrow}\alpha_3^*$. But we have the associativity for the sequential composition of the programs, that is, $P_m(\circ P_k\circ P_n)=(P_m\circ P_k)\circ P_n=P_m\circ P_k\circ P_n$. Consequently, the associativity for the composition of the arrows in $Int(G(\mathcal{A}))\subseteq \mathbf{DB}^{\mathbf{Sch}(G(\mathcal{A}))}$ holds as well. □

A system (a machine $\mathbf{M}_R$) that supports *transaction processing* guarantees that if the transaction executes some updates and then a failure occurs (for whatever reason) before the transaction reaches its normal termination, then those updates will be undone. Thus, the transaction either executes in its entirety or is totally canceled. In this way, a sequence of operations that is fundamentally not atomic can be made to look as if it really were atomic from an external point of view (and hence $\mathbf{M}_R$ encapsulates transactional atomicity).

The COMMIT and ROLLBACK operations are the key to the way it works:

- The COMMIT operation signals successful end-of-transaction, if its execution (of P_1) in $\mathbf{M}_R$ returns without errors to $\mathbf{M}_U$. It tells to transaction manager that a logical unit of work has been successfully completed, the database is in consistent state again, and all updates made by that unit of work can now be "committed" or made permanent. These states st' in $\mathbf{M}_R$ are distinguished by the fact that $t_2(st') * (t_3(st') + 1) = 1$ (i.e., $x_3 = 1$ and no errors, $x_2 = 0$).

- The ROLLBACK operation, by contrast, signals unsuccessful end-of-transaction: it tells the transaction manager that something has gone wrong when the execution of COMMIT was previously attempted and $\mathbf{M}_R$ responded by an error condition in the list $(0, n, \bot)$, with $n > 0$, to $\mathbf{M}_U$, the database might be in an inconsistent state. In that case, after unsuccessful COMMIT operation, the host program in $\mathbf{M}_U$ may send the call for ROLLBACK to $\mathbf{M}_R$ (i.e., calls the Application Plan P_2) and all the updates made by the logical unit of work must be "rolled back" or undone.

Executing either a COMMIT or a ROLLBACK operation establishes what is called a synchronization point (abbreviated synchpoint), which represents the boundary between two consecutive transactions. Thus, the consecutive transactions in $\mathbf{M}_R$ are defined by their states where $t_2(st'(k)) * (t_3(st'(k)) + 1) = 1$, $k > 0$. The only operations that establish a synchpoint are COMMIT, ROLLBACK and the machine initiation. When a synchpoint is established then:

- All updates by the program since the previous synchpoint are committed (COMMIT) or undone (ROLLBACK).
- All open cursors (in the host program P_U) are closed and all database positioning is lost.
- All record locks are released.

Transactions are indeed the unit of recovery and the unit of concurrency also. In general, a single program execution will consist of a sequence of several transactions, running one after another, with each COMMIT or ROLLBACK operation terminating one transaction and starting the next. Now we are ready to give a formal definition of an *atomic transaction* for a categorial RDB machine $\mathbf{M}_R$.

Definition 42 A concept of an atomic transaction for a Categorial RDB machine $\mathbf{M}_R$ and for a source program $P \in Obj_{\mathbf{P}_\mathcal{A}}$ is represented in the category $R_P = R_M(P)$ (by Proposition 29 and Corollary 20) as an arrow composed of all arrows between the state $st'(k - 1 - \mu y(t_2(st'(k-1-y)) * (t_3(st'(k-1-y)) + 1) - 1 = 0))$ and the state $st'(k)$, $k \geq 1$, such that $t_2(st'(k)) * (t_3(st'(k)) + 1) = 1$.

Remark If schema $\mathcal{A}$ has a nonempty set Σ_A of integrity constraints then an atomic arrow in $Int(G(\mathcal{A})) \subseteq \mathbf{DB}^{\mathbf{Sch}(G(\mathcal{A}))}$ is not necessarily a transition between two *models* of schema $\mathcal{A}$ (it is not guaranteed during the execution of each program P_n) in $\mathbf{M}_R$. Thus, only the strict subset of the arrows in $Int(G(\mathcal{A}))$ corresponds to the consistent transitions in $\mathbf{M}_R$.

Each atomic transaction is composed of a number of consecutive **RA** arrows P_{n_i}, $1 \leq i \leq k$, between the two consecutive synchpoints, not all of them must be consistent: what is important is that the resulting composition $P_{n_k} \circ \cdots \circ P_{n_1}$ is consistent.

The categorial semantics for a consistent database evolution-in-time can be given by the following "consistent" natural transformation and its derived "transaction" functors:

Theorem 11 *Let us define, for a given database schema $\mathcal{A}$ and its category of programs $\mathbf{P}_{\mathcal{A}}$, a constant functor $K_{\mathcal{A}} : \mathbf{P}_{\mathcal{A}} \to \mathbf{Cat}$ such that for every program $P \in Ob_{\mathbf{P}_{\mathcal{A}}}$, $K^0_{\mathcal{A}}(P) = \mathbf{DB}^{\mathbf{Sch}(G(\mathcal{A}))}$ and for each arrow $P \circ \_ \in Mor_{\mathbf{P}_{\mathcal{A}}}$, $K^1_{\mathcal{A}}(P \circ \_)$ is equal to the identity functor $Id : \mathbf{DB}^{\mathbf{Sch}(G(\mathcal{A}))} \to \mathbf{DB}^{\mathbf{Sch}(G(\mathcal{A}))}$.*

Then there is a "transaction" natural transformation $\tau_T : R_M \to K_A$ such that for each program P, we obtain a functor $T_P = \tau_T(P) : \mathbf{R}_P \to \mathbf{DB}^{\mathbf{Sch}(G(\mathcal{A}))}$ with:

1. *For any object $st'(k)$ in the category $\mathbf{R}_P$, $T^0_P(st'(k)) = \alpha^*$ such that*

$$\alpha^*(\mathcal{A}) = Out_{DB}\big(st'\big(k - \mu y\big(t_2\big(st'(k-y)\big) * \big(t_3\big(st'(k-y)\big)+1\big) - 1 = 0\big)\big)\big).$$

2. *For any arrow $g : st'(k) \to st'(m)$ in the category $\mathbf{R}_P$,*

$$T^1_P(g) = \big(T^0_P\big(st'(k)\big) \stackrel{\mathcal{M}}{\Longrightarrow} T^0_P\big(st'(m)\big)\big)$$

such that

$$\mathcal{M} = \begin{cases} \mathcal{M}_{P_{\text{NOP}}} & \textit{if } T^0_P(st'(k)) = T^0_P(st'(m)); \\ \mathcal{M}_{P_{i_n} \circ \cdots \circ P_{i_1}} & \textit{otherwise}, \end{cases}$$

where $P_{i_n} \circ \cdots \circ P_{i_1}$ is the sequential composition of ordered list of **RA***-arrows, considered as the programs of $\mathbf{M}_R$, between the states $st'(k_1)$ and $st'(m_1)$, for $k_1 = k - \mu y(t_2(st'(k-y)) * (t_3(st'(k-y)) + 1) - 1 = 0)$ and $m_1 = m - \mu y(t_2(st'(m-y)) * (t_3(st'(m-y)) + 1) - 1 = 0)$, such that the* **RA***-arrows between any two consecutive synchpoints $st'(j_1)$ and $st'(j_2)$, $k_1 \le j_1 < j_2 \le m_1$, with $T^0_P(st'(j_2)) = T^0_P(st'(j_1))$ are not considered.*

This functor is such that for its image, $Im(T_P) \subseteq Mod(G(\mathcal{A})) \subseteq \mathbf{DB}^{\mathbf{Sch}(G(\mathcal{A}))}$ is the set of all transactions between the models of the database schema $\mathcal{A}$.

Proof Notice that if $t_2(st'(k)) * (t_3(st'(k)) + 1) - 1 = 0$ then $j = \mu y(t_2(st'(k - y)) * (t_3(st'(k-y)) + 1) - 1 = 0) = 0$ ($j \ge 1$ otherwise). From the definition in point 1, for every $st'(k)$, $\alpha^* = T^0_P(st'(k))$ is an object (a functor) in $\mathbf{DB}^{\mathbf{Sch}(G(\mathcal{A}))}$ such that it is a model of schema $\mathcal{A}$ (i.e., it satisfies all integrity constraints in $\mathcal{A}$). For any arrow $g : st'(k) \to st'(m)$, $T^1_P(g)$ is a composed transaction, in which the intermediate identity transactions (and all arrows between two synchpoints $st'(j_1)$ and $st'(j_2)$, with $k_1 \le j_1 < j_2 \le m_1$, where in the second synchpoint $st'(j_2)$ the ROLLBACK operation was executed) are eliminated between two models of $\mathcal{A}$. Thus such a composed transaction, obtained by the sequential composition $P_{i_n} \circ \cdots \circ P_{i_1}$, corresponds to an arrow in $\mathbf{DB}^{\mathbf{Sch}(G(\mathcal{A}))}$ between two models of $\mathcal{A}$. Thus, $Im(T_P) \subseteq\subseteq Mod(G(\mathcal{A})) \subseteq \mathbf{DB}^{\mathbf{Sch}(G(\mathcal{A}))}$.

Let us show that T_P is a well-defined functor from $\mathbf{R}_P$ into $\mathbf{DB}^{\mathbf{Sch}(G(\mathcal{A}))}$:

1. For each identity arrow NOP : $st'(k) \to st'(k)$ we obtain the identity arrow $\alpha^* \stackrel{\mathcal{M}_{P_{\text{NOP}}}}{\Longrightarrow} \alpha^*$ where $\alpha^* = T^0_P(st'(k))$.

2. For any two arrows $g : st'(k) \to st'(m)$, $f : st'(m) \to st'(l)$, with $T_P^1(f) = (T_P^0(st'(m)) \xRightarrow{\mathcal{M}_{P_{jN}\circ\cdots\circ P_{j1}}} T_P^0(st'(l)))$ we obtain that

$$\begin{aligned}
T_P^1(f \circ g) &= \bigl(T_P^0\bigl(st'(k)\bigr) \xRightarrow{\mathcal{M}_{P_{jN}\circ\cdots\circ P_{j1}\circ P_{in}\circ\cdots\circ P_{i1}}} T_P^0\bigl(st'(l)\bigr)\bigr) \\
&= \bigl(T_P^0\bigl(st'(m)\bigr) \xRightarrow{\mathcal{M}_{P_{jN}\circ\cdots\circ P_{j1}}} T_P^0\bigl(st'(l)\bigr)\bigr) \circ \bigl(T_P^0\bigl(st'(k)\bigr) \\
&\qquad \xRightarrow{\mathcal{M}_{P_{in}\circ\cdots\circ P_{i1}}} T_P^0\bigl(st'(m)\bigr)\bigr) \\
&= T_P^1(f) \circ T_P^1(g).
\end{aligned}$$

The following commutative diagram demonstrates that τ_T is a natural transformation, for any arrow $P_1 \circ \_ : P \to P_2$ in P_A, where $P_2 = P_1 \circ P$ is sequential composition of the programs P and P_1:

$$\begin{array}{ccccc}
P & & \mathbf{R}_P = R_M(P) & \xrightarrow{T_P = \tau_T(P)} & \mathbf{DB}^{\mathbf{Sch}(G(\mathcal{A}))} = K_A(P) \\
P_1 \circ _ \downarrow & & in = R_M(P_1 \circ _) \downarrow & & \downarrow Id = K_A(P_1 \circ _) \\
P_2 & & \mathbf{R}_{P_2} = R_M(P_2) & \xrightarrow{T_{P_2} = \tau_T(P_2)} & \mathbf{DB}^{\mathbf{Sch}(G(\mathcal{A}))} = K_A(P_2)
\end{array}$$

Category $\mathbf{P}_{\mathcal{A}}$ Category **Cat**

In fact, from point 1, for any $st'(k) \in Ob_{\mathbf{R}_P} \subseteq Ob_{\mathbf{R}_{P_2}}$,

$$\begin{aligned}
T_{P_2}^0\bigl(st'(k)\bigr) &= Out_{DB}\bigl(st'\bigl(k - \mu y\bigl(t_2\bigl(st'(k-y)\bigr) \\
&\qquad * \bigl(t_3\bigl(st'(k-y)\bigr) + 1\bigr) - 1 = 0\bigr)\bigr)\bigr) \\
&= T_P^0\bigl(st'(k)\bigr).
\end{aligned}$$

And for any arrow $(g : st'(k) \to st'(m)) \in Mor_{\mathbf{R}_P} \subseteq Mor_{\mathbf{R}_{P_2}}$, $T_{P_2}^1(g) = T_P^1(g)$, and hence the diagram above commutes. □

Consequently, we have the following vertical composition $\tau_T \bullet \tau_E : U_M \stackrel{\bullet}{\longrightarrow} K_A$ of the "embedding" natural transformation $\tau_E : U_M \stackrel{\bullet}{\longrightarrow} R_M$ and of the "transaction" natural transformation $\tau_T : R_M \stackrel{\bullet}{\longrightarrow} K_A$,

$$\begin{array}{ccc}
 & U_M & \\
 & \xrightarrow{\qquad\qquad\qquad} & \\
 & \downarrow \tau_E & \\
\mathbf{P}_{\mathcal{A}} & \xrightarrow{\qquad R_M \qquad} & \mathbf{Cat} \\
 & \downarrow \tau_T & \\
 & \xrightarrow{\qquad\qquad\qquad} & \\
 & K_A &
\end{array}$$

where τ_E represents the embedding of the SQL statements from the host universal machine $\mathbf{M}_U$ into the categorial RDB machine $\mathbf{M}_R$ (by "forgetting" the parts of the program which are not the embedded SQL statements), while τ_T maps the embedded transitions in $\mathbf{M}_R$ into the RDBMS transactions.

Let, for a given source program P, the arrow in $\mathbf{U}_P = U_M(P)$, $f : st(j_1) \to st(j_2)$, be the sequence of atomic transitions of the compiled host program P_U between two consecutive operations 'COMMIT'. That is, with

$$\begin{aligned} &\big((i_1, st(j_1 - 1)), (i_2, \text{COMMIT}(st(j_1 - 1)))\big) \in \to_P \\ &\text{with } st(j_1) = \text{COMMIT}\big(st(j_1 - 1)\big), \end{aligned} \tag{6.6}$$

and

$$\begin{aligned} &\big((i_3, st(j_2 - 1)), (i_3, \text{COMMIT}(st(j_2 - 1)))\big) \in \to_P \\ &\text{with } st(j_1) = \text{COMMIT}\big(st(j_1 - 1)\big), \end{aligned} \tag{6.7}$$

in the computation system of $\mathbf{M}_U$. Then the states

$$st'(m) = F_P\big(st(j_1)\big) = \tau_E(P)\big(st(j_1)\big) \quad \text{and}$$
$$st'(k) = F_P\big(st(j_2)\big) = \tau_E(P)\big(st(j_2)\big)$$

are the two consecutive synchpoints in the RDB categorial machine $\mathbf{M}_R$ (because $t_2(st'(m)) * (t_3(st'(m)) + 1) = 1$ and $t_2(st'(k)) * (t_3(st'(k)) + 1) = 1$, with $st'(m) = st'(k - 1 - \mu y(t_2(st'(k - 1 - y)) * (t_3(st'(k - 1 - y)) + 1) - 1 = 0))$ and hence, from Definition 42, the transition from $st'(m)$ to $st'(k)$ is an *atomic transaction*.

Hence, the vertical composition of natural transformations above, for this source program P, is represented by the following diagram of functors in **Cat** on the left, and with the result of these functors on the right:

$$\begin{array}{lcc}
\mathbf{U}_P = U_M(P) & & st(j_1) \xrightarrow{\quad f \quad} st(j_2) \\
\quad \downarrow F_P = \tau_E(P) & & \downarrow F_P \\
\mathbf{R}_P = R_M(P) & & st'(m) = F_P(st(j_1)) \xrightarrow{F_P(f)} st'(k) = F_P(st(j_2)) \\
\quad \downarrow T_P = \tau_T(P) & & \downarrow T_P \\
\mathbf{DB}^{\mathbf{Sch}(G(\mathcal{A}))} = K_A(P) & & T_P(st'(m)) \xRightarrow{\mathcal{M}_{P_{i_n} \circ \cdots \circ P_{i_1}}} T_P(st'(m)) \\
\text{Category } \mathbf{Cat} & & \downarrow \\
 & \text{in } \mathbf{DB}, & A = \alpha^*(\mathcal{A}) \xrightarrow{\quad g \quad} A_1 = \alpha_1^*(\mathcal{A})
\end{array}$$

where $P_{i_n} \circ \cdots \circ P_{i_1}$ is the sequential composition of **RA**-arrows from $st'(m)$ to $st'(k)$ (as specified by point 2 of Theorem 11), and $\alpha^* = T_P(st'(m))$ and $\alpha_1^* = T_P(st'(k))$ and hence $(\alpha^* \xrightarrow{\mathcal{M}_{P_{i_n} \circ \cdots \circ P_{i_1}}} \alpha_1^*) = T_P(F_P(f))$.

The arrow $F_P(f)$ in $\mathbf{R}_P$ is an *atomic transaction*, while the bottom arrow $g : A \rightarrow A_1$ is a morphism in the **DB** category from the instance-database $A = Out_{DB}(st'(m))$ into the updated database $A_1 = Out_{DB}(st'(m))$ after this atomic transaction.

This morphism $g = g_n \circ \cdots \circ g_1$, where each g_j, $1 \leq j \leq n$, is a morphism obtained by Corollary 17 for a given schema mapping $\mathcal{M}_i = \{\Phi_i\} : \mathcal{A} \rightarrow \mathcal{A}$ and obtained by Proposition 26 for a given **RA**-algebra arrow P_{i_j} (in Sect. 5.3).

As a result, we obtain that each atomic transaction produces one morphism in the **DB** category, between the (consistent) database instances (that are the *models*) of a schema $\mathcal{A}$, before and after such an atomic transaction:

Corollary 21 *For a given database schema $\mathcal{A}$ and a source program P with SQL statements, each atomic transaction of this program P will result in a morphism in the* **DB** *category, between the models (i.e., consistent database instances) of a schema $\mathcal{A}$, before and after such an atomic transaction.*

This morphism is the result of a specific schema mapping from $\mathcal{A}$ into $\mathcal{A}$, thus it is representable by a categorial semantics of a database mapping system in Sect. 4.1.2.

Notice that this corollary is valid not only for all SQL program transactions (the data manipulation language DML), but also for the DDL functions, as CREATE TABLE and DROP TABLE, which change the database schemas as well:

1. In the case of a creation of a new table, for a relation $r \notin \mathcal{A} = (S_A, \Sigma_A)$, we my define a new schema $\mathcal{B} = (S_A \cup \{r\}, \Sigma_A)$ with the schema mapping $\mathcal{M}_{AB} = \{\Phi\} : \mathcal{A} \rightarrow \mathcal{B}$, where Φ is the SOtgd $\bigwedge_{r_i \in S_A} \forall \mathbf{x}_i (r_i^A(\mathbf{x}_i) \Rightarrow r_i^B(\mathbf{x}_i))$.
2. In the case of dropping of a table, i.e., of a relation $r \in \mathcal{A} = (S_A, \Sigma_A)$, we my define a new schema $\mathcal{B} = (S_A \backslash \{r\}, \Sigma_{A \backslash \{r\}})$ with the schema mapping $\mathcal{M}_{AB} = \{\Phi\} : \mathcal{A} \rightarrow \mathcal{B}$, where Φ is the SOtgd $\bigwedge_{r_i \in S_B} \forall \mathbf{x}_i (r_i^A(\mathbf{x}_i) \Rightarrow r_i^B(\mathbf{x}_i))$.

   Consequently, the whole story of the changes of a given database my be represented by the schema mappings, and their functorial denotational semantics based on **DB** category (in Sect. 4.1.2).

6.3 The Concurrent-Categorial RBD Machine

Most DBMSs are multiple-user (program) systems, that is, they are systems that allow any number of transactions to access the same database at the same time. We need the universal machine to provide $\mathbf{M}_U$ time-sharing features: each contemporary execution of programs is divided into a number of time-sharing windows where resource of machine is exclusively given to an particular application program, with the constraint that each conversation with the RDB machine $\mathbf{M}_R$ is completely closed inside such a time-window (atomic conversations in the time-sharing scenario). Hence each communication data between a universal and an RDB machine must contain also the identifier of the transaction ID_T. Moreover, each update of the database in the RDB machine $\mathbf{M}_R$ is marked by the transaction ID_T in the way that, when COMMIT or ROLLBACK is requested by some transaction ID_T, then only the updates made by that transaction are committed. As we will see, these new requirements will change the Input/Output functions of the universal and RDB machines, with respect to their previous definitions.

In such a system, some kind of *concurrency control mechanism* is needed to ensure that concurrent transactions do not interfere with each other's operation. Without such a mechanism, there are numerous problems that can arise; to avoid these problems, as most commonly encountered in practice, a concurrency control mechanism known as *locking* is used. The basic idea of locking is simple: when a transaction needs an assurance that some object that it is interested in—typically a database tuple—will not change in some unpredictable manner while its back is turned (as it were), it *acquires a lock* on that object. The effect of the lock is to "lock other transactions out of" the object, and thus in particular to prevent them from changing it. The first transaction is thus able to carry out its processing in the certain knowledge that the object in question will remain in an idle state for as long as that transaction wishes it to.

We assume throughout this section that X-locks (*exclusive locks*) and S-locks (*shared locks*) are the only kinds of locks available for categorial RDB machine $\mathbf{M}_R$.

If transaction $ID_T = N_a$ holds an X-lock on tuple $\mathbf{t}$, then a request from transaction $ID_T = N_b$ for a lock of either type on $\mathbf{t}$ will cause N_b to go into a wait state. N_b will wait until N_a's lock is released. The mechanism for such synchronization is as follows: The transaction N_b (of a program in a universal machine) makes CALLs for some embedded SQL statement and waits for response data from $\mathbf{M}_R$. If N_a's lock is not released, then the error flag which returns from $\mathbf{M}_R$ is different from 0 (in the same time-sharing window), and transaction N_b has to invoke again (possibly in a subsequent time-sharing window) the same CALLs (wait state). Only when N_a's lock is released (into some different time-sharing window) then the response for CALLs of the transaction N_b will return with the error flag equal to 0 and with requested data.

If transaction $ID_T = N_a$ holds (in the machine $\mathbf{M}_R$) an S-lock on tuple $\mathbf{t}$, then:

- A request from transaction $ID_T = N_B$ for an X-lock on $\mathbf{t}$ will cause N_b to go into a wait state (and N_b will wait until N_a's lock is released) as described above (by subsequent CALLs of the same embedded SQL statement, until response from $\mathbf{M}_R$ does not return with the error flag equal to 0);
- A request from transaction N_b for an S-lock on $\mathbf{t}$ will be granted (that is, N_b will now also hold an S-lock on $\mathbf{t}$).

When, in the categorial RDB machine $\mathbf{M}_R$, a transaction successfully retrieves a tuple, it automatically acquires an S-lock on that tuple. When a transaction successfully updates a tuple, it automatically acquires an X-lock on that tuple. In the categorial RDB machine $\mathbf{M}_R$, the X-locks of an transaction ID_T are held until the next synchpoint of the same transaction ID_T. S-locks are also normally held until that time.

Deadlock Locking can introduce problems of its own, principally the problem of deadlock. Deadlock is a situation in which two or more transactions are in simultaneous wait state, each one waiting for one of the others to release a lock before it can proceed. It is a responsibility of the universal machine program to detect it and break it. Breaking the deadlock involves choosing one of the deadlock transactions as the victim ID_T and rolling it back: invoking CALL of ROLLBACK of this ID_T toward the RDB machine $\mathbf{M}_R$, thereby releasing (in $\mathbf{M}_R$) its locks and so allowing some other transaction to proceed. Now we are in a position to see how locking is implemented in our categorial setting. Let us define a specialization of the universal (specified in Definition 39) and categorial RDB machines for the concurrent multiple-user/programs systems:

Definition 43 A concurrent multiple-user universal machine $\mathbf{M}_{UC}$ is a universal general purpose programming language machine $\mathbf{M}_U$ with the following specific features, where ID_T is the transaction identifier of the current transaction:

1. The extended input function In^*_{EU} satisfies the following properties:

$$\lambda In^*_{EU}(z) = \begin{cases} f : S_U \to S_U & \text{if } z = (n, m, ID_T, R) \in INP; \\ id : S_U \to S_U \,(\text{identity function}) & \text{otherwise,} \end{cases}$$

   where f is a mapping such that for a state $st \in S_U$, $f(st) \in S_U$ is obtained from st by the following setting: $x_1 = n, x_2 = m$, and the relation R is loaded into the subset of registers of S_U used as a dynamic data memory ('System/data/IO memory' of $\mathbf{M}_{UC}$ in Fig. 6.2).
2. The output function Out_U satisfies the following properties:

$$Out_U(st) = \begin{cases} \emptyset \text{ (empty set)} & \text{if } t_1(st) = 1; \\ (n, 0, ID_T, L) & \text{otherwise,} \end{cases}$$

   where L is a list of values in I/O memory of $\mathbf{M}_{UC}$ (used for 'INSERT...' and 'UPDATE...' SQL statements in the source program P) and n is the value of x_1. The output of this function is then transmitted to the concurrent RDB machine $\mathbf{M}_{RC}$.
3. The initial input value for the initial input function In_E contains also all programs (written for a database with a given schema) to be executed concurrently before a turn-off of the machine. They are loaded together with a Main program for time-shared Supervisor into its program memory.

A concurrent categorial RDB machine $\mathbf{M}_{RC}$ is a more complex version of the categorial RDB machine $\mathbf{M}_R$ (specified in Definition 40), with the new additional features:

Its Buffer Manager dedicates, for each existing transaction identifier $ID_T = N_a$ generated during the execution of some source program $P(i)$ in the concurrent universal machine $\mathbf{M}_{UC}$, a dynamic data memory where all **RA** arrows (obtained after variable bindings for the Application Plans $P_n, n \geq 3$, obtained by compilation from

SQL statements of this program $P(i)$ which are executed during this transaction ID_T) and the subset of updated relations of the current consistent database instance in data memory (obtained by $Out_{DB}(st'(k))$ in the current machine state $st'(k)$) are saved.

Consequently, the execution of these **RA** arrows obtained after the variable bindings use these "locally" updated relations by respecting X-locks on their tuples, while the rest of relations are taken in $Out_{DB}(st'(k))$.

Hence, the Buffer Manager manages the $n \geq 1$ "local" partitions of the data memory, one for each concurrent transaction, plus the last version of (committed) instance-database in $Out_{DB}(st'(k))$. In this way, each independent transaction (initiated by different time-shared source programs $P(i), i = 1, 2, 3, \dots$) saves its "local changes" (still non-committed, thus *possibly inconsistent*) independently from other transactions.

Each transaction ends with COMMIT or ROLLBACK statements, while the new transactions begin with CALL $n, n \geq 3$ statements by receiving a list $(n, 0, ID_T, L)$ from $\mathbf{M}_{UC}$ when there is still no partition for this ID_T, as follows:

- When a list $(n, 0, ID_T, L)$ is received from $\mathbf{M}_{UC}$:
  1. If there is still no partition for this ID_T, then the Buffer Manager creates a new partition for this new transaction ID_T, by binding variables for the Application Plan P_n and saving the obtained **RA** arrow in an internal list of this partition. Then this **RA** arrow is executed by using (and by respecting X-locks) the relations of the current consistent database in $Out_{DB}(st'(k))$. However, the resulting relation is saved only in this local partition, without changing the database in $Out_{DB}(st'(k))$.
  2. Otherwise, if there is a partition for this transaction, then a new **RA** arrow is generated by variable binding and it is saved at the end of the existing list of the previously generated **RA** arrows during this transaction. Then this **RA** arrow is executed by using (and by respecting X-locks) the locally updated relations in this partition and the rest of relations from $Out_{DB}(st'(k))$. The generated new relation will be saved locally in this partition, or will substitute the previous version of the same relation in this partition. Again, the data the database in $Out_{DB}(st'(k))$ is not changed.
- When a transaction $ID_T = N_a$ ends with a COMMIT, in the current machine state $st'(k)$, then all "locally" saved relations of this transaction are deleted and the ordered saved list of **RA** arrows are executed *again* but the resulting relations of the execution of each single **RA** arrow are directly saved in the current database instance in $Out_{DB}(st'(k))$. These executions guarantee that all updates during this transaction ID_T are done on the last consistent version of the database (because during this transaction $ID_T = N_a$ this instance-database may be changed by the COMMIT of some different transaction N_b of another concurrent program $P(j)$). At the end, all saved RA arrows of this transaction are deleted and this partition in data memory is eliminated.
- When a transaction $ID_T = N_a$ ends with a ROLLBACK then simply the partition of this transaction is eliminated (with all updated relations and **RA** arrows in it) and hence the current consistent database in $Out_{DB}(st'(k))$ remains unchanged.

Thus, differently from $\mathbf{M}_R$, in $\mathbf{M}_{RC}$, the current instance-database in $Out_{DB}(st'(k))$ is always a previously committed (thus, *consistent*) database.

Based on this more complex management of the concurrent transactions, the semantics for some operations of $\mathbf{M}_{RC}$ are modified w.r.t. their semantics for $\mathbf{M}_R$, as follows:

Definition 44 A concurrent-categorial RDB machine $\mathbf{M}_{RC}$ is a categorial RDB machine $\mathbf{M}_R$ with the following specific features:

1. – x_5 is the memory register used to save the number of the current transaction ID_T, whose continuation of execution is requested to $\mathbf{M}_{RC}$ by a CALL n.
2. VCON is the function which assigns 0 to x_2 if the database in current state $st'(k)$, obtained from $Out_{DB}(st'(k))$ by replacing its relations with locally updated relations in the partition with the transaction number in x_5, is consistent; 1 otherwise.
3. COM is the function which assigns 1 to x_3 and eliminates all local relation from the partition determined by transaction number in x_5. Hence, each **RA** arrow in the local list of this partition is executed in a given ordering by using the relations in the current database in $Out_{DB}(st'(k))$ and by modifying it with the obtained resulting relations. At the end of this process, we obtain a new state $st'(m)$ such that $Out_{DB}(st'(k))$ is the new consistent database instance resulting from this transaction, and it is written by the Buffer Manager in the physical storage memory as well. At the end, all locks are released and the partition with transaction number in x_5 is eliminated.
4. RBACK assigns 1 to x_3, all locks are released and the partition with transaction number in x_5 is eliminated.
5. The extended input function In^*_{ER} satisfies the following properties:

$$\lambda In^*_{ER}(z) = \begin{cases} f : S_R \to S_R & \text{if } z \text{ is } (n, m, ID_T, L) \text{ or a database;} \\ id : S_R \to S_R \text{ (identity function)} & \text{otherwise,} \end{cases}$$

   with function f such that for a state $st'(k) \in S_R$, $st'(k+1) = f(st'(k)) \in S_R$ is obtained from $st'(k)$ by the following setting: $x_1 = x_4 = n, x_2 = m, x_5 = ID_T$, and the list of values in L is loaded into the subset of registers of S_R used for the I/O memory if $z = (n, m, ID_T, L)$. Otherwise, if z is a database then it is loaded into the dynamic data-memory.
6. The output function Out_R satisfies the following properties:

$$Out_R\big(st'(k)\big) = \begin{cases} \emptyset \text{ (empty set)} & \text{if } t_1(st'(k)) = 0; \\ (0, n, ID_T, R) & \text{otherwise,} \end{cases}$$

   where ID_T is the value of x_5, R is a relation in I/O memory of $\mathbf{M}_{RC}$ (obtained by the last execution of a program P_k in the Application Plane) and n is the value of its register x_2, with $R = \bot$ if $n > 0$ or P_k is not obtained from an SQL statement 'SELECT...' in the source program P. If the output is a tuple $(0, n, ID_T, R)$ then it is transmitted to $\mathbf{M}_{UC}$.

6.3.1 Time-Shared DBMS Components

The synchronization process is represented by the same Fig. 6.2, while the DBMS components in Fig. 6.1 have some modified features, as follows:

The Bind compiler for $\mathbf{M}_{RC}$ is enriched with controls, to be inserted into compiled Application Plans, for embedded SQL statements—each CALL from a current transaction has to verify if there are lock-constraints caused by other transactions: if there is a blocking constraint then that application plan will not be executed but x_2 will be only imposed to 1 and a RETURN will be executed (when $\mathbf{M}_{UC}$ receives this error condition, it must wait, by repeating the same CALLs, until locks become released by some other transaction (in some subsequent time-shared window)).

Each execution of some program $P(m)$ on $\mathbf{M}_{UC}$ generates a number of transactions in one of the following cases: when a program begins the execution and for any execution of the statements COMMIT or ROLLBACK (in that case, the precedent transaction ID_T is closed, a new transaction ID'_T is open, and $Pr(ID_T) = Pr(ID'_T) = P(m)$ is imposed). At the beginning (initial state $st(0)$ of $\mathbf{M}_{UC}$) the function $Pr : N \rightarrow Ob_{\mathbf{P}_\mathcal{A}}$ for the given schema $\mathcal{A}$ supported by RDB machine is defined by $Pr(i) = P_\emptyset, i = 1, 2, 3, \ldots$, where $P_\emptyset$ is the empty program. We assume that each transaction identifier (during concurrent execution of a number of programs) is a unique number for a universal machine $\mathbf{M}_{UC}$: the first transaction has $ID_T = 1$ with $Pr(1)$ the first program to start the execution, and each next generated transaction increments this value:

- The main program of $\mathbf{M}_{UC}$ contains a Time-sharing Supervisor (in the basic time-shared window) that decides which application program

$$P(m) \in \{Pr(i) \mid 1 \leq i \leq K\},$$

  for $K = \min\{k \geq 1 \mid \forall j \geq k,\ Pr(j) = P_\emptyset\}$ (which did not reach the stop state) has to continue its execution in the next time-shared window (the next program instruction) and is also responsible for a management of the transaction identifiers and for a changing of the function $Pr : N \rightarrow Ob_{\mathbf{P}_\mathcal{A}}$ during run-time: when a transaction ID_T of the program $P(m) = Pr(ID_T)$ is finished then $Pr(ID_T) = P_\emptyset$ is imposed (it is imposed also when this program ends its execution) and hence there are no two different ID_T and ID'_T such that $Pr(ID_T) = Pr(ID'_T) \neq P_\emptyset$. The changing of the internal states of $\mathbf{M}_{UC}$, caused by this time-sharing supervisor, will not be considered (it is forgotten because it is not relevant) in the rest of this chapter.
- The management of the time-shared windows is as follows: each time that a new execution of some application program $P(m)$ begins, the supervisor opens a new window for such a program execution and sets $Pr(K) = P(m)$, $ID_T = K$ with $K = \min\{k \geq 1 \mid \forall j \geq k,\ Pr(j) = P_\emptyset\}$; when a program in a some time-shared window stops its execution (i.e., reaches a stop state of a program) this window is closed. If there is no application program execution then only the basic window for the main program of the Operation Multi-user system (mainly for a Time-shared Supervisor) remains open. The stop state of the main program corresponds to the turn-off for the universal machine $\mathbf{M}_{UC}$.

- By convention, each turn-on/turn-off of the $\mathbf{M}_{UC}$ determines a corresponding turn-on/turn-off of the RDB machine $\mathbf{M}_{RC}$.

Let us now introduce the operator of concurrent ("parallel", that is, time-shared) composition of application programs $P = P(1)|P(2)$ such that $P = P(1) \cup P(2)^*$ (where $P(2)^*$ is obtained from $P(2)$ by incrementing each address in its instructions by the integer $N + 1$, where N is the biggest address in $P(1)$), in the time-shared mode of the concurrent multiple-user universal machine $\mathbf{M}_{UC}$, by the following definition:

Definition 45 A *concurrent execution of a program* $P = P(1)|\cdots|P(n)$ on a machine $\mathbf{M}_{UC}$ is a computation system $CS = (TrS, Ins_S, Outs_S, Stop_S)$, where $TrS = (S_{PU}, \to_P)$ is a deterministic transition system with:

1. The set of configurations (obtained during the concurrent execution of programs): $S_{PU} = \{(i, st, ID_T) \mid i$ is the address of some instruction of the program $P(m) = Pr(ID_T)$, $1 \le m \le n$, executed during the transaction $ID_T \ge 1$, $st \in S_U$ is the state of $\mathbf{M}_{UC}$ before the execution of this ith instruction$\}$.
2. The $\to_P$ is a function defined for any time-shared window during the execution of a program $P(m)$ in the state $st \in S_U$ such that there is a transaction ID_T with $Pr(ID_T) = P(m)$ (if $P(m) \notin \{Pr(i) \mid 1 \le i \le K\}$, for $K = \min\{k \ge 1 \mid \forall j \ge k,\ Pr(j) = P_\emptyset\}$, then in this time-window the execution of this program $P(m)$ just begins and we set $Pr(K) = P(m)$ and $ID_T = K$), as follows:
   2.1. $((i, st, ID_T), (j, \lambda In^*_{EU}(z)(st), ID_T)) \in \to_P$ if $(i, In^*_{EU}, j) \in Pr(ID_T)$ is executed in this time-window, where $z \in INP$ is the input value taken by $\mathbf{M}_{UC}$ in this current state $st \in S_U$ and λ is the abstraction operator (currying) of λ-calculus such that $\lambda In^*_{EU}(z)(st) = I^*_{EU}(z, st)$.
   2.2. $((i, st, ID_T), (j, f(st), ID_T)) \in \to_P$, if $(i, f, j) \in Pr(ID_T)$ is executed in this time-window and $f \notin \{\text{COMMIT}, \text{ROLLBACK}\}$.
   If there is no instruction with the address j (we reached the end of this program) then $((j, f(st), ID_T), (j, f(st), ID_T)) \in \to_P$, and this time-window is closed by imposing $Pr(ID_T) = P_\emptyset$.
   2.3.
$$\big((i, st, ID_T), \big(j, f(st), ID_T\big)\big) \in \to_P$$
$$\big((j, f(st), ID_T), \big(\big(k, Out_U\big(f(st)\big)\big), ID_T\big)\big) \in \to_P,$$
   if $(i, f, j), (j, Out_U, k) \in Pr(ID_T)$ are executed in this time-window and $f \in \{\text{COMMIT}, \text{ROLLBACK}\}$. Set $Pr(K) = Pr(ID_T)$ with $K = \min\{k \ge 1 \mid \forall j \ge k,\ Pr(j) = P_\emptyset\}$, and after that set $Pr(ID_T) = P_\emptyset$ and $ID_T = K$.
   2.4. $((i, st, ID_T), (j, st, ID_T)) \in \to_P$, if $(i, t, j, k) \in Pr(ID_T)$ is executed in this time-window and $t(sk) = 1$,
   2.5. $((i, st, ID_T), (k, st, ID_T)) \in \to_P$, if $(i, t, j, k) \in Pr(ID_T)$ is executed in this time-window and $t(sk) = 0$.
3. $Ins_S(x) = (0, In_U(x), 1)$.
4. $Outs_S(i, st, ID_T) = Out_U(st)$.

5. $Stop_S : S_P \to \{0, 1\}$ is a total stop-criteria function such that

$$Stop_S(i, st, ID_T) = 0 \text{ if } i \text{ is not the address of any instruction of } Pr(ID_T).$$

6.3.2 The Concurrent Categorial Transaction Recovery

In what follows, the concurrent-categorial RDB machine $\mathbf{M}_{RC}$ will be considered as a computation model for the operational semantics for the database mappings, developed in the next chapter. In fact, in this section, we will demonstrate such a property by the final Corollary 22.

We may abstract a deterministic transition system $TrS = (S_{PU}, \to_P)$ in Definition 45, obtained by a concurrent execution of a program $P = P(1)|\cdots|P(n)$ on a machine $\mathbf{M}_{UC}$, by forgetting the instruction addresses and Transaction identifiers, that is, by substituting the transitions $((j, st(i), ID_T), (k, f(st(i)), ID_T)) \in \to_P$ by simple abstracted state-transitions $(st(i), f(st(i))) \in \to_P$. Let $ATrS_U = (\pi_2(S_{PU}), \to_P)$ be the abstracted deterministic transitions system obtained from the execution of the concurrent program $P = P(1)|\cdots|P(n)$ on a host machine $\mathbf{M}_{UC}$, with $\pi_2(S_{PU}) = \{st(i)|0 \le i \le N\} \subseteq S_U$ being the set of its internal states progressively numbered from 0 (initial state) to the last Nth state (end of this program).

Consequently, we are able to use the same synchronization mapping algorithm $Smap(ATrS_U, ATrS_R)$ in Sect. 6.2.1 in order to define the mapping

$$F_P^0 : \pi_2(S_{PU}) \to \pi_2(S_{PU}).$$

Hence, also for the concurrent programs $P = P(1)|\cdots|P(n), n \ge 1$, we can use Proposition 29 in order to define the categories $\mathbf{U}_P$ and $\mathbf{R}_P$, and Fig. 6.4 in Sect. 6.2.1 is valid also in this case for the "forgetful functor" $F_P : \mathbf{U}_P \to \mathbf{R}_P$ because each conversation between $\mathbf{M}_{UC}$ and $\mathbf{M}_{RC}$ is completely contained in a given time-window (the atomic conversations in the time-sharing scenario).

Consequently, we are able to define a category with the sequential composition of concurrent-programs, with the embedded SQL statements over a given database schema, by the following generalization of Definition 41:

Definition 46 Let $\mathcal{A}$ be a database schema. We define a category $\mathbf{P}_{\mathcal{A}}$ where the objects are the sequential compositions of the source programs with SQL statements over this schema $\mathcal{A}$, while the arrows are the actions $P_0 \circ \_ : P_1 \to P_2$ such that $P_2 = P_0 \circ P_1$ is the program obtained by sequential composition of $P_1 = (P_1(1)|\cdots|P_1(n)), n \ge 1$, and $P_0 = (P_0(1)|\cdots|P_0(m)), m \ge 1$.

Each identity arrow in $\mathbf{P}_{\mathcal{A}}$ is equal to $P_{\text{NOP}} \circ \_$ where P_{NOP} is the program composed of only the instruction NOP. The associativity of composition of the arrows is guaranteed by the associativity of the sequential composition of the programs, thus $\mathbf{P}_{\mathcal{A}}$ is well defined.

In the case when $n = m = 1$, we obtain the case described by Definition 41 for the sequential composition on non-concurrent programs.

The sequential composition $P_2 = (P_0(1)|\cdots|P_0(m)) \circ (P_1(1)|\cdots|P_1(n))$ means that the execution of all concurrent programs $P_1(i), 1 \leq i \leq n$ has to be finished before the beginning of the execution of the set of concurrent programs $P_0(i), 1 \leq i \leq m$.

Consequently, the definitions of the functors $U_M : \mathbf{P}_{\mathcal{A}} \to \mathbf{Cat}$ and $R_M : \mathbf{P}_{\mathcal{A}} \to \mathbf{Cat}$ with the "embedding" natural transformation $\tau_E : U_M \stackrel{\centerdot}{\longrightarrow} R_M$ in Corollary 20 are valid also for the concurrent programs $P = P(1)|\cdots|P(n), n \geq 1$.

In fact, for any arrow $P_0 \circ \_ : P_1 \to P_2$, with $P_1 = (P_1(1)|\cdots|P_1(n)), n \geq 1$, and $P_0 = (P_0(1)|\cdots|P_0(m)), m \geq 1$, in $\mathbf{P}_{\mathcal{A}}$, with $P_2 = P \circ P_1$, we have the following commutative diagram of functors in $\mathbf{Cat}$:

$$\begin{array}{ccc}
U_M(P_1) = \mathbf{U}_{P_1} & \xrightarrow{\;in = U_M(P_0 \circ _)\;} & U_M(P_2) = \mathbf{U}_{P_2} \\
\tau_E(P_1) = F_{P_1} \downarrow & & \downarrow \tau_E(P_2) = F_{P_2} \\
R_M(P_1) = \mathbf{R}_{P_1} & \xrightarrow{\;in = R_M(P_0 \circ _)\;} & R_M(P_2) = \mathbf{R}_{P_2}
\end{array}$$

Notice that the concept of the atomic transaction for the concurrent RDB machine $\mathbf{M}_{RC}$ is equal to that defined for the machine $\mathbf{M}_R$, given by Definition 42. The only difference is that such atomic transactions can be generated by different concurrent programs, while for $\mathbf{M}_R$ they are generated by only one host program.

Consequently, we have the same definition of the constant functor $K_A : \mathbf{P}_{\mathcal{A}} \to \mathbf{Cat}$ and the same "transaction" natural transformation $\tau_T : R_M \to K_A$ such that for each program $P = P(1)|\cdots|P(n), n \geq 1$, $K_A^0(P) = \mathbf{DB}^{\mathbf{Sch}(G(\mathcal{A}))}$. Thus, we obtain a functor $T_P = \tau_T(P) : \mathbf{R}_P \to \mathbf{DB}^{\mathbf{Sch}(G(\mathcal{A}))}$ as it was specified by Theorem 11 and hence, the image of this functor $im(T_P) \subseteq Mod(G(\mathcal{A})) \subseteq \mathbf{DB}^{\mathbf{Sch}(G(\mathcal{A}))}$ is the set of all transactions generated by the concurrent programs, between the models of database schema $\mathcal{A}$ (the object component T_P^0 generates all instance-databases modified by the concurrent programs, while the arrow component T_P^1 generates the transactions between them).

Thus, also for the concurrent time-sharing execution of the programs in $\mathbf{M}_{UC}$ and $\mathbf{M}_{RC}$, we obtain the following vertical composition $\tau_T \bullet \tau_E : U_M \stackrel{\centerdot}{\longrightarrow} K_A$ of the "embedding" natural transformation $\tau_E : U_M \stackrel{\centerdot}{\longrightarrow} R_M$ and of the "transaction" natural transformation $\tau_T : R_M \stackrel{\centerdot}{\longrightarrow} K_A$,

$$\begin{array}{ccc}
 & U_M & \\
 & \downarrow \tau_E & \\
\mathbf{P}_{\mathcal{A}} & \longrightarrow R_M \longrightarrow & \mathbf{Cat} \\
 & \downarrow \tau_T & \\
 & K_A &
\end{array}$$

where τ_E represents the embedding of the SQL statements from the host universal concurrent machine $\mathbf{M}_{UC}$ into the concurrent categorial RDB machine $\mathbf{M}_{RC}$ (by "forgetting" the parts of the program which are not the embedded SQL statements), while τ_T maps the embedded transitions in $\mathbf{M}_{RC}$ into the RDBMS transactions.

The vertical composition of natural transformations above, for a given concurrent program $P = P(1)|\cdots|P(N)$, $N \geq 1$, is represented as in non-concurrent case in Sect. 6.2.2 by the following diagram of functors in **Cat** on the left and with the result of these functors on the right:

$$
\begin{array}{lll}
\mathbf{U}_P = U_M(P) & & st(j_1) \xrightarrow{\;f\;} st(j_2) \\
F_P = \tau_E(P) \;\Big\downarrow & & \downarrow F_P \\
\mathbf{R}_P = R_M(P) & & st'(m) = F_P(st(j_1)) \xrightarrow{F_P(f)} st'(k) = F_P(st(j_2)) \\
T_P = \tau_T(P) \;\Big\downarrow & & \downarrow T_P \\
\mathbf{DB}^{\mathbf{Sch}(G(\mathcal{A}))} = K_A(P) & & T_P(st'(m)) \xRightarrow{\mathcal{M}_{P_{i_n}\circ\cdots\circ P_{i_1}}} T_P(st'(m)) \\
\text{Category } \mathbf{Cat} & & \downarrow \\
& \text{in } \mathbf{DB}, & A = \alpha^*(\mathcal{A}) \xrightarrow{\;g\;} A_1 = \alpha_1^*(\mathcal{A})
\end{array}
$$

where $P_{i_n} \circ \cdots \circ P_{i_1}$ is the sequential composition of **RA**-arrows (contained in one atomic transaction) from $st'(m)$ to $st'(k)$ (as specified by point 2 of Theorem 11) and $\alpha^* = T_P(st'(m))$ and $\alpha_1^* = T_P(st'(k))$, and hence $(\alpha^* \xRightarrow{\mathcal{M}_{P_{i_n}\circ\cdots\circ P_{i_1}}} \alpha_1^*) = T_P(F_P(f))$.

The arrow $F_P(f)$ in $\mathbf{R}_P$ is an *atomic transaction*, while the bottom arrow $g : A \to A_1$ is a morphism in the **DB** category from the instance-database $A = Out_{DB}(st'(m))$ into the updated database $A_1 = Out_{DB}(st'(m))$ after this atomic transaction.

This morphism $g = g_n \circ \cdots \circ g_1$, where each g_j, $1 \leq j \leq n$, is a morphism obtained by Corollary 17 for a given schema mapping $\mathcal{M}_i = \{\Phi_i\} : \mathcal{A} \to \mathcal{A}$ and obtained by Proposition 26 for a given **RA**-algebra arrow P_{i_j} (in Sect. 5.3).

As a result, we obtain that each atomic transaction produce one morphism in **DB** category, between the (consistent) database instances (that are models) of a schema $\mathcal{A}$, before and after such an atomic transaction:

Corollary 22 *For a given database schema $\mathcal{A}$ and a concurrent program*

$$P = P(1)|\cdots|P(n), \quad n \geq 1,$$

with SQL statements, each atomic transaction of this program P will result in a morphism in the **DB** *category, between the models (i.e., consistent database instances) of a schema $\mathcal{A}$, before and after such an atomic transaction.*

This morphism is the result of a specific schema mapping from $\mathcal{A}$ into $\mathcal{A}$, thus it is representable by a categorial semantics of a database mapping system in Sect. 4.1.2.

Notice that this corollary is valid not only for all SQL program transactions (the data manipulation language DML), but also for the DDL functions, as CREATE TABLE and DROP TABLE, which change database schemas as well. Consequently, the whole story of changes of a given database, in the concurrent-programs framework, my be represented by the schema mappings, and their functorial denotational semantics based on the **DB** category (in Sect. 4.1.2).

6.4 Review Questions

1. Definition 35 provides an input-extended abstract computation machine **MI** by introducing the (partial) input function In_E. Why is it important for the programs with embedded Σ_{RE} algebra terms, used as a database-sublanguage in order to be able to manipulate the data in a relational form and update an instance database? Why have we extended this (partial) input function into the total function In_E^*, from an algebraic point of view for computation machines and/or from the point of view of Computation Systems?
2. Why are we using the class of μ-recursive functions defined inductively by Kleene for the "universal machines"? What is your explanation for why we preferred a class of the input-extended Register machines **RMI** in Definition 38 (for which we needed to demonstrate that it is a universal machine as well), instead of the well known (in the theory of computations) Turing machines? An exercise: If you chose alternatively the Turing machines, show that the rest of the results in this chapter will be preserved as well.
3. In the rest of this chapter, we separated the universal machine $\mathbf{M}_U$, able to execute any computation program, from the Categorial Database Machine $\mathbf{M}_R$ dedicated to maintaining and updating the database instances. What is exactly the computation power of this DB machine (see the last section of the previous chapter)? Can you define a less powerful abstract categorial DB machine for the only Codd's SPJRU algebra and the corresponding subcategory of the **DB** category (consider only the first row of the Table 6.1 after Fig. 6.1 of the proposed relational DBMS)?
4. The Bind component in Fig. 6.1 produces an application plan, by transforming the original Σ_{RE} terms into the morphisms of the **RA** category, and by substituting the original algebra operators used in the previous chapter with the variable-binding operators. Why is it necessary? Provide an example and explain how it works, for example, for the term t_R which was used in Examples 28 and 33.
5. Why is the Categorial RDB machine in Definition 40 able to provide any kind of updating of a given instance database? Can it be used for the concurrent database updating from a number of concurrent programs and, if not, what else do we

need to satisfy these requirements? An exercise: reduce this machine in order to be able to execute only the terms of the Codd's SPJRU algebra Σ_R, and not to make any update of a given database instance.

6. Why is Theorem 10 a fundamental result for the database mapping denotational semantics based on the **DB** category? Would it be possible to define the class of abstract database machines in order to support the operational semantics for the Big Data Integration (the operational semantics will be formally provided in next chapter) without it? If yes, provide an alternative class and explain what will be the difference from this one.
7. Is the categorial approach to SQL embedding (based on the abstraction of the synchronization process in Fig. 6.3 by using the algorithm *Smap*) sufficient to represent the fundamental phases of the general database updating, with two basic processes: variable binding and execution? Take Example 34 (where the three basic atomic update operations are presented: updating (i), inserting (ii) and deleting (iii) of tuples in a given instance database $\alpha(\mathcal{A})$ of a schema $\mathcal{A}$) for the variable binding and execution in the Categorial Database Machine $\mathbf{M}_R$. How is this database updating synchrony process represented by the "forgetful" functorial semantics? What is the meaning of the "embedding" natural transformation τ_E?
8. For a given database schema $\mathcal{A}$, expressed by its sketch category $\mathbf{Sch}(G(\mathcal{A}))$ and its category of interpretations $Int(G(\mathcal{A}))$, explain what is a *transition* from one to another instance of this database, w.r.t. Example 34. How is the concept of an atomic transaction for a Categorial RDB machine $\mathbf{M}_R$ represented? Explain the meaning of the vertical composition of the "embedding" and "transaction" natural transformations, provided by this diagram

$$\begin{array}{ccc} & \xrightarrow{\quad U_M \quad} & \\ & \downarrow \tau_E & \\ \mathbf{P}_{\mathcal{A}} & \xrightarrow{\quad R_M \quad} & \mathbf{Cat} \\ & \downarrow \tau_T & \\ & \xrightarrow[\quad K_A \quad]{} & \end{array}$$

and how it determines the morphisms in the **DB** category, between two consistent instance databases of a given updated schema, as a basic transition which can be used for a definition of the Operational semantics of the database mapping systems.
9. Which kind of the concurrent control mechanism is used in order to define the concurrent-categorial RDB machines $\mathbf{M}_{RC}$? What are the principal modifications of the non-concurrent RDB machines in order to obtain these concurrent machines? Why do we need a different definition for the composition of user programs in this concurrent framework, and how is its semantics represented during a concurrent execution of a number of programs? Explain the semantics of the vertical composition of the "embedding" and "transaction" natural transformations, provided by the diagram above, but in this new concurrent setting.

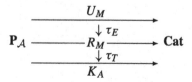

Reference

1. E. Börger, D. Rödding, *Berechenbarkeit, komlexität, logik. eine einfuhrung in algorithmen, sprachen und kalküle unter besonderer berücksichtigung ihrer komplexität* (F. Vieweg, Wiesbaden, 1986)

Operational Semantics for Database Mappings 7

7.1 Introduction to Semantics of Process-Programming Languages

In computer science, the process calculi (or process algebras) are a diverse family of related approaches to formally modeling concurrent systems. Process calculi provide a tool for the high-level description of interactions, communications, and synchronization between a collection of independent processes. They also provide algebraic laws that allow process descriptions to be manipulated and analyzed, and permit formal reasoning about equivalences between processes (e.g., using bisimulation). Leading examples of process calculi include Milner's Calculus of Communicating Systems (CCS) [34] and Hoare's theory of Communicating Sequential Processes (CCP) [18]. Both of them share the premise that the meaning of a process is fully determined by a synchronization tree, namely, a rooted, unordered tree whose edges are labeled with symbols denoting basic actions (or events). These trees are typically specified by a Structural Operational Semantics (SOS) [35] and so are, in fact, recursively enumerated trees. Two CCS process are distinguished according to an interactive game-like protocol called bisimulation, so that indistinguishable CCS processes are said to be bisimilar. Both CCS and CSP have the so-called silent actions, which in our case do not exist (basic actions are a set of relations obtained by computation of queries over non-empty models of database schemas), so that in the absence of silent action, and in the finitely branching processes (in database mapping systems, the number of inter-schema mappings from a given database schema toward another database schemas is always finite) we obtain the so-called Structural Operational Semantics with Guarded recursion (GSOS) in [7]. In GSOS, we have that bisimulation corresponds to the congruence w.r.t. all operations of CCS (here we will use only the parallel binary composition '$\|$' and action prefixing unary operations '$a.\_$' described in what follows), and Milner has argued extensively that in this case bisimulation yields the finest appropriate notion of the behavior of concurrent processes based on synchronization trees.

Z. Majkić, *Big Data Integration Theory*, Texts in Computer Science, DOI 10.1007/978-3-319-04156-8_7,

The semantics of a programming language is called *compositional* when the meaning of compound programs can be derived from the meaning of their subcomponents. A complete account of the meaning of programming language requires both an operational and denotational semantics:

- *Denotational semantics* is the mathematical interpretation which is better suited for reasoning about *programs* because of its *modularity*. The programs from the initial Σ_P-algebra, where Σ_P is the signature of the language corresponding to the basic constructs, and the corresponding unique homomorphism from the programs to the *denotational model* is called *initial algebra semantics* [17].
- *Operational semantics* of a programming language accounts for a formal description of its executable *behaviors* which should be *observable*. The *semantic domain*, carrier of the denotational model, is the final $\mathcal{B}_P$-coalgebra of the 'behavior' endofunctor $\mathcal{B}_P$ (final solution of a domain equation $X \cong \mathcal{B}_P(X)$). *Intended operational model* of a language based on the transition relation may also be seen as a $\mathcal{B}_P$-coalgebra: the unique homomorphism from this intended operational model ($\mathcal{B}_P$-coalgebra) to the semantic domain (final $\mathcal{B}_P$-coalgebra) is called the *final coalgebra semantics* [2, 36].
- *Semantic adequateness*. Both kinds of semantics should be related in such a way that one should be able to infer from the denotational semantics the operational behavior of the programs when the initial algebra and the final coalgebra semantics coincide [36]. The mathematical *structural* approach to operational semantics [35] is that every structural operational semantics (SOS) coinduces a denotational semantics, i.e., every SOS is *compositional*: This is achieved by defining operational semantics in terms of abstract, mathematical notions of *syntax* and *behavior*.

Bridging the gap between the operational and the denotational aspects of semantics is based on a suitable interplay between the standard *induction* principle and the dual *coinduction* principle (note that both principles are supported by the **DB** category, as demonstrated in the next chapter, Sects. 8.3.1 and 8.3.2, respectively).

The idea of coupling the initial algebra with coalgebra semantics was used in [36] to systematically derive denotational models from structural operational semantics [35]: the notion of observational equivalence (strong bisimulation) is a *congruence* if suitable restrictions are imposed on the syntactic format of the rules: among the rules in this format, 'abstract GSOS' rules and (negative) tree rules are known best. It was proven [39] that the structural rules correspond to the germ of an inductive functorial semantics, that is, they can be seen as an *action* of the syntax on the composite functor $\mathcal{B}_P \circ \mathcal{T}_P$, for abstract notions of *syntax* $\mathcal{T}_P$ (a monad freely generated by signature Σ_P) and *behavior* $\mathcal{B}_P$.

The fundamental results of this functorial operational semantics (FOS), created for the base **Set** category, are used to develop a new abstract FODS (functorial operational database-mapping semantics) concepts based on the denotational semantics of the database category **DB**.

Any given nonempty set of states in S and a given set of atomic actions in *Act* can be used to define a labeled transition system (LTS), used in structural operational semantics, as follows:

Definition 47 A transition is a triple $s \xrightarrow{a} s'$ with $a \in Act$, where $s, s' \in S$. An LTS is a (possibly infinite) set of transitions.

An LTS is finitely branching if each of its states has only finitely many outgoing transitions.

General Preliminary Considerations In this section, we consider the programs for execution of the schema-database mappings in a given schema database-mapping system given by a graph G as defined in Definition 14 of Sect. 2.6. Obviously, the programming language for the mapping system in a graph G is different from the low-level programming "select–project–join+union" language (SPJRU language [1], with the signature Σ_R used in order to define the queries $q(\mathbf{x})$ in the implications of an SOtgd Φ of each schema-database mapping $\mathcal{M}_{AB} = \{\Phi\} : \mathcal{A} \to \mathcal{B}$ (see Proposition 27 and Corollaries 18, 19 in Sect. 5.4).

Consequently, the programming language signature Σ_P here is different from the relational algebra signatures Σ_R and Σ_{RE} (in Definition 31) and used for algebraic and coalgebraic considerations in Sect. 8.3. The signature Σ_P will have, as we will see in what follows, a number of constants (i.e., actions in *Act*), only one unary operation to stop execution of programs, and two binary operators for the sequential and parallel execution of the database mapping programs. Now we can define a process graph, as follows:

Definition 48 A process graph P is an LTS in which one state $s \in S$ is elected to be the root. If the LTS contains a transition $A \xrightarrow{a} B$, then $P \xrightarrow{a} P'$ where P' has a root state s'.

- A process P_0 is finite if there are only finitely many sequences

$$P_0 \xrightarrow{a_1} P_1 \xrightarrow{a_2} \cdots \xrightarrow{a_k} P_k.$$

- A process P_0 is regular if there are only finitely many processes P_k such that $P_0 \xrightarrow{a_1} P_1 \xrightarrow{a_2} \cdots \xrightarrow{a_k} P_k$.

The signature of a basic framework for a database-mapping *process algebra* consists of a given set of states S and the following operators:

- First of all, we assume a finite nonempty set *Act* of atomic actions, representing indivisible behavior. Each action $a \in Act$ is constant (nullary operation) that can execute itself, after which it terminates successively.
- We assume a binary operator '$\|$' called *branching* (parallel) composition. If two terms t_1 and t_2 represent processes P_1 and P_2, respectively, then the term $t_1 \| t_2$ (the infix notation for $\|(t_1, t_2)$) represents the process that executes in parallel both processes P_1 and P_2. In other words, the process graph of $t_1 \| t_2$ is obtained by joining P_1 and P_2 at their root states. It is a commutative and associative operator.
- We assume a set of unary operators '$a.\_$' called *action prefixing* composition with $a \in Act$. If a term t_1 represents a process P_1 and $a \in Act$ is an action then the term $a.t_1$ represents the process that executes the action a and then the process

P_1. In other words, the process graph of $a.t_1$ is obtained by joining the edge a with the tree P_2. It is an associative but not a commutative operator, with right distributivity $a.(t_1 \| t_2) = a.t_1 \| a.t_2$.

In what follows, we can alternatively consider only one binary two sorts *sequential composition* operator '.', with the first argument in *Act* and the second argument being a term and the infix notation '$.(a, t_1)$'.

- Finally, we assume the nullary operator 'nil', that ends the process at a given state.

Consequently, in what follows, the program language for a data mapping system of a graph G is given by a general GSOS_{DB} grammar (with terms t), with the set *Act* of actions that can be *infinite* as well

$$(\mathrm{GSOS}_{DB}) \quad t := p|a|nil|a.t_1|t_1 \| t_2,$$

where 'nil' is a basic inert program nullary symbol (stop of execution constant), 'a.' is a unary action-prefixing operator symbol of programs, '$\|$' is a binary parallel composition symbol of programs, $a \in Act$ is an atomic action, and $p \in X$ is a program variable.

Notice that, differently from CCS, we do not use the 'alternative composition' but only the parallel composition '$\|$'. In the database mapping processes, the term $t\|t$ is not equal to t (in DB-denotational semantics, we will have that ground terms (without free variables) are instance-databases and '$\|$' denotes the coproduct in the **DB** category while '.' denotes the matching operator $\otimes$: two identical instance-databases with two mutually independent DBMS (i.e., $A + A$) are not equal (isomorphic in the **DB** denotational semantics) to the single database A with its DBMS), but the associative and commutative axioms are valid, and $t\|nil = t$ is an axiom as well.

Notice that $\|$ is a commutative and associative operator, so that $t_1 \| t_2 = t_2 \| t_1$ and $t_1 \| (t_2 \| t_3) = (t_1 \| t_2) \| t_3 = t_1 \| t_2 \| t_3$. Consequently, we can introduce the finite n-ary, $n \geq 1$, generalization '$\|^n$' for it (with '$\|^2$' equal to '$\|$' and '$\|^1$' being the identity operator), in the prefix forms $\|^1(t) = t$ and for $n \geq 2$, $\|^n(t_1, \dots, t_n) = t_1 \| \cdots \| t_n$.

In what follows, we will define a specific interpretation of this simple grammar for our case of database-mapping programs.

7.2 Updates Through Views

Given an update on a view of a relational database, how should it be translated to updates on the underlying database? The problem of updates through relational views has been studied for more than 20 years by the database community [5, 13, 20, 38].

Ideally, view updates must be translated to updates on the base tables (relations) in a way that the views state after the update is the same we would have gotten if the update has been applied to a materialized view instance.

Attention has been focused on updates through select–project–join views since they represent a common form of a view that can be easily reasoned about using key and foreign key information.

There are several choices of techniques that could be used to translate from updates on relational views to updates on the underlying relational database. Some consider a translation to be correct if it does not affect any part of the database that is outside of the view [5, 22]. Others consider a translation to be correct as long as it corresponds exactly to the specified update, and does not affect anything else in the view [14] (*side-effect free* relational view update).

The users specify an update *Upd* which may be an insertion or deletion (update is a combination of both of them) on the view instance $\|q_{B_i}(\mathbf{x}_i)\|_B$ of an instance-database $B = \alpha^*(\mathcal{B})$ of a database schema $\mathcal{B}$, where the R-algebra α^* is an mapping-interpretation (obtained from a given Tarski's interpretation by Definition 11) that is a model of $\mathcal{B}$. The goal is to find an update $\mathcal{W}$ on the base relational tables of the underlying database $\mathcal{B}$ that results in a new instance-database $B' = \alpha_1^*(\mathcal{B})$ (where α_1^* is a new mapping-interpretation which is a new model of $\mathcal{B}$), with a view $\|q_{B_i}(\mathbf{x}_i)\|_{B'}$ that implements the view update.

The update is said to be implemented without side-effects (i.e., side-effect free) in the view if $Upd(\|q_{B_i}(\mathbf{x}_i)\|_B) = \|q_{B_i}(\mathbf{x}_i)\|_{B'}$, that is, no other view tuple should be affected by the base tables modification apart from the one specified in the view update command, and no additional tuple should appear in the view after the base tables have been modified.

Generally, we have the following cases for the side-effect free updates:

- *Insertions*. An insertion over a nested relational algebra (NRA) view is a side-effect free when the corresponding relational view of a query $q_{B_i}(\mathbf{x}_i)$ is a select–project–join view, the primary and foreign keys of the source relations of $q_{B_i}(\mathbf{x}_i)$ are in the view and the joins are made only through foreign keys.
- *Deletions*. Deletions over an NRA view of a query $q_{B_i}(\mathbf{x}_i)$ are side-effect free when $q_{B_i}(\mathbf{x}_i)$ is well-nested. By well-nested we mean that the source relations in $q_{B_i}(\mathbf{x}_i)$ must be nested according to key–foreign key constraints in the underlying relations.

The algorithms for the side-effect free updates through views can be found in [20], for example.

Unfortunately, existing literature on *updates through views* has shown that for many common cases there may be no side-effect free translation [14, 20, 37].

This led researchers to permit side-effects, to develop algorithms to detect them [33], or to restrict the kind of updates that can be performed on a view [20].

If we cannot find a unique exact update (i.e., side-effect free) to underlying database instance $B = \alpha^*(\mathcal{B})$, an alternative is to find a *minimal* update to B that will cause the specified update to the view V of a query $q_{B_i}(\mathbf{x}_i)$.

In [8], two ways of measuring minimality have been considered: the first is by the number of changes in the tables in B; the second is by the number of side effects that changes in B cause in the view of the query $q_{B_i}(\mathbf{x}_i)$ (in addition to the required update).

In what follows, for our needs we will consider the second kind of the minimality for the side-effects.

Here, as follows, we will permit the general case of updating with minimal side-effects as well in the following two cases: Deletion and Insertion trough views.

The Update trough views is a combination of Deletion and, successively, of Insertion of updated information. Consequently, it will be considered successively in the section on operational semantics for database mapping systems. In that section, we will consider the trees of LTS that have a root in a modified database of a given database mapping graph G. Such process trees are of two kinds: one is based on forward propagation in the graph G that begins in the root database of LTS in which new tuples (such trees can be infinite as well) are inserted; another is based on the backward propagation from the root database of LTS in which a number of tuples are deleted.

The process tree for updates of a given root database is then the composition of the deletion (backward) tree (updated tuples are deleted) and of the insertion tree (new updated tuples are inserted in this root database).

7.2.1 Deletion by Minimal Side-Effects

In a deletion minimization, the input is a database $B = \alpha^*(\mathcal{B})$ of a schema $\mathcal{B}$ with its model α^*, a query $q_{B_i}(\mathbf{x}_i)$, a view $V = \|q_{B_i}(\mathbf{x}_i)\|_B$, and a set of tuples to be deleted $t \subseteq V$.

The view side-effect minimization is an algorithm for providing a set $\Delta B \subseteq B$ in order to minimize $|\Delta V|$ such that for the same query executed over updated database $B' = B \backslash \Delta B$ (which is another model of database schema $\mathcal{B}$) the obtained view is $V \backslash (\Delta V \cup t)$.

In other words, we wish to find a set of tuples in B whose removal will delete t while minimizing the number of other tuples deleted from the view V.

We can define this *view deletion translation* (VDT) as follows:

Definition 49 Given a view $V = \|q_B(\mathbf{x})\|_B$ of a database $B = \alpha^*(\mathcal{B})$ (i.e., a model of the schema $\mathcal{B}$) and a deletion request $-t$, where $t \subseteq V$, we say that a database deletion ΔB is a translation for $-t$ if ΔB causes the deletion of t from V when applied to database B. More formally, let $V' = \|q_B(\mathbf{x})\|_{B'}$ be the new view, obtained from the minimally updated database $B' = B \backslash \Delta B = \alpha_1^*(\mathcal{B})$ where α_1^* is the new updated model of $\mathcal{B}$, and let $\Delta V = V \backslash V'$ be the actual deleted set of tuples (the *view deletions induced by* ΔB). We say that ΔB is a VDT for $-t$ if $t \subseteq \Delta V$.

We denote the update of a model of $\mathcal{B}$ induced by VDT for $-t$ over a single view V by a transition $\alpha^* \overset{\mathcal{M}_{P_n}}{\Longrightarrow} \alpha_1^*$, where P_n is a sequential composition of **RA** arrows in the Application Plan considered as a program of categorial machine $\mathbf{M}_{RC}$ which executes the deleting of tuples in ΔB of the schema $\mathcal{B}$. From Proposition 30 in Sect. 6.2.2, this transition corresponds to the natural transformation $\eta_{P_n} : \alpha^* \overset{\bullet}{\longrightarrow} \alpha_1^*$ (an arrow in $Int(G(\mathcal{B})) \subseteq \mathbf{DB}^{\mathbf{Sch}(G(\mathcal{B}))}$) such that $f_D = \eta_{P_n}(\mathcal{B}) : \alpha^*(\mathcal{B}) \to \alpha_1^*(\mathcal{B})$, defined by Corollary 17 and specified in Example 34 as an arrow in **DB**, is obtained from 'DELETE...' SQL statements used to delete the tuples in ΔB.

If $\Delta V = t$ then ΔB is an *exact* translation (i.e., side-effect free) for $-t$. Otherwise ΔB is inexact, that is, with side-effects (or extra deletions) $E = \Delta V \backslash t$.

The problem of translating general select–project–join (SPJ) view deletions to underlying database deletions has been studied in [11] which gives an algorithm that exploits lineage information to find an exact (i.e., side-effect free) deletion-to-deletion translation whenever possible.

The lineage information is used as a starting point to enumerate all candidate witnesses for a deletion. In [8], it was shown that this minimization is NP-hard in general for monotonic select–project–join–union (SPJU) fragment of relational queries.

Proposition 31 *Let us consider a database mapping system (graph) $G = (V_G, E_G)$ with a current mapping-interpretation $\alpha^* : \mathbf{Sch}(G) \to \mathbf{DB}$ (i.e., from Definition 11 and Proposition 16, $\alpha^* \in Int(G) \subseteq \mathbf{DB}^{\mathbf{Sch}(G)}$) which is a model of each database schema in V_G (but not necessarily satisfies all inter-schema mappings in G) with $A = \alpha^*(\mathcal{A})$ obtained by deleting some tuples in $\mathcal{A}$ and there exists $\mathcal{M}_{BA} = \{\Phi\} \in E_G$, where Φ is a normalized SOtgd formula*

$$\exists \mathbf{f}\big(\forall \mathbf{x}_1\big(q_{B_1}(\mathbf{x}_1) \Rightarrow r_{A,1}(\mathbf{t}_1)\big) \wedge \cdots \wedge \forall \mathbf{x}_m\big(q_{B_m}(\mathbf{x}_m) \Rightarrow r_{A,m}(\mathbf{t}_m)\big)\big)$$

with $\mathbf{M}_{BA} = MakeOperads(\mathcal{M}_{BA}) = \{v_1 \cdot q_{B,1}, \ldots, v_m \cdot q_{B,m}, 1_{r_\emptyset}\} : \mathcal{B} \to \mathcal{A}$. Let (here S is defined in the algorithm 'Cmp' in Definition 11 for the relational symbols $r_{i_1}, \ldots, r_{i_k}$ in $q_{B_i}(\mathbf{x}_i)$)

$$\begin{aligned}\mathbb{S}_{\mathcal{B}} = \big\{(V_i, t_i) \mid V_i = \|q_{B_i}(\mathbf{x}_i)\|_{\alpha^*(\mathcal{B})},\ t_i = \big\{Cmp(S, \mathbf{d}_1, \ldots, \mathbf{d}_k) \mid \mathbf{d}_m \in \alpha(r_{i_m}),\\ m = 1, \ldots, k,\ \mathbf{b} = \alpha(q_{B,i})(\mathbf{d}_1, \ldots, \mathbf{d}_k) \neq \langle\rangle,\ \alpha(v_i)(\mathbf{b}) = \langle\rangle\big\} \neq \emptyset,\\ \textit{for } v_i \cdot q_{B,i} \in MakeOperads(\mathcal{M}_{BA}),\\ q_{B,i} \in O(r_{i_1}, \ldots, r_{i_k}, r_q), v_i \in O(r_q, r_{A,i})\big\}\end{aligned}$$

be a nonempty set where for each element $(V_i, t_i) \in \mathbb{S}_{\mathcal{B}}$, $1 \leq i \leq n = |\mathbb{S}_{\mathcal{B}}|$, $t_i' \subseteq t_i$ is a subset of tuples accepted to be consecutively deleted from the view V_i of a database $\mathcal{B}$. Then the concatenation of updates transitions (in Definition 49) of the instance-databases of $\mathcal{B}$

$$\alpha^* \stackrel{\mathcal{M}_{P_n}}{\Longrightarrow} \alpha_n^* \stackrel{\mathcal{M}_{P_{n-1}}}{\Longrightarrow} \alpha_{n-1}^* \cdots \stackrel{\mathcal{M}_{P_1}}{\Longrightarrow} \alpha_1^*$$

*where P_i is a sequential composition of **RA** arrows in the Application Plan considered as a program of categorial machine $\mathbf{M}_{RC}$ which executes the deleting of the tuples $t_i' \subseteq t_i$ in $\mathcal{B}$ for each $(V_i, t_i) \in \mathbb{S}_{\mathcal{B}}$, will be simply denoted by $\alpha^* \stackrel{\mathcal{M}_{BA}^{OP}}{\Longrightarrow} \alpha_1^*$, where $\mathcal{M}_{BA}^{OP} = \mathcal{M}_{P_n \circ \cdots \circ P_1}$ is associated to the inverted arrow w.r.t. $\mathcal{M}_{BA}$, so that $\alpha_1^*(\mathcal{B})$ is the final updated database instance of $\mathcal{B}$ with minimal side-effects. Hence, we extend these changes to the mapping-interpretation (a functor) $\alpha_1^* : \mathbf{Sch}(G) \to \mathbf{DB}$ such that for each $\mathcal{C} \neq \mathcal{B}$ in V_G, $\alpha_1^*(\mathcal{C}) \triangleq \alpha^*(\mathcal{C})$.*

This new mapping-interpretation $\alpha_1^* \in Int(G)$ satisfies $\mathcal{M}_{BA} : \mathcal{B} \to \mathcal{A}$ if for each $(V_i, t_i) \in \mathbb{S}_{\mathcal{B}}$, $t_i' = t_i$, but not necessarily of all mapping system G.

Proof Note that if $\mathbb{S}_{\mathcal{B}}$ is empty then there is no transition $\alpha^* \xrightarrow{\mathcal{M}_{BA}^{OP}} \alpha_1^*$; otherwise, for each $(V_i, t_i) \in \mathbb{S}_{\mathcal{B}}$, if $t_i' = t_i$ then this set of tuples in t_i for a given mapping $\mathcal{M}_{BA}$ is just the set of tuples for which this mapping is not satisfied. Thus, by eliminating them from the view V_i this mapping will be satisfied and hence after all deleting in $\alpha^* \xrightarrow{\mathcal{M}_{BA}^{OP}} \alpha_1^*$, the α_1^* will satisfy $\mathcal{M}_{BA}$. However, if $t_i' \subset t_i$ for at least one $(V_i, t_i) \in \mathbb{S}_{\mathcal{B}}$ then $\mathcal{M}_{BA}$ is not satisfied. This process is done for each ingoing arrow into the vertex $\mathcal{A}$. The interpretation α_1^* is a model of $\mathcal{A}$ and $\mathcal{B}$, but not necessarily of all mapping system G. □

Note that, based on this proposition, the transition $\alpha^* \xrightarrow{\mathcal{M}_{BA}^{OP}} \alpha_1^*$ (caused by a number of deletions of tuples in tables of a database $\mathcal{A}$) denotes the transition from a mapping-interpretation $\alpha^* \in Int(G) \subseteq \mathbf{DB}^{\mathbf{Sch}(G)}$ (which is a model for all vertices (i.e., the database schemas) in V_G, but it does not necessarily satisfy all inter-schema mappings in E_G) into a new mapping-interpretation $\alpha_1^* \in Int(G) \subseteq \mathbf{DB}^{\mathbf{Sch}(G)}$ which is a model for all vertices (or database schemas) in V_G.

Remark Notice that the accepted tuples t_i' to be deleted can be a strict subset of the "exact" set t_i because the local DBMS of the schema $\mathcal{B}$ can decide to protect from deleting its own tuples managed by its local legacy application during such a backward chaining. Obviously, if $t_i' \subset t_i$ than after the deleting of t_i' from $\mathcal{B}$, the mapping $\mathcal{M}_{BA}$ will remain unsatisfied.

Hence, this transition $\alpha^* \xrightarrow{\mathcal{M}_{BA}^{OP}} \alpha_1^*$ does not necessarily satisfy all inter-schema mappings in E_G. Consequently, the process of the updates propagates backward (w.r.t. the directed graph G of a database mapping system) from $\mathcal{A}$ into $\mathcal{B}$, and then can continue to propagate backward from $\mathcal{B}$ to another database schemas in E_G.

This backward propagation in G can be equivalently seen as a forward propagation in the inverted graph G^{OP}, and this is a reason that we are using the label $\mathcal{M}_{BA}^{OP}$ in the "atomic" transition $\alpha^* \xrightarrow{\mathcal{M}_{BA}^{OP}} \alpha_1^*$, where the mapping $\mathcal{M}_{BA} : \mathcal{B} \to \mathcal{A}$ is represented by the edge $\mathcal{M}_{BA}^{OP} : \mathcal{A} \to \mathcal{B}$.

Hence, we may see the transition $\alpha^* \xrightarrow{\mathcal{M}_{BA}^{OP}} \alpha_1^*$ as a *DB-atomic transition* of an LTS, based on the opposite graph G^{OP}, which represents the complete process of a backward chaining. In the case when the initial model of database mapping system G contains only finite extensions of all database schemas in G, the complete process will always produce a finite LTS tree (the branching is caused by a number of ingoing inter-schema-mappings in a given schema), also when G is not an acyclic graph: it hold from the fact that process of deleting tuples from finite databases cannot be infinite.

It is easy to show, based on the case 3 of Example 34 and Definition 49, that this atomic transition $\alpha^* \stackrel{\mathcal{M}_{BA}^{OP}}{\Longrightarrow} \alpha_1^*$, which represents the deleting of tuples $t_i' \subseteq t_i$ in the database $\mathcal{B}$ (based on deletion of n of its views, $(V_i, t_i) \in \mathbb{S}_{\mathcal{B}}$ for $i = 1, \dots, n$) of k different relations $r_1, \dots, r_k \in \mathcal{B}$ by the deleting operations 'DELETE FROM r_i WHERE C_i', $i = 1, \dots, k$, can be represented in the **DB** category as the morphisms $f_D = \eta_{P_n \circ \cdots \circ P_1}(\mathcal{B}) : \alpha^*(\mathcal{B}) \to \alpha_1^*(\mathcal{B})$.

This morphism in **DB** is represented as a horizontal arrow from the instance database $B = \alpha^*(\mathcal{B})$ in the diagram on the left ellipse, which represents the image of the data mapping graph G for the initial interpretation α^*, into the new updated instance $\alpha_1^*(\mathcal{B})$ of a schema $\mathcal{B}$ in the diagram on the right ellipse (which represents the image of the data mapping graph G for the resulting interpretation α_1^*).

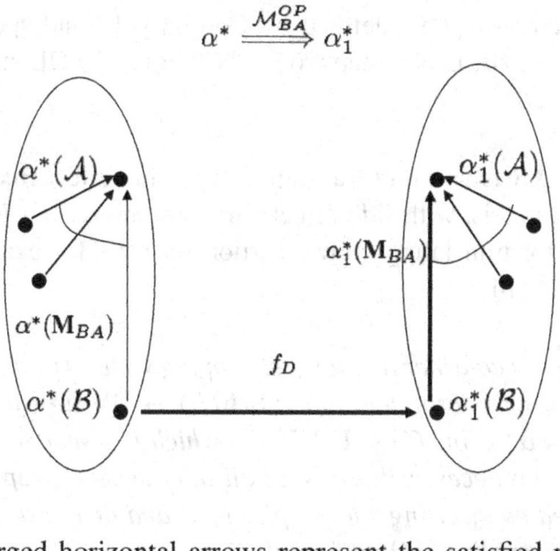

Note that enlarged horizontal arrows represent the satisfied morphisms in **DB** while the enlarged vertical arrow is a satisfied arrow of a database mapping graph G if $t_i' = t_i$ for all $(V_i, t_i) \in \mathbb{S}_{\mathcal{B}}$ for $i = 1, \dots, n$. This *backward propagation* is denoted by the cone of ingoing arrows into the node (schema) $\mathcal{A}$ in the diagram above).

7.2.2 Insertion by Minimal Side-Effects

In an insertion minimization, the input is a database $B = \alpha^*(\mathcal{B})$, a model of a schema $\mathcal{B}$, a query $q_{B_i}(\mathbf{x}_i)$, a view $V = \|q_{B_i}(\mathbf{x}_i)\|_B$, and a set of tuples to be inserted t.

The view side-effect minimization is an algorithm for providing a set $\Delta B \subseteq B$ in order to minimize $|\Delta V|$ such that for the same query executed over updated database $B' = B \cup \Delta B$ (which is another model of database schema $\mathcal{B}$) the obtained view is $V \cup (\Delta V \cup t)$. In other words, we wish to find a set of tuples in B whose addition will insert t while minimizing the number of other tuples inserted into the view V. We can define this *view insertion translation* (VIT) as follows:

Definition 50 Given a view $V = \|q_B(\mathbf{x})\|_B$ of a database $B = \alpha^*(\mathcal{B})$, a model of the schema $\mathcal{B}$, and an insertion request $+t$, we say that database insertion ΔB is a translation for $+t$ if ΔB causes the insertion of t into V when applied to B. More formally, let $V' = \|q_B(\mathbf{x})\|_{B'}$ be the new view, obtained from the minimally updated database $B' = B \cup \Delta B = \alpha_1^*(\mathcal{B})$, where α_1^* is the new updated model of $\mathcal{B}$, and let $\Delta V = V' \backslash V$ be the actual inserted view tuples (the *view insertions induced by* ΔB). We say that ΔB is a VIT for $+t$ if $t \subseteq \Delta V$.

We denote the update of a model of $\mathcal{B}$ induced by VIT for $+t$ over a single view V by a transition $\alpha^* \stackrel{\mathcal{M}_{P_n}}{\Longrightarrow} \alpha_1^*$, where P_n is a sequential composition of **RA** arrows in the Application Plan considered as a program of categorial machine $\mathbf{M}_{RC}$ which executes the inserting of tuples in ΔB of the schema $\mathcal{B}$. From Proposition 30 in Sect. 6.2.2, this transition corresponds to the natural transformation $\eta_{P_n} : \alpha^* \stackrel{\bullet}{\longrightarrow} \alpha_1^*$ (an arrow in $Int(G(\mathcal{B})) \subseteq \mathbf{DB}^{\mathbf{Sch}(G(\mathcal{B}))}$) such that $f_I = \eta_{P_n}(\mathcal{B}) : \alpha^*(\mathcal{B}) \to \alpha_1^*(\mathcal{B})$, defined by Corollary 17 and specified in Example 34 as an arrow in **DB**, is obtained from 'INSERT…' SQL statements used to insert the tuples in ΔB.

If $\Delta V = t$ then ΔB is an *exact* translation (i.e., side-effect free) for $+t$. Otherwise ΔB is inexact, that is, with side-effects (or extra insertions) $E = \Delta V \backslash t$.

The algorithms for translating view insertions can be, for example, used from operations defined in [14].

Proposition 32 *Let us consider a database mapping system (graph)* $G = (V_G, E_G)$ *with a current mapping-interpretation* $\alpha^* : \mathbf{Sch}(B) \to \mathbf{DB}$ (*i.e., from Definition* 11 *and Proposition* 16, $\alpha^* \in Int(G) \subseteq \mathbf{DB}^{\mathbf{Sch}(G)}$) *which is a model of each database schema in* V_G (*but not necessarily satisfies all inter-schema mappings in* G) *with* $A = \alpha^*(\mathcal{A})$ *obtained by inserting some tuples in* $\mathcal{A}$ *and there exists* $\mathcal{M}_{AB} = \{\Phi\} \in E_G$, *where* Φ *is a normalized SOtgd formula*

$$\exists \mathbf{f}\big(\forall \mathbf{x}_1\big(q_{A_1}(\mathbf{x}_1) \Rightarrow r_{B,1}(\mathbf{t}_1)\big) \wedge \cdots \wedge \forall \mathbf{x}_m\big(q_{A_m}(\mathbf{x}_m) \Rightarrow r_{B,m}(\mathbf{t}_m)\big)\big)$$

with $\mathbf{M}_{AB} = MakeOperads(\mathcal{M}_{AB}) = \{v_1 \cdot q_{A,1}, \ldots, v_m \cdot q_{A,m}, 1_{r_\emptyset}\} : \mathcal{A} \to \mathcal{B}$. *Let*

$$\begin{aligned}\mathbb{S}_{\mathcal{B}} = \big\{(V_i, t_i) \mid V_i = \alpha\big(\partial_1(v_i)\big), t_i = \big\{\mathbf{b} \mid \mathbf{d}_m \in \alpha(r_{i_m}) \in \alpha^*(\mathcal{A}), m = 1, \ldots, k,\\ \mathbf{b} = \alpha(q_{A,i})(\mathbf{d}_1, \ldots, \mathbf{d}_k) \neq \langle\rangle,\ \alpha(v_i)(\mathbf{b}) = \langle\rangle\big\} \neq \emptyset,\\ \text{for } v_i \cdot q_{A,i} \in MakeOperads(\mathcal{M}_{AB}),\\ q_{A,i} \in O(r_{i_1}, \ldots, r_{i_k}, r_q), v_i \in O(r_q, r_i)\big\}\end{aligned}$$

be a nonempty set where for each element $(V_i, t_i) \in \mathbb{S}_{\mathcal{B}}$, $1 \le i \le n = |\mathbb{S}_{\mathcal{B}}|$, $t_i' \subseteq t_i$ *is a subset of tuples accepted to be consecutively inserted into the view* V_i (*here just a relation* $\alpha(r_i)$ *of a database* $\mathcal{B}$ *and hence if* $t_i' = t_i$ *then all transitions are exact, i.e., side-effect free*), *then the concatenation of update transitions* (*in Definition* 50) *of the instance-databases of $\mathcal{B}$*

$$\alpha^* \xrightarrow{\mathcal{M}_{P_n}} \alpha_n^* \xrightarrow{\mathcal{M}_{P_{n-1}}} \alpha_{n-1}^* \cdots \xrightarrow{\mathcal{M}_{P_1}} \alpha_1^*$$

where P_i is a sequential composition of **RA** *arrows in the Application Plan considered as a program of categorial machine* $\mathbf{M}_{RC}$ *which executes the inserting of the tuples $t_i' \subseteq t_i$ in $\mathcal{B}$ for each $(V_i, t_i) \in \mathbb{S}_{\mathcal{B}}$ will be simply denoted by $\alpha^* \xrightarrow{\mathcal{M}_{AB}} \alpha_1^*$, where $\mathcal{M}_{BA} = \mathcal{M}_{P_n \circ \cdots \circ P_1}$ is associated to the arrow $\mathcal{M}_{BA}$, so that $\alpha_1^*(\mathcal{B})$ is the final updated database instance of $\mathcal{B}$. Hence, we extend these changes to the mapping-interpretation (a functor) $\alpha_1^* : \mathbf{Sch}(G) \to \mathbf{DB}$ such that for each $\mathcal{C} \neq \mathcal{B}$ in V_G, $\alpha_1^*(\mathcal{C}) \triangleq \alpha^*(\mathcal{C})$.*

This new mapping-interpretation $\alpha_1^* \in Int(G)$ satisfies a mapping $\mathcal{M}_{AB} : \mathcal{A} \to \mathcal{B}$ if for each $(V_i, t_i) \in \mathbb{S}_{\mathcal{B}}$, $t_i' = t_i$, but not necessarily of all mapping system G.

Proof Note that if $\mathbb{S}_{\mathcal{B}}$ is empty then there is no a transition $\alpha^* \xrightarrow{\mathcal{M}_{AB}} \alpha_1^*$; otherwise, for each $(V_i, t_i) = (\alpha(r_i), t_i) \in \mathbb{S}_{\mathcal{B}}$, if $t_i' = t_i$ then this set of tuples in t_i for a given mapping $\mathcal{M}_{AB}$ is just the set of tuples for which this mapping is not satisfied. Thus, by inserting them into the relation $\alpha(r_i)$ (i.e., view V_i), this mapping will be satisfied, and hence after all insertions in $\alpha^* \xrightarrow{\mathcal{M}_{AB}} \alpha_1^*$, α_1^* will satisfy $\mathcal{M}_{AB}$. This process is done for each outgoing arrow from the vertex $\mathcal{B}$, so that α_1^* will satisfy all outgoing arrows from the vertex $\mathcal{A}$ (the arrow $\mathcal{M}_{AB}$ is one of them). However, if $t_i' \subset t_i$ for at least one $(V_i, t_i) \in \mathbb{S}_{\mathcal{B}}$ (for example, the primary-key constraints for the relations can be broken by inserting all tuples in t_i) then $\mathcal{M}_{AB}$ is not satisfied, and hence the interpretation α_1^* is not necessarily a model of $\mathcal{B}$. □

Note that, based on this proposition, the transition $\alpha^* \xrightarrow{\mathcal{M}_{AB}} \alpha_1^*$ (caused by a number of insertions of tuples in tables of a database $\mathcal{A}$) denotes the transition from a mapping-interpretation $\alpha^* \in Int(G) \subseteq \mathbf{DB}^{\mathbf{Sch}(G)}$ (which is a model for all vertices (i.e., the database schemas) in V_G, but it does not necessarily satisfy all inter-schema mappings in E_G) into a new mapping-interpretation $\alpha_1^* \in Int(G) \subseteq \mathbf{DB}^{\mathbf{Sch}(G)}$. Hence, this transition $\alpha^* \xrightarrow{\mathcal{M}_{AB}} \alpha_1^*$ does not necessarily satisfy *all* inter-schema mappings in E_G. In order to guarantee that the insertion of the new tuples in the relation of $\mathcal{B}$ does not make its integrity constraints inconsistent, we can adopt a strategy to divide such relations into two relations (with the same attributes: the first one can be "locally" updated by the legacy software for this schema by applying to it all necessary integrity constraints; the second "imported" relations will be used only to collect the tuples from another databases in G and for this relation we do not apply any integrity constraint, so that the insertion of the new tuples provided by the inter-schema mapping into this database will not render it inconsistent.

In this way, each schema in a database-mapping system can have two partitions: one original (or "local") with all integrity constraints and updated only by the local legacy programs that fully respect the integrity constraints, and another "imported" partition composed by only relations used to receive the information from another databases without the integrity constraints which can make this schema inconsistent. In what follows, we consider that by using such strategies, we guarantee that the insertion of the information from another databases in G into a given schema always produces a new model (instance-database) of this schema.

Consequently, the process of the updates propagates from $\mathcal{A}$ into $\mathcal{B}$ and hence may continue to propagate from $\mathcal{B}$ to another database schemas in E_G (a forward chaining propagation in G).

$$\alpha^* \stackrel{\mathcal{M}_{AB}}{\Longrightarrow} \alpha_1^*$$

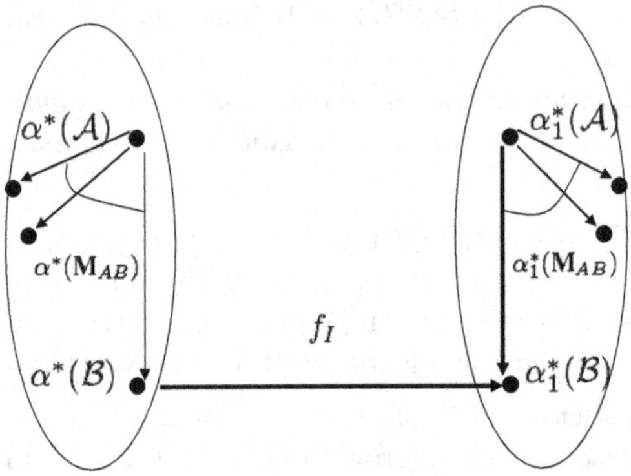

Hence, we may see this transition $\alpha^* \stackrel{\mathcal{M}_{AB}}{\Longrightarrow} \alpha_1^*$ as a *DB-atomic transition* of an LTS that represents the complete process of a forward chaining. In the case when G is not an acyclic graph, the tree of this LTS can be infinite as well (in the case, for example, when we use the full tgds for inter-schema mappings, so that $(q_{A_i}(\mathbf{x}_i) \Rightarrow \exists \mathbf{y}_i q_{B_i}(\mathbf{x}_i, \mathbf{y}_i)) \in \mathcal{M}_{AB}$, and in the place of tuple of variables $\mathbf{y}_i$ we insert the new Skolem constants).

Let r_{I_i} be the relational symbol for the set of tuples in $t_i' \subseteq t_i$, for $i = 1, \ldots, n$, $(V_i, t_i) \in \mathbb{S}_{\mathcal{B}}$. It is easy to show, based on case 2 of Example 34 and Definition 50, that this atomic transition $\alpha^* \stackrel{\mathcal{M}_{AB}}{\Longrightarrow} \alpha_1^*$, which represents the inserting of tuples in the $k = n$ relations of the database $\mathcal{B}$, $r_1, \ldots, r_k \in \mathcal{B}$ by the inserting operations 'INSERT INTO $r_i \ldots$', with r_{I_i} a relation to be inserted, $i = 1, \ldots, k$, can be represented in the **DB** category as the morphisms $f_I = \eta_{P_n \circ \cdots \circ P_1}(\mathcal{B}) : \alpha^*(\mathcal{B}) \to \alpha_1^*(\mathcal{B})$. This morphism in **DB** is represented as a horizontal arrow from the instance database $B = \alpha^*(\mathcal{B})$ in the diagram on the left ellipse, which represents the image of the data mapping graph G for the initial interpretation α^*, into the new updated instance $\alpha_1^*(\mathcal{B})$ of a schema $\mathcal{B}$ in the diagram on the right ellipse (which represents the image of the data mapping graph G for the resulting interpretation α_1^*).

Note that the enlarged horizontal arrow represent the satisfied morphism in **DB** while the enlarged vertical arrows is satisfied arrow of a database mapping graph G if $t_i' = t_i$ for all $(V_i, t_i) \in \mathbb{S}_{\mathcal{B}}$.

This *forward propagation* is denoted by the cone of arrows from the node (schema) $\mathcal{A}$ in the diagram above).

7.3 Denotational Model (Database-Mapping Process) Algebra

First of all, we need to clarify the relationships between the *database-mapping programs* specified by a schema-mapping graph $G = (V_G, E_G)$, where the nodes in V_G are the database schemas and the edges in E_G are the SOtgds which logically specify the schema mappings, and the *process-algebra language* (or its grammar) able to specify the execution of the database-mapping programs by a kind of process-programs. From the algebraic point of view, this *database-mapping program* is represented by the sketch category $\mathbf{Sch}(G)$ (with the embedding $\gamma : G \mapsto \mathbf{Sch}(G)$) where the arrows are algebraic R-operads obtained by "algebraization" of the SOtgds with its models represented by the R-algebras (i.e., the functors) $\alpha^* : \mathbf{Sch}(G) \to \mathbf{DB}$ satisfying all sketch's arrows, that is, $\alpha^* \in Mod(G) \subseteq \mathbf{DB}^{\mathbf{Sch}(G)}$. The *input* for such a program specified by the graph G (or sketch $\mathbf{Sch}(G)$) is a model $\alpha^* : \mathbf{Sch}(G) \to \mathbf{DB}$ and a ΔA of initial updating of a node (schema) $\mathcal{A} \in V_G$, in one of two possible ways described in Sects. 7.2.1 and 7.2.2, so that after such an update with obtained instance-database $A = \alpha^*(\mathcal{A})$, this modified functor α is not generally a new model of the mapping system G. Thus, the execution of the database-mapping program in $\mathbf{Sch}(G)$ begins with this initial updating of $\mathcal{A}$ and propagates in backward-chaining (if ΔA is a deletion of tuples in $\mathcal{A}$) or in forward-chaining (if ΔA is an insertion of tuples in $\mathcal{A}$) up to the moment when is reached the end. Two semantic scenarios are possible:

- *Strong Data Integration.* In this version, we consider the schema mappings and the integrity constraints of the schemas at the same ontological level, without any epistemic distinction of them. Thus, the updates (deletions and insertions) in each "local" database (a schema vertex in a graph G) caused by the schema mappings are obligatory and has to be respected as every integrity constraint of the schemas. This renders inconsistent the whole logic of a database mapping system if every mapping in G is not satisfied, so that from the architectural point of view this is a strongly centralized omniscient system, as in the case of a data-exchange setting. The end of a backward/forward chaining described above, in this case, must result in *a new model* (i.e., functor) $\alpha_1^* \in Mod(G) \subseteq \mathbf{DB}^{\mathbf{Sch}(G)}$ of G of the whole mapping system G (i.e., a model of all schemas and all schema mappings as well).
- *Weak Data Integration.* In this case, we only preserve the requirements that each single instance-database in G must be a model of its schema (as in all practical applications of RDB systems with the legacy application over them, which preserves its integrity constraints), but do not require that at the end of such a backward/forward chaining all (inter)schema mappings have to be satisfied. Thus, the architecture of such a solution will have two levels: a strong consistency level for each independent database in G, and a higher epistemic level which uses the SOtgd's schema mappings in order to communicate to single databases in G the *recommendations* which information has to be inserted/deleted in them as a response to one local changing of information in one database in G. The epistemic independence of these two logical systems permits each single database to (partially) implement such recommendations (if they are consistent with their

local knowledge) or not (for example, when deleting some information of a local database is recommended and such information is locally managed by its own legacy applications; or when inserting the tuples which would make the local knowledge inconsistent is recommended). In such an architecture, we do not require that each schema mapping in G be satisfied in each stable state of a database mapping system in G (i.e., when any backward/forward chaining is not in action).

If we adopt a P2P database architecture for a database mapping system in G, where each schema (a vertex in G) is a peer database, then we can consider a number of papers [9, 10, 12, 15, 16, 21, 23] that describe the Strong Data Integration and, especially, the problems about the inconsistencies caused by a forced transfer of the information from other peers into a given peer with its own local knowledge of integrity constraints.

The second solution is a more sophisticated solution, and it makes explicit an epistemic difference from the local knowledge of each peer (which is consistently managed by its own legacy applications) and the imported information from other peers: each peer (partially) accepts the imported information from other peers and is independent in its own decision which one to accept and which one to ignore. This extra knowledge received from a network of other peers and filtered locally (based on local knowledge) is similar to a kind of the social database-networks, where this imported knowledge from other peers is used not to brutally substitute its local consistent knowledge but only to consistently enlarge it (thus, it is a much more flexible solution of the standard data-exchange settings). Such an epistemic solution, however, without a material transfer of the information between the peers, based on an intensional FOL is analyzed in a number of papers [24–32].

In such a P2P architecture, each database schema (or vertex in G) can be implemented as an independent peer database $\mathcal{A} \in V_G$ managed by a universal machine $\mathbf{M}_{U_\mathcal{A}}$ (with its DBMS machine $\mathbf{M}_{R_\mathcal{A}}$), and the schema mappings are used as a program of a universal machine $\mathbf{M}_{U_{DB}}$ which communicate by Out_U and In^*_{EU} functions (see Definition 38 in Sect. 6.1) with each single peer database machine $\mathbf{M}_{U_\mathcal{A}}$. In fact, for example, during a forward chaining process, for a given mapping $\mathcal{M}_{AB} : \mathcal{A} \to \mathcal{B}$ and its operads-operations $q_i \in \mathit{MakeOperad}(\mathcal{M}_{AB})$, $q_i \in O(r_{A_1}, \dots, r_{A_1}, r_{B_i})$, $\mathbf{M}_{U_{DB}}$ will ask $\mathbf{M}_{U_\mathcal{A}}$ for the extensions of the current relations r_{A_j}, $j = 1, \dots, k$, and $\mathbf{M}_{U_\mathcal{B}}$ for the extension of r_{B_i}, by using Out_U and In^*_{EU} functions. Hence, it will compute $\triangle B$ recommended to be inserted into $\mathcal{B}$ and hence communicate it with Out_U to the machine $\mathbf{M}_{U_\mathcal{B}}$. The machine $\mathbf{M}_{U_\mathcal{B}}$ will take $\triangle B$ by its In^*_{EU} function and then will decide which part of it to (consistently) insert into its database $\mathcal{B}$ by generation of a new model (instance-database) $B = \alpha^*_1(\mathcal{B})$.

We are interested in the operational semantics of such programs specified by a graph G (i.e., a *program*) in the machine $\mathbf{M}_{U_{DB}}$, that is, in their *processes* of execution provided by its Computation system (Definition 37 in Sect. 6.1), but abstracted (reduced) to the transitions of the types defined in Propositions 31 and 32, and then refined as follows.

In order to be able to provide this operational semantics, the first step is just to specify *which* algebra of processes is adequate for it, and to define its process-programming grammar (or language). Hence, in what follows in this chapter, we will use the word "programs" for this process-algebra language, denoted by $GSOS_{DB}$ and adequate to represent the operational semantics for the execution of the database-mapping programs specified by a mapping graph G (or its "algebraic" sketch $\mathbf{Sch}(G)$). Note that such processes are not specified by the users (i.e., programmers) as the original database-mapping programs specified by a graph G, but are dynamically generated during the execution (forward or backward chaining) of such programs.

Denotational model of a process-programming language, introduced in Sect. 7.1 and given by the abstract grammar ($GSOS_{DB}$), is a Σ_P-algebra, where

$$\Sigma_P = \{nil, Act, \{a.\_\}_{a \in Act}, \|\}$$

is the signature of the language corresponding to the basic syntax constructs in $GSOS_{DB}$. The processes themselves form the initial such Σ_P-algebra and the corresponding homomorphism from this initial Σ_P-algebra (syntax algebra of Σ_P) to the *denotational model* is called initial algebra semantics [17].

In what follows, we will define the denotational model Σ_P-algebra in our framework of database mapping systems where the DB-denotational semantics is based on the category **DB**.

In the database-mappings context, a *visible action* $a \in Act$ (a nullary operator of the signature Σ_P) of a database-mapping program (a graph) G is the *kernel* of views that define the information-flux of a given schema mapping $\mathcal{M}_{AB} : \mathcal{A} \to \mathcal{B}$, for a given $\alpha^* \in Int(G) \subseteq \mathbf{DB}^{\mathbf{Sch}(G)}$. That is, based on point 2 of Definition 13,

$$a = \Delta(\alpha, MakeOperads(\mathcal{M}_{AB})) \in \mathcal{P}_{\text{fin}}(\underline{\Upsilon}) \subset Ob_{DB},$$

where $\mathcal{P}_{\text{fin}}$ is the finite powerset operator.

Consequently, the set of all possible visible actions is defined by $Act = \mathcal{P}_{\text{fin}}(\underline{\Upsilon})$, where $\underline{\Upsilon}$ is the total object in the subcategory of only simple objects $\underline{\mathbf{DB}}$ and each visible action $a \in Act$ is a finite set of views (i.e., relations), i.e., a simple object in **DB**.

Note that the low-level programming "select–project–join + union" language (SPJRU language [1], with the signature Σ_R, is encapsulated into this set of possible visible actions *Act*.

Proposition 33 *For any schema-database mapping program $G = (V_G, E_G)$, where V_G is the set of vertices (nodes) in G and E_G is the set of edges in G, the graph obtained by inverting orientation only of inter-schema mappings $\mathcal{M}_{AB} : \mathcal{A} \to \mathcal{B}$ by G^{OP}.*

There exists an LTS derived from G (or G^{OP}) and from a given initial mapping-interpretation $\alpha^* \in Int(G)$ of the sketch $\mathbf{Sch}(G)$, *which is a model of the root database schema in this LTS, such that all states of this LTS are the models of schema databases (vertices) in V_G. The set of visible atomic actions for the set of all LTS of database-mapping systems is equal to $Act = \mathcal{P}_{\text{fin}}(\underline{\Upsilon})$, with $\underline{\Upsilon}$ introduced in Definition* 26.

Proof Let us consider the root database schema $\mathcal{A} \in V_G$ for a given mapping-interpretation $\alpha^*_A \in Int(G) \subseteq \mathbf{DB}^{\mathbf{Sch}(G)}$ (not necessarily a model) of the sketch $\mathbf{Sch}(G)$ such that $A = \alpha^*(\mathcal{A})$ is a model of $\mathcal{A}$ after the insertion of a number of tuples in this database. The fact that A is a model of $\mathcal{A}$ can be verified in the **DB** category by the fact that the integrity-constraint arrow $\mathbf{T}_{\mathcal{A}\mathcal{A}_\top} = \{v_1 \cdot q_{A,1}, \ldots, v_n \cdot q_{A,n}, 1_{r_\emptyset}\} : \mathcal{A} \to \mathcal{A}_\top$ in $\mathbf{Sch}(G)$ is satisfied by α^*, that is, $\alpha(v_i)$ for $i = 1, \ldots, n$ are the injective functions (see Corollary 4 and Example 12 in Sect. 2.4.1).

That is, $A = \alpha^*(\mathcal{A}) \in Ob_{DB}$ (denoted alternatively by a state-pair $s = (\mathcal{A}, \alpha^*)$) is a model of the schema $\mathcal{A}$ and, consequently, the root state in this LTS. Then, an atomic action for a mapping $\mathcal{M}_{AB} : \mathcal{A} \to \mathcal{B}$ in E_G is given by

$$a = \Delta\bigl(\alpha, MakeOperads(\mathcal{M}_{AB})\bigr) \in \mathcal{P}_{\text{fin}}(\underline{\Upsilon}),$$

for the insertion transition $\alpha^* \overset{\mathcal{M}_{AB}}{\Longrightarrow} \alpha^*_1$ (in Proposition 32). The final transition step state of this action a is $s' = B = (\mathcal{B}, \alpha^*_1)$, where $\alpha^*_1 \in Int(G) \subseteq \mathbf{DB}^{\mathbf{Sch}(G)}$ is a new interpretation obtained from α^*_1 by changing only an extension of $\mathcal{B}$ in the way that B is a new obtained model of the database schema $\mathcal{B}$ after the transferring of the information flux $Flux(\alpha, MakeOperads(\mathcal{M}_{AB})) = T(\Delta(\alpha, MakeOperads(\mathcal{M}_{AB}))) = Ta$ from the instance database A into this database B. In the case of Strong Data Integration, α^*_1 satisfies the inter-schema mapping $\mathcal{M}_{AB}$ after the transition $\alpha^* \overset{\mathcal{M}_{AB}}{\Longrightarrow} \alpha^*_1$ (see the vertical arrow in the right ellipse of the diagram in Sect. 7.2.2).

Consequently, we have the atomic state-transition $(\mathcal{A}, \alpha^*) \xrightarrow{a} (\mathcal{B}, \alpha^*_1)$ or, equivalently, $A \xrightarrow{a} B$, caused by the insertion-transition $\alpha^* \overset{\mathcal{M}_{AB}}{\Longrightarrow} \alpha^*_1$, which corresponds in **DB** to the *atomic* morphism $f = \alpha^*(MakeOperads(\mathcal{M}_{AB})) : A \to B$ with the information flux $\widetilde{f} = Ta$ (in the case of Strong Data Integration, the inter-schema mapping $\mathcal{M}_{AB} : \mathcal{A} \to \mathcal{B}$ is satisfied by the final interpretation α^*_1).

By following this *forward-chaining* process, from the initial root and its interpretation α^*, for a given graph G, we obtain that an LTS is a tree where the states are the models of the database schemas and transitions are labeled by actions that are the kernels of the information fluxes of the inter-schema mappings.

We consider the class of cyclic database mapping graphs G as well, and hence such a tree can be also infinite.

It is easy to verify that there is a correspondence between atomic actions in Act and atomic morphisms in **DB** category, given by mapping

$$T : Act \to \bigl\{\widetilde{f} = B_T(f) \mid f \text{ is an atomic morphism in } \mathbf{DB}\bigr\}$$

such that for any $a = \Delta(\alpha, \mathbf{M}_{AB}) \in Act$, $Ta = Flux(\alpha, \mathbf{M}_{AB}) = \widetilde{f}$, where $f = \alpha^*(\mathbf{M}_{AB}) : \alpha^*(\mathcal{A}) \to \alpha^*(\mathcal{B})$ is a mapping morphism in **DB**.

Analogous considerations can be provided in the case when some tuples in the root database $\mathcal{A}$ are *deleted*. In this case, we will have the *backward-chaining* process that will propagate in the inverted graph G^{OP} and hence will delete the tuples in another databases in G as well (as explained in Sect. 7.2.1 for the transitions $\alpha^* \overset{\mathcal{M}^{OP}_{BA}}{\Longrightarrow} \alpha^*_1$). □

Remark The empty database $\bot^0 \in Act$ is an action as well: it is the '*nil*' action which stops the execution of a path in the LTS of a given process graph P in Definition 48.

Definition 51 Based on Proposition 33 and on the correspondence between the visible actions and atomic morphisms in **DB**, the process language for a database-mapping program of a graph $G = (V_G, E_G)$ is given by a general grammar (with the terms t and the program variables p):

$$(\mathcal{G}r) \quad t := p \mid f_a \mid \bot^1_A : A \to \bot^0 \mid t_1 \circ f_a \mid \langle t'_1, t'_2 \rangle,$$

where

- $f_a : A \to B$ is an atomic action such that $f_a = \alpha^*(MakeOperads(\mathcal{M}_{AB}))$ where α^* is a mapping-interpretation of **Sch**(G) and
$$a = \Delta\big(\alpha, MakeOperads(\mathcal{M}_{AB})\big),$$
and hence $\widetilde{f_a} = Ta$.
- $\bot^1_A$ is a unique arrow from $A = \alpha^*(\mathcal{A})$, $\mathcal{A} \in V_G$, into a terminal object $\bot^0$ in **DB** corresponding to '*nil*' operation, a basic inert program in GSOS$_{DB}$ (stop of execution).
- '$\circ$' is a binary sequential composition of arrows (morphisms) in **DB**. That is, the DB-denotational semantics of a sequential action-prefixing operator '$a.\_$' in GSOS$_{DB}$ here is given by '$\_ \circ f_a$'.
- For any two morphisms $t_1 : A \to B$ and $t_2 : A \to C$, their parallel composition (execution) is given by the unique arrow of product diagram in **DB**, $\langle t_1, t_2 \rangle : A \to B + C$. That is, the coproduct pairing '$\langle \_, \_ \rangle$' in **DB** represents the DB-denotational semantics of the syntax operator '$\|$' in GSOS$_{DB}$.

The weak point of this denotational grammar for abstract GSOS$_{DB}$ grammar is that here we possibly have an infinite number of different 'nil' operators $\bot^1_A : A \to \bot^0$ that are terminal arrows in **DB**. What is common for all of them is that their information flux is unique and equal to $\bot^0$ (i.e., zero object in **DB**), and hence it will be more appropriate to pass to the *objects* in **DB** (thanks to its symmetry properties) as terms of the denotational semantics of the abstract GSOS$_{DB}$ grammar.

Hence, in order to pass from the atomic morphisms in **DB** into the *visible* actions, we have to use the categorial symmetry of **DB** with the 'conceptualizing' operator $B_T : Mor_{DB} \to Ob_{DB}$ (in Definition 4 and Theorem 4). It is easy to see that this category symmetry transformation corresponds to the algebraic homomorphism $B_T : (Mor_{DB}, \circ, \langle \_, \_ \rangle) \to (Ob_{DB}, \otimes, \ddagger)$ (an extension of the homomorphism in Sect. 5.4 with a dual composition of the objects $* : Ob^2_{DB} \to Ob_{DB}$ equal to tensorial product '$\otimes$', i.e., to a matching operator such that for any two simple objects $A, B \in Ob_{DB}$, $A \otimes B = TA \cap TB$; it will be extended to a bifunctor in Sect. 8.1.1), such that the image of B_T is the subset of closed objects in **DB** and '$\ddagger$' is the data-separation operator (isomorphic to the disjoint union '$\uplus$', i.e., coproduct '$+$') in **DB** (introduced in Sect. 3.3.2). Consequently, we have:

- A visible atomic action $a \in Act = \mathcal{P}_{fin}(\underline{\Upsilon})$, where $\underline{\Upsilon}$ is the union of all simple objects in **DB**, for an atomic morphism $f_a : A \to B$, such that $B_T(f_a) = Ta$;
- A unique '*nil*' operator $B_T(\bot_A^1) = \bot^0 \in Act$;
- $B_T(t_2 \circ t_1) = t_2' \otimes t_1'$, where $t_1' = B_T(t_1) = \widetilde{t_1}, t_2' = B_T(t_2) = \widetilde{t_2}$ are terms of this 'dual' grammar (i.e., of closed objects in **DB**);
- From Definition 20, $B_T\langle t_1, t_2\rangle = B_T(t_1 + t_2) = B_T(t_1) \uplus B_T(t_2) = t_1' \uplus t_2' = t_1' + t_2' \simeq$ (from Lemma 12) $\simeq t_1' \ddagger t_2'$, and hence we will use '+' instead of '$\ddagger$' in what follows, where $t_1' = B_T(t_1) = \widetilde{t_1}, t_2' = B_T(t_2) = \widetilde{t_2}$ are terms of this 'dual' grammar (i.e., closed objects in **DB**).

Consequently, the denotation of the syntax operators in Σ_P for the processes of the database-mapping programs, and its language, can be provided by the **DB** operators (functions) $\Sigma_P = \{Act, \bot^0, \otimes, +\}$ and hence:

Definition 52 The *Database-mapping processes* or *terms t* induced by the above signature Σ_P and the process variables $p \in X$ (for objects in **DB**) with assignments $\sigma : X \to Ob_{DB}$, are given by the denotational DB-grammar:

$$(\mathcal{G}r_{DB}) \quad t ::= p \,|\, a \,|\, \bot^0 \,|\, a \otimes t_1 \,|\, t_1 + t_2,$$

where $a \in Act$ and $Act = \mathcal{P}_{fin}(\underline{\Upsilon})$ is the set of kernels of the information fluxes of (only) atomic morphisms in **DB**, $\otimes$ is the matching operator introduced in Definition 19 and $+$ is the coproduct in **DB**.

From the fact that the complex arrows in **DB** are composed of the simple ptp arrows between simple objects, it is enough for the operational semantics to consider the transitions based on the simple arrows only, so that *Act* is well defined.

7.3.1 Initial Algebra Semantics for Database-Mapping Programs

As we will show, the GSOS$_{DB}$ signature freely generates, for every set X of process variables $p \in X$, the set $\mathcal{T}_P X$ of process's terms t. This set $\mathcal{T}_P X$ is the carrier set of the *free* Σ_P-algebra over X where, in general, a Σ_P-algebra is given by a (carrier) set Z. Thus, in the case when $Z \subseteq Ob_{DB}$, the mapping $h : \Sigma_P Z \to Z$ is the Σ_P-algebra, that is, the *DB-denotational semantics* for the abstract syntax free Σ_P-algebra, based on the objects and basic operators in the **DB** category. The *syntax* of a database-mapping process language, given by the signature $\Sigma_P = \{nil, Act, \{a.\_\}_{a \in Act}, \|\}$ of the GSOS$_{DB}$ grammar, can be identified with monad, the *syntactical monad* $\mathcal{T}_P$ freely generated by the process constructors Σ_P, as presented in Sect. 5.1.1.

In what follows, we will use the generalized parallel execution operation symbols $\|^n \in \Sigma_P, n \geq 1$ so that '$\|^1$' is the identity, '$\|^2$' is equal to '$\|$', and the term $\|^n(p_1, \dots, p_n)$ is equivalent to $p_1\|(\dots(p_{n-1}\|p_n)\dots)$. The assumption that the signature $\Sigma_P = \{nil, Act, \{a.\_\}_{a \in Act}, \{\|^n\}_{n \geq 2}\}$ is finite is not essential for the correspondence between models of Σ_P and algebras of $\mathcal{T}_P$. If Σ_P is infinite, one can define $\mathcal{T}_P$ via an infinite coproduct (in **Set** it is the disjoint union $\biguplus$), commonly written for a set of process variables in X (in **Set** the '$\times$' denotes the Cartesian product)

$$\Sigma_P(X) = \biguplus_{\sigma \in \Sigma_P} X^{ar(\sigma)} = Act \cup \{nil\} \bigcup_{a_i \in Act} \{X \times \{i\}\} \bigcup_{n \geq 2} X^n \tag{7.1}$$

and $X \uplus \Sigma_P(X) = X \cup Act \cup \{nil\} \cup \{X \times \{i\} | a_i \in Act, i = 1, 2, \ldots\} \bigcup_{n \geq 2} X^n$, where X represents the set of variables, $X \times \{i\}$ is the domain for the ith action-prefixing unary operator $a_i.\_$ in Σ_P, while the singleton $\{nil\}$ is isomorphic to the terminal object **1** in **Set** and the finite Cartesian products $X^n = \overbrace{X \times \cdots \times X}^{n}$ is used as a domain for the finitary n-ary operator $\|^n$. More formally, for the signature Σ_P we define the endofunctor (denoted by the same symbol) $\Sigma_P : \mathbf{Set} \longrightarrow \mathbf{Set}$ such that for any object B in **Set**, $\Sigma_P(B) \triangleq \biguplus_{\sigma \in \Sigma_P} B^{ar(\sigma)}$ defined by (7.1), and any arrow in **Set** (a function) $f : B \longrightarrow C$, $\Sigma_P(f) \triangleq \biguplus_{\sigma \in \Sigma_P} f^{ar(\sigma)}$.

Moreover, for each set of variables X (an object in **Set**) we define the endofunctor $(X \uplus \Sigma_P) : \mathbf{Set} \longrightarrow \mathbf{Set}$ such that for any object B in **Set**, $(X \uplus \Sigma_P)(B) \triangleq X \biguplus_{\sigma \in \Sigma_P} B^{ar(\sigma)}$, and any arrow in **Set** (a function) $f : B \longrightarrow C$, $(X \uplus \Sigma_P)(f) \triangleq \{id_X\} \biguplus_{\sigma \in \Sigma_P} f^{ar(\sigma)}$, where id_X is the identity function for the set X.

For the signature endofunctor Σ_P and every set of process variables X, the initial Σ_P-algebra (Lambek's Lemma) is an isomorphism (for the carrier set $\mathcal{T}_P X$),

$$(X \uplus \Sigma_P)(\mathcal{T}_P X) \simeq_I \mathcal{T}_P X$$

with $\simeq_I = [inl_X, inr_X] : X \uplus \Sigma_P(\mathcal{T}_P X) \longrightarrow \mathcal{T}_P X$, where $inl_X : X \hookrightarrow \mathcal{T}_P X$ and $inr_X : \Sigma_P(\mathcal{T}_P X) \longrightarrow \mathcal{T}_P X$ are two injective arrows of the coproduct $X \uplus \Sigma_P(\mathcal{T}_P X)$ in **Set**. The free Σ_P-algebra $\mathcal{T}_P X$ on a set of variables X may be computed as the union of the chain obtained by iterating the functor $X \uplus \Sigma_P$ on the empty set $\emptyset$

$$\begin{aligned} \emptyset \subseteq X \uplus \Sigma_P(\emptyset) &\subseteq X \uplus \Sigma_P\big(X \uplus \Sigma_P(\emptyset)\big) \\ &\subseteq X \uplus \Sigma_P\big(X \uplus \Sigma_P\big(X \uplus \Sigma_P(\emptyset)\big)\big) \subseteq \cdots \end{aligned}$$

with $\Sigma_P(\emptyset) = Act \cup \{nil\}$ (so that the initial $X \uplus \Sigma_P$-algebra $(\mathcal{T}_P X, \simeq_I)$ is the *least fixed point* of the functor $X \uplus \Sigma_P : \mathbf{Set} \to \mathbf{Set}$).

Consequently, the injection $inr_X : \Sigma_P(\mathcal{T}_P X) \longrightarrow \mathcal{T}_P X$ represents the composition of terms in the carrier set $\mathcal{T}_P X$ (it is identity for the subset $Act \cup \{nil\}$), for example:

1. For an element $(p_i, 4) \in \{\mathcal{T}_P X \times \{4\}\} \subseteq \Sigma_P(\mathcal{T}_P X)$ where $p_i \in X \subseteq \mathcal{T}_P X$ is a given variable, $inr_X(p_i, 4) = a_4.p_i \in \mathcal{T}_P X$ is a term;
2. For $(t_1, 3) \in \{\mathcal{T}_P X \times \{3\}\} \subseteq \Sigma_P(\mathcal{T}_P X)$ where t_1 is a term with variables in $\mathcal{T}_P X$, $inr_X(t_1, 3) = a_3.t_1 \in \mathcal{T}_P X$ is a term;
3. For $(t_1, t_2) \in \mathcal{T}_P X^2$, $inr_X(t_1, t_2) = t_1 \| t_2 \in \mathcal{T}_P X$ is a term.

Thus, for any other $(X \uplus \Sigma_P)$-algebra, $l = [f, h] : X \uplus \Sigma_P(Z) \longrightarrow Z$ (where $f : X \to Z$ and $h : \Sigma_P(Z) \to Z$, there is a unique arrow $f_\#$ to it from this initial $(X \uplus \Sigma_P)$-algebra such that $l \circ (X \uplus \Sigma_P)(f_\#) = f_\# \circ [inl_X, inr_X]$, where

$$\Sigma_P(f_\#) = id_{Act \cup \{nil\}} \biguplus_{a_i \in Act} f_{\#,i} \biguplus_{n \geq 2} \overbrace{f_\# \times \cdots \times f_\#}^{n}$$

with $f_{\#,i} : \mathcal{T}_P X \times \{i\} \to Z \times \{i\}$ for each action $a_i \in Act$ such that $f_{\#,i}(t, i) = (f_\#(t), i)$, and $f_\# \times \cdots \times f_\#(t_1, \ldots, t_n) = (f_\#(t_1), \ldots, f_\#(t_n))$ such that the left diagram in **Set** bellow commutes (note that the diagram (I1) corresponds to the unique homomorphism $f_\# : (\mathcal{T}_P X, \simeq_I) \to (Z, l)$, from the initial "syntax" $(X \uplus \Sigma_P)$-algebra to another $(X \uplus \Sigma_P)$-algebra)):

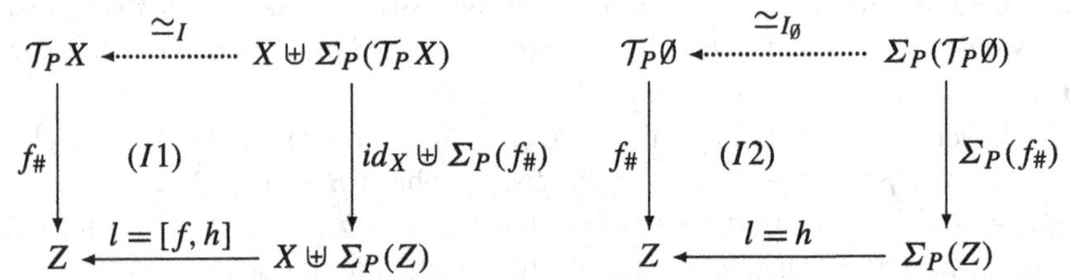

If the set of variables X is the empty set $\emptyset$ (thus f is the empty function, with empty graph), with $\emptyset \uplus S \simeq S$ in **Set**, we obtain the simpler commutative diagram above on the right, where the unique homomorphism $f_\# : \mathcal{T}_P \emptyset \to Z$ represents the *meaning* of the ground terms in $\mathcal{T}_P \emptyset$. Note that the right commutative diagram (I2) represents the unique homomorphisms $f_\#$ from the initial Σ_P-algebra of *ground terms* $(\mathcal{T}_P \emptyset, \simeq_{I_\emptyset})$ where $\simeq_{I_\emptyset}$ is an isomorphism equal to into inr_X for $X = \emptyset$, into the Σ_P-algebra (Z, h).

We can see in which way the mapping $f_\# : \mathcal{T}_P X \to Z$ is the unique inductive extension of the assignment $f : X \to Z$ (such that for any variable $p_k \in X$, $f_\#(p_k) = f(p_k)$) to all terms with variables in $\mathcal{T}_P X$, by these examples:

1. Let $(p_k) \in X$ be a free variable in X and hence from the left commutative diagram above $f_\#(\simeq_I (p_k)) = f_\#(p_k) = f(id_X(p_k)) = f(p_k)$. Thus, the reduction of $f_\#$ to $X \subseteq \mathcal{T}_P X$ is equal to the function f.
2. Let $(p_k, 4) \in X \times \{4\}$. Then, $\simeq_I(p_k, 4) = a_4.p_k \in \mathcal{T}_P X$, and from the commutativity $f_\#(\simeq_I(p_k, 4)) = f_\#(a_4.p_k) = h(f_{\#,4}(p_k, 4)) = h(f_\#(p_k), 4) = h(f(p_k), 4) =$ (when Z is a DB-denotational semantics, see bellow) $= a_4 \otimes f(p_k)$.
3. Let $(a_4.p_k, t_2) \in \mathcal{T}_P X^2$. Then, $\simeq_I(a_4.p_k, t_2) = a_4.p_k \parallel t_2 \in \mathcal{T}_P X$, and from the commutativity we have that $f_\#(\simeq_I(a_4.p_k, t_2)) = f_\#(a_4.p_k \parallel t_2) = h((f_\# \times f_\#)(a_4.p_k, t_2)) = h(f_\#(a_4.p_k), f_\#(t_2)) =$ (when Z is a DB-denotational semantics) $= (a_4 \otimes f(p_k)) + f_\#(t_2)$.

The inductive principle can be used to show that the operator $\mathcal{T}_P$ inductively extends to the *syntax* endofunctor (in Sect. 5.1.1) $\mathcal{T}_P : \mathbf{Set} \to \mathbf{Set}$ (the *syntax* monad freely generated by signature Σ_P).

The most interesting concrete syntax is an interpretation of the abstract $\mathcal{T}_P$ syntax, where '*nil*' is interpreted by $\bot^0$, any unary operation '$a_i.\_$' by the unary database operation $a_i \otimes \_$ and the binary operation '$\_\|\_$' by the **DB** coproduct $\_ + \_$ (and other finitary n-ary operations $\|^n$ by the corresponding **DB** n-ary coproducts, denoted by $+^n$). Thus:

Definition 53 The concrete interpretation of the abstract $\mathcal{T}_P$ syntax is defined by the following algebra isomorphism (which is identity for elements in *Act*):

$$\simeq_T : \left(\mathcal{T}_P X, Act, nil, \{a_i.\_\}_{a_i \in Act}, \left\{\|^n\right\}_{n \geq 2}\right)$$
$$\rightarrow \left(\widehat{\mathcal{T}}_P X, Act, \bot^0, \{a_i \otimes \_\}_{a_i \in Act}, \left\{+^n\right\}_{n \geq 2}\right)$$

between the abstract algebra and its concrete DB-interpretation.

This concrete DB-syntax will be denoted by $\widehat{\mathcal{T}}_P$, so that for any X there is a bijection between the set of abstract and these concrete terms, $\simeq_T : \mathcal{T}_P(X) \rightarrow \widehat{\mathcal{T}}_P(X)$, such that $\simeq_T(nil) = \bot^0$, $\simeq_T(a_i.t) = a_i \otimes (\simeq_T(t))$ and $\simeq_T(t_1 \| t_2) = (\simeq_T(t_1)) + (\simeq_T(t_2))$ (and analogously for all derived finitary operators '$\|^n$', $n \geq 3$).

In the case of the database-mapping DB-denotational semantics where $Z \subseteq Ob_{DB}$, given by Definition 52, with this concrete interpretation of the operators in Definition 53, the mapping $h : \Sigma_P Z \rightarrow Z$ represents the composition of objects in **DB** as ground terms of the grammar $(\mathcal{G}r_{DB})$ in Definition 52. In this case, each path $a_1 \otimes a_2 \otimes \cdots \otimes a_n$ of a given process $p \in X$ (which is an tree) is just the intersection of all information fluxes of the corresponding composition of arrows in **DB**, $f_n \circ \cdots \circ f_2 \circ f_1 : A \rightarrow B$, where A is the root state (a database) of this process and B is its final state (a database): thus, the *meaning of each path of such a process* is represented by the information transferred from the initial state A into the final state B of this process-path.

Note that, when the Σ_P-algebra $h : \Sigma_P(Z) \rightarrow Z$ is the *denotational semantics* for the initial (free) syntax algebra $\mathcal{T}_P X$ (when $Z \subseteq Ob_{DB}$ ($Act \subset Ob_{DB}$ as well), then $f : X \rightarrow Z$ is an assignment function which assigns to process variables in X the database-instances in Ob_{DB}, and $f_\# : \mathcal{T}_P X \rightarrow Z$ is its unique homomorphic extension to all terms with variables, called *initial algebra semantics* [17].

In what follows, we will investigate which assignment $f : X \rightarrow Z$ will be *adequate* in order to infer the operational behavior of database-mapping programs from the DB-denotational semantics.

7.3.2 Database-Mapping Processes and DB-Denotational Semantics

In order to define the database-mapping processes, for a given database-mapping program specified by a graph G, when the extension of a particular database schema is changed (by Insertion, Deletion or Updating (considered as deleting of old updated tuples and then by Insertion of new updated tuples)), we will consider the local modifications of the database schemas, caused by the propagation of the given initial modification of this particular database schema which becomes the root of an LTS.

Definition 54 Let $G = (V_G, E_G)$ be a graph of a given schema-database mapping program (from Definition 14 in Sect. 2.6 and Proposition 16 in Sect. 4.1.2). Then,

- For a given vertex $\mathcal{A} \in V_G$ we define a subgraph $G_{\mathcal{A}} \subseteq G$ composed of arrows $\mathcal{M}_{A\_} \in E_G$ such that $dom(\mathcal{M}_{A\_}) = \mathcal{A}$ and put them in the list $L_{\mathcal{A}}$ with $n \geq 0$ elements. Then, for a state (A, α^*), i.e., an instance-database $A = \alpha^*(\mathcal{A})$, we define the set $S_{(A,\alpha^*)}$ as follows:
  1. Set S to be the empty set, the pointer $i = 0$ to elements in $L_{\mathcal{A}}$, and $\alpha_1 = \alpha$.
  2. Set $\alpha = \alpha_1$ and $i = i + 1$. If $i > n$, goto 4.
  3. Take the ith arrow $\mathcal{M}_{AB}$ from the list $L_{\mathcal{A}}$. If there exists a transition $\alpha^* \stackrel{\mathcal{M}_{AB}}{\Longrightarrow} \alpha_1^*$ (from Proposition 32) such that $\alpha_1^* \neq \alpha^*$ then insert into S the tuple $\langle a, (\mathcal{B}, \alpha_1^*)\rangle$, where $a = \Delta(\alpha, MakeOperads(\mathcal{M}_{AB}))$. Goto 2.
  4. $S_{(A,\alpha^*)} = S$.
- For a given vertex $\mathcal{A} \in V_G$ we define a subgraph $G_{\mathcal{A}}^{OP} \subseteq G^{OP}$ composed of arrows $\mathcal{M}_{\_A}^{OP} \in G^{OP}$ such that $dom(\mathcal{M}_{\_A}^{OP}) = \mathcal{A}$ and put them in the list $L_{\mathcal{A}}$ with $n \geq 0$ elements. Then, for each state (A, α^*), i.e., an instance-database $A = \alpha^*(\mathcal{A})$, we define the set $S_{(A,\alpha^*)}^{OP}$ as follows:
  1. Set S to be the empty set, the pointer $i = 0$ to elements in $L_{\mathcal{A}}$, and $\alpha_1 = \alpha$.
  2. Set $\alpha = \alpha_1$ and $i = i + 1$. If $i > n$, goto 4.
  3. Take the ith arrow $\mathcal{M}_{BA}^{OP}$ from the list $L_{\mathcal{A}}$. If there exists a transition $\alpha^* \stackrel{\mathcal{M}_{BA}^{OP}}{\Longrightarrow} \alpha_1^*$ (from Proposition 31) such that $\alpha_1^* \neq \alpha^*$ then insert into S the tuple $\langle a, (\mathcal{B}, \alpha_1^*)\rangle$, where $a = \Delta(\alpha, MakeOperads(\mathcal{M}_{BA}))$. Goto 2.
  4. $S_{(A,\alpha^*)}^{OP} = S$.

Note that if $S_{(A,\alpha^*)}$ is not empty (i.e., $|S_{(A,\alpha^*)}| \geq 1$) then for each element

$$\langle a, (\mathcal{B}, \alpha_1^*)\rangle \in S_{(A,\alpha^*)} \quad \left(\text{or } \langle a, (\mathcal{B}, \alpha_1^*)\rangle \in S_{(A,\alpha^*)}^{OP}\right)$$

the mapping-interpretation $\alpha_1^* \in \mathbf{DB}^{\mathbf{Sch}(G_{\mathcal{A}})}$ (or $\alpha_1^* \in \mathbf{DB}^{\mathbf{Sch}(G_{\mathcal{A}}^{OP})}$) is a model of a database $\mathcal{B}$ and $\alpha_1^*(\mathcal{A}) = \alpha^*(\mathcal{A})$. We recall that in the definition of $S_{(A,\alpha^*)}$, all mapping arrows in $G_{\mathcal{A}}$ are satisfied (from Proposition 32) w.r.t. the extensions of all database schemas in $G_{\mathcal{A}}$.

An analogous result holds for $G_{\mathcal{A}}^{OP}$ as well (from Proposition 31). Based on this definition, we can define the semantics for the database-mapping processes by the equations, as follows:

DB-process algorithm $DBprog(G, \mathcal{A}, \alpha^*, \Delta A)$

Input. A graph $G = (V_G, E_G)$ of a database mapping program, a current model α^* of this system, the vertex $\mathcal{A} \in V_G$ (a database schema) where ΔA is a modification of its instance-database equal to A after such initial update which causes a modification of α^* by setting $\alpha * (\mathcal{A}) = A$. This process happens during an "insertion" processing described in Sect. 7.2.2 for the list $Y = S_{(A,\alpha^*)}$, or during "deletion" processing described in Sect. 7.2.1 for the list $Y = S^{OP}_{(A,\alpha^*)}$ given by Definition 54.

Output. A database-mapping process-program P composed of a set of flattened guarded equations with the set of process variables $p_k \in X$ and an assignment $ass : X \to \mathcal{S}$ where $\mathcal{S} \subseteq Ob_{DB}$ is the set of states.

1. Set $X = \{p_0, p_1\}$, $\mathcal{S}_{Gr} = \{(p_0, \bot^0, (\mathcal{A}_\emptyset, \alpha^*)), (p_1, A, (\mathcal{A}, \alpha^*))\}$, the set of equations $E = \{(p_0 = nil)\}$, the pointer to the variables $i = 1$.
2. Let $n = |Y|$. If $n = 0$ then insert the equation $(p_i = A)$ in E and goto 4.
   If $n \geq 2$ then set $N = j = 0$, L to be the empty list, and goto 3.
   Let $\langle a, (\mathcal{B}, \alpha_1^*)\rangle \in Y$.
   Control loop. If there exists an equation $(p_k = a.p_m) \in E$ with $(p_k, A) \in \pi_{\langle 1,2\rangle}\mathcal{S}_{Gr}$ and $(p_m, B) \in \pi_{\langle 1,2\rangle}\mathcal{S}_{Gr}$ such that $B = \alpha_1^*(\mathcal{B})$ then insert $(p_i = A)$ into E (i.e., interrupt a loop and close this path) and goto 4.
   Otherwise, insert new process variable p_{i+1} in X, $(p_{i+1}, \alpha_1^*(\mathcal{B}), (\mathcal{B}, \alpha_1^*))$ in $\mathcal{S}_{Gr}$ and the equation $(p_i = a.p_{i+1})$ in E. Goto 4.
3. $N = N + 1$. If $N \leq n$ goto 3.1.
   Let $j = |L|$ be the set of elements in the list L. If $j = 0$ (L is empty) then insert $(p_i = A)$ into E; Otherwise insert $(p_i = \|^j L)$. Goto 4.
   3.1. *Control loop*. If there exists an equation $(p_k = a.p_m) \in E$ with $(p_k, A) \in \pi_{\langle 1,2\rangle}\mathcal{S}_{Gr}$ and $(p_m, B) \in \pi_{\langle 1,2\rangle}\mathcal{S}_{Gr}$ such that $B = \alpha_1^*(\mathcal{B})$ then (i.e., interrupt a loop and close this path) and goto 3.
   Otherwise, $j = j + 1$. Insert p_{i+j} into L, p_{i+j}, p_{i+j+n} in X, $(p_{i+j} = a.p_{i+j+n})$ into E, and $(p_{i+j}, A, (\mathcal{B}, \alpha_1^*)), (p_{i+j+n}, \alpha_1^*(\mathcal{B}), (\mathcal{B}, \alpha_1^*))$ into $\mathcal{S}_{Gr}$. Goto 3.
4. If this process is interrupted by the administrator with an execution of the operation '*nil*', then set $X = \{p_0\}$, $\mathcal{S}_{Gr} = \{(p_0, \bot^0, (\mathcal{A}_\emptyset, \alpha^*))\}$, the set of equations $E = \{(p_0 = nil)\}$ and goto 6.
5. Continue the *forward* insertion (or *backward* deletion) chaining (a propagation) for the variable p_m, with the minimal index m such that it is not on the left side of any equation in E. If there is no such variable, goto 6.
   Set $(p_m, A, (\mathcal{A}, \alpha^*)) \in \mathcal{S}_{Gr}$, with $A = \alpha^*(\mathcal{A})$ and $Y = S_{(A,\alpha^*)}$ (or $Y = S^{OP}_{(A,\alpha^*)}$ during the deletion processing) given by Definition 54.
   If $|Y| \geq 1$ then set the index i to the value of this index of the *last* created program-variable and goto 2. Otherwise insert the equation $(p_m = A)$ into E and goto 5.
6. Stop. This process of generation of the new equations in this program P ends.
   Return the program P with equations in E and the assignment $ass : X \to \mathcal{S}$ (its graph is obtained by the projection on the first two columns in $\mathcal{S}_{Gr}$).

Note that the *Control loop* is an additional filtering w.r.t. the filtering provided by Definition 54 for computation of the sets $Y = S_{(A,\alpha^*)}$ (or $Y = S^{OP}_{(A,\alpha^*)}$ during the deletion processing) which controls that the target database of a transition is modified: by this control loop, we additionally control that such a transition is not a repetition of an equal previously executed transition, in order to interrupt the infinite *repetitive* loops. This control is affected by the universal machine $\mathbf{M}_{U_{DB}}$ during the execution of a program specified by a graph G. Note also that each path of the process ends with a leaf A equal to the instance-database which is the target of the last action a_n (a kernel of information flux transferred into A) so that $a_n \otimes A = Ta_n \cap TA = Ta_n$ and hence it preserves the meaning of the paths of each process described after Definition 53.

Remark The set of program equations E in P with the process variables in X, provided by the algorithm above, composes a *guarded system* of equations that has the unique (maximal fixed point) final coalgebra solution. In this guarded system, the term $\|^n(p_1, \ldots, p_n)$ (equivalent to $p_1 \| (\ldots (p_{n-1} \| p_n) \ldots))$ is considered as a *flattened* one-depth term with variables in X. Note that this process stops the execution when the administrator stops explicitly the execution by the operation '*nil*' (and resets all modifications produced by this program), or (for default) when the execution reaches the end or by the fact that the backward/forward process does not modify the databases further, or because the infinite loops are interrupted (it can happen only if the graph G is cyclic). Each equation in the system of equations E, obtained from this algorithm, has only one depth term on the right sides and the constants (leafs). Thus, such equations are 'flattened'. Notice that in the case of a node with $n \geq 2$ outgoing branches (described in point 3 of this algorithm), the direct transformation of transitions in Y into the equations would produce the non-flattened equation $p_i = \|(a_1.p_{i+1}, \ldots, a_n.p_{i+n})$, so, in order to obtain the flattened system of equations (in point 3 of this algorithm), we introduced the set of extra idle process transitions $p_i \xrightarrow{ass(p_{i+k})} p_{i+k}$ for $k = 1, \ldots, n$, that *do not change* the states (i.e., with $ass(p_{i+k}) = ass(p_i)$ for all $k = 1, \ldots, n$).

Let us consider a simple example:

Example 36 Let us consider the simple graph (a database-mapping program) G with three edges, $\mathcal{M}_{AB_1} : \mathcal{A} \to \mathcal{B}_1, \mathcal{M}_{AB_2} : \mathcal{A} \to \mathcal{B}_2$ and $\mathcal{M}_{AB_3} : \mathcal{A} \to \mathcal{B}_3$, and with a model α^* of $\mathbf{Sch}(G)$ in the case of a Strong Data Integration semantics. Let us insert a number of tuples into an instance-database $A' = \alpha^*(\mathcal{A})$ such that the updated database-instance is equal to $A = \alpha_1^*(\mathcal{A})$ that satisfy all three inter-schema mappings above (and hence α_1^* is a new model of the database mapping system in G), with the new updated databases for the schemas $\mathcal{B}_i$, $B_i = \alpha_1^*(\mathcal{B}_i), i = 1, 2, 3$, and $a_1 = \Delta(\alpha, MakeOperads(\mathcal{M}_{AB_1})) \in Act$ (and analogously, for a_2 and a_3).

Thus, our database process-program P, obtained from the *DBprog*$(G, \mathcal{A}, \alpha^*, \Delta A)$ algorithm has the set of process variables $p_0, p_1, \ldots, p_7 \in X$, with the graph of the mapping $ass : X \to \mathcal{S}$ equal to the set

$$\{(p_0, \bot^0), (p_1, A), (p_2, A), (p_3, A), (p_4, A), (p_5, B_1), (p_6, B_2), (p_7, B_3)\},$$

so that $ass(p_i) = A$ for $i = 1, \ldots, 4$, and $ass(p_{i+4}) = B_i$ for $i = 1, 2, 3$.

The system of guarded *flattened* equations of the program P provided by this algorithm is:

$$E = \{(p_0 = \bot^0), (p_1 = \|^3(p_2, p_3, p_4)), (p_2 = a_1.p_5),$$
$$(p_3 = a_2.p_6), (p_4 = a_3.p_7), (p_5 = B_1), (p_6 = B_2), (p_7 = B_3)\}.$$

Let us consider the non-flattened (direct) transition system on the left, and the flattened system (obtained from the algorithm '*DBprog*') on the right in the diagram bellow. Notice that the branching in the (direct) diagram on the left is expressed by the non-flattened equation $(p_1 = \|^3(a_1.p_2, a_2.p_3, a_3.p_4))$ while in the flattened system on the right by the equation $(p_1 = \|^3(ass(s_2).p_2, ass(s_3).p_3, ass(s_4).p_4))$ which is substituted by the *flattened* equation $(p_1 = \|^3(p_2, p_3, p_4)) \in E$.

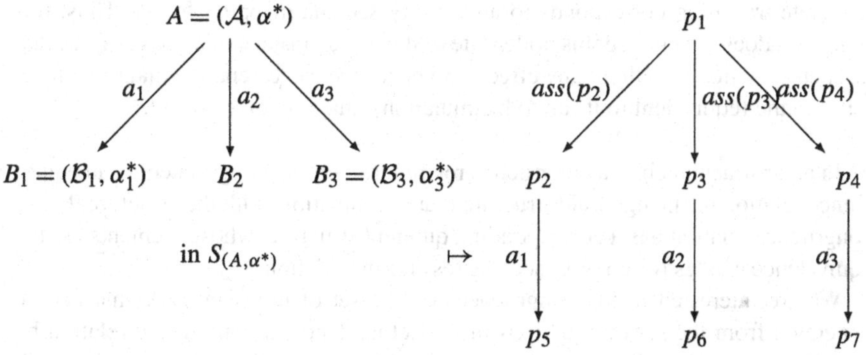

Let us show that this replacement (i.e., flattening) is valid: we denote the solution of the set of equations E for a variable p_i by $\mathbf{s}_E(p_i)$ and its value (i.e., denotation) by $\simeq_T(\mathbf{s}_E(p_i))$. For generality, we will demonstrate it for any branching node p_i with $n \geq 2$ outgoing branches (in our case $i = 1, n = 3$):

$$\begin{aligned}
&\simeq_T(\mathbf{s}_E(p_i)) \\
&= \simeq_T(\mathbf{s}_E(\|^n(ass(p_{i+1}).p_{i+1}, \dots, ass(p_{i+n}).p_{i+n}))) \\
&= +^n(ass(p_{i+1}) \otimes \simeq_T(\mathbf{s}_E(p_{i+n})), \dots, ass(p_{i+n}) \otimes \simeq_T(\mathbf{s}_E(p_{i+n}))) \\
&= +^n(ass(p_{i+1}) \otimes \simeq_T(\mathbf{s}_E(a_1.p_{i+n+1})), \dots, ass(p_{i+n}) \otimes \simeq_T(\mathbf{s}_E(a_n.p_{i+2n}))) \\
&= +^n(ass(p_{i+1}) \otimes a_1 \otimes \simeq_T(\mathbf{s}_E(p_{i+n+1})), \dots, ass(p_{i+n}) \otimes a_n \otimes \simeq_T(\mathbf{s}_E(p_{i+2n})))
\end{aligned}$$

(from $ass(p_{i+k}) \otimes a_k = Tass(p_{i+k}) \cap Ta_k = Ta_k$, $k = 1, \dots, n$ because Ta_k is the information flux from the instance-database $ass(p_{i+k}) = ass(p_i)$ for all $k = 1, \dots, n$, and hence $Ta_k \subseteq Tass(p_{i+k})$. Here, in our example $ass(p_{i+k}) = ass(p_i) = A$)

$$\begin{aligned}
&= +^n(a_1 \otimes \simeq_T(\mathbf{s}_E(p_{i+n+1})), \dots, a_n \otimes \simeq_T(\mathbf{s}_E(p_{i+2n}))) \\
&= +^n(\simeq_T(a_1.\mathbf{s}_E(p_{i+n+1})), \dots, \simeq_T(a_n.\mathbf{s}_E(p_{i+2n}))) \\
&= +^n(\simeq_T(\mathbf{s}_E(a_1.p_{i+n+1})), \dots, \simeq_T(\mathbf{s}_E(a_n.p_{i+2n})))
\end{aligned}$$

$$= +^n\big(\simeq_T\big(\mathbf{s}_E(p_{i+1})\big), \ldots, \simeq_T\big(\mathbf{s}_E(p_{i+n})\big)\big)$$
$$= \simeq_T\big(\mathbf{s}_E\big(\|^n(p_{i+1}, \ldots, p_{i+n})\big)\big).$$

Thus, the replacing of the non-flattened equation

$$\big(p_i = \|^n\big(ass(p_{i+1}).p_{i+1}, \ldots, ass(p_{i+n}).p_{i+n}\big)\big)$$

with the flattened equation $(p_1 = \|^n(p_{i+1}, \ldots, p_{i+n})) \in E$ is valid.

Consider that, from the fact that $ass(p_{i+k}) = ass(p_i)$ for $k = 1, \ldots, n$, each state-transition $p_i \xrightarrow{ass(p_{i+k})} p_{i+k}$ is the instance-database of a schema $\mathcal{A}$ in G and hence this state-transition corresponds to an identity schema mapping for $\mathcal{A}$. Thus, this transition does not change this node state $ass(p_i)$ (i.e., instance-database of $\mathcal{A}$), that is, such an action is without any effect, and hence we can extend the diagram on the left into the (equivalent to it (up to bisimulation)) diagram on the right.

In an abstract algebra, a congruence relation (or simply congruence) is an equivalence relation on an algebraic structure that is compatible with the structure. Every congruence relation has a corresponding quotient structure, whose elements are the equivalence classes (or congruence classes) for the relation.

We are interested in the congruences on the set of terms in $\mathcal{T}_P X$ that can be generated from the set of equations in E. Let us denote a congruence relation by $\equiv \subseteq \mathcal{T}_P X \times \mathcal{T}_P X$, which is an equivalence relation on $\mathcal{T}_P X$, so that: if $(t_1, t_2) \in E$ then $(t_1, t_2) \in \equiv$ (written also as $t_1 \equiv t_2$), and for each n-ary algebraic operation σ in the signature Σ_P, $t_i \equiv t'_i$ implies $\sigma(t_1, \ldots, t_n) \equiv \sigma(t'_1, \ldots, t'_n)$.

Thus, $\equiv$ is a congruence relation if it is an equivalence relation (which includes equations in E) and a subalgebra of the product $\mathcal{T}_P X \times \mathcal{T}_P X$.

The set of congruences of an algebraic system is a complete lattice. The meet is the usual set intersection. The join (of an arbitrary number of congruences) is the join of the underlying equivalence relations. This join corresponds to the subalgebra (of $\mathcal{T}_P X \times \mathcal{T}_P X$) generated by the union of the underlying sets of the congruences.

In what follows, we are interested in a particular congruence based on the guarded set of equations E on $\mathcal{T}_P X$ and the equivalence classes with representation elements in X.

Given a guarded equation $(p_k = t) \in E$, where $p_k \in X \subseteq \mathcal{T}_P X$ and $t \in \mathcal{T}_P X$, we denote the equivalence class equal to the set of terms $\{t \in \mathcal{T}_P X \mid p_k = t\}$ by $[p_k]_E$, so that $p_k \in [p_k]_E$, as follows:

Definition 55 For a given guarded set of equations E on $\mathcal{T}_P X$, of a program P obtained from the algorithm *DBprog*, we define the following set of equivalence classes in $\mathcal{T}_P X$:

1. The equivalence class $[p_k]_E$ of $p_k \in \mathcal{T}_P X$ determined by the guarded set of equations in E, for each variable $p_k \in X \subseteq \mathcal{T}_P X$, with $k \geq 1$.
2. The last equivalence class $[p_0]_E = \mathcal{T}_P X \backslash D$, for the variable $p_0 \in X$, where $D = \bigcup_{p_k \in X, k \geq 1} [p_k]_E$.

3. We introduce the bijection $[\_]_E : X \to \mathcal{T}_P X/_\equiv$, where $\mathcal{T}_P X/_\equiv$ is the quotient Σ_P algebra of terms with the congruence $\equiv$ defined for any two terms $t_1, t_2 \in \mathcal{T}_P X$ by $t_1 \equiv t_2$ iff $[t_1]_E = [t_2]_E$. Furthermore, for each n-ary operator $\sigma \in \Sigma$, in this quotient algebra of terms $\mathcal{T}_P X/_\equiv$ we have the corresponding n-ary operator $\sigma_\equiv$ such that $\sigma_\equiv([t_1]_E, \dots, [t_n]_E) = [\sigma(t_1, \dots, t_n)]_E$.

From this definition, for any two terms $t_1, t_2 \in \mathcal{T}_P X$, $t_1 \equiv t_2$ iff $t_1, t_2 \in [p_k]_E$ for some $p_k \in X$.

The following properties for a flattened system of equations obtained by the algorithm *DBprog* are valid:

Proposition 34 *For any flattened system of (also infinite) equations E with the assignment ass* $: X \to \mathcal{S}$*, obtained by the algorithm DBprog from a database-mapping program specified by a graph G, there exist the following mappings:*

- *A* $\mathcal{T}_P$*-coalgebra* $\mathbf{g}_E = inr_X \circ \Sigma_P(inl_X) \circ f_T : X \to \mathcal{T}_P X$*, where* $f_T : X \to \Sigma_P(X)$ *is a flattening mapping such that for any variable* $p_i \in X$,

$$f_T(p_i) = \begin{cases} a & \text{if } (p_i = a) \in E,\ a \in Act \cup \{nil\}; \\ (p_k, m) & \text{if } (p_i = a_m.p_k) \in E; \\ (p_{k_1}, \dots, p_{k_n}) & \text{if } (p_i = \|^n(p_{k_1}, \dots, p_{k_n})) \in E. \end{cases}$$

  Thus, $\mathbf{g}_E(p_i) = t \in \mathcal{T}_P X$ *if* $(p_i = t) \in E$.
- *A* $\mathcal{T}_P$*-algebra* $\mathbf{g}^E : \mathcal{T}_P X \to X$ *such that for any term* $t \in \mathcal{T}_P X$, $\mathbf{g}^E(t) = p_i$ *if* $t \in [p_i]_E$*, where* $[\_]_E : X \to \mathcal{T}_P X/_\equiv$ *is a bijection in Definition* 55.

  Hence, $\mathbf{g}^E \circ \mathbf{g}_E = id_X : X \to X$.
- *A mapping* $[\_] = \langle ass, \wp\rangle : X \to \mathcal{S} \times \mathcal{P}_{\text{fin}}(Act \times X)$*, where ass* $: X \to \mathcal{S}$ *is the assignment for variables in the equations in E, and* $\wp : X \to \mathcal{P}_{\text{fin}}(Act \times X)$ *is defined by: for any* $p_k \in X$,

$$\wp(p_k) = \begin{cases} \bigcup_{1 \le i \le n}\{(ass(p_{k+i}), p_{k+i})\} & \text{if } \mathbf{g}_E(p_k) = \|^n(p_{k+1}, \dots, p_{k+n}); \\ \{(a, p_{k+1})\} & \text{if } \mathbf{g}_E(p_k) = a.p_{k+1}; \\ \emptyset & \text{otherwise,} \end{cases}$$

  where $\mathcal{P}_{\text{fin}}$ *is the powerset operator for all* finite *subsets* ($\emptyset \in \mathcal{P}_{\text{fin}}(Z)$ *for any, including the empty, set Z*).

Proof From the algorithm *DBprog*, we have no two different equations with the same program variable on the left side, so that $\mathbf{g}_E$ is a function, and also $\mathbf{g}^E \circ \mathbf{g}_E = id_X$: for any $p_i \in X$, $\mathbf{g}_E(p_i) = t$ if $(p_i = t) \in E \subseteq \equiv$, so that $t \in [p_i]_E$, and $\mathbf{g}^E(t) = p_i$. □

Notice that from the algorithm *DBprog*, we always have a variable $p_0 \in X$ such that $\mathbf{g}_E(p_0) = nil$ and $\mathbf{g}^E(nil) = p_0$ because $(p_0 = nil) \in E$ always.

Proposition 35 *Given a flattened guarded system of equations E of a database-mapping program represented by a graph G, there is a unique $(X \uplus \Sigma_P)$-homomorphism $id_{X\#} : (\mathcal{T}_P X, \simeq_I) \rightarrow (X, [id_X, h_X])$ such that $id_{X\#} = \mathbf{g}^E$, where the mapping $h_X : \Sigma_P X \rightarrow X$ is defined by:*

1. *$nil \mapsto p_0$, and for any $A \in Act$, $A \mapsto p_k$ if $(p_k = A) \in E$; p_0 otherwise;*
2. *$(p_k, i) \mapsto p_m$ if $(p_m = a_i.p_k) \in E$; p_0 otherwise;*
3. *$(p_{i_1}, \ldots, p_{i_n}) \mapsto p_m$ if $(p_m = \|^n(p_{i_1}, \ldots, p_{i_n})) \in E$; p_0 otherwise,*

so that the following initial algebra semantics diagram commutes:

$$\begin{array}{ccccc}
X & \xrightarrow{inl_X} & \mathcal{T}_P X & \xleftarrow{inr_X} & \Sigma_P(\mathcal{T}_P X) \\
 & \searrow^{id_X} & \downarrow{\scriptstyle id_{X\#} = \mathbf{g}^E} & & \downarrow{\scriptstyle \Sigma_P(id_{X\#})} \\
 & & X & \xleftarrow{h_X} & \Sigma_P(X)
\end{array}$$

Proof It is easy to verify that the square diagram above commutes for $\mathbf{g}^E$: for each $p_k \in X$, $inl_X(p_k) = p_k$, with $\mathbf{g}^E(p_k) = p_k$, because $p_k \in [p_k]_E$, so that the left diagram commutes.

1. For $nil \in \Sigma_P(\mathcal{T}_P X)$, $\mathbf{g}^E(inr_X(nil)) = \mathbf{g}^E(nil) = p_0$ because from $(p_0 = nil) \in E$ we have $nil \in [p_0]_E$. At the same time, $h_X(\Sigma_P(\mathbf{g}^E(nil))) = h_X(nil) = p_0$. Similarly, for each $A \in Act$ (if $A \notin [p_i]_E$ for each $p_i \in X, i \geq 1$, then $A \in [p_0]_E$).
2. For $(t, i) \in \Sigma_P(\mathcal{T}_P X)$ we have two cases:
   2.1. $\mathbf{g}^E(t) = p_0$, i.e., $t \in [p_0]_E$, so that $h_X(\Sigma_P(\mathbf{g}^E(t, i))) = h_X(p_0, i) = p_0$ (there is no equation $(p_0 = a_i.p_0)$ in E). While $\mathbf{g}^E(inr_X(t, i)) = \mathbf{g}^E(a_i.t) = p_0$ because if $t \in [p_0]_E$ then also $a_i.t \in [p_0]_E$.
   2.2. $\mathbf{g}^E(t) = p_k$, i.e., $t \in [p_k]_E$ or $t \equiv p_k$, $k \geq 1$. This time $h_X(\Sigma_P(\mathbf{g}^E(t, i))) = h_X(p_k, i) = p_m$ if $(p_m = a_i.p_k) \in E$ (or if $a_i.p_k \in [p_m]_E$, i.e., from $t \equiv p_k$, $a_i.t \in [p_m]_E$); p_0 otherwise. While, $\mathbf{g}^E(inr_X(t, i)) = \mathbf{g}^E(a_i.t) = p_m$ if $a_i.t \in [p_m]_E$); p_0 otherwise. Thus, the commutativity is satisfied.
3. For $(t_1, \ldots, t_n) \in \Sigma_P(\mathcal{T}_P X)$, we have two cases:
   3.1. If $p_{i_k} \equiv t_k$ with $i_k > 0$ for all $k = 1, \ldots, n$ then
   $$h_X\big(\Sigma_P\big(\mathbf{g}^E(t_1, \ldots, t_n)\big)\big) = h_X(p_{i_1}, \ldots, p_{i_n}) = p_m$$
   if $(p_m = \|^n(p_{i_1}, \ldots, p_{i_n})) \in E$, i.e., if $\|^n(p_{i_1}, \ldots, p_{i_n}) \in [p_m]_E$; p_0 otherwise. While, $\mathbf{g}^E(inr_X(t_1, \ldots, t_n)) = \mathbf{g}^E(\|^n(t_1, \ldots, t_n)) = p_m$ if $\|^n(t_1, \ldots, t_n) \in [p_m]_E$ and, from $p_{i_k} \equiv t_k$, $\|^n(p_{i_1}, \ldots, p_{i_n}) \in [p_m]_E$; p_0 otherwise. Thus, the commutativity is satisfied.
   3.2. If there exists $1 \leq k \leq n$ such that $t_k \equiv p_0$ then
   $$h_X\big(\Sigma_P\big(\mathbf{g}^E(t_1, \ldots, t_k, \ldots, t_n)\big)\big) = h_X(p_{i_1}, \ldots, p_0, \ldots, p_{i_n}) = p_0$$
   because there does not exist the equation $(p_m = \|^n(p_{i_1}, \ldots, p_0, \ldots, p_{i_n}))$ in E. At the same time,

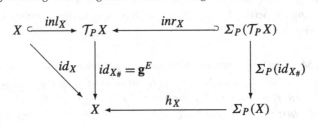

$$\mathbf{g}^E\left(inr_X(t_1,\ldots,t_k,\ldots,t_n)\right) = \mathbf{g}^E\left(\|^n(t_1,\ldots,t_k,\ldots,t_n)\right)$$
$$\text{(from } p_{i_k} \equiv t_k\text{)}$$
$$= \mathbf{g}^E\,\|^n(p_{i_1},\ldots,p_0,\ldots,p_{i_n}) \in [p_m]_E$$

for $m = 0$ (because if $m \geq 1$ then we would have the equation ($p_m = \|^n(p_{i_1},\ldots,p_0,\ldots,p_{i_n})$) in E, which is impossible). Thus, the commutativity is satisfied. From the fact that this is a (unique) homomorphism from the initial $(X \uplus \Sigma_P)$-algebra into $(X \uplus \Sigma_P)$-algebra $(X, [id_X, h_X])$ we conclude that $id_{X_\#} = \mathbf{g}^E$. □

Note that this flattened system of equations with program variables can be given by a function $\mathbf{g}_E : X \to \mathcal{T}_P(X)$ which is a *coalgebra* of the endofunctor (and the syntax monad) $\mathcal{T}_P(\_) : \mathbf{Set} \to \mathbf{Set}$ of the signature Σ_P.

Proposition 36 *Let $\mathbf{g}_E : X \to \mathcal{T}_P(X)$ be the guarded system of flattened equations for a database mapping process-program P. There is a* unique *solution for such a system of equations, given by the function $\mathbf{s}_E : X \to \mathcal{T}_\infty$, where $\mathcal{T}_\infty$ denotes the Σ_P algebra of all finite and infinite Σ_P-labeled trees (of ground terms).*

That is, $\mathcal{T}_\infty$ is the greatest fixed point of the endofunctor $\mathcal{T}_P : \mathbf{Set} \to \mathbf{Set}$, denoted by $gfp(\mathcal{T}_P)$ (i.e., $\mathcal{T}_P(gfp(\mathcal{T}_P)) = gfp(\mathcal{T}_P)$), so that the set of finite ground terms $\mathcal{T}_P(\emptyset) \subset \mathcal{T}_\infty$ is its strict subset (here $\emptyset$ denotes the empty set X of variables).

Proof We have seen that the algebra signature Σ_P of the database-mapping process has only one nullary operator *nil* (the 'stop of execution' operation), the finitary n-ary ($n \geq 2$) operator $\|^n$) and the set of unary operators (as $a_i.(\_)$ for example, for any $a_i \in Act$). Thus, the syntax monad (endofunctor) $\mathcal{T}_P : \mathbf{Set} \to \mathbf{Set}$ is polynomial. It is known [4] that generalized polynomial endofunctors of **Set** are iteratable and hence $\mathcal{T}_P$ has a final coalgebra $\simeq_\Sigma : \mathcal{T}_\infty \to \mathcal{T}_P(\mathcal{T}_\infty)$ which is (by Lambek's theorem) an isomorphism. Consequently, from the system of flattened equations given by $\mathbf{g}_E : X \to \mathcal{T}_P(X)$, which is a $\mathcal{T}_P(\_)$-coalgebra, we obtain the unique homomorphism $\mathbf{s}_E : X \to \mathcal{T}_\infty$ into this final coalgebra: this homomorphism assigns to each process variable in X a ground term which corresponds to the solution for this process variable. The solutions for the set of variables in X satisfy the flattened set of equations in Proposition 34. □

Note that $\mathcal{T}_\infty$ is computed by iterating the functor $\mathcal{T}_P$ on the empty set of variables $\emptyset$ up to the greatest fixed point (gfp), i.e., by the union of the chain

$$\mathcal{T}_P\emptyset \subseteq \mathcal{T}_P(\mathcal{T}_P\emptyset) \subseteq \cdots$$

of the least fixed points $\mathcal{T}_P\emptyset = lfp(\emptyset \uplus \Sigma_P)$, $\mathcal{T}_P(\mathcal{T}_P\emptyset) = lfp(\mathcal{T}_P\emptyset \uplus \Sigma_P)$, etc.

The most interesting concrete syntax is an interpretation of the abstract $\mathcal{T}_P$ syntax, given by Definition 53, where '*nil*' is interpreted by $\bot^0$, any unary operation '$a_i.\_$' by the unary database operation $a_i \otimes \_$ and the finitary n-ary operations $\|^n$,

≥ 2 by the corresponding **DB** n-ary coproducts $+^n$. This concrete DB-syntax in Definition 53 is provided by a bijection between the set of abstract and these concrete terms, $\simeq_T : \mathcal{T}_P(X) \to \widehat{\mathcal{T}}_P(X)$, such that $\simeq_T(nil) = \bot^0$, $\simeq_T(a_i.t) = a_i \otimes (\simeq_T(t))$, and $\simeq_T(\|^n(t_1,\dots,t_n)) = +^n((\simeq_T(t_1)),\dots,(\simeq_T(t_2)))$.

Consequently, for the set of abstract trees $\mathcal{T}_\infty$ we define its corresponding set of concrete trees $\mathbb{T}_\infty$, with the bijection $\simeq_T : \mathcal{T}_P(X) \to \widehat{\mathcal{T}}_P(X)$ and (for the ground terms only) $\simeq_T : \mathcal{T}_\infty \to \mathbb{T}_\infty$. Now we are able to define the semantics of process-programs based on the coalgebras, both for the abstract $\mathcal{T}_P$ and concrete DB-syntax $\widehat{\mathcal{T}}_P$:

Proposition 37 *Let* $\mathbf{g}_E : X \to \mathcal{T}_P(X)$ *be a guarded system of equations for a database mapping process-program* P *(in Proposition* 34*). Thus, the coalgebraic semantics for the database mapping process-programs can be given by the following commutative diagram in* **Set**:

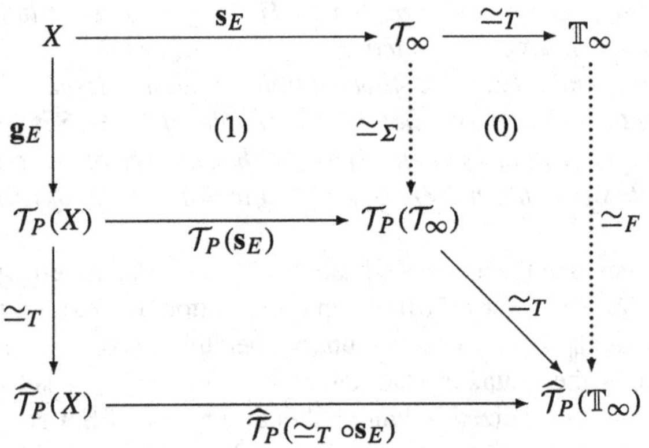

where $\simeq_T$ *is the isomorphism between the abstract and DB-terms in Definition* 53*, so that the terms in* $\mathbb{T}_\infty$ *are the objects in* **DB**, *that is, the instance-databases. The isomorphism* $\simeq_F$ *is the concrete interpretation of the Lambek's final coalgebra isomorphism* $\simeq_\Sigma$. *Consequently, we are able to use* $\mathbb{T}_\infty \subseteq Ob_{DB}$ *as a DB-denotational semantics for the abstract* Σ_P*-algebras.*

Proof From [4], the inner commutative diagram (1) represents the final coalgebra semantics for Σ_P, while the diagram (0) is composed of only isomorphisms, so that $\mathbb{T}_\infty \subseteq Ob_{DB}$ can be used as an adequate DB-denotational semantics algebra $h : \Sigma_P(\mathbb{T}_\infty) \to \mathbb{T}_\infty$, represented by the (syntax) initial algebra commutative diagram in Sect. 7.3.1, where $Z = \mathbb{T}_\infty$, such that $\Sigma_P(\mathbb{T}_\infty) = Act \cup \{nil\} \cup \{\mathbb{T}_\infty \times \{i\} | a_i \in Act\} \bigcup_{n\geq 2} \mathbb{T}^n_\infty$, with h being the identity for Act, $h(nil) = \bot^0$, and for $t_1, t_2 \in \mathbb{T}_\infty$, $h(t_1, i) = a_i \otimes t_1$, $h(t_1,\dots,t_n) = +^n(t_1,\dots,t_n)$.

So we obtain the following initial algebra semantics commutative diagram:

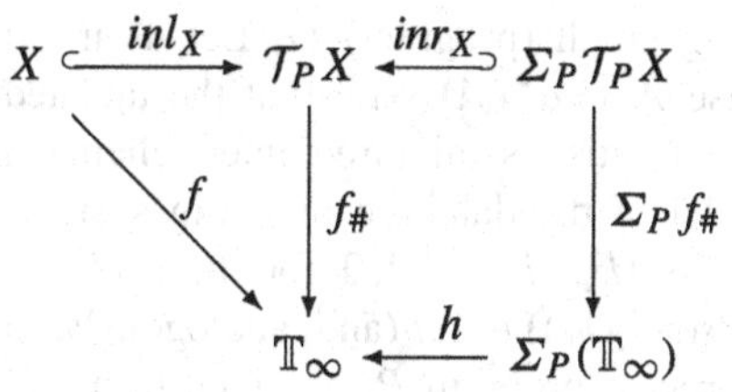

Note that the set of infinite ground terms $\mathcal{T}_\infty$ cannot be obtained in the inductive way we defined the set $\mathcal{T}_P(X)$ (when $X = \emptyset$).

The external commutative diagram of this proposition represents the final coalgebra semantics diagram analog to (1), but for its concrete DB-interpretation.

The isomorphism $\simeq_T : \mathcal{T}_\infty \to \mathbb{T}_\infty$ satisfies the homomorphic properties. Indeed, for any two terms $t, t' \in \mathcal{T}_\infty$,

$$\simeq_T(nil) = \bot^0, \ \simeq_T(a.t) = \simeq_T(a.\_)\bigl(\simeq_T(t)\bigr) = a \otimes \bigl(\simeq_T(t)\bigr),$$
$$\simeq_T\bigl(\|^n(t_1, \dots, t_n)\bigr) = \simeq_T\bigl(\|^n\bigr)\bigl(\simeq_T(t_1), \dots, \simeq_T(t_n)\bigr) = +^n\bigl(\simeq_T(t_1), \dots, \simeq_T(t_n)\bigr).$$

Thus, the commutative diagram (0) is composed of the isomorphic arrows (i.e., the bijections) in **Set**. □

Consequently, the trees in $\mathbb{T}_\infty$ are obtained from the trees in $\mathcal{T}_\infty$ by replacing the nodes $a \in Act, nil, a.\_$ and $\|^n$ with $a \in Act, \bot^0, a \otimes \_$ and $+^n$, respectively. Therefore, $\simeq_F : \mathbb{T}_\infty \to \widehat{\mathcal{T}_P}(\mathbb{T}_\infty)$ is a 'final' isomorphism as well, between the set of infinite trees that are possible solutions of the database-mapping process-programs.

For any given assignment $f : X \to \mathbb{T}_\infty$, the $(X \uplus \Sigma_P)$-algebra,

$$[f, h] : X \uplus \Sigma_P(\mathbb{T}_\infty) \to \mathbb{T}_\infty$$

of the diagram in the proof of Proposition 37 will be called as *DB-denotational semantics* for the initial $\mathcal{T}_P X$ algebra of the database-mapping process language with the set X of process variables.

Only for a particular assignment f this DB-denotational algebra is *adequate* for the operational semantics, as it will be explained in what follows. Notice that two terms in $\mathbb{T}_\infty$ (or, equivalently, in $\mathcal{T}_\infty$) are equal iff they are congruent (w.r.t. the set of equations of the operator $\|$, that is, its commutativity and associativity equations).

The final coalgebra semantics presented above is not observational, as is characteristic property of the behavior-based operational semantics. Such a behavior-based observational semantics will be presented in the next Sect. 7.4. Two precesses in the operational semantics (i.e., LTS trees) are equal if they are *bisimilar*, consequently, from the result above, if the terms (corresponding to these LTS trees) are congruent.

Example 37 Let us consider the continuation of Strong Data Integration semantics in Example 36, of the graph G with three edges, $\mathcal{M}_{AB_1} : \mathcal{A} \to \mathcal{B}_1, \mathcal{M}_{AB_2} : \mathcal{A} \to \mathcal{B}_2$ and $\mathcal{M}_{AB_3} : \mathcal{A} \to \mathcal{B}_3$, with a model (functor) $\alpha^* \in Mod(G) \subseteq \mathbf{DB}^{\mathbf{Sch}(G)}$

of this database-mapping graph (program) G. Let us insert a number of tuples into an instance-database $A' = \alpha^*(\mathcal{A})$ such that the updated database-instance is equal to $A = \alpha_1^*(\mathcal{A})$ and satisfies all three inter-schema mappings above (and hence α_1^* is a new model of the database mapping system in G), with the new updated databases $B_i = \alpha_1^*(\mathcal{B}_i), i = 1, 2, 3$ for the schemas $\mathcal{B}_1, \mathcal{B}_2$ and $\mathcal{B}_3$, and $a_1 = \Delta(\alpha, MakeOperads(\mathcal{M}_{AB_1})) \in Act$ (and, analogously, for a_2 and a_3).

Thus, our database process-program P, obtained from the algorithm '*DBprog*', with set of process variables $p_0, p_1, \ldots, p_7 \in X$, is the system of guarded *flattened* equations of the program P specified by algorithm *DBprog*:

$$E = \{(p_0 = \bot^0), (p_1 = \|^3(p_2, p_3, p_4)), (p_2 = a_1.p_5), \\ (p_3 = a_2.p_6), (p4 = a_3.p_7), (p_5 = B_1), (p_6 = B_2), (p_7 = B_3)\}.$$

Hence, for $i = 1, 2, 3$,

$$\mathbf{g}_E(p_1) = \|^3(p_2, p_3, p_4) = \|(p_2, \|(p_3, p_4)) \in \mathcal{T}_P(X), \\ \mathbf{g}_E(p_{i+1}) = a_i.p_{i+4}, \mathbf{g}_E(p_{i+4}) = B_i \in \mathcal{P}(\underline{\Upsilon}) = Act \subseteq \mathcal{T}_P(\emptyset) \subseteq \mathcal{T}_P(X)$$

as follows from the algorithm '*DBprog*' and the fact that $\alpha_1^* \in Mod(\mathbf{Sch}(G))$.

The solution of this set of equation is given by:

$$\simeq_T(\mathbf{s}_E(p_0)) = \bot^0, \\ \simeq_T(\mathbf{s}_E(p_{i+4})) = B_i \in \mathbb{T}_\infty, \quad i = 1, 2, 3, \\ \simeq_T(\mathbf{s}_E(p_{i+1})) = a_i \otimes B_i \in \mathbb{T}_\infty, \quad i = 1, 2, 3, \\ \simeq_T(\mathbf{s}_E(p_1)) = +(a_1 \otimes B_1, +(a_2 \otimes B_2, a_3 \otimes B_3)) \in \mathbb{T}_\infty.$$

We have the following equivalence classes:

$$[p_{i+4}]_E = \{p_{i+4}, B_i\}, \qquad [p_{i+1}]_E = \{p_{i+1}, a_i \otimes B_i\}, \quad \text{for } i = 1, 2, 3.$$

And

$$[p_1]_E = \{\|^3(p_2, p_3, p_4), \|^3(a_1 \otimes B_1, p_3, p_4), \|^3(p_2, a_2 \otimes B_2, p_4), \\ \|^3(p_2, p_3, a_3 \otimes B_3), \|^3(a_1 \otimes B_1, a_2 \otimes B_2, p_4), \|^3(a_1 \otimes B_1, p_3, a_3 \otimes B_3), \\ \|^3(p_2, a_2 \otimes B_2, a_3 \otimes B_3), \|^3(a_1 \otimes B_1, a_2 \otimes B_2, a_3 \otimes B_3), \\ \text{and all permutations of these tuples}\}.$$

At the same time,

$$[p_0]_E = \{p_0, nil, \bot^0\} \\ \cup \{\text{all other terms that are not in the above defined equivalence classes}\}.$$

It is easy to verify that for each p_i, $\{p_i, \mathbf{s}_E(p_i)\} \subseteq [p_i]_E$.

The solution of p_1 is the following tree (note that here we are using only the binary operators '$\|^2$', that is, '$\|$', instead of the unique operator '$\|^3$'):

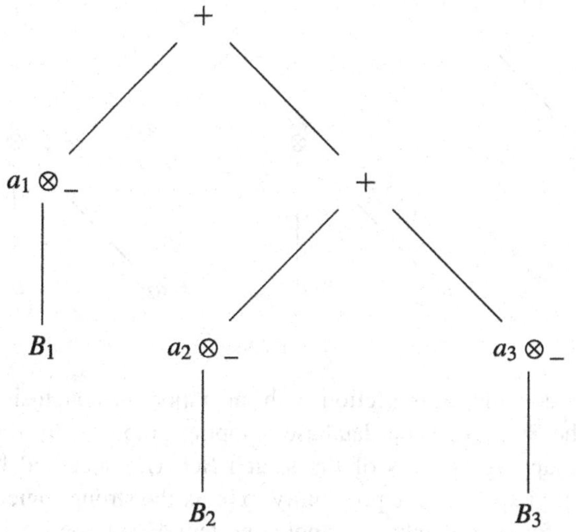

Alternatively, if we consider '$+^3$' as a finite 3-ary coproduct operator (analogous to '$\|^3$' in the guarded system of equations) then we obtain a very simple tree

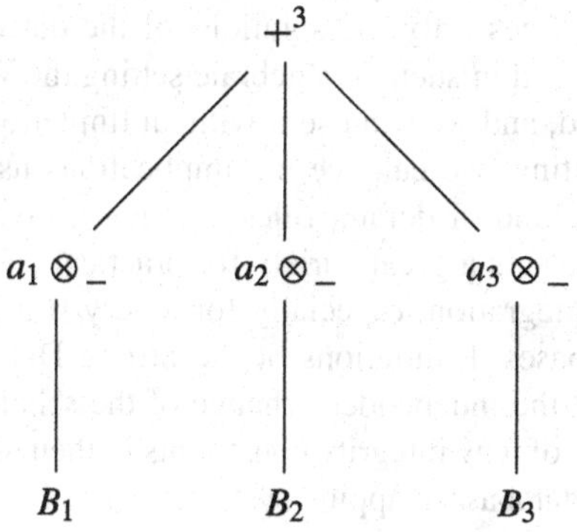

that is similar to the transition tree if instead of the nodes '$+^n$', $n \geq 2$, we use the states of a transition system, and instead of the unary nodes '$a \otimes \_$' we use the a-labeled edges that finish with another state, as we will see in the next section for the operational semantics of database-mapping process-programs. Note that if we simplify the family of action-prefixing unary operators '$a.\_$' by the binary connective '.' then we would obtain the following tree with leafs in *Act*:

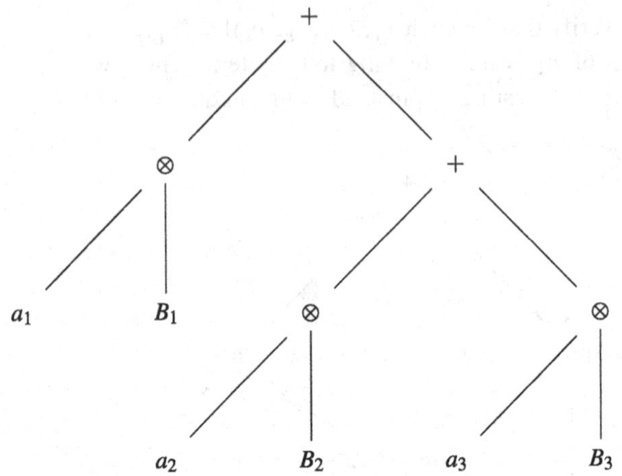

Remark We can conclude this section with an important conclusion: The "algebraization" of the SOtgds in the database-mapping graph G, by transforming its edges into the mapping-operads of the sketch $\mathbf{Sch}(G)$, specified by embedding $\gamma : G \mapsto \mathbf{Sch}(G)$, gives to us the possibility to relax the strong logical requirement that the solutions for such database-mapping system have to be only the *models* of its logical theory based on SOtgds (requirement which is satisfied by the Strong Data Integration semantics). In fact, in the algebraic setting for a given mapping-operad $\mathbf{M}_{AB} : \mathcal{A} \to \mathcal{B}$, for each $v_i \cdot q_i \in \mathbf{M}_{AB}$ and a given mapping interpretation (a functor) $\alpha^* \in Int(G) \subseteq \mathbf{DB}^{\mathbf{Sch}}(G)$, a logical model is obtained if each $\alpha(v_i)$ is an injective function. But in such an algebraic setting, all works well also if it is not an injection. Thus, not necessarily the solutions of the database-mapping system must always be models, and in such an algebraic setting the Weak Data Integration semantics is well defined, and we can use it without limitations of logical theories. In such a weakened setting, we can see the implications used in SOtgds for the database-mappings as a kind of default rules, and not as axioms of a logical theory. Such a framework offers a great variety for practical employment, differently from the Strong Data Integration, especially for a very Big Data Integration with thousands of peer databases. Limitations of the Strong Data Integration are very clear: it does not permit the independent change of the schemas of peer databases because the introduction of new integrity constraints in their schemas can invalidate the model of the whole database-mapping system.

In the following example, we will demonstrate a significant difference between Strong and Weak Data Integration semantics:

Example 38 Let us consider the case of Example 27 in Sect. 4.2.4, used for the two binary relations in a global schema $\mathcal{G}$, and consider now the case when each of these two relations belongs to two different schemas (peers) $\mathcal{A} = (\{r\}, \emptyset)$ and $\mathcal{B} = (\{r\}, \emptyset)$, with the schema mappings $\mathcal{M}_{AB} = \{\Phi\} : \mathcal{A} \to \mathcal{B}$ and $\mathcal{M}_{BA} = \{\Psi\}$:

$\mathcal{B} \to \mathcal{A}$, where Φ and Ψ are the SOtgds $\exists f_1 \forall x \forall y(r(x, y) \Rightarrow s(y, f_1(x, y)))$ and $\exists f_2 \forall x \forall z(s(x, z) \Rightarrow yr(x, f_2(x, z)))$, respectively, with two Skolem functions f_1 and f_2 which returns with fresh new Skolem marked values ω_i, $i = 0, 1, 2, \ldots$.

Thus, we obtain a cyclic graph $G = (V_G, E_G)$ with $V_G = \{\mathcal{A}, \mathcal{B}\}$ and $E_G = \{\mathcal{M}_{AB}, \mathcal{M}_{BA}\}$. The sketch **Sch**$(G)$ obtained by this graph will have the objects equal to set V_G, two identity arrows for them and mapping-operads

$$\mathbf{M}_{AB} = MakeOperads(\mathcal{M}_{AB}) = \{q_\emptyset\} \cup \{v_1 \cdot q_1\}$$

where $q_1 \in O(r, r_q)$, $v_1 \in O(r_q, s)$, and

$$\mathbf{M}_{BA} = MakeOperads(\mathcal{M}_{BA}) = \{q_\emptyset\} \cup \{v_2 \cdot q_2\}$$

where $q_2 \in O(s, s_q)$, $v_1 \in O(s_q, r)$.

Let us consider an initial situation when both r and s are the empty relations, and when a local legacy application inserts a tuple (a, b) into the relation r, so that we obtain a mapping-interpretation (a functor) $\alpha^* \in Int(G) \subseteq \mathbf{DB}^{\mathbf{Sch}}(G)$ such that $A = \alpha^*(\mathcal{A})$ has a single relation $\alpha(r) = \{(a, b), \langle\rangle\}$, so that a forward insertion chaining process will begin, and we will consider both Strong and Weak Data Integration semantics for this database-program specified by sketch **Sch**(G), as follows:

1. **Strong Data Integration semantics.** In this case the SOtgds in the database-mappings in G are considered as the axioms.

Transition	$\Delta r \in \mathcal{A}$	$\Delta s \in \mathcal{B}$	Equations	$\alpha_i^* \in Int(G)$
1		b, ω_0	$(p_0 = nil), (p_1 = \{\{b\}, \bot\}.p_2)$	α_1^*
2	b, ω_1		$(p_2 = \{\{b\}, \bot\}.p_3)$	α_2^*
3		ω_1, ω_2	$(p_3 = \{\{b, \omega_1\}, \bot\}.p_4)$	α_3^*
4	ω_1, ω_3		$(p_4 = \{\{b, \omega_1\}, \bot\}.p_5)$	α_4^*
5		ω_3, ω_4	$(p_5 = \{\{b, \omega_1, \omega_3\}, \bot\}.p_6)$	α_5^*
...	...	...	...	...

Let us consider the mapping-interpretations $\alpha_i^* \in Int(G)$ in each transition step:

Step 1. Here $f = \alpha_1(q_1) : \{(a, b)\} \to \{(b, f_1(a, b))\}$ where $f_1(a, b) = \omega_0$, $\alpha_1(r_q) = \{(b, \omega_0)\}$ and $\alpha_1(v_1) : \{(b, \omega_0)\} \to \{(b, \omega_0)\}$ is an injection, and hence $\mathcal{M}_{AB}$ is satisfied. However, $\alpha_1(s_q) = \{(b, f_2(b, \omega_0))\}$ with $f_2(b, \omega_0) = \omega_1$ and $\alpha_1(v_2) : \{(b, \omega_1)\} \to \{(a, b)\}$ is not an injection, and hence $\mathcal{M}_{BA}$ is not satisfied. Consequently, α_1^* is not a model of the database-mapping program in G.

Step 2. Here $g = \alpha_2(q_2) : \{(b, \omega_0)\} \to \{(a, b), (b, f_2(b, \omega_0))\}$ where $f_2(b, \omega_0) = \omega_1$, $\alpha_2(s_q) = \{(b, \omega_1)\}$ and $\alpha_2(v_2) : \{(b, \omega_1)\} \to \{(a, b), (b, \omega_1)\}$ is an injection, and hence $\mathcal{M}_{AB}$ is satisfied. However,

$$\alpha_2(r_q) = \{(b, \omega_0), (\omega_1, f_1(b, \omega_1))\}$$

with $f_1(b,\omega_1)=\omega_2$ and $\alpha_2(v_1):\{(b,\omega_0),(\omega_1,\omega_2)\}\rightarrow\{(b,\omega_0)\}$ is not an injection, and hence $\mathcal{M}_{AB}$ is not satisfied. Consequently, α_2^* is not a model of the database-mapping program in G.

Step 3. Here $f=\alpha_3(q_1):\{(a,b),(b,\omega_1)\}\rightarrow\{(b,\omega_0),(\omega_1,f_1(b,\omega_1))\}$ where $f_1(b,\omega_1)=\omega_2$, $\alpha_3(r_q)=\{(b,\omega_0),(\omega_1,\omega_2)\}$ and

$$\alpha_3(v_1):\{(b,\omega_0),(\omega_1,\omega_2)\}\rightarrow\{(b,\omega_0),(\omega_1,\omega_2)\}$$

is an injection. and hence $\mathcal{M}_{AB}$ is satisfied. However,

$$\alpha_3(s_q)=\{(b,\omega_1),(\omega_1,f_1(\omega_1,\omega_2))\}$$

with $f_1(\omega_1,\omega_2)=\omega_3$ and $\alpha_3(v_2):\{(b,\omega_1),(\omega_1,\omega_3)\}\rightarrow\{(a,b),(b,\omega_1)\}$ is not an injection, and hence $\mathcal{M}_{BA}$ is not satisfied. Consequently, α_3^* is not a model of the database-mapping program in G.

...

It is easy to verify that no mapping interpretation (i.e., a functor) $\alpha_i^*\in Int(G)\subseteq\mathbf{DB}^{\mathbf{Sch}}(G)$ for a finite $i=1,2,\ldots$ is a model of G, and each one of them satisfies only one of the two schema mappings in G. We will see the unique solution of this Strong Data Integration semantics for the database mapping program in the sketch $\mathbf{Sch}(G)$, based on the unique final colagebra semantics in the next section. From the fact that the solution of G in the Strong Data Integration Semantics has to be a model of G, clearly, the databases $\mathcal{A}$ and $\mathcal{B}$ of such a solution cannot be finite, so that this semantics is not applicable in practice.

2. **Weak Data Integration semantics.** In this case, the SOtgds in the database-mappings in G are considered as the default rules used to generate the recommendations for an updating of the peer databases. Thus it is not necessary that at the end of the forward chaining process all schema SOtgds mappings be satisfied.

   For example, we will consider the case when the peer databases do not accept the inserting in their local knowledge the tuples composed by only the Skolem marked null values. In fact, in this case, the forward-chaining will be finite:

Transition	$\Delta r\in\mathcal{A}$	$\Delta s\in\mathcal{B}$	Equations	$\alpha_i^*\in Int(G)$
1		b,ω_0	$(p_0=nil),(p_1=\{\{b\},\bot\}.p_2)$	α_1^*
2	b,ω_1		$(p_2=\{\{b\},\bot\}.p_3)$	α_2^*
3		$\emptyset$	$(p_3=\alpha_3^*(\mathcal{A}))$	$\alpha_3^*=\alpha_2^*$

The first two steps are equal to those in the previous case for Strong Data Integration semantics. In the third step, the tuple (ω_1,ω_2) is refused by peer $\mathcal{B}$, so that the transition (in Definition 54) $\alpha_2^*\overset{\mathcal{M}_{AB}}{\Longrightarrow}\alpha_3^*$ will have $\alpha_3^*=\alpha_2^*$ and hence $Y=S_{(A,\alpha^*)}=\emptyset$, so that the algorithm '*DBprog*' will reach the end with the final equation $(p_3=\alpha_3^*(\mathcal{A}))$, where $\alpha_3^*(\mathcal{A})=\alpha_2^*(\mathcal{A})=\{\bot,\alpha_2(r)\}$

with $\alpha_2(r) = \{(a, b), (b, \omega_1)\}$, while $\alpha_2^*(\mathcal{B}) = \{\perp, \alpha_2(s)\}$ with $\alpha_2(s) = \{(b, \omega_0)\}$. Hence, $ass(p_0) = \perp^0$, $ass(p_1) = \alpha_1^*(\mathcal{A}) = \{\perp, \alpha_1(r)\}$ with $\alpha_1(r) = \{(a, b)\}$, $ass(p_2) = \alpha_2^*(\mathcal{B}) = \{\perp, \alpha_2(s)\}$ with $\alpha_2(s) = \{(b, \omega_0)\}$, and $ass(p_3) = \alpha_3^*(\mathcal{A})$ defined above. It is easy to see that the final extensions of the relations r and s are equal to those in finite canonical solution $can_F(\mathcal{I}, D)$ of a global schema discussed in Sect. 4.2.4, which was, for the query-answering purposes, equivalent to the infinite canonical model of the global schema.

Notice that the meaning of this path-process $a_1.a_2$ where

$$a_1 = a_2 = \{\{b\}, \perp\} \in Act,$$

described after Definition 53, is $a_1 \otimes a_2 = Ta_1 \cap Ta_2 = T\{b\} \cup \perp$, which is equal to the meaning of the path-process for the Strong Data Integration semantics.

7.4 Operational Semantics for Database-Mapping Programs

In computer science, coalgebra has emerged as a convenient and suitably general way of specifying the reactive behavior of systems, including classes in object-oriented programming. While algebraic specification deals with functional behavior, typically using inductive datatypes generated by constructors, coalgebraic specification is concerned with reactive behavior modeled by coinductive process types that are observable by selectors, much in the spirit of automata theory. An important role is played here by final coalgebras, which are complete sets of possibly infinite behaviors.

As we have seen, the *syntax* of a programming language with a signature Σ_P is formally defined by *initial* Σ_P*-algebras* (and its denotational model) which inductively defines the *syntax-monad* $\mathcal{T}_P$, as we have specified in Sect. 7.3.1. The *semantics* of a programming language is dual to its *syntax*, so we will see in this section that it is based on the coinduction and *final coalgebras* and the *observation-comonad* $\mathcal{D}_P$.

The operational semantics of a language defines *how* programs are to be executed and what their observable effect is. Let us consider any *DB-mapping system* given by a graph $G = (V_G, E_G)$ in Definition 54, composed of a number of database schemas $\mathcal{A} \in V_G$ (and inter-schema mappings between them $\mathcal{M}_{AB} : \mathcal{A} \longrightarrow \mathcal{B}$, i.e., the edges in E_G), whose current extension is determined by a mapping-interpretation $\alpha_0^* \in Int(G) \subseteq \mathbf{DB}^{\mathbf{Sch}(G)}$.

The operational semantics is based on the process which begins with an updating of the extension (instance-database) of a given database schema $\mathcal{A} \in V_G$, from the previous extension $\alpha_0^*(\mathcal{A})$ into this updated extension $A = \alpha^*(\mathcal{A})$, where $\alpha^* \in \mathbf{DB}^{\mathbf{Sch}(G)}$, such that for any other database schema $\mathcal{B} \in V_G$ different from $\mathcal{A}$, $\alpha^*(\mathcal{B}) = \alpha_0^*(\mathcal{B})$, but $\alpha^* \notin Mod(\mathbf{Sch}(G))$ (i.e., it does not satisfy all inter-schema mappings in E_G). This process P begins by updating of the extension of the database $\mathcal{A}$ (that is the root of P) propagates in the graph G by creation of an

LTS (in Definitions 48 and the algorithm *DBprog*) up to the end of such a backward/forward chaining, or up to the moment when the execution is stopped (by a 'nil' operation) by a DB administrator of this database mapping system.

From Proposition 35, for a given database mapping graph G (a program), and the a database-mapping *process-program* P, we have assignment of the process-variables into the *states*, $ass : X \to \mathcal{S}$ (the 'states' in $\mathcal{S}$ are the coalgebraic (observational) duals to the algebraic process variables in X of the guarded system of equations E. For example, for a process $p_k \in X$, i.e., with a given state (instance database) $A = ass(p_k)$), and with its unique solution $\simeq_T(\mathbf{s}_E(p_k))$ (from Proposition 37), p_k represents the 'process in the state A'.

The operational *meaning* of this process p_k at the state $A = ass(p_k)$, denoted by $[p_k]$, from above is to execute the set of actions $a \in Act$ by passing then into the corresponding set of processes $p_{k+1}, p_{k+2}, \ldots$ (specified by the algorithm *DBprog* and based on the set $S_{(A,\alpha^*)}$ during a forward propagation of database insertions, and on the set $S^{OP}_{(A,\alpha^*)}$ during a backward propagation of database deletions).

Consequently, the meaning of a process p_k is that of a finitely branching LTS with the root in the process p_k (at the state (instance-database) $A = ass(s_k)$ of the schema $\mathcal{A} \in V_G$, that is an object in **DB**), and with the set of transitions $\{p_k \xrightarrow{a_i} p_{k+i} | 1 \leq i \leq n\}$, expressed equivalently by $(p_k, \{(a_1, p_{k+1}), \ldots (a_n, p_{k+n})\}) \in \mathcal{S} \times \mathcal{P}_{\text{fin}}(Act \times X)$, where $\mathcal{P}_{\text{fin}}$ is a finite powerset operation (such that for any set Y, $\emptyset \in \mathcal{P}_{\text{fin}}(Y)$, with $\mathcal{P}_{\text{fin}}(\emptyset) = \{\emptyset\}$) and '$\times$' is the Cartesian product (such that for any set Y, $Y \times \emptyset = \emptyset$).

The meaning of the process denoted by the variable p_k at the state (instance-database) $A = ass(p_k)$ should abstract from the name of the process involved in the transitions and focus to the actions which can be performed. It should be the following 'coinductively' defined set of ordered pairs in the following two basic cases (from Definition 54 and the algorithm *DBprog*), for the assignment $ass : X \to \mathcal{S}$, and branching $\wp : X \to \mathcal{P}_{\text{fin}}(Act \times X)$, that is, $[\_] = \langle ass, \wp \rangle$ (introduced by Proposition 34):

1. The meaning of an 'Insertion' process.

$$[p_k] = \big(ass(p_k), \wp(p_k)\big) = \big(ass(p_k), \big\{(A, p_{k+1}), \ldots (A, p_{k+n})\big\}\big),$$

   with $ass(p_i) = A = \alpha^*(\mathcal{A})$ for $i = k, \ldots, k+n$, where $n = |S_{(A,\alpha^*)}| \geq 2$, and $[p_{k+i}] = (ass(p_{k+i}), \{(a_i, p_{k+n+i})\})$ for $i = 1, \ldots, n$;
2. The meaning of a 'Deletion' process (analogous).

$$[p_k] = \big(ass(p_k), \wp(p_k)\big) = \big(ass(p_k), \big\{(A, p_{k+1}), \ldots (A, p_{k+n})\big\}\big),$$

   with $ass(p_i) = A = \alpha^*(\mathcal{A})$ for $i = k, \ldots, k+n$, where $n = |S^{OP}_{(A,\alpha^*)}| \geq 2$, and $[p_{k+i}] = (ass(p_{k+i}), \{(a_i, p_{k+n+i})\})$ for $i = 1, \ldots, n$.

Note this direct connection of each possible action a_i in the state A with the inter-schema mapping $\mathcal{A} \xrightarrow{\mathcal{M}_{AB_i}} \mathcal{B}_i$, $i = 1, \ldots, n$, in the database system given by a graph G (a program) in the insertion case, and with the inter-schema mapping $\mathcal{A} \xrightarrow{\mathcal{M}^{OP}_{B_i A}} \mathcal{B}_i$

in G^{OP} in the deletion case, respectively. Notice that $[p_k] = (ass(p_k), \emptyset) \in \mathcal{S} \times \mathcal{P}_{\text{fin}}(Act \times X)$ if $S_{(A,\alpha^*)}$ is the empty set (there is no branch from this state).

In general, this transition relation is not well-founded since, for instance, the cyclic database mapping programs P (that is, graphs G) are allowed.

The dual of foundation amounts to postulating that "the universe $V = \mathcal{P}_S(V)$ is a final $\mathcal{P}_S$-coalgebra", which is equivalent to Peter Aczel's "anti-foundation axiom" yielding non-well-founded sets [2]. Aczel's theory of non-well founded sets was driven by the quest for a set-theoretic foundation for the abstract semantics of Milner's CCS [34].

But one does not need to resort to non-standard foundation: as already clear in [2], coinductive definitions can be founded on final coalgebras and these exist also in the standard category **Set** of ordinary sets. What the anti-foundation axiom gives is the non-standard fact that the greatest (strict) fixed point $gfp(F) = F(gfp(F))$ of an endofunctor F on **SET** is a final F-coalgebra, provided F satisfies some mild conditions. In particular, the special final coalgebra theorem holds for the endofunctor mapping a class Y to the class $\mathcal{P}_S(A \times Y)$ having as elements (small) sets of pairs (a, y) with $a \in A$ and $y \in Y$. The behavior of CCS programs can be seen as a coalgebra of this endofunctor by taking for A the set *Act* of actions performed by programs.

If the anti-foundation axiom is not assumed then one cannot apply the special final coalgebra theorem of Aczel in order to obtain final coalgebras from greatest (strict) fixed points. The solution, adopted here, consists in taking the *finite* powerset endofunctor $\mathcal{P}_{\text{fin}} : \mathbf{Set} \to \mathbf{Set}$ mapping a set Y to the set $\mathcal{P}_{\text{fin}}(Y)$ of its finite subsets.

In particular, since each database mapping program (a graph G) has only finitely many inter-schema mappings from a given database schema, $[p_k] \in \mathcal{S} \times \mathcal{P}_{\text{fin}}(Act \times X)$, and it yields a semantics in the ordinary category **Set** of sets which is "almost" the same as Aczel's one by *abstract* behavior endofunctor $\mathcal{B}_{\mathcal{P}} = \mathcal{P}_{\text{fin}}(Act \times \_) : \mathbf{Set} \longrightarrow \mathbf{Set}$ (when we abstract from the names (states) of the programs involved in the transitions and focus to the actions which can be performed), and the general behavior endofunctor $\overline{B}_P = \mathcal{S} \times \mathcal{B}_{\mathcal{P}} : \mathbf{Set} \longrightarrow \mathbf{Set}$, so that $[p_k] \in \mathcal{S} \times \mathcal{B}_{\mathcal{P}}(X)$.

The difference is in the fact that the final coalgebra is an isomorphism (bijection) in **Set**, denoted by $\simeq_B : gfp(\overline{B}_P) \to \overline{B}_P(gfp(\overline{B}_P))$, rather than equality $gfp(\overline{B}_P) = \overline{B}_P(gfp(\overline{B}_P))$ in **Set**.

Concretely, the final coalgebra for the behavior $\mathcal{S} \times \mathcal{B}_P$ is the set of rooted, finitely branching trees, i.e., LTS with the *states* in $\mathcal{S}$ and labeled by the actions $a \in Act$, factored by the largest bisimulation relation. These (equivalence classes of) trees can be seen as the *abstract* global behaviors corresponding, for a given set X, to $\mathcal{B}_P(X) = \mathcal{P}_{\text{fin}}(Act \times X)$: the root $A = ass(p_k)$ (of a process p_k) of a tree $\mathcal{L}_A \in gfp(\mathcal{S} \times \mathcal{B}_P)$ is the starting point of an abstract computation with behavior $\mathcal{B}_P$. The quotient modulo bisimulation is needed in order to identify trees with *the same paths*. (Notice that the bisimulations may be understood as coalgebraic dual of 'algebraic' congruences, and they are relations which are closed under algebraic operations).

The fact that the nodes have no names reflects the *abstractness* of these global behaviors. Consequently, we are free to assign the names to nodes, and we will use a name p_k (for this state $ass(p_k) \in \mathcal{S}$, an instance of a schema $\mathcal{A}$) for a node name instead of the 'process p_k at state $ass(p_k) \in \mathcal{S}$'.

Consequently, for any DB-mapping system represented by a given graph G (a program) with the set of the processes p_k in the states $ass(p_k) \in \mathcal{S}$ (i.e., in different instance-databases $ass(p_k)$ for the database schemas in G), a single total meaning function is given in Proposition 34 by $[\_] = \langle ass, \wp \rangle : \mathcal{S} \longrightarrow \mathcal{S} \times \mathcal{P}_{\text{fin}}(Act \times X)$. It defines a one-to-one correspondence between the transition systems and co-algebras of the endofunctor $\overline{B}_P = \mathcal{S} \times B_{\mathcal{P}} = \mathcal{S} \times \mathcal{P}_{\text{fin}}(Act \times \_) : \textbf{Set} \longrightarrow \textbf{Set}$, so that the diagram bellow commutes.

$$
\begin{array}{ccc}
X & \xrightarrow{\quad [\_]^{@} \quad} & gfp(\overline{B}_P) \\[2pt]
\Big\downarrow {\scriptstyle [\_] = \langle ass, \wp \rangle} & \quad (4) \quad & \Big\downarrow {\scriptstyle \simeq_B} \\[2pt]
\overline{B}_P(X) = \mathcal{S} \times B_P(X) & \xrightarrow{\quad \overline{B}_P([\_]^{@}) \quad} & \overline{B}_P(gfp(\overline{B}_P))
\end{array}
$$

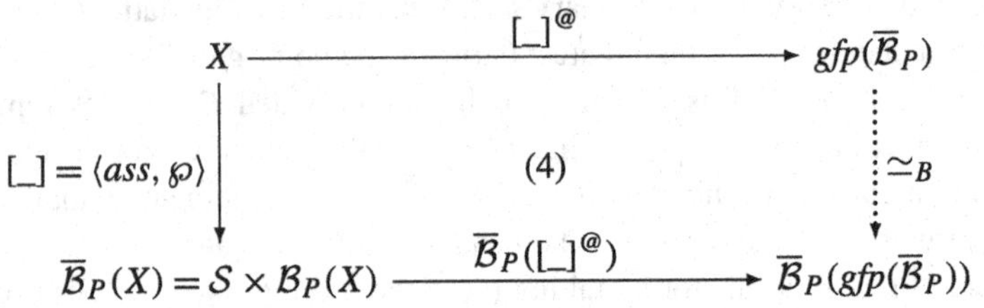

Final $(\mathcal{S} \times B_P)$-coalgebra diagram in **Set** ($\simeq_B$ is an isomorphism-bijection)
The coalgebras of the Power-set functor $\mathcal{P}$ are the same as the directed 'locally-small' graphs (the tree $\mathcal{L}_A$ is a pointed (by A) accessible graph) and, by anti-foundation axiom ($V = \mathcal{P}(V)$, for the universe V of all (also non-well-founded) sets, is a final $\mathcal{P}$-algebra), the *final coalgebra* is the class of rooted 'locally small' trees (possibly of infinite depth) factored by $\mathcal{P}$-bisimulation. A (possibly large) graph is *locally small* if the (possibly infinite) collection of children of every node is a (small) set.

The coalgebras are suitable to modeling the operational behavior of the programs of a language. The corresponding endofunctors B_P are called *behavior endofunctors*.

The final $(\mathcal{S} \times B_P)$-coalgebra $(gfp(\mathcal{S} \times B_P), \simeq_B)$, where $gfp(\mathcal{S} \times B_P) = \mathcal{S} \times B_P(gfp(\mathcal{S} \times B_P))$, is the greatest fixpoint of the behavior endofunctor, i.e., the set of rooted 'locally-small' (finitely branching) trees (of possibly infinite depth) factored by a strong bisimulation: for any database A, a labeled tree $\mathcal{L}_A$ with nodes in $\mathcal{S}$ and branches in Act is an element of the set $\mathcal{D}_P\mathcal{S} = gfp(\mathcal{S} \times B_P)$, obtained by the union of the chain (here $\mathbf{1}$ is a singleton $\mathcal{P}_{\text{fin}}(\emptyset) = \{\emptyset\}$, that is, terminal object in **Set**)

$$\emptyset \subseteq \mathcal{S} \times B_P(\emptyset) = \mathcal{S} \times \mathcal{P}_{\text{fin}}(\emptyset)$$

$$= \mathcal{S} \times \mathbf{1} \simeq \mathcal{S} \subseteq \mathcal{S} \times B_P(\mathcal{S}) \subseteq \mathcal{S} \times B_P\big(\mathcal{S} \times B_P(\mathcal{S})\big) \subseteq \cdots,$$

i.e., obtained by iterating the functor $\mathcal{S} \times B_P$ on the empty set of states $\emptyset$, up to the greatest fixed point (with $\mathcal{S} \subseteq \mathcal{D}_P\mathcal{S}$, so that for each single-node tree $A \in \mathcal{S} \subseteq \mathcal{D}_P\mathcal{S}$, $\simeq_B(A) = (A, \emptyset) \in \mathcal{S} \times B_P(\mathcal{D}_P\mathcal{S})$).

The final coalgebra isomorphism $\simeq_B : \mathcal{D}_P\mathcal{S} \rightarrow \mathcal{S} \times \mathcal{B}_P(\mathcal{D}_P\mathcal{S})$ splits into two epic projections (i.e., surjective functions) $\simeq_B = \langle fst_S, snd_S \rangle$, where for a given LTS tree $t = [p_k]^@$ with the root state $ass(p_k)$, fst_S restitutes its root state, i.e., $fst_S(t) = ass(p_k)$, while snd_S restitutes the trees obtained by division of all subtrees of $[A]^@$, $snd_S(t) = \{\xrightarrow{a_1} t_1, \ldots, \xrightarrow{a_n} t_n\} = \{(a_1, t_1), \ldots, (a_n, t_n)\} \in \mathcal{P}_{\text{fin}}(Act \times \mathcal{D}_P\mathcal{S}) = \mathcal{B}_P(\mathcal{D}_P\mathcal{S})$ where $t_i \in \mathcal{D}_P\mathcal{S}$, $1 \le i \le n$ are the subtrees of t if we eliminate its root state $ass(p_k)$, and $n \ge 1$ is the number of branches that come from the root $ass(p_k)$.

Consequently, for a set X and $(\mathcal{S} \times \mathcal{B})$-coalgebra $(X, [\_])$ in the category $\mathbf{Set}_{\mathcal{B}}$ of all $(\mathcal{S} \times \mathcal{B}_P)$-coalgebras, there is a unique coalgebra homomorphism $[\_]^@ : (X, [\_]) \longrightarrow (gfp(\mathcal{S} \times \mathcal{B}_P), \simeq_B)$ from this $(\mathcal{S} \times \mathcal{B}_P)$-coalgebra to the final $(\mathcal{S} \times \mathcal{B}_P)$-coalgebra such that the diagram above in **Set** commutes, where $[\_]^@ : X \longrightarrow gfp(\mathcal{S} \times \mathcal{B}_P)$ is a 'coinductive' extension of the arrow $[\_] : X \longrightarrow \mathcal{S} \times \mathcal{B}_P(X)$ such that for any process $p_k \in X$ at the state $ass(p_k) \in \mathcal{S}$, we obtain

- $[p_k]^@ \triangleq (ass(p_k), \{(a_i, [p_{k+i}]^@) \mid (a_i, p_{k+i}) \in \wp_S(s_k)\})$ if $\wp(p_k) \neq \emptyset$; $ass(p_k) \in \mathcal{S} \subseteq \mathcal{D}_P\mathcal{S}$, otherwise,

which is a labeled (possibly infinite) tree with the root in the state (instance database) $A = ass(p_k) \in \mathcal{S} \subseteq Ob_{DB}$. That is, $[p_k]^@$ is an LTS tree $\mathcal{L}_A$ with the root state $A = ass(p_k)$ composed of states in $\mathcal{S}$ and transitions in $Act \subseteq Ob_{DB}$.

One of the properties of Aczel's coinductive semantics for CCS (similar to this one) is that it maps two processes p_k and p_m to the same set if they are observationally equivalent (it is denoted by $p_k \sim p_m$):

$$(BSM) \quad p_k \sim p_m \quad \text{iff} \quad [p_k]^@ = [p_m]^@.$$

That is, the coinductive extension of the operational model $[\_] : X \longrightarrow \mathcal{S} \times \mathcal{B}_P(X)$ does preserve $\mathcal{B}_P$ bisimulation and, conversely, it can be "pulled back" to form the largest $\mathcal{B}_P$-bisimulation relation.

The above is a property which holds in general for every coinductive extension of coalgebras of endofunctors $\mathcal{S} \times \mathcal{B}_P$ preserving categorical (weak) pullbacks. Thus, one does not need to work with non-well founded sets, as we did.

A condition which well-behaved operational semantics should satisfy is *compositionality*: To every behavior $\mathcal{B}_P$ there corresponds a notion of observational equivalence called $\mathcal{B}_P$-bisimulation [3] (which for the behavior $\mathcal{B}_P = \mathcal{P}_{\text{fin}}(Act \times \_)$ corresponds to Milner's strong bisimulation—the finest notion of observational equivalence for transition relations). If this observational equivalence is a congruence (as in our case) w.r.t. the construct of the syntax, then the operational semantics is compositional. This means that programs with the same observable behavior can be interchanged in any context without affecting the overall observable behavior.

The construction of coalgebraic operational semantics can be done by using the duality property between algebras and coalgebras: instead of the set of variables X, by duality we use the corresponding set of states $\mathcal{S}$; instead of Σ_P syntax endofunctor, we use the behavior endofunctor $\mathcal{B}_P$; instead of coproduct (disjoint union) $\uplus$, we use the Cartesian product $\times$; instead of the empty set $\emptyset$ (zero object in **Set**), the singleton set $\mathbf{1} = \{\bot^0\}$ (terminal object in **Set**); and instead of the carrier set $\mathcal{T}_P X$ for the initial $X \uplus \Sigma_P$-algebra, we use the carrier set $\mathcal{D}_P S = gfp(\mathcal{S} \times \mathcal{B}_P)$ for the final $\mathcal{S} \times \mathcal{B}_P$-coalgebra, by inverting all arrows used in diagrams for the algebras, so that for the final coalgebra of the operational semantics we have the dual (w.r.t. the initial algebra semantics diagrams (I1) and (I2), respectively, in Sect. 7.3.1) commutative diagrams:

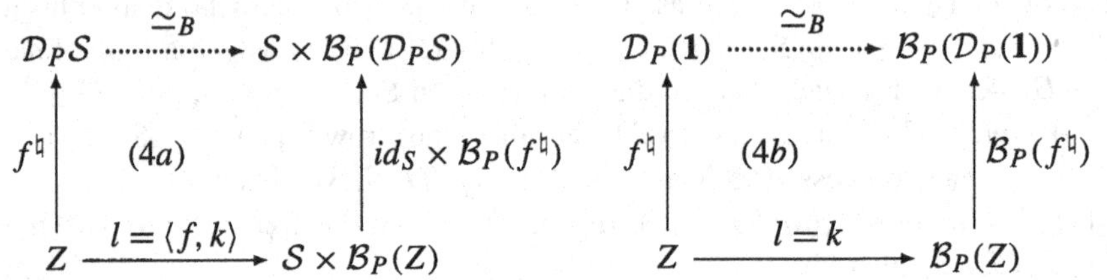

If the set of states $\mathcal{S}$ is the singleton set $\mathbf{1} = \{\bot^0\}$ (dual to the $\emptyset$ which is a zero object) and hence f is the constant function that maps all states into the label '$\bot^0$', with $\mathbf{1} \times Y \simeq Y$ in **Set**, then we obtain the simpler commutative diagram (4b) above on the right. The unique homomorphism $f^\natural : Z \to \mathcal{D}_P\mathcal{S}$ represents the unique coinductive extension of $k : Z \to \mathcal{B}_P Z$ along the covaluation ('renaming' of states) function $f : Z \to \mathcal{S}$.

Let us show how the final $(S \times \mathcal{B}_P)$-coalgebra in the diagram (4a) can be reduced to the final coalgebra semantics in the diagram (4), in the case when $Z = X$, f is the assignment function $ass : X \to \mathcal{S}$ and $k = \wp : X \to \mathcal{B}_P(X)$. First of all, $\mathcal{D}_P\mathcal{S}$ is equal to the $gfp(\mathcal{S} \times \mathcal{B}_P)$, and in this case $l = \langle f, k\rangle = \langle ass, \wp\rangle = [\_]$. From the fact that $ass^\natural$ is the unique arrow from the coalgebra $[\_] : X \to \mathcal{S} \times \mathcal{B}_P(X)$ to the final coalgebra $\simeq_B : gfp(\mathcal{S} \times \mathcal{B}_P) \to \mathcal{S} \times \mathcal{B}_P(gfp(\mathcal{S} \times \mathcal{B}_P))$, $ass^\natural$ is equal to $[\_]^@$, so that the diagram (4a) is equal to the diagram (4).

7.4.1 Observational Comonad

We will show that $\langle \mathcal{D}_P, \varepsilon, \delta\rangle$ is a cofreely generated observational comonad (dual of the freely generated syntax monad $\langle \mathcal{T}_P, \eta, \mu\rangle$) by the behavior endofunctor $\mathcal{B}_P = \mathcal{P}_{\text{fin}}(1 \uplus Act \times \_) : \mathbf{Set} \to \mathbf{Set}$.

For every set $\mathcal{S}$ of states, the value $\mathcal{D}_P$ at $\mathcal{S}$ is the set $gfp(\mathcal{S} \times \mathcal{B}_P)$ of all finite and infinite LTS trees with states in $\mathcal{S}$ and transitions in Act. In particular, $\mathcal{D}_P$ at the singleton $\{*\}$ is the set of all finite and infinite *abstract* LTS trees where all states are labeled by $*$ (that is, equivalently, by states with no names).

Consequently, from the fact that the final coalgebra isomorphism $\simeq_B : \mathcal{D}_P\mathcal{S} \to \mathcal{S} \times \mathcal{B}_P(\mathcal{D}_P\mathcal{S})$ splits into two epic projections (i.e., surjective functions) $\simeq_B =$

$\langle fst_S, snd_S \rangle$, the final coalgebra diagram (4a) above is equivalent to the following diagram:

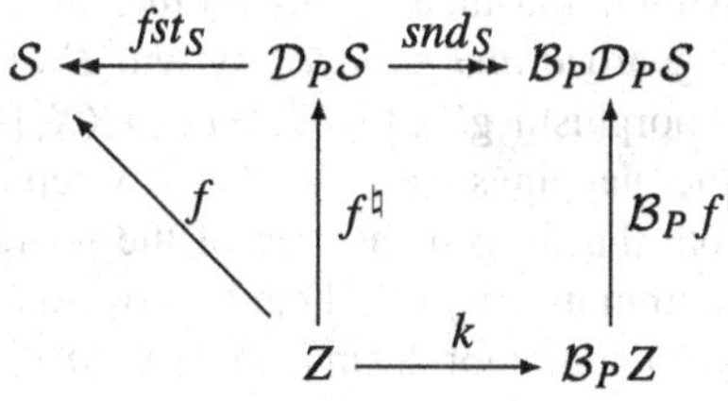

Hence, we obtain the endofunctor $\mathcal{D}_P : \mathbf{Set} \to \mathbf{Set}$, as follows:

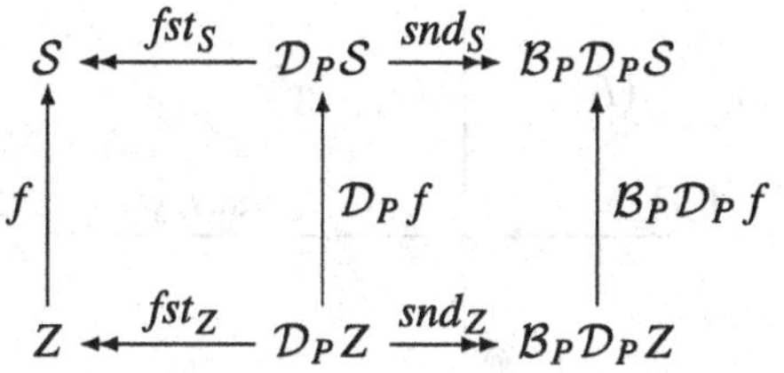

This function $\mathcal{D}_P(f) : \mathcal{D}_P Z \to \mathcal{D}_P\mathcal{S}$, applied to the LTS tree t of the global behavior, substitutes each state s in t by the 'renamed' state $f(s)$.

This coinductive principle can be used to show that the operator $\mathcal{D}_P$ inductively extends to the *observational* comonad $\mathcal{D}_P : \mathbf{Set} \to \mathbf{Set}$ (comonad freely generated by behavior $\mathcal{B}_P$), so that its coaction $\mathcal{D}_P f$ on arrow (i.e., a function) $f : Z \longrightarrow \mathcal{S}$ takes the coinductive extension of $snd_Z : \mathcal{D}_P Z \twoheadrightarrow \mathcal{B}_P(\mathcal{D}_P Z)$ along the composite $f \circ fst_Z$, i.e., $D_P(f) \triangleq (f \circ fst_Z)^\natural$ and hence the second diagram above commutes.

Consequently, we obtain the *observational* comonad $(\mathcal{D}_P, \varepsilon, \delta))$ with $\varepsilon_S = fst_S$, while the comultiplication $\delta_S = (id_{\mathcal{D}_P\mathcal{S}})^\natural$ is the unique coinductive extension of $snd_S : \mathcal{D}_P\mathcal{S} \twoheadrightarrow \mathcal{B}_P(\mathcal{D}_P\mathcal{S})$ along the identity on $\mathcal{D}_P\mathcal{S}$, and maps an LTS tree t into the same tree structure (with the same transitions) where all states are substituted by the tree t, as represented in the following commutative diagram:

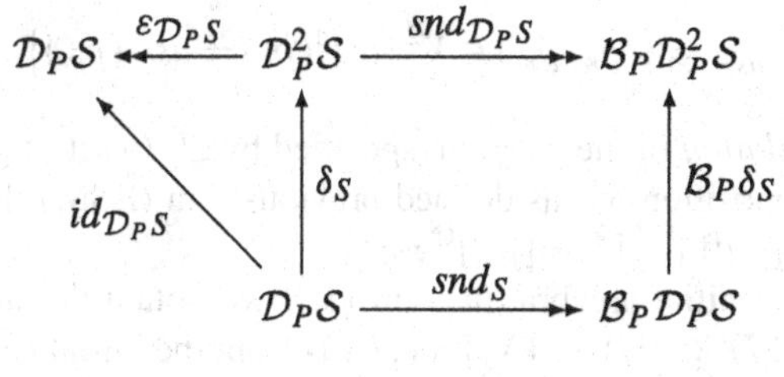

7.4.2 Duality and Database-Mapping Programs: Specification Versus Solution

Let us consider the following initial-final duality for the database-mapping process-programs expressed by a guarded *and flattened* system E of equations: the initial $(X \uplus \Sigma_P)$-algebra homomorphism $\mathbf{g}^E : (\mathcal{T}_P X, \simeq_I) \to (X, [id_X, h_X])$ (defined in the Proposition 35 with the mappings $h_X : \Sigma_P(X) \to X$ represented by the initial-algebra diagram (I_1). Note that in general case of the guarded, but not flattened system of equations, the commutativity of (I1) in Proposition 35 does not hold.

Dual $(\mathcal{S} \times \mathcal{B}_P)$-coalgebra homomorphism $[\_]^@ : (X, \langle ass, \wp\rangle) \to (\mathcal{D}_P\mathcal{S}, \simeq_B)$ is represented by the final-coalgebra diagram (4),

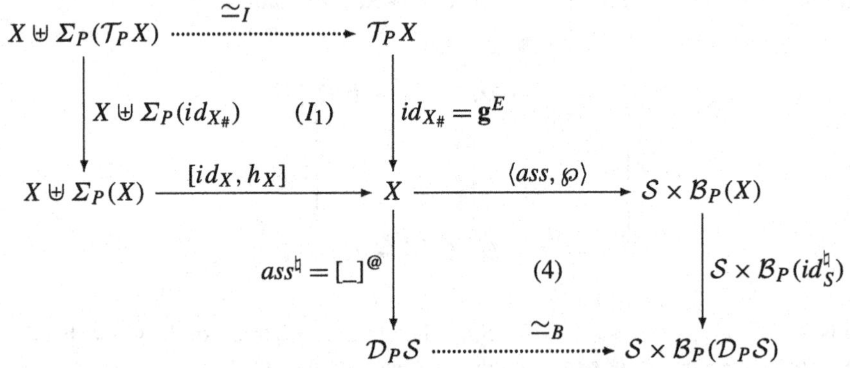

The commutative diagram (I_1) of the unique initial algebra homomorphism $id_{X_\#} = \mathbf{g}^E : (\mathcal{T}_P X, \simeq_I) \to (X, [id_X, h_X])$ represents the database-mapping process-program *specification* by a set of flattened guarded equations E with

$$E \subseteq \{(p_k = t) \mid t \in \mathcal{T}_P X \text{ with } p_k = \mathbf{g}^E(t)\},$$

while the commutative diagram (4) of the unique final coalgebra homomorphism $[\_]^@ : (X, [\_]) \to (\mathcal{B}_P\mathcal{S}, \simeq_B)$ represents the *solution* of this previously specified database-mapping process-program.

From the final-coalgebra diagram (4), we obtain $ass^\natural = [\_]^@ : X \to \mathcal{D}_P\mathcal{S}$.

It is easy to verify that the *kernel* of this unique coalgebra homomorphism into the final $(\mathcal{S} \times \mathcal{B}_P)$-coalgebra, $ass^\natural : (X, \langle ass, \wp\rangle) \to (\mathcal{D}_P\mathcal{S}, \simeq_B)$, defined by

$$K_{ass^\natural} = \{(p_k, p_m) \in X^2 \mid ass^\natural(p_k) = ass^\natural(p_m)\}$$

is a *bisimulation relation* of the program specified by $\mathbf{g}^E$ (a set of guarded flattened equations E in Proposition 34) as defined previously in (BSM). In fact, $p_k \sim p_m$, i.e., $(p_k, p_m) \in K_{ass^\natural}$ iff $[p_k]^@ = [p_m]^@$.

Dually, from the initial-algebra diagram (I_1), we obtain the unique homomorphism $id_{X_\#} = \mathbf{g}^E : (\mathcal{T}_P X, \simeq_I) \to (X, [id_X, h_X])$ from the initial $(X \uplus \Sigma_P)$-algebra. It is easy to verify that the kernel of the unique $(X \uplus \Sigma_P)$-algebra homomorphism

$id_{X_\#}$,

$$K_{id_{X_\#}} = \{(t_1, t_2) \in \mathcal{T}_P X \times \mathcal{T}_P X \mid id_{X_\#}(t_1) = id_{X_\#}(t_2)\}$$
$$= \{(t_1, t_2) \in \mathcal{T}_P X \times \mathcal{T}_P X \mid \mathbf{g}^E(t_1) = \mathbf{g}^E(t_2)\},$$

is the *congruence relation*, defined by $t_1 \equiv t_2$ iff $(t_1, t_2) \in K_{id_{X_\#}}$, of the same process-program (specified by $\mathbf{g}^E$) with the following equivalence classes, such that for a given variable $p_k \in X \subseteq \mathcal{T}_P X$ (from Proposition 34),

$$[p_k] = \{t \in \mathcal{T}_P X \mid \mathbf{g}^E(t) = \mathbf{g}^E(p_k) = p_k\} = \{t \in \mathcal{T}_P X \mid (p_k, t) \in K_{id_{X_\#}}\}$$
$$= \{t \in \mathcal{T}_P X \mid t \equiv p_k\}.$$

This congruence relation $\equiv$ has a corresponding quotient structure, denoted by $\mathcal{T}_P X_\equiv$ as it was specified in Definition 55, whose elements are the equivalence classes (or congruence classes).

The general notion of a congruence relation can be given a formal definition in the context of *universal algebra*, a field which studies ideas common to all algebraic structures. In this setting, a congruence relation is an equivalence relation $\equiv$ on an algebraic structure that satisfies for each n-ary operator $o_i \in \Sigma_P$, $o_i(t_1, \dots, t_n) = o_i(t'_1, \dots, t'_n)$ for any n-ary operation $o_i \in \Sigma_P$ and all terms $t_1, \dots, t_n, t'_1, \dots, t'_n$ satisfying $t_i \equiv t'_i$ for each $1 \le i \le n$.

Generally, this duality between the specification and the solution of the programs can be represented by the following table:

initial	final	description
X	$\mathcal{S}$	program variables—states
$\uplus$	$\times$	coproduct—product in **Set**
$\emptyset$	$\mathbf{1}$	initial—terminal object in **Set**
Σ_P	$\mathcal{B}_P$	signature—behavior
$\mathcal{T}_P$	$\mathcal{D}_P$	syntax monad—observational comonad
$\mathcal{T}_P X$	$\mathcal{D}_P \mathcal{S}$	lfp of $X \uplus \Sigma_P$—gfp of $\mathcal{S} \times \mathcal{B}_P$ functors
$(\mathcal{T}_P X, \simeq_I)$	$(\mathcal{D}_P, \simeq_B)$	initial algebra—final coalgebra
$\mathbf{g}^E = id_{X_\#}$	$[\_]^@ = ass^\natural$	program specification—solution homomorphism
$K_{id_{X_\#}}$	$K_{ass^\natural}$	congruence—bisimulation relation

7.5 Semantic Adequateness for the Operational Behavior

A denotational model $(X \uplus \Sigma_P)$-algebra $[f, h] : X \uplus \Sigma_P(Z) \rightarrowtail Z$ is *adequate* w.r.t. an operational semantics if it determines the operational behavior of the program up to observational equivalence.

Let us show that for $Z = \mathbb{T}_\infty$ the DB-denotational semantics Σ_P-algebra, $[f,h]: X \uplus \Sigma_P(\mathbb{T}_\infty) \rightarrowtail \mathbb{T}_\infty$ (in the commutative diagram represented in the proof of Proposition 37 in Sect. 7.3.2 is defined a mapping h), for $f = (\simeq_T \circ \mathbf{s}_E): X \to \mathbb{T}_\infty$, is adequate for the final coalgebra semantics of the database-mapping programs, given by the following commutative diagrams (1) and (2):

Proposition 38 *The DB-denotational semantics $(X \uplus \Sigma_P)$-algebra with the carrier set $\mathbb{T}_\infty$ is* adequate *for the final coalgebra semantics (with* the same *carrier set) of the database-mapping process-programs expressed by a guarded system of equations $\mathbf{g}_E$ (flattened or not) when $f = \simeq_T \circ \mathbf{s}_E$, so that the following diagram commutes*

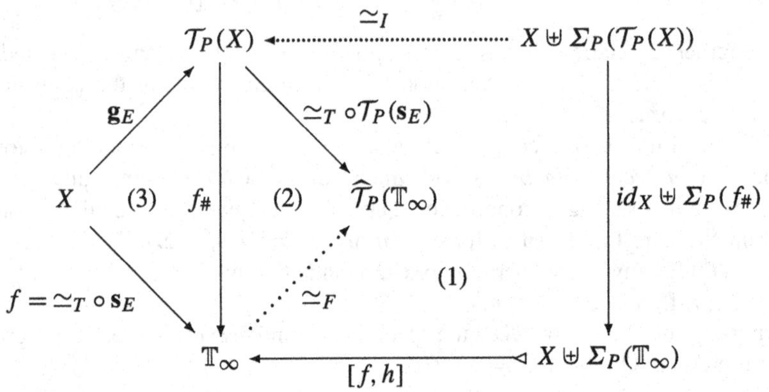

The commutative square diagram (1) *is obtained from the* initial algebra *semantics where $\simeq_I : X \uplus \Sigma_P(\mathcal{T}_P(X)) \to \mathcal{T}_P(X)$ is the* initial algebra *isomorphism for the Σ_P abstract signature, with the DB-denotational semantics algebra*

$$[f,h]: X \uplus \Sigma_P(\mathbb{T}_\infty) \rightarrowtail \mathbb{T}_\infty.$$

The commutative square diagram (composed of two triangles (3) *and* (2)) *is obtained from the* final coalgebra *semantics (in Proposition* 37).

Proof The assignment to process variables $f = \simeq_T \circ \mathbf{s}_E$ is just the *solution* for the system of flattened equations for a database mapping process-program P (Proposition 37, where the mapping h is specified in its proof).

Let us demonstrate that if $f = \simeq_T \circ \mathbf{s}_E$ then diagram (3) commutes: It is enough to show that for each variable $p_A \in X$, $f_\# \circ \mathbf{g}_E(p_A) = f(p_A)$. In fact, from the definition of the set of flattened equations E of a database-mapping process-program, in Proposition 34, we have that X is just the set of variables in E, and for each equation $(p_A = t) \in E$, where $p_A, t \in \mathcal{T}_P X$, $\mathbf{g}_E(p_A) = t$ and hence $f_\# \circ \mathbf{g}_E(p_A) = f_\#(t) =$ (by replacing t with p_A (from the fact that they are equal) $= f_\#(p_A) = f(p_A)$ (because the restriction of $f_\#$ to the variables $X \subseteq \mathcal{T}_P X$ is *equal* to f (from the initial-algebra semantics (diagram (1) above), $f_\#$ is the unique inductive extension of h along the mapping f with $f = f_\# \circ inl_X$). Thus, we can replace inl_X with $\mathbf{g}_E$ in the initial-semantics diagram. Consequently, the unique extension to all terms with variables, $f_\# : \mathcal{T}_P X \to \mathbb{T}_\infty$ (of h, which is defined in the proof of Proposition 37, along $f : X \to \mathbb{T}_\infty$) is defined recursively by (for $t, t_1, .., t_n \in \mathcal{T}_P X$):

1. For each variable $p_A \in X$, $f_\#(p_A) = f(p_A) = \simeq_T(\mathbf{s}_E(p_A))$;
2. $nil \mapsto \bot^0$, and $A \mapsto A$;
3. $a.t \mapsto a \otimes f_\#(t)$;
4. $\|^n(t_1, \ldots, t_n) \mapsto +^n(f_\#(t_1), \ldots, f_\#(t_n))$. □

In order to represent the adequateness of the DB-denotational semantics of the initial $\mathcal{T}_P(X)$ algebra, for the behavioral operational semantics expressed in the diagram (4) (in Sect. 7.4 for the behavior functor $\mathcal{B}_P$), we need the embedding $\mathcal{E} : \mathcal{D}_P\mathcal{S} \to \mathcal{T}_\infty$ of (also infinite) LTS trees in $\mathcal{D}_P\mathcal{S} = gfp(\mathcal{S} \times \mathcal{B}_P)$ into the ground terms (trees) in $\mathcal{T}_\infty$ and the embedding $\mathcal{E}_X : \mathcal{B}_P(\mathcal{S}) \to \mathcal{T}_P(X)$:

Definition 56 We define inductively the following embedding between (also infinite) trees, $\mathcal{E} : \mathcal{D}_P\mathcal{S} \to \mathcal{T}_\infty$, such that:

- For a single-state LTS (without the outgoing transitions) $A \in \mathcal{D}_P\mathcal{S}$, $A \mapsto A$;
- For an LTS (tree) $\mathcal{L}_A \in \mathcal{D}_P\mathcal{S}$ with the root state $A = fst_{\mathcal{S}}(\mathcal{L}_A)$ and $snd_{\mathcal{S}}(\mathcal{L}_A) = \{\xrightarrow{a} \mathcal{L}_B\}$ where $\mathcal{L}_B \in \mathcal{D}_P(\mathcal{S})$, $\mathcal{L}_A \mapsto a.\mathcal{E}(\mathcal{L}_B)$;
- For an LTS (tree) $\mathcal{L}_A \in \mathcal{D}_P\mathcal{S}$ with the root state $A = fst_{\mathcal{S}}(\mathcal{L}_A)$ and $snd_{\mathcal{S}}(\mathcal{L}_A) = \{\xrightarrow{a_i} \mathcal{L}_{B_i} \mid 1 \le i \le n\}$ where $\mathcal{L}_{B_i} \in \mathcal{D}_P\mathcal{S}$,
$$\mathcal{L}_A \mapsto \|^n\{a_i.\mathcal{E}(\mathcal{L}_{B_i}) \mid 1 \le i \le n\}.$$

Moreover, we define the embedding of the behavior into the syntax, for each set X, $\nu_X : \mathcal{S} \times \mathcal{B}_P(X) \to \mathcal{T}_P(X)$, natural in X, as follows:

- For the empty set $\emptyset \in \mathcal{B}_P(X) = \mathcal{P}_{\text{fin}}(Act \times X)$, $(A, \emptyset) \mapsto A$;
- For the singleton $\{(a, p_B)\} \in \mathcal{B}_P(X)$, $(A, \{(a, p_B)\}) \mapsto a.p_B$;
- For $\{(a_i, p_{B(i)}) \mid 1 \le i \le n\} \in \mathcal{B}_P(X)$, $n \ge 2$,
$$\left(A, \{(a_i, p_{B(i)}) \mid 1 \le i \le n\}\right) \mapsto \|^n\{a_i.p_{B(i)} \mid 1 \le i \le n\}.$$

Thus, $\nu : \mathcal{S} \times \mathcal{B}_P \stackrel{\cdot}{\longrightarrow} \mathcal{T}_P$ is a natural transformation, with a component $\nu_X = \nu(X)$.

It is easy to show that for each renaming function $f : X \to X'$ of variables in X (possibly involving equating some of the variables, first renaming and then applying the embedding is the same as first applying the embedding and then renaming. Thus, $\nu : \mathcal{S} \times \mathcal{B}_P \stackrel{\cdot}{\longrightarrow} \mathcal{T}_P$ is a natural transformation between the behavior and syntax endofunctors on **Set**.

Proposition 39 *Based on the embedding in Definition* 56, *the following is valid*:
$$\mathbf{g}_E = \nu(X) \circ [\_] : X \to \mathcal{T}_P(X) \quad \textit{and} \quad \mathbf{s}_E = \mathcal{E} \circ [\_]^@ \circ : X \to \mathcal{T}_\infty.$$

Consequently, the denotational semantics $(X \uplus \Sigma_P)$*-algebra with the carrier set* $\mathbb{T}_\infty$ *is adequate for the operational semantics when the following diagram commutes*:

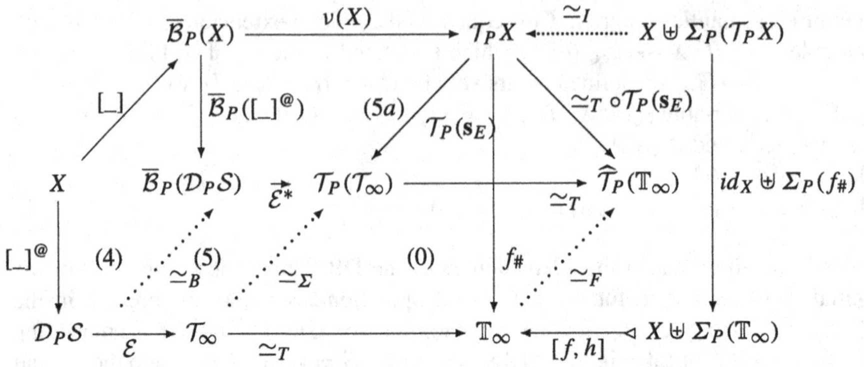

where $\mathcal{E}^* = \mathcal{T}_P(\mathcal{E}) \circ \nu(\mathcal{D}_P\mathcal{S})$, $f = \simeq_T \circ \mathbf{s}_E$ and $\overline{\mathcal{B}}_P = \mathcal{S} \times \mathcal{B}_P$.

Proof The commutativity of the diagram (4) is demonstrated in Sect. 7.4.2; the commutativity of (0) is proved in Proposition 37 for the final semantics of the system of flattened equations E specified by $\mathbf{g}_E : X \to \mathcal{T}_P X$ or, dually, by $\mathbf{g}^E : \mathcal{T}_P X \to X$ (in Proposition 34) and its solution $\mathbf{s}_E : X \to \mathbb{T}_\infty$.

Let us demonstrate the commutativity of the diagram (5) (the embedding of the LTS trees in $\mathcal{D}_P\mathcal{S}$ into the ground tree-terms in $\mathbb{T}_\infty$). We have the following cases:

1. A single-node tree $A \in \mathcal{D}_P\mathcal{S}$. Then $\simeq_B(A) = (A, \emptyset)$ and $\mathcal{E}^*(A, \emptyset) = \mathcal{T}_P(\mathcal{E}) \circ \nu(\mathcal{D}_P\mathcal{S})(A, \emptyset) = \mathcal{T}_P(\mathcal{E})(A) = \mathcal{E}(A) = A$. While, $\simeq_\Sigma \circ \mathcal{E}(A) = \simeq_\Sigma(A) = A$. Thus, $\simeq_\Sigma \circ \mathcal{E}(A) = \mathcal{E}^* \circ \simeq_B(A)$.
2. A tree $\mathcal{L}_A \in \mathcal{D}_P\mathcal{S}$ with a unique branch a from its root A. Then $\simeq_B(\mathcal{L}_A) = (A, \{(a, \mathcal{L}_B)\})$ and $\mathcal{E}^*(A, \{(a, \mathcal{L}_B)\}) = \mathcal{T}_P(\mathcal{E}) \circ \nu(\mathcal{D}_P\mathcal{S})(A, \{(a, \mathcal{L}_B)\}) = \mathcal{T}_P(\mathcal{E})(a.\mathcal{L}_B) = a.\mathcal{E}(\mathcal{L}_B)$. While, $\simeq_\Sigma \circ \mathcal{E}(\mathcal{L}_A) = \simeq_\Sigma(a.\mathcal{E}(\mathcal{L}_B)) = a.\mathcal{E}(\mathcal{L}_B)$. Thus, $\simeq_\Sigma \circ \mathcal{E}(\mathcal{L}_A) = \mathcal{E}^* \circ \simeq_B(\mathcal{L}_A)$.
3. A tree $\mathcal{L}_A \in \mathcal{D}_P\mathcal{S}$ with $n \geq 2$ branches from its root A. Then $\simeq_B(\mathcal{L}_A) = (A, \{(a_1, \mathcal{L}_{B_1}), \ldots, (a_n, \mathcal{L}_{B_n})\})$ and

$$\begin{aligned}
&\mathcal{E}^*\big(A, \big\{(a_1, \mathcal{L}_{B_1}), \ldots, (a_n, \mathcal{L}_{B_n})\big\}\big) \\
&= \mathcal{T}_P(\mathcal{E}) \circ \nu(\mathcal{D}_P\mathcal{S})\big(A, \big\{(a_1, \mathcal{L}_{B_1}), \ldots, (a_n, \mathcal{L}_{B_n})\big\}\big) \\
&= \mathcal{T}_P(\mathcal{E})\big(\|^n\{a_1.\mathcal{L}_{B_1}, \ldots, a_n.\mathcal{L}_{B_n}\}\big) = \|^n\big\{a_1.\mathcal{E}(\mathcal{L}_{B_1}), \ldots, a_n.\mathcal{E}(\mathcal{L}_{B_n})\big\}.
\end{aligned}$$

At the same time,

$$\begin{aligned}
\simeq_\Sigma \circ \mathcal{E}(\mathcal{L}_A) &= \simeq_\Sigma\big(\|^n\big\{a_1.\mathcal{E}(\mathcal{L}_{B_1}), \ldots, a_n.\mathcal{E}(\mathcal{L}_{B_n})\big\}\big) \\
&= \|^n\big\{a_1.\mathcal{E}(\mathcal{L}_{B_1}), \ldots, a_n.\mathcal{E}(\mathcal{L}_{B_n})\big\}.
\end{aligned}$$

Consequently, $\simeq_\Sigma \circ \mathcal{E}(\mathcal{L}_A) = \mathcal{E}^* \circ \simeq_B(\mathcal{L}_A)$.

From the fact that both $\mathbf{g}_E$ and $[\_]$ are based on the sets $S_{(A,\alpha^*)}$ (in Definition 54) for the outgoing arrows in G from a schema database $\mathcal{A}$, it is easy to verify that $\mathbf{g}_E = \nu(X) \circ [\_] : X \to \mathcal{T}_P(X)$. In fact, based on the algorithm *DBprog*, for each $p_k \in X$ we have one of the following two cases:

1. When $[p_k] = (ass(p_k), \{(ass(p_{k+1}), p_{k+1}), \ldots, (ass(p_{k+n}), p_{k+n})\})$, $n \geq 2$, we obtain $\nu(X)([p_k]) = \|^n(ass(p_{k+1}).p_{k+1}, \ldots, ass(p_{k+n}), p_{k+n})$ which is equivalent to the flattened term $\|^n(p_{k+1}, \ldots, p_{k+n})$ (see Example 36) with equation $(p_k = \|^n(p_{k+1}, \ldots, p_{k+n})) \in E$ and hence $\mathbf{g}_E(p_k) = \|^n(p_{k+1}, \ldots, p_{k+n})$.
2. When $[p_k] = (ass(p_k), \{(ass(p_{k+1}), p_{k+1})\})$, we obtain $\nu(X)([p_k]) = a.p_{k+1}$, and from the fact that $(p_k = a.p_{k+1}) \in E$, we obtain $\mathbf{g}_E(p_k) = a.p_{k+1}$.

Consequently, $\mathbf{s}_E = \mathcal{E} \circ [\_]^@ : X \to \mathcal{T}_\infty$, so that the final coalgebra semantics for the guarded system of equations of a given database mapping system is the result of the operational semantics based on the behavior endofunctor $\mathcal{B}_P$ and LTS of execution of these database programs. In fact, the commutative diagrams (4) (the operational semantics) and (5) (commutative diagram of the embedding of LTS trees into the ground term trees and (5a) compose the final coalgebra (bijection $\simeq_\Sigma$) semantics diagram (1) for the flattened system of equations in Proposition 37.

The commutative diagrams (4) (the operational semantics) and (5) (commutative diagram of the embedding of LTS trees into ground term trees and (0) (commutative diagram of isomorphisms in Proposition 37) define the commutative diagram (3) in Proposition 38 with arrows $\mathbf{g}_E$ and $\mathbf{s}_E$.

Consequently, the commutative diagram of this proposition extends the commutative diagram in Proposition 38 by the operational semantics and its final coalgebra isomorphism $\simeq_B$. Thus, the diagram above is composed of all fundamental (co)algebra isomorphisms: the initial Σ_P-algebra isomorphism $\simeq_I$ with DB-denotational semantics Σ_P-algebra, $[f, h] : X \uplus \Sigma_P(\mathbb{T}_\infty) \to \mathbb{T}_\infty$, where $f = \simeq_T \circ \mathbf{s}_E$ and both final coalgebra isomorphisms $\simeq_B$ and, derived from it, $\simeq_F$. □

In GSOS, only operations w.r.t. which bisimulation remains a congruence are admitted since bisimilar trees are easily seen to be *trace equivalent*, and hence in GSOS_{DB}, $a.(b\|c)$ and $a.b\|a.c$ are bisimilar or observationally equivalent and denoted by $a.(b\|c) = \|(a.b, a.c)$.

One of the properties of Aczel's coinductive semantics is that it maps two programs to the same set if and only if they are observationally equivalent:

$$[p_k]^@ = [p_m]^@ \quad \text{iff} \quad p_k \sim p_m.$$

That is, the coinductive extension of the operational model $[\_] : \mathcal{S} \longrightarrow \mathcal{S} \times \mathcal{B}_P(X)$ does preserve $\mathcal{B}_P$-bisimulation and, conversely, it can be 'pulled back' to form the largest $\mathcal{B}$-bisimulation relation. The above is a property which holds in general for every coinductive extension of coalgebras of endofunctors $\mathcal{B}_P$ preserving categorial (weak) pullbacks.

Proposition 40 *Let us define the set of the subcategories of the* **DB** *category,*

$$Sub(\mathbf{DB}) = \{D_{\mathcal{L}_A} | \mathcal{L}_A \in \mathcal{D}_P\mathcal{S}\},$$

where each tree $\mathcal{L}_A \in \mathcal{D}_P\mathcal{S}$ is an oriented graph with the root node A, so that $D_{\mathcal{L}_A}$ is a small category defined by this graph as follows: Each transition $B \xrightarrow{a} C$ in

$\mathcal{L}_A$, where $a \in Act \subseteq Ob_{DB}$, is substituted by an atomic **DB** *morphism (in Definition 18) $f_a = \alpha^*(MakeOperads(\mathcal{M}_{BC})) : B \to C$ where α is a mapping-interpretation (in Definition 11) such that $\alpha^*(\mathcal{B}) = B$ is a model of schema $\mathcal{B}$ in a G and, in the case of the forward propagation, $a = \Delta(\alpha, MakeOperads(\mathcal{M}_{BC}))$.*

Then we define the bijection $\simeq_{DB} : \mathcal{D}_P\mathcal{S} \to Sub(\mathbf{DB})$ such that for each LTS tree $\mathcal{L}_A \in \mathcal{D}_P\mathcal{S}, \simeq_D(\mathcal{L}_A) = D_{\mathcal{L}_A}$.

Consequently, $[p_k]^@ = [p_m]^@ \iff p_k \sim p_m$ iff $A = ass(p_k) \simeq B = ass(p_m)$ and the category $\simeq_{DB}(\mathcal{L}_A)$ is equal to the category $\simeq_{DB}(\mathcal{L}_B)$.

Proof For any LTS 'insertion' tree $\mathcal{L}_A \in \mathcal{D}_P\mathcal{S}$, a transition $A \xrightarrow{a} B$, derived from an inter-schema mapping $\mathcal{M}_{AB} : \mathcal{A} \to \mathcal{B}$, can be represented in **DB** by the arrow $f_a : A \to B$ such that $\widetilde{f_a} = Ta$. In fact, from the definition of $\simeq_{DB}$, $(A \xrightarrow{a} B)$ is mapped into an atomic morphism $f_a = \alpha^*(MakeOperads(\mathcal{M}_{AB}))$, where $MakeOperads(\mathcal{M}_{AB})$ is an arrow in the small category $\simeq_{DB}(\mathcal{L}_A)$, with $a = \Delta(\alpha, MakeOperads(\mathcal{M}_{AB}))$, so that $\widetilde{f_a} = Ta$.

Thus, we have the inclusion embedding $In : \simeq_{DB}(\mathcal{L}_A) \hookrightarrow \mathbf{DB}$. Analogous result holds for 'deletion' (backward propagation) trees as well.

Notice that this result can be obtained from Definition 51, too.

The equivalence of categories $\simeq_{DB}(\mathcal{L}_A)$ and $\simeq_{DB}(\mathcal{L}_B)$ is obtained from the fact that the trees $\mathcal{L}_A = [p_k]^@$ and $\mathcal{L}_B = [p_m]^@$ are bisimilar, i.e., *trace equivalent* with $A = ass(p_k) \simeq B = ass(p_m)$ (all corresponding arrows (transitions) in these categories that compose the traces have equal information fluxes). □

Consequently, any tree (i.e., oriented graph) in the $\mathcal{D}_P\mathcal{S}$ can be mapped by $\simeq_{DB}$ into a particular subcategory of **DB** and demonstrates that **DB** is an adequate category both for denotational and operational semantics for database mapping systems.

Example 39 For the database mapping process in Example 36 we obtain the following LTS tree $\mathcal{L}_A$ in $\mathcal{D}_P\mathcal{S}$ where $A = ass(p_1)$, and its small category $\simeq_{DB}(\mathcal{L}_A)$ in *Sub*(**DB**):

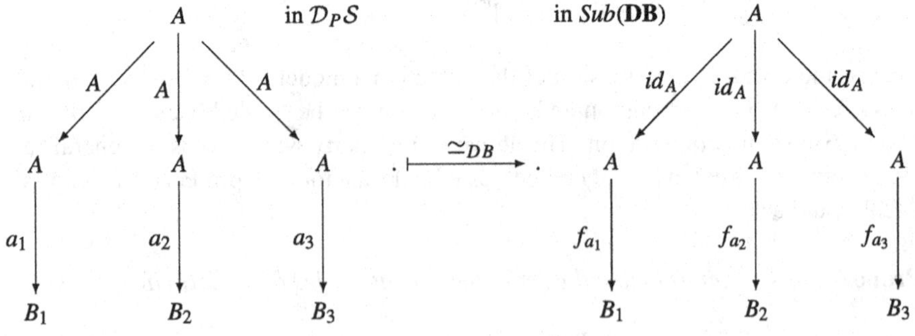

where $\widetilde{f_{a_i}} = Ta_i, i = 1, 2, 3$.

That is, it is similar to the transition tree in Example 36 (the tree with the process variables).

If in this LTS we substitute any label 'a' with the morphism 'f_a' such that $\widetilde{f}_a = Ta$, we obtain a diagram in **DB** category.

The subcategory $\simeq_{DB}(\mathcal{L}_A)$ is obtained from this graph (a tree) $\mathcal{L}_A$ by inserting the identity arrows for all vertices and all arrows that can be obtained from categorial composition of *atomic* arrows (i.e., transitions in LTS $\mathcal{L}_A$).

Consequently, from Proposition 40 and the final coalgebra diagram (4), we obtain the following final coalgebra diagram in **Set**, based on the **DB** category with $\mathcal{S} \subseteq Ob_{DB}$:

$$
\begin{array}{ccc}
X & \xrightarrow{\;\simeq_{DB}\,\circ[\_]^{@}\;} & Sub(\textbf{DB}) \\[1mm]
{\scriptstyle[\_]}\Big\downarrow & (6) & \Big\downarrow{\scriptstyle I_{Sub(\textbf{DB})}} \\[1mm]
\overline{\mathcal{B}}_P(X) = \mathcal{S} \times \mathcal{B}_P(X) & \xrightarrow{\;\simeq_{DB}\,\circ\overline{\mathcal{B}}_P([\_]^{@})\;} & Sub(\textbf{DB})
\end{array}
$$

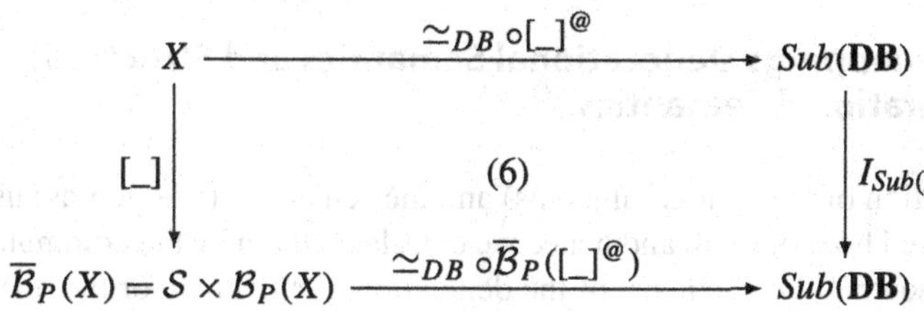

In fact, for each $p_k \in X$, the trees $[p_k]^{@} \in \mathcal{D}_P\mathcal{S} = gfp(\mathcal{S} \times \mathcal{B}_P)$ and $(\mathcal{S} \times \mathcal{B}_P)([[p_k]]^{@})$ have the same traces (from the final coalgebra semantics diagram (4)), so that $\simeq_{DB}([p_k]^{@})$ and $\simeq_{DB}((\mathcal{S} \times \mathcal{B}_P)([[p_k]]^{@}))$ are two equivalent subcategories of **DB**. Let $[p_k] = (A, \{(a_1, p_{k+1}), \dots, (a_n, p_{k+n})\})$, with $n \geq 2$ and $A = ass(p_k)$. Then the identity mapping $I_{Sub(\textbf{DB})} : Sub(\textbf{DB}) \to Sub(\textbf{DB})$ represents the mapping:

$$\simeq_{DB}\big([p_k]^{@}\big) \mapsto \simeq_{DB}\big(A, \{(a_1, [p_{k+1}]^{@}), \dots, (a_n, [p_{k+n}]^{@})\}\big),$$

by considering that the trees $\mathcal{L}_A = [p_k]^{@}$ and

$$\mathcal{L}'_A = \big(A, \{(a_1, [p_{k+1}]^{@}), \dots, (a_n, [p_{k+n}]^{@})\}\big)$$

are equal, but having only two different representations. Consequently, the diagram above commutes.

Proposition 41   *From the bijection* $\simeq_{DB} : \mathcal{D}_P\mathcal{S} \to Sub(\textbf{DB})$ *in Propositions* 39 *and* 40, *the following commutative diagram represents the adequate DB-denotational semantics for the operational semantics based on the* **DB** *category:*

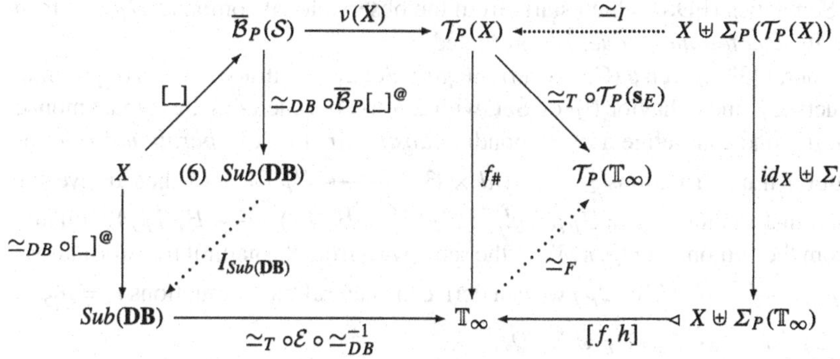

where, from Proposition 39, $\mathbf{g}_E = \nu(X) \circ [\_] : X \to \mathcal{T}_P(X)$, $\mathbf{s}_E = \mathcal{E} \circ [\_]^@ : X \to \mathcal{T}_\infty$, $f = \simeq_T \circ \mathbf{s}_E$ *and* $\overline{\mathcal{B}}_P = \mathcal{S} \times \mathcal{B}_P$.

Proof This commutative diagram is obtained from the commutative diagram in Proposition 39 where the square diagram (4) is substituted by equivalent square diagram (6), with an added bijection $\simeq_{DB}$ and the set of **DB** subcategories *Sub*(**DB**), and by considering that $\simeq_{DB}^{-1} \circ \simeq_{DB} = id_{\mathcal{D}_P\mathcal{S}}$, so that $f = \simeq_T \circ \mathcal{E} \simeq_{DB}^{-1} \circ \simeq_{DB} \circ [\_]^@ = \simeq_T \circ \mathcal{E} \circ [\_]^@ = \simeq_T \circ \mathbf{s}_E$. □

7.5.1 DB-Mappings Denotational Semantics and Structural Operational Semantics

In the description of the syntax (algebras) and the semantics (coalgebras) used previously, we used both of them and hence we correlated them by the commutative diagrams representing the adequacy of the denotational semantics for the operational semantics.

But we can consider the algebraic and coalgebraic structures in different hierarchic layers. For example, we can start with certain algebras describing one's application domain. On top of these one can have certain dynamic systems (processes) as coalgebras, involving such algebras (e.g., as codomains of attributes). And such coalgebraic systems may exist in an algebraic processes.

A concrete example of such layering of coalgebra on top of an algebra is given by Plotkin's so-called structural operational semantics (SOS) [35]. It involves a transition system (a coalgebra) describing the operational semantics of some language, by giving the transition rules by induction on the structure of the terms of the language. The later means that the set of terms of the language is used as an (initial) algebra.

Differently from the DB-denotational semantics where the carrier set of the denotational model (a $(X \uplus \Sigma)$-algebra) is the set of ground terms representing the objects (i.e., instance-databases) in base **DB** category, the carrier set in the SOS-denotational semantics is the $\mathcal{D}_P\mathcal{S} = gfp(\mathcal{S} \times \mathcal{B}_P)$ of the 'behavior functor $\mathcal{B}_P$', i.e., the set of LTS trees with the states in $\mathcal{S}$ and actions (i.e., transitions) in *Act*. In [39], both Functorial Operational semantics (FOS) (which starts with the syntactical monad $\mathcal{T}_P$ and then derives its *operational model*) and its dual Functorial Denotational Semantics (FDS) which starts from the observational comonad $\mathcal{D}_P$ and then derives its *denotational model* are presented.

In general [39], given a (Cartesian) category **Set** and arbitrary notions of program constructs Σ_P and behavior $\mathcal{B}_P$ on **Set**, with Σ_P freely generating the syntax monad $(\mathcal{T}_P, \eta, \mu)$, one can define a corresponding *abstract* notion of *operational rules* as the natural transformations $\varrho' : \Sigma_P(Id \times \mathcal{B}_P) \stackrel{\cdot}{\longrightarrow} \mathcal{B}_P\mathcal{T}_P$, and then derive the natural transformations $\varrho = \mathcal{B}_P\mu \circ \varrho'_{\mathcal{T}_P} : \Sigma_P(\mathcal{T}_P \times \mathcal{B}_P\mathcal{T}_P) \stackrel{\cdot}{\longrightarrow} \mathcal{B}_P\mathcal{T}_P$. By duality [39] from the comonad $\langle \mathcal{D}_P, \varepsilon, \delta \rangle$, in the same way, from the natural transformations $\rho' : \Sigma_P\mathcal{D}_P \stackrel{\cdot}{\longrightarrow} \mathcal{B}_P(Id \uplus \Sigma_P)$ we can derive the natural transformations $\rho = \rho'_{\mathcal{D}_P} \circ \Sigma_P\delta : \Sigma_P\mathcal{D}_P \stackrel{\cdot}{\longrightarrow} \mathcal{B}_P(\mathcal{D}_P \uplus \Sigma_P\mathcal{D}_P)$.

Let us show this property for our database-mapping programs (in [39], the authors considered the language where $\|$ is a non-deterministic choice, while here it is a parallel execution). The type $\mathcal{B}_P$ of the behavior of our database-mapping programming language is $\mathcal{B}_P(X) = (\mathcal{P}_{\text{fin}}(X))^{Act}$, the covariant functor mapping a set of process variables X to the set of functions from Act to the finite subsets of X. Let p_k and $p_{i+1}, \dots, p_{i+n}$ range over X, β range over $(\mathcal{P}_{\text{fin}}(X))^{Act}$ (a function $\beta : Act \rightarrow \mathcal{P}_{\text{fin}}(X))$), and let us write $a \rightsquigarrow \{p_{i+1}, \dots, p_{i+n}\}$ for the function β from Act to $\mathcal{P}_{\text{fin}}(X)$ mapping a to $\{p_{i+1}, \dots, p_{i+n}\}$.

For example, for $\wp : X \rightarrow \mathcal{P}_{\text{fin}}(Act \times X)$, let $\wp(p_k) = \{(a_1, p_{i+1}), \dots, (a_n, p_{i+n})\}$. Then β maps each action $a_m \in \pi_1(\wp(p_k))$, $1 \le m \le n$, into the subset of $\pi_2(\wp(p_k))$ (where π_1 and π_2 denote the first and second Cartesian projections) such that $(a_m, p_{i+m}) \in \wp(p_k)$ for each $p_{i+m} \in \pi_2(\wp(p_k))$; for each $a_m \notin \pi_1(\wp(p_k))$, $\beta(a_m) = \emptyset$ (empty set), that is, $\beta(a) = \{p_i \mid (a, p_i) \in \wp(p_k)\}$. Thus, the set $\wp(p_k)$ can be equivalently represented by this function $\beta : Act \rightarrow \mathcal{P}_{\text{fin}}(X)$, i.e., $\beta \in (\mathcal{P}_{\text{fin}}(X))^{Act}$.

Consequently, in what follows, we will use for the behaviors the functor $\mathcal{B}_P = \mathcal{P}_{\text{fin}}(Act \times \_) : \mathbf{Set} \rightarrow \mathbf{Set}$, instead of its equivalent functor $\mathcal{P}_{\text{fin}}(\_)^{Act} : \mathbf{Set} \rightarrow \mathbf{Set}$ used in [39]; this choice is justified specially when Act is infinite, an hence for the *finite* subsets in $\mathcal{P}_{\text{fin}}(Act \times X)$ we have no need to use the functions $\beta : Act \rightarrow X$ with the infinite domains.

The difference of our language where '$\|$' denotes the parallel executions, from the syntactically equal language in [39] where '$\|$' denotes the nondeterministic choice instead, is that in our case the function β can have more than one argument $a \in Act$ such that $\beta(a) \neq \emptyset$ (while in [39] β has exactly one such argument), so that β means just the parallel execution of all actions in the set $\{a \in Act \mid \beta(a) \neq \emptyset\}$. Moreover, we are using the finitary parallel operators '$\|^n$' for $n \ge 2$ as well.

Then, for example, for each operator σ of the signature $\Sigma_P = \{nil, Act, \{a.\_\}_{a \in Act}, \{\|^n\}_{n \ge 2}\}$, the corresponding rules can be modeled as a function

$$\Subset \sigma \Supset : \left(X \times \mathcal{P}_{\text{fin}}(Act \times X)\right)^{arity(\sigma)} \rightarrow \mathcal{P}_{\text{fin}}(Act \times \mathcal{T}_P X),$$

as follows:

1. For nullary operators, $\sigma = nil$ or $\sigma \in Act$,

$$\Subset \sigma \Supset = \emptyset \in \mathcal{P}_{\text{fin}}(Act \times \mathcal{T}_P X) \quad \text{a constant (nullary operator);}$$

2. For unary operators,

$$\Subset a.\_ \Supset \left(p_i, \left\{(a_1, p_{k_1}), \dots, (a_n, p_{k_n})\right\}\right)$$
$$= \left\{\left(a, \|^n(a_1.p_{k_1}, \dots, a_n.p_{k_n})\right)\right\} \in \mathcal{P}_{\text{fin}}(Act \times \mathcal{T}_P X);$$

3. For m-ary, $m \ge 2$, operators,

$$\Subset \|^m \Supset \left(\left(p_{i_1}, \left\{(a_{11}, p_{k_{1}1}), \dots, (a_{n_1 1}, p_{k_{n_1 1}})\right\}\right), \dots,\right.$$
$$\left.\left(p_{i_m}, \left\{(a_{1m}, p_{k_1 m}), \dots, (a_{n_m m}, p_{k_{n_m m}})\right\}\right)\right)$$

$$= \{(a_{11}, p_{k_1 1}), \ldots, (a_{n_1 1}, p_{k_{n_1 1}}), \ldots, (a_{1m}, p_{k_1 m}), \ldots, (a_{n_m m}, p_{k_{n_m m}})\}.$$

Notice that p_i and $p_{i_1}, i = 1, \ldots, m$ (the root variables) are forgotten during execution of these rules.

By using the universal property of coproducts $\uplus$ in **Set**, these functions can be glued into a single function (application of rules)

$$\varrho'_X : \{nil\} \cup Act\left(\biguplus_{a_i \in Act} (X \times \mathcal{B}_P X)\right) \biguplus_{n \geq 2} (X \times \mathcal{B}_P X)^n \to \mathcal{B}_P \mathcal{T}_P X,$$

i.e., into $\varrho'_X : \Sigma_P(X \times \mathcal{B}_P(X)) \to \mathcal{B}_P(\mathcal{T}_P X)$ with $\varrho' : \Sigma_P(Id \times \mathcal{B}_P) \stackrel{\cdot}{\longrightarrow} \mathcal{B}_P \mathcal{T}_P$, equal to that in [39].

Note one has a function ϱ'_X for each set X. Most importantly, the above definition of ϱ'_X is *natural* in X: for any renaming $f : X \to X'$ of variables in X (possibly involving equating some of the variables), first renaming and then applying the rules is the same as first applying the rules and then renaming, that is, the following diagram commutes

$$\begin{array}{ccc}
\Sigma_P(Id \times \mathcal{B}_P)X & \xrightarrow{\varrho'_X} & \mathcal{B}_P \mathcal{T}_P X \\
\Big\downarrow{\scriptstyle \Sigma_P(Id \times \mathcal{B}_P)f} & & \Big\downarrow{\scriptstyle \mathcal{B}_P \mathcal{T}_P f} \\
\Sigma_P(Id \times \mathcal{B}_P)X' & \xrightarrow{\varrho'_{X'}} & \mathcal{B}_P \mathcal{T}_P X'
\end{array}$$

Thus, we obtained the natural transformation $\varrho' : \Sigma_P(Id \times \mathcal{B}_P) \stackrel{\cdot}{\longrightarrow} \mathcal{B}_P \mathcal{T}_P$ and $\varrho = \mathcal{B}_P \mu \circ \varrho'_{\mathcal{T}_P} : \Sigma_P(\mathcal{T}_P \times \mathcal{B}_P \mathcal{T}_P) \stackrel{\cdot}{\longrightarrow} \mathcal{B}_P \mathcal{T}_P$ as in [39].

A Functorial Operational Semantics for a syntax monad $\mathcal{T}_P$ and a behavior $\mathcal{B}_P$ is the *operational monad* $\mathcal{T}_\varrho$ which 'lifts' the syntactical monad $(\mathcal{T}_P, \eta, \mu)$ (and inherits η and μ from it) to the coalgebras of the behavior endofunctor $\mathcal{B}_P$. It can be represented by the following commutative diagram in **Cat** (we denote the category of $\mathcal{B}_P$-coalgebras by $\mathcal{B}_{P\text{-cAlg}}$), where $U_{\mathcal{B}_P}$ denotes the forgetful functor:

$$\begin{array}{ccc}
\mathcal{B}_{P\text{-cAlg}} & \xrightarrow{\mathcal{T}_\varrho} & \mathcal{B}_{P\text{-cAlg}} \\
\Big\downarrow{\scriptstyle U_{\mathcal{B}_P}} & & \Big\downarrow{\scriptstyle U_{\mathcal{B}_P}} \\
\mathbf{Set} & \xrightarrow{\mathcal{T}_P} & \mathbf{Set}
\end{array}$$

That is, every natural transformation ϱ defines inductively a lifting (operational monad) $\mathcal{T}_\varrho$ of the monad $(\mathcal{T}_P, \eta, \mu)$ to the $\mathcal{B}_P$-coalgebras: given an operational

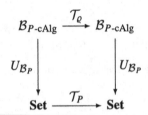

model $\mathcal{B}_P$-colagebra (i.e., 'assumption') $m : X \to \mathcal{B}_P(X)$, the structure ($\mathcal{B}_P$-colagebra) $\mathcal{T}_\varrho(m) : \mathcal{T}_P X \to \mathcal{B}_P(\mathcal{T}_P X)$ can be seen as the *operational model* on the set of terms (with the process variables) in $\mathcal{T}_P X$. In fact, for a given natural transformation ϱ and a set X, there is a morphism $\varrho_X : \Sigma_P(\mathcal{T}_P X \times \mathcal{B}_P \mathcal{T}_P X) \to \mathcal{B}_P \mathcal{T}_P X$ defined by:
1. $nil \mapsto \emptyset$, and $a \mapsto \emptyset$ for each $a \in Act$;
2. $((t, Y), i) \mapsto \{(a_i, a_k.t_k) \mid (a_k, t_k) \in Y\}$ if $|Y| \geq 1$; $\emptyset$ otherwise;
3. $((t_1, Y_1), \ldots, (t_n, Y_n)) \mapsto \bigcup_{1 \leq i \leq n} Y_i$.

For a given $\mathcal{B}_P$-coalgebra $m : X \to \mathcal{B}_P X$, there is a $(X \uplus \Sigma_P)$-algebra

$$\big(\!\big|\big[\eta_X, \mathcal{B}_P(\eta_X) \circ m\big], [h, \varrho_X]\big|\!\big) : X \uplus \Sigma_P(\mathcal{T}_P X \times \mathcal{T}_P X \times \mathcal{B}_P \mathcal{T}_P X) \to \mathcal{B}_P \mathcal{T}_P X,$$

with the unique arrow $[id_{\mathcal{T}_P X}, \mathcal{T}_\varrho(m)]$ from the initial 'syntax' $(X \uplus \Sigma_P)$-algebra, so that the following diagram commutes

$$\begin{array}{ccc}
\mathcal{T}_P X & \xleftarrow{\ \simeq_I = \langle \eta_X, inr_X \rangle\ } & X \uplus \Sigma_P(\mathcal{T}_P X) \\
\Big\downarrow {\scriptstyle f_\# = [\eta_X, \mathcal{B}_P(\eta_X) \circ m]_\# = [id_{\mathcal{T}_P X}, \mathcal{T}_\varrho(m)]} & & \Big\downarrow {\scriptstyle id_X \uplus \Sigma_P f_\#} \\
\mathcal{T}_P X \times \mathcal{B}_P \mathcal{T}_P X & \xleftarrow{\ \langle [\eta_X, \mathcal{B}_P(\eta_X) \circ m], [h, \varrho_X] \rangle\ } & X \uplus \Sigma_P(\mathcal{T}_P X \times \mathcal{B}_P \mathcal{T}_P X)
\end{array}$$

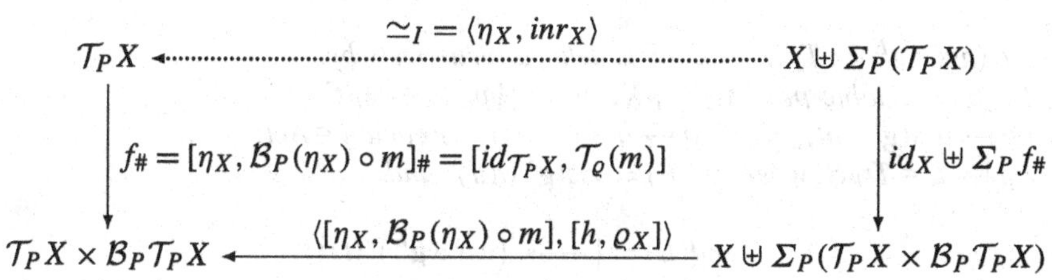

The unique parameter in this commutative diagram is the arrow $m : X \to \mathcal{B}_P X$ (all other arrows have fixed semantics). The mapping $\mathcal{T}_\varrho(m)$ is a component of the unique arrow from the initial $(X \uplus \Sigma_P)$-algebra, so that the diagram above can be reduced to the following interesting commutative component:

$$\begin{array}{ccccc}
X & \xhookrightarrow{\ inl_X = \eta_X\ } & \mathcal{T}_P X & \xhookleftarrow{\ inr_X\ } & \Sigma_P \mathcal{T}_P X \\
\Big\downarrow m & (8) & \Big\downarrow \mathcal{T}_\varrho(m) & & \Big\downarrow \Sigma_P f_\# \\
\mathcal{B}_P X & \xrightarrow{\ \mathcal{B}_P(\eta_X)\ } & \mathcal{B}_P \mathcal{T}_P X & \xleftarrow{\ \varrho_X\ } & \Sigma_P(\mathcal{T}_P X \times \mathcal{B}_P \mathcal{T}_P X)
\end{array}$$

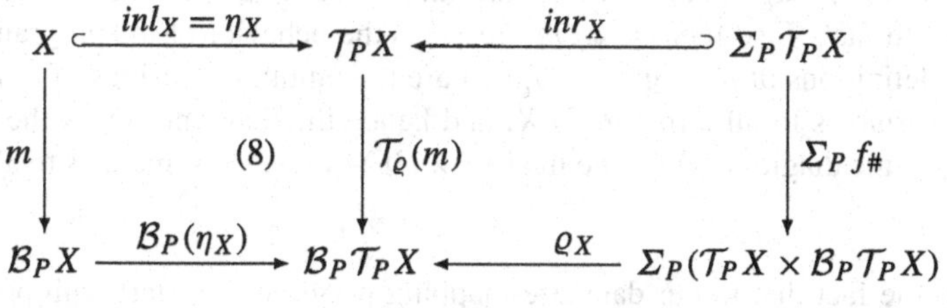

Thus, for any 'parameter' (coalgebra structure) $m : X \to \mathcal{B}_P X$, its *conservative extension* to all terms with variables $\mathcal{T}_\varrho(m) : \mathcal{T}_P X \to \mathcal{B}_P(\mathcal{T}_P X)$ is uniquely recursively determined (notice that for all variables $p_k \in X \subseteq \mathcal{T}_P X$, $\mathcal{T}_\varrho(m)(p_k) = m(p_k)$) by:
1. $p_k \mapsto m(p_k)$ (note that it can be the empty set $\emptyset$ as well);
2. $nil \mapsto \emptyset$, and $a \mapsto \emptyset$ for each $a \in Act$;
3. $a.t \mapsto \{(a, a_i.t_i) \mid (a_i, t_i) \in \mathcal{T}_\varrho(m)(t)\}$ if $|\mathcal{T}_\varrho(m)(t)| \geq 1$; $\emptyset$ otherwise;
4. $\|^n(t_1, \ldots, t_n) \mapsto \bigcup_{1 \leq i \leq n} \mathcal{T}_\varrho(m)(t_i)$.

Thus, the injective function $\eta_X : X \hookrightarrow \mathcal{T}_P X$ 'lifts' to the coalgebra mapping (diagram (8)) $\eta_X : (X, m) \to (\mathcal{T}_P X, \mathcal{T}_\varrho(m))$ for every coalgebra structure m on X.

Let us consider the following case when $m = \wp$, i.e., when $m : X \to \mathcal{B}_P X$ represents the behavior with $[\_] = \langle ass, \wp \rangle : X \to \mathcal{S} \times \mathcal{B}_P(X)$:

Proposition 42 *Let us consider the commutative diagram* (8) *in the case when* $m = \wp : X \to \mathcal{B}_P X$ *represents the behavior* $[\_] = \langle ass, \wp \rangle$ *of a program. Then, for* $Z = \mathcal{T}_P X$, $l = \mathcal{T}_Q(\wp)$ *and an extended assignment* $f = ass \circ \mathbf{g}^E : \mathcal{T}_P X \to \mathcal{S}$, *the commutative diagram* (4a) *in Sect.* 7.4 *becomes the following commutative diagram*

$$\begin{array}{ccc}
\mathcal{D}_P\mathcal{S} & \overset{\simeq_B}{\dashrightarrow} & \mathcal{S} \times \mathcal{B}_P(\mathcal{D}_P\mathcal{S}) \\
{\scriptstyle (ass \circ \mathbf{g}^E)^{\natural}} \uparrow & (4c) & \uparrow {\scriptstyle id_{\mathcal{S}} \times \mathcal{B}_P((ass \circ \mathbf{g}^E)^{\natural})} \\
\mathcal{T}_P X & \xrightarrow{\langle ass \circ \mathbf{g}^E,\, \mathcal{T}_Q(\wp) \rangle} & \mathcal{S} \times \mathcal{B}_P(\mathcal{T}_P X)
\end{array}$$

where $(ass \circ \mathbf{g}^E)^{\natural} : \mathcal{T}_P X \to \mathcal{D}_P\mathcal{S}$ *is defined recursively by:*
1. *For any variable* $p_k \in X \subseteq \mathcal{T}_P X$, $p_k \mapsto [p_k]^{@} \in \mathcal{D}_P\mathcal{S}$;
2. $nil \mapsto ass(\mathbf{g}^E(nil)) = \bot^0$, $a \mapsto ass(\mathbf{g}^E(a))$, *for each* $a \in Act$;
3. $a.t \mapsto \mathcal{L} \in \mathcal{D}_P\mathcal{S}$, *where* $fst(\mathcal{L}) = ass(\mathbf{g}^E(a.t))$ *and*

$$snd(\mathcal{L}) = \{\xrightarrow{a} (ass \circ \mathbf{g}^E)^{\natural}(t)\};$$

4. $\|^n(t_1, \ldots, t_n) \mapsto \mathcal{L} \in \mathcal{D}_P\mathcal{S}$, *where* $\mathcal{L}$ *is the union of all trees* $(ass \circ \mathbf{g}^E)^{\natural}(t_i)$, $1 \le i \le n$, *obtained by superposition of their roots in the unique root*

$$ass\big(\mathbf{g}^E\big(\|^n(t_1, \ldots, t_n)\big)\big).$$

Proof Now it is enough to demonstrate the validity of $\mathcal{T}_Q(\wp)^{\natural}$ for the variables $p_k \in X \subseteq \mathcal{T}_P X$. In fact, $\mathcal{T}_Q(\wp)(p_k) = \wp(p_k)$ respects the behavior of the programs. The recursive definitions of $ass \circ \mathbf{g}^E$ and $\mathcal{T}_Q(\wp)$ are the unique extensions of ass and $\wp$ from the variables to all terms in $\mathcal{T}_P X$, and hence the diagram (4c) is the unique extension of the diagram (4) for the final $(\mathcal{S} \times \mathcal{B}_P)$-coalgebra semantics to all terms in $\mathcal{T}_P X$. □

Due to the fact that in our database-mapping programs we start with processes that are LTS systems, from the observational comonad $\mathcal{D}_P$ described in Sect. 7.4, we will present also the second (equivalent) *dual* approach to the SOS.

A Functorial Denotational Semantics for a syntax monad $\mathcal{T}_P$ and a behavior endofunctor $\mathcal{B}_P$ is a *denotational comonad* $\mathcal{D}_\rho$ which 'lifts' the observational comonad $(\mathcal{D}_P, \varepsilon, \delta)$ (and inherits ε and δ from it) to the Σ_P-algebras of the syntactical monad $\mathcal{T}_P$. We have seen that Σ_P and its coalgebraic dual $\mathcal{B}_P$ are two endofunctors on a co-cartesian category **Set** and that $\mathcal{D}_P = (\mathcal{D}_P, \varepsilon, \delta)$ is the *cofree comonad generated* by $\mathcal{B}_P$ (dual to the algebraic syntax monad $\mathcal{T}_P = (\mathcal{T}_P, \eta, \mu)$). It can be represented by the following commutative diagram in **Cat** (we denote the category of Σ_P-algebras by $\Sigma_{P\text{-Alg}}$), where U^{Σ_P} denotes the forgetful functor:

$$\begin{array}{ccc}
\Sigma_P\text{-Alg} & \xrightarrow{\mathcal{D}_\rho} & \Sigma_P\text{-Alg} \\
\Big\downarrow U^{\Sigma_P} & & \Big\downarrow U^{\Sigma_P} \\
\mathbf{Set} & \xrightarrow{\mathcal{D}_P} & \mathbf{Set}
\end{array}$$

That is, every natural transformation $\rho : \Sigma_P\mathcal{D}_P \stackrel{\bullet}{\longrightarrow} \mathcal{B}_P(\mathcal{D}_P \uplus \Sigma_P\mathcal{D}_P)$ defines coinductively a lifting $\mathcal{D}_\rho$ of the comonad $\mathcal{D}_P$ to the Σ_P-algebras: given a Σ_P-algebra ('assumption') $k : \Sigma_P\mathcal{S} \to \mathcal{S}$, the structure ($\Sigma_P$-algebra) $\mathcal{D}_\rho(k) : \Sigma_P(\mathcal{D}_P\mathcal{S}) \to \mathcal{D}_P\mathcal{S}$ can be seen as the *denotational model* on the set of global behaviors (trees) $\mathcal{D}_P\mathcal{S}$. In fact, for a given natural transformation $\rho : \Sigma_P\mathcal{D}_P \stackrel{\bullet}{\longrightarrow} \mathcal{B}_P(\mathcal{D}_P \uplus \Sigma_P\mathcal{D}_P)$ and a set $\mathcal{S}$, there is a morphism $\rho_S : \Sigma_P\mathcal{D}_P(\mathcal{S}) \to \mathcal{B}_P(\mathcal{D}_P \uplus \Sigma_P\mathcal{D}_P)(\mathcal{S})$ and for a given Σ_P-algebra $k : \Sigma_P\mathcal{S} \to \mathcal{S}$, the $(\mathcal{S} \times \mathcal{B}_P)$-coalgebra $\langle[\varepsilon_s, k \circ \Sigma_P(\varepsilon_s)], [snd_S, \rho_S]\rangle : \mathcal{D}_P\mathcal{S} \uplus \Sigma_P\mathcal{D}_P\mathcal{S} \to \mathcal{S} \times \mathcal{B}_P(\mathcal{D}_P\mathcal{S} \uplus \Sigma_P\mathcal{D}_P\mathcal{S})$, with the unique arrow $f^\natural = [\varepsilon_s, k \circ \Sigma_P(\varepsilon_s)]^\natural$ to the final $(\mathcal{S} \times \mathcal{B}_P)$-coalgebra from a diagram (4a)

$$\begin{array}{ccc}
\mathcal{D}_P\mathcal{S} & \xrightarrow{\ \simeq_B = \langle\varepsilon_S, snd_S\rangle\ } & \mathcal{S} \times \mathcal{B}_P(\mathcal{D}_P\mathcal{S}) \\
\Big\uparrow f^\natural = [\varepsilon_S, k \circ \Sigma_P(\varepsilon_S)]^\natural = [id_{\mathcal{D}_P\mathcal{S}}, \mathcal{D}_\rho(k)] & (4d) & \Big\uparrow id_S \times \mathcal{B}_P(f^\natural) \\
\mathcal{D}_P\mathcal{S} \uplus \Sigma_P\mathcal{D}_P\mathcal{S} & \xrightarrow{\ \langle[\varepsilon_S, k \circ \Sigma_P(\varepsilon_S)], [snd_S, \rho_S]\rangle\ } & \mathcal{S} \times \mathcal{B}_P(\mathcal{D}_P\mathcal{S} \uplus \Sigma_P\mathcal{D}_P\mathcal{S})
\end{array}$$

The unique 'parameter' in this commutative diagram is the arrow $k : \Sigma_P\mathcal{S} \to \mathcal{S}$ (all other arrows have fixed semantics). The arrow $f^\natural$ (which is the unique coinductive extension of $[snd_S, \rho_S]$ along $[\varepsilon_S, k \circ \Sigma_P(\varepsilon_S)]$) is composed of two components, the identity $id_{\mathcal{D}_P\mathcal{S}} : \mathcal{D}_P\mathcal{S} \to \mathcal{D}_P\mathcal{S}$ and the component denoted by $\mathcal{D}_\rho(k) : \Sigma_P\mathcal{D}_P\mathcal{S} \to \mathcal{D}_P\mathcal{S}$, with $[id_{\mathcal{D}_P\mathcal{S}}, \mathcal{D}_\rho(k)] = f^\natural = [\varepsilon_S, k \circ \Sigma_P(\varepsilon_S)]^\natural$ that uniquely defines the arrow $\mathcal{D}_\rho(k)$ for any given arrow k, so that the diagram (4d) above can be reduced to the following commutative subdiagram

$$\begin{array}{ccccc}
\mathcal{S} & \xleftarrow{\ fst_S = \varepsilon_S\ } & \mathcal{D}_P\mathcal{S} & \xrightarrow{\ snd_S\ } & \mathcal{B}_P\mathcal{D}_P\mathcal{S} \\
\Big\uparrow k & (9) & \Big\uparrow \mathcal{D}_\rho(k) & & \Big\uparrow \mathcal{B}_P f^\natural \\
\Sigma_P\mathcal{S} & \xleftarrow{\ \Sigma_P(\varepsilon_S)\ } & \Sigma_P\mathcal{D}_P\mathcal{S} & \xrightarrow{\ \rho_S\ } & \mathcal{B}_P(\mathcal{D}_P\mathcal{S} \uplus \Sigma_P\mathcal{D}_P\mathcal{S})
\end{array}$$

Consequently, the 'lifting' comonad $\mathcal{D}_\rho$ maps a Σ_P-algebra $(\mathcal{S}, k : \Sigma_P\mathcal{S} \to \mathcal{S})$ into the Σ_P-algebra $(\mathcal{D}_P\mathcal{S}, \mathcal{D}_\rho(k) : \Sigma_P\mathcal{D}_P\mathcal{S} \to \mathcal{D}_P\mathcal{S})$ called SOS-denotational Σ_P-

algebra for the particular meaning of k. Thus, $\mathcal{D}_\rho(k)$ is a *conservative extension* of $k : \Sigma_P\mathcal{S} \to \mathcal{S}$ from the states (that are single-node trees) to all trees in $\mathcal{D}_P\mathcal{S}$, by:

- $\mathcal{D}_\rho(k)(y) = k(y)$, for $y \in \{nil\} \cup Act$;
- $\mathcal{D}_\rho(k)(i, \mathcal{L}_B)$ (where $\mathcal{L}_B$ is a tree with a root state B in $\mathcal{D}_P\mathcal{S}$) is the resulting tree $\mathcal{L}_A \in \mathcal{D}_P\mathcal{S}$, equal to $\langle a_i, \mathcal{L}_B\rangle$, i.e., with the branch a_i from the root $A = k(i, B)$ into the tree $\mathcal{L}_B$;
- $\mathcal{D}_\rho(k)(\mathcal{L}_{B_1}, \ldots, \mathcal{L}_{B_n})$ is the resulting tree $\mathcal{L}_A \in \mathcal{D}_P\mathcal{S}$, equal to the union of the trees $\mathcal{L}_{B_i}, i = 1, \ldots, n$, with the superposition of all roots B_i of these trees $i = 1, \ldots, n$ into the unique new root state equal to $A = k(B_1, \ldots, B_n)$.

Proposition 43 *Let us consider the commutative diagram* (9) *in the case when* $k = ass \circ \mathbf{g}^E \circ \kappa : \Sigma_P\mathcal{S} \to \mathcal{S}$ *represents the behavior of a program as in Proposition* 42. *Then, the commutative diagram in Proposition* 42 *can be extended to the following commutative diagram*:

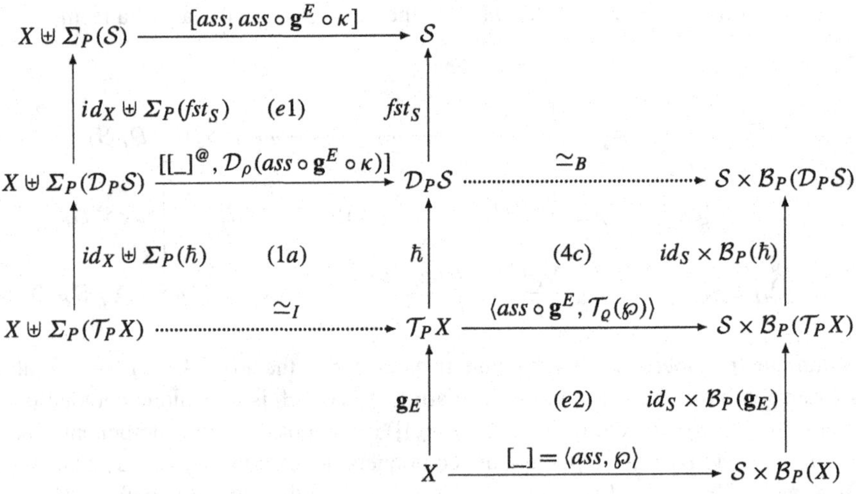

where $\hbar = (ass \circ \mathbf{g}^E)^\natural = [\_]^@_\#$ *is both a unique* $(\mathcal{S} \times \mathcal{B}_P)$*-coalgebra and* $(X \uplus \Sigma_P)$*-algebra homomorphism, and* $\kappa : \Sigma_P(\mathcal{S}) \to \mathcal{T}_P\emptyset \subset \mathcal{T}_P(X)$ *is defined by*:

1. $nil \mapsto nil$, $a \mapsto a$, *for each* $a \in Act$,
2. $(A, i) \mapsto a_i.A \in \mathcal{T}_P\emptyset$, *because* $A \in \mathcal{S} \subset Act$,
3. $(A_1, \ldots, A_n) \mapsto \|^n(A_1, \ldots, A_n) \in \mathcal{T}_P\emptyset$, *because for* $1 \le i \le n$, $A_i \in \mathcal{S} \subset Act$.

Proof It is enough to show that the diagram (1a) on the left commutes for $\hbar = (ass \circ \mathbf{g}^E)^\natural$ (defined in Proposition 42).

However, from the fact that $\hbar$ is the unique homomorphism from the initial $X \uplus \Sigma_P$ algebra into $(\mathcal{D}_P\mathcal{S}, [[\_]^@, \mathcal{D}_\rho(k)])$-algebra, also $\hbar = [\_]^@_\#$ must be satisfied, and hence $\hbar = (ass \circ \mathbf{g}^E)^\natural = [\_]^@_\# : \mathcal{T}_PX \to \mathcal{D}_P\mathcal{S}$. □

The diagram (e1) represents the diagram for naming the roots of the trees in $\mathcal{D}_P\mathcal{S}$ composed of the Σ_P signature. From this diagram we obtain the denotation relationship

$$fst_S \circ \mathcal{D}_\rho(k) = k \circ \Sigma_P(fst_S),$$

where $k = ass \circ \mathbf{g}^E \circ \kappa : \Sigma_P\mathcal{S} \to \mathcal{S}$ is corresponding to fst_S, in the way that determines the name of the root of composed trees by operations in Σ_P and the denotational model $\mathcal{D}_\rho(k)$.

From the fact that $fst_S \circ \hbar : \mathcal{T}_P X \to \mathcal{S}$ is the unique homomorphism (external square of the diagrams (1a) and (e1)), it must be equal to $ass_\#$, and from $ass = fst_S \circ \hbar \circ \mathbf{g}_E : X \to \mathcal{S}$, we obtain that $ass = ass_\# \circ \mathbf{g}_E$.

The diagram (e2) represents the generalization of the behavior from the *flattened* guarded behavior $[\_] = \langle ass, \wp\rangle : X \to \mathcal{S} \times \mathcal{B}_P(X)$ into the general guarded behavior $\langle ass, \mathcal{T}_\varrho(\wp) \circ \mathbf{g}_E\rangle = \langle ass \circ \mathbf{g}^E, \mathcal{T}_\varrho(\wp)\rangle \circ \mathbf{g}_E : X \to \mathcal{S} \times \mathcal{B}_P(\mathcal{T}_P X)$, which will be analyzed in next section.

Consequently, $\mathcal{T}_\varrho(\wp) \circ \mathbf{g}_E = \mathcal{B}_P(\mathbf{g}_E) \circ \wp$ represents the correspondence between the abstract behavior $\wp$ and the general behavior $\mathcal{T}_\varrho(\wp)$, and between the abstract behavior $\wp$ and the program equations defined by $\mathbf{g}_E$.

Consequently, the relationship between the abstract SOS-denotational semantics and the DB-denotational semantics is given by the following corollary:

Corollary 23 *For any given behavior* $[\_]$, *its unique coinductive extension* $[\_]^@$ *and derived from them* $\mathbf{g}_E = \nu(X) \circ [\_] : X \to \mathcal{T}_P(X)$ *(in Proposition* 39) *and* $\mathbf{g}^E$ *(in Proposition* 34), *the relationship between the* adequate *SOS-denotational semantics (with a denotational model* $\mathcal{D}_\rho(k)$ *in Proposition* 43, *where* $k = ass \circ \mathbf{g}^E \circ \kappa : \Sigma_P\mathcal{S} \to \mathcal{S}$) *and the* adequate *DB-denotational semantics are represented by the following commutative diagram between* $(X \uplus \Sigma_P)$*-algebras*

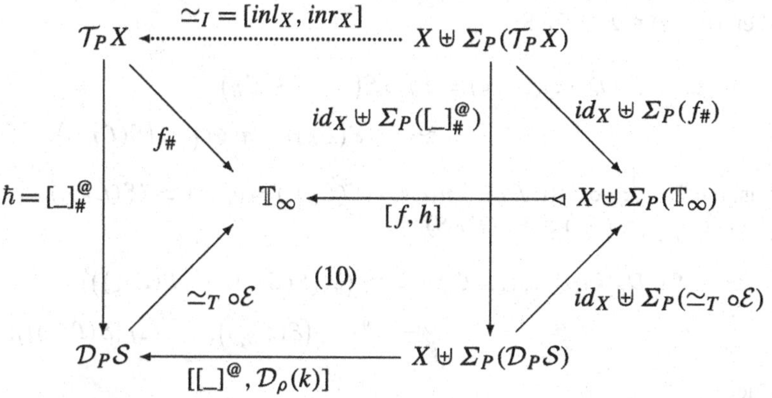

where $f = \simeq_T \circ \mathbf{s}_E = \simeq_T \circ \mathcal{E} \circ [\_]^@ : X \to \mathbb{T}_\infty$ *is the corresponding DB-mapping for the solution mapping* $[\_]^@ : X \to \mathcal{D}_P\mathcal{S} = gfp(\mathcal{S} \times \mathcal{B}_P)$.

Proof Notice that the component of the commutative diagram (10) is the two commutative diagrams bellow:

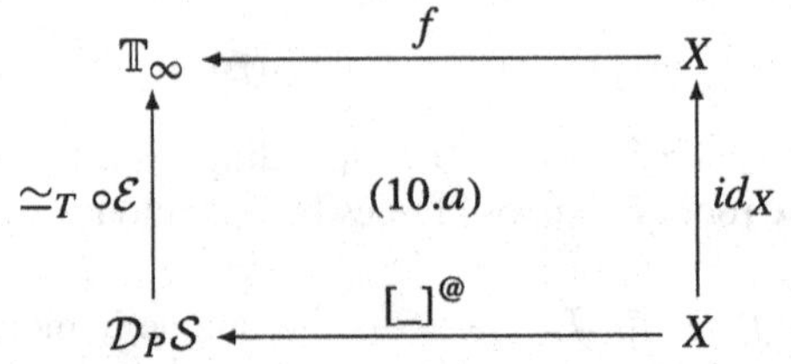

Thus, $f = {\simeq_T} \circ \mathcal{E} \circ [\_]^{@} : X \to \mathbb{T}_\infty(1)$ and, by Proposition 39, $f = {\simeq_T} \circ s_E$, so that $(\mathbb{T}_\infty, [f, h])$ is an adequate denotational semantics for the *syntax* initial free algebra $(\mathcal{T}_P X, \simeq_I)$ (in Proposition 38), and the commutative diagram

$$
\begin{array}{ccc}
\mathbb{T}_\infty & \xleftarrow{\ \ h\ \ } & \Sigma_P(\mathbb{T}_\infty) \\
{\scriptstyle \simeq_T \circ \mathcal{E}} \Big\uparrow & (10.b) & \Big\uparrow {\scriptstyle \Sigma_P(\simeq_T \circ \mathcal{E})} \\
\mathcal{D}_P\mathcal{S} & \xleftarrow{\ \mathcal{D}_\rho(k)\ } & \Sigma_P(\mathcal{D}_P\mathcal{S})
\end{array}
$$

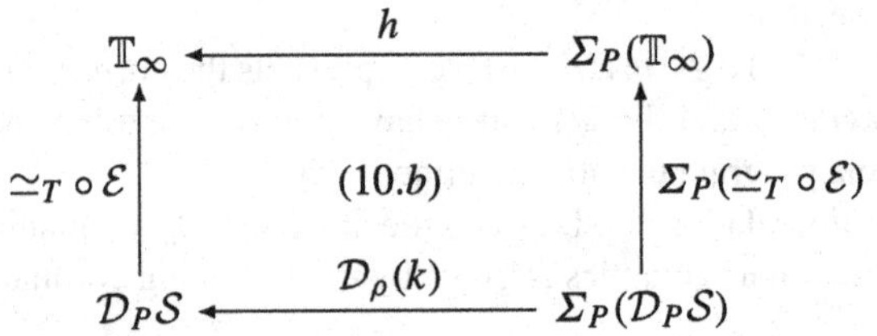

follows from the following possible cases (note that the mapping h is defined in the proof of Proposition 37):

1. When $nil \in \Sigma_P(\mathcal{D}_P\mathcal{S}) = \{nil\} \cup Act \bigcup_{a_i \in Act}\{\{i\} \times \mathcal{D}_P\mathcal{S}\} \bigcup_{n \geq 2}(\mathcal{D}_P\mathcal{S})^n$,

$$
{\simeq_T} \circ \mathcal{E} \circ \mathcal{D}_\rho(k)(nil) = {\simeq_T} \circ \mathcal{E}(\perp^0) = {\simeq_T}(nil) = \perp^0,
$$

and hence $h \circ \Sigma_P(\simeq_T \circ \mathcal{E})(nil) = h(nil) = \perp^0$.

2. When $(i, \mathcal{L}_B) \in \Sigma_P(\mathcal{D}_P\mathcal{S})$,

$$
{\simeq_T} \circ \mathcal{E} \circ \mathcal{D}_\rho(k)(i, \mathcal{L}_B) = {\simeq_T} \circ \mathcal{E}\left(A \xrightarrow{\ a_i\ } \mathcal{L}_B\right)
$$
$$
= {\simeq_T}\left(a_i.\mathcal{E}(\mathcal{L}_B)\right) = a_i \otimes \left({\simeq_T}(\mathcal{E}(\mathcal{L}_B))\right).
$$

Thus, $h \circ \Sigma_P(\simeq_T \circ \mathcal{E})(i, \mathcal{L}_B) = h(i, {\simeq_T}(\mathcal{E}(\mathcal{L}_B))) = a_i \otimes ({\simeq_T}(\mathcal{E}(\mathcal{L}_B)))$.

3. When $(\mathcal{L}_{B_1}, \ldots, \mathcal{L}_{B_n}) \in \Sigma_P(\mathcal{D}_P\mathcal{S})$,

$$
{\simeq_T} \circ \mathcal{E} \circ \mathcal{D}_\rho(k)(\mathcal{L}_{B_1}, \ldots, \mathcal{L}_{B_n}) = {\simeq_T}\left(\parallel^n(\mathcal{E}(\mathcal{L}_{B_1}), \ldots, \mathcal{E}(\mathcal{L}_{B_n}))\right)
$$
$$
= +^n\left({\simeq_T}(\mathcal{E}(\mathcal{L}_{B_1})), \ldots, {\simeq_T}(\mathcal{E}(\mathcal{L}_{B_n}))\right).
$$

Hence,

$$
h \circ \Sigma_P(\simeq_T \circ \mathcal{E})(\mathcal{L}_{B_1}, \ldots, \mathcal{L}_{B_n}) = h\left({\simeq_T}(\mathcal{E}(\mathcal{L}_{B_1})), \ldots, {\simeq_T}(\mathcal{E}(\mathcal{L}_{B_n}))\right)
$$
$$
= +^n\left({\simeq_T}(\mathcal{E}(\mathcal{L}_{B_1})), \ldots, {\simeq_T}(\mathcal{E}(\mathcal{L}_{B_n}))\right). \qquad \square
$$

Notice that this relationship, represented by the above commutative diagram, corresponds to the embedding of the SOS-denotational $(X \uplus \Sigma_P)$-algebra semantics into the DB-denotational $(X \uplus \Sigma_P)$-algebra semantics, that is, the embedding homomorphism

$$\simeq_T \circ \mathcal{E} : \big(\mathcal{D}_P\mathcal{S}, [[\_]^@, \mathcal{D}_\rho(k)]\big) \to \big(\mathbb{T}_\infty, [f, h]\big)$$

between these two adequate denotational semantics algebras for the *syntax* initial free algebra $(\mathcal{T}_P X, \simeq_I)$, in the commutative diagram in Corollary 23 between $(X \uplus \Sigma_P)$-algebras

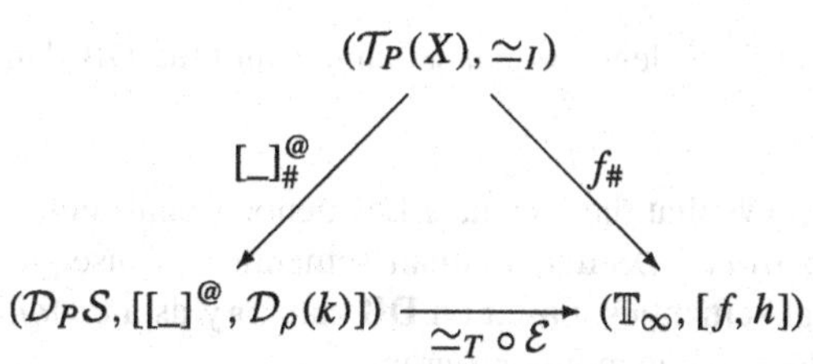

Moreover, we obtain a more specific relationship between the SOS and DB denotation semantics:

Corollary 24 *The DB-denotational semantics is isomorphic to the* abstract *SOS-denotational semantics (when $\mathcal{S} = \mathbf{1} = \{\bot^0\}$ is a terminal object in* **Set**, *so that all nodes of trees have the same name $\bot^0$). That is, the mapping $\simeq_T \circ \mathcal{E} : \mathcal{D}_P(\mathbf{1}) \to \mathbb{T}_\infty$ is an isomorphism, denoted by $\simeq_{Abs}$.*

Consequently, from the commutative diagram in Corollary 23, *we obtain that*

$$f_\# = \simeq_T \circ \mathcal{E} \circ [\_]^@_\# = \simeq_{Abs} \circ [\_]^@_\# .$$

Proof The mapping $\simeq_{Abs} = \simeq_T \circ \mathcal{E} : \mathcal{D}_P(\mathbf{1}) \to \mathbb{T}_\infty$ is defined in Definition 56 of the embedding $\mathcal{E} : \mathcal{D}_P\mathcal{S} \to \mathcal{T}_\infty$, when $\mathcal{S}$ is the singleton (a terminal object in **Set**) $\mathbf{1} = \{\bot^0\}$ and with the isomorphism $\simeq_T : \mathcal{T}_\infty \to \mathbb{T}_\infty$ in Definition 53, so that:

1. $\bot^0 \mapsto \bot^0$;
2. For any tree $\mathcal{L} = \{\bot^0 \xrightarrow{a} \mathcal{L}'\}$, $\mathcal{L} \mapsto a \otimes (\simeq_{Abs}(\mathcal{L}'))$;
3. For any tree $\mathcal{L}$ such that $\mathcal{L}$ is the union of the trees in $\{\mathcal{L}_1, \ldots, \mathcal{L}_n\}$ obtained by superposition of their roots in the unique root of $\mathcal{L}$, denoted by $\mathcal{L} = \underset{\mathbf{1}}{\uplus}\{\mathcal{L}_1, \ldots, \mathcal{L}_n\}$, $\mathcal{L} \mapsto +^n(\simeq_{Abs}(\mathcal{L}_1), \ldots, \simeq_{Abs}(\mathcal{L}_n))$.

In the case when $snd(\mathcal{L}_i) = \{\xrightarrow{a_1} \mathcal{L}'_1\}$ for each $1 \le i \le n$, we obtain a particular case $\mathcal{L} \mapsto +^n(a_1 \otimes (\simeq_{Abs}(\mathcal{L}_1)), \ldots, a_n \otimes (\simeq_{Abs}(\mathcal{L}_n)))$.

It is easy to verify that the inverse of '$\simeq_{Abs}$', denoted by '$\simeq^{-1}_{Abs}$', is defined recursively by:

1. $\bot^0 \mapsto \bot^0$,
2. $a \otimes t \mapsto \mathcal{L}$, such that $fst(\mathcal{L}) = \bot^0$ and $snd(\mathcal{L}) = \{\xrightarrow{a} (\simeq^{-1}_{Abs}(t))\}$,

3. $+^n(t_1, \dots, t_n) \mapsto \frac{\cup}{\mathbf{1}}\{\simeq_{Abs}^{-1}(t_1), \dots, \simeq_{Abs}^{-1}(t_n)\}$, thus, in the particular case

$$+^n(a_1 \otimes t_1, \dots, a_n \otimes t_n) \mapsto \mathcal{L} = \frac{\cup}{\mathbf{1}}\big\{\simeq_{Abs}^{-1}(a_1 \otimes t_1), \dots, \simeq_{Abs}^{-1}(a_n \otimes t_n)\big\},$$

so that $fst_{\mathcal{S}}(\mathcal{L}) = \bot^0$ and

$$snd_{\mathcal{S}}(\mathcal{L}) = \big\{\xrightarrow{a_1} \simeq_{Abs}^{-1}(t_1), \dots, \xrightarrow{a_n} \simeq_{Abs}^{-1}(t_n)\big\}.$$

Consequently, there exists the isomorphism

$$\simeq_{Abs} : \big(\mathcal{D}_P(\mathbf{1}), [\simeq_{Abs}^{-1} \circ f, \simeq_{Abs}^{-1} \circ h \circ \Sigma_P \simeq_{Abs}^{-1}]\big) \to \big(\mathbb{T}_\infty, [f, h]\big)$$

between the *abstract* SOS-denotational semantics and the DB-denotational semantics. □

This corollary shows that the adequate DB-denotational semantics is equivalent to the adequate *abstract* SOS-denotational semantics. Consequently, we can use the DB-denotational semantics, based on **DB** category as a *canonical* denotational semantics for the database-mapping programs.

Proposition 44 *The following commutative diagram represents the adequateness of the denotational semantics to the* abstract *operational semantics (when $\mathcal{S} = \mathbf{1} = \{\bot^0\}$) for the database-mapping programs with the final coalgebra isomorphism $\simeq_{DB} = \simeq_{Abs} \circ \simeq_B \circ \mathcal{B}_P(\simeq_{Abs}^{-1})$:*

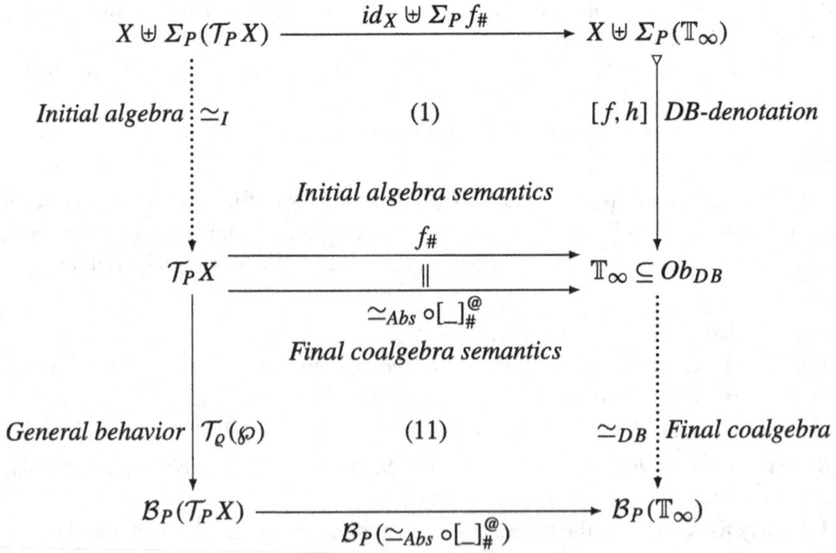

where, from Proposition 42, *the mapping $\wp : X \to \mathcal{B}_P X$ represents the abstract behavior $[\_] = \langle ass, \wp\rangle : X \to \{\bot^0\} \times \mathcal{B}_P X$ of a program, the* general *behavior*

$\mathcal{T}_\varrho(\wp)$ *is the unique extension of the behavior to all terms with variables and, by Proposition 39 and Corollary 23,* $f = \simeq_T \circ \mathcal{E} \circ [\_]^@ : X \to \mathbb{T}_\infty$ *is the* adequate *DB-denotational semantics, that is, the mapping whose results are represented as a composed objects in the* **DB** *category.*

Proof From Corollary 24, $f_\# = \simeq_{Abs} \circ [\_]^@_\#$.

From the commutative diagram (4c) in Proposition 43, in the case when $\mathcal{S} = 1$ is a terminal object in **Set**, so that $1 \times \mathcal{B}_P(\mathcal{T}_P X)$ can be substituted by (isomorphic to it) the set $\mathcal{B}_P(\mathcal{T}_P X)$, and $1 \times \mathcal{B}_P(\mathcal{D}_P(1))$ can be substituted by (isomorphic to it) the set $\mathcal{B}_P(\mathcal{D}_P(1))$. Thus, the constant mapping $ass \circ \mathbf{g}^E : \mathcal{T}_P X \to 1 = \{\perp^0\}$ can be eliminated, by extending this simplified (4c) diagram with the bijection $\simeq_{Abs} : \mathcal{D}_P(1) \to \mathbb{T}_\infty$, and we obtain the following commutative diagram

$$
\begin{array}{ccccc}
\mathcal{T}_P X & \xrightarrow{\hbar = [\_]^@_\#} & \mathcal{D}_P(1) & \xrightarrow{\simeq_{Abs}} & \mathbb{T}_\infty \subseteq Ob_{DB} \\
{\scriptstyle \mathcal{T}_\varrho(\wp)}\downarrow & (4c) & \vdots\,{\scriptstyle \simeq_B} & & \vdots\,{\scriptstyle \simeq_{DB}} \\
\mathcal{B}_P(\mathcal{T}_P X) & \xrightarrow[\mathcal{B}_P([\_]^@_\#)]{} & \mathcal{B}_P(\mathcal{D}_P(1)) & \xrightarrow[\mathcal{B}_P(\simeq_{Abs})]{} & \mathcal{B}_P(\mathbb{T}_\infty)
\end{array}
$$

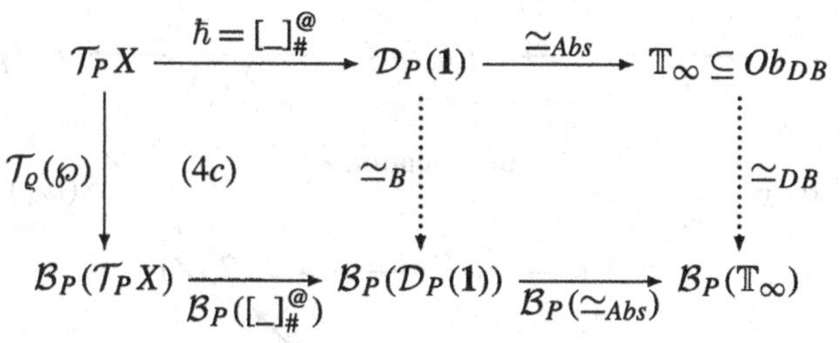

that composes the diagram (11) of this proposition.   □

In this setting, a denotational model is adequate w.r.t. an operational semantics when its $(X \uplus \Sigma_P)$-algebra homomorphism $f_\# : (\mathcal{T}_P X, \simeq_I) \to (\mathbb{T}_\infty, [f, h])$ is equal to the final $\mathcal{B}_P$-coalgebra homomorphism

$$\simeq_{Abs} \circ [\_]^@_\# : \left(\mathcal{T}_P X, \mathcal{T}_\varrho(\wp)\right) \to (\mathbb{T}_\infty, \simeq_{DB}).$$

Thus, in particular, the DB-semantic domain $\mathbb{T}_\infty$ (the objects in **DB** category) is the carrier of the final $\mathcal{B}_P$-coalgebra of the behavior $\mathcal{B}_P$.

Consequently, for each term $t \in \mathcal{T}_P X$ with the database-mapping process variables, $f_\#(t) = \simeq_{Abs}([\_]^@_\#(t))$.

7.5.2   Generalized Coinduction

The conventional coiteration schema, used in Sect. 7.4 for the operational semantics, assigns infinite behaviors to the processes of a set X by specifying for each element in X a direct observation and successor states (processes). Since the successors are taken from X again, the same specification applied to them reveals the second layer of the behavior, and so on.

A different approach is taken by the λ-coiteration schema [6], which is parameterized by the syntax monad $\mathcal{T}_P$ and a distributive law λ of $\mathcal{T}_P$ over general behavior endofunctor $\overline{\mathcal{B}}_P = \mathcal{S} \times \mathcal{B}_P$. It allows the successor states (processes) to be taken from $\mathcal{T}_P X$ instead of X, which may increase the expressiveness of representation. For the observations to be continued with these successors, the distributive law λ lifts the specification for X to $\mathcal{T}_P X$, so that we obtain the observables of the language $\mathcal{T}_P X$. This λ-coiteration schema needs two additional assumptions in order to uniquely define arrows into the final $\overline{\mathcal{B}}_P$-coalgebra $(\mathcal{D}_P \mathcal{S}, \simeq_B)$.

A major application of the notion of a distributive law λ in computer science has been given in [39].

The formal definition of the λ-coiteration schema [6] is based on the syntax monad $(\mathcal{T}_P, \eta, \mu)$ and behavior endofunctor F (here we will use the basic **Set** category, with $\emptyset \in F(X)$ for any set X) with the distributive law, expressed by the natural transformation $\lambda : \mathcal{T}_P F \to F\mathcal{T}_P$ such that the following diagrams (coherence unit and multiplication axioms) for each set X commute:

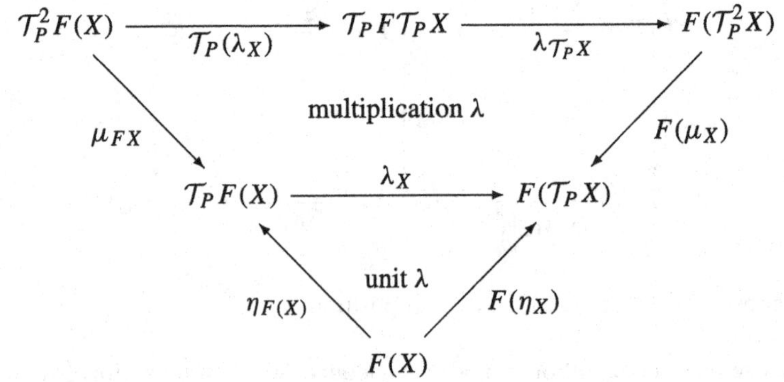

A distributive law λ assigns F-behaviors to the set of the language terms (composed programs) in $\mathcal{T}_P X$, given the behaviors of the elements from X. Intuitively, the presence of the distribution law tells us that the program terms and behaviors interact appropriately.

In the case of the database-mapping operational semantics, we have the following lemma:

Lemma 13 *In the case when* $F = \mathcal{B}_P = \mathcal{P}_{\text{fin}}(Act \times \_) :$ **Set** $\to$ **Set**, *we have a distributive law given by the natural transformation* $\lambda : \mathcal{T}_P \mathcal{B}_P \to \mathcal{B}_P \mathcal{T}_P$, *defined recursively for each set X by the mapping* $\lambda_X : \mathcal{T}_P(\mathcal{B}_P(X)) \to \mathcal{B}_P(\mathcal{T}_P X)$ *as follows:*

1. $nil \mapsto \emptyset$, *and for any* $a \in Act$, $a \mapsto \emptyset$;
2. $S_i \mapsto S_i$, *for any* $S_i \in \mathcal{B}_P(X) \subseteq T_P(\mathcal{B}_P(X))$;
3. $a.t \mapsto \{(a, a_k.t_k) \mid (a_k, t_k) \in \lambda_X(t)\}$ *if* $n = |\lambda_X(t)| \geq 1$; $\emptyset$ *otherwise*;
4. $\|^n(t_1, \dots, t_n) \mapsto \bigcup_{1 \leq i \leq n} \lambda_X(t_i)$.

Proof For any renaming of the variables $f : X \to Y$,

$$\lambda_Y \circ \mathcal{T}_P\big(\mathcal{B}_P(f)\big) = (\mathcal{B}_P)\big(\mathcal{T}_P(f)\big) \circ \lambda_X,$$

so that λ is a natural transformation. It is easy to verify that it satisfies the unit and multiplication axioms.

We have introduced the *abstract* notion of *operational rules* as the natural transformations $\varrho' : \Sigma_P(Id \times \mathcal{B}_P) \stackrel{\bullet}{\longrightarrow} \mathcal{B}_P\mathcal{T}_P$, and hence derive the natural transformations $\varrho = \mathcal{B}_P\mu \circ \varrho'_{\mathcal{T}_P} : \Sigma_P(\mathcal{T}_P \times \mathcal{B}_P\mathcal{T}_P) \stackrel{\bullet}{\longrightarrow} \mathcal{B}_P\mathcal{T}_P$. The natural transformation (i.e., *germ*) $\vartheta : \Sigma_P\mathcal{B}_P\mathcal{T}_P \stackrel{\bullet}{\longrightarrow} \mathcal{B}_P\mathcal{T}_P$ is a component of the natural transformation ϱ that represents the Σ_P-actions on the observables, so that for each set X its component ϑ_X is defined by:

1. $nil \mapsto \emptyset$, and for any $a \in Act$, $a \mapsto \emptyset$;
2. $(Y, i) \mapsto \{(a_i, a_k.t_k) \mid (a_k, t_k) \in Y\}$ if $n = |Y| \geq 1$; $\emptyset$ otherwise;
3. $(Y_1, \dots, Yt_n) \mapsto \bigcup_{1 \leq i \leq n} Y_i$.

Thus, there is a unique Σ_P-homomorphism

$$\lambda_X : (\mathcal{B}_P X, \simeq_P) \to \big(\mathcal{B}_P(\mathcal{T}_P X), [\mathcal{B}_P(\eta_X), \vartheta_X]\big)$$

from the initial algebra, with $\simeq_P = [inl_{\mathcal{B}_P X}, inr_{\mathcal{B}_P X}] = [\eta_{\mathcal{B}_P X}, inr_{\mathcal{B}_P X}]$ such that the following diagram commutes

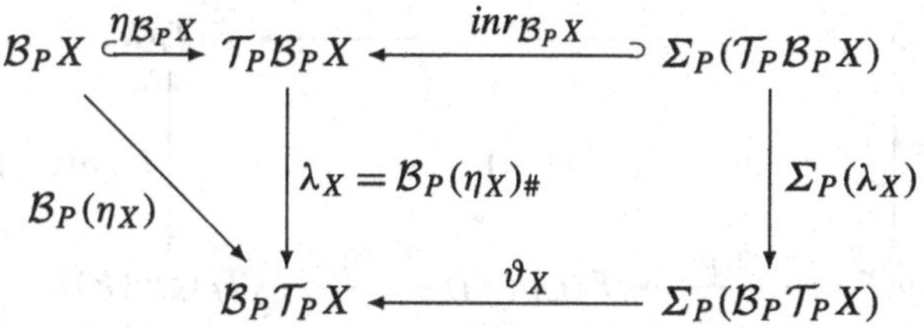

Consequently, it is easy to verify that the λ_X in this lemma is well defined. □

For example, for $S = \{(a_1, p_{k_1}), \dots, (a_n, p_{k_n})\} \in \mathcal{B}_P(X)$, from point 3 above, $\lambda_X(a.S) = \{(a, a_1.p_{k_1}), \dots, (a, a_n.p_{k_n})\} \in \mathcal{B}_P(\mathcal{T}_P X)$, with $\lambda_X(a.\{(a_1, p_{k_1})\}) = \{(a, a_1.p_{k_1})\}$, and $\lambda_X(a.\emptyset) = \emptyset$. Or, for $S_i \in \mathcal{B}_P(X)$, $i = 1, 2, 3, 4$ and $t_1 = S_1, t_2 = a_2.S_2, t_3 = a_3.S_3, t_4 = S_4$, from point 4,

$$\begin{aligned}
&\lambda_X\big(\|^4(t_1, t_2, t_3, t_4)\big)\\
&\quad = \lambda_X\big(\|^4(S_1, a_2.S_2, a_3.S_3, S_4)\big)\\
&\quad = \big\{(a_2, a_i.p_{k_i}) \mid (a_i, p_{k_i}) \in S_2\big\} \cup \big\{(a_3, a_i.p_{k_i}) \mid (a_i, p_{k_i}) \in S_3\big\} \cup S_1 \cup S_4.
\end{aligned}$$

The associate notion of a model is a so-called λ-bialgebra that consists of an object (a set) X with a pair of maps $\mathcal{T}_P X \stackrel{\beta}{\longrightarrow} X \stackrel{\gamma}{\longrightarrow} FX$, where β is a $\mathcal{T}_P$-algebra (Eilenberg–Moore monadic algebra such that $\beta \circ \eta_X = id_X$ and $\beta \circ \mu_x = \beta \circ \mathcal{T}_P(\beta)$), that is, β is a F-homomorphism from $(\mathcal{T}_P X, \lambda_X \circ \mathcal{T}_P(\beta))$ into F-coalgebra (X, γ) such that the following two diagrams commute:

$$\begin{array}{ccccccc}
\mathcal{T}_P X & \xrightarrow{\beta} & X & \qquad & \mathcal{T}_P(gfp(F)) & \xrightarrow{\beta_\lambda} & gfp(F) \\
\downarrow{\scriptstyle \mathcal{T}_P(\beta)} & & & & \downarrow{\scriptstyle \mathcal{T}_P(\beta_\lambda)} & & \\
\mathcal{T}_P F(X) & & \downarrow{\scriptstyle \gamma} & & \mathcal{T}_P F(gfp(F)) & (12) & \downarrow{\scriptstyle \simeq_F} \\
\downarrow{\scriptstyle \lambda_X} & & & & \downarrow{\scriptstyle \lambda_{gfp(F)}} & & \\
F(\mathcal{T}_P X) & \xrightarrow{F(\beta)} & F(X) & & F(\mathcal{T}_P(gfp(F))) & \xrightarrow{F(\beta_\lambda)} & F(gfp(F))
\end{array}$$

where the β_λ in diagram (12) is a unique F-homomorphism into the final F-coalgebra, and hence $\mathcal{T}_P(gfp(F)) \xrightarrow{\beta_\lambda} gfp(F) \xrightarrow{\simeq_F} F(gfp(F))$ is just the final λ-bialgebra. It was demonstrated [6] that for a given distributive law λ, for any guarded recursive mapping given by a $F\mathcal{T}_P$-coalgebra $\phi : X \to F(\mathcal{T}_P X)$, there is a unique "solution" $\phi^@ : X \to gfp(F)$ (so-called λ-coiterative arrow induced by ϕ) such that the following diagram commutes:

$$\begin{array}{ccccc}
X & & \xrightarrow{\phi} & & F(\mathcal{T}_P X) \\
\downarrow{\scriptstyle \phi^@} & & (12a) & & \downarrow{\scriptstyle F\mathcal{T}_P(\phi^@)} \\
gfp(F) & \xrightarrow{\simeq_F} & F(gfp(F)) & \xleftarrow{F(\beta_\lambda)} & F(\mathcal{T}_P(gfp(F)))
\end{array}$$

so that, $\phi^@ = \simeq_F^{-1} \circ F(\beta_\lambda) \circ F\mathcal{T}_P(\phi^@) \circ \phi : X \to gfp(F)$.

It is an interesting consideration in [19] that the uniqueness of the solution $\phi^@$, for a given guarded recursive mapping ϕ, is a based on the fact that there is a bijective representation of ϕ by the λ-bialgebra $\mathcal{T}_P^2 X \xrightarrow{\mu_X} \mathcal{T}_P X \xrightarrow{\phi_\#} F(\mathcal{T}_P X)$, where $\phi_\# = F(\mu_X) \circ \lambda_{\mathcal{T}_P X} \circ \mathcal{T}_P(\phi)$, (and, vice versa, for each λ-bialgebra

$$\mathcal{T}_P^2 X \xrightarrow{\mu_X} \mathcal{T}_P X \xrightarrow{g} F(\mathcal{T}_P X)$$

its corresponding guarded recursive mapping ϕ is equal to $g \circ \eta_X$), so that there is a unique homomorphism into the final λ-bialgebra

$$\mathcal{T}_P\big(gfp(F)\big) \xrightarrow{\beta_\lambda} gfp(F) \xrightarrow{\simeq_F} F\big(gfp(F)\big).$$

However, here we will present a more simple and direct representation, based on the standard final F-coalgebra semantics:

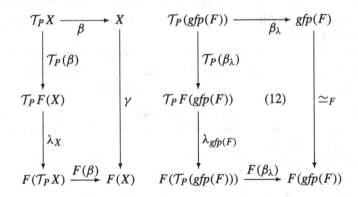

Proposition 45 *For each guarded recursive specification $\phi : X \to F(\mathcal{T}_P X)$, there exists a unique solution $\phi^@ = \phi_\#^@ \circ \eta_X : X \to gfp(F)$ such that $\phi_\#^@ : (\mathcal{T}_P X, \phi_\#) \to (gfp(F), \simeq_F)$ is the unique F-coalgebra homomorphism into the final F-coalgebra,*

where $\phi_{\#} \triangleq F(\mu_X) \circ \lambda_{\mathcal{T}_P X} \circ \mathcal{T}_P(\phi)$. *It is represented by the following commutative diagram:*

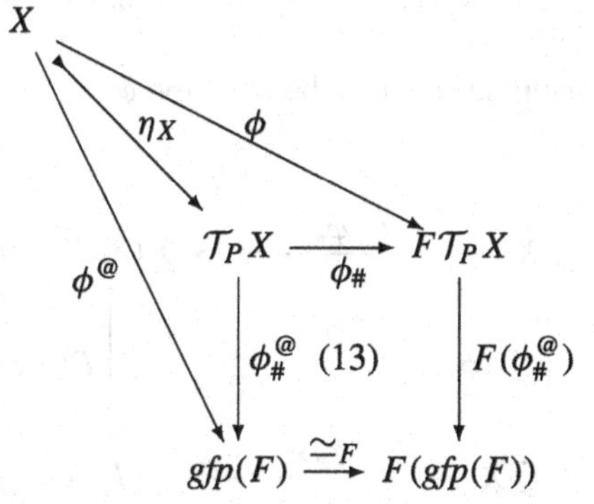

In particular, it is valid for the case when $F = \mathcal{B}_P = \mathcal{P}_{\text{fin}}(Act \times \_) : \mathbf{Set} \to \mathbf{Set}$.

Proof We have to show that for a general recursive specification $\phi_{\#} : \mathcal{T}_P X \to F(\mathcal{T}_P X)$, given by $\phi_{\#} \triangleq F(\mu_X) \circ \lambda_{\mathcal{T}_P X} \circ \mathcal{T}_P(\phi)$, its reduction to the variables $X \subseteq \mathcal{T}_P X$, expressed by the arrow $\phi_{\#} \circ \eta_X : X \to F(\mathcal{T}_P X)$, is *equal* to the given *guarded* recursive specification ϕ. In fact,

$$\phi_{\#} \circ \eta_X = F(\mu_X) \circ \lambda_{\mathcal{T}_P X} \circ \mathcal{T}_P(\phi) \circ \eta_X$$

$$\big(\text{from the natural transformation } \eta : Id \stackrel{\cdot}{\longrightarrow} \mathcal{T}_P \text{ applied to arrow } \lambda_X\big)$$

$$= F(\mu_X) \circ \lambda_{\mathcal{T}_P X} \circ \eta_{F\mathcal{T}_P X} \circ \phi$$

$$\big(\text{from the 'unit-}\lambda\text{' property when } X \text{ substituted by } \mathcal{T}_P X\big)$$

$$= F(\mu_X) \circ F(\eta_{\mathcal{T}_P X}) \circ \phi$$

$$(\text{from the monad property } I_d = \mu \circ \mathcal{T}_P \eta)$$

$$= \phi.$$

By using the natural transformation (a germ) $\vartheta : \Sigma_P F\mathcal{T}_P \stackrel{\cdot}{\longrightarrow} F\mathcal{T}_P$, in the proof of Lemma 13, it is easy to see by the following initial Σ_P-algebra commutative diagram that, for any $\phi : X \to F\mathcal{T}_P X$, the mapping $\phi_{\#}$ is its *unique* extension:

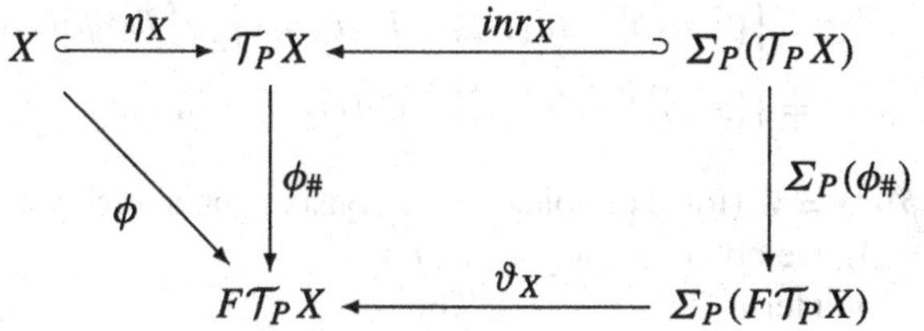

From this commutative diagram we obtain that this unique function $\phi_{\#}$ is defined recursively (based on the mapping ϑ_X defined in the proof of Lemma 13) by the following mapping:

1. $p_k \mapsto \phi(p_k)$ (note that it can be the empty set $\emptyset$ as well);
2. $nil \mapsto \emptyset$, and $a \mapsto \emptyset$ for each $a \in Act$;
3. $a.t \mapsto \{(a, a_i.t_i) \mid (a_i, t_i) \in \phi_\#(t)\}$ if $|\phi_\#(t)| \geq 1$; $\emptyset$ otherwise;
4. $\|^n(t_1, \ldots, t_n) \mapsto \bigcup_{1 \leq i \leq n} \phi_\#(t_i)$.

Let us show that this mapping satisfies the equation $\phi_\# = F(\mu_X) \circ \lambda_{\mathcal{T}_P X} \circ \mathcal{T}_P(\phi)$, i.e., the diagram

$$\begin{array}{ccc}
\mathcal{T}_P X & \xrightarrow{\phi_\#} & F(\mathcal{T}_P X) \\
\downarrow{\scriptstyle \mathcal{T}_P(\phi)} & & \uparrow{\scriptstyle F(\mu_X)} \\
\mathcal{T}_P F \mathcal{T}_P X & \xrightarrow{\lambda_{\mathcal{T}_P X}} & F(\mathcal{T}_P^2 X)
\end{array}$$

1. In the case $t = p_k \in X$, $F(\mu_X) \circ \lambda_{\mathcal{T}_P X} \circ \mathcal{T}_P(\phi)(p_k) = F(\mu_X) \circ \lambda_{\mathcal{T}_P X}(\phi(p_k))$ where $S = \phi(p_k) \in F(\mathcal{T}_P X)$, so that $F(\mu_X) \circ \lambda_{\mathcal{T}_P X}(S) = F(\mu_X)(S) = S$ because in S all terms are fattened (they are the variables). While $\phi_\#(p_k) = \phi(p_k) = S$.
2. For any $a \in Act$ (or *nil*),

$$F(\mu_X) \circ \lambda_{\mathcal{T}_P X} \circ \mathcal{T}_P(\phi)(a) = F(\mu_X) \circ \lambda_{\mathcal{T}_P X}(a) = F(\mu_X)(\emptyset) = \emptyset = \phi_\#(a).$$

Let us show that this equation is valid by structural induction on the length of terms (considered by a number of operators contained in a term), and assume that it holds for the terms $t, t_1, \ldots, t_n$, $n \geq 1$, of the length m. Then we consider a term with the length greater than m, that is, one of the following cases:

2.1. The case of a term $a.t$. Then

$$\begin{aligned}
& F(\mu_X) \circ \lambda_{\mathcal{T}_P X} \circ \mathcal{T}_P(\phi)(a.t) \\
&= F(\mu_X) \circ \lambda_{\mathcal{T}_P X}\big(a.\mathcal{T}_P(\phi)(t)\big) \\
&= F(\mu_X)\big(\{(a, a_k.t_k) \mid (a_k, t_k) \in \lambda_{\mathcal{T}_P X}\big(\mathcal{T}_P(\phi)(t)\big)\}\big) \\
&= \big\{\big(a, a_k.F(\mu_X)(t_k)\big) \mid (a_k, t_k) \in \lambda_{\mathcal{T}_P X}\big(\mathcal{T}_P(\phi)(t)\big)\big\} \\
&= \big\{\big(a, a_k.t'_k\big) \mid \big(a_k, t'_k\big) \in F(\mu_X)\big(\lambda_{\mathcal{T}_P X}\big(\mathcal{T}_P(\phi)(t)\big)\big)\big\} \\
&= \big\{\big(a, a_k.t'_k\big) \mid \big(a_k, t'_k\big) \in \phi_\#(t)\big\} = \phi_\#(a.t).
\end{aligned}$$

If $\mathcal{T}_P(\phi)(t) = \emptyset$ (for example, t is a constant or a variable p_k such that $\phi(p_k) = \emptyset$), we obtain analogous result.

2.2. The case of a term $\|^n(t_1, \ldots, t_n)$. Then

$$\begin{aligned}
& F(\mu_X) \circ \lambda_{\mathcal{T}_P X} \circ \mathcal{T}_P(\phi)\big(\|^n(t_1, \ldots, t_n)\big) \\
&\quad = F(\mu_X) \circ \lambda_{\mathcal{T}_P X}\big(\|^n\big(\mathcal{T}_P(\phi)(t_1), \ldots, \mathcal{T}_P(\phi)(t_n)\big)\big)
\end{aligned}$$

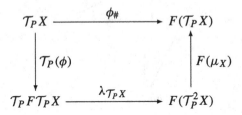

$$= F(\mu_X)\Big(\bigcup_{1\le i\le n} \lambda_{\mathcal{T}_P X}\big(\mathcal{T}_P(\phi)(t_i)\big)\Big)$$

$$= \bigcup_{1\le i\le n} F(\mu_X)\big(\lambda_{\mathcal{T}_P X}\big(\mathcal{T}_P(\phi)(t_i)\big)\big)$$

$$= \bigcup_{1\le i\le n} \phi_\#(t_i)$$

$$= \phi_\#\big(\|^n(t_1,\dots,t_n)\big).$$

Thus, $\phi_\# = F(\mu_X) \circ \lambda_{\mathcal{T}_P X} \circ \mathcal{T}_P(\phi)$. □

Note that in this representation of the (unique) solution of a guarded recursive specification $\phi : X \to F(\mathcal{T}_P X)$ we do not use the non-final semantics construction of the commutative diagram (12b) used in [6] and [19], so that we do not need to involve the mapping $\beta_\lambda : \mathcal{T}_P(gfp(F)) \to gfp(F)$. Moreover, we do not need to use λ-bialgebras in this representation: what we need is only the distribution law λ with its multiplication and unit axioms. Thus, this new minimalist approach is more rational and use the standard final semantics for F-coalgebras.

Lemma 14 *Let us consider Proposition* 45 *when* $F = \mathcal{B}_P$ *and* $\phi = \mathcal{B}_P(\eta_X) \circ \wp$. *Then we obtain that* $\phi_\#$ *is equal to the conservative extension (lifting) of* $\wp$, *that is,* $\phi_\# = \mathcal{T}_\varrho(\wp)$.

Proof Let as consider the definition of the mapping $\phi_\#$ given in the proof of Proposition 45. It is enough to show that $\phi_\# = \mathcal{T}_\varrho(\wp)$ for the variables $p_k \in X$. In fact, $\phi_\#(p_k) = \phi(p_k) = \mathcal{B}_P(\eta_X) \circ \wp(p_k) = \wp(p_k) = \mathcal{T}_\varrho(\wp)(p_k)$. □

Hence, from this lemma we obtain that the non-guarded specification of behavior $\phi_\# : \mathcal{T}_P \to \mathcal{B}_P(\mathcal{T}_P X)$ (the unique extension of the recursive guarded specification $\phi = \mathcal{B}_P(\eta_X) \circ \wp : X \to \mathcal{B}_P(\mathcal{T}_P X)$) is equal to conservative extension $\mathcal{T}_\varrho(\wp)$ of the non-recursive but guarded specification $\wp : X \to \mathcal{B}_P X$.

Based on Proposition 45 and Lemma 14, we obtain the following results:

Corollary 25 *The general Proposition* 45, *in the case when the endofunctor* F *is equal to the abstract behavior* $\mathcal{B}_P$, *can be applied to the final database-mapping semantics based on the general behavior endofunctor* $\overline{\mathcal{B}}_P = \mathcal{S} \times \mathcal{B}_P$ *with* $\mathcal{D}_P\mathcal{S} = gfp(\overline{\mathcal{B}}_P)$ *and a guarded recursive specification* $\phi = \mathcal{B}_P(\eta_X) \circ \wp$.

We obtain the following commutative diagram of the final $\overline{\mathcal{B}}_P$*-coalgebra semantics for the guarded non-recursive specification* $[\_] : X \to \overline{\mathcal{B}}_P(X)$ *(external square of this diagram), based on the* $\overline{\mathcal{B}}_P$*-coalgebra semantics for its derived non-guarded recursive specification* $\psi = \langle ass \circ \mathbf{g}^E, \phi_\#\rangle : \mathcal{T}_P X \to \overline{\mathcal{B}}_P(\mathcal{T}_P X)$ *(in internal square diagram* (14)) *where* $\phi_\# = \mathcal{T}_\varrho(\wp) = \mathcal{B}_P(\mu_X) \circ \lambda_{\mathcal{T}_P X} \circ \mathcal{T}_P(\phi)$,

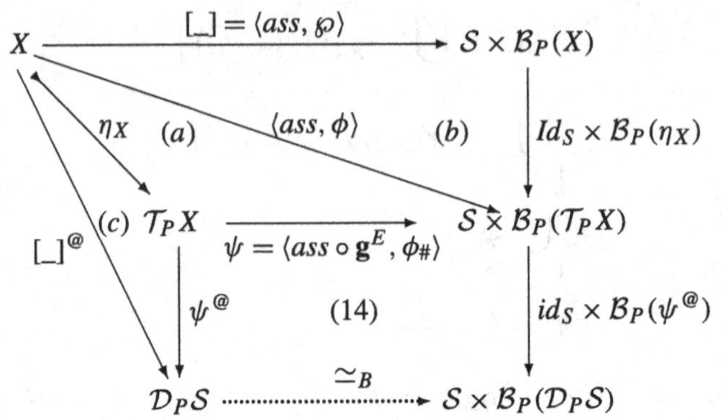

so that the unique non-recursive solution $[\_]^@ : X \to \mathcal{D}_P\mathcal{S}$ *is obtained from the corresponding unique non-guarded recursive solution (of general coinduction)* $\psi^@ : \mathcal{T}_P X \to \mathcal{D}_P\mathcal{S}$ *with* $[\_]^@ = \psi^@ \circ \eta_X$.

Proof The triangle (a) commutes from the fact that $\mathbf{g}^E \circ \eta_X = id_X$ (from the algorithm *DBprog* and Proposition 34), and $\phi = \phi_\# \circ \eta_X$ (from Proposition 45), while the triangle (b) corresponds to the particular definition of the guarded recursive specification in this proposition by $\phi = \mathcal{B}_P(\eta_X) \circ \wp$.

From the fact that there is only one $(S \times \mathcal{B}_P)$-homomorphism from the coalgebra $(X, [\_])$ into the final coalgebra $(\mathcal{D}_P\mathcal{S}, \simeq_B)$ it must be $[\_]^@ = \psi^@ \circ \eta_X : X \to \mathcal{D}_P\mathcal{S}$. □

7.6 Review Questions

1. Explain what the main differences between a denotational and an operational semantics are, and what the relationship between the behavior of the database mapping programs and observational semantics for RDB based on views and the power-view operator T is. Is this power-view operator T obtained from the initial semantics for the relational SPJRU algebra Σ_R in Sect. 5.1.1 a plausible factor which can determinate the semantic adequateness of these two semantics? Why did we obtain that the Structural Operational Semantics for the database mapping systems is a kind of the Guarded recursion (GSOS)? What can be the adequate interpretations for the branching (parallel) composition '$\|$' and the action prefixing composition '$a.\_$' of the GSOS grammar for the DB theory?
2. Why do we need the process of updates through views in database mapping systems (programs), that is, for a given update of a view of relational database how should it be translated to updates on the underlying database (consider the mapping-interpretation of the schema mappings)? What are the side-effect

free relational view updates? Why is it not possible to always have such side-effect free updates and which are the criteria to adopt in order to overcome such problems? How can we use the concepts of the transitions obtained from the **RA** arrows in the Application plan of the categorial RDB machines $\mathbf{M}_{RC}$ (defined in the previous chapter) for a formalization of the deletion (and insertion) by minimal side-effects?

3. Why do we obtain a backward chaining propagation (w.r.t. the schema mapping arrows) in the case of deletion and why do we obtain a forward chaining propagation in the case of insertion? In the presence of cyclic graphs of the schema mapping systems, infinite propagations are possible. For which kind of propagation does this happen? Can we use the methods of the fixpoint operators used in Sect. 4.2.4 in order to interrupt the infinite propagation (which generates the infinite extensions of the relations in the database) and useless insertion of the Skolem constants in our algebraic framework? (Why is it not possible in the pure logic framework?)
4. The problem of the infinite propagations, in cyclic database mapping systems with incomplete information, in Big Data Integration has a significant probability to happen. The logic used to specify the mappings based on SOtgds (for incomplete information) in order to be well defined needs the models of such logic theories, so that it will necessarily require the infinite processes and hence it will be inapplicable in practice. Explain the difference from the strong and weak Data Integration which is possible in the algebraic framework based on the **DB** category. What is an epistemic point of view in the more flexible weak integration? In which way can it be used for the P2P Data integrations?
5. The denotational (non-abstract) DB-grammar for the database-mapping precesses in Definition 52 is coherent with the observational operational semantics? Why? The database-mapping programs are well defined terms of the abstract $GSOS_{DB}$ signature Σ_P for a database-mapping processes presented in Sect. 7.1. Do this exercise: define an initial semantics for the denotational (non-abstract) DB grammar and its algebra, based on the syntax algebra of the $GSOS_{DB}$ signature Σ_P. Demonstrate that this initial semantics corresponds to the isomorphism in Definition 53.
6. The generation of the Labeled Transition Tree (LTS) of the database (process) updates is based on the atomic transitions (insertion or deletion of the tuples). How is such an atomic transition represented in the **DB** category? Why is it useful to represent such a process tree by an (also infinite) set of equations, and what are the variables of these equations? Why do we obtain a guarded system of equations? What would happen if we obtained an equation with an infinite number of variables w.r.t. the final semantics for the database processes, and why, fortunately, it cannot happen in our database mapping framework? Why do we need the flattening of a system of equations obtained by the DB-process algorithm *DBprog*? How can we define the equivalence classes

of the program's equations by using the initial algebra semantics? Why do we obtain a unique solution of a flattened guarded system of program's equations?

7. What is the operational meaning $[p_k]$ of a process (program) p_k at the state (an instance database) $A = ass(p_k)$? Why is the finite branching LTS of the obtained update processes fundamental w.r.t. the Peter Aczel's anti-foundation axiom, and which solution is adopted here? How are the abstract behavior functor and its final co-algebra semantics defined, and what is the greatest fixed point of it? What is a bisimulation relation and how is the observational equivalence for the processes defined? Explain the unique solution based on the final semantics and coinductive principle, and explain the meaning of the observational comonad.
8. What is the meaning of the "specification versus solution", that is, duality of the initial and final semantics? Define a table of dual concepts obtained from a combined diagram:

$$\begin{array}{ccccc}
X \uplus \Sigma_P(\mathcal{T}_P X) & \xrightarrow{\simeq_I} & \mathcal{T}_P X & & \\
\Big\downarrow {\scriptstyle X \uplus \Sigma_P(id_{X_\#})} & (I_1) & \Big\downarrow {\scriptstyle id_{X_\#} = \mathbf{g}^E} & & \\
X \uplus \Sigma_P(X) & \xrightarrow{[id_X, h_X]} & X & \xrightarrow{\langle ass, \wp \rangle} & \mathcal{S} \times \mathcal{B}_P(X) \\
 & & \Big\downarrow {\scriptstyle ass^\natural = [_]^@} & (4) & \Big\downarrow {\scriptstyle \mathcal{S} \times \mathcal{B}_P(id_{\mathcal{S}}^\natural)} \\
 & & \mathcal{D}_P \mathcal{S} & \xrightarrow{\simeq_B} & \mathcal{S} \times \mathcal{B}_P(\mathcal{D}_P \mathcal{S})
\end{array}$$

9. How are we able to combine the initial and final semantics diagrams in order to express the semantic adequateness of a denotational algebra (model) for the operational behavior? Can we use the final semantics of the abstract behavior endofunctor $\mathcal{B}_P$ instead of the final semantics of the flattened guarded set of equations for the processes, and how (explain the commutative diagram in Proposition 39)? What are the terms of the DB-denotational semantics for the abstract Σ_P-algebras, and their relationship with **DB** category? How can we represent the LTS trees in $\mathcal{D}_P \mathcal{S}$ (the greatest fixed point of the behavior endofunctor) by the subcategories of **DB**? Based on this representation, define the adequate DB-denotational semantics for the operational semantics based on the **DB** category.
10. What is the Structural Operational Semantics (SOS), the Functorial Operational Semantics (FOS) for a syntax monad $\mathcal{T}_P$ and a behavior $\mathcal{B}_P$, and what is its dual semantics? Explain the relationship between the abstract SOS-denotational semantics and the DB-denotational semantics given by the following commutative diagram

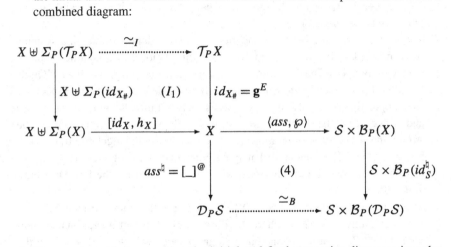

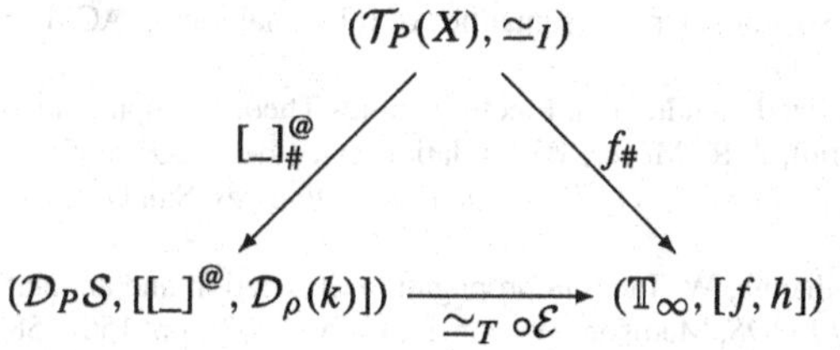

11. Demonstrate that the adequate DB-denotational semantics is equivalent to the adequate *abstract* SOS-denotational semantics, and explain why we can use the DB-denotational semantics, based on **DB** category, as a *canonical* denotational semantics for the database mapping programs.
12. Explain the final result: the equivalence of the initial algebra and final coalgebra semantics of the commutative diagram

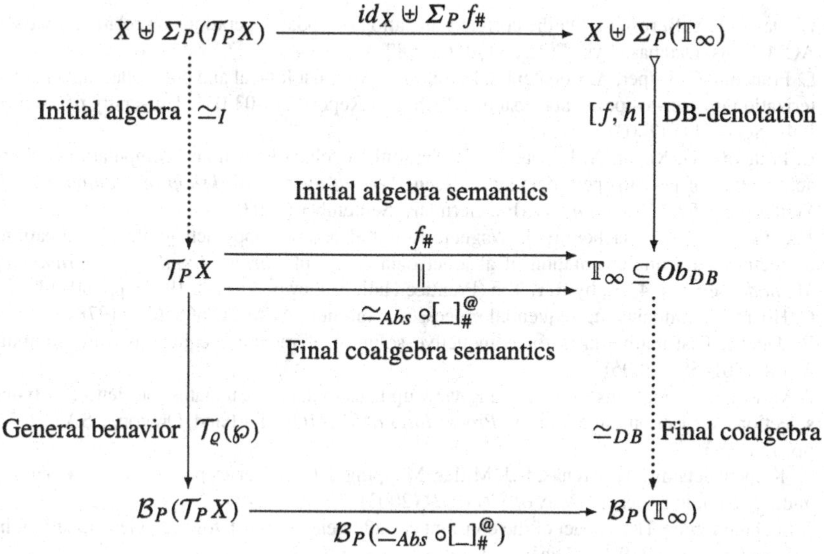

References

1. S. Abiteboul, R. Hull, V. Vianu, *Foundations of Databases* (Addison Wesley Publ. Co., Reading, 1995)
2. P. Aczel, *Non-well-Founded Sets*. Lecture Notes CSLI, vol. 14 (1988)
3. P. Aczel, N. Mendler, A final coalgebra theorem, in *Proc. Category Theory and Computer Science*. LNCS, vol. 389 (Springer, Berlin, 1989), pp. 357–365
4. P. Aczel, J. Adamek, S. Milius, J. Velebil, Infinite trees and completely iterative theories: a coalgebraic view. Theor. Comput. Sci. **300**, 1–45 (2003)

5. F. Bancilhon, N. Spyratos, Update semantics of relational views. ACM Trans. Database Syst. **6**(4), 557–575 (1981)
6. F. Bartels, Generalized coinduction. Electron. Notes Theor. Comput. Sci. **44**(1), 67–87 (2001)
7. B. Bloom, S. Istrail, A.R. Meyer, Bisimulation can't be traced: preliminary report, in *15th ACM Symposium on Principles of Programming Languages*, San Diego, CA, USA (1988), pp. 229–239
8. P. Buneman, S. Khanna, W. Tan, On propagation of deletion and annotations through views, in *Proceedings of PODS*, Madison, Wisconsin, USA (2002), pp. 150–158
9. D. Calvanese, E. Damaggio, G.D. Giacomo, M. Lenzerini, R. Rosati, Semantic data integration in P2P systems, in *Proc. of the Int. Workshop on Databases, Inf. Systems and P2P Computing*, Berlin, Germany, September (2003)
10. D. Calvanese, G.D. Giacomo, M. Lenzerini, R. Rosati, Logical foundations of peer-to- peer data integration, in *PODS 2004*, June 14–16, Paris, France (2004)
11. Y. Cui, J. Widom, Run-time translation of view tuple deletions using data lineage. Tech. Report, Stanford University (2001). http://dbpub.stanford.edu:8090/pub/2001-24
12. M. Dalal, Investigations into a theory of knowledge base revision, in *Proc. National Conference on Artificial Intelligence* (1988), pp. 475–479
13. U. Dayal, P.A. Bernstein, On the updatability of relational views, in *VLDB* (1978), pp. 368–377
14. U. Dayal, P.A. Bernstein, On the correct translation of update operations on relational views. ACM Trans. Database Syst. **7**(3), 381–416 (1982)
15. E. Franconi, G. Kuper, A. Lopatenko, L. Serafini, A robust logical and computational characterization of peer-to- peer data systems. Technical Report DIT-03-051, University of Trento, Italy, September (2003)
16. E. Franconi, G. Kuper, A. Lopatenko, L. Serafini, A robust logical and computational characterization of peer-to- peer data systems, in *Proc. of the Int. Workshop on Databases, Inf. Systems and P2P Computing*, Berlin, Germany, September (2003)
17. J.A. Gogen, J.W. Thacher, E.G. Wagner, A initial algebra approach to the specification, correctness and implementation of abstract data types, in *Current Trends in Programming Methodology*, vol. 4, ed. by R.Y. Yeh (Prentice Hall, Englewood Cliffs, 1978), pp. 80–149
18. C. Hoare, Communicating sequential processes. Commun. ACM **21**, 666–677 (1978)
19. B. Jacobs, Distributive laws for coinductive solution of recursive equations. Inf. Comput. **204**(4), 561–587 (2006)
20. A.M. Keller, Algorithms for translating view updates to database updates for views involving selections, projections, and joins, in *Proceedings of SIGMOD*, Portland, Oregon, USA (1985), pp. 154–163
21. A. Kementsietsidis, M. Arenas, R.J. Miller, Mapping data in peer-to-peer systems: semantics and algorithmic issues, in *Proc. of SIGMOD* (2003)
22. J. Lechtenbörger, The impact of the constant complement approach towards view updating, in *Proceedings of PODS*, San Diego, CA, USA (2003), pp. 49–55
23. D. Lembo, M. Lenzerini, R. Rosati, Source inconsistency and incompleteness in data integration, in *Proc. of the 9th Int. Workshop on Knowledge Representation Meets Databases (KRDB 2002)* (CEUR, Aachen, 2002)
24. Z. Majkić, Massive parallelism for query answering in weakly integrated P2P systems, in *Workshop GLOBE 04*, August 30–September 3, Zaragoza, Spain (2004)
25. Z. Majkić, Weakly-coupled ontology integration of P2P database systems, in *1st Int. Workshop on Peer-to-Peer Knowledge Management (P2PKM)*, August 22, Boston, USA (2004)
26. Z. Majkić, Weakly-coupled P2P system with a network repository, in *6th Workshop on Distributed Data and Structures (WDAS'04)*, July 5–7, Lausanne, Switzerland (2004)
27. Z. Majkić, Intensional logic and epistemic independency of intelligent database agents, in *2nd International Workshop on Philosophy and Informatics (WSPI 2005)*, April 10–13, Kaiserslautern, Germany (2005)
28. Z. Majkić, Intensional semantics for P2P data integration, in *Journal on Data Semantics (JoDS) VI, Special Issue on 'Emergent Semantics'*. LNCS, vol. 4090 (2006), pp. 47–66

29. Z. Majkić, Non omniscient intensional contextual reasoning for query-agents in P2P systems, in *3rd Indian International Conference on Artificial Intelligence (IICAI-07)*, December 17–19, Pune, India (2007)
30. Z. Majkić, Intensional first-order logic for P2P database systems, in *Journal of Data Semantics (JoDS XII)*. LNCS, vol. 5480 (Springer, Berlin, 2009), pp. 131–152
31. Z. Majkić, Sound and complete query-answering in intensional P2P data integration. arXiv:1103.0490v1 (2011), 02 March, pp. 1–27
32. Z. Majkić, B. Prasad, Soft query-answering computing in P2P systems with epistemically independent peers, in *Book on Soft Computing Applications in Industry, STUDFUZZ 226* (Springer, Berlin, 2008), pp. 331–356
33. C. Medeiros, F. Tompa, Understanding the implications of view update policies, in *VLDB* (1985)
34. R. Milner, *A Calculus of Communicating Systems*. LNCS, vol. 92 (Springer, Berlin, 1980)
35. G. Plotkin, A structural approach to operational semantics. Technical Report DAIMI FN-19, Computer Science Dept., Aarhus University (1981)
36. J. Rutten, D. Turi, Initial algebra and final coalgebra semantics for concurrency, in *LNCS*, vol. 803 (Springer, Berlin, 1994), pp. 530–582
37. A. Tomasic, Correct view update translation via containment. Technical note STAN-CS-TN-93-3, Stanford University, USA (1993), pp. 1–10
38. L. Tucherman, A.L. Furtado, M.A. Casanova, A pragmatic approach to structured database design, in *VLDB*, Florence, Italy (1983), pp. 219–231
39. D. Turi, G. Plotkin, Towards a mathematical operational semantics, in *Proc. 12th LICS Conf* (IEEE Computer Society Press, Washington, 1997), pp. 280–291

The Properties of DB Category 8

8.1 Expressive Power of the DB Category

This chapter is a direct continuation of Chap. 3, dedicated to the definition of the **DB** category and its basic properties. In fact, in Chap. 3, we demonstrated the duality and categorial symmetry properties of the **DB** category, the semantics for different types of (iso, mono)morphisms between its objects (databases), and basic set of operators: the power-view operator, the data separation (isomorphic to the (co)product and the data federation (connection). In this chapter, we will define the matching and, dual to it, the merging operator for databases. Then we will investigate another fundamental properties of the **DB** category and, especially, its (co)completeness, the lattice property for its objects, the enrichments, and also its topological properties.

First of all, we recall that **DB** is not a topos (from Proposition 5) because, generally, an arrow which is monic and epic is not an isomorphism in **DB** (the converse is valid in all categories, i.e., an isomorphism is always monic and epic). Let us explain the reason for that. In fact, for any *simple* morphism $f : A \to B$, so that both A and B are non-separation-composed objects, the following is true (from Corollary 9):

1. f is monic iff its information flux $\widetilde{f} = TA$;
2. f is epic iff its information flux $\widetilde{f} = TB$;
3. f is an isomorphism iff $\widetilde{f} = TA = TB$.

Consequently, each simple arrow, which is both monic and epic, is an isomorphism. But it does not hold for complex arrows, as used in Proposition 5, $k = [id_C, id_C] : A_1 \uplus A_2 \to B_1$, with $A_1 = A_2 = B_1 = C$ and $\frac{k}{\uplus} = \{id_C : A_1 \to B_1, id_C : A_2 \to B_1\}$ and hence $\widetilde{h} = \widetilde{id_C} \uplus \widetilde{id_C} = TC \uplus TC = T(C \uplus C)$. Thus, from Proposition 8, it is monic and, from Proposition 5, it is also epic, however, it is *not an isomorphism*.

Consequently, only for the complex arrows we have the situation that a given morphism which is both monic and epic is not also an isomorphism. Thus, **DB** is not a topos.

Now, in what follows, we will explain what exactly the particular condition is, when a complex arrow which is monic and epic is also an isomorphism in the **DB** category.

Z. Majkić, *Big Data Integration Theory*, Texts in Computer Science,
DOI 10.1007/978-3-319-04156-8_8,

Notice that an empty arrow $\bot^1 = \{q_\bot : \bot \to \bot\}$ between two databases, with $\widetilde{\bot^1} = \bot^0 = \{\bot\}$, is equivalent to the absence of the schema mapping between these two databases. In fact, because of that, we are always able to introduce the empty arrows between databases, as it was used in Lemma 8 for the construction of canonical complex morphisms in Sect. 3.2. In fact, an empty arrow is not a ptp arrow, so that the presence or absence of them in a given complex arrow does not change such an arrow (from Definition 23 two complex arrows are equal iff they have the same set of ptp arrows and the empty arrows are on ptp arrows). Moreover, note that each empty arrow $\bot^1 : A_j \to B_i$ is not monic if $A_j \neq \bot^0$ (to be monic it must satisfy $TA_j = \widetilde{\bot^1} = \bot^0$) and is not epic if $B_i \neq \bot^0$ (to be epic it must satisfy $TB_i = \widetilde{\bot^1} = \bot^0$), and is not an isomorphism if $A_j \neq \bot^0$ or $B_i \neq \bot^0$.

Notice also that, from the fact that $A \uplus \bot^0$ is isomorphic to A, we can replace any original complex-object B by a *strictly complex* object B' obtained by elimination of the empty objects $\bot^0$ from the disjoint union. Thus, any complex arrow $g : A \to B$ can be substituted by an equivalent arrow $f : A' \to B'$ between two strictly-complex objects A' and B' such that $\frac{f}{\uplus} = \frac{g}{\uplus}$ so that f can always be well-defined from the set $\frac{g}{\uplus}$ by using the canonical representation of g between strictly complex objects A' and B' (by Lemma 8). Consequently, any proof for the morphisms can be done by using its canonical morphism between the strictly-complex objects obtained from the original source and target objects (as we did in the last part of the proof of Proposition 8).

Lemma 15 *Each complex arrow $f : A \to B$ between the strictly-complex objects $A = \biguplus_{1 \le j \le m} A_j$, $B = \biguplus_{1 \le i \le k} B_i$, with $m, k \ge 2$, is an isomorphism iff it is both monic and epic and composed of $m = k$ point-to-point arrows.*

Proof From Proposition 6 and the fact that f is monic, for each A_j there must exist a ptp arrow $f_{ji} \in \frac{f}{\uplus}$ with $dom(f_{ji}) = A_j$, $|\frac{f}{\uplus}| \ge m$. Moreover, f is epic and hence for each B_i there must exist a ptp arrow $f_{li} \in \frac{f}{\uplus}$ with $cod(f_{li}) = B_i$, so that $|\frac{f}{\uplus}| \ge k$. Thus the fact that $|\frac{f}{\uplus}| = m = k$ is consistent with the facts that f is monic and epic, and for each A_j we have only one outgoing ptp arrow from it while for each B_i we have only one ingoing arrow into it.

Moreover, from the fact that f is both monic and epic, each of its ptp arrows $(f_{ji} : A_j \to B_i) \in \frac{f}{\uplus}$ must be epic and monic (by Proposition 6), and from the fact that it is a simple arrow, it must be an isomorphism with $\widetilde{f_{ji}} = TA_j = TB_i$ (i.e., $A_j \simeq B_i$), with its inverse $f_{ji}^{OP} : B_i \to A_j$. Consequently, from Definition 22, $\frac{f \circ f^{OP}}{\uplus} = \{f_{ji} \circ f_{ji}^{OP} | f_{ji} \in \frac{f}{\uplus}\} = \{id_{B_i} | f_{ji} \in \frac{f}{\uplus}\} =$ (from the fact that f is epic, for each B_i there must exist a ptp arrow f_{ji} with $cod(f_{ji}) = B_i) = \{id_{B_i} | 1 \le i \le m\} = \frac{id_B}{\uplus}$. Thus, from Definition 23, we obtain $f \circ f^{OP} = id_B$. Analogously, $\frac{f^{OP} \circ f}{\uplus} = \{f_{ji}^{OP} \circ f_{ji} | f_{ji} \in \frac{f}{\uplus}\} = \{id_{A_j} | f_{ji} \in \frac{f}{\uplus}\} =$ (from the fact that f is monic, for each A_j there must exist a ptp arrow f_{ji} with $dom(f_{ji}) = A_j) = \{id_{A_j} | 1 \le i \le m\} = \frac{id_A}{\uplus}$. Thus, from Definition 23, we obtain $f^{OP} \circ f = id_A$. Consequently, f is an isomorphism. □

Let us now generalize this result to all complex arrows:

Corollary 26 *Each complex arrow $f : A \to B$ is an isomorphism iff it is both monic and epic and the strictly-complex objects obtained from A and B are composed by $n = |\frac{f}{\uplus}|$ of nonempty simple databases.*

Proof Let A' and B' be the strictly-complex objects obtained from A and B, respectively, with the isomorphisms $is_1 : A \to A'$ and $is_2 : B \to B'$ (which are always monic and epic). If f is monic, then $f' = is_2 \circ f \circ is_1^{-1} : A' \to B'$ is monic from the fact that the composition of monic arrows is monic. Analogously, if f is epic then f' is epic. Consequently, if f is both monic and epic then f' is both monic and epic, so that if A' and B' are composed by the same number of non-empty objects (equal to $|\frac{f'}{\uplus}|$), from Lemma 15 we obtain that f' is an isomorphism. Thus, from the fact that $f = is_2^{-1} \circ f' \circ is_1$ is a composition of the three isomorphisms, is an isomorphism as well.

Notice that this is valid also when we obtain for the strictly complex objects $A' = B' = \bot^0$ of the objects A and B, and in such a case $n = 0$. □

Let us introduce the additional properties for the arrows in **DB** as principal morphisms (used for definition of "exponent" objects, Proposition 49), retraction pairs (for PO relation $\preceq$ and subcategory $\mathbf{DB}_I$, Theorem 6) and the idempotents (for category symmetry, Theorem 4):

Proposition 46 *The following properties for arrows in* **DB** *are valid*:

1. *For any two objects A and B, the arrow $h : A \longrightarrow B$ such that $\widetilde{h} \simeq A \otimes B$ is a principal morphism.*
2. *For any monomorphism $f : A \hookrightarrow B$ and its inverted arrow (an epimorphism) $f^{OP} : B \twoheadrightarrow A$, the pair (f, f^{OP}) is a retraction pair.*
3. *For any simple object A there is a category of idempotents on A, denoted by Ret_A and defined as follows*:
   - 3.1. *Objects of Ret_A form the set of all simple arrows from A into A, i.e., $Ob_{Ret_A} = \mathbf{DB}(A, A)$.*
   - 3.2. *For any two objects $f, g \in Ob_{Ret_A}$, each arrow between them is a pair $(h; h) : f \to g$ where $h : A \to A$ with $\widetilde{h} \subseteq \widetilde{f} \cap \widetilde{g}$, and hence there is the bijection $\sigma : Ret_A(f, g) \simeq \mathbf{DB}(\widetilde{f}, \widetilde{g})$ such that for any $(h; h) \in Ret_A(f, g)$, $\sigma(h; h) = T_e(h; h)$ for the category symmetry functor T_e in Theorem 4.*

Proof Claim 1. For any simple arrow $h : A \longrightarrow B$ with $\widetilde{h} = A \otimes B$, we have $\forall f : A \longrightarrow B, \exists g : A \longrightarrow A$ such that $f = h \circ g$. In fact, it is satisfied for $g \approx f$, i.e., $g = \{id_R : R \to R \mid R \in \widetilde{f}\} \cup \{q_\bot\}$. Let h be a complex arrow with a set of point-to-point arrows $h_{ji} : A_j \to B_i$ in $\frac{h}{\uplus}$. Hence, from the fact that $\widetilde{h} = A \otimes B$, we obtain

$\widetilde{h_{ji}} = A_j \otimes B_i = TA_j \cap TB_i$. For any arrow $f : A \to B$, from Proposition 7 we have that $\widetilde{f} \preceq A \otimes B$ and, consequently, for each of its point-to-point arrows $f_{li} : A_l \to B_i$, $\widetilde{f_{li}} \subseteq A_l \otimes B_i = TA_l \cap TB_i = \widetilde{h_{li}} \subseteq TA_l$. Thus, for this ptp arrow f_{li} we define a ptp arrow $g_{ll} = \{id_R : R \to R \mid R \in \widetilde{f_{li}} \subseteq TA_l\} \cup \{q_\bot\} : A_l \to A_l$, so that $\widetilde{g_{ll}} = \widetilde{f_{li}}$ and for composition $h_{li} \circ g_l : A_l \to B_i$, $\widetilde{h_{li} \circ g_{ll}} = \widetilde{h_{li}} \cap \widetilde{g_{ll}} = (TA_l \cap TB_i) \cap \widetilde{g_{ll}} = (A_l \otimes B_i) \cap \widetilde{f_{li}} = \widetilde{f_{li}}$. By this generation of ptp arrows g_{ll}, for each ptp arrow in $\frac{f}{\uplus}$ we define the canonical morphism $g : A \to A$ (in Lemma 8), such that $\frac{f}{\uplus} = \frac{h \circ g}{\uplus}$, that is, from Definition 23, $f = h \circ g : A \to B$.

Claim 2. For any simple arrow which is a monomorphism $f : A \hookrightarrow B$ and (by duality, Theorem 3) the epimorphism $f^{OP} : B \twoheadrightarrow A$ ($f^{OP} \approx f$), $f^{OP} \circ f = id_A$ (in fact, $\widetilde{f^{OP} \circ f} = \widetilde{f^{OP}} \cap \widetilde{f} = \widetilde{f} = TA = \widetilde{id_A}$).

For any complex monomorphism between two strictly-complex objects

$$f : \biguplus_{1 \le j \le m} A_j \hookrightarrow \biguplus_{1 \le i \le k} B_i,$$

from Proposition 6, $\frac{f}{\uplus} = \{f_{ji_j} : A_j \hookrightarrow B_{i_j} \mid 1 \le j \le m\}$, so that

$$\frac{f^{OP}}{\uplus} = \left\{f^{OP}_{ji_j} : B_{i_j} \twoheadrightarrow A_j \mid 1 \le j \le m\right\}$$

and hence $\frac{f^{OP} \circ f}{\uplus} = \{h_{jj} = f^{OP}_{ji_j} \circ f_{ji_j} = id_{A_j} : A_j \to A_j \mid 1 \le j \le m\}$, i.e., $f^{OP} \circ f = id_A$. If A and B are not strictly-complex, let A' and B' be their strictly-complex reductions with two isomorphisms $is_1 : A \to A'$ and $is_2 : B \to B'$ and the monomorphism $f_1 : A' \hookrightarrow B'$. Then, $f = is_2 \circ f_1 \circ is_1$ with $f^{OP} = is_1^{-1} \circ f_1^{OP} \circ is_2^{-1}$ and hence $f^{OP} \circ f = is_1^{-1} \circ f_1^{OP} \circ is_2^{-1} \circ is_2 \circ f_1 \circ is_1 = is_1^{-1} \circ f_1^{OP} \circ id_{B'} \circ f_1 \circ is_1 = is_1^{-1} \circ f_1^{OP} \circ f_1 \circ is_1 =$ (from the strictly-complex composition above) $= is_1^{-1} \circ id_{A'} \circ is_1 = is_1^{-1} \circ is_1 = id_A$.

Claim 3. For each $f : A \hookrightarrow A$, $f \circ f = f$ and hence it is idempotent and, consequently, an object in Ret_A. The identity arrow of an object f in Ret_A is the arrow $id_f = (f; f) : f \to f$. Given two arrows $(h; h) : f \to g$ and $(h_1; h_1) : g \to k$ and hence $\widetilde{h} \subseteq \widetilde{f} \cap \widetilde{g}$ and $\widetilde{h_1} \subseteq \widetilde{g} \cap \widetilde{k}$, their composition is an arrow $(h_1; h_1) \circ (h; h) = (h_1 \circ h; h_1 \circ h) : f \to k$ because $\widetilde{h_1 \circ h} = \widetilde{h_1} \cap \widetilde{h} \subseteq (\widetilde{f} \cap \widetilde{g}) \cap (\widetilde{g} \cap \widetilde{k}) \subseteq \widetilde{f} \cap \widetilde{k}$.

The mapping $\sigma : Ret_A(f, g) \simeq \mathbf{DB}(\widetilde{f}, \widetilde{g})$ is well defined. In fact, for any arrow in Ret_A, $(h; h) : f \to g$, from Theorem 4, $T_e(h; h) = in_{\widetilde{g}}^{OP} \circ T(h \circ f) \circ in_{\widetilde{g}} : \widetilde{f} \to \widetilde{g}$, so that $\widetilde{T_e(h; h)} = \widetilde{in_{\widetilde{g}}^{OP}} \cap \widetilde{T(h \circ f)} \cap \widetilde{in_{\widetilde{g}}} = \widetilde{g} \cap (\widetilde{f} \cap \widetilde{h}) \cap \widetilde{f} = \widetilde{h} \subseteq \widetilde{g} \cap \widetilde{f}$ is appropriate. The inverse of σ is given by $\sigma^{-1}(T_e(h; h)) = (k; k)$ where $k \triangleq \tau_{J(g)}^{-1} \circ h \circ \tau_{J(f)} : A \to A$ such that $\widetilde{k} = \widetilde{\tau_{J(g)}^{-1}} \cap \widetilde{h} \cap \widetilde{\tau_{J(f)}} = \widetilde{g} \cap \widetilde{h} \cap \widetilde{f} = \widetilde{h}$, that is, $k = h$. □

With this section we finished the investigation on the properties of the morphisms in the **DB** category, provided in Chap. 3, and now we are ready to investigate the properties of its objects (instance-databases) in the next sections, in order to define an algebraic database lattice, its enrichments and then investigate inductive/coinductive and computational properties.

8.1.1 Matching Tensor Product

Since the data residing in different databases may have inter-dependencies (they are based on the partial overlapping between databases, which is information about a common part of the world), we can define such an (partial) overlapping by morphisms of the category **DB**: information flux of each mapping between two objects A and B in **DB** is just a subset of this overlapping between these two databases, denoted by $A \otimes B$. It is "bidirectional", i.e., (by duality) for any mapping f from A into B there exists an *equivalent* mapping f^{OP} from B into A.

This overlapping represents the common matching between these two databases, and is equal to the maximal information flux which can be defined between these two databases. Consequently, we can formally introduce a denotational semantics for the database matching operation $\otimes$ as follows:

Theorem 12 **DB** *is a strictly symmetric idempotent monoidal category*

$$(\mathbf{DB}, \otimes, \Upsilon, \alpha, \beta, \gamma),$$

where Υ is the total object for a given universe of databases, with the "matching" tensor product $\otimes : \mathbf{DB} \times \mathbf{DB} \longrightarrow \mathbf{DB}$ defined as follows:

1. $\Upsilon \otimes A = A \otimes \Upsilon \triangleq TA$;

   Otherwise, for any two database instances (objects) $A = \biguplus_{1 \le j \le m} A_j$ and $B = \biguplus_{1 \le i \le k} B_i$, where all A_j and B_i are simple objects, with $k, m \ge 1$, the object $A \otimes B \equiv \otimes(A, B)$ is the overlapping (matching) between A and B, introduced by Definition 19:

$$A \otimes B \triangleq \begin{cases} TA \cap TB & \textit{if } m = k = 1,\ A = A_1,\ B = B_1; \\ \biguplus_{1 \le i \le k \& 1 \le j \le m} (A_j \otimes B_i) & \textit{otherwise.} \end{cases}$$

2. $id_\Upsilon \otimes f = f \otimes id_\Upsilon \triangleq Tf$;

   Otherwise, for any two arrows $f : A \to C$ and $g : B \to D$, the $f \otimes g \equiv \otimes(f, g) : A \otimes B \to C \otimes D$ is defined by canonical representation (Lemma 8) *of the set of ptp arrows*

$$\frac{f \otimes g}{\biguplus} \triangleq \begin{cases} \{\{id_R : R \to R \mid R \in (\widetilde{f} \cap \widetilde{g}) \neq \bot^0\}\} & \textit{if } \frac{f}{\biguplus} = \{f\},\ \frac{g}{\biguplus} = \{g\}; \\ \{h_{ij} = \{id_R \mid R \in \widetilde{f_i} \cap \widetilde{g_j}\} \neq \bot^1 \mid f_i \in \frac{f}{\biguplus}, g_j \in \frac{g}{\biguplus}\} & \textit{otherwise} \end{cases}$$

   such that for each point-to-point arrow

$$h_{ij} : T\big(dom(f_i)\big) \cap T\big(dom(g_j)\big) \to T\big(cod(f_i)\big) \cap T\big(cod(g_j)\big) \quad \textit{in } \frac{f \otimes g}{\biguplus},$$

   one has $\widetilde{h_{ij}} = \widetilde{f_i} \cap \widetilde{g_j} \neq \bot^0$.

Proof Let us show that for any two simple arrows $f : A \to C$ and $g : B \to D$ in the **DB** category, also the arrow $f \otimes g$ is a **DB** category arrow (that is, it satisfies Theorem 1).

From the fact that f and g are the arrows in **DB**, it holds from Theorem 1 that there exist the schema mappings $\mathcal{M}_{AC} : \mathcal{A} \to \mathcal{C}$, $\mathcal{M}_{BD} : \mathcal{B} \to \mathcal{D}$, and an R-algebra α such that $f = \alpha^*(MakeOperads(\mathcal{M}_{AC}))$ and $g = \alpha^*(MakeOperads(\mathcal{M}_{BD}))$ with $A = \alpha^*(\mathcal{A})$, $B = \alpha^*(\mathcal{B})$, $C = \alpha^*(\mathcal{C})$ and $D = \alpha^*(\mathcal{D})$.

Hence, we can specify the α-intersection schema mapping (from Definition 16),

$$\mathcal{M}_{AC} \underset{\alpha}{\overset{\cap}{-}} \mathcal{M}_{BD} = \{\Phi\} : \mathcal{A} \underset{\alpha}{\overset{\cap}{-}} \mathcal{B} \to \mathcal{C} \underset{\alpha}{\overset{\cap}{-}} \mathcal{D},$$

where Φ is a tgd

$$\bigwedge \Big\{ \forall \mathbf{x}\big(r(\mathbf{x}) \Rightarrow r(\mathbf{x})\big) \mid r \in \mathcal{A} \underset{\alpha}{\overset{\cap}{-}} \mathcal{B}, \alpha(r) \in Flux\big(\alpha, MakeOperads(\mathcal{M}_{AC})\big) \cap Flux\big(\alpha, MakeOperads(\mathcal{M}_{BD})\big) \Big\}.$$

Consequently,

$$\begin{aligned}
\mathbf{M} &= MakeOperads\Big(\mathcal{M}_{AC} \underset{\alpha}{\overset{\cap}{-}} \mathcal{M}_{BD}\Big) \\
&= \Big\{ 1_r \mid r \in \mathcal{A} \underset{\alpha}{\overset{\cap}{-}} \mathcal{B}, \alpha(r) \in Flux\big(\alpha, MakeOperads(\mathcal{M}_{AC})\big) \\
&\qquad \cap Flux\big(\alpha, MakeOperads(\mathcal{M}_{BD})\big) \Big\} \cup \{1_{r_\emptyset}\} \\
&= \Big\{ 1_r \mid r \in \mathcal{A} \underset{\alpha}{\overset{\cap}{-}} \mathcal{B}, \alpha(r) \in \tilde{f} \cap \tilde{g} \Big\} \cup \{1_{r_\emptyset}\}
\end{aligned}$$

is a sketch category arrow from the schema $\mathcal{A} \underset{\alpha}{\overset{\cap}{-}} \mathcal{B}$ into the schema $\mathcal{B} \underset{\alpha}{\overset{\cap}{-}} \mathcal{D}$, with

$$\alpha^*\Big(\mathcal{A} \underset{\alpha}{\overset{\cap}{-}} \mathcal{B}\Big) = TA \cap TB = A \otimes B, \qquad \alpha^*\Big(\mathcal{C} \underset{\alpha}{\overset{\cap}{-}} \mathcal{D}\Big) = TC \cap TD = C \otimes D,$$

so that

$$\begin{aligned}
\alpha^*(\mathbf{M}) &= \Big\{ \alpha(1_r) = id_R : R \to R \mid r \in \mathcal{A} \underset{\alpha}{\overset{\cap}{-}} \mathcal{B}, R = \alpha(r) \in \tilde{f} \cap \tilde{g} \Big\} \cup \{q_\perp\} \\
&= f \otimes g : A \otimes B \to C \otimes D.
\end{aligned}$$

Let us show that for any four simple arrows $f : A \to C$, $f' : C \to E$, $g : B \to D$ and $g' : B \to F$ (with, from Definition 22 in Sect. 3.2, $\frac{f}{\uplus} = \{f\}$, and analogously for all other arrows), where all A, B, C, D, E and F are simple objects,

$$\big(f' \circ f\big) \otimes \big(g' \circ g\big) = \big(f' \otimes g'\big) \circ (f \otimes g) : A \otimes B \to E \otimes F.$$

Thus,

$$
\begin{aligned}
\widetilde{(f' \circ f) \otimes (g' \circ g)} &= T(\widetilde{f' \circ f}) \cap T(\widetilde{g' \circ g}) = T(\widetilde{f}' \cap \widetilde{f}) \cap T(\widetilde{g}' \cap \widetilde{g}) \\
&= (\widetilde{f}' \cap \widetilde{f}) \cap (\widetilde{g}' \cap \widetilde{g}) = (\widetilde{f}' \cap \widetilde{g}') \cap (\widetilde{f} \cap \widetilde{g}) \\
&= T(\widetilde{f}' \cap \widetilde{g}') \cap T(\widetilde{f} \cap \widetilde{g}) = \widetilde{f' \otimes g'} \cap \widetilde{f \otimes g} \\
&= \widetilde{(f' \otimes g') \circ (f \otimes g)},
\end{aligned}
$$

so that from Definition 23,

$$
(f' \circ f) \otimes (g' \circ g) = (f' \otimes g') \circ (f \otimes g). \tag{8.1}
$$

Let us show that this property of composition is valid for complex morphisms $f : A \to C$, $f' : C \to E$, $g : B \to D$ and $g' : B \to F$ as well. Then,

$$
\begin{aligned}
&\frac{(f' \otimes g') \circ (f \otimes g)}{\uplus} \\
&= \Big\{ (f_i' \otimes g_j') \circ (f_i \otimes g_j) \mid f_i' \otimes g_j' \in \frac{f' \otimes g'}{\uplus},\ f_i \otimes g_j \in \frac{f \otimes g}{\uplus}, \\
&\qquad cod(f_i \otimes g_j) = dom(f_i' \otimes g_j') \Big\} \\
&= \Big\{ (f_i' \otimes g_j') \circ (f_i \otimes g_j) \mid f_i' \otimes g_j' \in \frac{f' \otimes g'}{\uplus},\ f_i \otimes g_j \in \frac{f \otimes g}{\uplus}, \\
&\qquad cod(f_i) = dom(f_i'),\ cod(g_j) = dom(g_j') \Big\} \\
&= \Big\{ (f_i' \otimes g_j') \circ (f_i \otimes g_j) \mid f_i' \circ f_i \in \frac{f' \circ f}{\uplus},\ g_j' \circ g_j \in \frac{g' \circ g_j}{\uplus} \Big\} \\
&\quad \text{(from the fact that all } f_i', f_i, g_j', g_j \text{ are simple arrows and (8.1))} \\
&= \Big\{ (f_i' \circ f_i) \otimes (g_j' \circ g_j) \mid f_i' \circ f_i \in \frac{f' \circ f}{\uplus},\ g_j' \circ g_j \in \frac{g' \circ g_j}{\uplus} \Big\} \\
&= \Big\{ \{ id_R \mid R \in \widetilde{f_i' \circ f_i} \cap \widetilde{g_j' \circ g_j} \} \mid f_i' \circ f_i \in \frac{f' \circ f}{\uplus},\ g_j' \circ g_j \in \frac{g' \circ g_j}{\uplus} \Big\} \\
&\quad \text{(by Definition 22)} \\
&= \frac{(f' \circ f) \otimes (g' \circ g)}{\uplus}.
\end{aligned}
$$

Consequently, $(f' \circ f) \otimes (g' \circ g) = (f' \otimes g') \circ (f \otimes g)$, and hence the compositional property for functor $\otimes$ is satisfied.

Let us show the functorial property for the identity arrows, and consider two objects $A = \biguplus_{1\le j\le m} A_j$ and $B = \biguplus_{1\le i\le k} B_i$, with their identity arrows $id_A = \biguplus_{1\le j\le m} id_{A_j}$ and $id_B = \biguplus_{1\le i\le k} id_{B_i}$ such that $\widetilde{id_A} = \biguplus_{1\le j\le m} \widetilde{id_{A_j}} = \biguplus_{1\le j\le m} TA_j = TA$, $\widetilde{id_B} = \biguplus_{1\le i\le k} \widetilde{id_{B_i}} = \biguplus_{1\le i\le k} TB_i = TB$, with $\frac{id_A}{\uplus} = \{id_{A_1}, \dots, id_{A_m}\}$ and $\frac{id_B}{\uplus} = \{id_{B_1}, \dots, id_{B_k}\}$. Then,

$$\begin{aligned}\frac{id_A \otimes id_B}{\uplus} &= \left\{\{q_\bot\} \cup \{id_R \mid R \in \widetilde{id_{A_j}} \cap \widetilde{id_{B_i}}\} \neq \bot^1 \mid id_{A_j} \in \frac{id_A}{\uplus}, id_{B_i} \in \frac{id_B}{\uplus}\right\}\\ &= \left\{\{q_\bot\} \cup \{id_R \mid R \in TA_j \cap TB_i\} \neq \bot^1 \mid 1 \le j \le m, 1 \le i \le k\right\}\\ &= \left\{id_{TA_j \cap TB_i} \mid TA_j \cap TB_i \neq \bot^0, 1 \le j \le m, 1 \le i \le k\right\}\\ &= \left\{id_{A_j \otimes B_i} \mid A_j \otimes B_i \neq \bot^0, 1 \le j \le m, 1 \le i \le k\right\}\\ &= \frac{id_{A\otimes B}}{\uplus}.\end{aligned}$$

Hence, from Definition 23, $id_A \otimes id_B = id_{A\otimes B} : A \otimes B \to A \otimes B$.

It is easy to verify that $\otimes$ is a monoidal bifunctor with natural isomorphic transformations (which generate an isomorphic arrow for each object in **DB**):

- $\alpha : (\_ \otimes \_) \otimes \_ \longrightarrow \_ \otimes (\_ \otimes \_)$, (associativity)
- $\beta : \Upsilon \otimes \_ \longrightarrow I_{DB}$, (left identity)
- $\gamma : \_ \otimes \Upsilon \longrightarrow I_{DB}$, (right identity)

such that $A \otimes B \simeq B \otimes A$, $A \otimes \Upsilon = \Upsilon \otimes A \simeq A$ and $A \otimes \bot^0 \simeq \bot^0$. For any morphism $f : A \longrightarrow B$, from Proposition 6 we obtain $\widetilde{f} \preceq A \otimes B$. □

A tensor product $\otimes$ of the monoidal category **DB** *is not* unique in contrast with the Cartesian product (we can have for simple databases, $A \otimes B = C \otimes B$ such that $C = A \cup A_1 \supset A$ with $TA_1 \cap TB = \bot^0$).

Notice that each $A \otimes B$ is a closed object (intersection of two closed objects TA and TB), and that the information flux of any morphism from A to B is a closed object included in this maximal information flux (i.e., overlapping) between A and B. A matching of two completely disjoint simple databases is equal to the empty zero object $\bot^0$.

Proposition 47 *Each object A with the monomorphism $\mu_A : A \otimes A \longrightarrow A$ and the epimorphism $\eta_A : \Upsilon \twoheadrightarrow A$ compose a monoid in the monoidal category* $(\mathbf{DB}, \otimes, \Upsilon, \alpha, \beta, \gamma)$.

Proof We have $A \otimes A \preceq A$ and hence $\mu_A = in : A \otimes A \longrightarrow A$ is a monomorphism (from Theorem 6 in Sect. 3.2.5). From the fact that $TA \subseteq T\Upsilon = \Upsilon$ and point 2 of Proposition 7, we obtain a monomorphism $in_{TA} : TA \hookrightarrow \Upsilon$ and hence a monomorphism (composition of two monic arrows) $in = in_{TA} \circ is_A : A \hookrightarrow \Upsilon$ with $\frac{in}{\uplus} = \frac{id_A}{\uplus}$. Consequently, by duality, its opposite arrow $\eta_A = in^{OP} : \Upsilon \twoheadrightarrow A$ is an epimorphism such that $\widetilde{\eta_A} = \widetilde{in^{OP}} = \widetilde{in} \simeq \biguplus \frac{in}{\uplus} = \biguplus \frac{id_A}{\uplus} = \widetilde{id_A} = TA$. It is easy to verify that

$\mu_A \circ (\mu_A \otimes id_A) \circ \alpha_{A,A,A} = \mu_A \circ (id_A \otimes \mu_A)$ is valid, and $\beta_A = \mu_A \circ (\eta_A \otimes id_A)$, $\gamma_A = \mu_A \circ (id_A \otimes \eta_A)$. □

We generalize the merging operator for objects for the set $S = \{A_1, \ldots, A_n\}$ of more than two databases by $\otimes S = A_1 \otimes \cdots \otimes A_n$.

8.1.2 Merging Operator

Merging of an instance-database A with another instance-database is a unary algebraic operation, similar to the binary operation Data Federation of two database instances (denoted by $\frac{\cup}{rn}$ in Sect. 3.3.1, i.e., the union $\cup$ with the renaming (*rn*) of the relational symbols in $\mathcal{A}$ and $\mathcal{B}$ that have originally the same names, by unifying these two databases, that are in two different DB machines, for example, under the common DBMS. Consequently, given a connection $\mathcal{A} \oplus \mathcal{B}$ of two schemas and an R-algebra α (which is their model), $A \frac{\cup}{rn} B = \alpha^*(\mathcal{A} \oplus \mathcal{B}))$.

In fact, the merging of two databases is defined as the database obtained by applying the power-view operator T to data federation of these two databases. That is, the data federation of two databases is isomorphic to the database obtained by their merging, from the behavioral point of view. Any view which can be obtained from the data federation (union with renaming) of two databases can also be obtained from the merging of these two databases, and vice versa.

In what follows, similarly to the matching tensor products which, for any two given databases, return a closed object, also the merging operator will return a closed object. As we will see, these two operators will provide the meet and joint operators of the complete algebraic database lattice, where $\perp^0$ and Υ are the bottom and top elements (i.e., the instance-databases), respectively.

Theorem 13 *For any fixed simple database $A \in Ob_{DB}$ we define the parameterized "merging with A" operator as an endofunctor $A \oplus \_ : \mathbf{DB} \longrightarrow \mathbf{DB}$, as follows:*

1. *For any object $B = \biguplus_{1 \le i \le k} B_i$, $k \ge 1$, the object $A \oplus B$ is a merging of A, defined by:*

$$A \oplus B \equiv \oplus(A, B) \triangleq \begin{cases} T(A \frac{\cup}{rn} B) = T(A \cup B) & \text{if } k = 1,\ B = B_1; \\ \biguplus_{1 \le i \le k}(A \oplus B_i) & \text{otherwise.} \end{cases}$$

2. *For any $f : B \to \biguplus_{1 \le j \le m} C_j$, $m \ge 1$, the arrow $A \oplus (f) : A \oplus B \to A \oplus C$ is defined by the canonical representation (Lemma 8) of*

$$\frac{A \oplus (f)}{\biguplus}$$
$$\triangleq \left\{ \left(id_A \frac{\cup}{rn} f_{ij} \right) : A \oplus B_i \to A \oplus C_j \mid f_{ij} : B_i \right.$$
$$\left. \to C_j \in \frac{f}{\biguplus}, \left(id_A \frac{\cup}{rn} f_{ij} \right) \neq \perp^1 \right\}.$$

Proof Let us show that for a given **DB** simple arrow $f : B \to C$ and hence, from Theorem 1, obtained from a schema mapping $\mathcal{M}_{BC} : \mathcal{B} \to \mathcal{C}$ and an R-algebra α, such that $f = \alpha^*(MakeOperads(\mathcal{M}_{BC}))$ with the simple databases $B = \alpha^*(\mathcal{B})$ and $C = \alpha^*(\mathcal{C})$, there is also an arrow $id_A \overset{\cup}{\underset{rn}{-}} f$ in **DB**. In fact, let us consider the identity schema mapping $\mathcal{M}_{AA} : \mathcal{A} \to \mathcal{A}$ and the sketch's mapping

$$\mathbf{M} = \left(MakeOperads(\mathcal{M}_{AA}) \overset{\cup}{\underset{rn}{-}} MakeOperads(\mathcal{M}_{BC}) \right) : \mathcal{A} \oplus \mathcal{B} \to \mathcal{A} \oplus \mathcal{C}.$$

Then,

$$\begin{aligned} &\alpha^* \left(MakeOperads(\mathcal{M}_{AA}) \overset{\cup}{\underset{rn}{-}} MakeOperads(\mathcal{M}_{BC}) \right) \\ &= \alpha^*\big(MakeOperads(\mathcal{M}_{AA})\big) \overset{\cup}{\underset{rn}{-}} \alpha\big(MakeOperads(\mathcal{M}_{BC})\big) \\ &= id_A \overset{\cup}{\underset{rn}{-}} f : A \overset{\cup}{\underset{rn}{-}} B \to A \overset{\cup}{\underset{rn}{-}} B \\ &= A \oplus (f) : A \overset{\cup}{\underset{rn}{-}} B \to A \overset{\cup}{\underset{rn}{-}} B. \end{aligned}$$

From the fact that $A \oplus B \simeq A \overset{\cup}{\underset{rn}{-}} B$ and $A \oplus C \simeq A \overset{\cup}{\underset{rn}{-}} C$, we obtain the **DB** arrow $id_A \overset{\cup}{\underset{rn}{-}} f : A \oplus B \to A \oplus C$.

It is easy to verify that for any two simple databases A and B the database $A \overset{\cup}{\underset{rn}{-}} B$ is different from $A \cup B$ only in the fact that by renaming we can have the copies of the same relations. Thus, the set of views (obtained by SPRJU queries) of both databases are equal. Consequently, $T(A \overset{\cup}{\underset{rn}{-}} B) = T(A \cup B) = T(TA \cup TB) = T(TA \overset{\cup}{\underset{rn}{-}} TB)$, that is, $A \oplus B = TA \oplus TB$, and $A \overset{\cup}{\underset{rn}{-}} B \simeq A \oplus B$.

Now we can verify that $A \oplus \_$ is an endofunctor. In fact, for any object $B = \biguplus_{1 \le i \le k} B_i$, $k \ge 1$ and its identity arrow $id_B = \biguplus_{1 \le i \le k} (id_{B_i} : B_i \to B_i)$,

$$\begin{aligned} \frac{A \oplus (id_B)}{\uplus} &= \left\{ \left(id_A \overset{\cup}{\underset{rn}{-}} id_{B_i} \right) : A \oplus B_i \to A \oplus B_i \mid id_{B_i} : B_i \to B_i \in \frac{id_B}{\uplus} \right\} \\ &= \left\{ A \oplus (id_{B_i}) : A \oplus B_i \to A \oplus B_i \mid id_{B_i} : B_i \to B_i \in \frac{id_B}{\uplus} \right\}. \end{aligned}$$

Thus, for each simple point-to-point arrow $A \oplus (id_{B_i}) : A \oplus B_i \to A \oplus B_i$, its flux is $\widetilde{A \oplus (id_{B_i})} = (\widetilde{id_A \overset{\cup}{\underset{rn}{-}} id_{B_i}}) = \widetilde{id_{A \overset{\cup}{\underset{rn}{-}} B_i}} = T(A \overset{\cup}{\underset{rn}{-}} B_i) = A \oplus B_i$. Consequently, for the identity arrows is valid the functorial property,

$$A \oplus (id_{B_i}) = id_{A \oplus B_i} : A \oplus B_i \to A \oplus B_i.$$

From the fact that for any simple object (i.e., a database) B, $A \subseteq A \oplus B$, each point-to-point arrow, resulting by application of this endofunctor, contains a sub-arrow id_A. Thus, based on composition of operads in point 4 of Definition 8 (Sect. 2.4), given two arrows $f : B \to C$ and $g : C \to D$, we have the compositional endofunctors property:

$$\frac{A \oplus (g) \circ A \oplus (f)}{\uplus} = \left\{ \left(id_A \frac{\cup}{rn} g_{jl} \right) \circ \left(id_A \frac{\cup}{rn} f_{ij} \right) \mid g_{jl} \circ f_{ij} \in \frac{g \circ f}{\uplus} \right\}$$
$$= \left\{ id_A \frac{\cup}{rn} (g_{jl} \circ f_{ij}) \mid g_{jl} \circ f_{ij} \in \frac{g \circ f}{\uplus} \right\}$$
$$= \frac{A \oplus (g \circ f)}{\uplus}.$$

Consequently, $A \oplus (g) \circ A \oplus (f) = A \oplus (g \circ f)$.

Moreover, $A \oplus B = B \oplus A$ (if B is simple database as well), $A \oplus \Upsilon = \Upsilon$, $A \oplus \bot^0 \simeq A$, $A \oplus TA = TA$ and $A \oplus A \simeq A$. □

'The merging with a complex-object A', when $A = \biguplus_{1 \le j \le m} A_j$, $m \ge 2$, can be obtained from the object-component of the functor defined in the theorem above, and it will be considered in more details in Sect. 8.1.5, by obtaining the symmetric operator '$\widehat{\oplus}$', such that for any two complex objects A and B the isomorphism $A \widehat{\oplus} B \simeq B \widehat{\oplus} A$ is valid. Matching '$\otimes$' and (generalized) merging '$\widehat{\oplus}$' operators are dual operators in the category **DB**: in fact, they are also two dual lattice operators (meet and join, respectively) w.r.t. the database ordering $\preceq$, as we will show in Sect. 8.1.5.

Remark Notice that $A \frac{\cup}{rn} B \simeq A \oplus B$, that is, from the behavioral point of view, the data federation is equivalent (isomorphic) to the data merging. That is, for any query over the data federation $A \frac{\cup}{rn} B$, which returns a view R, there exists a query over the data merging $A \oplus B$ which returns the same view R; and vice versa.

We generalize the merging operator for objects for the set $S = \{A_1, \dots, A_n\}$ of more than two simple databases by $\oplus S = T(\bigcup S) = T(A_1 \cup \dots \cup A_n)$.

8.1.3 (Co)Limits and Exponentiation

Previously, in Sect. 3.3, we introduced the (co)products in the **DB** category that are operators used in order to make the separation-composition of the databases. Here, instead, we will investigate the rest of the limits and colimits in the **DB** category in order to establish the general properties of this category, used as a base category for the functorial semantics of schema database mapping systems.

In order to explain these concepts in another way, we can see the limits and colimits as a left and a right adjunction for the diagonal functor $\Delta : \mathbf{DB} \longrightarrow \mathbf{DB}^{\mathbf{J}}$ for a small index category (i.e., a diagram) $\mathbf{J}$. For any colimit functor $F : \mathbf{DB}^{\mathbf{J}} \longrightarrow \mathbf{DB}$ we have a left adjunction to the diagonal functor, $(F, \Delta, \eta_C, \varepsilon_C) : \mathbf{DB}^{\mathbf{J}} \longrightarrow \mathbf{DB}$, with the colimit object $F(D)$ for any object (i.e., a diagram) $D \in \mathbf{DB}^{\mathbf{J}}$ and the universal (unit) cone (a natural transformation) $\eta_C : Id_{DB^J} \longrightarrow \Delta F$. Hence, by duality, the same functor F is also a right adjoint to the diagonal functor (i.e., the adjunction, $(\Delta, F, \eta, \varepsilon) : \mathbf{DB} \longrightarrow \mathbf{DB^J}$) with the limit object (equal to the colimit object above) $F(D)$ and the universal (counit) cone (a natural transformation) $\varepsilon : \Delta F \longrightarrow Id_{DB^J}$, such that $\varepsilon = \eta_C^{OP}$ and $\eta = \varepsilon_C^{OP}$.

Let us consider, for example, the coproducts ($F = + \equiv \uplus$) and the products ($F = \times \equiv +$). In this case, the diagram (a functor) $D = (A, B) \in \mathbf{DB^J}$ is just a diagram of two objects in **DB** without arrows between them (**J** is a discrete category with two index objects 1 and 2 and with their identity arrows, so that $\mathbf{DB^J} \simeq \mathbf{DB}^2 = \mathbf{DB} \times \mathbf{DB}$). We obtain for the universal cocone unit $\eta_C(A, B) : (A, B) \longrightarrow (A + B, A + B)$, one pair of the coproduct inclusion-monomorphisms $\eta_C(A, B) = (in_A, in_B)$ where $in_A = \langle id_A, \bot^1 \rangle : A \hookrightarrow A + B$ and $in_B = \langle \bot^1, id_B \rangle : B \hookrightarrow A + B$. Dually, for the products, the universal cone counit of the product $\varepsilon(A \times B, A \times B) : (A \times B, A \times B) \longrightarrow (A, B)$ is a pair of the product projection-epimorphisms $\varepsilon(A \times B, A \times B) = (p_A, p_B)$ where $p_A : A \times B \twoheadrightarrow A$ and $p_B : A \times B \twoheadrightarrow B$. Moreover, $A \times B = A + B$, $p_A = in_A^{OP} = [id_A, \bot^1]$ and $p_B = in_B^{OP} = [\bot^1, id_B]$, as represented in the following diagram:

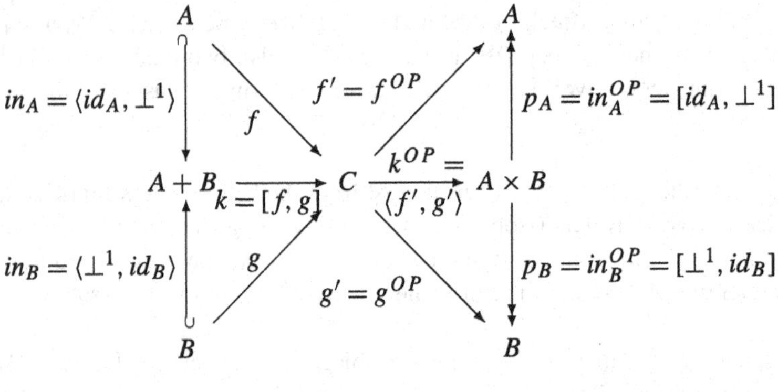

In the case when **J** is an index category with two objects and two parallel arrows from the first into the second object, each diagram (a functor) $D \in \mathbf{DB^J}$ is of the form $B \overset{f}{\underset{g}{\leftleftarrows}} A$. The limits of such diagrams are the equalizers, while their colimits are the coequalizers, as in the following example:

Example 40 Let us verify that each object in **DB** is a limit of some equalizer and a colimit of its dual coequalizer. In fact, for any simple object A, a "structure map" $h : TA \longrightarrow A$ of a *monadic* T-algebra (A, h) (from Proposition 57, $h : TA \rightarrow A$ is an isomorphism, thus epic and monic as well) derived from a monad (T, η, μ), where from Definition 25, in Sect. 3.2.1, $h \circ \eta_A = id_A$, and $h \circ Th = h \circ \mu_A$, we obtain the *coequalizer* (by Back's theorem, it is preserved by the endofunctor T, i.e., T creates a coequalizer) with a colimit A, and, by duality, we obtain the absolute equalizer with the limit A as well:

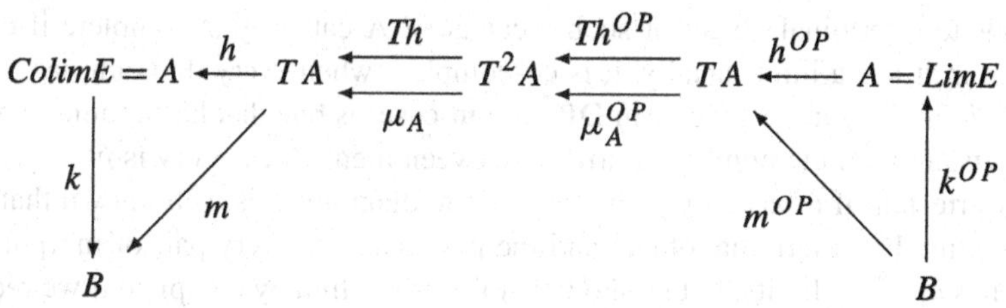

That is, whenever $m : TA \to B$ (with a simple object B) has $m \circ Th = m \circ \mu_A$ in **DB**, there is exactly one arrow $k : A \to B$ such that $m = k \circ h$. In fact, if $m \circ Th = m \circ \mu_A$ then $\widetilde{m} \cap \widetilde{Th} = \widetilde{m} \cap \widetilde{\mu_A}$, and from $\widetilde{m} \subseteq TA = \widetilde{\mu_A}$ and $\widetilde{Th} = \widetilde{h}$ we obtain $\widetilde{m} \cap \widetilde{h} = \widetilde{m}$, i.e., $\widetilde{m} \subseteq \widetilde{h}$. Consequently, in order to satisfy the commutativity $m = k \circ h$, that is, $\widetilde{m} = \widetilde{k} \cap \widetilde{h}$, we must have that $\widetilde{k} = \widetilde{m}$, so that k is uniquely determined by the morphism m.

Analogously, we obtain for the equalizer in the right part of the diagram above.

Lemma 16 **DB** *has the* (*co*)*equalizers for every pair of morphisms.*

Proof For each pair of simple morphisms $f, g : A \to B$ and hence $\widetilde{f} \subseteq TA$ and $\widetilde{g} \subseteq TA$, the $LimE = (\widetilde{f} \cap \widetilde{g}) \cup (TA \backslash (\widetilde{f} \cup \widetilde{g})) \subseteq TA$ is a limit object (up to isomorphism) of the diagram $B \underset{g}{\overset{f}{\leftleftarrows}} A$, so that the following diagram is an equalizer:

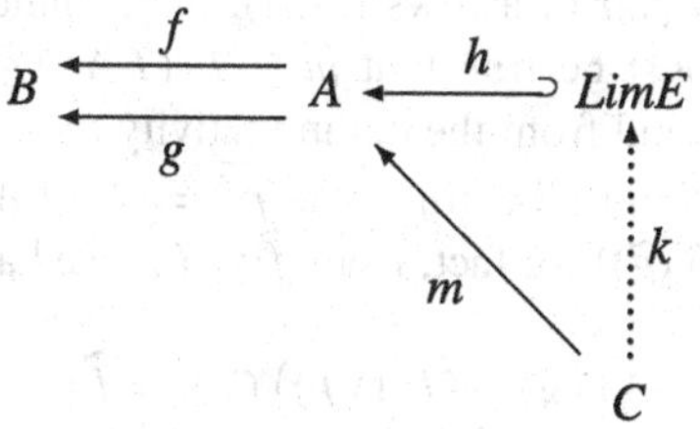

with (i) $f \circ h = g \circ h$ and (ii) $f \circ m = g \circ m$, where h is a monomorphism (every equalizer is monic). It is easy to verify that $\widetilde{h} = T(LimE) \subseteq TA$ is the maximal flux such that $\widetilde{f} \cap \widetilde{h} = \widetilde{g} \cap \widetilde{h}$ (equal to $\widetilde{f} \cap \widetilde{g}$). From (ii) we have $\widetilde{f} \cap \widetilde{m} = \widetilde{g} \cap \widetilde{m}$, so that it must be $\widetilde{m} \subseteq \widetilde{h}$ and hence for each $m : C \to A$ such that (ii) is true, there is a unique $k : C \to LimE$ with $\widetilde{k} = \widetilde{m}$ for which $m = h \circ k$. By duality, it is valid for coequalizers as well. The equalizers for the complex arrows are given as a particular case of the pullbacks after the proof of next Proposition 48. □

In the case when **J** is an index category with three objects and two arrows with the same codomain, each diagram (a functor) $D \in \mathbf{DB}^{\mathbf{J}}$ is of the form $A \xrightarrow{f} C \xleftarrow{g} B$. The limits of such diagrams are the pullbacks. In the case when **J** is an index category with one object only, the limit of such diagram corresponds to a terminal object of such a category. A category is complete if every diagram in it has a limit. Dually, it is co-complete when every diagram has a colimit. A *finite* diagram (a functor in $\mathbf{DB}^{\mathbf{J}}$ in our case) is one that has a finite number of objects, and a finite number of arrows between them. A category is *finitely complete* (Cartesian) if it has a limit for every finite diagram. It is well known that any category that has a terminal object and the pullbacks for every pair of morphisms, than it has *all* finite limits. Let us show that **DB** is not finitely complete if we reduce only to simple arrows and objects. For example, for the pullbacks. Let us suppose that the following diagram, where A, B and C are simple databases and *LimP* is a limit object of the diagram $A \xrightarrow{f} C \xleftarrow{g} B$, is a pullback (PB):

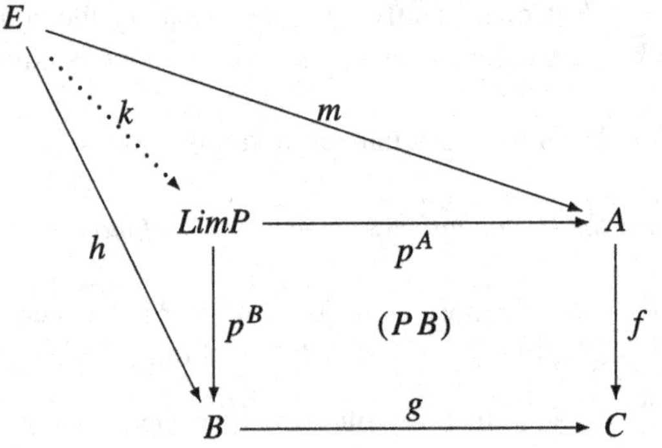

In order to have for every pair of arrows (m, h), the commutativity $f \circ m = g \circ h$, i.e., $\widetilde{f} \cap \widetilde{m} = \widetilde{g} \cap \widetilde{h}$ it must be true that $\widetilde{m} \subseteq T((TA \cap \widetilde{g}) \cup (TA \backslash \widetilde{f}))$ and $\widetilde{h} \subseteq T((TB \cap \widetilde{f}) \cup (TB \backslash \widetilde{g}))$, and from the commutativity $m = p^A \circ k$ and $h = p^B \circ k$, $\widetilde{m} \subseteq \widetilde{p^A}$ and $\widetilde{h} \subseteq \widetilde{p^B}$, it must be true that $\widetilde{p^A} = T((TA \cap \widetilde{g}) \cup (TA \backslash \widetilde{f}))$ and $\widetilde{p^B} = T((TB \cap \widetilde{f}) \cup (TB \backslash \widetilde{g}))$. In fact, from $\widetilde{f} \subseteq TA$ we have that

$$\big((TA \cap \widetilde{g}) \cup (TA \backslash \widetilde{f})\big) \cap \widetilde{f} = \widetilde{f} \cap \widetilde{g}$$

and hence for closed object $\widetilde{f} \cap \widetilde{g} = T(\widetilde{f} \cap \widetilde{g}) = T(((TA \cap \widetilde{g}) \cup (TA \backslash \widetilde{f})) \cap \widetilde{f}) = T((TA \cap \widetilde{g}) \cup (TA \backslash \widetilde{f})) \cap \widetilde{f} = \widetilde{p^A} \cap \widetilde{f}$. Similarly $\widetilde{p^B} \cap \widetilde{g} = \widetilde{f} \cap \widetilde{g}$ and hence $\widetilde{p^A} \cap \widetilde{f} = \widetilde{p^B} \cap \widetilde{g}$, i.e., $f \circ p^A = g \circ p^B$. However, if $\widetilde{m} \neq \widetilde{h}$ then there is no simple arrow $k : D \to LimP$ because we need $\widetilde{k} = \widetilde{h}$ (from $h = p^A \circ k$ and $\widetilde{h} \subseteq \widetilde{p^A}$) and similarly $\widetilde{k} = \widetilde{m}$ and hence $\widetilde{m} = \widetilde{h}$, which is a contradiction (the case when $\widetilde{m} = \widetilde{h}$ with $A = B$ is the case when it is a pullback with all simple objects, corresponding to equalizers in Lemma 16). Consequently, $LimP = ((TA \cap \widetilde{g}) \cup (TA \backslash \widetilde{f})) \times ((TB \cap \widetilde{f}) \cup (TB \backslash \widetilde{g}))$ and all arrows to and from it have to be *complex*, i.e., $k = \langle m, h \rangle$ and p^B and p^A substituted by $[\bot^1, p^B]$ and $[p^A, \bot^1]$, respectively. Consequently, we obtain:

Proposition 48 **DB** *category is* finitely (co)complete, *i.e.*, (co)*Cartesian*.

Proof Let us consider a pullback diagram (PB) above with $A = \biguplus_{1\leq j\leq M} A_j$, $B = \biguplus_{1\leq i\leq K} B_i$, $C = \biguplus_{1\leq l\leq N} C_l$ and $E = \biguplus_{1\leq x\leq N_1} E_x$ for $M, K, N, N_1 \geq 1$. Then a limit cone of this pullback is composed by limit object (see the ptp arrows in S_A and S_B in the diagram bellow) $LimP = \biguplus(S_E \cup S_A \cup S_B)$ where $S_E = \{D_{ilj} = D'_{jli} = (\widetilde{f_{jl}} \cap \widetilde{g_{il}}) \cup (TA_j \backslash (\widetilde{f_{jl}} \cup \widetilde{g_{il}})) \mid f_{jl} \in \frac{f}{\uplus}, g_{il} \in \frac{g}{\uplus}, i = j \text{ and } TB_i = TA_j\}$, $S_A = \{D_{ilj} = (TA_j \cap \widetilde{g_{il}}) \cup (TA_j \backslash \widetilde{f_{jl}}) \mid f_{jl} \in \frac{f}{\uplus}, g_{il} \in \frac{g}{\uplus}, TB_i \neq TA_j\}$ and $S_B = \{D'_{jli} = (TB_i \cap \widetilde{f_{jl}}) \cup (TB_i \backslash \widetilde{g_{il}}) \mid f_{jl} \in \frac{f}{\uplus}, g_{il} \in \frac{g}{\uplus}, TB_i \neq TA_j\}$, and with outgoing arrows

$$\frac{p^A}{\uplus} = \{p^A_{ilj} : D_{ilj} \hookrightarrow A_j \mid D_{ilj} \in S_E \cup S_A\} \quad \text{and}$$

$$\frac{p^B}{\uplus} = \{p^B_{jli} : D'_{jli} \hookrightarrow B_i \mid D'_{jli} \in S_E \cup S_B\}.$$

From the fact that

$$\frac{f \circ p^A}{\uplus} = \left\{ f_{jl} \circ p^A_{ijl} : D_{ilj} \to C_l \mid f_{jl} \in \frac{f}{\uplus}, g_{il} \in \frac{g}{\uplus} \right\} \quad \text{and}$$

$$\frac{g \circ p^B}{\uplus} = \left\{ g_{il} \circ p^B_{jli} \mid f_{jl} \in \frac{f}{\uplus}, g_{il} \in \frac{g}{\uplus} \right\},$$

we obtain the bijection $\sigma : \frac{f \circ p^A}{\uplus} \to \frac{g \circ p^B}{\uplus}$, such that for each $f_{ji} \circ p^A_{ijl} \in \frac{f}{\uplus}$, $\sigma(f_{ji} \circ p^A_{ijl}) = g_{il} \circ p^B_{jli}$ with

$$\sigma(\widetilde{f_{ji} \circ p^A_{ijl}}) = \widetilde{g_{il}} \cap \widetilde{p^B_{jli}} = \widetilde{g_{il}} \cap \widetilde{f_{jl}} = \widetilde{p^A_{ijl}} \cap \widetilde{f_{jl}} = \widetilde{f_{ji} \circ p^A_{ijl}}$$

and, consequently we have the same set of ptp arrows $\frac{f \circ p^A}{\uplus} = \frac{g \circ p^B}{\uplus}$, i.e., the commutativity of the square diagram (PB), that is, $f \circ p^A = g \circ p^B$.

Suppose that for another two arrows $m : E \to A$ and $h : E \to B$ the diagram (PB) commutes, i.e., $f \circ m = g \circ h$. For $\frac{m}{\uplus} = \frac{h}{\uplus} = \emptyset$ it is obvious, so we will consider nontrivial cases when both arrows have a nonempty set of ptp arrows. Let us define the sets (see the commutative diagram bellow composed of the ptp arrows of this pullback diagram), $S_1 = \{k_{xji} : E_x \to D_{ilj} \mid m_{xj} \in \frac{m}{\uplus}, p^A_{ilj} \in p^A \text{ with } \widetilde{k_{xji}} = \widetilde{m_{xj}} \subseteq \widetilde{p^A_{ilj}}\}$, and $S_2 = \{k'_{xji} : E_x \to D'_{jli} \mid h_{xi} \in \frac{h}{\uplus}, p^B_{jli} \in p^B \text{ with } \widetilde{k'_{xji}} = \widetilde{h_{xi}} \subseteq \widetilde{p^B_{jli}}\}$. Then we define the arrow $k : E \to LimP$ as the canonical representation (from Lemma 8) composed by the set of ptp arrows $\frac{k}{\uplus} \triangleq S_1 \cup S_2$.

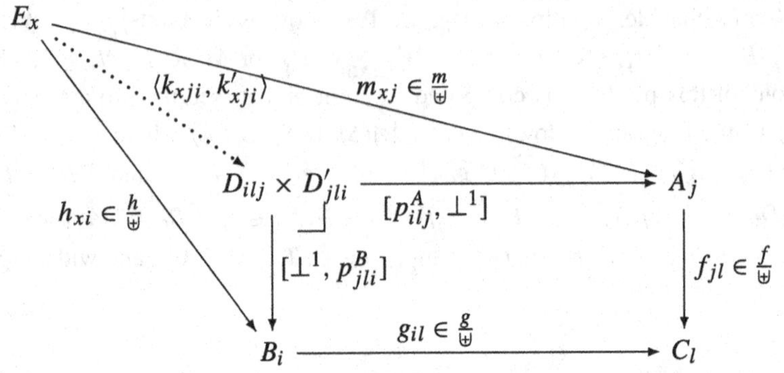

Let us show that the pullback diagram (PB) commutes for this arrow k, i.e., $p^B \circ k = h$ (and $p^A \circ k = m$). First of all, note that each ptp arrow $(s_{xi} : E_x \to B_i) \in \frac{p^B \circ k}{\uplus}$ is the union of the following set of compositions of the ptp arrows in $S_2 \subset \frac{k}{\uplus}$ and in $\frac{p^B}{\uplus}$, $s_{xi} = \bigcup\{p^B_{jli} \circ k'_{xji} : E_x \to B_i \mid k'_{xji} \in S_2, p^B_{jli} \in p^B$ and $1 \le j \le M\}$. From Lemma 7, the information flux of this ptp arrow is

$$
\begin{aligned}
\widetilde{s_{xi}} &= T\left(\bigcup\left\{\widetilde{p^B_{jli} \circ k'_{xji}} \mid p^B_{jli} \circ k'_{xji} \in \frac{s_{xi}}{\uplus}\right\}\right)\\
&= T\left(\bigcup\left\{\widetilde{p^B_{jli}} \cap \widetilde{k'_{xji}} \mid p^B_{jli} \circ k'_{xji} \in \frac{s_{xi}}{\uplus}\right\}\right)\\
&= T\left(\bigcup\left\{\widetilde{p^B_{jli}} \cap \widetilde{h_{xi}} \mid p^B_{jli} \circ k'_{xji} \in \frac{s_{xi}}{\uplus}\right\}\right)\\
&= T\left(\bigcup\left\{\widetilde{h_{xi}} \mid p^B_{jli} \circ k'_{xji} \in \frac{s_{xi}}{\uplus}\right\}\right)\\
&= T(\widetilde{h_{xi}})\\
&= \widetilde{h_{xi}},
\end{aligned}
$$

that is, $s_{x_i} = h_{xi}$. Consequently, $\frac{p^B \circ k}{\uplus} = \frac{h}{\uplus}$ and hence $p^B \circ k = h$. Analogously, $p^A \circ k = m$, so that k is this unique arrow which satisfies the pullback's requirements. □

Notice that in this definition of the pullbacks we considered only nonempty arrows f and g. Notice that when $\frac{f}{\uplus} = \frac{g}{\uplus} = \emptyset$ are empty morphisms, then the pullback diagram corresponds to the standard product diagram. In this case,

$$
\begin{aligned}
D_{ilj} &= (TA_j \cap \bot^0) \cup (TA_j \setminus \bot^0) = TA_j \simeq A_j,\\
D'_{jli} &= (TB_i \cap \bot^0) \cup (TB_i \setminus \bot^0) = TB_i \simeq B_i.
\end{aligned}
$$

Thus, $LimP = A \times B$ with the projections $p^A = [id_A, \bot^1]$, and $p^B = [\bot^1, id_B]$.

In fact, when $B = A$ and $S_A = S_B = \emptyset$, we obtain $p^A = p^B$ with limit object $LimP = \biguplus S_E$ corresponding to the equalizer.

From the fact that **DB** is an *extended symmetric* category, for any morphism $f : A \to B$ (i.e., each object $J(f) = \langle A, B, f\rangle$ in **DB** $\downarrow$ **DB**) the following diagram commutes (see the proof of Theorem 4 in Sect. 3.2.3):

$$\begin{array}{ccc} A = F_{st}(J(f)) & \xrightarrow{\quad f \quad} & B = S_{nd}(J(f)) \\ & {\scriptstyle \tau_{J(f)}} \searrow \quad \nearrow {\scriptstyle \tau^{-1}_{J(f)}} & \\ & \widetilde{f} = T_e(J(f)) & \end{array}$$

such that

$$T_e\big(J(f)\big) = \widetilde{f} \simeq \biguplus \frac{\widetilde{f}}{\uplus} = \biguplus_{f_{ji} \in \frac{f}{\uplus}} \widetilde{f_{ji}},$$

$$\frac{\tau_{J(f)}}{\uplus} = \left\{\tau_{J(f_{ji})} : A_j \twoheadrightarrow \widetilde{f_{ji}} \mid (f_{ji} : A_j \to B_i) \in \frac{f}{\uplus}\right\}, \quad \text{and}$$

$$\frac{\tau^{-1}_{J(f)}}{\uplus} = \left\{\tau^{-1}_{J(f_{ji}}: \widetilde{f_{ji}} \hookrightarrow B_i \mid (f_{ji} : A_j \to B_i) \in \frac{f}{\uplus}\right\}.$$

This decomposition of each morphism f into one epic $\tau_{J(f)} : A \twoheadrightarrow \widetilde{f}$ and one monic $\tau^{-1}_{J(f)} : \widetilde{f} \hookrightarrow B$ arrows will be used to demonstrate the existence of a limit in **DB** for each diagram D composed of a single morphism $f : A \to B$ in **DB**, by the pullback of the diagram $B \xrightarrow{id_B} B \xleftarrow{f} A$ (when $C = B$ and $g = id_B$). Here we will consider the following particular cases for f:

Corollary 27 *For any given morphism $f : A \to B$ such that for each its ptp arrow $f_{ji} : A_j \to B_i$, $TA_j \subseteq TB_i$, with $A = \biguplus_{1 \le j \le M} A_j, M \ge 1$ and $B = \biguplus_{1 \le j \le K} B_i, K \ge 1$, and a diagram $B \xrightarrow{id_B} B \xleftarrow{f} A$, the cone with the vertex $Lim(f) = \biguplus_{(f_{ji}:A_j \to B_i) \in \frac{f}{\uplus}} (A_j \times \widetilde{f_{ji}})$, and with outgoing arrows $\frac{p^B}{\uplus} = \{\tau^{-1}_{J(f_{ji})} : \widetilde{f_{ji}} \hookrightarrow B_i \mid f_{ji} \in \frac{f}{\uplus}\} = \frac{\tau^{-1}_{J(f)}}{\uplus}$ and $\frac{p^A}{\uplus} = \frac{id_A}{\uplus}$, defines a pullback diagram,*

$$\begin{array}{ccc} Lim(f) & \xrightarrow{\quad p^A \quad} & A \\ \Big\downarrow {\scriptstyle p^B} & (PB_f) & \Big\downarrow {\scriptstyle f} \\ B & \xrightarrow{\quad id_B \quad} & B \end{array}$$

This diagram can be simplified by substituting the limit object $Lim(f)$ with the object $A \times \widetilde{f}$ isomorphic to it.

Proof For a morphism f we have that for each of its ptp arrows $f_{ji} : A_j \to B_i$, $\widetilde{f_{ji}} \subseteq TA_j \subseteq TB_i$. The first part comes directly from the proof of Proposition 48 in the case when $C = B$ and $g = id_B$, with $l = i$,

$$D_{ilj} = (TA_j \cap \widetilde{g_{ii}}) \cup (TA_j \backslash \widetilde{f_{ji}}) = (TA_j \cap TB_i) \cup (TA_j \backslash \widetilde{f_{ji}}) = TA_j \simeq A_j \quad \text{and}$$

$$D'_{jli} = (TB_i \cap \widetilde{f_{ji}}) \cup (TB_i \backslash \widetilde{g_{ii}}) = \widetilde{f_{ji}} \cup (TB_i \backslash TB_i) = \widetilde{f_{ji}}.$$

From the fact that f is a complex monomorphism, we have a ptp arrow for every simple object A_j, $1 \leq j \leq M$, so from the commutativity we have that $Lim(f) \simeq A_1 \uplus \cdots \uplus A_M \uplus \frac{\widetilde{f}}{\uplus} \simeq A \times \widetilde{f}$ (we recall that the product '$\times$' is equal to $\uplus$). The rest comes directly from it. □

Remark Consequently, for each morphism $f : A \to B$ such that for each of its ptp arrows $f_{ji} : A_j \to B_i$, $TA_j \subseteq TB_i$, the following commutative diagram with the arrows p^B and p^A such that $\frac{p^B}{\uplus} = \frac{\tau^{-1}_{J(f)}}{\uplus}$ and $\frac{p^A}{\uplus} = \frac{id_A}{\uplus}$ is a pullback diagram:

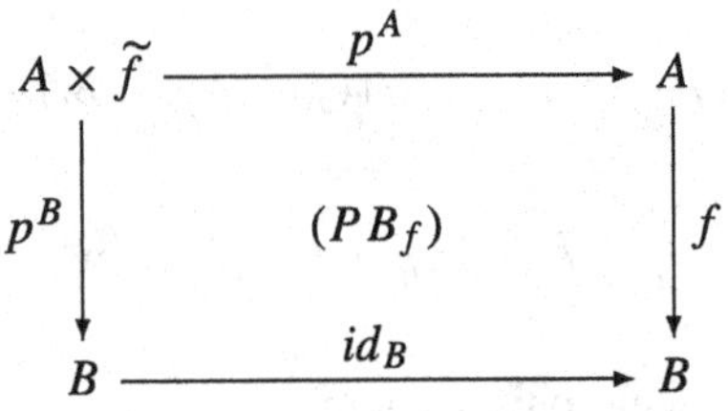

If the category is finitely complete and has the *exponentiation* as well than it is a Cartesian Closed Category (CCC). We say that a category has exponentiation if it has a product for any two objects, and if for any given objects B and C there is an object C^B and an arrow $eval_{B,C} : C^B \times B \to C$, called an *evolution* arrow, such that for any object A and arrow $f : A \times B \to C$, there is a unique arrow $\Lambda(f) : A \to C^B$ making the following diagram commute:

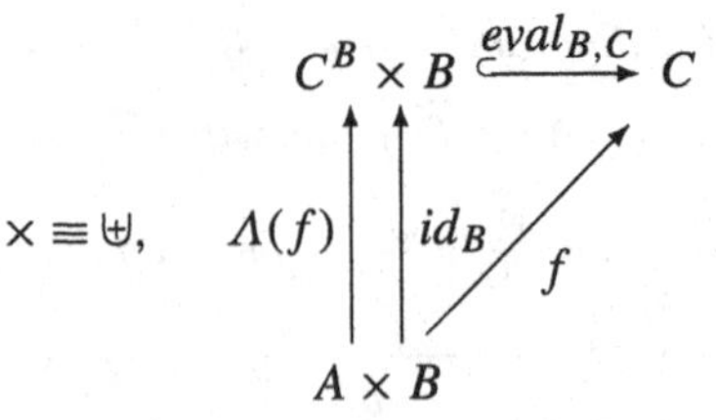

such that $eval_{B,C} \circ (\Lambda(f) \times id_B) = f$. The assignment of $\Lambda(f)$ to f establishes a bijection Λ from the arrows from $A \times B$ to C and the arrows from A to C^B.

It is well known that in a CCC with a zero object (i.e., when terminal and initial objects are isomorphic) such a category degenerates, i.e., *all* objects in such a category are isomorphic. The **DB** category has a zero object $\perp^0$ but does not degenerate:

this means that it cannot be a CCC. In what follows, we will demonstrate this fact by a demonstrating that it has no exponentiation.

Proposition 49 **DB** *is not a CCC.*

Proof Let C^B be the exponent object of the hom-set $\mathbf{DB}(C, B)$ which, as a set of the morphisms between C and D, cannot be an object of **DB** (differently from **Set** category). Hence, we have to construct this object $C^B \in \mathbf{DB}$ (to "internalize" the hom-set $\mathbf{DB}(C, B)$). For any two simple objects $B = \alpha^*(\mathcal{B})$ and $C = \alpha^*(\mathcal{C})$, the set of all arrows $\{f_1, f_2, \dots\} : B \to C$ where each simple arrow is $f_i = \alpha^*(\mathbf{M}^{(i)}_{BC})$, can be represented by a unique arrow $g = (\bigcup_{f_i \in \mathbf{DB}(B,C)} f_i) : B \to C$ (note that g is a unique ptp arrow of the complex arrow $\lceil f_1, f_2, \dots \rfloor : B \to C$ representing the set of all arrows from B into C), so we define the object C^B to be the information flux (from Definition 13) of this arrow

$$\begin{aligned} g &= \alpha^*\Big(\bigcup\big\{\mathbf{M}^{(i)}_{BC} = MakeOperads\big(\mathcal{M}^{(i)}_{BC}\big) \mid \mathcal{M}^{(i)}_{BC} : \mathcal{B} \to \mathcal{C}\big\}\Big) \\ &= \bigcup_{\mathcal{M}^{(i)}_{BC}:\mathcal{B}\to\mathcal{C}} \alpha^*\big(\mathbf{M}^{(i)}_{BC}\big) = \bigcup_{f_i \in \mathbf{DB}(B,C)} f_i. \end{aligned}$$

From point 2 of Definition 13, $\Delta(\alpha, \bigcup_{\mathcal{M}^{(i)}_{BC}:\mathcal{B}\to\mathcal{C}} \mathbf{M}^{(i)}_{BC}) = \bigcup_{\mathcal{M}^{(i)}_{BC}:\mathcal{B}\to\mathcal{C}} \Delta(\alpha, \mathbf{M}^{(i)}_{BC})$. Consequently,

$$\begin{aligned} \widetilde{g} &= Flux\Big(\alpha, \bigcup_{\mathcal{M}^{(i)}_{BC}:\mathcal{B}\to\mathcal{C}} \mathbf{M}^{(i)}_{BC}\Big) = T\Big(\Delta\Big(\alpha, \bigcup_{\mathcal{M}^{(i)}_{BC}:\mathcal{B}\to\mathcal{C}} \mathbf{M}^{(i)}_{BC}\Big)\Big) \\ &= T\Big(\bigcup_{\mathcal{M}^{(i)}_{BC}:\mathcal{B}\to\mathcal{C}} \Delta\big(\alpha, \mathbf{M}^{(i)}_{BC}\big)\Big) = T\Big(\bigcup_{\mathcal{M}^{(i)}_{BC}:\mathcal{B}\to\mathcal{C}} T\big(\Delta\big(\alpha, \mathbf{M}^{(i)}_{BC}\big)\big)\Big) \\ &= T\Big(\bigcup_{\mathcal{M}^{(i)}_{BC}:\mathcal{B}\to\mathcal{C}} Flux\big(\alpha, \mathbf{M}^{(i)}_{BC}\big)\Big) = T\Big(\bigcup_{f_i \in \mathbf{DB}(B,C)} \widetilde{f}_i\Big) \\ &= \bigoplus\big\{\widetilde{f}_i \mid f_i \in \mathbf{DB}(B, C)\big\}. \end{aligned}$$

Thus, we define the hom-object $C^B \triangleq \bigoplus\{\widetilde{f}_i \mid f_i \in \mathbf{DB}(B, C)\}$ which "internalize" the hom-sets (i.e., the *merging* of all closed objects obtained from a hom-set of arrows from B to C), i.e., the merging of all compact elements A such that $A \preceq B \otimes C$.

Hence, we obtain that

$$C^B = T\Big(\bigcup\big\{\widetilde{f}_i \mid f_i \in \mathbf{DB}(B, C)\big\}\Big)$$

(from Proposition 7)

$$= T\left(\bigcup\{\widetilde{f_i} \mid \widetilde{f_i} \subseteq B \otimes C\}\right)$$
$$= T\left(\bigcup\{\widetilde{f_i} \mid \widetilde{f_i} \preceq B \otimes C\}\right)$$

(from algebraic lattice property, Sect. 8.1.5)

$$= T(B \otimes C) = B \otimes C.$$

To have a Cartesian closed **DB** category (from Theorem 5 in Sect. 3.2.4, $\times \equiv \uplus$), for any complex arrow $f = [f_1, f_2] : A \uplus B \longrightarrow C$ in **DB**, we must have the exponential diagram commutativity, i.e., $f = eval_{B,C} \circ (\Lambda(f) \uplus id_B)$, where $\Lambda(f) : A \longrightarrow C^B$ is a simple derived morphism by "abstraction" Λ and the complex arrow $eval_{B,C} = [e_1, e_2] : C^B \times B \longrightarrow C$ is a universal "application" morphism. Thus, from the commutativity in **DB** of such an exponential diagram, we obtain $f_2 = e_2 : B \to C$ and $f_1 = e_1 \circ \Lambda(f) : A \to C$, thus $\widetilde{f_1} = \widetilde{e_1} \cap \widetilde{\Lambda(f)}$. However, $\widetilde{f_1} \subseteq TA \cap TC$ while $\widetilde{e_1} \cap \widetilde{\Lambda(f)} \subseteq (TA \cap TC) \cap TB$, so that there are the morphisms f_1 for which $\widetilde{f_1} \subset \widetilde{e_1} \cap \widetilde{\Lambda(f)}$ and hence the exponential diagram is not commutative. That is, **DB** has no exponentiation and hence is not a CCC. □

Remark From point 1 of Proposition 46 and the proof of this proposition, the "exponential" object C^B which "internalizes" in **DB** the hom-set of its morphisms **DB**(B, C) is equal to the information flux of the *principal morphism* from B into C.

Thus, from the fact that **DB** is not a CCC, we deduce that it is not a standard topos. But it still can be a kind of a *weak topos*, if we substitute the exponentiation with a kind of weak exponentiation (by replacing a (strong) product with a kind of (weak) tensorial product) and define a proper *subobject classifier* in **DB** (which has to be a kind of special pullback diagram, specified in dedicated Sect. 9.1.2).

8.1.4 Universal Algebra Considerations

In what follows, we will consider the signature Σ_R of the SPRJU Codd's relational algebra (in Definition 31, Sect. 5.1) which is used to define the views in the SOtgd implications for the schema database mappings. In order to explore the universal algebra properties of the category **DB** [18], where a morphism is not a function but a nonempty *set* of functions (this fact significatively complicates a definition of the mappings from **DB**-morphisms into the homomorphisms of the category of Σ_R-algebras), we will use an, equivalent to **DB**, a "sets-like" category $\mathbf{DB}_{sk}$ such that its arrows can be seen as total functions.

Proposition 50 *Let us denote the full skeletal subcategory of* **DB** *composed of closed objects only by* $\mathbf{DB}_{sk}$.

Such a category is equivalent to the category **DB**, *i.e., there exist an adjunction of a surjective functor* $T_{sk} : \mathbf{DB} \longrightarrow \mathbf{DB}_{sk}$ *and an inclusion functor* $In_{sk} : \mathbf{DB}_{sk} \longrightarrow \mathbf{DB}$ *such that* $T_{sk} In_{sk} = Id_{DB_{sk}}$ *and* $In_{sk} T_{sk} \simeq Id_{DB}$.

Proof Let us define $T^0_{sk} = T^0$ and $T^1_{sk} = T^1$, while In^0_{sk} and In^1_{sk} are two identity functions. It is easy to verify that these two categories are equivalent. In fact, there exists an adjunction $(T_{sk}, In_{sk}, \eta_{sk}, \varepsilon_{sk}) : \mathbf{DB} \longrightarrow \mathbf{DB}_{sk}$ based on the bijection $\sigma : \mathbf{DB}_{sk}(T_{sk}A, B) \to \mathbf{DB}(A, In_{sk}B)$ which is natural in simple objects $A \in \mathbf{DB}$ and $B \in \mathbf{DB}_{sk}$ (thus, B is closed, i.e., $B = TB$). In fact, for each $f \in \mathbf{DB}_{sk}(T_{sk}A, B)$, which is an arrow $f : TA \to B$ (from $T_{sk}A = TA$), we define $\sigma(f) = f \circ is_A : A \to B$; conversely, for any $g \in \mathbf{DB}(A, In_{sk}B)$, which is an arrow $g : A \to B$ (from $In_{sk}B = B$), we define $\sigma^{-1}(g) = g \circ is_A^{-1}$. Thus, $\sigma^{-1}(\sigma(f)) = \sigma(f) \circ is_A^{-1} = f \circ is_A \circ is_A^{-1} = f \circ id_{TA} = f$, and $\sigma(\sigma^{-1}(g)) = \sigma(g \circ is_A^{-1}) = g \circ is_A^{-1} \circ is_A = g \circ id_A = g$. □

In a given inductive definition, one defines a value of a function on all (algebraic) constructors (i.e., relational operators), in our example the endofunctor T as explained in Sect. 5.1.1, by the initial relational $(X \uplus \Sigma_R)$-algebra semantics with the syntax monad $(\mathcal{T}_P, \eta, \mu)$. It is provided for a given assignment (R-algebra) $\alpha : X \to TA$, where X is a set of relational variables (symbols) in a database schema $\mathcal{A}$, such that $A = \alpha^*(\mathcal{A})$. This inductive definition is expressed by the following commutative diagram in *Set*:

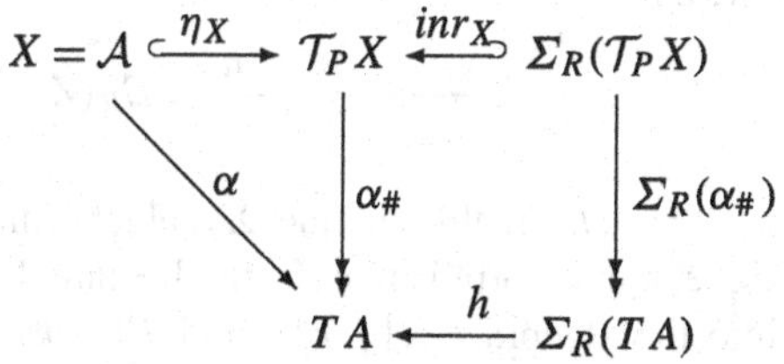

As we see, the *induction* for each schema $\mathcal{A}$ and R-algebra α, which is an interpretation of $\mathcal{A}$, defines the *closed objects* in **DB**.

The following is based on the fundamental results of the Universal algebra [4].

Based on Definition 31 of the Cod's SPRJU relational algebra, Σ_R is a finitary signature (in the usual algebraic sense: a collection Σ_R of function *symbols* with a function $ar : \Sigma_R \longrightarrow \mathcal{N}$ giving the finite arity of each function symbol) for a single-sorted (sort of relations) relational algebra. Here we will recall some Universal algebra concepts from Sect. 1.2:

We can speak of Σ_R-equations E and their satisfaction in a Σ_R-algebra, obtaining the notion of a (Σ_R, E)-algebra theory. In a special case when E is empty, we obtain a pure syntax version of Universal algebra, where $\mathbf{K}$ is a category of all Σ_R-algebras and the quotient-term algebras are simply term algebras $\mathcal{T}_P X$ for a given set of relational variables X.

An *algebra* for the algebraic theory (or type) (Σ_R, E) is given by a set X, called the *carrier* of the algebra, together with interpretations for each of the function symbols in Σ_R. A function symbol $o_i \in \Sigma_R$ of arity k must be interpreted by a function $\widehat{o_i} : X^k \longrightarrow X$. Given this, a term containing n distinct variables gives rise to a function $X^n \longrightarrow X$ defined by induction on the structure of the term. An algebra must also satisfy the equations given in E in the sense that equal *terms* give rise to identical functions (with obvious adjustments where the equated terms do not contain exactly the same variables). A *homomorphism* of algebras from an algebra X to an algebra Y is given by a function $g : X \longrightarrow Y$ which commutes with operations of the algebra $g(\widehat{o_i}(x_1, \dots, x_k)) = \widehat{o_i}(g(x_1), \dots, g(x_k))$.

This generates a variety category **K** of all relational algebras. Consequently, there is a bifunctor $E : \mathbf{DB}_{sk}^{OP} \times \mathbf{K} \longrightarrow \mathbf{Set}$ such that for any *simple* database instance A in $\mathbf{DB}_{sk}$ there exists the functor $E(A, \_) : \mathbf{K} \longrightarrow \mathbf{Set}$ with an universal element $(U(A), \varrho)$, where $\varrho \in E(A, U(A))$, i.e., $\varrho : A \longrightarrow U(A)$ is an inclusion function. $U(A)$ is a free algebra over A (i.e., the quotient-term algebra generated by a carrier database instance A) such that for any function $f \in E(A, Z)$, $f : A \rightarrow Z$, there is a unique homomorphism k (in **K**) from the free algebra $U(A)$ into an algebra Z, with $f = E(A, k) \circ \varrho$, represented by the following initial algebra semantics diagram in **Set**:

$$
\begin{array}{ccccc}
A & \xrightarrow{\varrho = inl_A} & U(A) & \xleftarrow{inr_A} & \Sigma_R(U(A)) \\
 & \searrow{\scriptstyle f \in E(A,Z)} & \downarrow{\scriptstyle E(A,k) = f_\#} & & \downarrow{\scriptstyle \Sigma_R(f_\#)} \\
 & & Z & \xleftarrow{h} & \Sigma_R(Z)
\end{array}
$$

where $k : (U(A), inr_A) \rightarrow (Z, h)$ is this unique Σ_R-algebra homomorphism corresponding to the unique homomorphism k from the initial $(A \uplus \Sigma_R)$-algebra $(U(A), [inl_A, inr_A])$ into the algebra $(Z, [f, k])$, and $E(A, k) = f_\#$ is a function equal to k but without the structural properties of this homomorphism (to commute with operations of the algebra).

Consequently, $U(A)$ is a quotient-term algebra, where a carrier is a set of equivalence classes of closed terms of a well defined formulae of a relational algebra, "constructed" by Σ_R-constructors (relational operators in SPJRU algebra: select, project, join and union) and elements (relations) of a database instance A.

From the so-called "parameter-theorem" we obtain that there exist:

- A unique universal functor $U : \mathbf{DB}_{sk} \longrightarrow \mathbf{K}$ such that for any simple database instance A in $\mathbf{DB}_{sk}$ it returns the free Σ_R-algebra $U(A)$. It is a quotient-term algebra where a carrier is a set $U(A)$ of the equivalence classes $[t_i]$ of closed terms $t_i \in \mathcal{T}_P A$ (in Definition 31) of well defined formulae of a relational algebra, "constructed" by Σ_R-constructors (i.e., the relational operators $o_i \in \Sigma_R$ in SPJRU algebra: select, project, join and union) and relational symbols of a database instance A. That is, $U(A) = \{[t_i] \mid t_i \in \mathcal{T}_P A\}$. An alternative for $U(A)$ is given by considering A as a set of variables rather than a set of constants. In either case, each k-ary operation $o_i \in \Sigma_R$, with the function $\widehat{o_i} : U(A)^k \rightarrow U(A)$ (i.e., a component of $inr_A : \Sigma_R(U(A)) \rightarrow U(A)$ where $\Sigma_R(U(A)) = \biguplus_{o_i \in \Sigma_R} U(A)^{ar(o_i)}$, so that $inr_A = \biguplus_{o_i \in \Sigma_R} \widehat{o_i}$ is the interpretation of all operations in the signature Σ_R) and hence is interpreted syntactically $\widehat{o_i}([t_1], \dots, [t_k]) = [o_i(t_1, \dots, t_k)]$, where, as usual, the square brackets denote the equivalence classes for terms t_i).

This universal functor U, for any simple morphism $f : A \longrightarrow B$ in $\mathbf{DB}_{sk}$ returns the homomorphism $f_H = U^1(f)$ from the Σ_R-algebra $U(A)$ into the Σ_R-algebra $U(B)$ such that for any term $o_i([t_1], \dots, [t_n]) \in U(A)$, $o_i \in \Sigma_R$, we obtain $f_H(o_i([t_1], \dots, [t_n])) = o_i(f_H([t_1]), \dots, f_H([t_n])) \in U(B)$.

- Its adjoint forgetful functor $F : \mathbf{K} \longrightarrow \mathbf{DB}_{sk}$ such that, for any free algebra $U(A)$ in **K**, the object $F(U(A))$ in $\mathbf{DB}_{sk}$ is equal to its carrier-set A (each term $o_i([t_1], \dots, [t]_n) \in U(A)$ is evaluated into a relation that is an element of this closed object A) and for each arrow $U^1(f)$, $F^1(U^1(f)) = f$, i.e., $FU = Id_{DB_{sk}}$ and $UF = Id_K$.

Remark In this universal algebra framework, we preferred to use the skeletal category $\mathbf{DB}_{sk}$ (which will be used for the algebraic considerations in what follows), instead of the complete category **DB**, equivalent to it. Note that if we use the **DB** then the unique universal functor $U : \mathbf{DB} \to \mathbf{K}$ is equal to the previous one, and the adjoint forgetful functor $F : \mathbf{K} \to \mathbf{DB}$ is defined by $F(U(A)) = TA$ (when A is a closed object then $A = TA$), with $F^1(U(f)) = Tf$ and hence TA can be substituted by A because they are isomorphic ($A \simeq TA$ in **DB**), and $Tf : TA \to TB$ is equivalent to $f : A \to B$ (i.e., $f \approx Tf$) because they have the same information fluxes, $\widetilde{f} = \widetilde{Tf}$.

It is immediate from the universal property that the mapping $A \mapsto U(A)$ extends to the endofunctor $F \circ U : \mathbf{DB}_{sk} \longrightarrow \mathbf{DB}_{sk}$. This functor carries a *monad structure* $(F \circ U, \eta, \mu)$ with $F \circ U$ an equivalent version of T but for this skeletal database category $\mathbf{DB}_{sk}$ (while for the equivalent version for **DB** we obtain the monad T, as demonstrated in Proposition 58 in Sect. 8.3). The natural transformation η is given by the obvious "inclusion" of A into $(F \circ U)(A) = F(U(A)) : R \longrightarrow [R]$ (each view (relation) R in a closed object A is an equivalence class of all algebra terms which produce this view). The natural transformation η is the unit of this adjunction of U and F, and hence it corresponds to an injective function $\varrho : A \longrightarrow U(A)$ in **Set**, given previously. The interpretation of μ is almost equally simple. An element of $(F \circ U)^2(A)$ is an equivalence class of terms built up from elements of $(F \circ U)(A)$, so that instead of $t(x_1, \dots, x_k)$, a typical element of $(F \circ U)^2(A)$ is given by the equivalence class of a term $t([t_1], \dots, [t_k])$. The transformation μ is defined by 'flattening-terms' map $[t([t_1], \dots, [t_k])] \mapsto [t(t_1, \dots, t_k)]$. This makes sense because a substitution of provably equal expressions into the same term results in provably equal terms.

The monadic properties of the endofunctor $T = F \circ U$ will be explored deeper in Sect. 8.3, after examination of the algebraic lattice, lfp (locally finite presentable properties of **DB** category and its enrichments.

Remark Consider a generalization to complex objects $A = \biguplus_{1 \le j \le m} A_j$, $m \ge 2$. A bifunctor $E : \mathbf{DB}_{sk}^{OP} \times \mathbf{K} \longrightarrow \mathbf{Set}$ for any *complex* database instance A in $\mathbf{DB}_{sk}$ generates a functor $E(A, \_) : \mathbf{K} \longrightarrow \mathbf{Set}$ with a universal element $(U(A), \varrho)$, where $U(A) = \biguplus_{1\leq j\leq m} U(A_j)$ and $\varrho = (\biguplus_{1\leq j\leq m} \varrho_j) \in E(A, U(A))$, i.e., $\varrho : \biguplus_{1\leq j\leq m} A_j \longrightarrow \biguplus_{1\leq j\leq m} U(A_j)$ is an inclusion function and $U(A)$ is a free algebra over A (i.e., a quotient-term algebra generated by a carrier database instance A). It is a disjoint union of the free algebras $U(A_j)$ over A_j, $j = 1, \dots, m$, such that for any function $f \in E(A, Z)$, $f : A \to Z$, with a set of point-to-point functions $f_{ji} : A_j \to Z_i \in \frac{f}{\uplus}$ for a complex object $Z = \biguplus_{1\leq j\leq k} Z_i, k \geq 1$, for each f_{ji} there is a unique homomorphism k_{ji} from the free algebra $U(A_j)$ into an algebra Z_i, with $f_{ji} = E(A_j, k) \circ \varrho_j$, represented by the following initial algebra semantics diagram in **Set**:

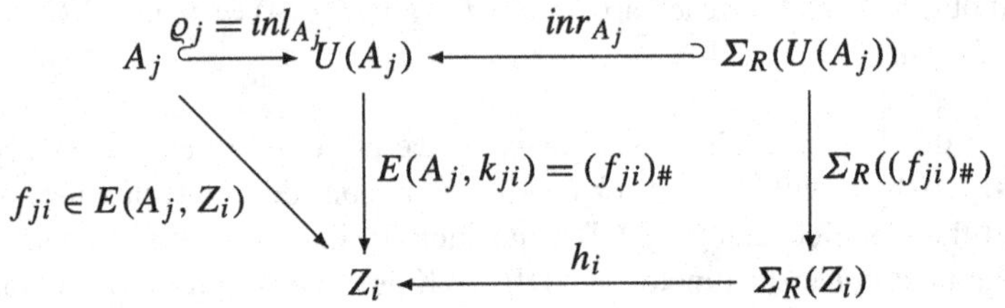

The set of such unique point-to point homomorphisms $k_{ji} : (U(A_j), inr_{A_j}) \to (Z_i, h_i)$, for each $f_{ji} \in \frac{f}{\uplus}$ define $\frac{k}{\uplus} = \{k_{ji} \mid f_{ji} \in \frac{f}{\uplus}\}$, that is, a unique Σ_R-algebra homomorphism $k : (U(A), inr_A) \to (Z, h)$ which corresponds to the unique homomorphism k from the initial $(A \uplus \Sigma_R)$-algebra $(U(A), [inl_A, inr_A])$ into the algebra $(Z, [f, k])$. Thus, $E(A, k) = f_\#$ is a function equal to k but without the structural properties of the homomorphism (to commute with operations of the algebra).

Let us define now the faithful forgetful functor from **DB** into **Set** category:

Theorem 14 *There exists a faithful forgetful functor* $F_{sk} = (F^0_{sk}, F^1_{sk}) : \mathbf{DB}_{sk} \longrightarrow$ **Set** *such that its object-component* F^0_{sk} *is equal to the powerset operator* $\mathcal{P}$, *while the arrow-component* F^1_{sk}, *for a given morphism* $f : A \to B$, *where* $A = \biguplus_{1\leq j\leq m} A_j$, $m \geq 1$, *and* $B = \biguplus_{1\leq i\leq k} B_i$, $k \geq 1$, *is defined as follows: for any set* $S \in F^0_{sk}(A) = \mathcal{P}(\biguplus_{1\leq j\leq m} A_j)$,

$$F^1_{sk}(f)(S)$$

$$\triangleq \begin{cases} \{R \mid R \in S \cap \widetilde{f_1}, (f_1 : A \to B) \in \frac{f}{\uplus}\} & \textit{if } m = k = 1; \\ \{R \mid (j, R) \in S \textit{ and } R \in \widetilde{f_{j1}}, (f_{j1} : A_j \to B_1) \in \frac{f}{\uplus})\} & \textit{if } m \geq 2, k = 1; \\ \{(i, R) \mid R \in S \cap \widetilde{f_{1i}}, (f_{1i} : A_1 \to B_i) \in \frac{f}{\uplus}\} & \textit{if } m = 1, k \geq 2; \\ \{(i, R) \mid (j, R) \in S, R \in \widetilde{f_{ji}}, (f_{ji} : A_j \to B_i) \in \frac{f}{\uplus}\} & \textit{otherwise.} \end{cases}$$

We define the faithful forgetful functor $F_{DB} = F_{sk} \circ T : \mathbf{DB} \longrightarrow$ **Set** *as well.*
Consequently, $\mathbf{DB}_{sk}$ *and* **DB** *are concrete categories.*

Proof The skeletal category $\mathbf{DB}_{sk}$ has the closed objects only, and hence for any two closed objects (that are sets of relations) $A = TA$ and $B = TB$, each *simple* arrow in **DB** between them $f : TA \longrightarrow TB$ (where A and B are simple databases) can be expressed in the **Set** category by the following (total) function from the sets TA to TB such that for any relation $R \in TA$, $f_T(R) = F_{sk}(f)(R) = R$ if $R \in \widetilde{f}$, $\bot$ otherwise. An analogous property is valid for its reversed equivalent morphism $f_T^{OP} : TB \longrightarrow TA$ in **DB**, too.

Let us demonstrate that the functor $F_{sk} : \mathbf{DB}_{sk} \longrightarrow \mathbf{Set}$ is well defined (we recall that any object A in $\mathbf{DB}_{sk}$ is closed, i.e., $A = TA$):

1. For any identity morphism in **DB**, we have the following two possible cases:
   1.1. $m = k = 1$ (case 1 of the definition of F_{sk}). Thus, for $f = id_A : TA \rightarrow TA$ ($A = TA$), with $\frac{id_A}{\uplus} = \{id_A\}$ and $\widetilde{id_A} = TA$, for any $S \in \mathcal{P}(A) = \mathcal{P}(TA)$ (thus, $S \subseteq TA$), $F_{sk}(id_A)(S) = \{R \mid R \in S \cap \widetilde{f_l}, f_l : A \rightarrow B \in \frac{id_A}{\uplus}\} = \{R \mid R \in S \cap \widetilde{id_A}\} = \{R \mid R \in S \cap TA\} = \{R \mid R \in S\} = S$ and hence $F_{sk}(id_A) = id_{\mathcal{P}(A)}$ is the identity function for the powerset set $\mathcal{P}(A)$ in **Set**.
   1.2. $m = k \geq 2$ (case 4 of the definition of F_{sk}). Thus, for
   $$f = id_A = \biguplus_{1 \leq j \leq m} id_{TA_j} : TA_j \rightarrow TA_j \quad with\ \widetilde{id_{TA_j}} = TA_j$$
   (hence $\frac{id_A}{\uplus} = \{id_{A_1}, \ldots, id_{A_m}\}$), for any $S \in \mathcal{P}(\biguplus_{1 \leq j \leq m} A_j)$,
   $$\begin{aligned} F_{sk}(id_A)(S) &= \left\{(i, R) \mid (j, R) \in S, R \in \widetilde{f_l}, f_l : A_j \rightarrow B_i \in \frac{id_A}{\uplus}\right\} \\ &= \left\{(j, R) \mid (j, R) \in S, R \in \widetilde{id_{A_j}}, id_{A_j} : A_j \rightarrow A_j \in \frac{id_A}{\uplus}\right\} \\ &= \{(j, R) \mid (j, R) \in S, R \in TA_j = A_j\} = S, \end{aligned}$$
   and hence $F_{sk}(id_A) = id_{\mathcal{P}(A)}$ is the identity function for the powerset set $\mathcal{P}(A)$ in **Set**.
2. For any $g : B \rightarrow C$ in $\mathbf{DB}_{sk}$, with $C = \biguplus_{1 \leq l \leq n} C_l$, $n \geq 1$, and a composition $g \circ f : A \rightarrow C$, let us show that $F_{sk}(g \circ f)(S) = F_{sk}(g)(F_{sk}(f)(S))$ for any $S \in \mathcal{P}(A)$. In the case when $n = 1$, it can be directly reduced to one of the cases 1 and 2 of the definition of $F_{sk}(f)$ above. Thus, let us show this property for two most complex cases:
   2.1. $m = 1, k, n \geq 2$. Let
   $$S_1 = F_{sk}(f)(S) = \left\{(i, R) \mid R \in S \cap \widetilde{f_{1i}}, (f_{1i} : A_1 \rightarrow B_i) \in \frac{f}{\uplus}\right\}.$$
   Then, $F_{sk}(g)(F_{sk}(f)(S)) = F_{sk}(g)(S_1) = \{(l, R) \mid (i, R) \in S_1, R \in \widetilde{g_{il}}, (g_{il} : B_i \rightarrow C_l) \in \frac{f}{\uplus}\} = \{(l, R) \mid R \in S \cap \widetilde{f_{1i}}, (f_{1i} : A_1 \rightarrow B_i) \in \frac{f}{\uplus}, R \in \widetilde{g_{il}}, (g_{il} : B_i \rightarrow C_l) \in \frac{f}{\uplus}\} = \{(l, R) \mid R \in S \cap \widetilde{f_{1i}} \cap \widetilde{g_{il}}, (g_{il} \circ f_{1i} : A_1 \rightarrow C_l) \in \frac{g \circ f}{\uplus}\} = \{(l, R) \mid R \in S \cap \widetilde{g_{il} \circ f_{1i}}, (g_{il} \circ f_{1i} : A_1 \rightarrow C_l) \in \frac{g \circ f}{\uplus}\} = F_{sk}(g \circ f)(S)$.
   2.2. $m, k, n \geq 2$. Let
   $$S_1 = F_{sk}(f)(S) = \left\{(i, R) \mid (j, R) \in S, R \in \widetilde{f_{ji}}, (f_{ji} : A_j \rightarrow B_i) \in \frac{f}{\uplus}\right\}.$$

Then, $F_{sk}(g)(F_{sk}(f)(S)) = F_{sk}(g)(S_1) = \{(l, R) \mid (i, R) \in S_1, R \in \widetilde{g_{il}}, (g_{il} : B_i \to C_l) \in \frac{f}{\uplus}\} = \{(l, R) \mid (j, R) \in S, R \in \widetilde{f_{ji}}, (f_{ji} : A_j \to B_i) \in \frac{f}{\uplus}, R \in \widetilde{g_{il}}, (g_{il} : B_i \to C_l) \in \frac{f}{\uplus}\} = \{(l, R) \mid (j, R) \in S, R \in \widetilde{f_{ji}} \cap \widetilde{g_{il}}, (g_{il} \circ f_{ji} : A_j \to C_l) \in \frac{g \circ f}{\uplus}\} = \{(l, R) \mid (j, R) \in S, R \in \widetilde{g_{il} \circ f_{ji}}, (g_{il} \circ f_{ji} : A_j \to C_l) \in \frac{g \circ f}{\uplus}\} = F_{sk}(g \circ f)(S)$.

Consequently, the compositional property of this functor is valid for the arrows, $F_{sk}(g \circ f) = F_{sk}(g) \circ F_{sk}(f)$.

3. Let us show that $F^1_{sk}(f) = F^1_{sk}(h)$ implies $f = h$, i.e., F_{sk} is a faithful functor, as follows:
   3.1. $m = k = 1$. Thus $f = \lceil f_1, \ldots, f_n \rfloor : A_1 \to B_1, n \geq 1$ and $g = \lceil g_1, \ldots, g_{n_1} \rfloor : A_1 \to B_1, n_1 \geq 1$. Then $\frac{f}{\uplus}$ is a singleton set composed of a unique ptp arrow $f_{11} : A_1 \to B_1$ which is obtained by fusing (Lemma 7) all arrows $f_l, l = 1, \ldots, n$, and, analogously, $\frac{g}{\uplus}$ is a singleton set composed of a unique ptp arrow $g_{11} : A_1 \to B_1$, obtained by fusing all arrows $g_l, l = 1, \ldots, n_1$. They are defined by their information fluxes, $\widetilde{f_{11}} = \{R \mid R \in A_1, F_{sk}(f)(\{R\}) \neq \emptyset\} = \{R \mid R \in A_1, F_{sk}(g)(\{R\}) \neq \emptyset\} = \widetilde{g_{11}}$, and hence from Definition 23, $f_{11} = g_{11}$, thus $\frac{f}{\uplus} = \{f_{11}\} = \{g_{11}\} = \frac{g}{\uplus}$ and hence $f = g$.
   3.2. $m \geq 2, k = 1$. Thus each ptp arrow $f_{j1} : A_j \to B_1 \in \frac{f}{\uplus}$ is defined by its information flux, $\widetilde{f_{j1}} = \{R \mid R \in A_j, F_{sk}(f)(\{(j, R)\}) \neq \emptyset\} = \{R \mid R \in A_j, F_{sk}(g)(\{(j, R)\}) \neq \emptyset\} = \widetilde{g_{j1}}$, and hence from Definition 23, $f_{j1} = g_{j1} : A_j \to B_1$. It holds for all ptp arrows with $1 \leq j \leq m$, thus $\frac{f}{\uplus} = \frac{g}{\uplus}$ and hence $f = g$.
   3.3. $m = 1, k \geq 2$. Thus each ptp arrow $f_{1i} : A_1 \to B_i \in \frac{f}{\uplus}$ is defined by its information flux, $\widetilde{f_{1i}} = \{R \mid R \in A_1, (i, R) \in F_{sk}(f)(\{R\})\} = \{R \mid R \in A_1, (i, R) \in F_{sk}(g)(\{R\})\} = \widetilde{g_{1i}}$, and hence from Definition 23, $f_{1i} = g_{1i} : A_1 \to B_i$. It holds for all ptp arrows with $1 \leq i \leq k$, thus $\frac{f}{\uplus} = \frac{g}{\uplus}$ and hence $f = g$.
   3.4. $m, k \geq 2$. Thus each ptp arrow $f_{ji} : A_j \to B_i \in \frac{f}{\uplus}$ is defined by its information flux, $\widetilde{f_{ji}} = \{R \mid R \in A_j, (i, R) \in F_{sk}(f)(\{(j, R)\})\} = \{R \mid R \in A_j, (i, R) \in F_{sk}(g)(\{(j, R)\})\} = \widetilde{g_{ji}}$, and hence from Definition 23, $f_{ji} = g_{ji} : A_j \to B_i$. It holds for all ptp arrows with $1 \leq j \leq m$ and $1 \leq i \leq k$, thus $\frac{f}{\uplus} = \frac{g}{\uplus}$ and hence $f = g$.

   Consequently, also $F_{DB} = F_{sk} \circ T$ is a faithful functor. Thus, $\mathbf{DB}_{sk}$ and $\mathbf{DB}$ are concrete categories. □

8.1.5 Algebraic Database Lattice

This section is an extension of Sect. 3.2.5 dedicated to the partial ordering '$\preceq$' for databases with an introduction of the PO subcategory $\mathbf{DB}_I$.

We recall that the symbol '$=$' is used to express the fact that both sides name the same objects, whereas '$\approx$' is used to build equations which may or may not be true for particular elements. Note that in the case of equations between arrows in the **DB** category (i.e., the commutative diagrams) we use the symbol '$=$' instead of '$\approx$', and '$=$' is used for set equality relation as well.

We recall (see [4]) the fact that a complete lattice $\mathcal{C}$ is a poset P such that for any subset S both $\inf(S)$ (greatest lower bound) and $\sup(S)$ (least upper bound) exist in P (with $0 = \inf(P)$, $1 = \sup(P)$): for any $A, B, C \in \mathcal{C}$, and binary operations "join" $\bigvee$ and "meet" $\bigwedge$ (in the case of the **Set** category, these operators are the set-union and set-intersection, respectively, while for the **DB** category we will show that they are merging and matching operators, respectively), the following identities are satisfied (see the first 5 equations (laws) for complete lattices in Sect. 1.2)

1. $A \vee A \approx A,\ A \wedge A \approx A$; (idempotency)
2. $A \vee B \approx B \vee A,\ A \wedge B \approx B \wedge A$; (commutativity)
3. $A \vee (B \vee C) \approx (A \vee B) \vee A,\ A \wedge (B \wedge C) \approx (A \wedge B) \wedge C$; (associativity)
4. $A \vee 0 \approx A,\ A \wedge 1 \approx A$;
5. $A \approx A \vee (A \wedge B),\ A \approx A \wedge (A \vee B)$. (absorption)

Let us denote the restriction of the closure endofunctor $T : \mathbf{DB} \longrightarrow \mathbf{DB}$ to PO category $\mathbf{DB}_I$ by $T_I : \mathbf{DB}_I \longrightarrow \mathbf{DB}_I$. We have seen in Sect. 3.2.5 that $\mathbf{DB}_I$ is a PO category where each arrow $in : A \hookrightarrow B$ is a monomorphism with $A \preceq B$ (as defined by Theorem 6 in Sect. 3.2.5). We recall that, by Definition 28 and Corollary 14 in Sect. 3.4.1, for the behavioral equivalence '$\approx$', $A \approx B$ iff ($A \preceq B$ and $B \preceq A$), and $A \simeq B$ implies $A \approx B$.

Thus, we obtained a partial order $(Ob_{DB}, \preceq)$. Let us show that it is a lattice ordered set, i.e., that every pair of objects $A, B \in Ob_{DB}$ has the least-upper-bound (sup) and the greatest-lower-bound (inf).

First of all, if we consider the subset of only simple objects then $A \preceq B$ is equal to $TA \subseteq TB$, $A \otimes B = TA \cap TB$ and $A \oplus B = T(A \cup B) = T(TA \cup TB)$. Hence, the matching and merging operators correspond to the meet and join lattice operators. For the complex objects we can continue to use the matching operator as meet operation, but we have to generalize the "merging with A" to complex objects as well, in the way that it can be used as a join operation:

Definition 57 Let $A = \biguplus_{1 \le j \le m} A_j, m \ge 1$, and $B = \biguplus_{1 \le i \le k} B_i, k \ge 1$, be two complex objects, then we define the following sets:

$$S_A = \{A_j \mid \exists B_i.A_j \succeq B_i, 1 \le j \le m \ne 1, 1 \le i \le k \ne 1\},$$

$$S_B = \{B_i \mid \exists A_j.A_j \preceq B_i, 1 \le j \le m \ne 1, 1 \le i \le k \ne 1\},$$

$$S_{AB} = \{A_j \oplus B_i \mid A_j \notin S_A, B_i \notin S_B, 1 \le j \le m, 1 \le i \le k\}.$$

We define the generalized merging operator '$\widehat{\oplus}$' as follows:

$$A \widehat{\oplus} B \equiv \widehat{\oplus}(A, B) \triangleq \begin{cases} A_1 \oplus B_1 = T(A_1 \cup B_1) \in S_{AB} & \text{if } m = k = 1; \\ \biguplus(S_A \cup S_B \cup S_{AB}) & \text{otherwise.} \end{cases}$$

We recall (see the remark after Definition 21) that for a given set of sets $S = \{S_1, \ldots, S_n\}$, $\biguplus S$ represents the disjoint union of sets in S by taking *any ordering* of them (because they are equal up to isomorphism, that is, $\biguplus S \simeq (S_1 \uplus S_2 \uplus \cdots \uplus S_n)$).

We have the following interesting cases:
1. If $S_A = A$ then $S_{AB} = \emptyset$ (empty set), and hence $A \preceq B$;
2. If $S_B = B$ then $S_{AB} = \emptyset$, and hence $A \succeq B$;
3. If $S_A = A$ and $S_B = B$ then $S_{AB} = \emptyset$, and hence $A \approx B$;
4. If $S_A = S_B = \emptyset$ then S_{AB} is the disjoint union of all possible mergings of the simple objects (i.e., databases) in A and B, with $A \npreceq B$ and $B \npreceq A$.

We have the following important properties for this generalization of the "merging with A" operator '$\widehat{\oplus}$':

Lemma 17 *The following properties for any two objects A and B in* **DB** *are valid*:
1. (*Commutativity*) $A\widehat{\oplus}B \simeq B\widehat{\oplus}A$, *thus*, $A\widehat{\oplus}B \approx B\widehat{\oplus}A$;
2. (*Generalized merging*) *If A is a simple object, then* $A\widehat{\oplus}B \simeq A \oplus B$. *Thus*, $A\widehat{\oplus}B \approx A \oplus B$;
3. $A\widehat{\oplus}B \succeq A$ *and* $A\widehat{\oplus}B \succeq B$;
4. *If* $A \preceq B$ *then* $A\widehat{\oplus}B \approx B$.

Proof Claim 1. From $A\widehat{\oplus}B = \biguplus(S_A \cup S_B \cup S_{AB})$ we obtain the isomorphism (from the commutativity of $\uplus$), $A\widehat{\oplus}B \simeq B\widehat{\oplus}A$ and hence $A\widehat{\oplus}B \approx B\widehat{\oplus}A$ (from Corollary 14 any two isomorphic objects are behaviorally equivalent), which represents the commutativity property of the generalized merging operator.

Claim 2. It is a generalization of the "merging with A" operator in Theorem 13, from the fact that when $m = 1$ (i.e., A is a simple object), we obtain $S_A = S_B = \emptyset$ and hence $A\widehat{\oplus}B = \biguplus S_{AB} = \biguplus\{A_1 \oplus B_i \mid 1 \leq i \leq k\} \simeq \biguplus_{1 \leq i \leq k} A_1 \oplus B_i =$ (from Theorem 13) $= A \oplus B$. Thus, $A\widehat{\oplus}B \approx A \oplus B$.

Claim 3. For each $1 \leq j \leq m$, the simple object A_j is in S_A or in S_{AB} in form $A_j \otimes B_i \supseteq TA_j$ and hence $A\widehat{\oplus}B \succeq A$. Analogously, for each $1 \leq i \leq k$, the simple object B_i is in S_B or in S_{AB} in form $A_j \otimes B_i \supseteq TB_i$, so that $A\widehat{\oplus}B \succeq B$.

Claim 4. From the fact that $A\widehat{\oplus}B \succeq B$, let us show that if $A \preceq B$ (with a mapping $\sigma : \{1, \ldots, m\} \to \{1, \ldots, k\}$ such that $\forall_{1 \leq j \leq m}.TA_j \subseteq TB_{\sigma(j)}$) then $A\widehat{\oplus}B \preceq B$ as well. In fact, if $A \preceq B$ then $S_B = B$ and $S_{AB} = \emptyset$. Thus, the elements in the disjoint union of $A\widehat{\oplus}B$ are all B_i, with $TB_i \subseteq TB_i$, $1 \leq i \leq k$, or some A_j, so that $TA_j \subseteq TB_{\sigma(j)}$ and hence $A\widehat{\oplus}B \preceq B$. □

We are now ready to introduce the following database lattice L_{DB}:

Proposition 51 *The set Ob_{DB} of all database instances (objects) of* **DB**, *both with the generalized merging and matching tensor products* $\widehat{\oplus}$ *and* $\otimes$ *(read "join" and "meet", respectively) is a lattice with partial ordering '$\preceq$' (introduced by Definition* 19).

This lattice $L_{DB} = (Ob_{DB}, \preceq, \otimes, \widehat{\oplus})$ *is a complete lattice with the top and bottom objects* Υ *and* $\bot^0$, *respectively.*

Proof First of all, if $A \preceq B$ with a mapping $\sigma : \{1, \ldots, m\} \to \{1, \ldots, k\}$ such that $\forall_{1 \leq j \leq m}.TA_j \subseteq TB_{\sigma(j)}$, we obtain, from points 3 and 4 of Lemma 17, that the generalized merging '$\widehat{\oplus}$' is equal to the join operator of this lattice w.r.t. the ordering '$\preceq$'. Let us show that the matching operator $\otimes$ is the meet operator of this lattice. In fact, from Theorem 12, $A \otimes B \preceq A$ and $A \otimes B \preceq B$ (for each disjoint component $A_j \otimes B_i$ in $A \otimes B$, $A_j \otimes B_i = TA_j \cap TB_i \subseteq TA_j$ and $A_j \otimes B_i = TA_j \cap TB_i \subseteq TB_i$). If $A \preceq B$ with a mapping $\sigma : \{1, \dots, m\} \to \{1, \dots, k\}$ such that $\forall_{1 \leq j \leq m}.TA_j \subseteq TB_{\sigma(j)}$, then $A \preceq A \otimes B$ as well (we define a mapping $\sigma' : \{1, \dots, m\} \to \{(1,1), \dots, (1,k), \dots, (m,1), \dots (m,k)\}$ such that $\sigma'(j) = (j, \sigma(j)), 1 \leq j \leq m$, with $TA_j \subseteq TA_j \otimes TB_{\sigma(j)} = TA_j \cap TB_{\sigma(j)} = TA_j$). Thus, if $A \preceq B$ then $A \otimes B \approx A$.

Hence, $\inf(A, B) = A \otimes B$ and $\sup(A, B) = A \widehat{\oplus} B$.

Consequently, the equivalence in this lattice is equal to the strong behavioral equivalence of the **DB** category (from Corollary 14 in Sect. 3.4.1). We have to show that the following laws hold:

1. $A \widehat{\oplus} A \approx A,\ A \otimes A \approx A$; (idempotency)
2. $A \widehat{\oplus} B \approx B \oplus A,\ A \otimes B \approx B \otimes A$; (commutativity)
3. $A \widehat{\oplus} (B \widehat{\oplus} C) \approx (A \widehat{\oplus} B) \widehat{\oplus} A,\ A \otimes (B \otimes C) \approx (A \otimes B) \otimes C$; (associativity)
4. $A \widehat{\oplus} \bot^0 \approx A,\ A \otimes \Upsilon \approx A$; (bottom-top laws, see Proposition 10)
5. $A \approx A \widehat{\oplus} (A \otimes B),\ A \approx A \otimes (A \widehat{\oplus} B)$. (absorption)

The commutative and idempotency laws hold directly from the definitions of $\widehat{\oplus}$ (with Lemma 17) and $\otimes$, and from the results obtained previously. The second associative law for '$\widehat{\oplus}$' holds from the transitive property of the PO relation '$\preceq$' used in definitions for the sets of simple objects S_A, S_B in Definition 57. The isomorphism (a) $A \otimes (B \otimes C) \approx (A \otimes B) \otimes C$ comes from definition of matching in Theorems 12, and the fact that disjoint union is commutative up to an isomorphism. Hence, from the fact that isomorphic objects are behaviorally equivalent, we obtain the associativity $A \otimes (B \otimes C) \approx (A \otimes B) \otimes C$ for the matching operator. Let us show absorbtion: We have $(A \otimes B) \preceq A$ and hence, from point 4 of Lemma 17, $A \widehat{\oplus} (A \otimes B) \approx A$. Analogously, from point 3 of Lemma 17, $A \preceq (A \widehat{\oplus} B)$ and from above, $A \otimes (A \widehat{\oplus} B) \approx A$.

For any subset $K \subseteq Ob_{DB}$, $\bot^0 \preceq \inf(K) = \bigotimes_{A_i \in K} A_i \in Ob_{DB}$ and $\Upsilon \succeq \sup(K) = \widehat{\bigoplus}_{A_i \in K} A_i \in Ob_{DB}$ (it holds also for the limit case when $K = Ob_{DB}$, with $A_i \subseteq \Upsilon$, i.e., $A_i \preceq \Upsilon$, so that $\sup(Ob_{DB}) = \bigoplus_{A_i \in Ob_{DB}} A_i = \Upsilon \in Ob_{DB}$). Consequently, the lattice $(Ob_{DB}, \preceq)$ for every subset K has the greatest-lower-bound and the least-upper-bound, and hence it is a complete lattice. □

An element B in a lattice $\mathcal{C}$, $B \in \mathcal{C}$, is *compact* iff, for every $X \subseteq \mathcal{C}$ such that $\sup(X)$ exists and $B \preceq \sup(X)$, then there exists a *finite* $X' \subseteq_\omega X$ such that $B \preceq \sup(X')$: the set of compact elements in $\mathcal{C}$ is denoted by $Comp\mathcal{C}$. For example, the lattice of subsets of a set is an algebraic lattice where the compact elements are finite sets.

A lattice is *algebraic* if it is *complete* and *compactly generated*: a lattice $(\mathcal{C}, \preceq)$ is compactly generated if every element of $\mathcal{C}$ is a sup of compact elements less then or equal to it, i.e., for every $A \in \mathcal{C}$, $A \approx \sup\{B \in Comp\mathcal{C} \mid B \preceq A\}$. The closed elements, obtained by a closure operator $\mathcal{J}$ from finite subsets, are compact elements. Clearly, finite lattices are algebraic. But in our case, the complete database lattice L_{DB} in Proposition 51 is infinite, thus we need to investigate if it is algebraic or not.

For example, the subset [0, 1] of the real numbers is a complete lattice, but it is not algebraic.

The *finite objects* in **DB** are the databases with a finite number of n-ary relations (n is a finite number $n \in \omega$, and the nullary relation $\perp$ is an element of each object in **DB** category); the *extension* of relations is not necessarily finite—in such a case, for a finite object A in **DB**, the object TA is composed of infinite number of relations, that is, TA is an infinite object.

We will demonstrate that the database lattice is an *algebraic lattice*. One way of producing, and recognizing, complete algebraic lattices is through *closure* operators. We can use this approach because the power-view operator $T : Ob_{DB} \rightarrow Ob_{DB}$ is a closure operator (by Theorem 6 in Sect. 3.2.5).

However, the complete lattice derived from a given closure operator must have the meet operator '$\wedge$' equal to the set-intersection operator, and, in our case of a complete database lattice $L_{DB} = (Ob_{DB}, \preceq, \otimes, \widehat{\oplus})$, the meet operator is the matching operator: only for the *simple objects* in **DB** it is equal to set-intersection '$\cap$'. Consequently, we are able to use the closure operator T in order to demonstrate that the sublattice of L_{DB} composed of only simple databases is algebraic.

It is easy to show that the principal properties of the objects and morphisms in the **DB** category are derivable using the properties of the simple objects and simple morphisms between them. For example, a complex object $A \uplus B$ represents the separation-composition of the simple objects (databases) A and B, and hence the algebraic properties of the complex objects are derivable from the algebraic properties of the simple objects that compose them. The same holds for complex arrows.

The complex arrow $h = [f, g] : A \uplus B \rightarrow C$, $h = f \uplus g : A \uplus B \rightarrow C \uplus D$, $h = \langle f, g \rangle : A \rightarrow C \uplus D$ and $h = \lceil f, g \rfloor : A \rightarrow C$ are representable in **DB** by the set of their simple ptp arrows f and g.

Consequently, in order to investigate the algebraic database lattice, we can begin from the full subcategory $\underline{\mathbf{DB}}$ of **DB** composed of *only simple* objects, introduced by Definition 26 in Sect. 3.2.5 with the object $\underline{\Upsilon}$ equal to the union of all simple objects (databases).

By definition, a closed-set system is *algebraic* if $\mathcal{C}$ is closed under unions of upward directed subsets, i.e., for every $S \subseteq \mathcal{C}$, $\bigvee S \in \mathcal{C}$. Equivalently, the closure operator $\mathcal{J}$ on a set Υ is algebraic if it satisfy the following "finitary" property: for any subset $X \subseteq \Upsilon$, $\mathcal{J}(X) = \bigcup\{\mathcal{J}(X') \mid X' \subseteq_\omega X\}$, where $X' \subseteq_\omega X$ means that X' is a *finite* subset of X.

Proposition 52 *Let $\mathcal{C} = Ob_{\underline{DB}_{sk}} \subset Ob_{\underline{DB}}$ be the set of all closed objects (w.r.t. the power-view closure operator T) of the $\underline{\mathbf{DB}}$ category. The following properties for a database closure are valid*:

- *A closed-set system $(\underline{\Upsilon}, \mathcal{C})$ consists of the "total" closed object (i.e., the top database instance) $\underline{\Upsilon} \in \mathcal{C}$ and the set $\mathcal{C}$ which is closed under intersections of arbitrary subsets. That is, for any $K \subseteq \mathcal{C}$, $\bigcap K \in \mathcal{C}$.*
- *The closure operator T is algebraic on the set $\underline{\Upsilon}$.*

- *The complete lattice* $(\mathcal{C}, \subseteq)$ *obtained by this closure operator* T *is an algebraic lattice with meet* $\otimes$ *and join* $\oplus$ *operators. The compact elements of* $(\mathcal{C}, \subseteq)$ *are closed objects of the* **DB** *category generated by* finite *simple objects* $A \subseteq_\omega \underline{\Upsilon}$.
- *The lattice* $(Ob_{\underline{DB}}, \preceq, \otimes, \oplus)$ *is an algebraic lattice such that the algebraic closure operator* T *is a surjective homomorphism from this lattice into the 'skeletal' algebraic lattice of closed databases* $(\mathcal{C}, \subseteq)$.

Proof From Theorem 6 in Sect. 3.2.5, $T : Ob_{DB} \longrightarrow Ob_{DB}$ is a closure operator over the subset $Ob_{\underline{\mathbf{DB}}}$ and hence the set of closed (and simple) objects $\mathcal{C}$ is closed under intersection. From Definition 26 in Sect. 3.2.5, the total object $\underline{\Upsilon}$ is the top object such that for any $A \in \mathcal{C}$ (thus, $A = TA$), $A \subseteq \underline{\Upsilon}$. In fact, each object $A \in Ob_{\underline{\mathbf{DB}}}$ is a subset of $\underline{\Upsilon}$ and vice versa, each subset of $\underline{\Upsilon}$ (that contains $\bot$) is a database instance and hence an object in **DB** category.

From the Universal algebra theory it holds that each closure operator and its equivalent closure-set system $(\underline{\Upsilon}, \mathcal{C})$ generate a complete lattice $(\mathcal{C}, \subseteq)$ such that for any subset $K \subseteq \mathcal{C}$ of *simple* closed objects $K = \{A_i = TA_i \mid i \in I, A_i$ is closed set of **DB**$\}$ we obtain:

- The greatest lower bound $\bigwedge K = \bigwedge_{i \in I} A_i = \bigcap_{i \in I} A_i = \bigcap K = \bigotimes_{i \in I} A_i$ (from Theorem 12 for the simple databases and from $A_i = TA_i$), that is, meet lattice operator $\bigwedge$ corresponds to the matching operation $\otimes$.
- The least upper bound $\bigvee K = \bigvee_{i \in I} A_i = T(\bigcup_{i \in I} A_i) = T(\bigcup_{i \in I} TA_i) = \bigoplus_{i \in I} A_i$ (from Theorem 13 for the simple databases), that is, the join lattice operator $\bigvee$ corresponds to the merging operation $\oplus$, and hence for $K = \mathcal{C}$ we obtain $\bigvee \mathcal{C} = T(\bigcup_{A_i \in Ob_{\underline{\mathbf{DB}}}} A_i) =$ (by Definition 26) $= T\underline{\Upsilon} = \underline{\Upsilon}$.

Let us demonstrate that T is algebraic, that is, $T(A) = \bigcup\{T(A') \mid A' \subseteq_\omega A\}$ also for the infinite simple objects $A \subseteq \underline{\Upsilon}$. Let $U : \underline{\mathbf{DB}} \longrightarrow \mathbf{K}$ be the unique universal functor (which is a restriction of the functor U to the subcategory $\underline{\mathbf{DB}}$ of only simple objects, described previously in Sect. 8.1.4). Then, for this infinite simple database A, $U(A) = \{[t_i] \mid t_i \in \mathcal{T}_P A\}$ is an infinite set of the equivalence classes $[t_i]$ of SPJRU algebra terms with an infinite carrier set A of relational tables. From the fact that for all Σ_R-algebra operators $o_i \in \Sigma_R$ with finite $n = ar(o_i) \leq 2$ and $R_1, \dots, R_n \in A$, $t_i = \widehat{o_i}(R_1, \dots, R_n) \in \mathcal{T}_P A$, we can define a finite database $A' = \{R_1, \dots, R_n\} \subseteq_\omega A$ such that $t_i = \widehat{o_i}(R_1, \dots, R_n) \in \mathcal{T}_P A'$, $U(A) = \{[t_i] \mid t_i \in \mathcal{T}_P A\} = \bigcup\{[t_i] \mid t_i \in \mathcal{T}_P A', A' \subseteq_\omega A\}$. Thus, from Definition 31 in Sect. 5.1, $TA = \{R = \|[t_i]\|_\# \mid t_i \in \mathcal{T}_P A\} = \bigcup\{R = \|[t_i]\|_\# \mid t_i \in \mathcal{T}_P A', A' \subseteq_\omega A\} = \bigcup\{\{R = \|[t_i]\|_\# \mid t_i \in \mathcal{T}_P A'\} \mid A' \subseteq_\omega A\} = \bigcup\{TA' \mid A' \subseteq_\omega A\}$. Hence, T is an algebraic closure operator and, consequently, the lattice $(\mathcal{C}, \subseteq)$ obtained from this closure operator and the closed-set system $(\underline{\Upsilon}, \mathcal{C})$ is algebraic. It is well known that the compact elements of such a lattice $(\mathcal{C}, \subseteq)$, obtained from the algebraic closure operator T, are precisely the closed sets $T(A)$, where A is a finite subset of $\underline{\Upsilon}$.

The ordering in the lattice $(Ob_{\underline{DB}}, \preceq, \otimes, \oplus)$ is defined for any two simple databases $A, B \in Ob_{\underline{DB}}$ by $A \preceq B$ iff $TA \subseteq TB$. Thus, $A \approx B$ iff $TA = TB$, so that $A \approx TA$ with $A \otimes B = TA \cap TB$ and $A \oplus B = T(A \cup B) = T(TA \cup TB)$. From the fact that each non-closed element A in this lattice is equivalent to the closed element obtained from it ($A \approx TA$), this lattice has the same structure of the algebraic lattice of closed databases $(\mathcal{C}, \subseteq)$. Let us show now that it is also algebraic: in fact, for each $A \in Ob_{\underline{DB}}$ we have

$$\begin{aligned}
A &\approx TA \\
&\quad \text{(from the algebraic property of } T\text{)} \\
&= \bigcup\{TB \mid B \subseteq_\omega TA\} \approx T\Bigl(\bigcup\{TB \mid B \subseteq_\omega TA\}\Bigr) \\
&\quad \bigl(B \text{ is finite and hence } B' = TB \in Comp\,C \text{ is compact element}\bigr) \\
&= T\Bigl(\bigcup\bigl\{B' \in Comp\,C \mid B' \preceq A\bigr\}\Bigr) \\
&= \sup\bigl\{B' \in Comp\,C \mid B' \preceq A\bigr\},
\end{aligned}$$

so that this lattice is compactly generated and hence algebraic as well. Consequently, we obtain the surjective homomorphism $T : (Ob_{\underline{DB}}, \preceq, \otimes, \oplus) \to (\mathcal{C}, \subseteq, \cap, T\cup)$. □

Now we can extend the lattice $(\mathcal{C}, \subseteq)$ of only closed objects of **DB** into a lattice of all objects of **DB** category:

Corollary 28 *PO subcategory* $\mathbf{DB}_I \subseteq \mathbf{DB}$, *with* $\mathbf{DB}_I = L_{DB} = (Ob_{DB}, \preceq, \otimes, \widehat{\oplus})$, *is an algebraic lattice.*

Proof In Proposition 52, it was shown that T is an algebraic closure operator for all *simple* objects in **DB**. Let us show that this closure operator (for all objects in **DB**) is algebraic for the complex objects $A = \biguplus_{1 \le j \le m} A_j$, $j \ge 2$ as well. In fact,

$$\begin{aligned}
T(A) &= T\Bigl(\biguplus_{1\le j\le m} A_j\Bigr) \\
&= \biguplus_{1\le j\le m} T(A_j) \quad \text{(from the fact that each } A_j \text{ is a simple object)} \\
&= \biguplus_{1\le j\le m} \Bigl(\bigcup\{T(A'_j) \mid A'_j \subseteq_\omega A_j\}\Bigr) \\
&= \bigcup\{(j, T(A'_j)) \mid A'_j \subseteq_\omega A_j, 1 \le j \le m\} \\
&= \bigcup\Bigl(\biguplus_{1\le j\le m} \{T(A'_j) \,\big|\, A'_j \subseteq_\omega A_j\}\Bigr) \\
&= \bigcup\Bigl\{\biguplus_{1\le j\le m} T(A'_j) \,\Big|\, \biguplus_{1\le j\le m} A'_j \subseteq_\omega \biguplus_{1\le j\le m} A_j\Bigr\} \\
&= \bigcup\{T(A') \mid A' \subseteq_\omega A\}.
\end{aligned}$$

Thus, the closure operator T is algebraic for all objects in **DB** and, consequently, the complete lattice $L_{DB} = (Ob_{DB}, \preceq, \otimes, \widehat{\oplus})$ is algebraic. Thus, the PO subcategory $\mathbf{DB}_I = L_{DB}$ (from Theorem 6 in Sect. 3.2.5) is algebraic as well. □

The algebraic property is very useful in order to demonstrate the properties of the **DB** category: in order to demonstrate the theorems in general, we need to extend the inductive process of a proof beyond ω steps to the transfinite.

Zorn's lemma (equivalent to the Axiom of Choice of the set theory) allows us to do this. The database lattice $L_{DB} = (Ob_{DB}, \preceq, \otimes, \widehat{\oplus})$ is a (nonempty) poset with the property that every chain $K \subseteq Ob_{DB}$ (i.e., linearly ordered subset) has an upper bound $\bigvee K = \bigcup K$ (because this poset is *algebraic*) in Ob_{DB}. Hence, we can apply Zorn's lemma which asserts that L_{DB} has a maximal element.

Remark From the fact that $\mathbf{DB}_I = L_{DB} = (Ob_{DB}, \preceq, \otimes, \widehat{\oplus})$ is an algebraic lattice, we obtain for the total object $\Upsilon = T\Upsilon = \bigcup\{TA \mid A \subseteq_\omega \Upsilon\}$, i.e., it is the union of all closed objects generated by only *finite* objects of **DB** and hence the union of all *compact* elements of $(\mathcal{C}, \subseteq)$.

Let ω be the category of natural numbers with the arrows $\leq : j \longrightarrow k$ corresponding to the total order relation $j \leq k$, i.e., $\omega = \{0 \rightarrow 1 \rightarrow 2 \rightarrow \cdots\}$. An endofunctor $H : C \longrightarrow D$ is *ω-cocontinuous* if it preserves the colimits of functors $J : \omega \longrightarrow C$, that is, when $H Colim J \simeq Colim HJ$ (the categories C and D are thus supposed to have these colimits).

Notice that a functor $J : \omega \longrightarrow C$ is a diagram in C of the form $\{C_0 \rightarrow C_1 \rightarrow C_2 \rightarrow \cdots\}$. For ω-cocontinuous endofunctors the construction of the *initial algebra* is inductive [24].

Proposition 53 *For each simple object A in the category* **DB**, *the "merging with A" endofunctor $\Sigma_A = A \oplus \_ : \mathbf{DB} \longrightarrow \mathbf{DB}$ is ω-cocontinuous.*

Proof Let us consider any chain in **DB** (all arrows are monomorphisms (i.e., "$\preceq$") in a corresponding chain of the $\langle Ob_{DB}, \preceq\rangle$ algebraic lattice), that is a diagram $\mathcal{D}$,

$$\bot^0 \preceq_0 (\Sigma_A \bot^0) \preceq_1 (\Sigma_A^2 \bot^0) \preceq_2 \cdots \Sigma_A^\omega,$$

where $\bot^0 = \{\bot\}$ (with the empty relation $\bot \in A$, so that $\Sigma_A \bot^0 = A \oplus \bot^0 = T(A \cup \bot^0) = TA$) is the initial object in **DB**, with unique monic arrow $\bot^1 = \preceq_0 : \bot^0 \hookrightarrow (\Sigma_A \bot^0)$ with $\widetilde{\bot^1} = \bot^0$ and the consecutive arrows $\preceq_n = \Sigma_A^n \bot^1 : (\Sigma_A^n \bot^0) \hookrightarrow (\Sigma_A^{n+1} \bot^0)$ with $\widetilde{\Sigma_A^n \bot^1} = TA$, for all $n \geq 1$, as representation of a functor $J : \omega \longrightarrow \mathbf{DB}$. The endofunctor Σ_A preserves colimits because it is monotone and $\Sigma_A^\omega = TA$ is its fixed point, i.e., $\Sigma_A^\omega = TA = T(A \cup TA) = T(A \cup \Sigma_A^\omega) = \Sigma_A(\Sigma_A^\omega)$. Thus, the colimit $Colim J = \Sigma_A^\omega$ of the base diagram $\mathcal{D}$, given by the functor $J : \omega \longrightarrow \mathbf{DB}$, is equal to $Colim J = (A \oplus \_)^\omega \bot^0 = TA$. Thus $\sum_{A Colim J} = T(A \cup Colim J) = T(A \cup TA) = T(TA) = TA = Colim \Sigma_A J$ (where $Colim \Sigma_A J$ is a colimit of the diagram $\Sigma_A \mathcal{D}$).

The ω-cocompleteness amounts to chain-completeness, i.e., to the existence of least upper bound of ω-chains. Consequently, Σ_A is ω-cocontinuous endofunctor, i.e., a monotone function which preserves l.u.bs of ω-chains. □

In what follows, we will pass from lattice-based concepts, as l.u.bs of directed subsets, compact subsets, and algebraic lattices, to the categorially generalized concepts as directed colimits, finitely presentable (fp) objects, and locally finitely presentable (lfp) categories, respectively:

- A *directed colimit* in **DB** is a colimit of the functor $F : (J, \preceq) \longrightarrow \mathbf{DB}$, where $(J, \preceq)$ is a directed partially ordered set, such that for any two objects $j, k \in J$ there is an object $l \in J$ such that $j \preceq l, k \preceq l$, considered as a category. For example, when $J = Ob_{DB}$ we obtain the algebraic (complete and compact) lattice which is an directed PO-set such that for any two objects $A, B \in J$ there is an object $C \in J$ with $A \preceq C$ and $B \preceq C$ (when $C = \sup(A, B) \in J$).
- An object A is said to be *finitely presentable* (fp), or finitary, if the functor $\mathbf{DB}(A, \_) : \mathbf{DB} \to \mathbf{Set}$ preserves directed colimits (or, equivalently, if it preserves filtered colimits). We write $\mathbf{DB}_{fp}$ for the full subcategory of **DB** on the finitely presentable objects: it is essentially small. Intuitively, fp objects are "finite objects", and a category is lfp if it can be generated from its finite objects: a *strong generator* **M** of a category is its small full subcategory such that $f : A \longrightarrow B$ is an isomorphism *iff* for all objects C of this subcategory, given a hom-functor $\mathbf{M}(C, \_) : \mathbf{M} \longrightarrow \mathbf{Set}$, the following isomorphism of hom-setts $\mathbf{M}(C, f) : \mathbf{M}(C, A) \longrightarrow \mathbf{M}(C, B)$ in **Set** is valid.

From Theorem 1.11 [2], a category is *locally finitely presentable* (lfp) iff it is cocomplete and has a strong generator.

Corollary 29 **DB** *and* $\mathbf{DB}_{sk}$ *are concrete, locally small, and locally finitely presentable categories* (*lfp*).

Proof Given any two objects A and B in **DB**, the hom-set $\mathbf{DB}(A, B)$ of all arrows $f : A \longrightarrow B$ corresponds to the directed subset (see Proposition 6) $K = \{\widetilde{f} \mid \bot^0 \preceq \widetilde{f} \preceq A \otimes B\} \subseteq Ob_{\mathbf{DB}_{sk}}$, which is bounded algebraic (complete and compact) sublatice of $\mathcal{C}$. Thus, the set of all arrows $f : \Upsilon \longrightarrow \Upsilon$ corresponds to the directed set $K = \{\widetilde{f} \mid \bot^0 \preceq \widetilde{f} \preceq \Upsilon \otimes \Upsilon = \Upsilon\}$, which is equal to the lattice $(Ob_{\mathbf{DB}_{sk}}, \preceq)$ (which is a sublattice of the algebraic lattice L_{DB} in Proposition 51, composed of only closed objects in **DB**). Thus, **DB** is *locally small* (has small hom-sets), and, since $\mathbf{DB} \supseteq \mathbf{DB}_{sk}$, also $\mathbf{DB}_{sk}$ is locally small.

Let us show that the full subcategory $\mathbf{DB}_{\text{fin}}$, composed of closed objects obtained from *finite* database objects, is a strong generator of **DB**: in fact, if $A \simeq B$, where $A = \biguplus_{1 \le j \le m} A_j, m \ge 1$ and $B = \biguplus_{1 \le i \le k} B_i, k \ge 1$ are two *finite* databases (so that from Lemma 15 and Corollary 26 in Sect. 8.1, $k = m$ with a bijection $\sigma : \{1, \dots, m\} \to \{1, \dots, k\}$ of simple databases $A_j \simeq B_{\sigma(j)}, 1 \le j \le m$) then for all $C = \biguplus_{1 \le l \le n} C_l, m \ge 1$ in $\mathbf{DB}_{\text{fin}}$, $|\mathbf{DB}_{\text{fin}}(C, A)|$ is the rank of the complete sublattice of $\langle Ob_{\mathbf{DB}_{sk}}, \preceq\rangle$ bounded by $\bot^0 \preceq D \preceq C \otimes A$, while $|\mathbf{DB}_{\text{fin}}(C, B)|$ is the rank of the complete sublattice of $\langle Ob_{\mathbf{DB}_{sk}}, \preceq\rangle$ bounded by $\bot^0 \preceq D_1 \preceq C \otimes B$. From $A \simeq B$ we deduce $C \otimes A \simeq C \otimes B$, thus $|\mathbf{DB}_{\text{fin}}(C, A)| = |\mathbf{DB}_{\text{fin}}(C, B)|$, i.e., there is a bijection $\upsilon : \mathbf{DB}_{\text{fin}}(C, A) \to \mathbf{DB}_{\text{fin}}(C, B)$ which is an isomorphism in **Set**. Thus, **DB**, which is cocomplete and has this strong generator $\mathbf{DB}_{\text{fin}}$, is an lfp. □

We define a representable functor $\mathbf{DB}(A, \_) : \mathbf{DB} \longrightarrow \mathbf{Set}$ such that $\mathbf{DB}(A, B)$ is the hom-set of all morphisms in **DB** from A into B, and for any arrow $g : B \longrightarrow C$, $\mathbf{DB}(A, g)$ is the function such that for any morphism $f \in \mathbf{DB}(A, B)$ we obtain the morphism $h = \mathbf{DB}(A, g)(f) \triangleq g \circ f \in \mathbf{DB}(A, C)$.

We say that a functor $H : \mathbf{DB} \longrightarrow \mathbf{Set}$ preserves colimits if the image $H\nu : HF \longrightarrow HColimF$ for the colimit $(\nu, ColimF)$ of a functor $F \in \mathbf{DB}^{\mathbf{J}}$ is a colimiting cone (or cocone) for HF. In this case, we are interested in $H = \mathbf{DB}(\Upsilon, \_)$.

Let us show, for example, that the object Υ is a finitely presentable (fp) (it was demonstrated previously by remark that $\Upsilon = T\Upsilon = \bigcup\{TA \mid A \subseteq_\omega \Upsilon\}$), i.e., the fact that its hom-functor $\mathbf{DB}(\Upsilon, \_) : \mathbf{DB} \longrightarrow \mathbf{Set}$ preserves directed colimits:

Proposition 54 *The total object, that is, the matching monoidal unit Υ is a finitely presentable (fp).*

Proof Let us consider the object *ColimF* in **DB** (a colimit of the functor $F \in \mathbf{DB}^{\mathbf{J}}$, where F can be seen as a base diagram for this *directed* colimit, composed of a finite number of (possibly complex) objects $B_1, \dots, B_n$ with PO-arrows "$\preceq$" between them), such that the arrows $h_i : B_i \hookrightarrow ColimF$ are the monomorphisms corresponding to partial ordering $B_i \preceq ColimF$ for directed colimits, as specified by Theorem 6 in Sect. 3.2.5. These monomorphisms are the components of the cocone $(\nu, ColimF)$ where $\nu : F \longrightarrow \Delta ColimF$ is a natural transformation (and Δ is a diagonal constant functor) such that its components are $\nu(B_i) = \nu_{B_i} = h_i$, $1 \le i \le n$. The translation of this directed colimit diagram from **DB** into **Set** by using the functor $\mathbf{DB}(\Upsilon, \_)$, for any cocone monomorphism h_i, can be represented by the following diagram:

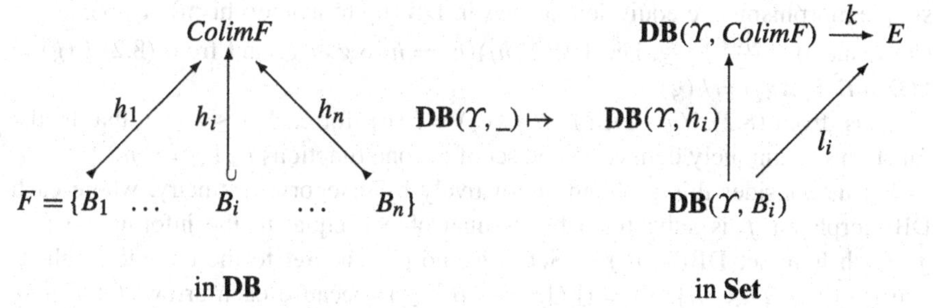

Thus, for this directed colimit and a set of all objects in the diagram (i.e., the functor) F, $S = \{B_i \mid B_i \in F\} \subseteq Ob_{DB}$, $ColimF = \sup(S) = \sup\{B_i \mid B_i \in F\} = \sum_{B_i \in F} B_i = \bigoplus_{B_i \in F} B_i$. Then for each monomorphism $h_i : B_i \preceq ColimF$, from Proposition 6, any of its simple ptp arrows $h_{ij} \in \frac{h_i}{\uplus}$ is monic and hence with $\widetilde{h_{ij}} = TB_i$.

In fact, for any other cocone diagram (ν', D) with a vertex D and with natural transformation $\nu' : F \to \Delta D$ that provides the components (ptp arrows) $h'_i = \nu'(B_i) = \nu'_{B_i}$, there is a unique arrow $k : ColimF \to D$ such that $h'_i = k \circ h_i$, $1 \le i \le n$.

This colimit diagram is then translated by the functor $H = \mathbf{DB}(\Upsilon, \_)$ in an analogous colimit diagram in **Set** with the cocone-base HF, as presented by the right-hand side of the diagram above for any fixed cocone monic arrow $h_i : B_i \hookrightarrow ColimF$. Let us show that for any other cocone E in **Set**, for the same cocone-base HF (where $H = \mathbf{DB}(\Upsilon, \_) : \mathbf{DB} \to \mathbf{Set}$) there is a unique arrow (a function) k from $\mathbf{DB}(\Upsilon, ColimF)$ to the set E (vertex of a cocone E) such that the diagram composed of these two cocones commutes. That is, for each object $\mathbf{DB}(\Upsilon, B_i)$ in the cocone-base HF, we obtain a function l_i which is a component of cocone E in **Set**:

$$l_i = k \circ \mathbf{DB}(\Upsilon, h_i) : \mathbf{DB}(\Upsilon, B_i) \to E, \tag{8.2}$$

for each $1 \le i \le n$.

From the functorial property of $H = \mathbf{DB}(\Upsilon, \_) : \mathbf{DB} \to \mathbf{Set}$, we have that for the function $\mathbf{DB}(\Upsilon, h_i) : \mathbf{DB}(\Upsilon, B_i) \to \mathbf{DB}(\Upsilon, ColimF)$ it holds that for each DB-morphism $(g : \Upsilon \to B_i) \in \mathbf{DB}(\Upsilon, B_i)$, $\mathbf{DB}(\Upsilon, h_i)(g) = h_i \circ g$ is a DB-morphism with the set of ptp (simple) arrows

$$\frac{h_i \circ g}{\uplus} = \left\{ h_{ij} \circ g_l \mid cod(g_l) = dom(h_{ij}), g_l \in \frac{g}{\uplus}, h_{ij} \in \frac{h_i}{\uplus} \right\}$$

where for each monic simple arrow h_{ij}, $\widetilde{h_{ij}} = T(dom(h_{ij}))$ and

$$\widetilde{g_l} \subseteq T\big(dom(g_l)\big) \cap T\big(cod(g_l)\big) \subseteq T\big(cod(g_l)\big) = T\big(dom(h_{ij})\big) = \widetilde{h_{ij}}.$$

That is, $\widetilde{h_{ij} \circ g_l} = \widetilde{h_{ij}} \cap \widetilde{g_l} = \widetilde{g_l}$ and hence, by Definition 28 in Sect. 3.4.1, these two simple morphisms are equivalent arrows in **DB** (up to isomorphism), $h_{ij} \circ g_l \approx g_l$. Consequently, $\frac{h_i \circ g}{\uplus} \approx \frac{g}{\uplus}$, i.e., $\mathbf{DB}(\Upsilon, h_i)(g) = h_i \circ g \approx g$, and from (8.2) $l_i(g) = k(\mathbf{DB}(\Upsilon, B_i)(g)) = k(g)$.

Thus, from (8.2), $\forall g \in \mathbf{DB}(\Upsilon, B_i).k(g) = l_i(g)$, for each $1 \le i \le n$, that is, the function k is uniquely defined by the set of cocone functions l_i, $1 \le i \le n$.

Let us consider this problem alternatively by category symmetry, where each DB-morphism f is substituted by its dual object, equal to the information flux $\widetilde{f}$: Each hom-set $\mathbf{DB}(\Upsilon, B_i)$ in **Set** is isomorphic in **Set** to the complete sublattice $(\{\widetilde{f} \mid \widetilde{f} \preceq \Upsilon \otimes B_i\}, \preceq) = (\{\widetilde{f} \mid \widetilde{f} \preceq TB_i\}, \preceq)$ (because each arrow $f : \Upsilon \longrightarrow B_i$ corresponds to the closed object $\widetilde{f} \preceq \Upsilon \otimes B_i = TB_i$). On the other hand, $HColimF = \mathbf{DB}(\Upsilon, ColimF)$ is isomorphic to the complete sublattice $(S, \preceq)$, where $S = \{\widetilde{f} \mid \widetilde{f} \preceq \sum_{B_i \in F} TB_i = \bigoplus_{B_i \in F} TB_i\}$. Hence, all arrows of the cocone $H\nu$, $\mathbf{DB}(\Upsilon, h_i) : \mathbf{DB}(\Upsilon, B_i) \hookrightarrow \mathbf{DB}(\Upsilon, ColimF)$ are injective functions $(\{\widetilde{f} \mid \widetilde{f} \preceq TB_i\}, \preceq) \subseteq (S, \preceq)$ (also all arrows in the base diagram HF in **Set** are injective functions of the form $(\{\widetilde{f} \mid \widetilde{f} \preceq TB_j\}, \preceq) \subseteq (\{\widetilde{f} \mid \widetilde{f} \preceq TB_k\}, \preceq)$).

Thus, the PO arrows (the monomorphisms) '$\preceq$' in **DB** are translated into injective functions (i.e., the monomorphisms in **Set**) and hence the directed-colimit diagram in **DB** with *ColimF* equal to the l.u.b. of all objects in its cocone-base F is translated by the functor $H = \mathbf{DB}(\Upsilon, \_)$ into the direct-colimit diagram (with respect to inclusion-PO '$\subseteq$') in **Set** where the $H(\mathit{ColimF})$ is the l.u.b. with respect to partial ordering $\subseteq$. Consequently, any other E that makes this diagram commutative with the same cocone base HF (composed of all injective functions) from $H(B_i)$ into E for all $1 \leq i \leq n$ must be a bigger set then l.u.b. $H(\mathit{ColimF})$ and with unique injective function $k : H(\mathit{ColimF}) \hookrightarrow E$. All arrows of the cocone E, $l_i : (\{\widetilde{f} | \widetilde{f} \preceq TB_i\}, \preceq) \longrightarrow E$, must be injective functions (however, with different domains) in order to preserve the commutativity of this colimiting cocone E. Thus, the function $k : (S, \preceq) \longrightarrow E$ is a unique function such that, for any $v \in \langle S, \preceq \rangle$, $k(v) = l_i(v)$ for some $l_i : (\{\widetilde{f} | \widetilde{f} \preceq TB_i\}, \preceq) \longrightarrow E$ and $v \in (\{\widetilde{f} | \widetilde{f} \preceq TB_i\}, \preceq)$. From $\mathit{HColim} = \mathbf{DB}(\Upsilon, \mathit{ColimF}) \simeq S$ we can conclude that there is a unique arrow in **Set** from $\mathit{HColimF}$ into E. Consequently, HColim is a colimit in **Set**, i.e., $H = \mathbf{DB}(\Upsilon, \_)$ preserves directed colimits, and hence Υ is finitely presentable. □

Remark We emphasize the fact that Υ is an fp object for a more general considerations of the theory of enriched categories, which will be elaborated in Sect. 8.2 as, for example, the demonstration that the monad based on the power-view endofunctor $T : \mathbf{DB} \to \mathbf{DB}$ is an enriched monad. The Kelly–Power theory applies in the case of a symmetric monoidal closed category, which is an lfp and a closed category: it is equivalent to demanding that the underlying ordinary category is lfp and that the monoidal structure on this ordinary category restricts to one on its fp objects. For details see [10, 11], and consider particularly that the unit Υ must be finitely presentable.

A locally finitely presentable category [7] is the category of models for an essentially *algebraic* theory, which allows operations whose domain is an equationally defined subset of some product of the previously defined domains (the canonical example is a composition in a category, which is defined only on composable, not arbitrary pairs of arrows).

In fact, we deduce from the algebraic (complete and compact) lattice $(Ob_{DB}, \preceq)$ that for any simple object A, $A \simeq TA = \bigoplus\{TB \mid B \subseteq_\omega A\} = \bigoplus S$ (we recall that $\oplus$ is a generalization in **DB** of the union operation $\cup$ for sets and $X \oplus Y = TX \oplus TY$), where the set $S = \{B \mid B \subseteq_\omega A\}$ is upward directed, i.e., for any two finite $B_1, B_2 \subseteq_\omega A$ there is $C = B_1 \cup B_2 \in S$ such that $B_1 \preceq C$ and $B_2 \preceq C$, with $TC = B_1 \oplus B_2$. That is, any object in **DB** is generated from finite objects and this generated object is just a directed colimit of these fp objects.

An important consequence of this freedom is that we can express *conditional* equations in the logic for databases.

Another important result, from the fact that **DB** is a complete and cocomplete lfp category, is that it can be used as the category of models for essentially algebraic theory [15, 23] as a relational database theory. Thus, it is a category of models for a finite limit sketch, where sketches are called graph-based logic [3, 12], and it is well known that a relational database *scheme* can readily be viewed, with some inessential abstraction involved, as a sketch. By Liar's theorem, a category **DB** is accessible [13, 20] because it is sketchable.

Remark A standard application of sketches is used to define a theory of a *single* database scheme, so that objects of this graph-based logic theory are single relations of such a database and arrows between them are used to define the common database functional dependencies, inclusion dependencies and other database constraints. Here, in the case of inter-database schema mappings, we need to use the whole databases as objects in this lfp **DB** category: the price for this higher level of abstraction is that arrows in **DB** are much more complex than in a standard setting, and hence are not (simple) functions but the *sets of functions*.

Let us show that the top closed object $\underline{\Upsilon}$ for the simple databases in **DB**, introduced by Definition 26, is a maximal database instance of a particular database schema $\widehat{\Upsilon}$, which defines the minimal cardinality for a set of relational symbols (variables) $\mathbb{R}$ and its R-algebras α introduced for operads in Sect. 2.4.1 by Definition 10 (obtained as the extensions of Tarski's semantics of FOL).

Proposition 55 *Let us define the minimal set of relational symbols by a bijection* $\alpha_0 : \mathbb{R} \to \underline{\Upsilon}$, *so that the schema database for a closed object* $\underline{\Upsilon}$ *in* **DB** *is equal to* $\widehat{\Upsilon} = (\mathbb{R}, \emptyset)$ *and hence* α_0 *is an R-algebra with* $\underline{\Upsilon} = \alpha_0^*(\widehat{\Upsilon})$. *Thus, for any other R-algebra* $\alpha \neq \alpha_0$, $\alpha^*(\widehat{\Upsilon}) \subseteq \underline{\Upsilon}$, *so that the following initial* Σ_R*-algebra semantics is valid for* `all` *R-algebras* α

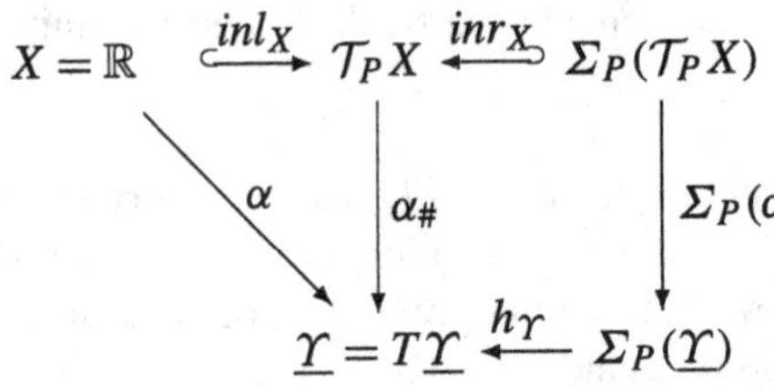
$X = \mathbb{R}$
inl_X
$T_P X$
inr_X
$\Sigma_P(T_P X)$
α
$\alpha_\#$
$\Sigma_P(\alpha_\#)$
$\underline{\Upsilon} = T\underline{\Upsilon}$
h_Υ
$\Sigma_P(\underline{\Upsilon})$

That is, $\bigcup_\alpha \alpha^*(\widehat{\Upsilon}) = \bigcup\{A \mid A$ *is simple object in* **DB**$\} = \bigcup_{A \in Ob_{\underline{\mathbf{DB}}}} A = \underline{\Upsilon}$.

The total object in **DB** *is* $\Upsilon = \underline{\Upsilon} \cup (\biguplus_\omega \underline{\Upsilon})$.

Proof From the bijection $\alpha_0 : \mathbb{R} \to \Upsilon$, the minimal cardinality for the set of relational symbols $\mathbb{R}$ in our operads semantics for database mappings is determined by the cardinality of the closed object $\underline{\Upsilon}$ defined by Definition 26.

So, for a defined schema $\widehat{\Upsilon} = (\mathbb{R}, \emptyset)$ and this R-algebra α_0,

$$\alpha_0^*(\widehat{\Upsilon}) = \{\alpha_0(r) \mid r \in \widehat{\Upsilon}\} = \{\alpha_0(r) \mid r \in \mathbb{R}\} = \underline{\Upsilon}.$$

The Σ_R-algebra generated by the carrier set $\underline{\Upsilon}$, $\Upsilon_\Sigma = (\underline{\Upsilon}, h_\Upsilon)$, is the bottom horizontal arrow of the commutative diagram in **Set** above such that the function h_Υ, for each unary (or binary) operator $o_i \in \Sigma_P = \Sigma_R$ and the relations $\alpha_\#(t), \alpha_\#(t_1), \alpha_\#(t_2) \in T\underline{\Upsilon} = \underline{\Upsilon}$, is defined by:

1. $h_{\Upsilon}(\bot) = \bot \in \Upsilon \in Ob_{DB}$ (for each object in **DB**, the empty relation is an element of it, with $\bot^0 = \{\bot\}$).
2. $h_{\Upsilon}(\alpha_{\#}(t), i) = \alpha_{\#}(o_i(t))$, where $o_i(t) \in \mathcal{T}_P X$.
3. $h_{\Upsilon}((\alpha_{\#}(t_1), \alpha_{\#}(t_2)), -1) = \alpha_{\#}(t_1 \text{ TIMES } t_2)$, and $h_{\Upsilon}((\alpha_{\#}(t_1), \alpha_{\#}(t_2)), 0) = \alpha_{\#}(t_1 \text{ UNION } t_2)$ if $\alpha_{\#}(t_1)$ and $\alpha_{\#}(t_2)$ are union compatible; $\bot$ otherwise.

Thus, $\alpha_{\#}$ is the unique epic inductive extension of h_{Υ} along the assignment $\alpha : X \to \underline{\Upsilon}$, to all terms with variables in $\mathcal{T}_P X$ (the mapping $\alpha_{\#}$ is equal to α for the variables in $X \subseteq \mathcal{T}_P X$). (The injections inl_X and inr_X are defined at the end of Sect. 5.1.1.)

Heaving defined this schema $\widehat{\Upsilon} = (\mathbb{R}, \emptyset)$, the instance database $\alpha^*(\widehat{\Upsilon})$ is an object in **DB** for each R-algebra α. Let us show that $\alpha^*(\widehat{\Upsilon}) \subseteq \underline{\Upsilon}$.

From the fact that $\underline{\Upsilon}$ is defined as a set of all n-ary relations (for finite $n \geq 0$) with values in a fixed universe $\mathcal{U}$, it means that, for any n-ary relational symbol $r \in \mathbb{R}$, the relation $\alpha(r)$ is an element in $\underline{\Upsilon}$. Thus, $\alpha^*(\widehat{\Upsilon}) = \{\alpha(r) \mid r \in \widehat{\Upsilon}\} = \{\alpha(r) \mid r \in \mathbb{R}\} \subseteq \underline{\Upsilon}$. Consequently,

$$\underline{\Upsilon} = \underline{\Upsilon} \bigcup_{\alpha \neq \alpha_0} \alpha^*(\widehat{\Upsilon}) = \alpha_0^*(\widehat{\Upsilon}) \bigcup_{\alpha \neq \alpha_0} \alpha^*(\widehat{\Upsilon}) = \bigcup_{\alpha} \alpha^*(\widehat{\Upsilon}) = \bigcup_{A \in S \subseteq Ob_{DB}} A$$

where $S = \{\alpha^*(\widehat{\Upsilon}) \in Ob_{DB} \mid \alpha \neq \alpha_0\} \subseteq \{A \mid A \text{ is simple object in } \mathbf{DB}\} = \underline{\Upsilon}$. □

Remark Notice that the top simple object $\underline{\Upsilon}$ is the set of all finitary relations (i.e., with a finite number of columns $n = ar(R)$, but some of them can have also infinite sets of tuples) that can be obtained from a fixed universe $\mathcal{U} = \mathbf{dom} \cup SK$, where $SK = \{\omega_0, \omega_1, \dots\}$ is an infinite set of indexed Skolem constants and **dom** is a fixed domain of values.

As specified in the introduction (Sect. 1.4) for any relation $R \in \underline{\Upsilon}$ with an infinite number of tuples, $SK \subseteq val(\{R\})$ and $val(\{R\}) \cap \mathbf{dom}$ is always *finite* (because a domain $dom(a) \subset \mathbf{dom}$ of any attribute $a \in \mathbf{att}$ of a relation is finite). Consequently, the infinite relations can have only finite sets of values of **dom** and become infinite only because they have *all* Skolem constants: it happens only when we have the cyclic tgds with existentially quantified right-hand side of implication (the case of database mappings with incomplete information, as explained in Example 27 in Sect. 4.2.4):

$R_1 = \|r\|_{can(\mathcal{I},\mathcal{D})}$	$R_2 = \|s\|_{can(\mathcal{I},\mathcal{D})}$
a, b	b, ω_0
b, ω_1	ω_1, ω_2
ω_1, ω_3	ω_3, ω_4
ω_3, ω_5	ω_5, ω_6
...	...

with $\pi_2(R_2) = \{\langle w_i \rangle \mid w_i \in SK\}$ the infinite unary relations composed of the infinite set of all Skolem constants.

The number of Skolem constants must be infinite also when **dom** is a finite set because in the foreign key constraint $\pi_2(R_1) \subseteq \pi_1(R_2)$ and $\pi_1(R_2) \subseteq \pi_1(R_1)$ in Example 27, the non-key attributes are formed by means of fresh (new) marked values in SK (in the place of existentially quantified predicate's variables (relational attributes)) and we cannot insert the Skolem values that are previously used because in that case we would introduce the equivalences in the extensions of relations (for their quantified attributes) which are not defined in the database schemas.

Consequently, for each fixed domain **dom** we will obtain a particular simple-top object $\underline{\Upsilon}$ and hence a specific **DB** category.

Hence, we have to deal with a class of database categories **DB** for the class of all possible domains **dom** and, consequently, a class of the complete algebraic lattices $(\mathcal{C}, \subseteq) = (Ob_{\underline{\mathbf{DB}}_{sk}}, \subseteq, \otimes, \oplus, \perp^0, \underline{\Upsilon})$ in Proposition 52, which will be used to define a class of Heyting algebras and a particular intermediate (superintuitionistic) propositional logic in Sect. 9.1.3.

8.2 Enrichment

It is not misleading, at least initially, to think of an enriched category as being a category in which the hom-sets carry some extra structure (partial order $\preceq$ of algebraic sublattice $(Ob_{DB}, \preceq)$ in our case) and in which that structure is preserved by composition. The notion of enriched category [10] is more general and allows for the hom-objects ("hom-sets") of the enriched category to be objects of some monoidal category, traditional called V.

Let us demonstrate that **DB** is a *monoidal closed* category:

Definition 58 For any two objects $B = \biguplus_{1 \le i \le k} B_i, k \ge 1$ and $C = \biguplus_{1 \le l \le n} C_l, n \ge 1$, the set of all arrows $\mathbf{DB}(B, C)$ from B into C can be represented by unique principal arrow $h : B \to C$ (Proposition 46) such that its set of ptp arrows is

$$\frac{h}{\uplus} \triangleq \left\{ h_{il} = \bigcup \left\{ f_{il} : B_i \to C_l \mid f_{il} \in \frac{f}{\uplus}, f \in \mathbf{DB}(B, C) \right\} \neq \emptyset \mid 1 \le i \le k, \right.$$
$$\left. 1 \le l \le n \right\}.$$

The object C^B is equal to the information flux of this principal arrow. Thus, we define the hom-object

$$C^B \triangleq \widetilde{h} \simeq \biguplus \frac{\widetilde{h}}{\uplus}.$$

Remark For any two *simple* objects B and C (when $n = k = 1$), the set of all arrows $S = \{f_1, f_2, \ldots\} : B \to C$, from B into C, can be represented by a unique complex arrow, $\lceil S \rfloor : B \to C$. Thus, from Definition 22 and Lemma 7 in Sect. 3.2, we obtain the hom-object $C^B \triangleq \widetilde{\lceil S \rfloor} = \bigoplus \{\widetilde{f_j} \mid f_j \in \mathbf{DB}(B, C)\}$. From Proposition 7, for each simple arrow $f_j \in \mathbf{DB}(B,C)$, $\widetilde{f_j} \subseteq B \otimes C$. Thus, based on the complete lattice $L_{DB} = (Ob_{DB}, \preceq, \otimes, \widehat{\oplus})$ in Proposition 51, the set $\{\widetilde{f_j} \mid f_j \in \mathbf{DB}(B,C)\}$ is a sublattice with the l.u.b. equal to $B \otimes C$, so that the merging (join operator of this lattice) of all elements in this set is equal to its l.u.b. That is, we obtain that $C^B = \bigoplus\{\widetilde{f_j} \mid f_j \in \mathbf{DB}(B,C)\} = T(\bigcup\{\widetilde{f_j} \mid f_j \in \mathbf{DB}(B,C)\}) = T(\bigcup\{\widetilde{f_j} \mid \widetilde{f_j} \subseteq B \otimes C = TB \cap TC\}) = T(TB \cap TC) = T(TB) \cap T(TC) = TB \cap TC = B \otimes C$.

Consequently, the hom-set $\mathbf{DB}(B,C)$ in **Set** of all morphisms $B \to C$ can be 'internalized' into the hom-object C^B in DB category by merging of compact elements $A \preceq B \otimes C$ (where $B \otimes C$ is the "distance" between B and C, following Lawvere's idea, as follows from the definition of metric space for the **DB** category in Sect. 9.1.1).

The aim to 'internalize' the hom-sets is justified by the necessity to substitute the base category **Set** by the base database category **DB** in the cases as Yoneda embeddings for categories by the contravariant functor $H : \mathbf{C}^{OP} \to \mathbf{Set}^{\mathbf{C}}$ introduced in technical preliminaries (Sect. 1.5).

Lemma 18 *For any two objects B and C, $C^B \simeq B^C \simeq B \otimes C$.*

Proof From Definition 58,

$$
\begin{aligned}
C^B &= \biguplus \frac{\widetilde{h}}{\uplus} \\
&\quad \text{(see the remark above for simple objects)} \\
&\simeq \biguplus_{1 \le i \le k, 1 \le l \le n} \left(\bigoplus \{\widetilde{f_{il}} \mid f_{il} \in \mathbf{DB}(B_i, C_l)\} \right) \\
&= \biguplus_{1 \le i \le k, 1 \le l \le n} B_i \otimes C_l \\
&\quad \text{(from Definition 19)} \\
&= B \otimes C.
\end{aligned}
$$

□

Remark From duality, for any two objects $A = \biguplus_{1 \le j \le m} A_j, m \ge 1$ and $B = \biguplus_{1 \le i \le k} B_i, k \ge 1$ there is the bijection $\_^{OP} : \mathbf{DB}(A,B) \to \mathbf{DB}(B,A)$ such that for any morphism $f : A \to B$, $f^{OP} = \_^{OP}(f) : B \to A$, where

$$
\frac{f^{OP}}{\uplus} = \left\{ f_{ji}^{OP} \mid f_{ji} : A_j \to B_i \in \frac{f}{\uplus} \right\},
$$

with equal information fluxes $\widetilde{f_{ji}^{OP}} = \widetilde{f_{ji}}$ (the flux transmitted from simple objects A_j into B_i can be reflexed from B_i to A_j as well). Consequently, by Lemma 18,

$$A^B \simeq \biguplus_{1\le j\le m, 1\le i\le k} \left(\bigoplus \{\widetilde{f_{ji}} \mid f_{ji} \in \mathbf{DB}(A_j, B_i)\} \right)$$

$$\simeq \biguplus_{1\le i\le k, 1\le j\le m} \left(\bigoplus \{\widetilde{f'_{ij}} \mid f'_{ij} = f_{ji}^{OP} \in \mathbf{DB}(B_i, A_j)\} \right) \simeq B^A \simeq A \otimes B,$$

i.e., $B \otimes A \simeq A^B \simeq B^A \simeq A \otimes B$.

That is, the *cotensor* B^A (i.e., hom-object) of any two objects A and B, which is a particular limit in **DB**, is equal to the corresponding colimit, that is, *tensor product* $A \otimes B$, of these two objects; this fact is based on the duality property of **DB** category.

8.2.1 DB Is a V-Category Enriched over Itself

Generally, a monoid M acting on the set Ob_{DB} may be seen as a general metric space where for any $B \in Ob_{DB}$ the distance C^B is a set of $R \in S$ (relations, or views, in our case) whose action sends B to C (this action provides a possibility to pass from the "state" B to "state" C of the database "system" of objects in **DB**), as we investigated in the chapter dedicated to operational semantics for the database mapping systems.

A monoidal category is closed if the functor $\_ \otimes B : \mathbf{DB} \longrightarrow \mathbf{DB}$ has a right adjoint $(\_)^B : \mathbf{DB} \longrightarrow \mathbf{DB}$ for every object B, $((\_)^B, \_ \otimes B, \eta_\otimes, \varepsilon_\otimes) : \mathbf{DB} \longrightarrow \mathbf{DB}$, with the counit $\varepsilon_C : C^B \otimes B \longrightarrow C$ called the evaluation at C (denoted by $eval_{B,C}$).

Theorem 15 *Strictly symmetric idempotent monoidal category* $(\mathbf{DB}, \otimes, \Upsilon)$ *is a monoidal bi-closed: for every object* B*, there exists in* **Set** *an isomorphism (a bijection)* $\Lambda : \mathbf{DB}(A \otimes B, C) \cong \mathbf{DB}(A, C^B)$ *such that for any* $f \in \mathbf{DB}(A \otimes B, C)$, $\Lambda(f) \approx f$, *the hom-object* C^B *together with a monomorphism* $eval_{B,C} : C^B \otimes B \hookrightarrow C$, *the following "exponent" diagram*

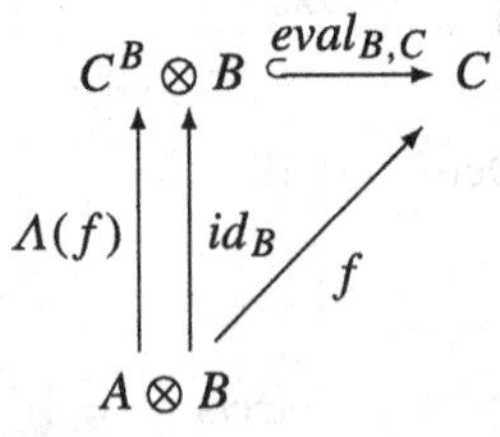

commutes, with $f = eval_{B,C} \circ (\Lambda(f) \otimes id_B)$.

Proof Let us consider the simple objects A, B, C and a morphism

$$f = \alpha^*(MakeOperads(\mathcal{M})) : A \otimes B \to C,$$

obtained from a sketch's mapping $\mathbf{M} : \mathcal{A} \overset{\cap}{\underset{\alpha}{-}} \mathcal{B} \to \mathcal{C}$ with $A = \alpha^*(\mathcal{A})$, $B = \alpha^*(\mathcal{B})$ and $C = \alpha^*(\mathcal{C})$ obtained from the schemas $\mathcal{A}$, $\mathcal{B}$ and $\mathcal{C}$. Thus. $\widetilde{f} \subseteq T(A \otimes B) \cap TC = TA \cap TB \cap TC$.

Then we define the schema mapping $\mathcal{M}_1 = \{\Phi\} : \mathcal{A} \underset{\alpha}{\cap} \mathcal{A} \to \mathcal{B} \underset{\alpha}{\cap} \mathcal{C}$ with SOtgd Φ equal to conjunction $\bigwedge\{\forall\mathbf{x}(r(\mathbf{x}) \Rightarrow r(\mathbf{x})) \mid r \in \mathcal{A}\underset{\alpha}{\cap}\mathcal{A}$ and $\alpha(r) \in \widetilde{f} \subseteq \alpha^*(\widehat{\mathcal{A}}) = TA\}$ with the sketch's mapping $\mathbf{M}_1 = \{1_r \in O(r,r) \mid \alpha(r) \in \widetilde{f} \subseteq TA\}$, where $\mathcal{A}\underset{\alpha}{\cap}\mathcal{B}$ is the α-intersection schema (defined in Definition 16 in Sect. 3.1) with $\alpha^*(\mathcal{A}\underset{\alpha}{\cap}\mathcal{A}) = TA$ and hence its flux is $\widetilde{\alpha^*(\mathbf{M}_1)} = \mathit{Flux}(\alpha, \mathbf{M}_1) = \widetilde{f}$.

Then we define the **DB** morphism

$$\Lambda(f) = \alpha^*(\mathbf{M}_1) \circ is_A : A \to \alpha\left(\mathcal{B} \underset{\alpha}{\cap} \mathcal{C}\right) = B \otimes C = C^B.$$

This morphism from A into C^B is uniquely defined by the morphism $f : A \otimes B \to C$: in fact, its flux is uniquely determined by

$$\widetilde{\Lambda(f)} = \widetilde{\alpha^*(\mathbf{M}_1) \circ is_A} = \widetilde{\alpha^*(\mathbf{M}_1)} \cap \widetilde{is_A} = \widetilde{f} \cap TA = \widetilde{f}.$$

Analogously, we define the schema mapping $\mathcal{M}_2 = \{\Psi\} : \mathcal{B}\underset{\alpha}{\cap}\mathcal{C} \to \mathcal{C}$ with SOtgd Ψ equal to the conjunction $\bigwedge\{\forall\mathbf{x}(r(\mathbf{x}) \Rightarrow r(\mathbf{x})) \mid r \in \mathcal{B}\underset{\alpha}{\cap}\mathcal{C}$ and $\alpha(r) \in \alpha^*(\mathcal{B}\underset{\alpha}{\cap}\mathcal{C}) = TB \cap TC\}$ with the sketch's mapping

$$\mathbf{M}_2 = \mathit{MakeOperads}\bigl\{1_r \in O(r,r) \mid \alpha(r) \in TB \cap TC\bigr\}.$$

Hence, we define the **DB** morphism $eval_{B,C} = \alpha^*(\mathbf{M}_2) : C^B \otimes B = \alpha^*(\mathcal{B}\underset{\alpha}{\cap}\mathcal{C}) \to C$ which is a monomorphism from the fact that $\widetilde{eval_{B,C}} = TB \cap TC = C^B \otimes B$.

Consequently, we obtain for the commutativity of this exponential diagram, that is, $\widetilde{f} = \widetilde{eval_{B,C} \circ (\Lambda(f) \otimes id_B)} = \widetilde{eval_{B,C}} \cap \widetilde{(\Lambda(f) \otimes id_B)} = \widetilde{eval_{B,C}} \cap (\widetilde{\Lambda(f)} \otimes \widetilde{id_B}) = B \otimes C \cap \widetilde{\Lambda(f)} \cap TB = TB \cap TC \cap \widetilde{\Lambda(f)} = \widetilde{\Lambda(f)}$ (from the fact that $\widetilde{\Lambda(f)} \subseteq C^B = TB \cap TC$). Thus, $f = eval_{B,C} \circ (\Lambda(f) \times id_B)$ iff $\Lambda(f) \approx f$.

Λ is a bijection because $\mathbf{DB}(A \otimes B, C) = \{g \mid \bot^0 \subseteq \widetilde{g} \subseteq A \otimes B \otimes C\} \cong \{\widetilde{g} \mid \widetilde{g} \in K\}$, where K is a bounded algebraic sublattice (of closed objects) of the lattice $(\mathcal{C}, \subseteq)$ in Proposition 52 and $\cong$ denotes a bijection, i.e., $K = \{a \mid a \in \mathcal{C}$ and $a \subseteq TA \cap TB \cap TC\}$. Also $\mathbf{DB}(A, C^B) = \mathbf{DB}(A, B \otimes C) \cong K$, thus $|\mathbf{DB}(A \otimes B, C)| = |\mathbf{DB}(A, C^B)| = |K|$. Consequently, Λ is a bijection such that for any $f \in \mathbf{DB}(A \otimes B, C)$, $\widetilde{\Lambda(f)} = \widetilde{f} \in K$, i.e., $\Lambda(f) \approx f$.

Let us consider now the complex objects $A = \biguplus_{1 \le j \le m} A_j, m \ge 1$, $B = \biguplus_{1 \le i \le k} B_i$, $k \ge 1$ and $C = \biguplus_{1 \le l \le n} C_l, n \ge 1$. Then $eval_{B,C}$ is the monomorphism $in : C \otimes B \hookrightarrow C$, obtained by Theorem 6 in Sect. 3.2.5 from the PO relation (from the isomorphism $C^B \otimes B \simeq C \otimes B \otimes B$), $C^B \otimes B \approx C \otimes B \otimes B$ (from $C \otimes B \preceq B$) $\approx C \otimes B \preceq C$ in the complete lattice $L_{DB} = (Ob_{DB}, \preceq, \otimes, \widehat{\oplus})$ (in Proposition 51), with $\frac{in}{\uplus} = \{in_{li} : C_l \otimes B_i \hookrightarrow C_l \mid 1 \le l \le n, 1 \le i \le k\}$.

For a given complex arrow $f : A \otimes B \to C$, we define the mapping Λ such that $\frac{f}{\uplus} \mapsto \{f'_{jli} : A_j \to C_l \otimes B_i \mid (f_{jil} : A_j \otimes B_i \to C_l) \in \frac{f}{\uplus}$ and $\widetilde{f'_{jli}} = \widetilde{f_{jil}}\}$, and its inverse Λ^{-1} such that

$$\frac{\Lambda(f)}{\uplus} \mapsto \left\{ f_{jil} : A_j \otimes B_i \to C_l \mid \left(f'_{jli} : A_j \to C_l \otimes B_i\right) \in \frac{\Lambda(f)}{\uplus} \text{ and } \widetilde{f_{jil}} = \widetilde{f'_{jli}} \right\},$$

and hence Λ is a bijection.

Let us show the commutativity $f = in \circ (\Lambda(f) \otimes id_B)$ of the diagram, i.e., that $\frac{f}{\uplus} = \frac{in \circ (\Lambda(f) \otimes id_B)}{\uplus}$. In fact, for each $(in_{li} \circ (f'_{jli} \otimes id_{B_i}) : A_j \otimes B_i \to C_l) \in \frac{in \circ (\Lambda(f) \otimes id_B)}{\uplus}$, $\widetilde{in_{li} \circ (f'_{jli} \otimes id_{B_i})} = (TC_l \cap TB_i) \cap (\widetilde{f'_{jli}} \cap TB_i) =$ (from $\widetilde{f'_{jli}} \subseteq TA_j \cap TB_i \cap TC_l) = \widetilde{f'_{jli}} = \widetilde{f_{jil}}$, for the corresponding $(f_{jil} : A_j \otimes B_i \to C_l) \in \frac{f}{\uplus}$. Consequently, $in_{li} \circ (f'_{jli} \otimes id_{B_i}) = f_{jil}$ for each ptp arrow in $in \circ (\Lambda(f) \otimes id_B)$ and vice versa. Hence, $f = in \circ (\Lambda(f) \otimes id_B) = eval_{B,C} \circ (\Lambda(f) \otimes id_B)$.

Consequently, **DB** is closed and symmetric, that is, a biclosed category. □

We have seen that all objects in **DB** are finitely representable (from Corollary 29 and Proposition 54). Let us denote the representable functor $\mathbf{DB}(\Upsilon, \_)$ by $V = \mathbf{DB}(\Upsilon, \_) : \mathbf{DB} \longrightarrow \mathbf{Set}$. By putting $A = \Upsilon$ in $\Lambda(f) : A \to C^B$ of the 'exponent' diagram in Theorem 15, and by using the isomorphism $\beta : \Upsilon \otimes B \simeq B$, we get a natural isomorphism in **Set** (a bijection Λ), $\mathbf{DB}(B, C) \simeq V(C^B) = \mathbf{DB}(\Upsilon, C^B)$.

Then C^B is exhibited as a lifting through V of the hom-set $\mathbf{DB}(B, C)$ (i.e., in the case when B and C are the simple objects, the hom-object C^B is a set of all views which gives a possibility to pass from a "state" B to a "state" C). It is called the *internal hom* of B and C.

By putting $B = \Upsilon$ in $\Lambda(f) : A \to C^B$ of the 'exponent' diagram in Theorem 15, and by using the isomorphism $\gamma : A \otimes \Upsilon \simeq A$, we deduce a natural isomorphism $i : C \simeq C^\Upsilon$ (it is obvious, from the fact that $C^\Upsilon \simeq C \otimes \Upsilon \simeq C$).

The fact that a monoidal structure is closed means that we have an internal Hom functor, $(\_)^{(\_)} : \mathbf{DB}^{OP} \times \mathbf{DB} \to \mathbf{DB}$, which 'internalizes' the external Hom functor, $Hom : \mathbf{DB}^{OP} \times \mathbf{DB} \to \mathbf{Set}$, such that for any two objects A and B, the hom-object $B^A = (\_)^{(\_)}(A, B)$ represents the hom-set $Hom(A, B)$ (i.e., the set of all morphisms from A to B).

Monoidal closed categories generalize the Cartesian closed ones in that they also possess exponent objects B^A which "internalize" the hom-sets. One may then ask if there is a way to "internally" describe the behavior of functors on morphisms. That is, given a monoidal closed category C and a functor $F : C \longrightarrow C$, consider, say, $f \in C(A, B)$, then $F(f) \in C(F(A), F(B))$. Since the hom-object B^A and $F(B)^{F(A)}$ represent hom-sets $C(A, B)$ and $C(F(A), F(B))$ in C, one may study the conditions under which F is "represented" by a morphism in $C(B^A, F(B)^{F(A)})$, for each A and B.

Theorem 16 *The endofunctor* $T : \mathbf{DB} \longrightarrow \mathbf{DB}$ *is closed.*

DB *is a V-category enriched over itself, with the composition law monomorphism* $m_{A,B,C} : C^B \otimes B^A \hookrightarrow C^A$ *and identity element (epimorphism)* $j_A : \Upsilon \twoheadrightarrow A^A$ *which "picks up" the identity in* A^A.

The monad (T, η, μ) *is an enriched monad, and hence* **DB** *is an object of V-cat and endofunctor* $T : \mathbf{DB} \to \mathbf{DB}$ *is an arrow of V-cat.*

Proof Let us show that the arrows introduced by this proposition satisfy the morphisms in the definition of the **DB** category; see Theorem 1 in Sect. 3.2. For a given database schemas $\mathcal{A}$, $\mathcal{B}$ and $\mathcal{C}$ and an R-algebra α such that $A = \alpha^*(\mathcal{A})$, $B = \alpha^*(\mathcal{B})$, $C = \alpha^*(\mathcal{C})$, we have that there exist the α-intersection schemas $\mathcal{A}\underset{\alpha}{\overset{\cap}{-}}\mathcal{C}$ and $(\mathcal{A}\underset{\alpha}{\overset{\cap}{-}}\mathcal{C})\underset{\alpha}{\overset{\cap}{-}}\mathcal{B}$ so that $\alpha^*(\mathcal{A}\underset{\alpha}{\overset{\cap}{-}}\mathcal{C}) = A \otimes C = C^A$ and

$$\alpha^*\left(\left(\mathcal{A}\underset{\alpha}{\overset{\cap}{-}}\mathcal{C}\right)\underset{\alpha}{\overset{\cap}{-}}\mathcal{B}\right) = TA \cap TC \cap TB = A \otimes C \otimes B = C^B \otimes B^A.$$

Thus, we may define a schema mapping $\mathcal{M} = \{\Phi\} : (\mathcal{A}\underset{\alpha}{\overset{\cap}{-}}\mathcal{C})\underset{\alpha}{\overset{\cap}{-}}\mathcal{B} \to \mathcal{A}\underset{\alpha}{\overset{\cap}{-}}\mathcal{C}$, where Ψ is a SOtgd $\bigwedge\{\forall \mathbf{x}(r(\mathbf{x}) \Rightarrow r(\mathbf{x})) \mid r \in (\mathcal{A}\underset{\alpha}{\overset{\cap}{-}}\mathcal{C})\underset{\alpha}{\overset{\cap}{-}}\mathcal{B}\}$, and

$$\mathbf{M} = \mathit{MakeOperads}(\mathcal{M}) = \left\{1_r \mid r \in \left(\mathcal{A}\underset{\alpha}{\overset{\cap}{-}}\mathcal{C}\right)\underset{\alpha}{\overset{\cap}{-}}\mathcal{B}\right\} \cup \{1_{r_\emptyset}\}$$

and hence we obtain a monomorphism

$$\begin{aligned} m_{A,B,C} &= \alpha^*(\mathbf{M}) \\ &= \{1_R : R \to R \mid R \in TA \cap TC \cap TB\} \cup \{q_\perp\} : C^B \otimes B^A \hookrightarrow C^A. \end{aligned}$$

In the case of complex objects A, B and C, this monomorphism $m_{A,B,C}$ is uniquely determined by the PO relation $C^B \otimes B^A \preceq C^A$ as specified by Theorem 6 in Sect. 3.2.5. In fact, from $C^B \otimes B^A \simeq (B \otimes C) \otimes (A \otimes B)$, $C^B \otimes B^A \approx (B \otimes C) \otimes (A \otimes B) \approx (C \otimes B) \otimes (B \otimes A) \approx$ (from $B \otimes B \approx B$) $\approx C \otimes B \otimes A \approx B \otimes (A \otimes C) \preceq$ (from the complete lattice $L_{DB} = (Ob_{DB}, \preceq, \otimes, \widehat{\oplus})$ in Proposition 51) $\preceq A \otimes C$ (from $A \otimes C \simeq C^A$) $\approx C^A$.

The identity element j_A is defined by a schema mapping $\mathcal{M}_1 = \{\Psi\} : \widehat{\Upsilon} \to \mathcal{A}\underset{\alpha}{\overset{\cap}{-}}\mathcal{A}$, where Ψ is a SOtgd $\bigwedge\{\forall \mathbf{x}(r(\mathbf{x}) \Rightarrow r(\mathbf{x})) \mid r \in \mathcal{A}\underset{\alpha}{\overset{\cap}{-}}\mathcal{A}\}$ and hence

$$j_A = \alpha^*\big(\mathit{MakeOperads}(\mathcal{M}_1)\big) = \{1_R : R \to R \mid R \in TA\} \cup \{q_\perp\} : \Upsilon \twoheadrightarrow A^A.$$

In the case when $A = \biguplus_{1 \le j \le m} A_j$, $m \ge 2$ is a complex object, with $A^A = A \otimes A$ so that j_A is defined by the set of ptp arrows, $\frac{j_A}{\uplus} = \frac{id_A \otimes id_A}{\uplus}$ with $\widetilde{j_A} \simeq \biguplus \frac{j_A}{\uplus} = \biguplus \frac{id_A \otimes id_A}{\uplus} = A \otimes A$, and consequently, by Proposition 8 in Sect. 3.2, j_A is an epimorphism.

Analogously, for all other morphisms used in this proof a schema mapping can be provided from which they are derived by an R-algebra α.

It is easy to verify that for each pair of objects (i.e., databases) A and B in **DB** there exists $f_{AB} \in \mathbf{DB}(B^A, (TB)^{TA})$, called an *action* of T on B^A, such that for all $g \in \mathbf{DB}(A, B)$ the following is valid:

$$f_{AB} \circ \Lambda\big(g \circ \beta(A)\big) = \Lambda\big(T(g) \circ \beta(TA)\big) : \Upsilon \longrightarrow (TB)^{TA},$$

where $\beta : \Upsilon \otimes \_ \longrightarrow I_{DB}$ is a left identity natural transformation of a monoid $(\mathbf{DB}, \otimes, \Upsilon, \alpha, \beta, \gamma)$, thus $\beta(A) = is_A^{-1} : \Upsilon \otimes A \to A$ (where $\Upsilon \otimes A = TA$ from Theorem 12) and $\beta(TA) = id_{TA}$. In fact, we take $\widetilde{f_{AB}} = id_{B^A}$ (the action of T on $f : A \to B$ is equal to $Tf : TA \to TB$ with $\widetilde{Tf} = \widetilde{f}$), which is an identity action; moreover, $B^A \simeq A \otimes B = TA \otimes TB \simeq TB^{TA}$. Hence, by considering that for $A = \biguplus_{1 \le j \le m}, m \ge 1$, $B = \biguplus_{1 \le i \le k}, k \ge 1$ and $is_A^{-1} : TA \to A$ with $\frac{is_A^{-1}}{\uplus} = \{is_{A_j}^{-1} : TA_j \to A_j \mid 1 \le j \le m\}$, we obtain (see the proof of Theorem 15):

$$\frac{\Lambda(T(g))}{\uplus} = \left\{T(g_{ji}) \mid (g_{ji} : A_j \to B_i) \in \frac{g}{\uplus}\right\}$$

and

$$\begin{aligned}
\frac{\Lambda(g \circ is_A^{-1})}{\uplus} &= \left\{g'_{ji} \mid \big(g_{ji} \circ is_{A_j}^{-1} : TA_j \to B_i\big) \in \frac{g \circ is_A^{-1}}{\uplus}, \widetilde{g'_{ji}} = \widetilde{g_{ji} \circ is_{A_j}^{-1}}\right\} \\
&= \left\{g'_{ji} \mid g_{ji} \in \frac{g}{\uplus}, \widetilde{g'_{ji}} = \widetilde{g_{ji}} \cap \widetilde{is_{A_j}^{-1}}\right\} \\
&= \left\{g'_{ji} \mid g_{ji} \in \frac{g}{\uplus}, \widetilde{g'_{ji}} = \widetilde{g_{ji}} \cap TA_j\right\} \\
&\quad (\text{from } \widetilde{g_{ji}} \subseteq A_j \otimes B_i = TA_j \cap TB_i \subseteq TA_j) \\
&= \left\{g'_{ji} \mid g_{ji} \in \frac{g}{\uplus}, \widetilde{g'_{ji}} = \widetilde{g_{ji}}\right\}.
\end{aligned}$$

Thus, for each $g_{ji} \in \frac{g}{\uplus}$, $\widetilde{g'_{ji}} = \widetilde{g_{ji}} = T(g_{ji})$, i.e., $g'_{ji} = T(g_{ji})$, and consequently,

$$\Lambda\big(g \circ is_A^{-1}\big) = \Lambda\big(T(g)\big). \tag{8.3}$$

Hence, $f_{AB} \circ \Lambda(g \circ \beta(A)) = id_{B^A} \circ \Lambda(g \circ is_A^{-1}) = \Lambda(g \circ is_A^{-1}) =^{(1)} \Lambda(T(g)) = \Lambda(T(g) \circ id_{TA}) = \Lambda(T(g) \circ \beta(TA))$.

Consequently, T is a closed endofunctor.

The composition law $m_{A,B,C}$ may be equivalently represented by a natural transformation $m : (B \otimes \_) \otimes (\_ \otimes B) \longrightarrow \otimes$ and an identity element j_A by natural transformation $j : Y \longrightarrow \otimes \circ \Delta$, where $\Delta : \mathbf{DB} \longrightarrow \mathbf{DB} \times \mathbf{DB}$ is a diagonal functor, while $Y : \mathbf{DB} \longrightarrow \mathbf{DB}$ is a constant endofunctor, $Y(A) \triangleq \Upsilon$ for any A and $Y(f) \triangleq id_\Upsilon$ for any arrow f in **DB**. It is easy to verify that two coherent diagrams (associativity and unit axioms) commute, and hence **DB** is enriched over itself V-category (as, for example, **Set** category).

T is a V-functor: for each pair of objects A and B there exists an identity map (see above) $f_{AB} : B^A \longrightarrow (TB)^{TA}$, subject to the compatibility with composition m and with the identities expressed by the commutativity of the diagram

$$\begin{array}{ccc}
C^B \otimes B^A & \xhookrightarrow{\quad m_{A,B,C} \quad} & C^A \\
{\scriptstyle f_{BC}} \Big\downarrow \otimes \Big\downarrow {\scriptstyle f_{AB}} & & \Big\downarrow {\scriptstyle f_{AC}} \\
TC^{TB} \otimes TB^{TA} & \xhookrightarrow{\quad m_{TA,TB,TC} \quad} & TC^{TA}
\end{array}$$

that is, $f_{AC} \circ m_{A,B,C} = m_{TA,TB,TC} \circ (f_{BC} \otimes f_{AB})$ and $j_{TA} = f_{AB} \circ j_A$. From the fact that f_{BC} and f_{AB} are the identity arrows, $(f_{BC} \otimes f_{AB})$ is identity as well, so it is enough to show that $m_{TA,TB,TC} = m_{A,B,C}$. In fact, for any $A = \biguplus_{1 \le j \le m}, m \ge 1$, $B = \biguplus_{1 \le i \le k}, k \ge 1$ and $C = \biguplus_{1 \le l \le n}, n \ge 1$, and from $C^B \simeq B \otimes C$, $B^A \simeq A \otimes B$, we obtain

$$\begin{aligned}
&\frac{m_{A,B,C}}{\biguplus} \\
&= \big\{in_{ilji_1} : (B_i \otimes C_l) \otimes (A_j \otimes B_{i_1}) \hookrightarrow A_j \otimes C_l \\
&\qquad |1 \le j \le m, 1 \le i, i_1 \le k, 1 \le l \le n\big\} \\
&= \big\{in_{ilji_1} : (TB_i \cap TC_l) \cap (TA_j \cap TB_{i_1}) \hookrightarrow TA_j \cap TC_l \\
&\qquad |1 \le j \le m, 1 \le i, i_1 \le k, 1 \le l \le n\big\} \\
&= \big\{in_{ilji_1} : \big(T(TB_i) \cap T(TC_l)\big) \cap \big(T(TA_j) \cap T(TB_{i_1})\big) \hookrightarrow T(TA_j) \cap T(TC_l) \\
&\qquad |1 \le j \le m, 1 \le i, i_1 \le k, 1 \le l \le n\big\} \\
&= \big\{in_{ilji_1} : (TB_i \otimes TC_l) \otimes (TA_j \otimes TB_{i_1}) \hookrightarrow TA_j \otimes TC_l \\
&\qquad |1 \le j \le m, 1 \le i, i_1 \le k, 1 \le l \le n\big\} \\
&= \frac{m_{TA,TB,TC}}{\biguplus},
\end{aligned}$$

that is, $m_{TA,TB,TC} = m_{A,B,C}$.

It is easy to verify that also natural transformations

$$\eta : I_{DB} \longrightarrow T \quad \text{and} \quad \mu : T^2 \longrightarrow T$$

satisfy the V-naturality condition (V-natural transformation η and μ are an Ob_{DB}-indexed family of components $\delta_A : \Upsilon \twoheadrightarrow TA$ in **DB** (for η, $\delta_A : \Upsilon \twoheadrightarrow (TA)^A$, $(TA)^A \approx TA$; while for μ, $\delta_A : \Upsilon \twoheadrightarrow TA^{T^2A}$, $(TA)^{T^2A} \approx TA$). This map f_{AB} is equal also for the endofunctor identity I_{DB}, and for the endofunctor T^2, because $B^A \simeq A \otimes B \simeq (TB)^{TA} = (T^2B)^{T^2A}$. □

In fact, each monoidal closed category is itself a V-category: hom-sets from A to B are defined as "internalized" hom-objects (cotensors) B^A. The composition is given by the image of the bijection $\Lambda : \mathbf{DB}(D \otimes A, C) \simeq \mathbf{DB}(D, C^A)$, where $D = C^B \otimes B^A$, of the arrow $\varepsilon_B \circ (id_{C^B} \otimes \varepsilon_A) \circ \alpha_{C^B,B^A,A}$ and hence $m_{A,B,C} = \Lambda(\varepsilon_B \circ (id_{C^B} \otimes \varepsilon_A) \circ \alpha_{C^B,B^A,A}) = \Lambda(eval_{B,C} \circ (id_{C^B} \otimes eval_{A,B}) \circ \alpha_{C^B,B^A,A})$, that is, a monomorphism. The identities are given by the image of the isomorphism $\beta_A : \Upsilon \otimes A \longrightarrow A$, under the bijection $\Lambda : \mathbf{DB}(\Upsilon \otimes A, A) \simeq \mathbf{DB}(\Upsilon, A^A)$ and hence $j_A = \Lambda(\beta_A) : \Upsilon \twoheadrightarrow A^A$ (j_A that is an epimorphism).

Moreover, for a V-category **DB**, the following isomorphism (which extends the tensor–cotensor isomorphism Λ of exponential diagram in Theorem 15) is valid in all enriched Lawvere theories [23], $\mathbf{DB}(D \otimes A, C) \simeq \mathbf{DB}(D, C^A) \simeq \mathbf{DB}(A, \mathbf{DB}(D, C))$.

Finally, from the fact that **DB** is an lfp category enriched over the lfp symmetric monoidal closed category with a tensor product $\otimes$ (i.e., the matching operator for databases), and the fact that T is a finitary enriched monad on **DB**, from Kelly–Power theorem we have that **DB** admits a presentation by operations and equations, what was implicitly assumed in the definition of this power-view operator (see also [16, 19]).

8.2.2 Internalized Yoneda Embedding

As we have seen in Sect. 8.2.1, the fact that a monoidal structure is closed means that we have an internal hom-functor $(\_)^{(\_)} : \mathbf{DB}^{OP} \times \mathbf{DB} \rightarrow \mathbf{DB}$ which "internalizes" the external hom-functor, $hom : \mathbf{DB}^{OP} \times \mathbf{DB} \rightarrow \mathbf{Set}$, such that for any two objects A and B, the hom-object $B^A = (\_)^{(\_)}(A, B)$ represents the hom-set $hom(A, B) = \mathbf{DB}(A, B)$ (the set of all morphisms from A to B). We may apply the Yoneda embedding to the **DB** category by the contravariant functor $H : \mathbf{DB}^{OP} \rightarrow \mathbf{Set}^{\mathbf{DB}}$, such that for any object A we obtain a functor $H_A = \mathbf{DB}(A, \_) : \mathbf{DB} \rightarrow \mathbf{Set}$ and hence for any object B in **DB**, $H_A(B) = hom(A, B) = \mathbf{DB}(A, B)$ is a hom-set of all morphisms in **DB** between A and B, and for any morphism $f : B \rightarrow C$ in **DB** the function $H_A(f) = \mathbf{DB}(A, f) : \mathbf{DB}(A, B) \rightarrow \mathbf{DB}(A, C)$ is a composition $H_A(f) = f \circ \_$.

The internalized Yoneda embedding can be obtained by replacing the base **Set** category with the base database category **DB** and hence this embedding is provided by the contravariant functor $\overline{H} : \mathbf{DB}^{OP} \rightarrow \mathbf{DB}^{\mathbf{DB}}$. So, we internalize a hom-set $H_A(B) = hom(A, B) = \mathbf{DB}(A, B)$ in **Set** by the (co)tensor $\overline{H}_A(B) = A \otimes B$ in **DB**, and we internalize a composition function $H_A(f) = f \circ \_$ in **Set** by the (co)tensor composition of morphisms in **DB**, $\overline{H}_A(f) = id_A \otimes f : A \otimes B \rightarrow A \otimes C$.

This internalized Yoneda embedding for **DB** category satisfies the following property:

Proposition 56 *For any object A in* **DB** *with a monomorphisms* $in_A : A \hookrightarrow \Upsilon$, *the natural transformation* $\phi : \overline{H}_A \rightarrow T$ *is monomorphic and such that* $\phi = in_A \otimes id_{\{\_\}}$.

Proof For any object A, from the internalized Yoneda embedding $\overline{H} : \mathbf{DB}^{OP} \to \mathbf{DB}^{\mathbf{DB}}$, we obtain the covariant functor $\overline{H}_A = \overline{H}(A)$ such that $\overline{H}_A = A \otimes \_ : \mathbf{DB} \to \mathbf{DB}$. Consequently, for the covariant power-view functor $T : \mathbf{DB} \to \mathbf{DB}$, the natural transformation $\phi : \overline{H}_A \to T$ is well defined and monomorphic if, for *every* arrow $f : B \to C$ in $\mathbf{DB}$, the following diagram commutes

$$\begin{array}{ccc}
\overline{H}_A(B) = A \otimes B & \xrightarrow{\overline{H}_A(f) = id_A \otimes f} & A \otimes C = \overline{H}_A(C) \\
\phi_B \big\downarrow & & \big\downarrow \phi_C \\
\Upsilon \otimes B = TB & \xrightarrow{Tf} & TC = \Upsilon \otimes C
\end{array}$$

Notice that, from the fact that for each $A \in Ob_{DB}$, $A \preceq \Upsilon$ and $\overline{H}_{\Upsilon} = \Upsilon \otimes \_ = T$. Thus, $\phi : \overline{H}_A \to \overline{H}_{\Upsilon}$, that is, for any object B, $\phi_B = in_A \otimes id_B : A \otimes B \hookrightarrow \Upsilon \otimes B$ and $\phi_C = in_A \otimes id_c$, where $in_A : A \hookrightarrow \Upsilon$ is the unique monomorphism (in this case the inclusion) of A into the total object Υ, so that the diagram above can be reduced to the following commutative diagram in $\mathbf{DB}$ for any $f \in \mathbf{DB}(B, C)$),

$$\begin{array}{ccc}
\overline{H}_A(B) = A \otimes B & \xrightarrow{\overline{H}_A(f) = id_A \otimes f} & A \otimes C = \overline{H}_A(C) \\
\phi_B = in_A \otimes id_B \big\downarrow & & \big\downarrow \phi_C = in_A \otimes id_C \\
TB = \overline{H}_{\Upsilon}(B) = \Upsilon \otimes B & \xrightarrow{id_{\Upsilon} \otimes f = Tf} & \Upsilon \otimes C = \overline{H}_{\Upsilon}(C) = TC
\end{array}$$

so that $Tf \circ \Phi_B = (id_{\Upsilon} \otimes f) \circ (in_A \otimes id_B) =$ (from functorial property of $\otimes$) $= (id_{\Upsilon} \circ in_A) \otimes (f \circ id_B) = in_A \otimes f = (in_A \circ id_A) \otimes (id_C \circ f) = (in_A \otimes id_C) \circ (id_A \otimes f) = \phi_C \circ \overline{H}_A(f)$. Consequently, $\phi = id_{\{\_\}} \otimes in_A$. □

This proposition provides the idea in which way we are able, by using the contravariant functor $\overline{H} : \mathbf{DB}^{OP} \to \mathbf{DB}^{\mathbf{DB}}$, to define the internalized Yoneda representation of the arrows in $\mathbf{DB}^{OP}$ by the natural transformations in the category of functors $\mathbf{DB}^{\mathbf{DB}}$: by assuming that in Proposition 56, the natural transformation $\phi = \overline{H}(in_A^{OP})$ is just the Yoneda representation of the epimorphism $in_A^{OP} : \Upsilon \twoheadrightarrow A$ in $\mathbf{DB}^{OP}$ (i.e., of the monomorphism $in_A : A \hookrightarrow \Upsilon$ in $\mathbf{DB}$).

Corollary 30 *The internalized Yoneda representation is provided by the contravariant functor* $\overline{H} : \mathbf{DB}^{OP} \to \mathbf{DB}^{\mathbf{DB}}$*, defined as follows:*

- *For each object A, we define the endofunctor* $\overline{H}_A = \overline{H}(A) : \mathbf{DB} \to \mathbf{DB}$ *such that for any* $f : B \to C$, $\overline{H}_A(f) = \overline{H}(A)(f) = id_A \otimes f : \overline{H}_A(B) \to \overline{H}_A(C)$, *where* $\overline{H}_A(B) = B \otimes A \simeq A^B$.

- *For each arrow* $g^{OP} : D \rightarrow A$ *in* $\mathbf{DB}^{OP}$, *we define the natural transformation* $\phi = \overline{H}(g^{OP}) = g \otimes id_{\{\_\}} : \overline{H}(A) \rightarrow \overline{H}(D)$.

In fact, this corollary can be represented by the following commutative diagram in **DB**:

$$\begin{array}{ccc} \overline{H}_A(B) = A \otimes B & \xrightarrow{\overline{H}_A(f) = id_A \otimes f} & A \otimes C = \overline{H}_A(C) \\ \phi_B = g \otimes id_B \downarrow & & \downarrow \phi_C = g \otimes id_C \\ \overline{H}_D(B) = D \otimes B & \xrightarrow{\overline{H}_D(f) = id_D \otimes f} & D \otimes C = \overline{H}_D(C) \end{array}$$

8.3 Database Mappings and (Co)monads: (Co)induction

The notion of a monad is one of the most general mathematical notions for *every* algebraic theory, that is, every set of operations satisfying equational laws can be seen as a monad (which is also a *monoid* in a category of endofunctors of a given category: the "operation" μ being the associative multiplication of this monoid and η its unit). In Sect. 5.1.1, we considered in details the syntax monads $(\mathcal{T}_P, \eta, \mu)$ in the **Set** category for the relational algebras and, for the particular Σ_R (Codd's SPRJU relational algebra), the relationship with the power-view monad T of the DB category, which will be here analyzed in more details.

We used monads to provide the *denotational semantics to database mappings*, as it was explained in Sect. 5.1.1: in order to interpret database mappings (i.e., morphisms) in the category **DB**, we distinguish the object A (a database instance of type (or schema) $\mathcal{A}$) from the object TA of observations (i.e., the computations of type $\mathcal{A}$ without side-effects), and take as a denotation of (view) mappings the elements of TA (which are the views of (type) $\mathcal{A}$). In particular, we identify the A with the object of values (of type $\mathcal{A}$), and hence we obtain the object of observations by applying the unary type-constructor T (power-view operator introduced in Sect. 3.2.1) to A.

It is well known that each endofunctor defines the algebras and coalgebras (the left and right commutative diagrams bellow), based on Definition 24 in Sect. 3.2.1:

$$\begin{array}{ccc} TA & \xrightarrow{Tf} & TB \\ h \downarrow & & \downarrow k \\ A & \xrightarrow{f} & B \end{array} \qquad\qquad \begin{array}{ccc} TA & \xrightarrow{Tf_1} & TB \\ h_1 \uparrow & & \uparrow k_1 \\ A & \xrightarrow{f_1} & B \end{array}$$

We recall that a monad (T, η, μ) (from Proposition 9 in Sect. 3.2.1, the endofunctor $T : \mathbf{DB} \to \mathbf{DB}$ is a monad), given by commutative diagrams

$$\begin{array}{ccc} T^3A & \xrightarrow{T\mu_A} & T^2A \\ {\scriptstyle \mu_{TA}}\downarrow & & \downarrow{\scriptstyle \mu_A} \\ T^2A & \xrightarrow{\mu_A} & TA \end{array} \qquad \begin{array}{ccccc} TA & \xrightarrow{\eta_{TA}} & T^2A & \xleftarrow{T\eta_A} & TA \\ & {\scriptstyle id_A}\searrow & \downarrow{\scriptstyle \mu_A} & \swarrow{\scriptstyle id_A} & \\ & & TA & & \end{array}$$

defines an adjunction $(F^T, G^T, \eta^T, \epsilon^T) : \mathbf{DB} \longrightarrow CT_{\text{alg}}$ such that $G^T \circ F^T = T : \mathbf{DB} \longrightarrow \mathbf{DB}$, $\eta^T = \eta$, $\epsilon^T = \eta^{OP}$ and $\mu = G^T \epsilon^T F^T$. The functors $F^T : \mathbf{DB} \to CT_{\text{alg}}$ (into the Eilenberg–Moore category of *monadic* T-algebras, associated to the monad (T, η, μ)) and forgetful functor $G^T : CT_{\text{alg}} \to \mathbf{DB}$ are defined as follows: for any object (database) A, $F^T(A) = (TA, \mu_A : T^2A \to TA)$, $F^T(f) = Tf$ while $G^T(A, g : TA \to A) = A$ and $G^T(f) = f$.

Let us consider the following properties for monadic algebras/coalgebras in **DB** (introduced by Definition 24 and 25 in Sect. 3.2.1):

Proposition 57 *The following properties for the monad (T, η, μ) and the comonad (T, η^C, μ^C) hold:*

- *The Eilenberg–Moore categories CT_{alg} and CT_{coalg} of the monad $T : \mathbf{DB} \longrightarrow \mathbf{DB}$ (from Proposition 9) are isomorphic ($CT_{\text{coalg}} = CT_{\text{alg}}^{OP}$), complete and cocomplete. The object $(\bot^0, id_{\bot^0} : \bot^0 \longrightarrow \bot^0)$ is an initial T-algebra in CT_{alg} and a terminal T-coalgebra in CT_{coalg}.*
- *For each object A in **DB** category there exist the unique monadic T-algebra $(A, \eta_A^C : TA \longrightarrow A)$ and the unique comonadic T-coalgebra $(A, \eta_A : A \longrightarrow TA)$ with $\eta_A^C = \eta_A^{OP}$ (i.e., $\eta_A^C = is_A^{-1} \approx \eta_A = is_A \approx id_A$).*
- *The free monadic T-algebra $(TA, \mu_A : T^2A \longrightarrow TA)$ is dual (and equal) to the cofree monadic T-coalgebra $(TA, \mu_A^C : TA \longrightarrow T^2A)$ with $\mu_A^C = \mu_A^{OP}$ (i.e., $\mu_A^C = \mu_A = id_{TA}$).*

Proof Let us define the functor $F : CT_{\text{alg}} \longrightarrow CT_{\text{coalg}}$ such that for any T-algebra $(A, h : TA \longrightarrow A)$ we obtain the dual T-coalgebra $F^0(A, h) = (A, h^{OP} : A \longrightarrow TA)$, where the component F^1 for the arrows is an identity function; and the functor $F : CT_{\text{coalg}} \longrightarrow CT_{\text{alg}}$ such that for any T-coalgebra $(A, k : A \longrightarrow TA)$ we obtain the dual T-algebra $G^0(A, k) = (A, k^{OP} : TA \longrightarrow A)$, where the component G^1 for the arrows is an identity function. Hence, $FG = I_{T_{\text{coalg}}}$ and $GF = I_{T_{\text{alg}}}$. The categories CT_{alg} and CT_{coalg} are complete and cocomplete as **DB** category ($CT_{\text{coalg}} = CT_{\text{alg}}^{OP}$). The rest is easy to verify (by Proposition 9 and Corollary 11 in Sect. 3.2.1): each monadic T-algebra/coalgebra is an isomorphism. The free monadic T-algebra and the cofree monadic T-coalgebra are equal because $TA = T^2A$, thus, μ_A, μ_A^C are identity arrows (from duality property of **DB**). □

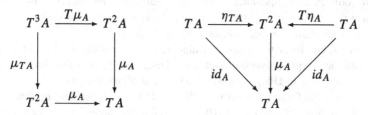

As we can see, each monadic T-coalgebra is an *equivalent inverted* arrow in **DB** of some monadic T-algebra and vice versa: the fundamental duality property of **DB** introduces the equivalence of monadic T-algebras and monadic T-coalgebras and hence the equivalence of the dichotomy "*construction* versus *observation*" or duality between induction and coinduction principles [9].

We have seen (from the Universal algebra considerations in Sect. 8.1.4) that there exists the unique universal functor $U : \mathbf{DB}_{sk} \longrightarrow \mathbf{K}$ such that for any simple instance-database A in $\mathbf{DB}_{sk}$ it returns the free Σ_R-algebra $U(A)$.

Its adjoint is the forgetful functor $F : \mathbf{K} \longrightarrow \mathbf{DB}_{sk}$ such that for any free algebra $U(A)$ in **K** the object $F \circ U(A)$ in $\mathbf{DB}_{sk}$ is equal to its carrier-set A (for each n-ary operation $o_i \in \Sigma_R$, the term $o_i(R_1, \dots, R_n) \in U(A)$ with $R_i \in A$, $i = 1, \dots, n$, is evaluated into some view (*computed* relation) of this closed object A in $\mathbf{DB}_{sk}$).

Finitariness. In a locally finitely presentable (lfp) category, every object can be given as the directed (or filtered) colimit of the finitely presentable (fp) objects. Hence, if the action of a monad preserves this particular kind of colimits, its action on any object will be determined by its action on the fp objects; such a monad is called finitary.

Let us verify that the power-view closure 2-endofunctor $T : \mathbf{DB} \longrightarrow \mathbf{DB}$ is a *finitary* monad.

Proposition 58 *The power-view closure 2-endofunctor* $T : \mathbf{DB} \longrightarrow \mathbf{DB}$ *is immediate from the universal property of composed adjunction* $(UT_{sk}, In_{sk}F, In_{sk}\eta_U T_{sk} \cdot \eta_{sk}, \varepsilon_U \cdot U\varepsilon_{sk}F) : \mathbf{DB} \longrightarrow \mathbf{K}$, *i.e.,* $T = In_{sk}FUT_{sk} \simeq Id_{DB}$. *It is finitary.*

The category **DB** *is equivalent to the (Eilenberg–Moore) category* T_{alg} *of all monadic T-algebras and is equivalent to the category* T_{coalg} *of all monadic T-coalgebras.*

Its equivalent skeletal category $\mathbf{DB}_{sk}$ *is, instead, isomorphic to* T_{alg} *and* T_{coalg}.

Proof For any object A in **DB**, $In_{sk}FUT_{sk}(A) = In_{sk}T_{sk}(A) = TA$, and for any morphism $f : A \longrightarrow B$ in **DB**, $In_{sk}FUT_{sk}(f) = In_{sk}K_{sk}(f) = In_{sk}(f_T) = Tf$ (where, from Theorem 14, $f_T = T^1_{sk}(f)$ and $\widetilde{f_T} = \widetilde{Tf} = \widetilde{f}$).

The adjunction-equivalence $(T_{sk}, In_{sk}, \eta_{sk}, \varepsilon_{sk})$ between **DB** and $\mathbf{DB}_{sk}$ and the adjunction-isomorphism $(U, F, \eta_U, \varepsilon_U)$ of $\mathbf{DB}_{sk} \simeq \mathbf{K}$ produce a composed adjunction $(UT_{sk}, In_{sk}F, In_{sk}\eta_U T_{sk} \cdot \eta_{sk}, \varepsilon_U \cdot U\varepsilon_{sk}F) : \mathbf{DB} \longrightarrow \mathbf{K}$, which is an equivalence.

Hence, $\mathbf{K} \simeq \mathbf{DB}_{sk}$ and, from universal algebra (Back's theorem) theory, $\mathbf{K} \simeq T_{\text{alg}}$. Thus, $\mathbf{DB}_{sk} \simeq T_{\text{alg}}$. From this fact and the fact that **DB** is equivalent to $\mathbf{DB}_{sk}$, we obtain that **DB** is equivalent to T_{alg}. Similarly, by duality, we obtain that **DB** is equivalent to T_{coalg}.

In order to show the finitarity condition, let us consider the term algebra $U(A)$ over a simple database A with *infinite* set of relations. Since every operation $\rho \in \Sigma_R$ can only take finitely many arguments, every term $t \in U(A)$ can only contain finitely many variables from A. Hence, instead of building the term algebra over the *infinite* database A, we can also build the term algebras over *all finite subsets*(of relations) A_0 of A and take union of these: $U(A) = \bigcup\{U(A_0) \mid A_0 \subseteq_\omega A\}$. This result comes directly from Universal algebra because the closure operator T is *algebraic* and $(\mathcal{C}, \subseteq)$, where $\mathcal{C}$ is a set of all closed simple objects in $\underline{\mathbf{DB}}$, is an algebraic (complete+compact) lattice. □

The notion of T-algebra subsumes the notion of a Σ_R-algebra (Σ_R-algebras can be understood as algebras in which the operators (of the signature) are not subject to any law, i.e., with empty set of equations). In particular, the monad $\mathcal{T}_P$ freely generated by a signature Σ_R is such that $\mathcal{T}_{\mathrm{alg}}$ is isomorphic to the category of Σ_R-algebras. Therefore, the *syntax* of a programming language can be identified with monad, the *syntactical monad* $\mathcal{T}_P$ freely generated by the program constructors Σ_R.

More formally, as it was presented in Sect. 5.1.1, for the signature Σ_R there is an endofunctor $\Sigma_R : \mathbf{Set} \longrightarrow \mathbf{Set}$ such that for any object B, $\Sigma_R(B) \triangleq \biguplus_{o_i \in \Sigma_R} B^{ar(o_i)}$, and for any arrow in $\mathbf{Set}$ (a function) $f : B \longrightarrow C$, $\Sigma_R(f) \triangleq \biguplus_{o_i \in \Sigma_R} f^{ar(o_i)}$.

Thus, also for any object A in $\mathbf{Set}$ we have an endofunctor $\Sigma_{R_A} = A \uplus \Sigma_R : \mathbf{Set} \longrightarrow \mathbf{Set}$ such that for any object B in $\mathbf{Set}$,

$$\Sigma_{R_A}(B) = (\Sigma_R \uplus A)(B) = A \uplus \Sigma_R B \triangleq A \biguplus_{\sigma \in \Sigma_R} B^{ar(\sigma)},$$

and for any arrow $f : B \longrightarrow C$, $\Sigma_{R_A}(f) \triangleq id_A \biguplus_{\sigma \in \Sigma_R} f^{ar(\sigma)}$.

We define an *iteratable* endofunctor H of a category $\mathbf{D}$ if for every object X of $\mathbf{D}$ the endofunctor $X \uplus H(\_)$ has an initial algebra. It is well known that the signature endofunctor Σ_R in $\mathbf{Set}$ category is ω-cocontinuous and iteratable.

The problem which will be analyzed in details in what follows is: What are the concepts in the $\mathbf{DB}$ category (and relative commutative diagrams) corresponding to the initial Σ_R-algebra concepts and syntax monad $\mathcal{T}_P X$ for the SPRJU relational algebra terms expressed by the commutative diagrams in the $\mathbf{Set}$ category? For example, the set of terms $\mathcal{T}_P X$ in the initial algebra semantics diagram is not an object in the $\mathbf{DB}$ category, but $\alpha_\#$ image which is a set TA of the relations (views) obtained by evaluation of terms in $\mathcal{T}_P X$ is a power-view object in $\mathbf{DB}$. Hence, the power-view monad $T : \mathbf{DB} \to \mathbf{DB}$ is a natural correspondence to the syntax monad $\mathcal{T}_P : \mathbf{Set} \to \mathbf{Set}$.

This process of translation of (co)inductive principles from standard $\mathbf{Set}$ into $\mathbf{DB}$ category will be elaborated in the next sections.

We have to take in mind the difference between $\mathbf{Set}$, as an elementary topos used for ordinary Lawvere theories, that is, to study equational theories (varieties in Universal algebra), and finitary monads on $\mathbf{Set}$ (for example, Moggi's leading examples of monads in the setting of call-by-value λ-calculus, with examples drawn primarily from the programming language ML and adopted in functional programming, in particular in the development of the language Haskell), from our monoidal closed V-category $\mathbf{DB}$ in the setting of enriched Lawvere theories.

More specifically, the idea of a monad as a model of computations, where the programs take values in A, say, to computations in TB, Kleisli categories provide the right setting to describe this approach and hence will be presented in a dedicated Sect. 8.4.

8.3.1 DB Inductive Principle and DB Objects

We have shown in Sect. 5.1.1 that for any schema $\mathcal{A}$ and an R-algebra α which is a model of $\mathcal{A}$, such that the simple instance-database $A = \alpha^*(\mathcal{A})$ is an object in **DB**, the set of views (relations) TA is determined from the *unique epic* $(X \uplus \Sigma_R)$-homomorphism $\alpha_\#$, from the initial $(X \uplus \Sigma_R)$-algebra $(\mathcal{T}_P X, \simeq)$ into this $(X \uplus \Sigma_R)$-algebra $(TA, [\alpha, h_A])$. This is represented by the commutative diagram in **Set** (here X is the set of relational symbols in database schema $\mathcal{A}$)

$$
(\mathcal{A},\alpha) \quad \mapsto \quad
\begin{array}{ccccc}
X = \mathcal{A} & \overset{inl_X}{\hookrightarrow} & \mathcal{T}_P X & \overset{inr_X}{\hookleftarrow} & \Sigma_R(\mathcal{T}_P X) \\
 & \searrow^{\alpha} & \downarrow{\scriptstyle \alpha_\#} & & \downarrow{\scriptstyle \Sigma_R(\alpha_\#)} \\
 & & TA & \overset{h_A}{\longleftarrow} & \Sigma_R(TA)
\end{array}
$$

where Σ_R and $\mathcal{T}_P X$ are the SPRJU algebra and the set of its terms with variables in X, respectively, given by Definition 31 (see also Example 29 in Sect. 5.1.1), with

$$\Sigma_R(X) = \biguplus_{o_k \in \Sigma_R} X^{ar(o_k)} = \bigcup_{o_i \in \Sigma_R, ar(o_i)=1, i \geq 1} \{X \times \{i\}\} \bigcup_{i=-1,0} \{X^2 \times \{i\}\}.$$

Consequently, the *inductive principle*, used to construct unique $\alpha_\#$ from α, is in strict relationship with the properties of the *objects* in the **DB** category and especially with the construction of closed objects that are *least fixed points* of the monotonic power-view operator T. We will see that the dual *coinductive principle* is instead in a strict relationship with the *morphisms* in the **DB** category.

From the fact that **DB** is an lfp category (Corollary 29) enriched over the lfp symmetric monoidal closed category (Theorem 15) with a tensor product $\otimes$ (matching operator for databases), and the fact that T is a finitary enriched monad on **DB** (Theorem 16), by Kelly–Power theorem, **DB** admits a presentation by operations and equations, so that **DB** is the category of models [17] for an essentially algebraic theory.

In what follows, we will use the category $\mathbf{DB}_I$ (the "poset" subcategory of **DB** with the same objects and with only monic arrows, $in_B : B \hookrightarrow A$ iff $B \preceq A$ (i.e., $TB \subseteq TA$ for the simple objects), introduced by Theorem 6 in Sect. 3.2.5). Then we will introduce a functor $\Sigma_D : \mathbf{DB}_f \to \mathbf{DB}$, where $\mathbf{DB}_f$ is a full subcategory of $\mathbf{DB}_I$ composed of only *finite objects* (i.e., the databases), for the signature of relational algebra w.r.t. the lfp category **DB** enriched over itself but where an arrow is not a function (differently from standard algebra signature Σ_R defined over **Set** category where an arrow is a function) but *a set* of functions. This definition of Σ_D is correct because all sigma operations $o_i \in \Sigma_R$ of a relational algebra are finitary, i.e., with arity $n = ar(o_i)$ being a finite number. Thus, a view-mapping arrow $f_{o_i} : A \to TA$ in **DB**, equal (from Definition 17 in Sect. 3.1.2) to $\alpha^*(\mathbf{M}_{A\hat{A}}) = \{\alpha(q_i), q_\bot\}$

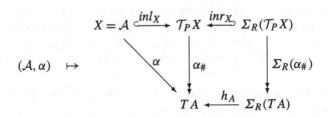

where $\mathbf{M}_{A\widehat{A}} = \{q_i, 1_\emptyset\} = MakeOperads(\{\forall \mathbf{x}_i(q_i(\mathbf{x}_i) \Rightarrow r_q(\mathbf{x}_i))\}) : \mathcal{A} \to \widehat{\mathcal{A}}$, which represents such an operation from a database A into the closed database TA will have a finite cardinality of $\partial_0(f_{o_i}) \subseteq_\omega A$, with cardinality $|\partial_0(f_{o_i})| = ar(o_i)$, so that we can restrict Σ_D to finite databases only. The extension of Σ_D to *all* databases, as infinite databases which are not closed objects (i.e., compact objects in **DB**), can be successively obtained by left Kan extension of this finite restriction as will be demonstrated in what follows.

First of all, we have to demonstrate the existence of the ω-cocontinuous endofunctors for the **DB** category which can be used for a construction of the initial algebra based on morphisms of the **DB** category. We have demonstrated in Proposition 53 that the "merging with A" endofunctor $\sum_A = A \oplus \_ : \mathbf{DB} \longrightarrow \mathbf{DB}$ is ω-cocontinuous, i.e., a monotone function which preserves l.u.bs of ω-chains.

Let us define another ω-cocontinuous endofunctor:

Proposition 59 *For each object A in the category* **DB** *with coproduct* $+$, *equal from Theorem 5 to the disjoint union, the endofunctor* $A + T\_ = A \uplus T\_ : \mathbf{DB} \longrightarrow \mathbf{DB}$ *is ω-cocontinuous.*

Proof Constant endofunctor $A : \mathbf{DB} \to \mathbf{DB}$ is an ω-cocontinuous endofunctor, identity endofunctors are ω-cocontinuous, colimit functors (thus coproduct $+$, which is, by Theorem 5 in Sect. 3.3.2, equal to disjoint union $\uplus$) are ω-cocontinuous (because of the standard "interchange of colimits"). Since ω-cocontinuity is preserved by functor composition $\circ$, for the second endofunctor $A + T\_ = (A + Id\_) \circ T$ it is enough to show that T is an ω-cocontinuous endofunctor. In fact, let us consider the following diagram obtained by an iterative application of the endofunctor T

$$\bot^0 \preceq_0 (T\bot^0) \preceq_1 (T^2\bot^0) \preceq_2 \cdots T^\omega \bot^0,$$

where $\bot^0$ is the initial object in **DB** and all objects $T^n\bot^0 = \bot^0$ and hence all arrows in this chain are identities. Thus, $ColimJ = T^\omega\bot^0 = \bot^0$, and $TColimJ = ColimTJ = \bot^0$, so that T is an ω-cocontinuous endofunctor. □

In what follows, we will make the translation of the inductive principle from the **Set** into **DB** category, for any given database schema $\mathcal{A}$ and R-algebra α, based on the following considerations:

- The object (i.e., a set of relational symbols of a database schema) $\mathcal{A}$ in **Set** is considered as set of variables $X = \mathcal{A}$, while an instance of this schema $A = \alpha^*(\mathcal{A})$, obtained by R-algebra $\alpha : X \to TA$, is an object in **DB**. Analogously, the set of terms with variables $\mathcal{T}_P X$ used in the **Set** category is translated (by the surjective function $\alpha_\# : \mathcal{T}_P X \twoheadrightarrow TA$, obtained as a unique homomorphism from the initial Σ_R-algebra of terms $\mathcal{T}_P X$ in **Set**) into the set TA of all views (which are relations obtained by computation of these terms with variables in $X = \mathcal{A}$).

  However, TA (in **DB** it is an *instance database* of a schema $\mathcal{A}$, while in **Set** it is considered as *a set* of relations) is not a carrier set for the initial $(A \uplus \Sigma_D)$-algebra in **DB** for the endofunctor $\Sigma_D = T : \mathbf{DB} \to \mathbf{DB}$ (see bellow) because, generally, TA is not isomorphic in **DB** to $A \uplus \Sigma_D(TA)$ (in fact, $T(TA) = TA \not\simeq TA \uplus TA = TA \uplus TTA = A \uplus \Sigma_D(TA) \simeq T(A \uplus \Sigma_D(TA))$, i.e., $T(TA) \neq T(A \uplus \Sigma_D(TA))$).

- Cartesian product $\times : \mathbf{Set} \to \mathbf{Set}$ is translated into the matching operation (tensor product) $\otimes : \mathbf{DB} \longrightarrow \mathbf{DB}$.

  This translation is based on observations that any n-ary algebraic operator $o_i \in \Sigma_R$ is represented as a function (an arrow in **Set**) $\widehat{o_i} : \mathcal{T}_P X^n \to \mathcal{T}_P X$ which using the n-fold Cartesian product $\mathcal{T}_P X \times \cdots \times \mathcal{T}_P X$ as domain, while such an operator in the **DB** category is represented by an atomic view-mapping $f_{o_i} = \{q_\perp\} \cup \{q_j : \alpha(r_{i1}) \times \cdots \times \alpha(r_{in}) \to \alpha_\#(o_i(r_{i1}, \ldots, r_{in})) \mid$ for each subset of relational symbols $r_{i1}, \ldots, r_{in} \in \mathcal{A} \underset{\alpha}{\overset{\cap}{-}} \mathcal{A}\} : TA \to TA$, where $\alpha_\# : \mathcal{T}_P X \to TA$ is the unique extension (of the initial algebra semantics in the diagram in the previous section) of an R-algebra (i.e., an assignment) $\alpha : X \to A \subseteq TA$ for the relational symbols in X.

  Thus, this algebraic operator o_i is translated into an arrow from TA into TA.

  In fact, if we replace $\times$ by $\otimes$ in the n-fold $\overbrace{\mathcal{T}_P X \times \cdots \times \mathcal{T}_P X}^{n}$, we obtain (for a simple database A) $\overbrace{TA \otimes \cdots \otimes TA}^{n} = \overbrace{T(TA) \cap \cdots \cap T(TA)}^{n} = T(TA) = TA$.
- Any disjoint union endofunctor $X \uplus \_ : \mathbf{Set} \to \mathbf{Set}$ (used for a construction of the syntax endofunctor $X \uplus \Sigma_R : \mathbf{Set} \to \mathbf{Set}$) is translated into the **DB** endofunctor $A \uplus \Sigma_D : \mathbf{DB} \longrightarrow \mathbf{DB}$ (the coproduct $\uplus$ in **Set** is translated by the coproduct $\uplus$ in **DB**).

  We have only finitary operators $o_i \in \Sigma_R$ and hence to model them we can use only the finite databases, i.e., finite simple objects $A \in Ob_{DB}$. Thus, the set of all functions obtained from the signature Σ_R, $\widehat{o_i} : A^{ar(o_i)} \to TA$, can be internalized in **DB** by a homomorphism $\bigoplus_{o_i \in \Sigma_R} : \Sigma_D(A) \to TA$, where, analogously as in **Set** where $\Sigma_R(X) = \biguplus_{o_i \in \Sigma_R} X^{ar(o_i)}$, here in **DB** we have $\Sigma_D(A) = \bigoplus_{o_i \in \Sigma_R} A^{ar(o_i)}$.

  Consequently, for a finite A we obtain

$$\Sigma_D(A) = \bigoplus_{o_i \in \Sigma_R} \big(\overbrace{A \otimes \cdots \otimes A}^{ar(o_i)}\big) = \bigoplus_{o_i \in \Sigma_R} TA = TA,$$

  and hence the endofunctor Σ_D is a restriction of the power-view endofunctor $T : \mathbf{DB} \to \mathbf{DB}$ to finite objects in **DB**.

It is well known [14] that for any monoidal category **C** with a monoidal product $\otimes$, any two functors $F_1 : \mathbf{P}^{OP} \to \mathbf{C}$ and $F_2 : \mathbf{P} \to \mathbf{C}$ have a tensor functorial product $F_1 \otimes_P F_2 = \int^{p \in \mathbf{P}} (F_1 p) \otimes (F_2 p)$.

In what follows, we will consider the *simple* databases only: these results then are extended to all complex databases, by considering that they are all finite (co)products (equivalent, up to isomorphism, to a disjoint union) of the simple databases.

In our case, we take for **C** the lfp *enriched (co)complete* category **DB** with the monoidal product corresponding to the matching database operation $\otimes$, with $F_2 = \Sigma_D : \mathbf{DB}_f \to \mathbf{DB}$ and F_1 being the hom-functor for a given simple database A (i.e., an object in **DB**), $\mathbf{DB}_I(\_, A) \circ K : \mathbf{DB}_f \to \mathbf{DB}$, where $K : \mathbf{DB}_f \hookrightarrow \mathbf{DB}_I$ is an inclusion functor. Notice that for any finite database, i.e., object $B \in \mathbf{DB}_f$, the $\mathbf{DB}_I(K(B), A)$ is a hom-object $A^{K(B)}$ of enriched database subcategory $\mathbf{DB}_I$. In this context, we obtain that for any object (also infinite) A in $\mathbf{DB}_I$ (that is, in **DB**), we have a tensor product

$$\begin{aligned}\big(\mathbf{DB}_I(\_, A) \circ K\big) \otimes_P \Sigma_D &= \int^{B \in \mathbf{DB}_f} \mathbf{DB}_I\big(K(B), A\big) \otimes \Sigma_D B \\ &= \int^{B \in \mathbf{DB}_f} \mathbf{DB}_I(B, A) \otimes \Sigma_D B.\end{aligned}$$

This tensorial product comes with a dinatural transformation [6], $\beta : S \to A$, where $S = \mathbf{DB}_I(\_, A) \otimes \Sigma_{D\_} : \mathbf{DB}_f^{OP} \times \mathbf{DB}_f \to \mathbf{DB}$ and A is a constant functor between the same categories of the functor S. Thus, for any given object A in **DB** we have a collection of arrows $\beta_B : \mathbf{DB}_I(B, A) \otimes \Sigma_D B \to A$ (for every object $B \in \mathbf{DB}_f$).

In the standard case of **Set**, which is a (co)complete lfp with the monoidal product $\otimes$ equal to the Cartesian product $\times$, such arrows are $\beta_B : \mathbf{Set}(B, A) \times \Sigma_R(B) \to A$, where B is a finite set with cardinality $n = |B|$, so that $\mathbf{Set}(B, A)$ is a set of all tuples of arity n composed of elements of the set A, while $\Sigma_R(B)$ here is *interpreted* as a set of all *basic* n-ary algebra operations. So that β_B is a specification for all basic algebra operations with arity n, and is a function such that for any n-ary operation $o_i \in \Sigma_R(B)$ and a tuple $(a_1, \ldots, a_n) \in A^n \simeq \mathbf{Set}(B, A)$ (where $\simeq$ is an isomorphism in **Set**) returns the result $\beta_B((a_1, \ldots, a_n), o_i) = \widehat{o_i}(a_1, \ldots, a_n) \in A$. If for a given B with $n = |B|$, $\Sigma_R(B)$ is the empty set (that is, we have no n-ary operators in Σ_R), then $\mathbf{Set}(B, A) \times \Sigma_R(B) = \mathbf{Set}(B, A) \times \emptyset = \emptyset$ ($\emptyset$ is the initial object in **Set**), so that $\int^{B \in \mathbf{DB}_f} \mathbf{DB}_I(B, A) \otimes \Sigma_R B \simeq \int^{B \in \mathbf{DB}_f \& \Sigma_R B \neq \emptyset} \mathbf{DB}_I(B, A) \otimes \Sigma_R B$. In this case, $\beta_B : \emptyset \to A$ is an empty function.

In the nonstandard case when, instead of the base category **Set**, another lfp-enriched (co)complete category is used, as the **DB** category in our case, the *interpretation* for this tensorial product and dinatural transformation β is obviously very different, as we will see in what follows. First of all, the hom-set $\mathbf{DB}_I(B_i, A)$ in the preorder DB category is internalized in a hom-object in $\mathbf{DB}_I$ as follows:

$\mathbf{DB}_I(B_i, A) = \widetilde{in_i} = TB$ if there exists a monomorphism $in_i : B_i \hookrightarrow A$ in $\mathbf{DB}_I$; $\bot^0$ otherwise.

Based on the considerations explained before, we obtain that the finitary signature functor $\Sigma_D : \mathbf{DB}_f \to \mathbf{DB}$ has a left Kan extension of Σ_D along the category inclusion $K : \mathbf{DB}_f \to \mathbf{DB}_I$ [5] in the enriched category **DB**, $Lan_K(\Sigma_D) : \mathbf{DB}_I \to \mathbf{DB}$, and left Kan extension $Lan_{J \circ K}(\Sigma_D) : \mathbf{DB} \to \mathbf{DB}$ for the inclusion functor $J : \mathbf{DB}_I \hookrightarrow \mathbf{DB}$ (this second extension is a direct consequence of the first one because J does not introduce extension for objects, differently from K, from the fact that $\mathbf{DB}_I$ and **DB** have the same objects). Thus it is enough to analyze only the first left Kan extension (left one) given by the following commutative diagrams:

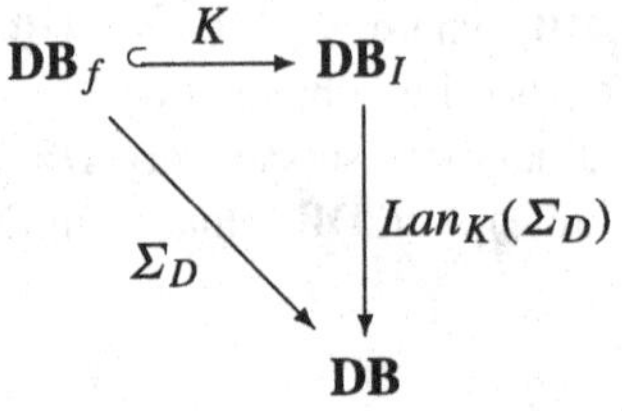

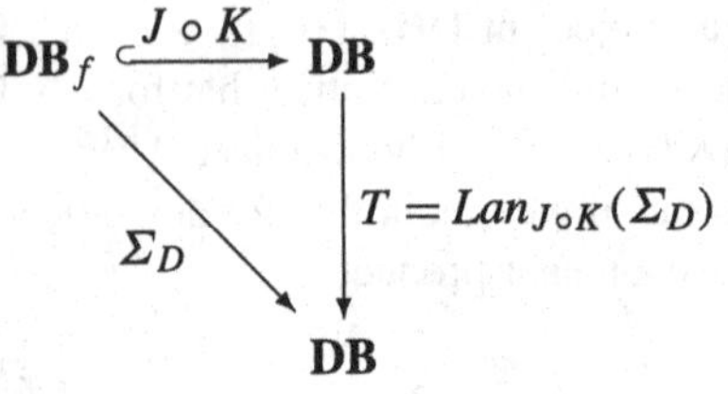

That is, we have the functor $Lan_K : \mathbf{DB}^{\mathbf{DB}_f} \to \mathbf{DB}^{\mathbf{DB}_I}$ which is a left adjoint to the functor $G = \_ \circ K : \mathbf{DB}^{\mathbf{DB}_I} \to \mathbf{DB}^{\mathbf{DB}_f}$, and hence the left Kan extension of $\Sigma_D \in \mathbf{DB}^{\mathbf{DB}_f}$ along K is given by functor $Lan_K(\Sigma_D) \in \mathbf{DB}^{\mathbf{DB}_I}$, and natural transformation $\eta : \Sigma_D \to Lan_K(\Sigma_D) \circ K$ is an universal arrow. That is, for any other functor $S : \mathbf{DB}_I \to \mathbf{DB}$ and a natural transformation $\alpha : \Sigma_D \to S \circ K$, there is a unique natural transformation $\sigma : Lan_K(\Sigma_D) \to S$ such that $\alpha = \sigma K \bullet \eta$ (where $\bullet$ is a vertical composition for natural transformations).

This adjunction is denoted by a tuple $(Lan_K, G, \varepsilon, \eta)$, where

$$\eta_C : C \to W\big(Lan_K(C)\big)$$

is a universal arrow for each object C (let us consider the case when $C = \Sigma_D$). In fact, for any object (i.e., a functor) S in $\mathbf{DB}^{\mathbf{DB}_I}$ and a morphism (i.e., a natural transformation) $\alpha : \Sigma_D \to G(S)$, there exists a unique morphism $\sigma : Lan_K(\Sigma_D) \to S$ such that the following two adjoint diagrams commute:

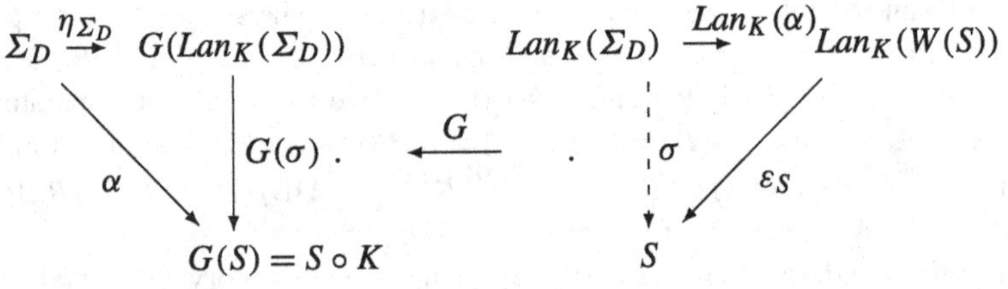

so that η_{Σ_D} is a universal arrow from Σ_D into G and, dually, ε_S is a couniversal arrow from Lan_K into S. Consequently, in this adjunction, the unit η generates a universal arrow for each object in $\mathbf{DB}^{\mathbf{DB}_f}$ and the counit ε generates a couniversal arrow for each object in $\mathbf{DB}^{\mathbf{DB}_I}$.

From the well know theorem for left Kan extensions, when we have a tensorial product $\int^{B \in \mathbf{DB}_f} \mathbf{DB}_I(K(B), A) \otimes \Sigma_D B$ for every $A \in \mathbf{DB}_I$, i.e., $A \in \mathbf{DB}$, the function for the objects of the functor $Lan_K(\Sigma_D)$ (here $B \preceq_\omega A$ means that $B \in \mathbf{DB}_f$ & $B \preceq A$, and we take the case when $A \notin \mathbf{DB}_f$ is an *infinite* database which is not a closed object in **DB**) is defined by:

$$Lan_K(\Sigma_D)(A) \triangleq \int^{B \in \mathbf{DB}_f} \mathbf{DB}_I\big(K(B), A\big) \otimes \Sigma_D B$$

$$= \int^{B \in \mathbf{DB}_f} \mathbf{DB}_I(B, A) \otimes \Sigma_D B \quad \big(\text{from } K(B) = B\big)$$

$$= \int^{B \preceq_\omega A} \mathbf{DB}_I(B, A) \otimes \Sigma_D B$$

$$+ \int^{B \in \mathbf{DB}_f \ \& \ B \not\preceq_\omega A} \mathbf{DB}_I(B, A) \otimes \Sigma_D B.$$

Since a hom-object $\mathbf{DB}_I(B, A)$ is an empty database $\bot^0$ (zero object in **DB**) if there is no (monic) arrow from B to A, we can continue as follows:

$$\begin{aligned}
&Lan_K(\Sigma_D)(A) \\
&= \int^{B \preceq_\omega A} \mathbf{DB}_I(B, A) \otimes \Sigma_D B + \bot^0 \\
&\overset{(*)}{\simeq} \int^{B \preceq_\omega A} \mathbf{DB}_I(B, A) \otimes \Sigma_D B \quad (\text{by Corollary 13}) \\
&= \int^{B \preceq_\omega A} \widetilde{in_B} \otimes \Sigma_D B \quad (\text{here } in_B : B \hookrightarrow A \text{ is a unique monic arrow into } A) \\
&= \int^{B \preceq_\omega A} TB \otimes \Sigma_D B \\
&= \int^{B \preceq_\omega A} TB \otimes TB \quad \big(\text{for finite } B,\ \Sigma_D(B) = T(B)\big) \\
&= \int^{B \preceq_\omega A} TB \\
&= \bigvee \big\{T(B) \mid B \preceq_\omega A\big\} \\
&\quad \big(\text{l.u.b. of compact elements of directed set } \{B \mid B \preceq_\omega A\}\big) \\
&= T(A),
\end{aligned}$$

from the fact that, by Corollary 28, the poset $\mathbf{DB}_I$ is a complete algebraic lattice [17] $(\mathbf{DB}_I, \preceq)$ with the meet and join operators $\otimes$ and $\oplus$, respectively, and with compact elements TB for each *finite* database B.

Remark Let us consider now which kind of interpretation can be associated to the tensor product (see (*) above): $\int^{B \in \mathbf{DB}_f} \mathbf{DB}_I(B, A) \otimes \Sigma_D B \simeq \int^{B \preceq_\omega A} \mathbf{DB}_I(B, A) \otimes \Sigma_D B$ and its B-components (for $B \preceq_\omega A$, that is, $B \subseteq_\omega TB \subseteq TA$), $\mathbf{DB}_I(B, A) \otimes \Sigma_D B$, in the enriched lfp database category **DB**:

The second component $\Sigma_D B$ cannot be a set of signature operators, just because an object in **DB** is not a set of functions, and it is not interesting in this interpretation: in fact, it can be omitted from B-component because it is equal to TB which is the l.u.b. of the first component, i.e., of the hom-object $\mathbf{DB}_I(B, A) = \mathbf{DB}_I(B, TA) = \widetilde{in_B}$, for an inclusion arrow $in_B : B \hookrightarrow TA$.

Hence, the component $\Sigma_D B$ in $\mathbf{DB}_I(B, A) \otimes \Sigma_D B$ means that the significant computation of the object $Lan_K(\Sigma_D)(A) \in \mathbf{DB}$ is done only for the *finite* objects B in **DB**, while the component $\mathbf{DB}_I(B, A)$ means that such objects (i.e., the instance databases) have to be the subobjects of the (also infinite) database A. That is, $B \preceq A$, and hence B has to satisfy the condition $B \preceq_\omega A$.

Thus, the meaning of the natural transformation β is to represent the monomorphism $\beta_B : B \hookrightarrow A$ in **DB** for each finite object $B \preceq_\omega A$. That is, β represents the set of all monomorphisms from *finite* objects into a given (also infinite) object A in $\mathbf{DB}_I$ category: it is *a cocone* of the object A in $\mathbf{DB}_I$ (with a monomorphism $\beta_B = in_\emptyset : \bot^0 \hookrightarrow A$, when $\mathbf{DB}_I(B, A)$ is the empty hom-set).

Notice that the same result can be obtained by considered the colimits in **DB** and hence by considering the object $Lan_K(\Sigma_D)(A) \in \mathbf{DB}$ as a colimit (point-to-point) such that the composition $(K \downarrow A) \xrightarrow{P} \mathbf{DB}_f \xrightarrow{\Sigma_D} \mathbf{DB}$ (where P is a functor-projection, mapping $\langle B, K(B) \to A\rangle \mapsto B$), for each (also infinite) object $A \in \mathbf{DB}_I$, is a colimit in **DB**, with cocone β, denoted by $Lan_K(A) = \overrightarrow{Colim}((K \downarrow A) \xrightarrow{P} \mathbf{DB}_f \xrightarrow{\Sigma_D} \mathbf{DB})$, which is the dinatural transformation β used for a computation of the tensor functor product. Here, instead, this cocone β (a dinatural transformation) is represented by the comma category $(K \downarrow A)$ (where the objects are the monomorphisms from the finite objects $B_i \in \mathbf{DB}_f$ with $K(B_i) = B_i \in \mathbf{DB}_I$ into a fixed object $A \in \mathbf{DB}_I$), that is, to the set β of monomorphisms $\{in_i : K(B_i) \hookrightarrow A \in \mathbf{DB}_I \mid B_i \in \mathbf{DB}_f\}$. The functor $Lan_K(\Sigma_D)$ maps this cocone in $\mathbf{DB}_I$ into the colimit cocone represented by the corresponding set of monomorphisms $\{\preceq_i : \Sigma_D(B_i) \hookrightarrow Lan_K(\Sigma_D)(A) \in \mathbf{DB} \mid B_i \in \mathbf{DB}_f$ so that $\Sigma_D(B_i) = T(B_i)\}$. Thus, the colimit object $Lan_K(\Sigma_D)(A)$ is just the union of all fluxes of the monomorphisms from the finite objects B_i into it, that is,

$$\begin{aligned}
Lan_K(\Sigma_D)(A) &= \bigcup\bigl\{\widetilde{in_i} \mid \bigl(in_i : K(B_i) \hookrightarrow A\bigr) \in \beta\bigr\} \\
&= \bigcup\bigl\{\widetilde{in_i} \mid (in_i : B_i) \hookrightarrow A) \in \beta\bigr\} \\
&= \bigcup\bigl\{TB_i \mid (in_i : B_i) \hookrightarrow A) \in \beta\bigr\} \\
&= \bigcup\{TB_i \mid B_i \preceq_\omega A\} \\
&= \bigvee\{B_i \mid B_i \preceq_\omega A\} \\
&= TA
\end{aligned}$$

in the complete *algebraic* lattice C of closed objects in **DB**, by Corollary 28.

Consequently, we obtain that $Lan_K(\Sigma_D)$, the conservative extension of Σ_D to all (also infinite non-closed) objects in **DB**, is equal (up to isomorphism) to the endofunctor T.

That is, formally we obtain:

Corollary 31 *The following strong relationship between the relational-algebra signature endofunctor Σ_D, translated into the database category* **DB**, *and the closure endofunctor T holds*: $T = Lan_{J \circ K}(\Sigma_D) : \mathbf{DB} \to \mathbf{DB}$.

As we can see the translation of the relational-algebra signature Σ_R is provided by the power-view endofunctor $Lan_{J\circ K}(\Sigma_D) = T : \mathbf{DB} \to \mathbf{DB}$, as informally presented in the introduction.

Consequently, the endofunctor $(A \uplus Lan_{J\circ K}(\Sigma_D)) = (A \uplus T) : \mathbf{DB} \to \mathbf{DB}$, from Proposition 59, is the ω-cocontinuous endofunctor $(A \uplus T) : \mathbf{DB} \to \mathbf{DB}$, with a chain

$$\perp^0 \preceq_0 ((A \uplus T) \perp^0) \preceq_1 ((A \uplus T)^2 \perp^0) \preceq_2 \cdots (A \uplus T)^\omega,$$

where

$$\begin{aligned}
(A \uplus T) \perp^0 &= \perp^0 \uplus A, \\
(A \uplus T)^2 \perp^0 &= (A \uplus T)(\perp^0 \uplus A) = \perp^0 \uplus A \uplus TA, \\
(A \uplus T)^3 \perp^0 &= (A \uplus T)(\perp^0 \uplus A \uplus TA) \\
&= \perp^0 \uplus A \uplus TA \uplus T^2 A = \perp^0 \uplus A \uplus TA \uplus TA,
\end{aligned}$$

and we obtain that the colimit of this diagram in **DB** is $(A \uplus T)^\omega = \perp^0 \uplus A \uplus \coprod_\omega TA$. From the fact that for the coproduct (and initial object $\perp^0$), $\perp^0 \uplus B \simeq B$ for any B, then we can take as the colimit $(A \uplus T)^\omega = A \uplus \coprod_\omega TA$. This colimit is a least fixpoint of the monotone operator $(A \uplus T)$ in the complete lattice of databases in **DB** (by Knaster–Tarski theorem for least fixpoints, $(A \uplus T)((A \uplus T)^\omega) \simeq (A \uplus T)^\omega$).

Notice that the coproduct of two databases A and B in the **DB** category [16, 19] corresponds to completely disjoint databases, in the way that it is not possible to use the relations from both these databases in the *same* query: hence $T(A \uplus B) = TA \uplus TB$, that is, the set of all views of a coproduct $A \uplus B$ is a disjoint union of views of A and views of B.

In fact,

$$\begin{aligned}
(A \uplus T)((A \uplus T)^\omega) &= A \uplus T\Big(A \uplus \coprod_\omega TA\Big) = A \uplus TA \uplus T \coprod_\omega TA \\
&= A \uplus TA \uplus \coprod_\omega TTA = A \uplus TA \uplus \coprod_\omega TA \\
&= A \uplus \coprod_\omega TA = (A \uplus T)^\omega.
\end{aligned}$$

We can denote this isomorphic arrow in the **DB** category, which is the initial $(A \uplus T)$-algebra, by $[inl_A, inr_A] : (A \uplus T)(A \uplus \coprod_\omega TA) \to (A \uplus \coprod_\omega TA)$.

Consequently, the variable injection $inl_A : A \hookrightarrow \mathcal{T}_P A$ in **Set** is translated into the corresponding monomorphism $inl_A = \langle id_A, \perp^1 \rangle : A \hookrightarrow (A \uplus \coprod_\omega TA)$ in the **DB** category, with the information flux $\widetilde{inl_A} = TA \uplus \perp^0 \simeq TA$. The right inclusion function $inr_A : \Sigma_R \mathcal{T}_P A \hookrightarrow \mathcal{T}_P A$ in **Set** is translated into an isomorphism (which is also a monomorphism) $inr_A : \Sigma_D(A \uplus \coprod_\omega TA) \simeq (A \uplus \coprod_\omega TA)$ in the **DB** category, based on the fact that

$$T\left(A \uplus \coprod_\omega TA\right) = TA \uplus T \coprod_\omega TA = TA \uplus \coprod_\omega TA \simeq A \uplus \coprod_\omega TA.$$

So that $\widetilde{inr_A} = TA \uplus \coprod_\omega TA = T(A \uplus \coprod_\omega TA) \cap T(\Sigma_D(A \uplus \coprod_\omega TA))$.

Moreover, by this translation, any Σ_R algebra $h : \Sigma_R B \to B$ in **Set** is translated into an isomorphism in **DB**, $h_D = is_B^{-1} : TB \to B$ with $\widetilde{h_D} = TB$.

Consequently, the initial algebra for a given simple database A with a set of views in TA, of the ω-cocontinuous endofunctor $(A \uplus T) : \mathbf{DB} \longrightarrow \mathbf{DB}$ comes with an induction principle, which we can rephrase follows: For every finite object B (for which $\Sigma_D = T$) and Σ_D-algebra structure $h_D : \Sigma_D B \longrightarrow B$ (which must be an isomorphism) and every simple mapping $f : A \longrightarrow B$ (between the simple objects A and B) there exists a unique arrow $f_\# : A \uplus \coprod_\omega TA \longrightarrow B$ such that the following diagram in **DB**

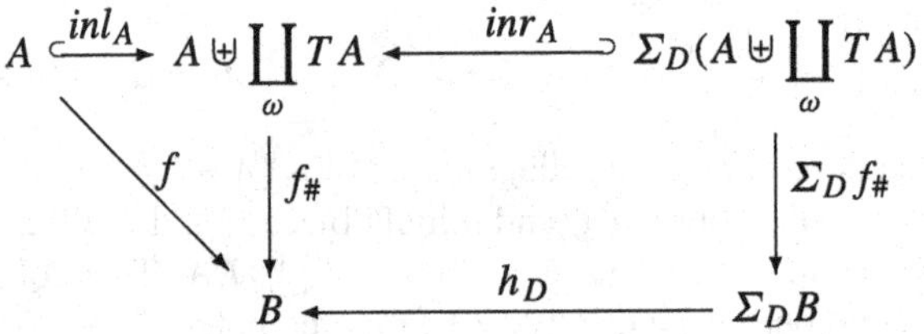

commutes, where $\Sigma_D = T$ and $f_\# = [f, h_D \circ \Sigma_D(f_\#) \circ inr_A^{OP}]$ is the unique *inductive extension of* h_D *along the mapping* f.

It is easy to verify that it is valid. From the fact that

$$\begin{aligned} f_\# \circ inl_A &= [f, h_D \circ \Sigma_D(f_\#) \circ inr_A^{OP}] \circ \langle id_A, \bot^1 \rangle \\ &= \lceil f \circ id_A, h_D \circ \Sigma_D(f_\#) \circ inr_A^{OP} \circ \bot^1 \rfloor = \lceil f, \bot^1 \rfloor, \end{aligned}$$

$$\widetilde{f_\# \circ inl_A} = \widetilde{\lceil f, \bot^1 \rfloor} = \widetilde{f} \uplus \bot^0 \text{ (by Lemma 9)} \simeq \widetilde{f}.$$

Thus, by Definition 23, the left triangle commutes, i.e., $f = f_\# \circ inl_A$.

From Proposition 7, $\widetilde{f_\#} \subseteq T(A \uplus \coprod_\omega TA) \cap TB = \widetilde{inr_A} \cap TB$, and $\widetilde{\Sigma_D f_\#} = T\widetilde{\widetilde{f_\#}} = \widetilde{f_\#}$, and $\widetilde{h_D} = is_B^{-1} = TB$. Consequently,

$$\widetilde{f_\# \circ inr_A} = \widetilde{f_\#} \cap \widetilde{inr_A} = \widetilde{f_\#} = \widetilde{f_\#} \cap TB = \widetilde{h_D \circ \Sigma_D(f_\#)}$$

and hence, form Definition 23, the commutativity of the right square holds, i.e., $h_D \circ \Sigma_D(f_\#) = f_\# \circ inr_A$.

The diagram above can be equivalently represented by the following unique morphism between initial $(A \uplus \Sigma_D)$-algebra and any other $(A \uplus \Sigma_D)$-algebra:

$$\begin{array}{ccc}
A \uplus \coprod_{\omega} TA & \xleftarrow{\;inl_A \uplus inr_A\;} & A \uplus \Sigma_D(A \uplus \coprod_{\omega} TA) \\
\Big\downarrow f_{\#} & & \Big\downarrow id_A \uplus \Sigma_D(f_{\#}) \\
B & \xleftarrow{\;[f, h_D]\;} & A \uplus \Sigma_D(B)
\end{array}$$

Thus, we obtain the following corollary:

Corollary 32 *For each simple object A in the* **DB** *category, there is the initial Σ_A-algebra, $\langle A \uplus \coprod_{\omega} TA,\ [inl_A, inr_A]\rangle$, where $inl_A : A \hookrightarrow (A \uplus \coprod_{\omega} TA)$ is a monomorphism, while $inr_A : \Sigma_D(A \uplus \coprod_{\omega} TA) \hookrightarrow (A \uplus \coprod_{\omega} TA)$ is an isomorphism.*

This inductive principle can be used also to show that the endofunctor T : **DB** $\longrightarrow$ **DB** inductively extends to the composed endofunctor $E = (I_{DB} \uplus \coprod_{\omega} \circ T)$: **DB** $\to$ **DB**, where I_{DB} is the identity endofunctor for **DB**, while the endofunctor $\coprod_{\omega}$: **DB** $\to$ **DB** is an ω-coproduct.

Indeed, to define its action Tf on an arrow $f : A \longrightarrow B$, take the inductive extension of $inr_B : \Sigma_D(B \uplus \coprod_{\omega} TB) \longrightarrow (B \uplus \coprod_{\omega} TB)$ (of the $(B \uplus \Sigma_D)$: **DB** $\longrightarrow$ **DB** endofunctor with initial $(B \uplus \Sigma_D)$-algebra structure

$$[inl_B, inr_B] : (B \uplus \Sigma_D)\Big(B \uplus \coprod_{\omega} TB\Big) \longrightarrow \Big(B \uplus \coprod_{\omega} TB\Big)$$

along the composite $inl_B \circ f$, i.e.,

$$(inl_B \circ f)_{\#} = \Big[inl_B \circ f, inr_B \circ \Sigma_D\Big(f \uplus \coprod_{\omega} Tf\Big) \circ inr_A^{OP}\Big] = f \uplus \coprod_{\omega} Tf.$$

The second diagram

$$\begin{array}{ccccc}
A & \xhookrightarrow{\;inl_A\;} & A \uplus \coprod_{\omega} TA & \xleftarrow{\;inr_A\;} & \Sigma_D(A \uplus \coprod_{\omega} TA) \\
\Big\downarrow f & & \Big\downarrow f \uplus \coprod_{\omega} Tf & & \Big\downarrow \Sigma_D(f \uplus \coprod_{\omega} Tf) \\
B & \xhookrightarrow{\;inl_B\;} & B \uplus \coprod_{\omega} TB & \xleftarrow{\;inr_B\;} & \Sigma_D(B \uplus \coprod_{\omega} TB)
\end{array}$$

commutes [16]. In fact, for the left square we have $(f \uplus \coprod_{\omega} Tf) \circ inl_A = (f \uplus \coprod_{\omega} Tf) \circ \langle id_A, \bot^1\rangle = \ulcorner f \circ id_A, (\coprod_{\omega} Tf) \circ \bot^1 \urcorner = \ulcorner f, \bot^1 \urcorner = \langle id_B, \bot^1\rangle \circ f = inl_B \circ f$.

The proof that the right square commutes as well is the same as the proof given above for the right square of the unique inductive extension of h_D along f.

It is easy to verify that $inl_A = \eta_A : A \hookrightarrow EA$, where $EA = A \uplus \coprod_\omega TA$, is obtained from the natural transformation $\eta : I_{DB} \longrightarrow E$. Another example is the definition of the operation $\mu_A : E^2A \longrightarrow EA$ by inductively extending $inr_A :$ $\Sigma_D EA \longrightarrow EA$ along the identity id_{EA} of the object EA (consider the first diagram, substituting A and B with the object $EA = A \uplus \coprod_\omega TA$, f with id_{EA} and h_D with inr_A). Inductively derived η_A (which is a monomorphism), μ_A (which is an identity, i.e., $\mu_A = id_{EA}$, because $E^2 = E$) and the endofunctor E, define the monad (E, η, μ), i.e., this monad is inductively extended in a *natural way* from the signature endofunctor $\Sigma_D = T : \mathbf{DB} \longrightarrow \mathbf{DB}$.

So the "DB-syntax" monad (E, η, μ), where $E = I_{DB} \uplus \coprod_\omega \circ T = I_{DB} \uplus T \circ \coprod_\omega$ is an inductive algebraic extension of the "coalgebraic" *observation based* power-view monad (T, η, μ).

With this result we internalized the initial **Set**-based SPRJU algebra semantics used for database-mappings, into the initial **DB**-based semantics of the same Σ_R algebra.

8.3.2 DB Coinductive Principle and DB Morphisms

As we have seen in the chapter dedicated to operational semantics of database-mappings, the coalgebras are suitable mathematical formalizations of reactive systems and their behavior, as in our case when we are considering database mappings based on tgds, and their observable information fluxes which are parts of the tuples generated from a source database by these mapping (tgds) and, successively, transferred into the target database.

This subsection provides an application of the corecursion, that is, a construction method which uses the final coalgebras [1]. In order to provide an intuitive understanding of the rest of this section, let us focus on the coalgebraic point of view of a database mappings.

In fact, we know that each tgd used in the schema mappings, or as an integrity constraint over a schema, produces the tuples. Hence, given a sketch's mapping $\mathbf{M}_{AB} = \{q_1, \ldots, q_k, 1_\emptyset\} : \mathcal{A} \to \mathcal{B}$ with $q_i = v_i \cdot q_{A,i} \in O(r_{i1}, \ldots, r_{i_k}, r_{B_i})$, an operad's expression $e \Rightarrow r_{B_i}(\mathbf{t}_i)$, where $v_i \in O(r_{q_i}, r_{B_i})$ (with a new introduced relational symbol r_{q_i} of the same type as r_{B_i}) and $q_{A,i} \in O(r_{i1}, \ldots, r_{i_k}, r_{q_i})$, is an operad's expression $e \Rightarrow r_{q_i}(\mathbf{t}_i)$ with tuple of terms in $\mathbf{t}_i$ composed of variables in e and Skolem functions f_m obtained by Skolemization of the existential quantifiers in a given tgd.

The functorial translation (Lawvere theories) by a mapping-interpretation α (i.e., an R-algebra in Definition 11 and Corollary 4 in Sect. 2.4.1), from the mapping sketch category into the **DB**, satisfies such a schema mapping $\mathbf{M}_{AB}$ if $\alpha(v_i) : \alpha(r_{q_i}) \to \alpha(r_{B_i})$ is an injection. The computation of the relation $\alpha(r_{q_i})$ is done by the function $\alpha(q_{A,i})$ (in Definition 11) obtained from the logical formula $e[(\_)m/r_{im}]_{1 \le m \le k}$, that can be equivalently represented by a composed term $t_e = E(e[(\_)m/r_{im}]_{1 \le m \le k})$ (where E is the algorithm of an equivalent translation of the logical formulae into the relational algebra terms) of the relational algebra with "parameters" in $\{r_{i1}, \dots, r_{i_k}\}$. The operators used for this term are those of the SPRJ algebra (for conjunctive queries). But we need the union operation as well, in the case when we have more than one $q_i \in S \subseteq \mathbf{M}_{AB}$ with the same relational symbol on the right-hand side of their operad's expression, that is, when q_i, q_j are $e \Rightarrow r_B$ and $e' \Rightarrow r_B$, respectively.

So, generally, $\alpha(r_{q_i}) = \bigcup_{q_i \in S} Im(\alpha(q_{A,i}))$ (an a union of the images of the functions $\alpha(q_{A,i})$) and hence provided by the algebraic equation

$$r_{q_i} \approx \text{UNION}\{E(e[(\_)m/r_{im}]_{1 \le m \le k}) \mid q_i = (e \Rightarrow r_B) \in S\}. \tag{8.4}$$

Consequently, we need the '_ UNION _' relational algebra operation as well. Moreover, if we have a function f with arity $m = ar(f)$ in the operad's expressions (i.e., in the SOtgd of considered schema mapping), then we need the unary relational algebra operators 'EXTEND _ ADD a, $name$ AS $name = f(name_1, \dots, name_m)$' as well. The complex equation (8.4) may be flattened, by adding to $X_{\mathbf{M}_{AB}} = \{r_{q_i} \mid q_i \in \mathbf{M}_{AB}\} \subseteq X$ a set of new fresh variables (relational symbols). This process has to be done for all operad's expressions in a schema mapping $\mathbf{M}_{AB}$, with all parameters that are the relations in a source schema $\mathcal{A}$.

Notice that, from the fact that for any mapping $\mathbf{M}_{AB}$, $(1_\emptyset : r_\emptyset \to r_\emptyset) \in \mathbf{M}_{AB}$, $r_\emptyset \in X$.

Thus, the computation of the extension $\alpha(r_{q_i})$ of the variables $r_{q_i} \in X$ is obtained as a solution of this system of equations, by considering the relations $\alpha(r_{i1}) \in A = \alpha^*(\mathcal{A})$ (used on the left side of operad's expressions) as the *parameters*. Thus, for a given instance database $A = \alpha^*(\mathcal{A})$ considered as a set of parameters, with such a flattened guarded system of equations, we compute the extensions of the relations in the target database schema $\mathcal{B}$: the satisfaction of the mapping $\mathbf{M}_{AB}$ means that each condition $\alpha(r_{q_i}) \subseteq \alpha(r_{B_i})$ where $r_{B_i} \in \mathcal{B}$ has to be satisfied.

For any inter-schema mapping $\mathbf{M}_{AB}$, we obtain a non-recursive flattened guarded system of equations. But in the case of the integrity schema constraints based on tgds, when $\mathcal{B} = \mathcal{A}_\top$ (in this case, from Corollary 6 in Sect. 2.4.3, the information flux into the target database $\mathcal{B} = \mathcal{A}_\top$ is always empty), generally we obtain a recursive system of equations (see Example 41 bellow), with (see Example 29 in Sect. 5.1.1)

$$\Sigma_R^+(X) = \biguplus_{o_k \in \Sigma_R^+} X^{ar(o_k)} = \{\bot\} \quad \bigcup_{o_i \in \Sigma_R^+, ar(o_i)=1, i \ge 1} \{X \times \{i\}\} \bigcup_{i=-1,0} \{X^2 \times \{i\}\}$$

so that $\Sigma_R^+(\emptyset) = \{\bot\} = \bot^0$ and $\Sigma_R^+(\bot^0) \simeq S = \{\bot\} \cup \{o_i(\bot) \mid o_i$ is an unary operator of type 'EXTEND...'$\}$, i.e., the set of all unary relations with only one tuple.

Hence, we obtained a recursive system of equations, which can be represented by using a syntax monad (a polynomial endofunctor of **Set**), $\mathcal{T}_P^+ : \mathbf{Set} \to \mathbf{Set}$, derived from the signature Σ_R^+ (which is the SPRJU algebra Σ_R extended by an additional nullary operator '$\bot$' (empty-relation constant) and the set of unary operator 'EXTEND _ ADD a, $name$ AS e') introduced by Definition 31 in Sect. 5.1.

The set $\mathcal{T}_P^+ X$ of its terms with variables is equal to the lfp of the endofunctor $X \uplus \Sigma_R^+ : \mathbf{Set} \to \mathbf{Set}$ and hence, when $X = \emptyset$ is the empty set, the least fixpoint (lfp) $\mathcal{T}_P^+(\emptyset)$ is obtained by the chain (note that in **Set**, the empty set $\emptyset$ is the initial object, so that $\emptyset \uplus Y = Y$):

$$\emptyset \subseteq \emptyset \uplus \Sigma_R^+(\emptyset) = \bot^0 \subseteq \emptyset \uplus \Sigma_R^+\big(\emptyset \uplus \Sigma_R^+(\emptyset)\big) = \Sigma_R^+\big(\Sigma_R^+(\emptyset)\big) = \Sigma_R^+\big(\bot^0\big)$$
$$\subseteq \emptyset \uplus \Sigma_R^+\big(\emptyset \uplus \Sigma_R^+\big(\emptyset \uplus \Sigma_R^+(\emptyset)\big)\big) = \Sigma_R^+\big(\Sigma_R^+\big(\bot^0\big)\big) \subseteq \cdots,$$

by iterating the functor $\emptyset \uplus \Sigma_R^+$ on the empty set $\emptyset$.

If this chain is infinite then we can generate the finitary relations with an infinite number of tuples, but not the relations with an infinite number of attributes (by an infinite number of applications of the operations of type 'EXTEND...' we do not produce the infinitary relations because this operation increments the number of attributes of only finitary relations, so that we always obtain a finitary relation). Hence, based on Definition 31 in Sect. 5.1, in this case we obtain that $\Upsilon \supseteq \{\|t_R\|_\# \mid t_R \in \mathcal{T}_P^+(\emptyset)\}$, because all relations in Υ (from Definition 26 in Sect. 8.1.5, Υ is the union of all *simple* objects (databases) in **DB** category) are finitary (i.e., have a finite arity).

Note that here we do not use the parameters in A as the set of nullary operations in Σ_R^+ (because the signature Σ_R^+ has only unary and binary operators), differently from the analog case of the set of flattened guarded system of equations used in operational semantics (Proposition 37) in Sect. 7.3.2.

Thus, the system of guarded equations above, which defines a mapping from a simple database A to a simple database B, my be expressed by the function $f_e : X \to A \uplus \mathcal{T}_P^+(X)$, such that for the left-hand side (a variable $r \in X$) of a flattened guarded equation $r \approx o_i(r_1, \dots, r_k)$ (with operation $o_i \in \Sigma_R^+$) it returns its right-hand side, i.e., $f_e(r) = o_i(r_1, \dots, r_k) \in \mathcal{T}_P^+(X)$, while for an equation with a parameter R, $r \approx R$, $f_e(r) = R \in A$. Consequently, this function f_e is just a coalgebra (X, f_e) of the polynomial **Set** endofunctor $A \uplus \mathcal{T}_P^+(\_) : \mathbf{Set} \to \mathbf{Set}$ with the signature Σ_R^+.

It is well known [1] that such polynomial endofunctors of **Set** have a *final coalgebra* which is the algebra of all finite and infinite Σ-labeled trees with leafs in A (which are the ground terms). Hence, $T_\infty(A) \simeq A \uplus \mathcal{T}_P^+(T_\infty(A))$, i.e., $T_\infty(A)$ is the greatest fixpoint (gfp) of the endofunctor $A \uplus \mathcal{T}_P^+(\_)$ obtained by union of the chain

$$\emptyset \subseteq A \uplus \mathcal{T}_P^+(\emptyset) \subseteq A \uplus \mathcal{T}_P^+\big(A \uplus \mathcal{T}_P^+(\emptyset)\big) \subseteq \cdots,$$

i.e., obtained by iterating the functor $A \uplus \mathcal{T}_P^+$ on the empty set, so that each parameter $R \in A \subseteq T_\infty(A)$ is a single node tree (tree composed of only the leaf R) in $T_\infty(A)$.

So, the flattened guarded system of equations defined by a mapping f_e has the *unique* solution $f_s : X \to T_\infty(A)$, which is the $A \uplus \mathcal{T}_P^+(\_)$-coalgebra homomorphism from the coalgebra (X, f_e) into the final coalgebra $(T_\infty(A), \simeq)$, as represented by the following commutative diagram in **Set** category:

$$\begin{array}{ccccc} X_{\mathbf{M}_{AB}} & \overset{in}{\hookrightarrow} & X & \xrightarrow{\quad f_s \quad} & T_\infty(A) = gfp(A \uplus \mathcal{T}_P^+) \\ (\alpha, \mathbf{M}_{AB} : \mathcal{A} \to \mathcal{B}) \mapsto & & f_e \big\downarrow & (c.1) & \big\downarrow \simeq_F \\ & & A \uplus \mathcal{T}_P^+(X) & \xrightarrow{A \uplus \mathcal{T}_P^+(f_s)} & A \uplus \mathcal{T}_P^+(T_\infty(A)) \end{array}$$

where the inclusion $in : X_{\mathbf{M}_{AB}} \hookrightarrow X$ represents the generation of new auxiliary variables in X from a given set of variables (defined by a mapping $\mathbf{M}_{AB} : \mathcal{A} \to \mathcal{B}$) during the process of flattening of the complex equations $\alpha(r_{q_i}) = \bigcup_{q_i \in S \subseteq \mathbf{M}_{AB}} Im(\alpha(q_{A,i}))$, for each $r_{q_i} \in X_{\mathbf{M}_{AB}}$.

The isomorphism (a bijection) $\simeq : T_\infty(A) \to A \uplus \mathcal{T}_P^+(T_\infty(A))$ is defined, for any ground-term tree $t \in T_\infty(A)$, by:

- $\simeq (t) = t \in A$ if t is a parameter; $t \in \mathcal{T}_P^+(T_\infty(A))$ if $t = o_i(t_1, t_2)$ or $t = o_i(t_1)$,

while its inverse $\simeq^{-1} : A \uplus \mathcal{T}_P^+(T_\infty(A)) \to T_\infty(A)$, by:

- $\simeq^{-1}(t) = t$ if $t \in A$ or $t \in \mathcal{T}_P^+(T_\infty(A))$.

Consequently, each pair $(\alpha, \mathbf{M}_{AB})$ which defines a morphism $f = \alpha^*(\mathbf{M}_{AB}) : A \to \alpha^*(\mathcal{B})$ in the **DB** category generates this coinductive process with final $(A \uplus \mathcal{T}_P^+)$-coalgebra semantics represented by commutative diagram above, and hence the coinduction is intimately linked with the morphisms in the **DB** category, dually to the induction which is linked with the (closed) objects in the **DB** category.

Notice that the final coalgebra semantics, which defines the unique solution for all variables in $r_{q_i} X_{\mathbf{M}_{AB}} \subseteq X$, will satisfy the schema mapping $\mathbf{M}_{AB}$ if each component function $\alpha(v_i) : f_s(r_{q_i}) \to \alpha(r_{B_i})$, where $\alpha(r_{B_i}) \in B = \alpha^*(\mathcal{B})$, of the morphism $f = \alpha^*(\mathbf{M}_{AB}) : A \to \alpha^*(\mathcal{B})$ in the **DB** category, is an injection. In such a case, the solution subset of relations in $\{f_s(r_{q_i}) \mid r_{q_i} \in X_{\mathbf{M}_{AB}} \subseteq X\}$ defines the information flux of this **DB** morphism f (as specified by the kernel of the information flux in Definition 13 in Sect. 2.4.3).

In Example 29 (Sect. 5.1.1), we have remarked that the nullary operator (empty-relation constant) $\bot \in \Sigma_R^+$ can be used to build the relations. In fact, we may build a tuple of a k-ary relation by the following set of equations with variables (relational symbols) in X (we recall that $r_\emptyset \in X$):

$$r_\emptyset \approx \bot; \quad \text{(it is always satisfied)}$$

$$r_{1,1} \approx \text{EXTEND } r_\emptyset \text{ ADD } at(1), name_1 \text{ AS } name_1 = d_{11};$$

$$r_{1,2} \approx \text{EXTEND } r_{1,1} \text{ ADD } at(1), name_1 \text{ AS } name_1 = d_{12};$$

$$\ldots$$

$$r_{1,k} \approx \text{EXTEND } r_{1,k-1} \text{ ADD } at(k), name_k \text{ AS } name_k = d_{1k},$$

so that $R_1 = f_s(r_{1,k}) = \{(d_{1k}, \ldots, d_{1k})\}$ is a k-ary relation with this tuple. Analogously, we are able to construct all other k-ary tuples and hence, by using the operation 'UNION', we obtain the final k-ary relation. Consequently, *every* relation R may be represented as a term of the Σ_R^+-algebra.

Corollary 33 *For any simple object (database) A,*

$$T_\infty(A) = gfp(A \uplus T_P^+) = gfp(T_P^+) = T_\infty(\bot^0),$$

and hence the diagram (c.1) *above can be substituted by*

$$\begin{array}{ccccc}
 & & T_\infty(\bot^0) = gfp(T_P^+) & \xrightarrow{\simeq_F} & T_P^+(T_\infty(\bot^0)) \\
 & & \uparrow f_s & (c.2) & \uparrow T_P^+(f_s) \\
X_{\mathbf{M}_{AB}} & \xrightarrow{in} & X & \xrightarrow{f_e} & T_P^+(X)
\end{array}$$

Note that $T_\infty(\bot^0)$ is the set of all (also infinite) term-trees with leafs equal to the empty relation $\bot \in \bot^0$, that is, it is a gfp obtained by the union of the following chain:

$$\emptyset \subseteq T_P^+(\emptyset) \subseteq T_P^+(T_P^+(\emptyset)) \subseteq T_P^+(T_P^+(T_P^+(\emptyset))) \subseteq \cdots,$$

and hence $T_P^+(\emptyset) \subseteq T_\infty(\bot^0)$.

Example 41 Let us consider a Data Integration framework, described in Sect. 4.2.3, with the foreign key constraints (FK) in the global schema $\mathcal{G}$ given by Example 27 (Sect. 4.2.4). The database mapping graph of this GAV data integration system, at the logical sketch's level, can be represented by the graph composed of two mapping-operad arrows, $\mathbf{M}_{S\mathcal{G}_T} = MakeOperads(TgdsToSOtgd(\mathcal{M})) : \mathcal{S} \longrightarrow \mathcal{G}_T$, where $\mathcal{G}_T = (S_G, \emptyset)$ is used for the retrieved (from the source database $\mathcal{S}$ only) data and, consequently, has the same global database schema as $\mathcal{G}$ but with empty set of integrity constraints, and the new inter-schema mapping-operad

$$\mathbf{M}_{G_T G} = \{q_i \mid q_i = ((\_)_1(\mathbf{x}_i) \Rightarrow (\_)(\mathbf{x}_i)) \in O(r_i, r_i')$$
$$\text{for each } r_i \in \mathcal{G}_T \text{ and the same relation } r_i' \in \mathcal{G}\} \cup \{1_{r_\emptyset}\} : \mathcal{G}_T \longrightarrow \mathcal{G}$$

that represents the inclusion between retrieved global database $ret(\mathcal{I}, D)$ and the complete canonical global database $can(\mathcal{I}, D)$ (obtained by satisfaction of the integrity constraints Σ_G for the foreign key constraints).

The integrity-constraints mapping operad

$$\mathbf{T}_{G\mathcal{A}_\top} = MakeOperads(\{EgdsToSOtgd(\Sigma_G^{\mathrm{egd}})$$
$$\wedge\, TgdsToCanSOtgd(\Sigma_G^{\mathrm{tgd}})\}) : \mathcal{G} \to \mathcal{A}_\top$$

is satisfied because we assumed that all primary key-integrity constraints for a global schema are satisfied by the retrieved database $ret(\mathcal{I}, D)$, and foreign-key constraints are satisfied by an insertion of the new tuples (with the Skolem constants) in the relations as explained in Sect. 4.2.4.

Based on Example 12, for each egd (PK) for a relation r in $\mathcal{G}$, we obtain the $n-k$ operators $\{q_1, \ldots, q_{n-k}\} \subseteq \mathbf{T}_{GA_\top}$ such that for any $1 \le i \le n-k$,

$$q_i = \big((\_)_1(\mathbf{y}) \wedge (\_)_2(\mathbf{z}) \wedge \big((y_{1_1} = z_{1_1}) \wedge \cdots \wedge (y_{1_k} = z_{1_k})\big)$$
$$\wedge \big((y_{j_1} \neq z_{l_1}) \vee \cdots \vee (y_{j_{n-k}} \neq z_{l_{n-k}})\big)\big) \Rightarrow (\_)(\overline{0}, \overline{1}).$$

For a given set of k foreign-key constraints, represented by tgds (FK)

$$\exists \mathbf{f}_i \big(\forall \mathbf{x}_i \big(r_{i,1}(\mathbf{x}_i) \Rightarrow r_{i,2}(\mathbf{t}_i)\big)\big), \quad i = 1, \ldots, k,$$

where $\mathbf{t}_i$ is a tuple of terms with variables in $\mathbf{x}_i$ and functional symbols in $\mathbf{f}_i$, we obtain $q_i \in O(r_{i,1}, r_{i,2}, r_\top)$, with $q_i = ((\_)_1(\mathbf{x}_i) \wedge \neg(\_)_2(\mathbf{t}_i)) \Rightarrow (\_)(\overline{0}, \overline{1})$.

Let us consider the foreign-key constraints (FK) in a global schema $\mathcal{G}$ given by Example 27 (Sect. 4.2.4). Suppose that we have two relations r and s in $\mathcal{G}$, both of arity 2 and having the first attribute as a key, and the following foreign key constraints in $\mathcal{G}$:

1. $\pi_2(r) \subseteq \pi_1(s)$, of the tgd $\forall x \forall y (r(x, y) \Rightarrow s(y, f_1(x, y)))$, with $\mathbf{K} = \{1\}$ key indexes of s, and $\mathbf{K}' = \{2\}$.
2. $\pi_1(s) \subseteq \pi_1(r)$, of the tgd $\forall x \forall z (s(x, z) \Rightarrow r(x, f_2(x, z)))$, with $\mathbf{K} = \{1\}$ key indexes of r, and $\mathbf{K}' = \{1\}$ and f_1 and f_2 introduced by Skolemization of the existential quantifications.

That is, $\mathcal{G} = (S_G, \Sigma_G)$ with $S_G = \{r, s\}$, $ar(r) = ar(s) = 2$ and

$$\Sigma_G^{\text{tgd}} = \big\{\forall x \forall y \big(r(x, y) \Rightarrow \exists z s(y, z)\big), \forall x \forall z \big(s(x, z) \Rightarrow \exists y r(x, y)\big)\big\}.$$

Thus, in this case we obtain the schema mapping

$$\mathbf{T}_{GA_\top} = MakeOperads\big(\{\Phi\}\big) : \mathcal{G} \to \mathcal{A}_\top,$$

with SOtgd Φ equal to

$$\exists f_1, f_2 \big(\forall x, y \big(r(x, y) \wedge \neg s\big(y, f_1(x, y)\big) \Rightarrow r_\top(\overline{0}, \overline{1})\big)$$
$$\wedge \forall x, y \big(s(x, z) \wedge \neg r\big(x, f_2(x, z)\big) \Rightarrow r_\top(\overline{0}, \overline{1})\big)\big).$$

Consequently, we have to find an R-algebra α such that $can(\mathcal{I}, D) = \alpha^*(\mathcal{G})$ is *a model* of this Data Integration system, that is, such α that satisfies the constraint **DB**-morphism $f_{\Sigma_G} = \alpha^*(\mathbf{T}_{GA_\top}) = \{\alpha(q_1), \alpha(q_2), q_\perp\}$, for

$$q_1 = v_1 \cdot q_{G,1} = (\_)_1(x, y) \wedge \neg(\_)_2\big(y, f_1(x, y)\big) \Rightarrow (\_)(\overline{0}, \overline{1}) \in O(r, s, r_\top),$$

with $q_{G,1} \in O(r, s, r_{q_1})$, $v_1 \in O(r_{q_1}, r_\top)$, and for

$$q_2 = v_2 \cdot q_{G,2} = (\_)_1(x, z) \wedge \neg(\_)_2\big(x, f_2(x, z)\big) \Rightarrow (\_)(\overline{0}, \overline{1}) \in O(s, r, r_\top),$$

with $q_{G,2} \in O(s, r, r_{q_2})$, $v_2 \in O(r_{q_2}, r_\top)$.

The condition that α is a model is satisfied if both functions $\alpha(v_1) : \alpha(r_{q_1}) \to R_=$ and $\alpha(v_2) : \alpha(r_{q_2}) \to R_=$ are injections (we have that for every α, $\alpha(r_\top) = R_=$).

This condition means that α satisfies the implications in Φ which are equivalent to the implications $\forall x, y(r(x, y) \Rightarrow s(y, f_1(x, y)))$ and $\forall x, z(s(x, z) \Rightarrow r(x, f_2(x, z)))$ (they are algebraically expressed by the relations in points 1 and 2 above). As we have examined in Sect. 4.2.4, the canonical model $can(\mathcal{I}, D)$ in this case is infinite. The sketch category derived by this data integration system is denoted by $\mathbf{Sch}(\mathcal{I})$.

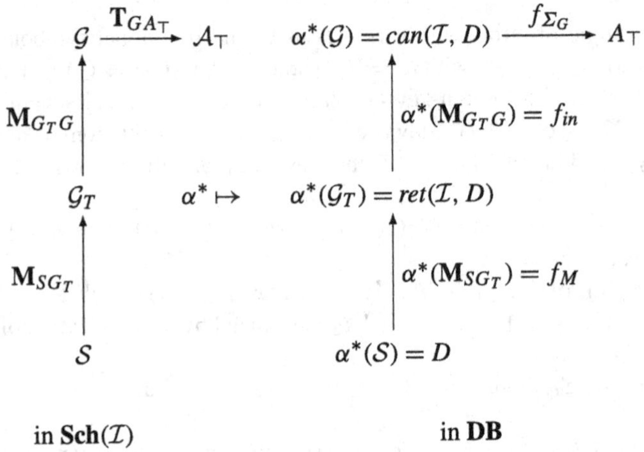

where is the "truth" database $A_\top = \alpha^*(\mathcal{A}_\top) = \{\alpha(r_\top)\}$ with the built-in relation $\alpha(r_\top) = R_=$.

Thus, in order to pass from the logic framework into the Σ_R-algebra framework, we need to transform the two tgds in points 1 and 2 above into the guarded set of equations with Σ_R terms. Let $R = \|r\|_{ret(\mathcal{I},D)}$ and $S = \|r\|_{ret(\mathcal{I},D)}$ be the extensions of r and s in the retrieved (from the source database D) database $ret(\mathcal{I}, D)$, respectively, which are imported into the global database by injective mapping $\mathbf{M}_{G_T G} : \mathcal{G}_T \to \mathcal{G}$. Hence, the final extensions of r and s are determined by the solution of the following flattened guarded system of equations with variables (relational symbols) in $X = \{r_{X,1}, \dots, r_{X,6}\}$, such that $ar(r_{X,3}) = ar(r_{X,5}) = 3$ and all other relational symbols have the arity equal to 2:

$r_{X,1} \approx r_{X,4}$ UNION $r_{X,7}$;

$r_{X,2} \approx r_{X,6}$ UNION $r_{X,8}$;

$r_{X,3} \approx$ EXTEND $r_{X,2}$ ADD $at(3), name_1$ AS $name_1 = f_1(nr(1), nr(2))$;

$r_{X,4} \approx r_{X,3}[nr(2), nr(3)]$;

$r_{X,5} \approx$ EXTEND $r_{X,1}$ADD $at(3), name_2$ AS $name_2 = f_2(nr(1), nr(2))$;

$r_{X,6} \approx r_{X,5}\big[nr(1), nr(3)\big];$

$r_{X,7} \approx R;$

$r_{X,8} \approx S.$

Thus, the flattened system of *recursive* guarded equations above may be expressed by the function $f_e : X \to A \uplus \mathcal{T}_P^+(X)$ (for example, $f_e(r_{X,1}) = r_{X,4}$ UNION $r_{X,7}$, $f_e(r_{X,4}) = r_{X,3}[nr(2), nr(3)]$, $f_e(r_{X,8}) = S \in A$), where the "parameters" are the relations of the database $A = \{R, S\} \in Ob_{DB}$. This system is just a coalgebra of the polynomial endofunctor $A \uplus \mathcal{T}_P^+(\_) : \mathbf{Set} \to \mathbf{Set}$ with the signature Σ_R^+. The right-hand sides of equations (except for two last equations) belong to $\uplus\mathcal{T}_P^+(X)$. The right-hand sides of the last two equations belong to "parameters" in the database $A = \{R, S\} \in Ob_{DB}$.

Hence, the flattened guarded system of equations defined by a mapping f_e has the *unique* solution $f_s : X \to \mathcal{T}_\infty(A)$, which is the $A \uplus \mathcal{T}_P^+(\_)$-coalgebra homomorphism from the coalgebra (X, f_e) into the final coalgebra $(\mathcal{T}_\infty(A), \simeq)$. In our case, when the mapping f_e defines the flattened guarded system of equations presented above, the unique solution of this system, provided by the mapping f_s, defines the R-algebra α which satisfies the constraint mapping $\mathbf{T}_{GA_\top} : \mathcal{G} \to \mathcal{A}_\top$, by:

$$\alpha(r) \triangleq f_s(r_{X,1}),$$

$$\alpha(s) \triangleq f_s(r_{X,2}),$$

so that $\alpha^*(\mathcal{G}) = can(\mathcal{I}, D)$ is a model of the global schema $\mathcal{G}$. In this case of the recursive system of equations with the Skolem functions f_1 and f_2 that introduce the new Skolem constants, both $f_s(r_{X,1})$ and $f_s(r_{X,2})$ are infinite trees (i.e., infinite ground terms with constants in $A = \{R, S\}$).

Let us consider the case when $R = \{(a, b)\}$ and S is an empty relation. Then we obtain the following infinite set of tuples for the relations $\alpha(r)$ and $\alpha(s)$:

$\alpha(r)$	$\alpha(s)$
a, b	b, ω_1
b, ω_2	ω_2, ω_3
ω_2, ω_4	ω_4, ω_5
...	...

where $\omega_1 = f_1(a, b)$, $\omega_2 = f_2(b, \omega_1)$, $\omega_3 = f_1(b, \omega_2)$, $\omega_4 = f_2(\omega_2, \omega_3)$, $\omega_5 = f_1(\omega_2, \omega_4)$, etc.

Diagram (c.2) in Corollary 33 is valid for a general guarded system of equations, and we need to specify the flattening of such a system. It may be provided by specifying the equations with a mapping $f_T : X \to \Sigma_R^+(X)$, so that diagram (c.2) above can be substituted by (notice that $\bot^0 = \{\bot\}$ is a singleton set, thus a terminal object in $\mathbf{Set}$) the following diagram:

$$T_\infty(\perp^0) \xrightarrow{\simeq_1} \perp^0 \times \Sigma_R^+(T_\infty(\perp^0)) \xrightarrow{\simeq} \Sigma_R^+(T_\infty(\perp^0)) \xrightarrow{\simeq_2} T_P^+(T_\infty(\perp^0))$$

$$f_s = f^\# \Big\uparrow \quad (c.3) \qquad \perp^0 \times \Sigma_R^+(f_s) \Big\uparrow \quad (c.4) \qquad \Sigma_R^+(f_s) \Big\uparrow \qquad T_P^+(f_s) \Big\uparrow$$

$$X \xrightarrow{\langle f, f_T \rangle} \perp^0 \times \Sigma_R^+(X) \xrightarrow{\simeq} \Sigma_R^+(X) \xrightarrow{k_X} T_P^+(X)$$

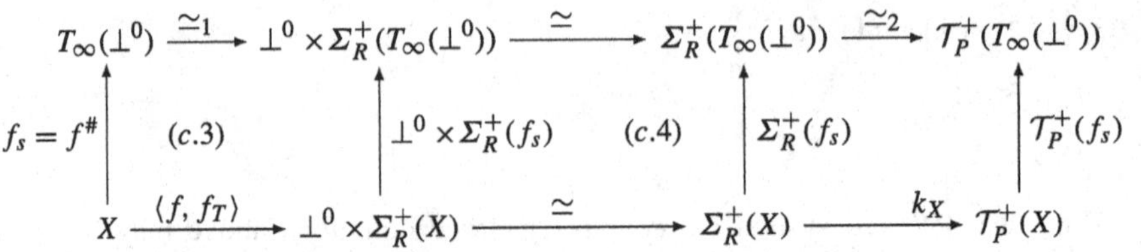

such that $\simeq_F = \simeq_2 \circ \simeq \circ \simeq_1$ and $f_e = k_X \circ \simeq \circ f_T$, where $k_X = inr_X \circ \Sigma_R^+(inl_X)$, with the bijections (isomorphisms in **Set**) $\simeq_1 = \langle p_l, p_r \rangle$ (p_l and p_r are the left and right projections, respectively, as in the diagram bellow), with the isomorphism

$$\simeq : \big(\perp, (t_R, i)\big) \mapsto (t_R, i); \qquad \big(\perp, \big((t_R, t_R'), i\big)\big) \mapsto \big((t_R, t_R'), i\big),$$

and the isomorphisms

$$\simeq_2 = inr_{T_\infty(\perp^0)} : (t_R, i) \mapsto o_i(t_R); \qquad \big((t_R, t_R'), i\big) \mapsto o_i\big(t_R, t_R'\big),$$

with its inverse,

$$\simeq_2^{-1} : o_i(t_R) \mapsto (t_R, i); \qquad o_i\big(t_R, t_R'\big) \mapsto \big((t_R, t_R'), i\big).$$

Diagram (c.3) represents the final coalgebra semantics of the $(\perp^0 \times \Sigma_R^+)$-coalgebras, where $f_s = f^\#$ is the unique homomorphism from the $(\perp^0 \times \Sigma_R^+)$-coalgebra $(X, \langle f, f_T \rangle)$ into the final $(\perp^0 \times \Sigma_R^+)$-coalgebra $(T_\infty(\perp^0), \simeq_1)$, and can be represented by the following commutative diagram in **Set**:

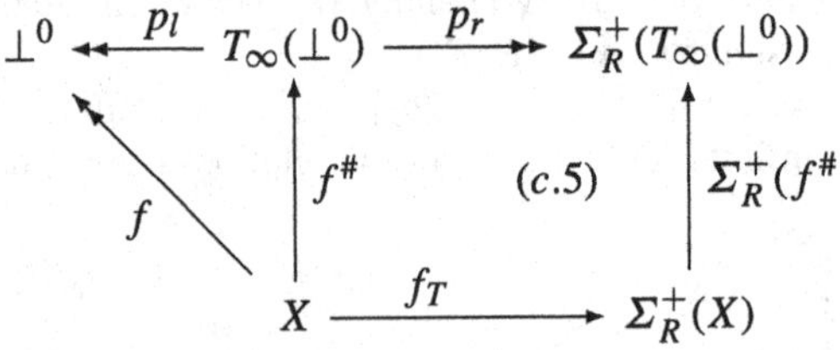

Thus, $f^\#$ is the unique *coinductive extension of f_T along the mapping f*, so that the commutative triangle on the left represents the initialization for the relational symbols (the variables) in X by initial extension equal to $\perp$ (empty relation), while the commutative square (c.5) on the right represents the final solution (extensions) of these relational symbols in X, specified by f_T (the set of *flattened* guarded equations).

The first projection p_l assigns to each tree-term the unique value $\perp \in \perp^0$, while the second projection p_r is a mapping defined by (for any tree-term $t_R, t_R' \in T_\infty(\perp^0)$):

$$\perp \mapsto \perp;$$

$$o_i(t_R) \mapsto (t_R, i), \qquad \text{(for each unary operator with } i \geq 1)$$

$o_i(t_R, t'_R) \mapsto ((t_R, t'_R), i)$, (for the binary unary operators,

'UNION' (with $i = 0$), 'TIMES' (with $i = -1$))

with its inverse $p_r^{OP} : \Sigma_R^+(T_\infty(\bot^0)) \to T_\infty(\bot^0)$ defined by:

$\bot \mapsto \bot$;

$(t_R, i) \mapsto o_i(t_R)$, (for each unary operator with $i \geq 1$)

$((t_R, t'_R), i) \mapsto o_i(t_R, t'_R)$,

so that $p_r \circ p_r^{OP}$ and $p_r^{OP} \circ p_r$ are the identity functions, thus p_r is an isomorphism (bijection), such that $p_r = \simeq \circ \simeq_1$.

Consequently, the diagram (c.3) + (c.4) is equal to the diagram (c.5) and represents the final semantics of the Σ_R^+-coalgebras, where $f_s = f^\#$ is the unique homomorphism from the Σ_R^+-coalgebra (X, f_T) into the final Σ_R^+-coalgebra $(T_\infty(\bot^0), p_r)$.

8.4 Kleisli Semantics for Database Mappings

The basic idea behind the semantic of programs [21] is that a program denotes a morphism from A (the object of *values* of type A) to TB (the object of *computations* of type B), according to the view of "programs as functions from values to computations", so that the natural category for interpreting programs (in our case, a particular equivalent "computation" database mappings of the form $f_1 \triangleq \eta_B \circ f : A \longrightarrow TB$, derived from a database mapping $f : A \longrightarrow B$, such that $f_1 \approx f$) is not the **DB** category, but the Kleisli category $\mathbf{DB}_T$.

Definition 59 Given a monad (T, η, μ) over a category **C**, we have [14]:

- A **Kleisli triple** is a triple $(T, \eta, \_^*)$, where for $f : A \longrightarrow TB$ we have $f^* : TA \longrightarrow TB$ such that the following equations hold: $\eta_A^* = id_{TA}$, $f^* \circ \eta_A = f$ and $g^* \circ f^* = (g^* \circ f)^*$, for $f : A \longrightarrow TB$ and $g : B \longrightarrow TC$.
  A Kleisli triple satisfies the *mono requirement* provided η_A is monic for each object A.
- A **Kleisli category** $\mathbf{C}_T$ has the same objects as **C** category. For any two objects A and B there is the bijection between arrows $\theta : \mathbf{C}(A, TB) \cong \mathbf{C}_T(A, B)$. For any two arrows $f : A \longrightarrow B$ and $g : B \longrightarrow C$ in $\mathbf{C}_T$, their composition is defined by $g \circ f \triangleq \theta(\mu_C \circ T\theta^{-1}(g) \circ \theta^{-1}(f))$.

The mono requirement for the monad (T, η, μ) [22] is satisfied because, by Proposition 9 in Sect. 3.2.1, for each object A the arrow $\eta_A : A \longrightarrow TA$ is an isomorphism $\eta_A = is_A$ (we denote its inverse by η_A^{-1}) and hence it is also monic. Consequently, the category **DB** is a *computational model* for the view-mappings (which are the programs) based on observations (i.e., views) with the typed operator T, so that:

- TA is a *type of computations* (i.e. observations of the object of values A (of a type A), which are the views of the database A).
- η_A is the *inclusion* of values into computations (i.e., inclusion of elements of the database A into the set of views of the database A). It is an isomorphism $\eta_A = is_A : A \longrightarrow TA$ in the **DB** category.
- f^* is the *equivalent extension* of a database mapping $f : A \longrightarrow TB$ "from values to computations" (programs correspond to call-by-value parameter passing) to a mapping "from computations to computations" (programs correspond to call-by-name) such that $f^* = Tf = f \circ \eta_A^{-1} = f \circ is_A^{-1}$ and, consequently, $f^* \approx f$. Thus, in the **DB** category, the call-by-value ($f : A \longrightarrow TB$) and call-by-name ($f^* : TA \longrightarrow TB$) paradigms of programs are represented by *equivalent* morphisms, $f \approx f^*$.

  Notice that in skeletal category $\mathbf{DB}_{sk}$ (which is equivalent to **DB**) all morphisms correspond to the call-by-name paradigm because each arrow is a mapping from computations into computations (which are closed objects).

However, in our case, the Kleisli category is a perfect model only for a subset of database mappings in **DB**: exactly for every view-mapping (i.e., query) $q_A : A \longrightarrow TA$ which is just an arrow in Kleisli category $\theta(q_A) : A \longrightarrow A$. For a general database mapping $f : A \longrightarrow B$ in **DB**, only its (equivalent to f) "computation extension" $\eta_B \circ f : A \longrightarrow TB$ is an arrow $\theta(\eta_B \circ f) : A \longrightarrow B$ in the Kleisli category.

Consequently, the Kleisli category is a model for the database mappings up to the equivalence "$\approx$".

It means that, generally, the database mappings are not simply programs from values into computations. In fact, the semantics of a database mapping, between any two objects A and B, is equal to telling that for some set of computations (i.e, query-mappings) over A we have the same equivalent (in the sense that these programs produce the same computed value, i.e., a view) set of computations (i.e., query-mappings) over B: it is fundamentally an *equivalence* of computations.

This is a consequence of the fact that each database mapping (which is *a set of functions*) from A into B is naturally bidirectional, i.e., a morphism $f : A \longrightarrow B$ is equivalent to its inverted morphism $f^{OP} : B \longrightarrow A$ (explained by the duality property $\mathbf{DB} = \mathbf{DB}^{OP}$ [16]).

Let us define this equivalence formally:

Definition 60 Each simple database mapping $h : A \longrightarrow B$ defines an equivalence of programs (**DB** epimorphisms), $h_A \triangleq \tau(J(h)) : A \twoheadrightarrow TH$ and $h_B \triangleq \tau^{-1}(J(h))^{OP} : B \twoheadrightarrow TH$, where τ and τ^{-1} are natural transformations of the categorial symmetry in Theorem 4, and H is the kernel (in Definition 13) of the information flux $\widetilde{h}$ (i.e., $TH = \widetilde{h}$).

Hence, $h_A \approx h \approx h_B$, such that the computations of these two programs (i.e., arrows of Kleisli category $\mathbf{DB}_T$) are equal, that is, $\partial_1(h_A) = \partial_1(h_B)$. We can also give an alternative model for the equivalent computational extensions of the database mappings in **DB** category:

Proposition 60 *Denotational semantics of each simple mapping f, between any two database instances A and B, is given by the unique equivalent "computation" arrow $f_1 \triangleq \eta_B \circ f$ in T_{coalg} from the monadic T-coalgebra (A, η_A) into the cofree monadic T-coalgebra (TB, μ_B^C), $f_1 : (A, \eta_A) \longrightarrow (TB, \mu_B^C)$ or, dually, by the unique equivalent arrow $f_1^{OP} \triangleq (\eta_B \circ f)^{OP} = f^{OP} \circ \eta_B^{OP}$ from the free monadic T-algebra (TB, μ_B) into the monadic T-algebra (A, η_A^{OP}).*

Proof We have to show that for any morphism $f : A \to B$ between an object $A = \biguplus_{1 \le j \le m} A_j, m \ge 1$ and $B = \biguplus_{1 \le i \le k} B_i, k \ge 1$, $\mu_B^C \circ \eta_B \circ f = T(\eta_B \circ f) \circ \eta_A$ is valid. In fact, by Proposition 57, η_B is the isomorphism $is_B : B \to TB$, so that $Tis_B = id_{TB} : TB \to TB$, due to the fact that

$$\begin{aligned}
\frac{Tis_B}{\uplus} &= \left\{ Tis_{B_i} : TB_i \to TB_i \mid is_{B_i} \in \frac{is_B}{\uplus} \right\} \\
&\quad (\text{from } \widetilde{Tis_{B_i}} = \widetilde{id_{TB_i}} = TB_i, Tis_{B_i} = id_{TB_i}) \\
&= \left\{ id_{TB_i} : TB_i \to TB_i \mid is_{B_i} \in \frac{is_B}{\uplus} \right\} \\
&= \{ id_{TB_i} \mid 1 \le i \le k \} \\
&= \frac{id_{TB}}{\uplus},
\end{aligned}$$

and $\mu_B^C = id_{TB}$. Consequently, $\mu_B^C \circ \eta_B \circ f = id_{TB} \circ is_B \circ f = is_B \circ f$, so that

$$\begin{aligned}
\widetilde{\mu_B^C \circ \eta_B \circ f} &\simeq \biguplus \frac{\widetilde{is_B \circ f}}{\uplus} = \biguplus_{f_{ji} \in \frac{f}{\uplus}} \widetilde{is_{B_i} \circ f_{ji}} \\
&= \biguplus_{f_{ji} \in \frac{f}{\uplus}} \widetilde{is_{B_i}} \cap \widetilde{f_{ji}} = \biguplus_{f_{ji} \in \frac{f}{\uplus}} TB_i \cap \widetilde{f_{ji}} = \biguplus_{f_{ji} \in \frac{f}{\uplus}} \widetilde{f_{ji}} \simeq \tilde{f}.
\end{aligned}$$

Analogously, $T(\eta_B \circ f) \circ \eta_A = T\eta_B \circ Tf \circ is_A = id_{TB} \circ Tf \circ is_A = Tf \circ is_A$. Thus,

$$\begin{aligned}
\widetilde{T(\eta_B \circ f) \circ \eta_A} &\simeq \biguplus \frac{\widetilde{Tf \circ is_A}}{\uplus} = \biguplus_{f_{ji} \in \frac{f}{\uplus}} \widetilde{Tf_{ji} \circ is_{A_j}} \\
&= \biguplus_{f_{ji} \in \frac{f}{\uplus}} \widetilde{Tf_{ji}} \cap \widetilde{is_{A_j}} = \biguplus_{f_{ji} \in \frac{f}{\uplus}} \widetilde{Tf_{ji}} \cap TA_j \\
&= \biguplus_{f_{ji} \in \frac{f}{\uplus}} \widetilde{Tf_{ji}} \simeq \widetilde{Tf} \overset{\text{(by Theorem 2)}}{\simeq} \simeq \tilde{f}.
\end{aligned}$$

Thus, $\widetilde{\mu_B^C \circ \eta_B \circ f} \simeq T(\widetilde{\eta_B \circ f}) \circ \eta_A$, so that $\mu_B^C \circ \eta_B \circ f = T(\eta_B \circ f) \circ \eta_A : A \to T^2 B$. □

Note that each view-map (i.e., a query) $q_A : A \longrightarrow TA$ is equal to its denotational semantics arrow in T_{coalg}, $q_A : (A, \eta_A) \longrightarrow (TA, \mu_A^C)$.

It is well known (see an analog case for Eilenberg–More category in Sect. 8.3) that for a Kleisli category there exists an adjunction $(F_T, G_T, \eta_T, \varepsilon_T)$ such that we obtain the same monad (T, η, μ) such that $T = G_T F_T$, $\mu = G_T \varepsilon_T F_T$, $\eta = \eta_T$ (for any arrow $h \in \mathbf{DB}_T(A, B)$, $G_T^0(A) = TA$, $G_T^1(h) = \mu_B \circ T\theta^{-1}(h)$, and for any arrow $f \in \mathbf{DB}(A, B)$, $F_T^0(A) = A$, $F_T^1(f) = \eta_B \circ f$). Let us see now how the Kleisli category $\mathbf{DB}_T$ is "internalized" into the $\mathbf{DB}$ category.

Theorem 17 *The Kleisli category* $\mathbf{DB}_T$ *of the monad* (T, η, μ) *is isomorphic to the* **DB** *category. That is, it may be "internalized" in* **DB** *by the faithful forgetful functor* $K = (K^0, K^1) : \mathbf{DB}_T \longrightarrow \mathbf{DB}$ *such that* K^0 *is an identity function and* $K^1 \triangleq \phi\theta^{-1}$ *where, for any two objects A and B, we have the following bijections of the hom-sets:*

- $\theta : \mathbf{DB}(A, TB) \cong \mathbf{DB}_T(A, B)$ *is a Kleisli bijection and*
- $\phi : \mathbf{DB}(A, TB) \cong \mathbf{DB}(A, B)$ *such that* $\phi(\_) = \eta_{cod(\_)}^{OP} \circ (\_)$ *is a* **DB** *category bijection, respectively.*

We can generalize a "representation" for the **DB** *category (instead of* **Set** *category):*

A "representation" of functor K is a pair (Υ, φ)*, where* Υ *is the total object and* $\varphi : \mathbf{DB}_T(\Upsilon, \_) \simeq K$ *is a natural isomorphism. The functor* $\mathbf{DB}_T(\Upsilon, \_) : \mathbf{DB}_T \longrightarrow \mathbf{DB}$ *defines the "internalized" hom-sets in* $\mathbf{DB}_T$*, i.e.,* $\mathbf{DB}_T^0(\Upsilon, B) \triangleq TB^{\Upsilon}$ *and* $\mathbf{DB}_T^1(\Upsilon, f_T) \triangleq id_\Upsilon \otimes T\theta^{-1}(f_T)$.

Proof Let us show that ϕ is really a bijection in **DB**. For any program morphism $f : A \longrightarrow TB$ we obtain $\phi(f) = \eta_B^{OP} \circ f : A \longrightarrow B$ and, vice versa, for any $g : A \longrightarrow B$ its inverse $\phi^{-1}(g) \triangleq \eta_B \circ g$, thus, $\phi\phi^{-1}(g) = \phi(\eta_B \circ g) = \eta_B^{OP} \circ (\eta_B \circ g) = (\eta_B^{OP} \circ \eta_B) \circ g = id_B \circ g = g$ (because η_B is an isomorphism) and hence $\phi\phi^{-1}$ is an identity function. Analogously, $\phi^{-1}\phi(f) = \phi^{-1}(\eta_B^{OP} \circ f) = \eta_B \circ (\eta_B^{OP} \circ f) = (\eta_B \circ \eta_B^{OP}) \circ f = id_{TB} \circ f = f$, i.e., $\phi^{-1}\phi$ is an identity function and hence ϕ is a bijection. Let us demonstrate that K is a functor.

For any identity arrow $id_T = \theta(\eta_A) : A \longrightarrow A$ in $\mathbf{DB}_T$ we obtain $K^1(id_T) = \phi\theta^{-1}(\theta(\eta_A)) = \phi(\eta_A) = \eta_A^{OP} \circ \eta_A = id_A$ (because η_A is an isomorphism). For any two arrows $g_T : B \longrightarrow C$ and $f_T : A \longrightarrow B$ in Kleisli category, we obtain

$$K^1(g_T \circ f_T) = K^1\big(\theta\big(\mu_C \circ T\theta^{-1}(g_T) \circ \theta^{-1}(f_T)\big)\big)$$

(from Definition 59 of Kleisli category;

we introduce $g \triangleq \theta^{-1}(g_T) : B \longrightarrow TC$ and $f \triangleq \theta^{-1}(f_T) : A \longrightarrow TB$)

$$= \phi\theta^{-1}\big(\theta(\mu_C \circ Tg \circ f)\big) = \phi\theta^{-1}\big(\theta(id_{TC} \circ Tg \circ f)\big) = \phi\theta^{-1}\big(\theta(Tg \circ f)\big)$$

$$\left(\text{it is easy to verify in } \mathbf{DB} \text{ that } Tg = g \circ \eta_B^{OP}, \text{ from } \widetilde{Tg} \simeq \widetilde{g} \simeq \biguplus \frac{\widetilde{g \circ \eta_B^{OP}}}{\uplus}\right)$$

$$= \phi\big(g \circ \eta_B^{OP} \circ f\big) = \eta_C^{OP} \circ g \circ \eta_B^{OP} \circ f = \phi(g) \circ \phi(f)$$

$$= \phi\theta^{-1}\big(\theta(g)\big) \circ \phi\theta^{-1}\big(\theta(f)\big)$$

$$= K^1\big(\theta\theta^{-1}(g_T)\big) \circ K^1\big(\theta\theta^{-1}(f_T)\big) = K^1(g_T) \circ K^1(f_T).$$

Thus, each arrow $f_T : A \longrightarrow B$ in $\mathbf{DB}_T$ is "internalized" in **DB** by its representation $f \triangleq K^1(f_T) = \phi\theta^{-1}(f_T) = \eta_B^{OP} \circ \theta^{-1}(f_T) : A \longrightarrow B$, where $\theta^{-1}(f_T) : A \longrightarrow TB$ is a program equivalent to the database mapping $f : A \longrightarrow B$, i.e., $\theta^{-1}(f_T) \approx f$.

The functor K is faithful. In fact, for any two arrows $f_T, h_T : A \longrightarrow B$ in $\mathbf{DB}_T$, $K^1(f_T) = K^1(h_T)$ implies $f_T = h_T$. Indeed, from $K^1(f_T) = K^1(h_T)$ we obtain $\phi\theta^{-1}(f_T) = \phi\theta^{-1}(h_T)$, and if we apply a bijection $\phi\theta^{-1}$, we obtain $\phi\theta^{-1}\phi\theta^{-1}(f_T) = \phi\theta^{-1}\phi\theta^{-1}(h_T)$, i.e., $\theta\theta^{-1}(f_T) = \theta\theta^{-1}(h_T)$, or equivalently, $f_T = h_T$ (the compositions $\theta\theta^{-1}$ and $\phi\phi^{-1}$ are the identity functions).

Let us show that K is an isomorphism: from the adjunction $(F_T, G_T, \eta_T, \varepsilon_T) : \mathbf{DB} \longrightarrow \mathbf{DB}_T$, where F_T^0 is identity, $F_T^1 \triangleq \theta\phi^{-1}$, we obtain that $F_T \circ K = I_{DB_T}$ and $K \circ F_T = I_{DB}$. Thus, the functor K is an isomorphism of **DB** and Kleisli category $\mathbf{DB}_T$.

Notice that from Theorem 12 in Sect. 8.1.1, $\mathbf{DB}_T^0(\Upsilon, A) \triangleq TA^\Upsilon = A^\Upsilon = TA$ and $\mathbf{DB}_T^1(\Upsilon, f_T) \triangleq id_\Upsilon \otimes T\theta^{-1}(f_T) = T\theta^{-1}(f_T)$.

Thus, $K^0(A) = A \simeq TA = \mathbf{DB}_T^0(\Upsilon, A)$, and for $f_T : A \to B$, $K^1(f_T) = \phi\theta^{-1}(f_T) = \eta^{OP} \circ \theta^{-1}(f_T) : A \to B$, and from the fact that $Tg \approx is_{cod(g)}^{-1} \circ g = \eta_{cod(g)}^{OP} \circ g$ for any arrow g in **DB**, we obtain $K^1(f_T) \approx \mathbf{DB}_T^1(\Upsilon, f_T)$. Consequently, we obtain the natural isomorphism $\varphi : \mathbf{DB}_T(\Upsilon, \_) \simeq K$. □

Remark It is easy to verify that a natural isomorphism $\eta : I_{DB} \longrightarrow T$ of the monad (T, η, μ) is equal to the natural transformation $\eta : K \longrightarrow G_T$ (by considering that $G_T : \mathbf{DB}_T \longrightarrow \mathbf{DB}$ is defined by $G_T^0 = T^0$ and for any $f_T : A \longrightarrow B$ in $\mathbf{DB}_T$, $G_T^1(f_T) \triangleq \mu_B \circ T\theta^{-1}(f_T) : TA \longrightarrow TB$).

Hence, the functor F_T has two different adjunctions: the universal adjunction $(F_T, G_T, \eta_T, \mu_T)$ which defines the same monad (T, η, μ), and this particular (for the **DB** category only) isomorphism's adjunction (F_T, K, η_I, μ_I) which defines the trivial identity monad.

We are now ready to define the semantics of queries in **DB** category and the categorial definition of *query equivalence*. This is important in the context of the Database integration/exchange and for the theory of *query-rewriting* [8].

When we define a mapping (i.e., an arrow or morphism) $f : A \longrightarrow B$ between two simple databases A and B, we define implicitly the information flux $\widetilde{f}$, i.e., the set of views of A transferred by this mapping into B. Thus, in the context of the query-rewriting, we consider only the queries (i.e., view-maps) whose resulting view (i.e., an observation) belongs to the information flux of this mapping. Consequently, given any two queries, $q_{A_i} : A \longrightarrow TA$ and $q_{B_j} : B \longrightarrow TB$, they have to satisfy (w.r.t. query rewriting constraints) the condition $\partial_1(q_{A_i}) \in \widetilde{f}$ (the $\partial_1(q_{A_i})$ is a singleton containing the resulting view of this query) and $\partial_1(q_{B_j}) \in \widetilde{f}$. So, a well-rewritten query over B, $q_{B_j} : B \longrightarrow TB$, such that it is equivalent to the original query, i.e., $q_{B_j} \approx q_{A_i}$, has to satisfy the condition $\partial_1(q_{B_j}) = \partial_1(q_{A_i}) \in \widetilde{f}$. Now we can provide the denotational semantics for a query-rewriting in a data integration/exchange environment:

Proposition 61 *Each database query is a (non-monadic) T-coalgebra. Any homomorphism between two T-coalgebras $f : (A, q_{A_i}) \longrightarrow (B, q_{B_j})$, where A and B are two simple objects (databases) in* **DB**, *defines the semantics for the relevant query-rewriting, when $\partial_1(q_{A_i}) \in \widetilde{f}$.*

Proof Let us consider the following commutative diagram, where vertical arrows are the T-coalgebras:

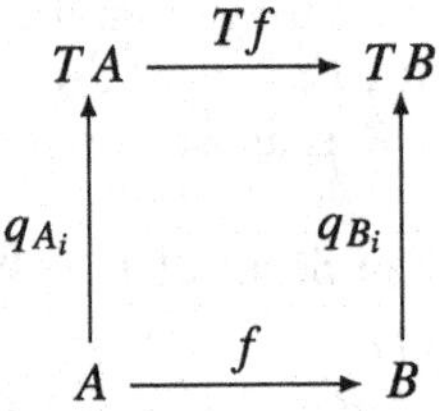

The homomorphism between two (non-monadic) T-coalgebras $f : (A, q_{A_i}) \longrightarrow (B, q_{B_j})$ means that it the commutativity $q_{B_j} \circ f = Tf \circ q_{A_i} : A \longrightarrow TB$ is valid and, from the duality property, we obtain $q_{B_i} = Tf \circ q_{A_i} \circ f^{OP}$. Consequently, for a given mapping $f : A \to B$ between databases A and B, for every query q_{A_i} such that $\partial_1(q_{A_i}) \in \widetilde{f}$ (i.e., $\widetilde{q_{A_i}} \subseteq \widetilde{f}$), we have an equivalent rewritten query q_{B_i} over a database B. In fact, $\widetilde{q_{B_i}} = \widetilde{Tf} \cap \widetilde{q_{A_i}} \cap \widetilde{f^{OP}} = \widetilde{q_{A_i}}$ because of the fact that $\widetilde{q_{A_i}} \subseteq \widetilde{f}$ and $\widetilde{f^{OP}} = \widetilde{Tf} = \widetilde{f}$. Consequently, we obtain the equivalence of these two queries $q_{B_j} \approx q_{A_i}$. □

8.5 Review Questions

1. Why is **DB** category not a topos? If the topos property for isomorphic arrows is valid for the simple arrows in **DB**, why is it not generally valid for the complex, if the complex arrows are representable by a set of simple ptp arrows? What would happen with the **DB** category if we would force the topos property for all arrows in **DB** and why can't it be accepted? Why can't the redundancy of the equal arrows be considered equal to a single arrow (mapping), from the DB mapping robustness point of view?

2. What is the way to reduce each complex object to a *strictly* complex object and why is this reduction invariant for the ptp arrows between two complex objects? Explain why it is important for the both epic and monic complex arrows to have the same number of ptp arrows in order to obtain the isomorphism? What is the sense of the "maximality" of the principal arrows (morphisms)? What are the canonical examples of the retraction pairs of the arrows in relationship with the categorial symmetry of the **DB** category? What is the relationship between the category of idempotents on a given simple object A and the categorial symmetry of the **DB** category?
3. Why does the matching between two instance database represent the maximal common information (the views) contained in both databases? Is the extension to complex databases appropriate? (If no, explain why, and try to define it.) Is there a natural sense for the matching operation what has to be a dual operation w.r.t. matching? The merging endofunctor is defined as a unary operator by prefixing with a given instance database. Is it a limitation to have a family of parameterized unary operators instead of a unique binary endofunctor? The generalized binary merging operator (for the objects only) is introduced later, in order to have a maximal mathematical flexibility, in order to elaborate on a database complete lattice. Do you have an idea how this generalization can be provided for the arrows as well, in order to obtain a binary merging endofunctor?
4. The **DB** category is finitely (co)complete, that is, it is (co)Cartesian. The fundamental property is that it has all pullbacks. Why is it not a Cartesian Closed Category (CCC)? We have demonstrated that it is not a topos, based on the properties of its morphisms (that both an epic and monic complex arrow is not necessarily an isomorphism), so the absence of the exponentiation is another proof. It is well known that the CCCs are the models for the typed λ-calculus, with the currying operator Λ. Is it reasonable to suspect that **DB** has the λ-calculus computation capability, if we demonstrated that the computation power of **DB** is equivalent to the full relational algebra Σ_{RE} with all update operations for the databases? How can the absence of this important property of the **DB** category be explained w.r.t. to the types of the logic formulae (i.e., SOtgds) used for the database mappings? Can you indicate which kind of the more powerful logic formula we would need for the database mappings in order to obtain the Cartesian Closed denotational **DB** category?
5. Explain the difference between the Universal algebra theory for a **DB** category and the initial algebra semantics for Σ_R relational algebras (i.e., the syntax monad $\mathcal{T}_P X$) provided in Sect. 5.1.1. Why is it possible to substitute in this setting the **DB** category with its skeletal subcategory $\mathbf{DB}_{sk}$ composed of only closed objects? What is a variety **K**, and why do we deal with the *quotient* term algebras? What is the meaning of the bifunctor $E : \mathbf{DB}_{sk}^{OP} \times \mathbf{K} \to \mathbf{Set}$ w.r.t. the initial algebra semantics and adjoint universal functor U and forgetful functor F, and their composition $F \circ U$? Show that the **DB** and $\mathbf{DB}_{sk}$ are concrete categories.
6. What is the reason to develop a lattice structure for the databases, from a (many) logic theory point of view (for example, Boolean algebras are based on the particular lattice structures)? The lattice $L_{DB} = (Ob_{DB}, \preceq, \otimes, \widehat{\oplus})$ is a complete infinite lattice with the top and bottom objects Υ and $\bot^0$, respectively. Why it is

important to demonstrate that a lattice of the *simple* databases in **DB** is an *algebraic* lattice and its closure power-view operator T is algebraic as well? What are the compact elements of such a lattice? Show that **DB** and $\mathbf{DB}_{sk}$ are locally finitely presentable categories with the finitely presentable total object Υ (which is the matching monoidal unit as well). Explain what it means for the theory of also *infinite databases*.

7. In which way do the monoidal closed categories generalize the Cartesian closed ones and "internalize" the hom-sets? Why is **DB** a V-category enriched over itself? Show that **DB** is a monoidal closed category with closed endofunctor (power-view operator) T. How are we able to demonstrate the property that **DB** admits a presentation by operations and equations? Show in which way we are able to "internalize" the Yoneda embedding, by replacing the base **Set** category with the database category **DB**.
8. The notion of T-algebra subsumes the notion of a Σ_R algebra (which can be understood as algebras in which the operators of the signature are not subject to any law (equation)). Demonstrate that the **DB** category is equivalent to the Eilenberg–Moore category of all monadic T-algebras and to the category of all monadic T-coalgebras as well. Why does the **DB** category belong to the setting of *enriched* Lawvere theories? How are we able to translate the inductive principle from **Set** (which is a CCC) into the monoidal (with matching tensor product $\otimes$) V-category **DB**? Why do we need the left Kan extension of the finitary signature functor Σ_D? What is the relationship between this relational-algebra signature Σ_D and the closure (power-view) endofunctor T? How can we define the "DB-syntax" monad E (where the initial semantics is expressed in **DB** instead in standard **Set** category) by an inductive algebraic extension of the "coalgebraic" observation based power-view monad T?
9. In which way is the DB coinductive principle in relationship with the morphisms (mapping-interpretations) of the **DB** category, different from the relationship of the inductive principle and objects (databases)? For which kind of mappings can we obtain a recursive system of equations derived from the relational algebra signature Σ_R^+ with the instructions of type 'EXTEND...' and the set of terms $\mathcal{T}_P^+(X)$? Explain the coinductive meaning of the following commutative diagram

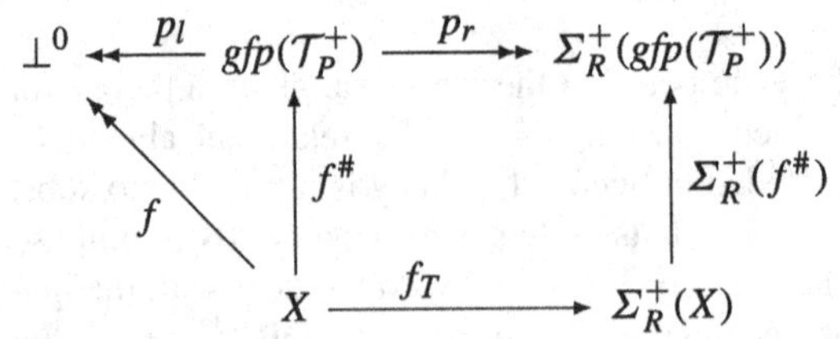

10. What is Kleisli semantics for the **DB** category and DB-mapping programs, as a paradigm of "programs as functions from values to computations"? What is the meaning in this setting of a closed object TA, the inclusion of values into computations and call-by-value and call-by-name paradigms? How are we able to express the denotational semantics of each simple mapping between two databases

by the, equivalent to it, "computation" morphism in **DB**? Why is the Kleisly category of the monad (power-view endofunctor) T isomorphic to the **DB** category, and in which way may this Kleisly category be internalized in **DB**? Show the categorial definition of the query equivalence.

References

1. P. Aczel, J. Adamek, S. Milius, J. Velebil, Infinite trees and completely iterative theories: a coalgebraic view. Theor. Comput. Sci. **300**, 1–45 (2003)
2. J. Adamek, J. Rosicky, *Locally Presentable and Accessible Categories*. London Mathematical Society Lecture Note Series, vol. 189 (Cambridge University Press, Cambridge, 1994)
3. M. Barr, C. Wells, *Toposes, Triples and Theories*. Grundelehren der math. Wissenschaften, vol. 278 (Springer, Berlin, 1985)
4. P.M. Cohn, *Universal Algebra* (Harper and Row, London, 1965)
5. E. Dubuc, Kan extensions in enriched category theory, in *Lecture Notes in Math.*, vol. 145 (Springer, Berlin, 1970), pp. 275–291
6. E. Dubuc, R. Street, Dinatural transformations, in *Lecture Notes in Math.*, vol. 137 (Springer, Berlin, 1970), pp. 126–138
7. P. Gabriel, F. Ulmer, *Lokal prasentierbare kategorien*. Lecture Notes in Mathematics, vol. 221 (Springer, Berlin, 1971)
8. A.Y. Halevy, Theory of answering queries using views. SIGMOD Rec. **29**(4), 40–47 (2000)
9. B. Jacobs, J. Rutten, A tutorial on (co)algebras and (co)induction. Bull. Eur. Assoc. Theor. Comput. Sci. **62**, 222–259 (1997)
10. G.M. Kelly, *Basic Concepts of Enriched Category Theory*. London Mathematical Society Lecture Note Series, vol. 64 (Cambridge University Press, Cambridge, 1982)
11. G.M. Kelly, A.J. Power, Adjunctions whose counits are coequalizers, and presentations of finitary enriched monads. J. Pure Appl. Algebra **89**, 163–179 (1993)
12. Y. Kinoshita, A.J. Power, M. Takeyama, Sketches. J. Pure Appl. Algebra **143**, 275–291 (1999)
13. C. Lair, Sur le genre d'esquissibilite des categories modelables (accessibles) possedant les produits de deux. Diagrammes **35**, 25–52 (1996)
14. S.M. Lane, *Categories for the Working Mathematician* (Springer, Berlin, 1971)
15. F.W. Lawvere, Functorial semantics of algebraic theories. Proc. Natl. Acad. Sci. **50**, 869–872 (1963)
16. Z. Majkić, Abstract database category based on relational-query observations, in *International Conference on Theoretical and Mathematical Foundations of Computer Science (TMFCS-08)*, Orlando FL, USA, July 7–9 (2008)
17. Z. Majkić, Algebraic operators for matching and merging of relational databases, in *International Conference in Artificial Intelligence and Pattern Recognition (AIPR-09)*, Orlando FL, USA, July 13–16 (2009)
18. Z. Majkić, Induction principle in relational database category, in *Int. Conference on Theoretical and Mathematical Foundations of Computer Science (TMFCS-09)*, Orlando FL, USA, July 13–16 (2009)
19. Z. Majkić, DB category: denotational semantics for view-based database mappings. arXiv:1103.0217v1 (2011), 24 February, pp. 1–40
20. M. Makkai, R. Pare, *Accessible Categories: The Foundations of Categorical Model Theory*. Contemporary Mathematics, vol. 104 (Am. Math. Soc., Providence, 1989)
21. E. Moggi, Computational lambda-calculus and monads, in *Proc. of the 4th IEEE Symp. on Logic in Computer Science (LICS'89)* (1989), pp. 14–23
22. E. Moggi, Notions of computation and monads. Inf. Comput. **93**(1), 55–92 (1991)
23. A.J. Power, Enriched Lawvere theories. Theory Appl. Categ. **6**, 83–93 (2000)
24. M. Smith, G. Plotkin, The category-theoretic solution of recursive domain equations. SIAM J. Comput. (1982)

Weak Monoidal DB Topos 9

9.1 Topological Properties

A traditional axiomatic foundation of mathematics is the set theory, in which all mathematical objects are ultimately represented by sets (even functions which map between sets). More recent work in category theory allows this foundation to be generalized using topoi; each topos completely defines its own mathematical framework. The category of sets forms a familiar topos, and working within this topos is equivalent to using traditional set-theoretic mathematics. But one could instead choose to work with many alternative topoi. A standard formulation of the axiom of choice makes sense in any topos, and there are topoi in which it is invalid. Constructivists will be interested to work in a topos without the law of excluded middle.

The definition of *topos* was originally proposed by Lawvere and Tierney, using terms which were available when they started the topos theory in 1969 [12, 24], it was required that a category **C** (see Sect. 8.1.3) satisfy the following properties:

1. **C** is finitely complete (has a terminal object and pullbacks);
2. **C** is finitely co-complete (has an initial object and pushouts);
3. **C** has exponentiation;
4. **C** has a subobject classifier (to be introduced in the next Sect. 9.1.2).

Subsequently, C. Juul Mikklsen discovered that condition 2 is implied by the combination of conditions 1, 3, and 4 [19].

Thus, a topos can be defined as a CCC (Cartesian Closed Category, see Sect. 8.1.3) with a subobject classifier.

The word "elementary" or "classic" (which from now on will be understood) has a special technical meaning having to do with the nature of the definition of topos.

The most known topoi are **Set** and, for any small category **C**, the functor category $\mathbf{Set}^{\mathbf{C}}$. The category **Grph** of graphs of the kind permitting multiple directed edges between two vertices is a topos. A graph $G = (V_G, E_G)$ consists of two sets, an edge set E_G and a vertex set V_G, and two functions s, t between those sets, assigning to every edge $e \in E_G$ its source $s(e) \in V_G$ and target $t(e) \in V_G$. **Grph** is thus equivalent to the functor category $\mathbf{Set}^{\mathbf{C}}$, where **C** is the category with two objects

Z. Majkić, *Big Data Integration Theory*, Texts in Computer Science,
DOI 10.1007/978-3-319-04156-8_9,

E_G and V_G and two morphisms $s, t : E_G \to V_G$ giving respectively the source and target of each edge.

The Yoneda Lemma asserts that $\mathbf{C}^{OP}$ embeds in $\mathbf{Set}^{\mathbf{C}}$ as a full subcategory. In the graph example, the embedding represents $\mathbf{C}^{OP}$ as the subcategory of $\mathbf{Set}^{\mathbf{C}}$ whose two objects are V' as the one-vertex no-edge graph and E' as the two-vertex one-edge graph (both as functors), and whose two nonidentity morphisms are the two graph homomorphisms from V' to E' (both as natural transformations). The natural transformations from V' to an arbitrary graph (functor) G constitute the vertices of G while those from E' to G constitute its edges. Although $\mathbf{Set}^{\mathbf{C}}$, which we can identify with **Grph**, is not made concrete by either V' or E' alone, the functor $U : \mathbf{Grph} \to \mathbf{Set}^2$ sending object G to the pair of sets $(\mathbf{Grph}(V', G), \mathbf{Grph}(E', G))$ and morphism $h : G \to H$ to the pair of functions $(\mathbf{Grph}(V', h), \mathbf{Grph}(E', h))$ is faithful. That is, a morphism of graphs can be understood as a pair of functions, one mapping the vertices and the other the edges, with application still realized as composition but now with multiple sorts of generalized elements. This shows that the traditional concept of a concrete category as one whose objects have an underlying set can be generalized to cater for a wider range of topoi by allowing an object to have multiple underlying sets, that is, to be multisorted.

An important result, obtained from Proposition 49 in Sect. 8.1.3, demonstrates that **DB** is not a CCC and, consequently, **DB** *is not* a standard topos. Thus, for each sketch category, obtained from a database-mapping graph G, $\mathbf{DB}^{\mathbf{Sch}(G)}$ (see Sect. 4.1.3 for the semantics of database-mapping systems) is not a topos as well.

Consequently, in the rest of this chapter, we will investigate some topological properties of the database category **DB**. That is, we will consider its metric, subobject classifier and topos properties. We will show that **DB** is a metric space, weak monoidal topos, and that there are some negative results, e.g., that **DB** is not well-pointed, has no power objects and pullbacks do not preserve epics.

9.1.1 Database Metric Space

In a metric space X, we denote by $X(A, B)$ the non-negative real quantity of X-distance from the point A to the point B. In a database context, for any two given databases A and B, their matching is inverse proportional to their distance: The maximal distance, ∞, between any two objects is equal to the minimal possible matching, i.e., ∞ is represented by the closed object $\bot^0$, while the minimal distance 0 is obtained for their maximal matching, i.e., when these two objects are isomorphic ($A \simeq B$).

Following this reasoning, we are able to formally define the concept of the database distance. In a first approach, we will consider only the simple databases and hence the subcategory $\underline{\mathbf{DB}}$:

Definition 61 If A and B are any two objects in $\underline{\mathbf{DB}}$ then their distance, denoted by $d(A, B)$, is defined as follows:

$$d(A,B) = \begin{cases} \underline{\Upsilon} & \text{if } A \simeq B; \\ A^B & \text{otherwise.} \end{cases}$$

The (binary) partial *distance relation* '$\sqsubseteq$' on *closed* database objects is defined as the inverse of the set inclusion relation '$\subseteq$'.

Notice that each *distance* is a *closed database object* (such that $A = TA$): the minimal distance is the total simple object $\underline{\Upsilon}$ which is closed, the maximal distance is the zero object $\bot^0$ which is closed, and any hom-object B^A ($B^A = TA \cap TB$) is the intersection of two closed objects, and hence a closed object as well.

Let us show that this definition of the distance for the databases satisfies the general metric space properties.

A categorical version of a metric space under the name enriched category, or V-category, is introduced in [5, 13], where the distances become the hom-objects. In this case, the definition of a database distance in the V-category **DB** (which is a strictly symmetric monoidal category ($\mathbf{DB}, \otimes, \Upsilon$)) is different, as we can see. For example, for every $A \in Ob_{\underline{\mathbf{DB}}}$, $d(A, A) = \underline{\Upsilon} \supset A^A$.

Proposition 62 *The transitivity law for the distance relation* $\sqsubseteq$ *and the triangle inequality, for any three simple databases* A, B *and* C, $d(A, B) \otimes d(B, C) \sqsupseteq d(A, C)$ *is valid in the database metric space. Moreover,*

- *There exists a strong link between the database PO-relation '$\preceq$' and the distance PO-relation '$\sqsubseteq$',*
  - $A \preceq B$ *iff* $\forall (C \not\simeq A)(d(A, C) \sqsupseteq d(B, C))$, *thus*
  - $A \simeq B$ *iff* $\forall C(d(A, C) = d(B, C))$.
- *The distances in* $\underline{\mathbf{DB}}$ *are locally closed. That is, for each object A there exists the bijection*

$$\phi : \{d(A, B) \mid B \in Ob_{\underline{\mathbf{DB}}}, B \not\simeq A\} \to \underline{\mathbf{DB}}(A, A)$$

where $\underline{\mathbf{DB}}(A, A)$ *is the hom-set of all endomorphisms of* A.

Proof The transitivity of $\sqsubseteq$ is valid because it is equal to the inverse set inclusion relation $\subseteq^{-1}$. Let us show the triangle inequality:

1. If $A \simeq C$ then $d(B, C) = d(B, A) = d(A, B)$, thus

$$d(A, B) \otimes d(B, C) = d(A, B) \sqsupseteq \underline{\Upsilon} = d(A, C).$$

2. If $A \not\simeq C$ then we have the following possibilities:
   2.1. If $A \simeq B$ then $d(A, B) \otimes d(B, C) = \underline{\Upsilon} \otimes d(B, C) = d(B, C) = d(A, C)$ (from $A \simeq B$), i.e., $d(A, B) \otimes d(B, C) \sqsupseteq d(A, C)$.
   2.2. The case $B \simeq C$ is analogous to the case 2.1.
   2.3. If $A \not\simeq B$ and $B \not\simeq C$ then $d(A, B) \otimes d(B, C) = TA \cap TB \cap TC \subseteq TA \cap TC = d(A, C)$, i.e., $d(A, B) \otimes d(B, C) \sqsupseteq d(A, C)$.

It is easy to show that ϕ is a bijection. Indeed, from the fact that

$$\{d(A,B) \mid B \in Ob_{\underline{\mathbf{DB}}}, B \not\approx A\} = \{S \mid S = TS \subseteq TA\},$$

we get:
1. $d(A, \bot^0) = \bot^0 = \widetilde{\bot^1} \mapsto \bot^1 : A \to A$;
2. $d(A,B) = TA \cap TB \mapsto f : A \to A$, such that $\widetilde{f} = TA \cap TB \subseteq TA$;
3. $d(A,\Upsilon) = TA \cap T\Upsilon = TA = \widetilde{id_A} \mapsto id_A : A \to A$. □

Notice that the locally closed property means that, for any distance $d(A, B)$ from a simple database A, we have a morphism $\varphi(d(A, B)) = f : A \to A$ such that $d(A, B) = \widetilde{f}$.

Now we can conservatively extend the metric space to all, also complex, objects in **DB**:

Proposition 63 *If A and B are any two objects in* **DB** *then their distance, denoted by* $d(A, B)$, *is defined as follows*:

$$d(A,B) = \begin{cases} \Upsilon & \textit{if } A \approx B; \\ A^B & \textit{otherwise.} \end{cases}$$

The (binary) partial distance relation *'$\sqsubseteq$' on closed database objects is defined as the inverse of the PO relation '$\preceq$'. This is a conservative extension of the metrics for simple objects in Definition* 61.

Proof From Definition 28 and Corollary 14 in Sect. 3.4.1, we obtain that the strong behavioral equivalence '$\approx$' is a strict generalization of the **DB** category isomorphism '$\simeq$', so that this extension of the metrics to all objects in **DB** is a generalization of the metrics for only simple objects (databases). Let us show that it is a conservative extension. In fact, $A \approx B$ iff ($A \preceq B$ and $A \succeq B$) iff ($A \sqsupseteq B$ and $A \sqsubseteq B$), where for the simple objects (by Theorem 6 in Sect. 3.2.5) $A \preceq B$ iff $TA \subseteq TB$. Consequently, for the simple objects $A \approx B$ iff $TA = TB$, i.e., $A \simeq B$.

Moreover, all distances are closed objects and for the simple objects the sets TS are closed as well, and hence for any set of simple objects A, B, C and D, $d(A, B) \preceq d(C, D)$ iff $d(A, B) \subseteq d(C, D)$, and so the metric ordering '$\sqsubseteq$' in this proposition is a conservative extension of that in Definition 61.

Let us show the metric property $d(A, B) \otimes d(B, C) \sqsupseteq d(A, C)$. In fact, $d(A, B) \otimes d(B, C) = (A \otimes B) \otimes (B \otimes C) \simeq$ (by commutativity of tensor product) $\simeq (B \otimes B) \otimes (A \otimes C)$. That is, from Corollary 14 in Sect. 3.4.1, $d(A, B) \otimes d(B,C) \approx (B \otimes B) \otimes (A \otimes C)$ and $(B \otimes B) \otimes (A \otimes C) \preceq A \otimes C = d(A, C)$, so that $d(A, B) \otimes d(B, C) \preceq d(A, C)$, i.e., $d(A, B) \otimes d(B, C) \sqsupseteq d(A, C)$.

In Proposition 51, Sect. 8.1.5, we demonstrated that the database lattice $L_{DB} = (Ob_D B, \preceq, \otimes, \widehat{\oplus})$ is a complete algebraic lattice with the top and bottom objects Υ and $\bot^0$, respectively. Thus, from the fact that each distance $d(A, B) = A \otimes B$ is a closed object in **DB**, $\Upsilon \succeq d(A, A)$ and $d(A, B) \succeq \bot^0$, that is, $\Upsilon \sqsubseteq d(A, A)$ and $d(A, B) \sqsubseteq \bot^0$. Thus, $d(A, B) = A \otimes B \simeq B \otimes A = d(B, A)$, i.e., $d(A, B) \simeq$

$d(B, A)$ and hence, from Corollary 14 in Sect. 3.4.1, we obtain the metric symmetry property $d(A, B) \approx d(B, A)$. □

From the definition of the distance we have that for the infinite distance $\bot^0$ (which is the terminal and initial object in the **DB** category; also called an infinite object) we obtain: $d(\bot^0, \bot^0) = \Upsilon$ (zero distance is the total object in the **DB** category), and for any other database $A \not\approx \bot^0$, $d(A, \bot^0) = \bot^0$, the distance from A to the infinite object (database) is *infinite*. Thus, the bottom element $\bot^0$ and the top element Υ in the database lattice are, for this database metric system, the infinite and zero distances (closed objects), respectively. Let us make a comparison between this database metric space in Proposition 63 and the general metric space (Frechet axioms):

Frechet axioms	**DB** metric space
$d(A, B) + d(B, C) \geq d(A, C)$	$d(A, B) \otimes d(B, C) \sqsupseteq d(A, C)$
$0 \leq d(A, A)$	$\Upsilon \sqsubseteq d(A, A)$
if $d(A, B) = 0$ then $A = B$	if $d(A, B) \approx \Upsilon$ then $A \approx B$
$d(A, B) < \infty$	$d(A, B) \sqsubseteq \bot^0$
$d(A, B) = d(B, A)$	$d(A, B) \approx d(B, A)$

Note that by Definition 19 in Sect. 3.2 of the PO relation '$\preceq$', $A \approx \Upsilon$ only when $A = \Upsilon$. Thus, $d(A, B) \approx \Upsilon$ implies $d(A, B) = \Upsilon$ and, from Proposition 63 above, $A \approx B$. Thus, a database metric space $\mathbf{DB}_{\text{met}}$, where the distances are the *closed* objects (databases), is a subcategory of $\mathbf{DB}_I^{OP}$ composed of only closed objects of **DB**: each arrow is an epimorphism $in^{OP} : A \twoheadrightarrow B$ (for a monomorphism $in : B \hookrightarrow A$ in $\mathbf{DB}_I$ defined by Theorem 6 in Sect. 3.2.5 from $B \preceq A$) which corresponds to the distance relation $A \sqsubseteq B$. Hence, we can say that a database metric space is embedded in the **DB** category, where the distances are the closed databases and the distance relations are the epimorphisms between closed databases.

9.1.2 Subobject Classifier

The concept of a *subobject* is the categorial version of the set-theoretical *subset*. The main idea is to regard a subset B of a given object A as a monomorphism from B into A. Every subset $B \subseteq A$ in the category **Set** can be described by its characteristic function $C_f : A \longrightarrow \Omega$, such that $C_f(x) = 1$ (True) if $x \in B$, 0 (False) otherwise, where $\Omega = \mathbf{2} = \{0, 1\}$ is the set of truth values (see Definition 1 in Sect. 1.3). The subobject classifier for **Set** is the object $\Omega = \mathbf{2}$ with the arrow $true : \{*\} \longrightarrow \Omega$ (with $true(*) = 1$), and a constant function $t_B : B \rightarrow \{*\}$ from B into a terminal object $\{*\}$ (a singleton) that satisfies the Ω-axiom:

For each monomorphism $in_B : B \hookrightarrow A$ there is one and only one characteristic arrow $C_{in_B} : A \longrightarrow \Omega$ such that the diagram on the left

$$\begin{array}{ccc} B & \overset{in_B}{\hookrightarrow} & A \\ {\scriptstyle t_B}\downarrow & & \downarrow{\scriptstyle C_{in_B}} \\ \{*\} & \xrightarrow{true} & \Omega \end{array} \qquad\qquad \begin{array}{ccc} B & \overset{in_B}{\hookrightarrow} & A \\ {\scriptstyle t_B}\downarrow & & \downarrow{\scriptstyle 1 - C_{in_B}} \\ \{*\} & \xrightarrow{false} & \Omega \end{array}$$

is a pullback square (the commutative diagram on the right is its logic dual with $false(*) = 0$). The PO relation in **Set** is a set-inclusion $B \subseteq A$ which is represented in this category by the injective function (a monomorphism) $in_B : B \hookrightarrow A$, where the characteristic function C_{in_B} represents the image of this injective function in_B.

In order to generalize this idea for **DB**, for any two database instances

$$A = \biguplus_{1 \le j \le m} A_j, \quad m \ge 1 \quad \text{and} \quad B = \biguplus_{1 \le i \le k} B_i, \quad k \ge 1,$$

we have to use the PO relation '$\preceq$' in the **DB** category (see Theorem 6 in Sect. 3.2.5) and hence B is a subobject of A iff $B \preceq A$, defined by Definition 19 with a mapping $\sigma : \{1, \ldots, k\} \to \{1, \ldots, m\}$ (if A and B are simple objects, i.e., $m = k = 1$, then the subobject relation reduces to the set inclusion $TB \subseteq TA$), which is represented by a monomorphism $f : B \hookrightarrow A$ with $\frac{f}{\uplus} = \{f_{i\sigma(i)} : B_i \hookrightarrow A_{\sigma(i)} \mid 1 \le i \le k\}$ (see Theorem 6).

The image of a monic arrow f in **DB** corresponds to its information flux $\widetilde{f}$ transferred into the target object A, that is, the characteristic mapping $C_f : A \to \Omega$ in **DB** must have exactly the information flux of the monomorphism f. Thus, for this subobject monomorphism, we obtain $\widetilde{f} = \biguplus \widetilde{\frac{f}{\uplus}} \simeq \biguplus_{1 \le i \le k} \widetilde{f_{i\sigma(i)}} = \biguplus_{1 \le i \le k} TB_i = TB$, so that from the fact that $TB \simeq B$, and from the fact that two isomorphic objects are also behaviorally equivalent, we obtain $\widetilde{f} \approx TB \approx B$ and hence the subobject relation $\widetilde{f} \approx B \preceq A$. That is, instead of the subobject B of A, we can use the equivalent characteristic-mapping subobject $\widetilde{f} \preceq A$ (in the case of simple database, we obtain $\widetilde{f} = TB \subseteq TA$). From the extreme case of the monomorphism $f = id_\Upsilon : \Upsilon \hookrightarrow \Upsilon$ we obtain that for its characteristic mapping $C_f : \Upsilon \to \Omega$, $\widetilde{C_f} \simeq \widetilde{f} = \widetilde{id_\Upsilon} = \Upsilon$. Consequently, we obtain $\Omega = \Upsilon$ which is a top object of the complete algebraic database lattice $L_{DB} = (Ob_{DB}, \preceq, \otimes, \widehat{\oplus})$ in Proposition 51 of Sect. 8.1.5, and hence we can use the bottom and top objects of this complete lattice for the false and true logical values in **DB** (as in the case of the many-valued logics with complete lattice of truth-values).

The constant "true" morphism in **DB** for the subobject classifier with the target object $\Omega = \Upsilon$ is its identity morphism $true = id_\Upsilon : \Upsilon \to \Omega$. The opposite "false" morphism in **DB** is the empty morphism $false = \bot^1 \cup (\biguplus_\omega \bot^1) : \Upsilon \to \Omega$. This interpretation for the true and false morphisms in **DB** is supported by this considerations: the false logical value has to be associated by the empty flux

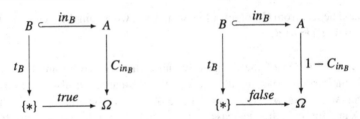

$\bot^0 \simeq \widetilde{\bot^1 \cup (\biguplus_\omega \bot^1)}$, and the true logical value by the maximal information flux $\Upsilon = \widetilde{id_\Upsilon} = \widetilde{true}$, that is, the bottom and the top objects of the lattice $L_{DB} = (Ob_{DB}, \preceq, \otimes, \widehat{\oplus})$.

Another analogous approach which can be used is based on modal logics (see, for example, Definition 10 in [18] for the Kripke semantics where the bottom element of the lattice is considered as a trivial world in which any logical formula is satisfied): if we consider each relation as a possible world, then the set with the empty relation $\bot^0 = \{\bot\}$ corresponds to the falsity, while the set of all relations (equal to Υ) to the truth. In this interpretation, the lattice L_{DB} is used as a PO set of the possible worlds, analogously to the Kripke interpretation of the intuitionistic propositional logic, i.e., for the topoi. And this closely relates our weak topos with the Kripke semantics of its propositional logic based on the subobject classifier in **DB**.

Let us show that in the **DB** category, differently from **Set**, the "characteristic function" C_f from A into Ω can be represented by a morphism $C_f = in_B \circ f^{OP} : A \rightarrow \Upsilon$, with $in_B = \biguplus_{1 \le i \le k}(in_{B_i} : B_i \hookrightarrow \Upsilon_i = \underline{\Upsilon})$ such that

$$\frac{C_f}{\uplus} = \left\{ C_{f_{ij}} = in_{B_i} \circ f_{ij}^{OP} : A_j \rightarrow \Upsilon_i = \underline{\Upsilon} \mid f_{ij} \in \frac{f}{\uplus} \right\}.$$

Hence

$$\widetilde{C_f} = \widetilde{\biguplus \frac{C_f}{\uplus}} \simeq \biguplus \left\{ \widetilde{in_{B_i} \circ f_{ij}^{OP}} \mid f_{ij} \in \frac{f}{\uplus} \right\} = \biguplus \left\{ TB_i \cap \widetilde{f_{ij}^{OP}} \mid f_{ij} \in \frac{f}{\uplus} \right\}$$

(by duality)

$$= \biguplus \left\{ TB_i \cap \widetilde{f_{ij}} \mid f_{ij} \in \frac{f}{\uplus} \right\} = \biguplus \left\{ \widetilde{f_{ij}} \mid f_{ij} \in \frac{f}{\uplus} \right\},$$

i.e., $\widetilde{C_f} \simeq \widetilde{f}$, so that $\widetilde{C_f} \approx \widetilde{f} \approx B \preceq A$ (characteristic arrow is well defined).

Consequently, we are ready for a formal definition of the subobject classifier in the **DB** category, based on the limit cone (see the pullback diagram in Proposition 27) of the diagram $\Upsilon \xrightarrow{true} \Omega \xleftarrow{C_f} A$, as follows:

Theorem 18 *The subobject classifier for* **DB** *is the object* $\Omega = \Upsilon$ *with the arrow* $true = id_\Upsilon : \Upsilon \longrightarrow \Omega$ *that satisfies the* Ω*-axiom:*

For any given subobject $B = \biguplus_{1 \le i \le k} B_i \preceq A = \biguplus_{1 \le j \le m} A_j$, *with* $k, m \ge 1$ *and* $\sigma : \{1, \ldots, k\} \rightarrow \{1, \ldots, m\}$ *(i.e., a monomorphism* $f : B \hookrightarrow A$ *with* $\frac{f}{\uplus} = \{f_{i\sigma(i)} : B_i \hookrightarrow A_{\sigma(i)} \mid 1 \le i \le k\}$*; see Theorem 6), there is one and only one characteristic morphism* $C_f = in_B \circ f^{OP} : A \rightarrow \Upsilon$, *where* $in_B = \biguplus_{1 \le i \le k}(in_{B_i} : B_i \hookrightarrow \Upsilon_i = \underline{\Upsilon})$, *with the vertex* $LimP = A \times B$ *and with outgoing arrows* p^Υ *and* p^A *with* $\frac{p^\Upsilon}{\uplus} = \frac{in_B}{\uplus}$ *and* $\frac{p_A}{\uplus} = \frac{id_A}{\uplus}$, *such that the diagram*

$$\begin{array}{ccc} A\times B & \xrightarrow{\quad p^A \quad} & A \\ \Big\downarrow p^{\Upsilon} & (\Omega) & \Big\downarrow C_f \\ \Upsilon & \xrightarrow{\quad true \quad} & \Omega \end{array}$$

is a pullback square. Thus, **DB** *is a kind of a monoidal elementary topos.*

Proof From Proposition 55 in Sect. 8.1.5 dedicated to algebraic database lattice, we have $\Omega = \Upsilon = \underline{\Upsilon} \cup (\biguplus_{\omega} \underline{\Upsilon}) = \underline{\Upsilon} \cup (\overbrace{\underline{\Upsilon} \uplus \underline{\Upsilon} \uplus \cdots}^{\omega})$. Diagram ($\Omega$) above is a special case of the class of pullbacks defined by Corollary 27 because $C_f : A \to \Omega$ satisfies the condition that for each its ptp arrow $C_{f_{ij}} : A_j \to \Upsilon_i = \underline{\Upsilon}$, $TA_j \subseteq T\Upsilon_i = T\underline{\Upsilon} = \underline{\Upsilon} = \bigcup_{A\in Ob_{\mathbf{DB}}} A \subset \Upsilon = \bigcup_{A\in Ob_{\mathbf{DB}}} A$ (from Proposition 26 in Sect. 3.2.5). Thus, $Lim(C_f) = A \times \widetilde{C_f}$ where $\widetilde{C_f} \simeq \widetilde{f} \simeq (f : B \hookrightarrow A$ is a monomorphism) $\simeq TB \simeq B$ and hence $Lim(C_f) \simeq A\times B$. For a monomorphism $f : B \hookrightarrow A$ we have that $\widetilde{f} \simeq \widetilde{TB} \simeq \widetilde{\tau^{-1}_{J(C_f)}} \simeq \widetilde{in_B}$ and hence, from Corollary 27, $\frac{p^{\Upsilon}}{\uplus} = \frac{\tau^{-1}_{J(C_f)}}{\uplus} = \frac{in_B}{\uplus}$.

Consequently, for each ptp monomorphism $f_{ij} : B_i \hookrightarrow A_j$ with $j = \sigma(i)$, we obtain the following pullback square where $\widetilde{C_{f_{ij}}} \simeq \widetilde{f_{ij}} \simeq TB_i \simeq B_i$:

$$\begin{array}{ccc} A_j\times B_i & \xrightarrow{\quad [id_{A_j}, \bot^1] \quad} & A_j \\ \Big\downarrow [\bot^1, in_{B_i}] & & \Big\downarrow C_{f_{ij}} \\ \underline{\Upsilon} & \xrightarrow{\quad id_{\underline{\Upsilon}} \quad} & \underline{\Upsilon} \end{array}$$

□

Note that the logic dual (width "false" arrow $false = \bot^1 \cup (\biguplus_{\omega} \bot^1) : \Upsilon \to \Omega$) and the complement-characteristic morphism $f' = (1 - C_f)$, such that $\frac{(1-C_f)}{\uplus} = \{f'_{ji} : A_j \to \Upsilon_i \mid (f_{ij} : B_i \to A_j) \in \frac{f}{\uplus}, 1 \le i \le k, j = \sigma(i)$ with $\widetilde{f'_{ji}} = TA_j \backslash \widetilde{f_{ij}}\}$, corresponds to the following pullback diagram

$$\begin{array}{ccc} \Upsilon\times B & \xrightarrow{\quad p^A \quad} & A \\ \Big\downarrow p^{\Upsilon} & & \Big\downarrow (1-C_f) \\ \Upsilon & \xrightarrow{\quad false \quad} & \Omega \end{array}$$

where for $\frac{p^A}{\uplus} = \frac{f}{\uplus}$ and $\frac{p^{\Upsilon}}{\uplus} = \frac{id_{\underline{\Upsilon}}}{\uplus}$.

We have that $\frac{false}{\uplus} = \emptyset$, and hence from the proof of Proposition 48 for the components of $LimP$ of the diagram $\Upsilon \xrightarrow{false} \Omega \xleftarrow{(1-C_f)} A$, for $l = i$, $D_{ilj} = (TA_j \cap \widetilde{\bot^1}) \cup (TA_j \backslash \widetilde{f'_{ji}}) = \bot^0 \cup \widetilde{f_{ij}} = \widetilde{f_{ij}}$ while $D'_{jli} = (T\Upsilon_i \cap \widetilde{f'_{ji}}) \cup (T\Upsilon_i \backslash \widetilde{\bot^1}) =$

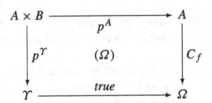

$T\Upsilon_i = \Upsilon_i = \underline{\Upsilon}$ for every $i \in \omega$. Consequently, $LimP \simeq (\widetilde{f} \times \Upsilon) \simeq B \times \Upsilon \simeq \Upsilon \times B$. Thus, from the proof of Proposition 48, $\frac{p^A}{\uplus} = \frac{f}{\uplus}$ and $\frac{p^{\Upsilon}}{\uplus} = \frac{id_{\Upsilon}}{\uplus}$ for this pullback diagram. It is easy to verify that $(1 - \widetilde{C_f}) \circ p^A = \bot^0$ so that this diagram commutes.

9.1.3 Weak Monoidal Topos

The standard topos is a finitely complete and cocomplete category with subobject classifier and with exponents. The well know example for the standard topos is the **Set** category, but also the arrow-category **Set** $\downarrow I$ of bundles (fibres) over a set I of indices (where an object (a bundle) is a function with codomain equal to I) used as a model for a Boolean algebraic logic. A bundle is a set-theoretic concept of the sheaf concept, and one of the primary sources of topos theory is algebraic geometry, in particular the study of sheaves [11], and the idea closely tied up with models of intuitionistic logic. The fundamental theorem of topoi demonstrates that if a category **D** is any topos and A is its object, then the arrow-category **D** $\downarrow A$ is also a topos.

In a standard topos, the following properties of the morphisms are valid:

- ISOMORPHIC $\equiv$ MONIC + EPIC;
- EQUALIZER $\equiv$ MONIC.

These properties are not valid in **DB** because it is not a standard topos. Recall that in every category, we have that ISOMORPHIC *implies* MONIC + EPIC and that EQUALIZER *only implies* MONIC, so this is valid in **DB** as well.

In the previous sections of this chapter, we demonstrated that the database category **DB** as a kind of *weak monoidal* topos, which differs from the standard topos by the fact that instead of the standard exponents (with the Cartesian product in the exponent diagrams) we have the hom-objects which satisfy the "exponent" diagrams where the Cartesian product '$\times$' is replaced by the monoidal tensor product '$\otimes$'. Let us consider some of the properties of this weak monoidal topos.

Proposition 64 *The weak monoidal topos* **DB** *preserves the epic–monic factorization of the standard topoi.*

Proof It is well known that in all topoi, each arrow has an epic–monic factorization. In fact, for any $f : A \to B$ for a diagram $B \xleftarrow{f} A \xrightarrow{f} B$ we have a pushout colimit (because a topos is cocomplete) with cocone $p : B \to ColimP$ and $q : B \to ColimP$, such that $p \circ f = q \circ f$. In this case, we denote by $im_f : f(A) \hookrightarrow B$ a monic arrow in a given topos, then it is an equalizer of this pair of arrows $p, q : B \to ColimP$, so that in this topos this equalizer is represented by the following commutative diagram

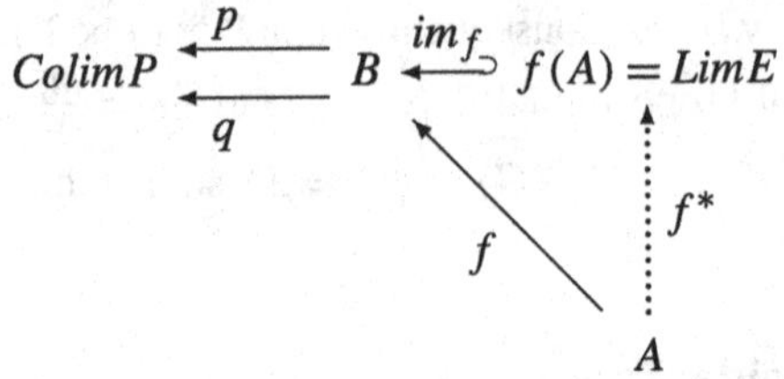

such that $f^* : A \to f(A)$ is an epic arrow, and consequently, we obtain the epic–monic factorization $f = f^* \circ im_f$. For example, in the topos **Set**, $f(A) = \{f(x) \mid x \in A\}$ is the image of the function $f : A \to B$, and for each $x \in A$, $f^*(x) = f(x)$.

Instead, in the weak monoidal topos **DB**, the epic–monic factorization is determined by its categorial symmetry, and the following diagram (see the proof of Theorem 4 in Sect. 3.2.3) commutes

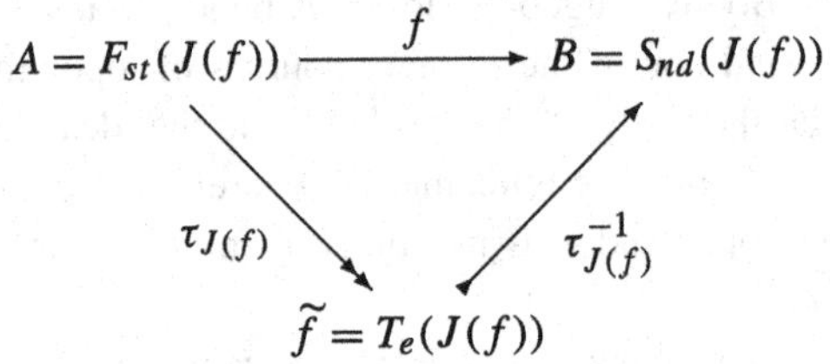

such that $T_e(J(f)) = \widetilde{f} \simeq \widetilde{\biguplus \frac{f}{\uplus}} = \biguplus_{f_{ji} \in \frac{f}{\uplus}} \widetilde{f_{ji}}$,

$$\frac{\tau_{J(f)}}{\uplus} = \left\{ \tau_{J(f_{ji})} : A_j \twoheadrightarrow \widetilde{f_{ji}} \mid (f_{ji} : A_j \to B_i) \in \frac{f}{\uplus} \right\}, \quad \text{and}$$

$$\frac{\tau^{-1}_{J(f)}}{\uplus} = \left\{ \tau^{-1}_{J(f_{ji})} : \widetilde{f_{ji}} \hookrightarrow B_i \mid (f_{ji} : A_j \to B_i) \in \frac{f}{\uplus} \right\},$$

with $\widetilde{f} \simeq \widetilde{\tau_{J(f)}} \simeq \widetilde{\tau^{-1}_{J(f_{ji})}}$, that is, from Definition 28 in Sect. 3.4.1, $f \approx \tau_{J(f_{ji})} \approx \tau^{-1}_{J(f_{ji})}$.

In fact, this equivalence is a general result of compositions with monic and epic arrows, as follows: □

Proposition 65 *The monomorphism $\tau^{-1}_{J(f)} : \widetilde{f} \hookrightarrow B$ is the smallest subobject of B through which $f : A \to B$ factors. That is, if $f = in \circ g$ for any $g : A \to C$ and a monic arrow $in : C \hookrightarrow B$ then there is a unique $k : \widetilde{f} \to C$ making*

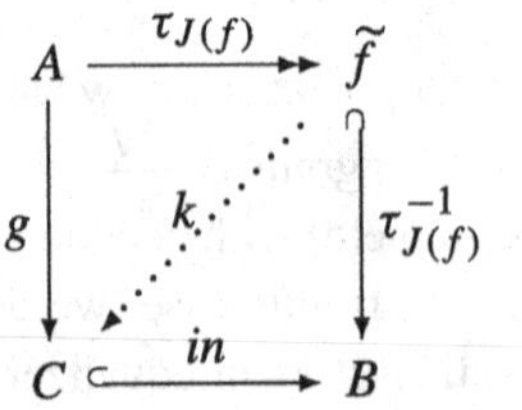

commute, and hence $\widetilde{\tau^{-1}_{J(f)}} \preceq \widetilde{in}$, as in the ordinary topoi.

Proof Let $A = \biguplus_{1\le j\le m} A_j$, $B = \biguplus_{1\le i\le k} B_i$ and $C = \biguplus_{1\le l\le n} C_l$ with $m, k, n \ge 1$.
Then, for any ptp arrow $(f_{ji} : A_j \to B_i) \in \frac{f}{\uplus}$ we define the set of indexes:

$$L_{ji} = \left\{ l \mid 1 \le l \le n, (g_{jl} : A_j \to C_l) \in \frac{g}{\uplus} \text{ and } (in_{li} : C_l \to B_i) \in \frac{in}{\uplus} \right\}.$$

Consequently, from the fact that $f = in \circ g$, for each ptp arrow $f_{ji} \in \frac{f}{\uplus}$ the following must hold:
(i) $\widetilde{f_{ji}} = T(\bigcup\{\widetilde{in_{li} \circ g_{jl}} \mid l \in L_{ij}\}) = T(\bigcup\{\widetilde{in_{li}} \cap \widetilde{g_{jl}} \mid l \in L_{ij}\}) = T(\bigcup\{TC_l \cap \widetilde{g_{jl}} \mid l \in L_{ij}\}) = T(\bigcup\{\widetilde{g_{jl}} \mid l \in L_{ij}\})$, and hence
(ii) $\widetilde{g_{jl}} \subseteq \widetilde{f_{ji}}$ for each $l \in L_{ij}$.
From the fact that it holds for every f_{ji}, we can define an arrow $k : \widetilde{f} \to C$ as follows:
for each $(f_{ji} : A_j \to B_i) \in \frac{f}{\uplus}$, we define a ptp arrow for each $l \in L_{ij}$, $k_{jil} : \widetilde{f_{ji}} \to C_l$ such that
(iii) $\widetilde{k_{jil}} = \widetilde{g_{jl}}$.
Let us show that $g = k \circ \tau_{J(f)}$, that is, for each ptp arrow $(g_{jl} : A_j \to C_l) \in \frac{g}{\uplus}$, we have that $\widetilde{g_{jl}} = T(\bigcup\{\widetilde{k_{jil} \circ \tau_{J(f_{ji})}} \mid 1 \le i \le k, (\tau_{J(f_{ji})} : A_j \to \widetilde{f_{ji}}) \in \frac{\tau_{J(f)}}{\uplus}$ and $(k_{jil} : \widetilde{f_{ji}} \to C_l) \in \frac{k}{\uplus}\})$. Indeed,

$$T\left(\bigcup\left\{\widetilde{k_{jil} \circ \tau_{J(f_{ji})}} \mid 1 \le i \le k, \tau_{J(f_{ji})} \in \frac{\tau_{J(f)}}{\uplus} \text{ and } k_{jil} \in \frac{k}{\uplus}\right\}\right)$$

$$= T\left(\bigcup\left\{\widetilde{k_{jil}} \cap \widetilde{\tau_{J(f_{ji})}} \mid 1 \le i \le k, \tau_{J(f_{ji})} \in \frac{\tau_{J(f)}}{\uplus} \text{ and } k_{jil} \in \frac{k}{\uplus}\right\}\right) \quad \text{(from (iii))}$$

$$= T\left(\bigcup\left\{\widetilde{g_{jl}} \cap \widetilde{\tau_{J(f_{ji})}} \mid 1 \le i \le k, \tau_{J(f_{ji})} \in \frac{\tau_{J(f)}}{\uplus} \text{ and } k_{jil} \in \frac{k}{\uplus}\right\}\right)$$

$$= T\left(\bigcup\left\{\widetilde{g_{jl}} \mid 1 \le i \le k, \tau_{J(f_{ji})} \in \frac{\tau_{J(f)}}{\uplus} \text{ and } k_{jil} \in \frac{k}{\uplus}\right\}\right)$$

(from $\widetilde{f_{ji}} = \widetilde{\tau_{J(f_{ji})}}$ and (ii))

$$= T\widetilde{g_{jl}} = \widetilde{g_{jl}}.$$

Finally, let us show that $\tau^{-1}_{J(f)} = in \circ k$, that is, for each ptp monic arrow $(\tau^{-1}_{J(f_{ji})} : \widetilde{f_{ji}} \hookrightarrow B_i) \in \frac{\tau^{-1}_{J(f)}}{\uplus}$, $\widetilde{\tau^{-1}_{J(f_{ji})}} = T(\bigcup\{\widetilde{in_{li} \circ k_{jil}} \mid 1 \le l \le n, (k_{jil} : \widetilde{f_{ji}} \to C_l) \in \frac{k}{\uplus}$ and $(in_{li} : C_l \to B_i) \in \frac{in}{\uplus}\})$. Indeed,

$$T\left(\bigcup\left\{\widetilde{in_{li} \circ k_{jil}} \mid 1 \le l \le n, k_{jil} \in \frac{k}{\uplus} \text{ and } in_{li} \in \frac{in}{\uplus}\right\}\right)$$

$$= T\left(\bigcup\left\{\widetilde{in_{li}} \cap \widetilde{k_{jil}} \mid 1 \le l \le n, k_{jil} \in \frac{k}{\uplus} \text{ and } in_{li} \in \frac{in}{\uplus}\right\}\right)$$

$$= T\Big(\bigcup\Big\{TC_l \cap \widetilde{k_{jil}} \mid 1 \le l \le n, k_{jil} \in \frac{k}{\uplus} \text{ and } in_{li} \in \frac{in}{\uplus}\Big\}\Big)$$

$$= T\Big(\bigcup\Big\{\widetilde{k_{jil}} \mid 1 \le l \le n, k_{jil} \in \frac{k}{\uplus} \text{ and } in_{li} \in \frac{in}{\uplus}\Big\}\Big)$$

$$= T\Big(\bigcup\{\widetilde{g_{jl}} \mid l \in L_{ji}\}\Big) \quad \text{(from (iii))}$$

$$= \widetilde{f_{ji}} \quad \text{(from (i))}$$

$$= \widetilde{\tau^{-1}_{J(f_{ji})}}.$$

□

There is another important property of this weak monoidal topos **DB** which is valid in any standard topos:

Proposition 66 *The following topos properties are valid in* **DB**:

- *Coproduct preserve pullbacks. That is, if*

$$\begin{array}{ccc} A & \xrightarrow{f} & C \\ {\scriptstyle g}\downarrow & (16.a) & \downarrow{\scriptstyle k} \\ B & \xrightarrow{h} & D \end{array} \qquad\qquad \begin{array}{ccc} A' & \xrightarrow{f'} & C \\ {\scriptstyle g'}\downarrow & (16.b) & \downarrow{\scriptstyle k} \\ B' & \xrightarrow{h'} & D \end{array}$$

are pullbacks in the **DB** *category, then so is the following diagram*

$$\begin{array}{ccc} A + A' & \xrightarrow{[f, f']} & C \\ {\scriptstyle g+g'}\downarrow & (16) & \downarrow{\scriptstyle k} \\ B + B' & \xrightarrow{[h, h']} & D \end{array}$$

- *Pullbacks preserve epics.*

Proof Claim 1. We recall that by duality $A + B$ is equal to $A \times B$ (equal to disjoint union $A \uplus B$). Let us show that for any two arrows $l_1 : E \to B + B'$ and $l_2 : E \to C$ it holds that $[h, h'] \circ l_1 = k \circ l_2$, then there is a unique arrow $e_1 : E \to A + A'$ such that the following diagram commutes:

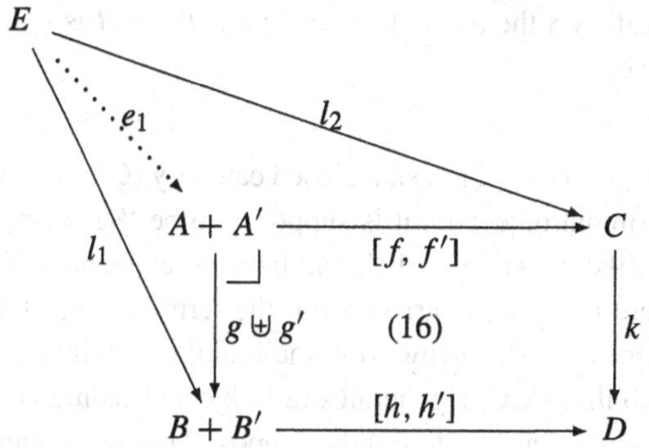

Let $A = \biguplus_{1 \le j \le m} A_j, m \ge 1$, $A' = \biguplus_{1 \le j \le m'} A'_j, m' \ge 1$, $B = \biguplus_{1 \le i \le k} B_i, k \ge 1$ and $B' = \biguplus_{1 \le i \le k'} B'_i, k' \ge 1$. Let $l : E \to B$ be the subset of ptp arrows in $\frac{l_1}{\uplus}$ such that the codomain of each of these ptp arrows is an object in $B_i, 1 \le i \le k$, and let $l' : E \to B'$ be the subset of ptp arrows in $\frac{l_1}{\uplus}$ such that the codomain of each of these ptp arrows is an object in $B'_i, 1 \le i \le k'$. Then $l_1 = \langle l, l' \rangle$, and hence we obtain two commutative diagrams:

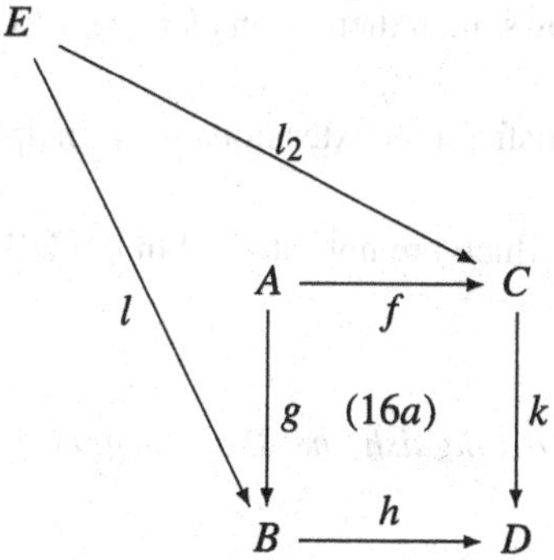

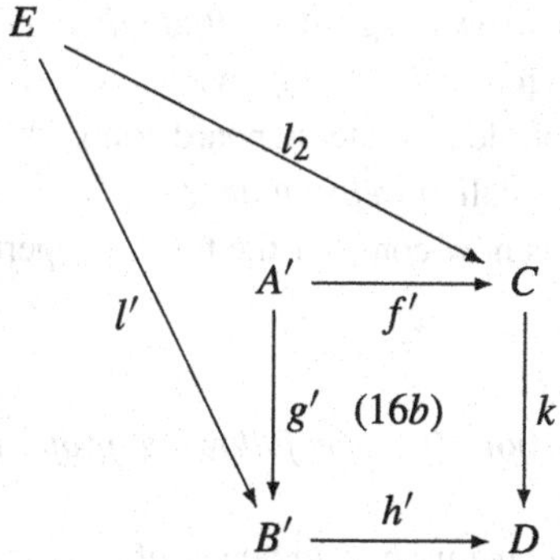

so that, from the fact that by hypothesis both diagrams (16a) and (16b) are the pullbacks diagrams, we have the unique arrows $e : E \to A$ such that $l = g \circ e$, $l_2 = f \circ e$, and $e' : E \to A'$ such that $l' = g' \circ e'$, $l_2 = f' \circ e'$. Consequently, there is a unique arrow $e_1 = \langle e, e' \rangle : E \to A + A'$ such that $l_1 = \langle l, l' \rangle = \langle g \circ e, g' \circ e' \rangle = (g \uplus g') \circ \langle e, e' \rangle = (g \uplus g') \circ e$, and $l_2 = \lceil f \circ e, f' \circ e' \rfloor$ (from the fact that $\lceil f \circ e, f' \circ e' \rfloor = T(\widetilde{f \circ e} \cup \widetilde{f' \circ e'}) = T(\widetilde{l_2} \cup \widetilde{l_2}) = T\widetilde{l_2} = \widetilde{l_2}$), so that $l_2 = \lceil f \circ e, f' \circ e' \rfloor = [f, f'] \circ \langle e, e' \rangle = [f, f'] \circ e$. Consequently, diagram (16) is a pullback diagram.

Claim 2. Let us show it for the pullback diagram $A \xrightarrow{f} C \xleftarrow{g} B$ composed by the simple arrows. Then, form the proof of Proposition 48 in Sect. 8.1.3, the pullback object is $LimP = ((TA \cap \widetilde{g}) \cup (TA \backslash \widetilde{f})) \times ((TB \cap \widetilde{f}) \cup (TB \backslash \widetilde{g}))$ with two monic cone arrows, $[p^A, \bot^1] : LimP \hookrightarrow A$ and $[\bot^1, p^B] : LimP \hookrightarrow B$ where $\widetilde{p^A} = T((TA \cap \widetilde{g}) \cup (TA \backslash \widetilde{f}))$ and $\widetilde{p^B} = T((TB \cap \widetilde{f}) \cup (TB \backslash \widetilde{g}))$. If $f : A \twoheadrightarrow C$ is epic then $\widetilde{f} = TC$. Thus, $\widetilde{p^B} = T((TB \cap \widetilde{f}) \cup (TB \backslash \widetilde{g})) = T((\underbrace{TB \cap TC}) \cup (TB \backslash \widetilde{g})) = TB$ (from the fact that $\widetilde{g} \subseteq (TB \cap TC)$). Consequently, $[\bot^1, p^B] \simeq \widetilde{p^B} = TB$, and hence from Proposition 8 the arrow $[\bot^1, p^B] : LimP \hookrightarrow B$ is epic as well, so that pullbacks preserve epics. □

Since a standard topos is a Cartesian closed category (CCC) with subobject classifier (and with terminal object **1**), it is supposed to be '**Set**-like', its initial object **0** ought to behave like the empty set $\emptyset$, and have no elements. Otherwise, if **0** has elements, then there is a unique arrow from the terminal object **1** into initial object **0** (and its unique opposite arrow from the initial object into **1**) which has to be an isomorphism, so that CCC degenerates (all objects become isomorphic, that is, 'equal'). Thus, in a non-degenerate standard topos, **0** has no elements.

In **DB**, terminal and initial objects are equal to the zero object $\bot^0 = \{\bot\}$, and from the fact that it is a non-degenerate category we can see that it is not a standard topos.

The question of the existence of *elements of objects* relates to the notion of *extensionality principle*, the principle that sets with the same elements are identical. For the arrows in **Set** (the functions), this principle takes the following form: if two parallel arrows $f, g : A \to B$ are *distinct* arrows, then there is an *element* $x : \mathbf{1} \to A$ of A such that $f \circ x \neq g \circ x$.

A non-degenerate standard topos that satisfies this extensionality principle for arrows is called *well-pointed.*

Let us now consider the topos properties which are not satisfied in the **DB** category.

Proposition 67 *The following properties distinguish the* **DB** *category from a topos*:

1. **DB** *category has no power objects.*
2. **DB** *category is not well-pointed.*

Proof Claim 1. By definition, the power object of A (if it exists) is an object $\mathcal{P}(A)$ which represents the contravariant functor $Sub(\_ \times A) : \mathbf{DB} \to \mathbf{Set}$, where for any object B, $Sub(B) = \{\widetilde{f} | f \text{ is a subobject of } A\} = \{\widetilde{f} | \widetilde{f} \subseteq TA\}$ is the set of all subobjects (monomorphic arrows with the target object B) of B. Let us show that for any simple object $A \not\simeq \bot^0$ there is no the power object $\mathcal{P}(A)$ such that the bijection $\mathbf{DB}(\_ , \mathcal{P}(A)) \simeq Sub(\_ \times A)$ holds in **Set**. In fact, $Sub(A \times A) = Sub(A + A) = \{\widetilde{f} \mid \widetilde{f} \preceq T(A + A) = TA + TA\} = \{\widetilde{\langle f_1, f_2 \rangle} \mid \widetilde{f_1} \uplus \widetilde{f_2} \preceq TA \uplus TA\}$. Hence, $|\mathbf{DB}(A, \mathcal{P}(A))| = |\{\widetilde{f} | \widetilde{f} \subseteq TA \cap T(\mathcal{P}(A))\}| \subseteq |\{\widetilde{f} | \widetilde{f} \subseteq TA\}| \subseteq |Sub(A \times A)|$.

Claim 2. The extensionality principle for arrows "if $f, g : A \to B$ is a pair of distinct parallel arrows, then there is an element $x : \bot^0 \to A$ of A such that $f \circ x \neq g \circ x$" does not hold because $\widetilde{f \circ x} = \widetilde{g \circ x} = \bot^0$ for the (unique) element (arrow) $x : \bot^0 \to A$ such that $\widetilde{x} = \bot^0$. □

9.2 Intuitionistic Logic and DB Weak Monoidal Topos

The classical rules of logic are representable in the topos **Set** by the operations on the set $\mathbf{2} = \{0, 1\}$, and then can be developed in any topos **C** by using Ω in place of **2**.

This gives the "logic" of **C**, which characterizes the behavior of subobjects in a standard topos **C**.

Let us briefly consider the logic of a classic topos **C** with a terminal object 1 (equal to the singleton $\{*\}$ when **C** is the **Set** category), where the truth-functions are the particular characteristic arrows of the subobject classifier pullback squares, for example,

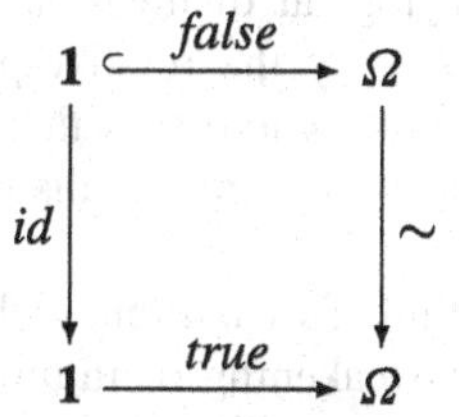

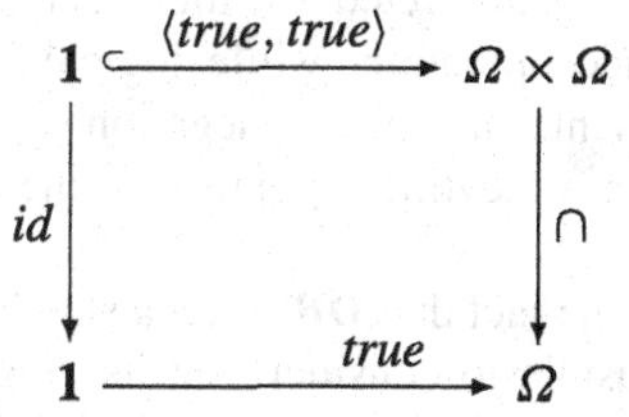

for the "negation truth-function" characteristic arrow $\sim : \Omega \to \Omega$, and "conjunction truth-function" characteristic arrow $\cap : \Omega \times \Omega \to \Omega$. Analogously, we denote the "disjunction truth-function" characteristic arrow by $\cup : \Omega \times \Omega \to \Omega$.

We consider a case of the propositional logic (CPL) with the set of propositional letters in PR, and logical connectives $\wedge, \vee$ and $\neg$ (conjunction, disjunction and negation, respectively). The language of such a logic is composed of a set of formulae as follows: All propositional letters are formulae; If ϕ and ψ are formulae, then $(\phi \wedge \psi)$, $(\phi \vee \psi)$ and $\neg\phi$ are formulae (the logical implication in classic logic $(\phi \Rightarrow \psi)$ is derived and equal to $\phi \vee \neg\psi$). In order to develop a theory of meaning, or semantics, for PL, we use the truth-functions. By a value assignment we shall consider any function $V : PR \to \Omega$, which assigns a truth-value to each propositional letter, and provides a "meaning" or "interpretation" to the members of PR. It is uniquely (inductively) extended to all PL formulae, trough recursive application of the rules:

- $V(\neg\phi) = \sim V(\phi)$;
- $V(\phi \wedge \psi) = V(\phi) \cap V(\psi)$;
- $V(\phi \vee \psi) = V(\phi) \cup V(\psi)$.

We are now able [15] to do propositional logic in any standard topos **C**. Recall that each standard topos is well-pointed, so that a truth value in **C** is an arrow $\mathbf{1} \to \Omega$ and that $\mathbf{C}(\mathbf{1}, \Omega)$ denotes the collection of such **C**-arrows.

Then, in a topos **C**, a **C**-valuation is a function $V : PR \to \mathbf{C}(\mathbf{1}, \Omega)$ assigning to each propositional letter $p \in PR$ a truth-value (arrow) $V(p) : \mathbf{1} \to \Omega$. This function is uniquely (inductively) extended to all formulae of PL by the following rules of composition of arrows in **C**:

- $V(\neg\phi) = \sim \circ V(\phi) : \mathbf{1} \to \Omega$;
- $V(\phi \wedge \psi) = \cap \circ \langle V(\phi), V(\psi)\rangle : \mathbf{1} \to \Omega$;
- $V(\phi \vee \psi) = \cup \circ \langle V(\phi), V(\psi)\rangle : \mathbf{1} \to \Omega$.

In this way, generally, the logic is internally represented in a standard topos **C**. We shall say that ϕ is a **C**-valid formula, denoted by $\mathbf{C} \models \phi$, iff for every **C**-valuation V, $V(\phi) = true : \mathbf{1} \to \Omega$.

Based on this translation, it was demonstrated that any **C**-valid formula is a theorem of classical propositional logic (CPL), but CPL is not sound for **C**-validity: in fact, the "law of excluded middle", $\phi \vee \neg\phi$, *is not* **C**-valid.

Moreover, it was demonstrated that the 'topos logic' based on the topos-valuation is *intuitionistic* propositional logic (see, for example, [8]), that is, from the algebraic point of view, it is a Heyting algebra (it was demonstrated that a collection of arrows (truth-values) in $\mathbf{C}(\mathbf{1}, \Omega)$ is a Heyting algebra).

A Heyting algebra is based on a complete *distributive* lattice where logical conjunction is interpreted by the meet lattice operation, logical disjunction by the join lattice operation, while logical implication $\phi \Rightarrow \psi$ by the relative-pseudo-complement, and logical negation $\neg\phi$ is equal to $\phi \Rightarrow 0$, where 0 is the logical falsity (in a Heyting algebra it is the bottom element of its complete distributive lattice).

From the fact that **DB** is not a standard topos but a kind of weak monoidal topos, it is plausible to consider that its "logic" is a kind of weakening of intuitionistic logic. In what follows, we will try to describe this new non-standard propositional logic.

Notice that the **DB** *is not* well-pointed (from Proposition 67), so that the way to "internalize the logic" in **DB** has to be very different w.r.t. to the case of standard topoi presented previously. The terminal object **1** in **DB** is the zero object $\bot^0 = \{\bot\}$, so that we have only one arrow $\bot^1 : \mathbf{1} \to \Omega$, and hence the collection of arrows $\mathbf{DB}(\mathbf{1}, \Omega)$ is a singleton and, consequently, cannot represent the collection of truth values of its logic. But, as in the case of the subobject classifier pullback diagram in **DB**, where the truth-arrow (*true*) is equal to the identity arrow $id_\Upsilon : \Upsilon \to \Omega$ (in **DB**, we have $\Omega = \Upsilon$), the set of truth values is the collection $\mathbf{DB}(\Upsilon, \Omega)$ of arrows. Consequently, we have to discover which kind of algebra is represented by this collection $\mathbf{DB}(\Upsilon, \Omega)$, and then to define a logic of such an algebra.

In Sect. 8.1.5, we defined the complete algebraic database lattice $L_{DB} = (Ob_{DB}, \preceq, \otimes, \widehat{\oplus})$ (see Proposition 51), where the meet operation is equal to the database-matching operator $\otimes$ and the join operator is equal to the generalized database-merging operator in Definition 57. In order to make a relationship between the collection $\mathbf{DB}(\Upsilon, \Omega)$ of truth-values (arrows) with this complete (but not distributive) lattice L_{DB}, we will use the following property (based on the symmetry property of arrows and objects in **DB**):

Lemma 19 *There is a bijection β between the truth-values and closed objects in* **DB**, *that is,* $\beta : \mathbf{DB}(\Upsilon, \Omega) \cong Ob_{\mathbf{DB}_{sk}}$, *such that for any truth-value* $f : \Upsilon \to \Omega$, $\beta(f) = \widetilde{f}$ *which is a closed object in the skeletal category* $\mathbf{DB}_{sk}$ *equivalent to* **DB** *(by Proposition* 50).

Proof For each truth-value $f : \Upsilon \to \Omega$ in $\mathbf{DB}(\Upsilon, \Omega)$, from Proposition 6, $\widetilde{f} \preceq \Upsilon \otimes \Omega = \Upsilon$ (recall that $\Omega = \Upsilon = \underline{\Upsilon} \cup (\biguplus_\omega \underline{\Upsilon})$), and from the fact that $\bot \in \widetilde{f}$, $\bot^0 = \{\bot\} \preceq \widetilde{f}$, so that β is a surjective mapping. Conversely, for each closed object $A =$

$\biguplus_{1\leq j\leq m} A_j \in Ob_{\mathbf{DB}_{sk}}, m \geq 1, A = TA$, and hence we can define a function $(f = \beta^{-1}(A) \triangleq \biguplus_{1\leq j\leq m} f_j) \in \mathbf{DB}(\Upsilon, \Omega)$ with $f_j = \{id_R : R \to R \mid R \in A_j\} : \underline{\Upsilon} \to \underline{\Upsilon}$, so that $\beta(f) = TA = A$. □

Based on this lemma, and from the fact that in the lattice L_{DB} we have that for any object $A \in Ob_{DB}$, $A \approx TA$ (where '$\approx$' is the equivalence relation in this lattice), we obtain that the lattice L_{DB} is equivalent to the lattice of only closed objects and, consequently, the lattice of truth-values in $\mathbf{DB}(\Upsilon, \Omega)$ is isomorphic to the lattice L_{DB}. Thus we can use the meet and join operators of the lattice L_{DB} as an algebraic counterparts of the logical operations for conjunction and disjunction, respectively. Hence, we introduce the following non-standard propositional logic for the weak monoidal topos **DB**:

Definition 62 In the weak monoidal database topos **DB**, a **DB**-valuation is a function $V : PR \to \mathbf{DB}(\Upsilon, \Omega)$ assigning to each propositional letter $p \in PR$ a truth-value (arrow) $V(p) : \Upsilon \to \Omega$. This function is uniquely (inductively) extended to all formulae of PL (where $\wedge$ and $\vee$ are the logical operators conjunction and disjunction, respectively) by the following rules:

- $V(\phi \wedge \psi) = V(\phi) \otimes V(\psi) = \beta^{-1}(\beta(V(\phi)) \otimes \beta(V(\psi))) : \Upsilon \to \Omega$;
- $V(\phi \vee \psi) = V(\phi) \widehat{\oplus} V(\psi) = \beta^{-1}(\beta(V(\phi)) \widehat{\oplus} \beta(V(\psi))) : \Upsilon \to \Omega$,

where 'β' is the bijection, $\beta : \mathbf{DB}(\Upsilon, \Omega) \cong Ob_{\mathbf{DB}_{sk}}$, in Lemma 19.

We say that ϕ is **DB**-valid formula, denoted by $\mathbf{DB} \models \phi$, iff for every **DB**-valuation V, $V(\phi) = true = id_\Upsilon : \Upsilon \to \Omega$.

We can consider the reduction of this logic to only simple databases (from the fact that complex databases are only disjoint union of the simple databases), by the class of "simple" **DB**-valuation:

Definition 63 Let us define the strict complete sublattice of truth-values

$$L_{\underline{DB}} = (Ob_{\underline{\mathbf{DB}}_{sk}}, \preceq, \otimes, \oplus) = (Ob_{\underline{\mathbf{DB}}_{sk}}, \subseteq, \cap, T\cup) \subset L_{DB},$$

with the bottom object $\bot^0$ and the unique complex object equal to the top simple object $\underline{\Upsilon}$. The "simple" **DB**-valuations for this complete sublattice is given by a subset of functions $V : PR \to \mathbf{DB}(\Upsilon, \Omega)$ assigning to each propositional letter $p \in PR$ a truth-value (arrow) $V(p) : \Upsilon \to \Omega$ such that $\frac{V(p)}{\uplus} = \{f : \underline{\Upsilon} \to \underline{\Omega}\} = \underline{\Upsilon}$ is a simple arrow in **DB**, and consequently, V may be equivalently substituted by $\underline{V} : PR \to \underline{\mathbf{DB}}(\underline{\Upsilon}, \underline{\Omega})$, such that $\frac{V(p)}{\uplus} = \{\underline{V}(p)\}$ with $true = id_{\underline{\Upsilon}} : \underline{\Upsilon} \to \underline{\Omega}$ and $false = \bot^1 : \underline{\Upsilon} \to \underline{\Omega}$. Then we obtain the unique extension to all formulae by:

- $\underline{V}(\phi \wedge \psi) = \beta^{-1}(\beta(\underline{V}(\phi)) \cap \beta(\underline{V}(\psi))) : \underline{\Upsilon} \to \underline{\Omega}$;
- $\underline{V}(\phi \vee \psi) = \beta^{-1}(T(\beta(\underline{V}(\phi)) \cup \beta(\underline{V}(\psi)))) : \underline{\Upsilon} \to \underline{\Omega}$,

where 'β' is a reduction of bijection in Lemma 19 to $\beta : \underline{\mathbf{DB}}(\underline{\Upsilon}, \underline{\Omega}) \cong Ob_{\underline{\mathbf{DB}}_{sk}}$ of only simple arrows, where $\underline{\mathbf{DB}}_{sk}$ is a full skeletal subcategory of $\underline{\mathbf{DB}}$ with only closed simple objects.

It is easy to verify that the lattice $L_{\underline{DB}}$ is well defined (it is an algebraic complete lattice from Proposition 52 in Sect. 8.1.5) by the restriction of matching and merging operators to simple *closed* objects A and B, given by $A \otimes B = TA \cap TB = A \cap B$ (by Theorem 12) $= T(A \cap B)$ and $A \widehat{\oplus} B = A \oplus B = T(A \cup B)$ (by Definition 57) and hence the PO relation '$\preceq$' reduces to the subset relation '$\subseteq$'.

We can consider a simple closed object $TA \in Ob_{\underline{\mathbf{DB}}_{sk}}$ as a representative element of an equivalence class of the instance databases (based on the equivalence relation corresponding to the isomorphism of objects in **DB** category), $[A] = \{B \in Ob_{\underline{\mathbf{DB}}} \mid TB = TA\}$, with $TA \in [A]$, and for any two $B, C \in [A]$, they are behaviorally equivalent (hence, isomorphic) $B \simeq C$ in **DB**. Thus, in what follows we will consider this "weak monoidal topos"-logic based on the DB-quotient (w.r.t. behavioral equivalence) lattice $L_{\underline{DB}}$, with the meet lattice operator equal to $T\cap = \cap$ and the join operator $T\cup$, for the logical conjunction and disjunction, respectively. The logical operators for negation and implication will be considered in the next sections.

The examination of the closure algebraic operators (the power-view operator T in our case) from a logical point of view and the investigation of the modal negation operators based on the complete lattice of truth-values, provided by the Birkhoff polarity, has been presented in [16]. In the next section, we will briefly present these results and, successively, consider if some of them can feature the power-view operator T in our **DB**-algebraic lattice $L_{\underline{DB}}$ in Definition 63.

9.2.1 Birkhoff Polarity over Complete Lattices

Generally, lattices arise concretely as the substructures of closure systems (intersection systems) where a closure system is a family $\mathcal{F}(W)$ of subsets of a set W such that $W \in \mathcal{F}(W)$ and if $A_i \in \mathcal{F}(W), i \in I$, then $\bigcap_{i \in I} A_i \in \mathcal{F}(W)$. Then the representation problem for general lattices is to establish that every lattice can be viewed, up to an isomorphism, as a collection of subsets in a closure system (on some set W), closed under the operations of the system.

Closure operators Γ are canonically obtained by the composition of two maps of Galois connections. The Galois connections can be obtained from any binary relation $R \subseteq W^2$ on a set W [2] (i.e., Birkhoff *polarity*) in a canonical way:

If (W, R) is a set with a particular relation on a set W, $R \subseteq W \times W$, with mappings $\lambda_R : \mathcal{P}(W) \to \mathcal{P}(W)^{OP}$, $\varrho_R : \mathcal{P}(W)^{OP} \to \mathcal{P}(W)$, such that for any $U, V \in \mathcal{P}(W)$, $\lambda_R U = \{a \in W \mid \forall u \in U.((u, a) \in R)\}$, $\rho_R V = \{a \in W \mid \forall v \in V.((a, v) \in R)\}$, where $\mathcal{P}(W)$ is the *powerset poset* with the bottom element being the empty set $\emptyset$ and the top element W, and $\mathcal{P}(W)^{OP}$ its dual (with $\subseteq^{OP}$ inverse of $\subseteq$), then we obtain the induced Galois connection $\lambda_R \dashv \rho_R$, i.e., $\lambda_R U \subseteq^{OP} V$ iff $U \subseteq \rho_R V$.

The following lemma is useful for the relationship of these set-based operators with the operation of negation in the complete lattices.

Lemma 20 (INCOMPATIBILITY RELATION [2]) *Let* $(W, \leq, \wedge, \vee, 0, 1)$ *be a complete lattice with '$\wedge$' and '$\vee$' the meet and join lattice operators, respectively, and with* 1 *and* 0 *the top and bottom elements in* W*. Then we can use a binary relation*

$R \subseteq W \times W$ as an `incompatibility` *relation for set-based negation operators λ and ρ, with the following properties: for any $U, V \subseteq W$,*
1. *(Additivity) $\lambda_R(U \cup V) = \lambda_R U \cup^{OP} \lambda_R V = \lambda_R U \cap \lambda_R V$, with $\lambda_R \emptyset = \emptyset^{OP} = W$;*
2. *(Multiplicativity) $\rho_R(U \cap^{OP} V) = \rho_R(U \cup V) = \rho_R U \cap \rho_R V$, with $\rho_R W^{OP} = \rho_R \emptyset = W$;*
3. *$\lambda_R(U \cap V) \supseteq \lambda_R U \cup \lambda_R V$, $\rho_R(U \cap V) \supseteq \rho_R U \cup \rho_R V$, and $\lambda_R \rho_R V \supseteq V$, $\rho_R \lambda_R U \supseteq U$.*

Setting $\Gamma_R = \rho_R \lambda_R : \mathcal{P}(W) \to \mathcal{P}(W)$, the operator Γ_R is a monotone mapping and a closure operator such that for any $U, V \in \mathcal{P}(W)$
1. $U \subseteq V$ implies $\Gamma_R(U) \subseteq \Gamma_R(V)$;
2. $U \subseteq \Gamma_R(U)$, for any $U \in \mathcal{P}(W)$;
3. $\Gamma_R \Gamma_R(U) \subseteq \Gamma_R(U)$, thus, from facts 1 and 2, $\Gamma_R \Gamma_R(U) = \Gamma_R(U)$.

The set $\{U \in \mathcal{P}(W) \mid U = \Gamma_R(U)\}$ is called the set of *stable* (i.e., closed) sets.

We will consider the case when $(W, R_{\leq})$ is a complete lattice with the binary relation $R_{\leq} = \{(a, b) \in W^2 \mid a \leq b\}$, equal to the partial order '$\leq$' in the lattice $(W, \leq)$. This choice is meaningful and based on the fact that we try to connect the weak monoidal topos with the algebraic complete lattice $L_{\underline{DB}} = (Ob_{\underline{\mathbf{DB}}_{sk}}, \subseteq, \cap, T\cup)$ (in Definition 63), with the standard topos intuitionistic logic and its Heyting algebra.

In what follows, we denote by R^{-1} the inverse relation of R, and by $\overline{R} = W^2 \backslash R$ the complement of R.

Example 42 Let us consider the Kripke semantics of the intuitionistic negation (point 5 in the introduction, Sect. 1.2), given by
5. $\mathcal{M} \models_a \neg\phi$ iff $\forall b(a \geq b$ implies not $\mathcal{M} \models_b \phi)$,

so that the set of worlds where $\neg\phi$ is satisfied is denoted by

$$\|\neg\phi\|_K = \{a \in W \mid \mathcal{M} \models_a \neg\phi\} \in H(W).$$

Let us show that this semantics is a particular case of additive negation operator λ_R of the Birkhoff polarity, when the incompatibility relation is $R = \{(a, b) \in W^2 \mid a \nleq b\} = \overline{R_{\leq}}$, and $R_{\leq}$ is a binary poset relation '$\leq$', i.e., $R_{\leq} = \{(a, b) \in W^2 \mid a \leq b\}$.

In fact,

$\mathcal{M} \models_a \neg\phi$

iff $\mathcal{M} \models_a \Box_{R_{\leq}^{-1}} \neg_c \phi$ (from S4 modal Kripke semantics for intuitionistic negation)

iff $\forall b\big((a, b) \in R_{\leq}^{-1}$ implies $\mathcal{M} \models_b \neg_c \phi\big)$

iff $\forall b(a \geq b$ does not imply $\mathcal{M} \models_b \phi)$ (point 5 in Sect. 1.2)

iff $\forall b\big(b \leq a$ implies $b \notin \|\phi\|_K\big)$

iff $\forall b\big(b \in \|\phi\|_K$ implies $b \nleq a\big)$

iff $\forall b\big(b \in \|\phi\|_K$ implies$(b, a) \in \overline{R_{\leq}}\big)$

iff $\forall b \in \|\phi\|_K.(b,a) \in \overline{R_{\leq}}$

iff $a \in \lambda_{\overline{R_{\leq}}}(\|\phi\|_K)$.

Thus, $\|\neg\phi\|_K = \lambda_{\overline{R_{\leq}}}(\|\phi\|_K)$.

In fact, as follows from this example, the negation operator in intuitionistic logic (i.e., Heyting algebra) has a Kripke semantics base on the accessibility relation which is a PO relation '$\leq$' between the possible worlds. This fact, together with the fact that in the autoreferential semantics we use the elements of W as possible worlds, explains why we have a special interest to consider the incompatibility relation equal to the PO relation used in a complete lattice $(W, \leq)$.

The resulting Galois connection on this partial order is the familiar Dedekind–McNeile Galois connection of antitonic mappings λ and ρ, with $\Gamma = \rho\lambda$ (by $\downarrow a$, we denote the ideal $\{b \in W \mid b \leq a\}$ and by $\uparrow a$, we denote $\{b \in W \mid a \leq b\}$):

$$\lambda U = \{a \in W \mid \forall u \in U.(u \leq a)\}, \qquad \rho V = \{a \in W \mid \forall v \in V.((a \leq v)\}.$$

It is easy to verify that for every subset $S \subseteq W$, we obtain $\Gamma(S) = \downarrow \bigvee(S)$ (the '$\bigvee$' is extension of the *binary* join operator $\vee$ of the lattice to any subset S of elements in W, that is, it is the l.u.b. of S in W, i.e., $\bigvee(S) \in W$).

In fact, for each singleton $U = \{b\}$, $\lambda U = \{a \in W \mid \forall u \in \{b\}.(u \leq a)\} = \uparrow b$. Thus, from the additivity (point 1 of the lemma above), we obtain $\lambda S = \bigcap\{\lambda\{b\} \mid b \in S\} = \bigcap\{\uparrow b \mid b \in S\} = \uparrow \bigvee(S)$. Analogously, from the multiplicativity (point 2 of the lemma above), we obtain $\rho S = \downarrow \bigwedge$ where '$\bigwedge$' is the extension of the *binary* meet operator $\wedge$ of the lattice to any subset S of elements in W, that is, it is the g.l.b. of S in W.

Consequently, $\lambda = \uparrow \bigvee$ is a kind of an additive modal negation and $\rho = \downarrow \bigwedge$ is a kind of a multiplicative modal negation, so that $\Gamma = \rho\lambda = \downarrow \bigwedge \uparrow \bigvee =$ (from $\bigwedge \uparrow = id_W$) $= \downarrow \bigvee$ from the logical point of view is a composition of these two modal negations.

Notice that $\Gamma(\emptyset) = \Gamma(\{0\}) = \{0\}$. Hence, from the monotone property of Γ and from $\emptyset \subseteq U$ (for any $U \in \mathcal{P}(W)$), we obtain $\{0\} = \Gamma(\emptyset) \subseteq \Gamma(U)$. That is, any stable (closed) set contains the bottom element $0 \in W$, and the minimal closed set is $\{0\}$ and not the empty set $\emptyset$ while, naturally, the maximal closed set is W. Notice that we have similar facts for the database complete lattice $L_{\underline{DB}} = (Ob_{\underline{\mathbf{DB}}_{sk}}, \subseteq, \cap, T\cup)$ with the bottom object (a singleton) $\bot^0 = \{\bot\} \neq \emptyset$ and with a similar property for the power-view closure operator, $T(\bot^0) = \bot^0$ and $T(\underline{\Upsilon}) = \underline{\Upsilon}$ for the bottom and top objects in $Ob_{\underline{\mathbf{DB}}_{sk}}$, respectively.

Moreover, each closed set is a downclosed ideal and there is the bijection between the set of closed sets and the set of algebraic truth-values in W, so that we are able to define the representation theorem for a complete lattice $(W, \leq)$ as follows:

Proposition 68 (AUTOREFERENTIAL REPRESENTATION FOR COMPLETE INFINITARY DISTRIBUTIVE LATTICES) *Let $(W, \leq)$ be a complete infinitary distributive lattice. We define the Dedekind–McNeile coalgebra $\downarrow : W \to \mathcal{P}(W)$, where*

$\downarrow = \Gamma in$, $\Gamma = \rho\lambda$ is a Dedekind–McNeile closure operator and $in : W \to \mathcal{P}(W)$ the inclusion map $a \mapsto \{a\}$. We denote by $\mathcal{F}(W) = \{\downarrow a \mid a \in W\}$ the closure system because for each $a \in W$, $\downarrow a$ is a closed set. Then, the $(\mathcal{F}(W), \subseteq)$ is a complete lattice of closed subsets, with the meet operator $\cap$ (set intersection), and the join operator $\sqcup = \Gamma\bigcup = \downarrow\bigvee\bigcup$, different *from the set union $\bigcup$, such that the following isomorphism is valid:*

$$\downarrow : (W, \leq, \wedge, \vee, 0, 1) \simeq (\mathcal{F}(W), \subseteq, \cap, \sqcup, \{0\}, W),$$

with $\bigvee\downarrow = id_W$ and $\downarrow\bigvee = id_{\mathcal{F}(W)}$.

Proof We will show that each closed set $U \in \mathcal{F}(W)$ is an ideal in $(W, \leq)$. In fact, for any $a \in W$, $\downarrow a = \Gamma in(a) = \rho\lambda(\{a\}) = \{a' \in W \mid a' \leq a\}$ is an ideal. Thus, for any two $a, b \in W$,
1. If $a \leq b$ then $\downarrow a \subseteq \downarrow b$;
2. $\downarrow a \cap \downarrow b = \{a' \in W \mid a' \leq a\} \cap \{b' \in W \mid b' \leq b\} = \{c \in W \mid c \leq a \wedge b\}$
$= \downarrow (a \wedge b) \in \mathcal{F}(W)$;
3. $\downarrow a \sqcup \downarrow b = \downarrow\bigvee\bigcup(\downarrow a, \downarrow b) = \downarrow\bigvee(\downarrow a \cup \downarrow b) = \downarrow\bigvee(\{a' \in W \mid a' \leq a\} \cup \{b' \in W \mid b' \leq b\}) = \downarrow (a \vee b) \in \mathcal{F}(W)$.

The operator $\sqcup$ is the join operator in $\mathcal{F}(W)$ and hence $\downarrow a \subseteq \downarrow b$ iff $\downarrow a \sqcup \downarrow b = \downarrow b$.

Thus, the original lattice $(W, \leq)$ is isomorphic to the lattice $(\mathcal{F}(W), \subseteq)$ via map $a \mapsto \downarrow a$. It is easy to verify that the inverse of $\downarrow$, i.e., $\downarrow^{-1} = (\Gamma in)^{-1} = in^{-1}\Gamma^{-1} : \mathcal{P}(W) \to W$ is equal to the supremum $\bigvee$, i.e., $\downarrow^{-1}(U) = \bigvee\{a \in U\}$, with $\bigvee\downarrow a = a$, and hence $\bigvee\downarrow = id_W$ and $\downarrow\bigvee = id_{\mathcal{F}(W)}$ are the identity functions for W and $\mathcal{F}(W)$, respectively. It is easy to verify that $\mathcal{F}(W)$ is a complete lattice, such that for every subset $S \subseteq \mathcal{F}(W)$, its l.u.b. is $\bigvee_{\mathcal{F}} S = \Gamma \bigcup S$ with the bottom element $\downarrow 0 = \{0\}$ and the top element $\downarrow 1 = W$, and for any $U = \downarrow a$, $V = \downarrow b \in \mathcal{F}(W)$, $U \cap V = \downarrow (a \wedge b) \in \mathcal{F}(W)$ and $U \sqcup V = \downarrow (a \vee b) \in \mathcal{F}(W)$. □

Remark What is important to notice is that the autoreferential representation, based only on the lattice ordering $(W, \leq)$, naturally introduces the modality in a lattice-based logic. In fact, a modal operator $\Gamma : \mathcal{P}(W) \to \mathcal{F}(W) \subset \mathcal{P}(W)$, based on the partial order of a complete lattice $(W, \leq)$, plays the fundamental role for the modal disjunction $\sqcup$ in the canonical autoreferential lattice $\mathcal{F}(W)$. We need it because, generally, for any two $a, b \in W$, $\downarrow a \bigcup \downarrow b \notin \mathcal{F}(W)$. Its restriction on $\mathcal{F}(W)$ is an *identity* function $id_{\mathcal{F}(W)}$, which is a selfadjoint (universal and existential) modal operator.

Notice that an analogous phenomenon is presented in our logic of weak monoidal topos, as considered at the end of the previous section. Notice also that in his approach Lewis distinguishes the standard, or *extensional,* disjunction '$\vee_c$' from the *intensional* disjunction $\vee$ "in that at least one of the disjoined propositions is "necessarily true" [14, p. 523]. In the same way, he defined also the "strict" logic implication $\phi \Rightarrow \psi$ equal to $\Box(\phi \Rightarrow_c \psi)$ where '$\Rightarrow_c$' is the classic extensional implication such that $\phi \Rightarrow_c \psi$ is equal to $\neg_c\phi \vee_c \psi$ (here '$\neg_c$' denotes the classic logical negation). In what follows, we will see that this property holds also for the implication defined as relative pseudo-complement in a complete distributive lattice W, that is, for the *intuitionistic* logic implication.

In fact, if we consider that each formula is true in the bottom element (a "contradiction" world) $0 \in W$ of the complete lattice W of "possible worlds", then the complex algebra of intuitionistic logic is a Heyting algebra,

$$\mathbf{H}(W) = \big(H(W), \subseteq, \cap, \cup, \Rightarrow_h, \neg_h\big),$$

based on r.p.c. lattice $(H(W), \subseteq, \cap, \cup, \{0\}, W)$, where $H(W)$ is not a set of closed subsets (as in L_{alg}) but a set of *hereditary* subsets S (with $0 \in S$) of a poset $(W, \leq)$ such that if $a \in S$ and $b \leq a$ then $b \in S$, and differently from L_{alg}, its meet operator is a simple set-union. The relative pseudo-complement for any two hereditary subsets $U, V \in H(W)$ is

$$U \Rightarrow_h V = \bigcup (Z \in H(W) \mid Z \cap U \subseteq V\} = W \setminus \bigcup_{b \in U \setminus V} \{a \in W \mid (b, a) \in R_{\leq}\},$$

while in L_{alg} we apply the closure power-view operator T on it, as well. Based on Kripke possible-worlds semantics, we obtain for $\mathbf{H}(W)$ the modal propositional logic with a universal modal logical operator of logic "necessity" '$\Box$' (for S4 frame system with a reflexive and transitive accessibility relation R corresponding to the partial order '$\leq$' between the possible worlds in W) such that $\phi \Rightarrow \psi$ is logically equivalent to $\Box(\phi \Rightarrow_c \psi)$ with classic propositional implication '$\Rightarrow_c$', and $\neg\phi$ is logically equivalent to $\Box\neg_c\phi$ with a classic propositional negation '$\neg_c$'.

Example 43 The smallest *nontrivial* bilattice is Belnap's 4-valued bilattice [1, 6, 17] $W = \{1, 0, \bot, \top\}$ where 1 is *true*, 0 is *false*, $\top$ is inconsistent (both true and false) or *possible*, and $\bot$ is *unknown*. As Belnap observed, these values can be given two natural orders: *truth* order, $\leq$, and *knowledge* order, $\leq_k$, such that $0 \leq \top \leq 1$, $0 \leq \bot \leq 1$, and $\bot \leq_k 0 \leq_k \top$, $\bot \leq_k 1 \leq_k \top$. The meet and join operators under partial ordering '$\leq$' are denoted '$\wedge$' and '$\vee$'; they are natural generalizations of the usual conjunction and disjunction notions. Meet and join under $\leq_k$ are denoted by '$\circledast$' and '$\circledcirc$', so that $0 \circledast 1 = \bot$, $0 \circledcirc 1 = \top$, $\top \wedge \bot = 0$ and $\top \vee \bot = 1$.

The *bilattice negation* [7] is defined by $\neg 0 = 1$, $\neg 1 = 0$, $\neg\bot = \bot$ and $\neg\top = \top$. It is easy to see that the De Morgan law is valid, $\neg(a \wedge b) = \neg a \vee \neg b$.

Thus, w.r.t. the truth ordering '$\leq$', the elements in $\mathcal{F}(W)$ are $\downarrow 0 = \{0\}$, $\downarrow \bot = \{0, \bot\}$, $\downarrow \top = \{0, \top\}$ and $\downarrow 1 = W$. That is, $\mathcal{F}(W) = \{\{0\}, \{0, \bot\}, \{0, \top\}, W\}$.

From the isomorphism in Proposition 68, we have the isomorphism

$$\downarrow : (W, \leq, \wedge, \vee, 0, 1) \simeq \big(\mathcal{F}(W), \subseteq, \cap, \sqcup, \{0\}, W\big).$$

So that, $\downarrow \bot \sqcup \downarrow \top = \downarrow \bigvee(\downarrow \bot \cup \downarrow \top) = \downarrow \bigvee\{0, \bot, \top\} = \downarrow 1 = W \neq \downarrow \bot \cup \downarrow \top$, that is, $\sqcup \neq \cup$.

Notice that the set of hereditary subsets of W (see the introduction in Sect. 1.2) is equal to $H(W) = \{\{0\}, \{0, \bot\}, \{0, \top\}, \{0, \bot, \top\}, W\}$, that is, to the lattice $(H(W), \subseteq, \cap, \cup, \{0\}, W)$ which is, differently from $\mathcal{F}(W)$, closed under the set-union (join

operator in $H(W)$). In fact, $\{0, \bot\} \cup \{0, \top\} = \{0, \bot, \top\} \in H(W)$. Consequently, we have the following surjective homomorphism between lattices,
(e) $\Gamma : (H(W), \subseteq, \cap, \cup, \{0\}, W) \to (\mathcal{F}(W), \subseteq, \cap, \sqcup, \{0\}, W)$,
such that $\Gamma(\{0\}) = \{0\}$ (bottom element in $\mathcal{F}(W)$), $\Gamma(\{0, \bot\}) = \{0, \bot\}$, $\Gamma(\{0, \top\}) = \{0, \top\}$, and $\Gamma(\{0, \bot, \top\}) = \Gamma(W) = W$. We have also that for any two hereditary sets $S_1, S_2 \in H(W)$,

- $\Gamma(\cap(S_1, S_2)) = (\Gamma\cap)(\Gamma(S_1), \Gamma(S_2)) = \cap(\Gamma(S_1), \Gamma(S_2))$, and
- $\Gamma(\cup(S_1, S_2)) = (\Gamma\cup)(\Gamma(S_1), \Gamma(S_2)) = \sqcup(\Gamma(S_1), \Gamma(S_2))$,

and hence Γ is a homomorphism between these two lattices.

It is easy to verify that the join operator $\sqcup$, in the compact set-based representation $\mathcal{F}(W)$, reduces to the standard set union $\cup$ when the complete lattice $(W, \leq)$ is a total ordering (as, for example, in the fuzzy logic with the closed interval of reals $W = [0, 1]$ for the set of truth values).

Let us extend Example 43 to relative pseudo-complements, so that for two elements (sets) in $H(W)$, $S_1 = W$ and $S_2 = \{0, \bot, \top\}$, $S_1 \Rightarrow_h S_2 = \bigcup\{Z \in H(W) \mid Z \cap S_1 \subseteq S_2\} = \bigcup\{Z \in H(W) \mid Z \subseteq S_2\} = \{0, \bot, \top\} = S_2$.

Let us show that $\Gamma(S_1 \Rightarrow_h S_2) = (\Gamma S_1) \widehat{\Rightarrow} (\Gamma S_2)$. In fact, $\Gamma(S_1 \Rightarrow_h S_2) = \Gamma\{0, \bot, \top\} = W$, while

$$\begin{aligned}(\Gamma S_1) \widehat{\Rightarrow} (\Gamma S_2) &= W \widehat{\Rightarrow} \big(\Gamma\{0, \bot, \top\}\big) = W \widehat{\Rightarrow} W = \Gamma(W \Rightarrow_h W) \\ &= \Gamma\Big(\bigcup\{Z \in H(W) \mid Z \cap W \subseteq W\}\Big) \\ &= \Gamma\bigcup\{Z \in H(W) \mid Z \subseteq W\} = \Gamma W = W.\end{aligned}$$

We can generalize these results as follows:

Corollary 34 *There is the following homomorphism between the distributive complete (r.p.c.) lattice $(H(W), \subseteq, \cap, \cup, \{0\}, W)$ of hereditary subsets in $(W, \leq)$, and its "closed" version complete lattice:*

1. $\Gamma : (H(W), \subseteq, \cap, \cup, \{0\}, W) \to (\mathcal{F}(W), \subseteq, \Gamma\cap, \Gamma\cup, \{0\}, W)$, *where* $\Gamma\cap = \cap$ *and* $\Gamma\cup = \sqcup$ *are the meet and join lattice operators in* $\mathcal{F}(W)$.
   Thus, we can define the following Heyting algebra of closed hereditary subsets:
2. $\mathbf{F}(W) = (\mathcal{F}(W), \subseteq, \cap, \Gamma\cup, \widehat{\Rightarrow}, \widehat{\neg}, \{0\}, W)$, *where* $\widehat{\Rightarrow}$ *and* $\widehat{\neg}$ *are (relative) pseudo-complements in this r.p.c. lattice, and hence there is the following homomorphisms of Heyting algebras* $\mathbf{H}(W) = (H(W), \subseteq, \cap, \cup, \Rightarrow_h, \neg_h, \{0\}, W)$ *and* $\mathbf{F}(W)$:
3. $\Gamma : \mathbf{H}(W) \to \mathbf{F}(W)$.

Proof By composition of the homomorphism $\bigvee : (H(W), \subseteq, \cap, \cup, \{0\}, W) \to (W, \leq, \wedge, \vee, 0, 1)$ and homomorphism $\downarrow$ in Proposition 68, we obtain the homomorphism

$$\Gamma = \downarrow \bigvee : \mathbf{H}(W) \to \big(\mathcal{F}(W), \subseteq, \Gamma\cap, \Gamma\cup, \{0\}, W\big),$$

with $\{0\} = \Gamma(\{0\})$, $W = \Gamma(W)$, and, for any two $S_1, S_2 \in H(W)$,

1. $\Gamma(S_1 \cup S_2) = \downarrow \bigvee(S_1 \cup S_2) = \downarrow (\bigvee S_1 \vee \bigvee S_2) =$ (from $\bigvee \downarrow = id_W) = \downarrow ((\bigvee \downarrow) \bigvee S_1 \vee (\bigvee \downarrow) \bigvee S_2) = \bigvee \downarrow (\downarrow \bigvee S_1 \cup \downarrow \bigvee S_2) = \Gamma(\Gamma S_1 \cup \Gamma S_2) \in \mathcal{F}(W)$.

Thus, Γ is a homomorphism for join operators, and, analogously,

2. $\Gamma(S_1 \cap S_2) = \Gamma(\Gamma S_1 \cap \Gamma S_2) =$ (from the fact that Γ is a closure operator) $= \Gamma S_1 \cap \Gamma S_2 \in \mathcal{F}(W)$.

Thus, Γ is a homomorphism for meet operators. For each $a \in W$, the set $\downarrow a$ is closed, so that $\Gamma(\downarrow a) = \downarrow a$, and for any $S \in H(W)$, $a, b \in W$,

3. $S \cap \downarrow a \subseteq \downarrow b$ iff $\Gamma S \cap \downarrow a \subseteq \downarrow b$.

 (the implication from right to left is obvious, based on the closure property of Γ, $S \subseteq \Gamma S$; for the implication from left to right, we can use the monotonicity property of the closure operator Γ, and hence if $S \cap \downarrow a \subseteq \downarrow b$ then $\Gamma(S \cap \downarrow a) =$ (from 2.) $= \Gamma S \cap \Gamma(\downarrow a) = \Gamma S \cap \downarrow a \subseteq \Gamma(\downarrow b) = \downarrow b$.

Let us show that the complete lattice $(\mathcal{F}(W), \subseteq, \Gamma\cap, \Gamma\cup, \{0\}, W)$ is also distributive and hence an r.p.c. lattice, so that we can define the (relative) pseudocomplements $\widehat{\Rightarrow}$ and $\widehat{\neg}$. In fact, for any $S_1 = \Gamma S_1 = \downarrow a$, $S_2 = \Gamma S_2 = \downarrow b$, $S_3 = \Gamma S_3 \in \mathcal{F}(W) \subseteq H(W)$,

4.

$$\begin{aligned}
\Gamma \cup (S_1 \cap S_2, S_3) &= \Gamma \cup (\Gamma S_1 \cap \Gamma S_2, \Gamma S_3) = \Gamma\big((\Gamma S_1 \cap \Gamma S_2) \cup \Gamma S_3\big) \\
&\quad \text{(by the distributivity of } \cap \text{ and } \cup \text{)} \\
&= \Gamma\big((\Gamma S_1 \cup \Gamma S_3) \cap (\Gamma S_2 \cup \Gamma S_3)\big) \\
&\quad \text{(from the homomorphism property 1.)} \\
&= \Gamma(\Gamma S_1 \cup \Gamma S_3) \cap \Gamma(\Gamma S_2 \cup \Gamma S_3) \\
&= \Gamma(S_1 \cup S_3) \cap \Gamma(S_2 \cup S_3) = \Gamma \cup (S_1, S_3) \cap \Gamma \cup (S_2, S_3).
\end{aligned}$$

5.

$$\begin{aligned}
\Gamma \cup (S_1, S_2) \cap S_3 &= \Gamma \cup (S_1, S_2) \cap \Gamma S_3 = \Gamma(S_1 \cup S_2) \cap \Gamma S_3 \\
&\quad \text{(from the homomorphism property 1.)} \\
&= \Gamma\big((S_1 \cup S_2) \cap S_3\big) \\
&\quad \text{(by the distributivity of } \cap \text{ and } \cup \text{)} \\
&= \Gamma\big((S_1 \cap S_3) \cup (S_2 \cap S_3)\big) = \Gamma \cap \big((S_1 \cap S_3), (S_2 \cap S_3)\big).
\end{aligned}$$

From the fact that the complete distributive lattice of hereditary subsets $\mathbf{H}(W)$ satisfies the infinite distributivity law (so it is a Heyting algebra), from this homomorphism (an analogous proof as for points 3 and 4), also the complete distributive lattice $(\mathcal{F}(W), \subseteq)$ of closed hereditary subsets satisfies the infinite distributive law, and, consequently, it is a Heyting algebra with:

6.

$$(\downarrow a) \widehat{\Rightarrow} (\downarrow b) = \Gamma \cup \big\{\downarrow c \in \mathcal{F}(W) \mid \downarrow c \cap \downarrow a \subseteq \downarrow b\big\}$$

$$= \Gamma \cup \Big\{ \bigcup \{ S \in H(W) \mid S \subseteq \downarrow c \in \mathcal{F}(W) \} \mid \downarrow c \cap \downarrow a \subseteq \downarrow b \Big\}$$

$$\overset{\text{(from 3)}}{=} \Gamma \cup \Big\{ \bigcup \{ S \in H(W) \mid S \subseteq \downarrow c, S \cap \downarrow a \subseteq \downarrow b \} \mid \downarrow c \in \mathcal{F}(W) \Big\}$$

$$= \Gamma \cup \{ S \in H(W) \mid S \cap \downarrow a \subseteq \downarrow b \}$$

$$= \Gamma(\downarrow a \Rightarrow_h \downarrow b),$$

thus, for any $S_1, S_2 \in H(W)$, $\Gamma(\Gamma(S_1) \Rightarrow_h \Gamma(S_2)) = (\Gamma S_1) \widehat{\Rightarrow} (\Gamma S_2)$, and

$$S_1 \Rightarrow_h S_2 = \bigcup \{ S \in H(W) \mid S \cap S_1 \subseteq S_2 \}$$

(monotonicity of Γ)

$$= \bigcup \{ S \in H(W) \mid \Gamma S \cap \Gamma S_1 \subseteq \Gamma S_2 \}$$

$$\overset{\text{(from 3)}}{=} \bigcup \{ S \in H(W) \mid S \cap \Gamma S_1 \subseteq \Gamma S_2 \} = (\Gamma S_1) \Rightarrow_h (\Gamma S_2).$$

Thus, $\Gamma(S_1 \Rightarrow_h S_2) = \Gamma(\Gamma(S_1) \Rightarrow_h \Gamma(S_2)) = (\Gamma S_1) \widehat{\Rightarrow} (\Gamma S_2)$, and hence Γ is a homomorphism for relative pseudo-complements as well (thus $\Gamma(\neg_h S_1) = \widehat{\neg}(\Gamma S_1)$). □

The Heyting algebra $\mathbf{F}(W)$ for closed hereditary subsets of a complete lattice $(W, \leq)$ will be used in what follows for the logic of the weak monoidal topos.

9.2.2 DB-Truth-Value Algebra and Birkhoff Polarity

It is well known that the semantics of intuitionistic logic of the standard topoi was provided by Samuel Kripke [10], by introducing the concept of the truth of a given propositional formula ϕ in a "possible world" $a \in W$ for a given model $\mathcal{M}$ of such a logic, denoted by $\mathcal{M} \models_a \phi$.

In such a Kripke semantics (extended to all propositional logics with *modal* logical operators as well) we denote the set of all possible worlds where this formula ϕ is true in a given model $\mathcal{M}$ by $\|\phi\| = \{a \in W \mid \mathcal{M} \models_a \phi\}$.

Remark Notice that in the original Kripke semantics, we can have $\|\phi\|$ be the empty set, while here, in our "lifted" Kripke semantics $\mathcal{M} \models_0 \phi$ (each formula is true in the "contradiction" world $a = 0 \in W$) is always satisfied $0 \in \|\phi\|$. This lifting is necessary in our case because we will use a logic for databases and every simple database $A \in Ob_{\mathbf{DB}}$ is not the empty set of relations (in fact, $\perp \in A$ for each simple object A in the **DB** category).

The complex algebra of such sets for the intuitionistic logic is a Heyting algebra, $\mathbf{H}(W) = (H(W), \subseteq, \cap, \cup, \Rightarrow_h, \neg_h)$ where $H(W)$ is not a set of closed subsets but a set of *hereditary* subsets S (with $0 \in S$) of a poset $(W, \leq)$ such that if $a \in S$ and $b \leq a$ then $b \in S$, and differently from $L_{\underline{DB}}$, its join operator is the simple set-union.

Based on Kripke possible-worlds semantics, we obtain for $\mathbf{H}(W)$ the modal propositional logic with a universal modal logical operator of logical "necessity" '$\Box_{R_{\leq}^{-1}}$' (with the S4 frame system, i.e., with a reflexive and transitive accessibility relation $R_{\leq}^{-1}$ corresponding to a partial order '$\geq$' between the possible worlds in W) such that $\phi \Rightarrow \psi$ is logically equivalent to $\Box_{R_{\leq}^{-1}}(\phi \Rightarrow_c \psi)$ with the standard propositional implication '$\Rightarrow_c$', and $\neg\phi$ is logically equivalent to $\Box_{R_{\leq}^{-1}}\neg_c\phi$ with the standard negation '$\neg_c$' (see Example 42).

It is clear that the logical interpretation of $L_{\underline{DB}}$ has to be different just because each of its join operators is preceded by the closure operation T for which we have to find the right logical interpretation. Taking in account these differences between the standard Kripke semantics (with its Heyting complex algebra $\mathbf{H}(W)$) and our DB lattice of truth-values $L_{\underline{DB}}$, we need to develop a new Kripke semantics for the weak monoidal topos logic.

Hence, in an analogous way, we will present a possible-worlds semantics for the algebraic logic of the weak monoidal topos **DB**. First of all, we have to establish what are the possible worlds in such a logic: from the fact that the sets $\|\phi\| = \{a \in W \mid \mathcal{M} \models_a \phi\}$ of this logic have to be the elements in $Ob_{\mathbf{DB}_{sk}}$, we conclude that $\|\phi\| \in Ob_{\mathbf{DB}_{sk}}$, i.e., each $\|\phi\|$ is a simple *closed* object (database), composed of finitary *relations*. Consequently, the possible worlds for the weak monoidal topos **DB** must contain all relations, i.e., $W \supseteq \underline{\Upsilon} = \bigcup_{A \in Ob_{\underline{DB}}} A$ (see Definition 26 in Sect. 3.2.5).

In such a possible-worlds semantics, each propositional formula ϕ represents an instance-database whose extension is given by $\|\phi\|$, and hence for a given relation R, $R \in \|\phi\|$ iff $\mathcal{M} \models_R \phi$, that is, iff ϕ is "true in the world $a = R \in W$".

Notice that every relation $R \in \underline{\Upsilon}$ is finitary (i.e., with finite value $ar(R)$), but can have an infinite number of tuples as well. From Theorem 1, each non-closed database $A \in Ob_{\underline{DB}}$ is a finite subset of $\underline{\Upsilon}$, that is, $A \subseteq_\omega \underline{\Upsilon}$ if $A \neq TA$. A closed database $TA \in Ob_{\underline{DB}}$ instead may have an infinite number of relations, if A have a relation with the infinite number of tuples.

An (observationally) equivalent relation R_A to a given database A, that is, such that $T(\{R_A\}) = TA$, may be given as a Cartesian product of all relations in A (because from R_A by finitary projections we can obtain all relations in A). However, in order to have $R_A \in \underline{\Upsilon}$, we must ensure the condition that $ar(R_A)$ is finite number, that is, that $|A|$ (the number of relations in A) is finite. In order to guarantee this finitary property of transformations of each instance-database into an equivalent single relation in $\underline{\Upsilon}$ (based on the algebraic property of the closure power-view operator T, Proposition 52 in Sect. 8.1.5), the first attempt is as follows:

Lemma 21 *Let us denote a vector of a relation R by $\overrightarrow{R} \triangleq \bigcup_{1 \leq i \leq ar(R)} \pi_i(R)$ if $ar(R) \geq 2$; R otherwise. Analogously, we define a vector relation $\overrightarrow{A} \triangleq \bigcup_{R \in A} \overrightarrow{R}$ obtained from a database-instance A. We denote the unary Skolem relation by $\overrightarrow{SK} \triangleq \{\langle \omega_i \rangle \mid \omega_i \in SK\}$. Based on this, we introduce the PO set $(\underline{\Upsilon}, \preceq_v)$ such that for any two relations, $R_1, R_2 \in \underline{\Upsilon}$, $R_1 \preceq_v R_2$ iff $\overrightarrow{R}_1 \subseteq \overrightarrow{R}_1$, and we denote the derived equality by $R_1 =_v R_2$ iff ($R_1 \preceq_v R_2$ and $R_2 \preceq_v R_1$).*

Hence, we obtain the following complete distributive lattice of relations
(e1) $L_v = (\underline{\Upsilon}, \preceq_v, \wedge_v, \vee_v, 0, 1)$, *where* 0 *equal to empty relation* $\perp$, *and* 1 *is equal to top relation* $\overrightarrow{\underline{\Upsilon}} = \{\langle a \rangle \mid a \in \mathcal{U}\}$, *and for any two relations* $R_1, R_2 \in \underline{\Upsilon}$, *the meet and join operators are defined by*
(e2) $R_1 \wedge_v R_2 \triangleq \overrightarrow{R}_1 \cap \overrightarrow{R}_2$, $R_1 \vee_v R_2 \triangleq \overrightarrow{R}_1 \cup \overrightarrow{R}_2$.
The vectorization operation $\overrightarrow{\_} : \mathcal{P}(\underline{\Upsilon}) \to \underline{\Upsilon}$ *is the l.u.b. ('$\bigvee$') operation of the lattice* L_v.

Proof Any vector relation is a unary relation and, by definition, $\overrightarrow{\perp} = \perp = \{\langle\rangle\}$ where $\langle\rangle$ is the empty tuple (Definition 10 in Sect. 2.4.1). Consequently, for any R, $\perp = \overrightarrow{\perp} = \{\langle\rangle\} \subseteq \overrightarrow{R}$, so that $\perp \preceq_v R$, that is, $\perp$ is the bottom element of this lattice. By definition, $\overrightarrow{\underline{\Upsilon}} = \bigcup_{R \in \underline{\Upsilon}} \overrightarrow{R} = \bigcup_{R \in \underline{\Upsilon}} \{\langle a \rangle \mid a \text{ is a value in } R\} = \{\langle a \rangle \mid a \in \mathcal{U}\}$ (where $\underline{\Upsilon}$ is generated by all values in $\mathcal{U} = \mathbf{dom} \cup SK$, where **dom** is a finite domain of values and $SK = \{\omega_0, \omega_1, \dots\}$ is the infinite number of indexed Skolem constants, i.e., marked null values). Analogously, for any $R \in \underline{\Upsilon}$, $\overrightarrow{R} \subseteq \{\langle a \rangle \mid a \in \mathcal{U}\} = \overrightarrow{\underline{\Upsilon}} \in \underline{\Upsilon}$, thus $R \preceq_v \overrightarrow{\underline{\Upsilon}}$ so that $\overrightarrow{\underline{\Upsilon}}$ is the top element of this lattice. It is easy to verify, form (e2) that this lattice is distributive. We have that for any (also infinite) subset $S \subseteq \underline{\Upsilon}$, $glb(S) = \bigwedge_{R \in S} R = \bigcap_{R \in S} \overrightarrow{R}$ and $lub(S) = \bigvee_{R \in S} R = \bigcup_{R \in S} \overrightarrow{R} = \overrightarrow{S}$, thus well defined, and lub operation is equal to the vectorization operation $\overrightarrow{\_}$; hence, this is a complete lattice. □

It is easy to see that the "vectorization" operation $\overrightarrow{\_}$ is a homomorphism $\overrightarrow{\_} : L_v \to (\{\overrightarrow{R} \mid R \in \underline{\Upsilon}\}, \subseteq, \cap, \cup, 0, 1)$ between the complete distributive lattices, where the right-hand side lattice is the quotient of the left-hand side lattice w.r.t. the equivalence $=_v$, that is, equal to $L_v / =_v$. If we denote by $[R]$ the equivalence class of relations w.r.t. the equivalence $=_v$, then $\overrightarrow{R} \in [R]$ is the representing element of this equivalence class, and the kernel of the homomorphism $\overrightarrow{\_}$ corresponds to the congruence relation defined by $=_v$. Notice that from the fact that L_v is a complete distributive lattice (satisfying the infinite distributivity (equation 10 in Sect. 1.2), it holds that it is also a complete Heyting algebra with relative pseudo-complements.

Sadly, we have the desired property $TA = \downarrow_v \overrightarrow{A} = \{R \in \underline{\Upsilon} \mid R \preceq_v \overrightarrow{A}\}$ only when the unary relation $\overrightarrow{A}$ *is finite*. However, generally it holds that $TA \subseteq \downarrow_v \overrightarrow{A}$ (in fact, if $R \in TA$ then $\overrightarrow{R} = \bigcup_{1 \le i \le ar(R)} \pi_i(R) \subseteq \overrightarrow{A} = \overrightarrow{(\overrightarrow{A})}$, thus $R \preceq_v \overrightarrow{A}$, i.e., $R \in \downarrow_v \overrightarrow{A}$). Let us show that inverse does not hold, i.e., $TA \not\supseteq \downarrow_v \overrightarrow{A}$:

Example 44 That inverse does not hold, that is, $TA \not\supseteq \downarrow_v \overrightarrow{A}$, we can see by considering Example 27 in Sect. 4.2.4, when $A = can(\mathcal{I}, D)$ has two infinite binary relations R_1 and R_2:

$R_1 = \|r\|_{can(\mathcal{I},\mathcal{D})}$	$R_2 = \|s\|_{can(\mathcal{I},\mathcal{D})}$
a, b	b, ω_0
b, ω_1	ω_1, ω_2
ω_1, ω_3	ω_3, ω_4
ω_3, ω_5	ω_5, ω_6
...	...

Let us consider a unary infinite relation

$$R = \{\langle \omega_{4i}\rangle \mid i = 0, 1, \ldots\} = \{\langle \omega_0\rangle, \langle \omega_4\rangle, \langle \omega_8\rangle, \ldots\}$$

such that $\overrightarrow{R} = R \preceq_v R_2 \preceq_v \overrightarrow{A} = \{\langle a\rangle, \langle b\rangle\} \cup \{\langle \omega_i\rangle \mid i = 0, 1, \ldots\} = \overrightarrow{SK} \cup \{\langle a\rangle, \langle b\rangle\}$, that is, $\overrightarrow{R} \in \downarrow_v \overrightarrow{A}$. However, there is no *finite* term t of Σ_R (SPRJU) algebra such that $\|t\|_A = R$, so that $R \notin TA$. In fact, we can obtain R only as an *infinite* union of terms π_2 (R_2 *WHERE* $name_2 = \omega_i$), for $i = 0, 4, 8, \ldots$, and not by π_2 (R_2 *WHERE* C) because there is no unary select operation _ *WHERE* C in Σ_R with the condition C equal to an infinite string '($name_2 = \omega_0$) $\vee$ ($name_2 = \omega_4$) $\vee$ ($name_2 = \omega_8$) $\vee \cdots$', and by using negation (complements) we cannot obtain a finite string for a condition C as well. Consequently, with finite-length Σ_R-algebra terms we cannot obtain R and hence $\downarrow_v \overrightarrow{A} \nsubseteq TA$.

Instead, for the complete lattice in the next Lemma 22, we have the behavioral equivalence $TA = \downarrow (R_1$ *TIMES* $R_2)$. Notice that in this case, differently from the case above, $R \npreceq R_2$ because $T\{R\} \nsubseteq T\{R_2\}$. In fact, the view '$R$ *WHERE* $name_1 \neq \omega_0$' in $T\{R\}$ equal to $\{\langle \omega_{4i}\rangle \mid i = 1, 2, \ldots\}$ is not a view in $T\{R_2\}$ (analogously to the case above, it can't be obtained by finite-length Σ_R-term from R_2).

Consequently, we need an alternative method of an *equivalent* transformation of a database-instance A into a single finitary relation R_A (with a finite $n = ar(R_A)$), with possibly an infinite set of tuples, in order to guarantee that $TA = \downarrow R_A$, as follows:

Proposition 69 *There exists a function $\digamma : Ob_{\underline{DB}} \to \underline{\Upsilon}$ such that for each $A \in Ob_{\underline{DB}}$, $\digamma(TA) \triangleq \digamma(A)$ with $\digamma(\bot^0) \triangleq \bot$ and:*
1. *$\digamma(A) \in T(\{\digamma(A)\}) = T(A)$, with finite $k = ar(\digamma(A))$ and $R_{\Upsilon} \triangleq \digamma(\underline{\Upsilon})$.*
2. *If A has infinite relations then $\overrightarrow{SK} \subseteq \overrightarrow{A}$ with $\{\overrightarrow{SK}, \overrightarrow{A}\} \subseteq T(A)$.*

Proof It is easy to see that for every finite set (a database) $A \in Ob_{\underline{DB}}$, that is, when $n = |A|$ is finite, so that the relation $R = \times_C(A)$ (the Cartesian product of relations in A) is finitary (i.e., $ar(R) = \sum_{R_i \in A} ar(R_i)$ is finite), we can define $\digamma(A) \triangleq R$ (an alternative is to define $\digamma(A) \triangleq \overrightarrow{A}$).

Thus, we have only to show that $F(A)$ exists also when A is an infinite set of relations (it can happen for a closed database, $A = TB$, where B is a non-closed simple database (thus with finitely many relations), with at least one infinite finitary relation, or in the case $A = B \otimes C = TB \cap TB$ when both TA and TB are infinite sets of relations) because in such a case the Cartesian product of the infinite set of relations in A will result in an infinitary relation, while all relations in $\underline{\Upsilon}$ are finitary relations.

Let us consider such a closed infinite database $A = TA$. Notice that each relation R in this infinite set A has a finite $n = ar(R)$, but $|R|$ may be infinite. If a simple database A has an infinite relation R then $\overrightarrow{SK} \subseteq \overrightarrow{A}$, where $\overrightarrow{SK} = \{\langle\omega_k\rangle \mid k = 0, 1, 2, \dots\}$ is an infinite unary relation with all Skolem constants. Consequently, all infinite relations are infinite because they have an *infinite subset* of Skolem constants. So, let us consider the subset A_S of all infinite relations $R \in A_S \subseteq TA$ composed of only Skolem constants (i.e., with infinite set $val(\{R\}) \subseteq SK$), and we define its quotient subset $[A_S]$ as follows:

(i) For each pair $R, R' \in [A_S] \subseteq A_S$, $T(\{R\}) \neq T(\{R'\})$;

(ii) For each $R \in A_S$, we have a unique $R' \in [A_S]$ with $T(\{R\}) = T(\{R'\})$ and $ar(R') \leq ar(R)$.

Hence, $[A_S]$ is the set of all distinct infinite relations (composed of Skolem constants only) with minimal arities. There is a maximal relation $R^M_A \in [A_S]$ such that for every $R \in [A_S]$, $T(\{R\}) \subset T(\{R_M\})$. In fact, for each pair $R_1, R_2 \in [A_S]$, R_1 TIMES $R_2 \in A_S$, thus there exists $R \in [A_S]$ such that $T(\{R\}) = T(R_1 \text{ TIMES } R_2)$, with $T(\{R_1\}) \subset T(\{R\})$ and $T(\{R_2\}) \subset T(\{R\})$. Consequently, there is a maximal relation R^M_A in $[A_S]$ with finite arity $n = ar(R^M_A)$ such that any (infinite) relation $R \in A_S$ is also in $T(\{R^M_A\})$. It is unique because all relations in $[A_S]$ are distinct, and hence

(a) $\{R^M_A\} \subseteq A_S \subseteq TA$, and, consequently, $T(\{R^M_A\}) \subseteq T(TA) = TA$.

Let us define the unary relation $\overrightarrow{A} \setminus \overrightarrow{R}{}^M_A$ composed of the values which are not in R^M_A; thus, it is a finite relation, and consequently,

(b) $\overrightarrow{A} \setminus \overrightarrow{R}{}^M_A \in TA$, so that $T(\{\overrightarrow{A} \setminus \overrightarrow{R}{}^M_A\}) \subseteq TA$.

Thus we can define $F(A) \triangleq (\overrightarrow{A} \setminus \overrightarrow{R}{}^M_A)$ TIMES R^M_A, with the finite arity equal to $n + 1$ and hence $F(A) \in \underline{\Upsilon}$.

Thus, from (a), $R^M_A \in TA$, and from (b), $\overrightarrow{A} \setminus \overrightarrow{R}{}^M_A \in TA$, so we obtain that $F(A) \in TA$, and from (a) and (b) also that $T(\{F(A)\}) \subseteq TA$.

Let us also show that $T(\{F(A)\}) \supseteq TA$. From the fact that $\overrightarrow{F(A)} = \overrightarrow{A}$, each finite relation in TA can be generated from the relation $F(A)$: if we consider the infinite set of "atomic" relations, $\mathfrak{J}_A = \{R = \{\langle a\rangle\} \mid \langle a\rangle \in \overrightarrow{A}\} \subseteq T(\{F(A)\}) \subseteq TA$, any finite relation $R \in TA$ can be obtained by a finite-length term over a finite subset of relations in $\mathfrak{J}_A$ and hence $R \in T(\{F(A)\})$. Any infinite relation in TA which is not in $T(\{R^M_A\})$ can be obtained as a union of an infinite relation in $T(\{R^M_A\})$ (composed of only Skolem constants) and a finite relation which is in $T(\{F(A)\})$. Thus, $TA \subseteq T(\{F(A)\})$.

The relation $R_{\underline{\Upsilon}} = F(\underline{\Upsilon})$ is finitary, and hence it is a relation in $\underline{\Upsilon}$ (where all relations are finitary). Thus we have that $T(F(\underline{\Upsilon})) = T(\{R_{\underline{\Upsilon}}\}) = T\underline{\Upsilon} = \underline{\Upsilon}$.

Let us prove Claim 2. From the fact that $F(A) \in TA$ and $\overrightarrow{A} = \overrightarrow{TA} = \overrightarrow{T(\{F(A)\})} = \overrightarrow{F(A)} = \bigcup\{\pi_i(F(A)) | 1 \leq i \leq n = ar(F(A))\} \in TA$ (because n is finite, so that we have a finite union, which can be represented by a finite-length term of Σ_R algebra). From the fact that A has infinite relations, we have that $SK \subseteq val(A)$ and hence $\overrightarrow{SK} \subseteq \overrightarrow{A}$. Let $S = \{a_1, \ldots, a_m\} = val(A) \backslash SK \subseteq \mathbf{dom}$, thus a finite subset of values in the finite set **dom**, then $\overrightarrow{SK}$ can be obtained by a term $\overrightarrow{A}$ WHERE $((name \neq a_1) \wedge \cdots \wedge (name \neq a_m))$ of Σ_R algebra and, consequently, $\overrightarrow{SK} \in TA$, and hence from $R_A^M \in TA$, we must have that

(c) $T(\{\overrightarrow{SK}\}) \subseteq T(\{R_A^M\})$ and hence $\overrightarrow{R_A^M} = \overrightarrow{SK}$.

Consequently, for an infinite A (thus with infinite relations), we set $F(A) = (\overrightarrow{A} \backslash \overrightarrow{SK})$ TIMES R_A^M. □

Notice that from Proposition 69 above we obtain that, for any $A \in Ob_{\underline{DB}}$ (that is, $A \subseteq \underline{\Upsilon}$), the set of views which can be obtained from A is equal to the set of view that can be obtained from a single relation $R_A = F(A)$, and $R_A \in \underline{\Upsilon}$ (i.e., $n = ar(R_A)$ is finite).

We can have a number of different functions F that satisfy the constraint 1, but for any two such functions F, F' and every database $A \in Ob_{\underline{DB}}$, $T(F(A)) = TA = T(F'(A))$, that is, $F(A)$ and $F'(A)$ are observationally equivalent (hence isomorphic objects in the **DB** category). Consequently, it is not important which one of these functions will be chosen for a transformation of a database-instance $A \in Ob_{\underline{DB}}$ into a single, equivalent to it finitary relation $F(A) \in \underline{\Upsilon}$.

Example 45 Let us consider the following continuation of Example 44, when $A = T(can(\mathcal{I}, D))$ is an infinite database, with two infinite binary relations $\{\bot, R_1, R_2\} = can(\mathcal{I}, D) \subseteq T(can(\mathcal{I}, D))$ represented in the following table:

$R_1 = \lVert r \rVert_{can(\mathcal{I},\mathcal{D})}$	$R_2 = \lVert s \rVert_{can(\mathcal{I},\mathcal{D})}$	R_{11}	R_{21}
a, b	b, ω_0	ω_1, ω_3	ω_1, ω_2
b, ω_1	ω_1, ω_2	ω_3, ω_5	ω_3, ω_4
ω_1, ω_3	ω_3, ω_4	ω_5, ω_7	ω_5, ω_6
ω_3, ω_5	ω_5, ω_6	ω_7, ω_9	ω_7, ω_8
...	...	...	...

From point 2 of Proposition 69 above, $\overrightarrow{A} \in TA$, and for each $a \in val(A)$, $R = \{\langle a \rangle\}$ can be obtained by the term of the Σ_R algebra $\overrightarrow{A}$ WHERE $(name = a)$, so that $R \in TA$ and, consequently, $R' = \{\langle \omega_0, \omega_0, \omega_0, \omega_0 \rangle\}$ is obtained by the Σ_R-term $\{\langle \omega_0 \rangle\}$ TIMES $(\{\langle \omega_0 \rangle\}$ TIMES $(\{\langle \omega_0 \rangle\}$ TIMES $\{\langle \omega_0 \rangle\}))$, thus $R' \in TA$ and hence,

$$R_{11} \triangleq \big(R_1 \text{ WHERE } (nome_1 \neq a) \wedge (name_1 \neq b)\big) \in TA;$$

$$R_{21} \triangleq \big(R_2 \text{ WHERE } (nome_1 \neq b)\big) \in TA.$$

Consequently, we define

$$R_A^M \triangleq (R_{11} \text{ TIMES } R_{21}) \text{ UNION } R' \in TA,$$

so that $\overrightarrow{R_A^M} = \overrightarrow{SK}$, with $\overrightarrow{SK} \in T(\{R_A^M\})$, that is, $T(\{\overrightarrow{SK}\}) \subseteq T(\{R_A^M\})$ (in fact, $\overrightarrow{SK} = (\pi_1(R_A^M))$ UNION $\pi_4(R_A^M))$. But R_A^M cannot be obtained by a finite-length term over the unary relation $\overrightarrow{SK}$ and hence $R_A^M \notin T(\{\overrightarrow{SK}\})$.

Thus, $\overrightarrow{A} \backslash \overrightarrow{R_A^M} = \overrightarrow{A} \backslash \overrightarrow{SK} = \{\langle a\rangle, \langle b\rangle\} \in TA$ is a finite unary relation with only two tuples, so that

$$F(A) = \{\langle a\rangle, \langle b\rangle\} \text{ TIMES } R_A^M$$

is a finitary relation with $ar(F(A)) = 5$ and with an infinite set of tuples.

In order to show that $T(\{F(A)\}) = TA = T(T(can(\mathcal{I}, D))) = T(can(\mathcal{I}, D)) = A$, it is enough to show that $\{R_1, R_2\} \subseteq T(\{F(A)\})$:

1. $R_{31} = \{\langle a, b\rangle\} = ((\pi_1(F(A))$ WHERE $(name = a))$ TIMES $(\pi_1(F(A))$ WHERE $(name = a))) \in T(\{F(A)\})$;
2. $R_{32} = \{\langle b, \omega_1\rangle\} = ((\pi_1(F(A))$ WHERE $(name = b))$ TIMES $(\pi_2(F(A))$ WHERE $(name = \omega_1))) \in T(\{F(A)\})$;
3. $R_{33} = \{\langle b, \omega_0\rangle\} = ((\pi_1(F(A))$ WHERE $(name = b))$ TIMES $(\pi_2(F(A))$ WHERE $(name = \omega_0))) \in T(\{F(A)\})$.

Thus,

$$R_1 = (\pi_{\langle 2,3\rangle}(F(A)) \text{ UNION } (R_{31} \text{ UNION } R_{31})) \in T(\{F(A)\}),$$
$$R_2 = (\pi_{\langle 4,5\rangle}(F(A)) \text{ UNION } R_{33}) \in T(\{F(A)\}).$$

From the fact that for each $R_A = F(A) \in \underline{\Upsilon}$, $T(\{R_A\}) = TA \subseteq \underline{\Upsilon} = T(\{R_{\underline{\Upsilon}}\})$, we are able to introduce a partial ordering between the relations in $\underline{\Upsilon}$.

In order to introduce this partial ordering between the relations, and a complete lattice of them, that is, in order to obtain the possible worlds with the partial ordering (as it was developed by Kripke for intuitionistic logic), we do as follows:

Lemma 22 *We define the PO set* $(\underline{\Upsilon}, \leq)$ *such that for any two relations,* $R_1, R_2 \in \underline{\Upsilon}$,

$$R_1 \leq R_2 \quad \textit{iff} \quad T(\{R_1\}) \subseteq T(\{R_1\}),$$

so that $\bot$ *is the bottom element in* $(\underline{\Upsilon}, \leq)$.

We denote $R_1 = R_2$ *iff* $R_1 \leq R_2$ *and* $R_2 \leq R_2$. *Consequently, we obtain the following complete lattice of relations:*

(i) $L_{\underline{\Upsilon}} = (\underline{\Upsilon}, \leq, \wedge, \vee, 0, 1)$, *where* 0 *equal to empty relation* $\bot$, *and* 1 *is equal to top relation* $R_{\underline{\Upsilon}} = F(\underline{\Upsilon})$, *and for any two relations* $R_1, R_2 \in \underline{\Upsilon}$ *we define the meet and join operators by:*

(ii) $R_1 \wedge R_2 \triangleq F(T\{R_1\} \cap T\{R_2\})$, $R_1 \vee R_2 \triangleq F(T\{R_1\} \cup T\{R_2\})$, *where* $F : Ob_{\underline{DB}} \to \underline{\Upsilon}$ *is a function introduced in Proposition 69. Consequently, the composed operation '*F*' in this lattice corresponds to the lattice l.u.b. (supremum) operator '*$\bigvee$*' of this lattice, and for any instance-database* $A \in Ob_{\mathbf{DB}}$,

(iii) $TA = \downarrow R_A$, *where* $R_A = \digamma(A) \in \underline{\Upsilon}$.

Proof This PO is well defined since a singleton $\{R\}$ is equivalent to a database, and $T(\{R\})$ is the set of all views that can be obtained from it in SPRJU relational algebra, with $\bot \in T(\{R\}) \in Ob_{\underline{\mathbf{DB}}}$. $T\{\bot\} = \{\bot\} = \bot^0 \subseteq A \subseteq TA$ for each database A in **DB**, so that $T\{\bot\} \subseteq T\{R\}$ for any $R \in \underline{\Upsilon}$, i.e., $\bot \leq R$.

From point 2 of Proposition 69, for any simple database A, $TA = T(\{\digamma(A)\})$, which is valid when A is a singleton set of relations as well. Let $R_1 \leq R_2$, i.e., $T\{R_1\} \subseteq T\{R_1\}$. Then,

$$T(R_1 \wedge R_2) = T\big(\big\{\digamma\big(T\{R_1\} \cap T\{R_2\}\big)\big\}\big) = T\big(\big\{\digamma\big(T\{R_1\}\big)\big\}\big) = T\{R_1\},$$

so that we obtain the lattice equality $R_1 \wedge R_2 = R_1$. While

$$T(R_1 \vee R_2) = T\big(\big\{\digamma\big(T\{R_1\} \cup T\{R_2\}\big)\big\}\big) = T\big(\big\{\digamma\big(T\{R_2\}\big)\big\}\big) = T\{R_2\},$$

so that we obtain the lattice equality $R_1 \vee R_2 = R_2$. Consequently, the meet and join operations of this lattice are well defined. The finitary relation $\digamma(A)$ is the l.u.b. of all elements in TA. In fact, for each view (relation) $v \in TA$, $T\{v\} \subseteq TA = T(\{\digamma(A)\})$, so that $v \leq \digamma(A)$ and, consequently, $\digamma(A) \in TA$ is the l.u.b. of relations in TA. Thus, for each closed database $TA \in Ob_{\underline{\mathbf{DB}}_{sk}}$, $\digamma(TA) = \bigvee(TA)$ and $TA = \downarrow(\digamma(A))$.

Consequently, from the fact that $\underline{\Upsilon} = T\underline{\Upsilon}$ is a closed object in **DB**, we obtain that the top element of this complete lattice is $1 = R_{\underline{\Upsilon}} = \digamma(\underline{\Upsilon}) = \digamma(T\underline{\Upsilon}) = \bigvee(T\underline{\Upsilon}) = \bigvee(\underline{\Upsilon})$.

From the fact that $\vee(R_1, R_2) \triangleq \digamma(\bigcup\{T\{R_1\}, T\{R_2\}\})$, we obtain for the supremum

$$\bigvee A \triangleq \digamma\Big(\bigcup\big\{T\{R\} \mid R \in A\big\}\Big) = \digamma(A),$$

(from the set-equality $T(\{\digamma(\bigcup\{T\{R\} \mid R \in A\})\}) = T(\{\digamma(A)\}))$ and hence the supremum in this lattice corresponds to the relation $R_{\underline{\Upsilon}} = \digamma(\underline{\Upsilon}) = \bigvee(\underline{\Upsilon})$. □

We recall that a Boolean algebra is a distributive lattice (which satisfies the equations of points 1–6 in Sect. 1.2) with additional negation operator $\neg$, such that it satisfies the following additional equations:

$$a \wedge 0 = 0, \qquad a \vee 1 = 1, \qquad a \wedge \neg a = 0, \qquad a \vee \neg a = 1.$$

The lattice of relations $L_{\underline{\Upsilon}}$ in Lemma 22 and the lattice L_v in Example 44 above have the following properties:

Proposition 70 *The complete distributive lattice* L_v *in Lemma* 21 *is a Boolean algebra* $(\underline{\Upsilon}, \wedge_v, \vee_v, \neg_v, \bot, \overrightarrow{\underline{\Upsilon}})$ *with negation operation defined by*:

$$\neg_v R = \overrightarrow{R}_{\neg} \triangleq \big\{\langle a\rangle \mid a \in \mathcal{U}, \langle a\rangle \notin \overrightarrow{R}\big\}$$

for any $R \in \underline{\Upsilon}$. *If* $\underline{\Upsilon}$ *is finite, then the lattice* $L_{\underline{\Upsilon}}$ *in Lemma 22 is a Boolean algebra* $(\underline{\Upsilon}, \wedge, \vee, \neg, \bot, \times_c(\underline{\Upsilon}))$ *as well, with negation* $\neg$ *equal to the pseudo-complement.*

Proof For any $R \in \underline{\Upsilon}$, $R \wedge_v \bot = \overrightarrow{R} \cap \bot = \bot$, $R \vee \overrightarrow{\underline{\Upsilon}} = \overrightarrow{R} \cup \overrightarrow{\underline{\Upsilon}} = \overrightarrow{\underline{\Upsilon}}$, and $R \wedge_v \neg_v R = \overrightarrow{R} \cap \overrightarrow{R}_{\neg} = \bot$, $R \vee_v \neg_v R = \overrightarrow{R} \cup \overrightarrow{R}_{\neg} = \overrightarrow{\underline{\Upsilon}}$, thus the lattice $L_{\underline{\Upsilon}}$ is a Boolean algebra.

Let us show that, when $\underline{\Upsilon}$ is finite, the lattices $L_{\underline{\Upsilon}}$ and L_v are *equal*, that is, the binary relations (ordering) $\leq$ and $\leq_v$ are equal. Thus, we have to show that for any two relations $R_1, R_2 \in \underline{\Upsilon}$, $T\{R_1\} \subseteq T\{R_2\}$ iff $\overrightarrow{R}_1 \subseteq \overrightarrow{R}_2$. In fact, from the fact that $\underline{\Upsilon}$ is finite (thus $\mathcal{U} = \mathbf{dom}$, without Skolem constants), every relation $R \in \underline{\Upsilon}$ is finite, so that (a) $T\{R\} = T\overrightarrow{R}$:

1. (From right to left) If $\overrightarrow{R}_1 \subseteq \overrightarrow{R}_2$ then $T\overrightarrow{R}_1 \subseteq T\overrightarrow{R}_2$, thus from (a), $T\{R_1\} \subseteq T\{R_2\}$.
2. (From left to right) If $T\{R_1\} \subseteq T\{R_2\}$ then from (a), $T\overrightarrow{R}_1 \subseteq T\overrightarrow{R}_2$. Thus for any $R = \{\langle a \rangle\} \in T\overrightarrow{R}_1$, thus, $\langle a \rangle \in \overrightarrow{R}_1$, we have $R = \{\langle a \rangle\} \in T\overrightarrow{R}_2$, so that $\langle a \rangle \in \overrightarrow{R}_1$ and hence $\overrightarrow{R}_1 \subseteq \overrightarrow{R}_2$.

Consequently, we obtain $R_1 \leq R_2$ iff $R_1 \leq_v R_2$, that is, we obtain the equal ordering, thus the same lattices, both with the bottom element $\bot$ (empty relation), and with the top element $F(\underline{\Upsilon}) = \overrightarrow{\underline{\Upsilon}}$. Let us show that the Boolean negation in the lattice $L_{\underline{\Upsilon}}$ is the pseudo-complement. In fact, for any $R \in \underline{\Upsilon}$, for its pseudo-complement in $L_{\underline{\Upsilon}}$, we have (for *finite* $\underline{\Upsilon}$):

$$\begin{aligned}
\neg R &= \bigvee\{R_1 \in \underline{\Upsilon} \mid R_1 \wedge R \leq \bot\} = F\Big(\bigcup\{T\{R_1\} \mid R_1 \wedge R \leq \bot\}\Big) \\
&= F\Big(\bigcup\{T\{R_1\} \mid R_1 \wedge R = \bot\}\Big) \\
&\quad \text{(from (a) and equal meet operations } \wedge \text{ and } \wedge_v) \\
&= F\Big(\bigcup\{T\{\overrightarrow{R}_1\} \mid R_1 \wedge_v R = \bot\}\Big) = F\Big(\bigcup\{T\{\overrightarrow{R}_1\} \mid \overrightarrow{R}_1 \cap \overrightarrow{R} = \bot\}\Big) \\
&= F\Big(\bigcup\{T\{\overrightarrow{R}_1\} \mid \overrightarrow{R}_1 \subseteq \overrightarrow{R}_{\neg}\}\Big) = F(T\{\overrightarrow{R}_{\neg}\}) = F(\{\overrightarrow{R}_{\neg}\}) = \overrightarrow{R}_{\neg} = \neg_v R.
\end{aligned}$$

□

From these results we may conclude that the important properties of the lattice of finitary relations have to be investigated when $\mathcal{U}$ is infinite (thus with an infinite number of Skolem constants), that is, when the set of all finitary relations (but which can have infinite number of tuples) in $\underline{\Upsilon}$ is *infinite*.

We are ready now to establish the relationship between the Galois connection of Birkhoff polarity (its special case when the incompatibility binary relation is a PO-set) and the power-view operator T for databases:

Corollary 35 *The power view operator T is equal to the closure operator $\Gamma = \rho\lambda$ of Dedekind–McNeile Galois connection of λ and ρ in the complete lattice $(\underline{\Upsilon}, \preceq)$ defined in Lemma 22.*

Proof From Lemma 22 we have that '$\digamma$' is an l.u.b. ('$\bigvee$') in the complete r.p.c. lattice $(\underline{\Upsilon}, \preceq)$, so that $\Gamma = \rho\lambda = \downarrow \bigvee = \downarrow \digamma$.

For any simple object (i.e., instance-database) $A \in Ob_{\mathbf{DB}}$, $TA = \{R \in \underline{\Upsilon} \mid R \preceq \digamma(A)\} = \downarrow \digamma(A)$. Consequently, $T = \downarrow \digamma = \downarrow \bigvee = \Gamma : \mathcal{P}_{\perp}(\underline{\Upsilon}) \to \mathcal{P}_{\perp}(\underline{\Upsilon})$, where $\mathcal{P}_{\perp}(\underline{\Upsilon}) = \{S \cup \{\perp\} \mid S \subseteq \underline{\Upsilon}\} = Ob_{\mathbf{DB}}$, from the fact that $\Gamma(\{0\}) = \Gamma(\{\perp\}) = \{\perp\} = \perp^0$ is the bottom object in $Ob_{\underline{\mathbf{DB}}_{sk}}$, and $\Gamma(\underline{\Upsilon}) = \underline{\Upsilon}$ is the top object in $Ob_{\underline{\mathbf{DB}}_{sk}}$. Notice that for any $A \subseteq \underline{\Upsilon}$, $\Gamma(A \cup \{\perp\}) = \rho(\lambda(A \cup \{\perp\})) = \rho(\lambda(A) \cap \lambda(\{\perp\})) = \rho(\lambda(A) \cap \underline{\Upsilon}) = \rho(\lambda(A)) = \Gamma(A)$ and hence $T = \Gamma : \mathcal{P}_{\perp}(\underline{\Upsilon}) \to \mathcal{P}_{\perp}(\underline{\Upsilon})$ is equal to $\Gamma : \mathcal{P}(\underline{\Upsilon}) \to \mathcal{P}(\underline{\Upsilon})$ applied to nonempty subsets of $\underline{\Upsilon}$. □

Based on the definition of the **DB**-algebraic lattice $L_{\underline{DB}} = (Ob_{\underline{\mathbf{DB}}_{sk}}, \preceq, \otimes, \oplus, \{\perp\}, \underline{\Upsilon}) = (Ob_{\underline{\mathbf{DB}}_{sk}}, \subseteq, \cap, T\cup, \perp^0, \underline{\Upsilon})$ in Definition 63, and on the definition of the complete lattice of relations $(\underline{\Upsilon}, \preceq, \wedge, \vee, \perp, R_{\underline{\Upsilon}})$ in Lemma 22, we obtain the following representation for **DB**-algebraic logic:

Proposition 71 *The algebra of truth-values*

$$L_{\mathrm{alg}} = (Ob_{\underline{\mathbf{DB}}_{sk}}, \preceq, \otimes, \oplus, \widehat{\Rightarrow}, \widehat{\neg}) = (Ob_{\underline{\mathbf{DB}}_{sk}}, \subseteq, \cap, T\cup, \widehat{\Rightarrow}, \widehat{\neg})$$

*is a Heyting algebra, obtained as extension of the lattice $L_{\underline{DB}}$ in Definition 63, with the "simple" **DB**-valuations $\mathbb{V} = \beta\underline{V} : PR \to Ob_{\underline{\mathbf{DB}}_{sk}}$, and with the following extension to all formulae:*

$$\begin{aligned}
\mathbb{V}(\phi \wedge \psi) &= \mathbb{V}(\phi) \otimes \mathbb{V}(\psi) = \mathbb{V}(\phi) \cap \mathbb{V}(\psi) = \bigcap\big(\mathbb{V}(\phi), \mathbb{V}(\psi)\big) \\
&= T\cap\big(\mathbb{V}(\phi), \mathbb{V}(\psi)\big); \\
\mathbb{V}(\phi \vee \psi) &= \mathbb{V}(\phi) \oplus \mathbb{V}(\psi) = T\big(\mathbb{V}(\phi) \cup \mathbb{V}(\psi)\big) = T\cup\big(\mathbb{V}(\phi), \mathbb{V}(\psi)\big); \\
\mathbb{V}(\phi \Rightarrow \psi) &= \mathbb{V}(\phi) \widehat{\Rightarrow} \mathbb{V}(\psi) = T\Big(\bigcup\{Z \in Ob_{\underline{\mathbf{DB}}_{sk}} \mid Z \cap \mathbb{V}(\phi) \subseteq \mathbb{V}(\psi)\}\Big); \\
\mathbb{V}(\neg\phi) &= \widehat{\neg}\mathbb{V}(\phi) = T\Big(\bigcup\{Z \in Ob_{\underline{\mathbf{DB}}_{sk}} \mid Z \cap \mathbb{V}(\phi) = \perp^0\}\Big).
\end{aligned}$$

*We say that ϕ is **DB**-valid formula, denoted by $\mathbf{DB} \models_s \phi$, iff for every **DB**-valuation $\mathbb{V}$, $\mathbb{V}(\phi) = \underline{\Upsilon}$ (dual to $\underline{V}(\phi) = id_{\underline{\Upsilon}} : \underline{\Upsilon} \to \underline{\Omega}$ where $id_{\underline{\Upsilon}} = \{id_R : R \to R \mid R \in \underline{\Upsilon}\}$).*

Proof It is easy to verify that the restriction of the matching and merging operators to simple *closed* objects A and B are given by $A \otimes B = TA \cap TB = A \cap B$ (from Theorem 12) $= T(A \cap B)$ and $A \oplus B = T(A \cup B)$ (from Definition 57), and hence PO relation '$\preceq$' reduces to the subset relation '$\subseteq$'.

From Corollary 35, $T = \Gamma$ for the poset $W = \underline{\Upsilon}$, so that $\mathcal{F}(W) = \mathcal{F}(\underline{\Upsilon}) = \{\downarrow R \mid R \in \underline{\Upsilon}\} = \{T\{R\} \mid R \in \underline{\Upsilon}\} = \{TA \mid R = F(A), A \subseteq \underline{\Upsilon}\} = \{TA \mid A \in Ob_{\mathbf{DB}}\} = Ob_{\mathbf{DB}_{sk}}$. Consequently, we obtain that

$$L_{\mathrm{alg}} = \big(\mathcal{F}(\underline{\Upsilon}), \subseteq, \Gamma\cap, \Gamma\cup, \widehat{\Rightarrow}, \widehat{\neg}, \bot^0, \underline{\Upsilon}\big)$$

is equal to the Heyting algebra $\mathbf{F}(W)$ (for $W = \underline{\Upsilon}$) in Corollary 34. Notice that the operations of negation and disjunction are not standard (set-complement and set-union). Instead, they are modified by the power-view operator T and hence are (*multi*) *modal* operators. □

Notice that we can consider each closed object $TA \in Ob_{\mathbf{DB}_{sk}}$ as a representative element of the equivalence class $[A] = \{B \in Ob_{\mathbf{DB}} \mid TB = TA\}$, with $A \simeq TA \simeq \downarrow F(TA) \simeq \downarrow F(A) \in \mathcal{F}(\underline{\Upsilon})$, that is, $A, TA, \{F(TA)\}, \downarrow F(A) \in [A]$. Because of that, the **DB** algebraic logic L_{alg} (of the weak monoidal topos) is factored by the equivalence relation equal to the isomorphism of objects in **DB** category.

From the fact that the bottom element 0 is equal to the empty relation $\bot$, in Kripke semantics of this logic for each formula ϕ, from the fact that $\mathcal{M} \models_{\bot} \phi$, for the set $A = \|\phi\| = \{a \in \underline{\Upsilon} \mid \mathcal{M} \models_a \phi\} \in \mathcal{F}(\underline{\Upsilon})$ of this logic, we obtain a standard result $\bot \in A \simeq TA \in Ob_{\mathbf{DB}_{sk}}$.

Consequently, the logic of the weak monoidal topos determines the databases up to the behavioral (observational) equivalence (which is an isomorphism of the simple objects in the **DB** category), that is, as a set of all possible views (all possible observations) of a given database (associated to a given logical formula ϕ). From this point of view, this weak monoidal topos logic is an abstract-behavior logic which does not determinate a model of a given database schema, but the abstract behavior of such a model, defined by the set of all observations (views) which can be obtained by using a Σ_R (SPJRU) Codd's relational algebra.

This result is coherent with the fact that for each model (instance-database) $A = \alpha^*(\mathcal{A})$ of a database schema $\mathcal{A}$, the closed object TA is an "abstract behavioral model" of a database A, represented by the set of all observations (i.e., views) which can be obtained from this database. From the fact that both matching and merging database-lattice operators are based on the power-view closure operator T, we have seen that the logical connectives of the weak monoidal topos are closed by $\Gamma = T$, so that the obtained database logic is abstract in the same way.

From the fact that for each different database set of individual symbols **dom** (and universe $\mathcal{U} = \mathbf{dom} \cup SK$ with infinite set of Skolem constants $SK = \{\omega_0, \omega_1, \dots\}$ for marked null values, used in Data Integration with incomplete information as presented in Sect. 4.2.4) we obtain a different **DB** category with different total objects $\underline{\Upsilon}$, the weak monoidal topos logic corresponds to this class of all **DB** categories:

Definition 64 We define the WMTL (weak monoidal topos database logic) by the following class of complete Heyting algebras in Proposition 71:

$$\mathcal{HA}_{DB} = \{L_{\mathrm{alg}} \mid \text{for every } \mathbf{dom}\} \subset \mathcal{HA}.$$

The following property for the closed object (instance-database) in each (determined by **dom**) **DB** category is valid:

Lemma 23 *The set* $\mathfrak{J}_A$ *of join-irreducible elements in any closed object (instance database)* $A = TA \in Ob_{\underline{\mathbf{DB}}_{sk}}$ *is equal to the set of its atoms,*

$$\mathfrak{J}_A = \{R = \{\langle d_i \rangle\} \mid d_i \in val(A)\} = \{R = \{\langle d_i \rangle\} \mid \langle d_i \rangle \in \vec{A}\},$$

so that $\mathfrak{J}_A \subseteq A$, *and* $A \simeq \mathfrak{J}_A$ *iff* $\mathfrak{J}_A$ *is a finite nonempty set.*

Proof The 'atomic database' $\mathfrak{J}_A$ was introduced in Sect. 1.4. Let us consider the sublattice $(TA, \leq) = (\downarrow (F(A)), \leq) \subseteq (\underline{\Upsilon}, \leq)$ with the bottom relation $\bot$ and the top relation $F(A)$. Then for each $R \in \mathfrak{J}_A$, $\bot \leq R$, so that R is an atom in this lattice $(TA, \leq)$. Each relation $R_1 \in (\downarrow F(A), \leq) = (TA, \leq)$ different from atoms is composed of at least two distinct values in $S \subseteq \mathcal{U}$ and hence it is a join of a subset of atoms $\{R = \{\langle d \rangle\} \mid d \in S\}$, i.e., $R_1 = \bigvee\{R = \{\langle d \rangle\} \mid d \in S\}$, and consequently, the set of join irreducible elements in $(TA, \leq)$ is equal to the set of its atoms.

If $\mathfrak{J}_A$ is finite, then $A = TA$ is finite as well with all finite relations, so that $A = T\mathfrak{J}_A$, i.e., $A \simeq \mathfrak{J}_A$. If $\mathfrak{J}_A$ is *infinite*, for a database $A = TA$ with at least one relation $R \in A$ with infinite number of tuples, then $T\mathfrak{J}_A$ is a set of all *finite* relations (because each query is a finite-length SPJRU term, thus it can generate from $\mathfrak{J}_A$ only a finite relation) with $T\mathfrak{J}_A \subset TA$ (thus, $\mathfrak{J}_A \not\simeq A$). □

Thus, if $\mathfrak{J}_A$ is finite then it is a set of generators of the relational algebra (A, Σ_R) (hence, this algebra (A, Σ_R) is *finitely generated* by a finite set $\mathfrak{J}_A$ because any element (relation) in A can be expressed as a finite-length term in $\mathfrak{J}_A$), that is, of the database A. Also the lattice $(TA, \leq)$ is then finitely generated by $\mathfrak{J}_A$.

Notice that in the case when $A = \underline{\Upsilon}$ (for a given universe $\mathcal{U} = \mathbf{dom} \cup SK$), we obtain that $\mathfrak{J}_{\underline{\Upsilon}} = \{R = \{\langle d \rangle\} \mid d \in \mathcal{U}\}$ is an infinite set, but it is only a strict subsetset of generators of $\underline{\Upsilon}$ (the relational algebra $(\underline{\Upsilon}, \Sigma_R)$ cannot be generated by $\mathfrak{J}_{\underline{\Upsilon}}$ only).

A special case of a database with a relation with an infinite number of tuples may appear in Data Integration with cyclic tgds with the existentially quantified right-hand side of material implication, as explained in Sect. 4.2.4.

9.2.3 Embedding of WMTL (Weak Monoidal Topos Logic) into Intuitionistic Bimodal Logics

Intuitionistic propositional logic IL and its extensions, known as intermediate or superintuitionistic logics, in many respects can be regarded just as fragments of classical modal logics containing S4. At the syntactical level, the Gödel translation embeds every intermediate logic into modal logics. Semantically, this is reflected by the fact that Heyting algebras are precisely the algebras of open elements of topological Boolean algebras. This embedding is a powerful tool for transferring various kinds of results from intermediate logics to modal ones and back via preservation theorems.

That the Gödel translation can be extended to an embedding of at least a few particular intuitionistic modal systems into some classical multi-modal logics was observed by several authors [22, 23, 27].

The propositional intuitionistic logic with connectives $\wedge, \vee, \Rightarrow, \mathbf{0}$ ($\neg\phi$ is defined as $\phi \Rightarrow \mathbf{0}$ where $\mathbf{0}$ is contradiction) and the modal operators $\odot_i$, $i = 1, \ldots, n$, is an IM-logic (Intuitionistic Modal logic [26]) if it is a set of formulae containing intuitionistic logic IL (with only first four connectives above) and closed under substitution, Modus Ponens and the congruence rules $\frac{\phi \Leftrightarrow \psi}{\odot_i \phi \Leftrightarrow \odot_i \psi}$, for all $i = 1, \ldots, n$.

The way how, for a given negation logical operator of intuitionistic logic, we can find its algebraic counterpart in the Birkhoff polarity was shown previously in Example 42. Now, in this section, we will represent the algebraic operators in the quotient algebra L_{alg} by the corresponding logical operations. Notice that in L_{alg} all operators are preceded by the closure operator Γ (equal to power-view closure operator T for databases), so that the first step is to obtain the logical operator corresponding to this algebraic closure operator.

Remark First of all, we have to verify if the monotonic closure operator Γ can be interpreted by a unique modal operator (in this case it has to be an existential modal operator from the fact that it is monotonic and satisfies the first additive modal property for the bottom element $\{0\} \in \mathcal{F}(W)$, $\Gamma(\{0\}) = \{0\}$). The answer is negative because the second additive property does not hold, and from Lemma 20, instead of equality we have for any two $S, S' \in \mathcal{F}(W)$ only that $\Gamma(S \cup S') \supseteq (\Gamma(S) \cup \Gamma(S'))$.

However, the closure operator $\Gamma = \rho\lambda$ is a monotonic operator, obtained by composition of two negation operators, multiplicative operator ρ and additive operator λ, both of which are modal negation operators (as we have seen in Example 42, the intuitionistic negation $\neg$ is a modal, $\neg = \Box_{R_{\leq}^{-1}} \neg_c$, where $\neg_c$ is classic (standard) negation and $\Box_{R_{\leq}^{-1}}$ universal modal operator for accessibility relation $R_{\leq}^{-1}$, we obtained the algebraic additive operator $\lambda_{\overline{R}_{\leq}^{-1}}$). Consequently, Γ in a logical representation has to be a *composed modal* logical operator, as follows:

Lemma 24 *The closure operator $\Gamma = \rho\lambda$, obtained from Dedekind–McNeile Galois connection, corresponds to the composition of two modal operators $\Box_{\overline{R}_{\leq}} \Diamond_{\overline{R}_{\leq}^{-1}}$, where $\Box_{\overline{R}_{\leq}}$ is a universal modal operator for the accessibility relation $\overline{R}_{\leq}$, and $\Diamond_{\overline{R}_{\leq}^{-1}}$ is an existential modal operator for the accessibility relation $\overline{R}_{\leq}^{-1}$.*

Both these two operators are normal modal operators, and we denote by IL_{Γ} the IM-logic with $\odot_1 = \Box_{\overline{R}_{\leq}}$ and $\odot_2 = \Diamond_{\overline{R}_{\leq}^{-1}}$.

Proof For a given formula ϕ of the DB weak topos logic with the algebra L_{alg}, let $S = \|\phi\|_K$ be a set of worlds where ϕ is true. Then, for the incompatibility binary relation $R_{\leq}$ (equal to the poset $\leq$), we have the Dedekind–McNeile Galois connection with

$$\begin{aligned}
\lambda(S) &= \{a \in W \mid \forall b \in S.(b,a) \in R_{\leq}\} \\
&= \{a \in W \mid \forall b(b \in S \text{ implies } (a,b) \in R_{\leq}^{-1})\} \\
&= \{a \mid \forall b((a,b) \notin R_{\leq}^{-1} \text{ implies } b \notin S)\} \\
&= \{a \mid \forall b((a,b) \in \overline{R_{\leq}^{-1}} \text{ does not imply } b \in \|\phi\|_K)\} \\
&= \{a \mid \forall b((a,b) \in \overline{R_{\leq}^{-1}} \text{ does not imply } \mathcal{M} \models_b \phi)\} \\
&= \{a \mid \mathcal{M} \models_a \Box_{\overline{R_{\leq}^{-1}}} \neg_c \phi\} = \|\Box_{\overline{R_{\leq}^{-1}}} \neg_c \phi\|_K.
\end{aligned}$$

Thus, the modal additive algebraic negation λ is represented by a composition $\Box_{\overline{R_{\leq}^{-1}}} \neg_c$ of the universal modal operator $\Box_{\overline{R_{\leq}^{-1}}}$ and classic (standard) logical negation $\neg_c$. Analogously,

$$\begin{aligned}
\rho(S) &= \{a \in W \mid \forall b \in S.(a,b) \in R_{\leq}\} \\
&= \{a \in W \mid \forall b(b \in S \text{ implies } (a,b) \in R_{\leq})\} \\
&= \{a \mid \forall b((a,b) \notin R_{\leq} \text{ implies } b \notin S)\} \\
&= \{a \mid \forall b((a,b) \in \overline{R_{\leq}} \text{ does not imply } b \in \|\phi\|_K)\} \\
&= \{a \mid \forall b((a,b) \in \overline{R_{\leq}} \text{ does not imply } \mathcal{M} \models_b \phi)\} \\
&= \{a \mid \mathcal{M} \models_a \Box_{\overline{R_{\leq}}} \neg_c \phi\} \\
&= \|\Box_{\overline{R_{\leq}}} \neg_c \phi\|_K.
\end{aligned}$$

Thus, the modal multiplicative algebraic negation ρ is represented by a composition $\Box_{\overline{R_{\leq}}} \neg_c$ of the universal modal operator $\Box_{\overline{R_{\leq}}}$ and classic (standard) logical negation $\neg_c$.

Consequently, the closure operator $\Gamma = \rho\lambda$ is represented by the composition

$$(\Box_{\overline{R_{\leq}}} \neg_c)(\Box_{\overline{R_{\leq}^{-1}}} \neg_c).$$

That is, from the fact that $\neg_c \Box_{\overline{R_{\leq}^{-1}}} \neg_c$ is equal to the existential modal operator $\Diamond_{\overline{R_{\leq}^{-1}}}$, we obtain that Γ is represented by modal composition $\Box_{\overline{R_{\leq}}} \Diamond_{\overline{R_{\leq}^{-1}}}$. □

In fact, it is easy to verify that the Kripke semantics of the closure operator Γ corresponds to

$$\mathcal{M} \models_a \Box_{\overline{R_{\leq}}} \Diamond_{\overline{R_{\leq}^{-1}}} \phi \quad \text{iff} \quad \forall b(\forall c(\mathcal{M} \models_c \phi \text{ implies } c \leq b) \text{ implies } a \leq b),$$

such that for any down-closed hereditary set $\|\phi\|_K$,

$$\|\Box_{\overline{R}_\leq}\Diamond_{\overline{R}_\leq^{-1}}\phi\|_K = \Gamma(\|\phi\|_K) = \downarrow\bigvee(\|\phi\|_K).$$

Definition 65 Based on definition of propositional logic WMTL in Definition 64, we are able to define the following Gödel-like translation from WMTL into this S4 extended bimodal logic:

1. $p^M = \Box_{\overline{R}_\leq}\Diamond_{\overline{R}_\leq^{-1}}\Box_{R_\leq^{-1}}p$;
2. $(\phi \wedge \psi)^M = \phi^M \wedge_c \psi^M$;
3. $(\phi \vee \psi)^M = \Box_{\overline{R}_\leq}\Diamond_{\overline{R}_\leq^{-1}}(\phi^M \vee_c \psi^M)$;
4.

$$\begin{aligned}(\phi \Rightarrow \psi)^M &= \Box_{\overline{R}_\leq}\Diamond_{\overline{R}_\leq^{-1}}\Box_{R_\leq^{-1}}(\phi^M \Rightarrow_c \psi^M)\\ &= \Box_{\overline{R}_\leq}\Diamond_{\overline{R}_\leq^{-1}}\Box_{R_\leq^{-1}}(\neg_c\phi^M \vee_c \psi^M),\end{aligned}$$

where $\wedge_c, \vee_c, \Rightarrow_c$ and $\neg_c$ on the right-hand sides are the logical connectives of classic propositional logic.

Notice that only for disjunction we obtained different translation of the Gödel translation for IL.

Thus, based on Corollary 34 which gives us the relationships between the logical operators of an ordinary Heyting algebra $\mathbf{H}(W)$ and the "closed" version $\mathbf{F}(W)$ with a set of closed hereditary subsets in $\mathcal{F}(W)$, by a homomorphism $\Gamma : \mathbf{H}(W) \rightarrow \mathbf{F}(W)$, we are able to define a new autoreferential Kripke semantics for this multimodal logic (for the three accessibility relations, $\overline{R}_\leq, \overline{R}_\leq^{-1}$ and $R_\leq^{-1}$) of weak monoidal topos (its quotient algebra L_{alg}, when $W = \underline{\Upsilon}$ and, by Corollary 35, Γ is equal to power-view operator T).

Notice that from the fact that all three accessibility relations $\overline{R}_\leq, \overline{R}_\leq^{-1}$ and $R_\leq^{-1}$ are derived from the basic PO-relation $\leq$ of the lattice $(W, \leq)$, in what follows, we will use this simplification (by using only the relation $\leq$) for this autoreferential Kripke semantics:

Proposition 72 *An autoreferential Kripke model of the "closed" algebraic logic* $\mathbf{F}(W)$ *in Corollary* 34, *based on the set of possible worlds in* W, *is a pair* $\mathcal{M} = ((W, \leq), V)$, *where* $(W, \leq)$ *is a Kripke frame and* $V : PR \rightarrow \mathcal{F}(W)$, *for a set* PR *of propositional symbols, is a mapping which assigns to each propositional symbol* $p \in PR$ *a closed hereditary subset* $V(p) \in \mathcal{F}(W)$, *so that* $V(p)$ *is the set of possible worlds at which* p *is true (with* $0 \in V(p)$*).*

We now extend the notion of truth at a particular possible world $a \in W$ *to all formulae, by introducing the expression* $\mathcal{M} \models_a \phi$, *to be read "a formula* ϕ *is true in* $\mathcal{M}$ *at* a*", defined inductively as follows:*

1. $\mathcal{M} \models_a p$ *iff* $a \in V(p)$, *and* $\mathcal{M} \models_0 \phi$ *for every formula* ϕ;
2. $\mathcal{M} \models_a \phi \wedge \psi$ *iff* $\mathcal{M} \models_a \phi$ *and* $\mathcal{M} \models_a \psi$;

3. $\mathcal{M} \models_a \phi \vee \psi$ *iff* $\forall b(a \nleq b$ *implies* $\exists c(c \nleq b$ *and* $(\mathcal{M} \models_c \phi$ *or* $\mathcal{M} \models_c \psi)))$;
4. $\mathcal{M} \models_a \phi \Rightarrow \psi$ *iff* $\forall b(a \nleq b$ *implies* $\exists c(c \nleq b$ *implies* $\forall w(w \leq c$ *implies* $(\mathcal{M} \models_w \phi$ *implies* $\mathcal{M} \models_w \psi))))$;
5. $\mathcal{M} \models_a \neg\phi$ *iff* $\mathcal{M} \models_a \phi \Rightarrow \mathbf{0}$ *iff* $\forall b(a \nleq b$ *implies* $\exists c(c \nleq b$ *implies* $\forall w(w \leq c$ *implies not* $\mathcal{M} \models_w \phi)))$.

Proof Based on Corollary 34, we obtain:
1. $\phi \vee \psi$ is equal to $\Box_{\overline{R}_\leq} \Diamond_{\overline{R}_\leq^{-1}}(\phi \vee_c \psi)$ (different from standard Kripke representation);
2. $\phi \Rightarrow \psi$ is equal to $\Box_{\overline{R}_\leq} \Diamond_{\overline{R}_\leq^{-1}} \Box_{R_\leq^{-1}}(\phi \Rightarrow_c \psi)$, where $\Box_{R_\leq^{-1}}(\phi \Rightarrow_c \psi)$ is a standard Kripke representation for intuitionistic implication (given by point 4 in the introduction, Sect. 1.2);
3. $\neg\phi$ is equal to $\Box_{\overline{R}_\leq} \Diamond_{\overline{R}_\leq^{-1}} \Box_{R_\leq^{-1}} \neg_c \phi$, where $\Box_{R_\leq^{-1}} \neg_c \phi$ is a standard Kripke representation for intuitionistic negation (given by point 5 in the introduction, Sect. 1.2);

Consequently, the complex algebra for this modal Kripke semantics is equal to the "closed" algebra $\mathbf{F}(W) = (\mathcal{F}(W), \subseteq, \cap, \Gamma\cup, \widehat{\Rightarrow}, \widehat{\neg})$, where for any two $V_1, V_2 \in \mathcal{F}(W)$:

$$\begin{aligned}
V_1 \sqcup V_2 &= \big(\Gamma \cup (V_1, V_2)\big) \\
&= \big\{a \in W \mid \forall b\big((a,b) \in \overline{R}_\leq \text{ implies } \exists c\big((b,c) \in \overline{R}_\leq^{-1} \text{ and } (c \in V_1 \cup V_2)\big)\big)\big\} \\
&= \big\{a \in W \mid \forall b\big((a,b) \notin \overline{R}_\leq \text{ or } \exists c\big((b,c) \in \overline{R}_\leq^{-1} \text{ and } (c \in V_1 \cup V_2)\big)\big)\big\} \\
&= \big\{a \in W \mid \nexists b\big((a,b) \in \overline{R}_\leq \text{ and } \nexists c\big((b,c) \in \overline{R}_\leq^{-1} \text{ and } (c \in V_1 \cup V_2)\big)\big)\big\} \\
&= W \backslash \big\{a \in W \mid \exists b\big((a,b) \in \overline{R}_\leq \text{ and } \nexists c\big((b,c) \in \overline{R}_\leq^{-1} \text{ and } (c \in V_1 \cup V_2)\big)\big)\big\} \\
&= W \backslash \Bigg(\bigcup_{b \in W \text{ and } \nexists c((b,c) \in \overline{R}_\leq^{-1} \text{ and } (c \in V_1 \cup V_2))} \{a \in W \mid (a,b) \in \overline{R}_\leq\} \Bigg) \\
&= W \backslash \Bigg(\bigcup_{b \in W \backslash \bigcup_{c \in V_1 \cup V_2} \{b \in W \mid (b,c) \in \overline{R}_\leq^{-1}\}} \{a \in W \mid (a,b) \in \overline{R}_\leq\} \Bigg) \\
&= W \backslash \Bigg(\bigcup_{b \in W \backslash \bigcup_{c \in V_1 \cup V_2} \{b \in W \mid c \nleq b\}} \{a \in W \mid a \nleq b\} \Bigg)
\end{aligned}$$

$$V_1 \widehat{\Rightarrow} V_2 = W \backslash \Bigg(\bigcup_{b \in W \backslash \bigcup_{c \in V_3} \{b \in W \mid c \nleq b\}} \{a \in W \mid a \nleq b\} \Bigg),$$

where $V_3 = W \backslash (\bigcup_{c \in V_1 \backslash V_2} \uparrow c)$. □

For the weak monoidal topos, we have the Kripke frame $(W, \leq) = (\underline{\Upsilon}, \leq)$ and $\Gamma = T$, so that the complex algebra is equal to the factor algebra

$$L_{\text{alg}} = \left(Ob_{\underline{\mathbf{DB}}_{sk}}, \preceq, \otimes, \oplus, \widehat{\Rightarrow}, \widehat{\neg}, \{\bot\}, \underline{\Upsilon}\right)$$
$$= \left(Ob_{\underline{\mathbf{DB}}_{sk}}, \subseteq, \cap, T\cup, \widehat{\Rightarrow}, \widehat{\neg}, \bot^0, \underline{\Upsilon}\right) = \mathbf{F}(\underline{\Upsilon})$$

in Proposition 71, with $T\cup = \Gamma\cup = \sqcup$, and this autoreferential semantics is a new Kripke semantics for the logic of the weak monoidal topos.

In fact, from Proposition 68, $\mathcal{F}(\underline{\Upsilon}) = \{\downarrow R \mid R \in \underline{\Upsilon}\} = \{\Gamma in(R) \mid R \in \underline{\Upsilon}\} = \{\Gamma\{R\} \mid R \in \underline{\Upsilon}\} = \{T\{R\} \mid R \in \underline{\Upsilon}\} \subseteq Ob_{\underline{\mathbf{DB}}_{sk}}$. Conversely, for each closed object (instance-database) $A = TA \in Ob_{\underline{\mathbf{DB}}_{sk}}$ we have from Lemma 22 $A = TA = \downarrow R \in \mathcal{F}(\underline{\Upsilon})$ where $R = \digamma(A) \in \underline{\Upsilon}$, that is, $Ob_{\underline{\mathbf{DB}}_{sk}} \subseteq \mathcal{F}(\underline{\Upsilon})$ as well. Thus, $Ob_{\underline{\mathbf{DB}}_{sk}} = \mathcal{F}(\underline{\Upsilon})$. In this frame, the Kripke valuation $V : PR \to \mathcal{F}(\underline{\Upsilon})$ is exactly the algebraic **DB**-valuation $\mathbb{V} = \beta\underline{V} : PR \to Ob_{\underline{\mathbf{DB}}sk}$ in Proposition 71.

Notice that, differently from the standard Kripke semantics of $\mathbb{IL}$ given in Sect. 1.2, the logical disjunction here is a modal operation as well.

In fact, the frame semantics of WMTL is a particular kind of a *general frame* $\mathfrak{F} = (W, \leq, \mathcal{F}(W))$ where $(W, \leq)$ is a Kripke frame and $\mathcal{F}(W)$ is a set of closed hereditary subsets of W closed under operations of Heyting algebra $\mathbf{F}(W)$ (that is, under $\cap, T\cup, \widehat{\Rightarrow}$ and $\widehat{\neg}$) and used to restrict the allowed valuations in the frame: a model $\mathcal{M}$ based on Kripke frame $(W, \leq)$ is *admissible* in the general frame $\mathfrak{F}$ if $\{a \in W \mid \mathcal{M} \models_a p\} \in \mathcal{F}(W)$ for every propositional variable $p \in PR$. The closure conditions on $\mathcal{F}(W)$ then ensure that $\{a \in W \mid \mathcal{M} \models_a \phi\} \in \mathcal{F}(W)$ for every formula ϕ (not only variable).

From this point of view, a Kripke frame (W, R) may be identified with a general frame in which *all* valuations are admissible, that is, a general frame $(W, R, \mathcal{P}(W))$, where $\mathcal{P}(W)$ denotes the power set of W.

The Kripke semantics of the propositional WMTL (Weak Monoidal Topos Logic), given by proposition above, is different from the Kripke semantics of the $\mathbb{IL}$ given in Sect. 1.2 (with a general frame $\mathfrak{F} = (W, \leq, H(W))$ where $H(W)$ is the set of all hereditary subsets of W closed under $\cap, \cup$ and $\Rightarrow_h$), so that we have to show that WMTL is not CPL, but satisfies all axioms (from 1 to 11 in Sect. 1.2) of $\mathbb{IL}$ (intuitionistic propositional logic), and consequently, is an intermediate logic.

The translation from WMTL into this S4 extended bimodal logic, given by Definition 65, can be expressed dually by algebraic embedding of the Heyting algebra $\mathbf{H}(\underline{\Upsilon}) = (H(\underline{\Upsilon}), \cap, \cup, \Rightarrow_h, \bot^0)$ into an extended Heyting algebra by the two new modal-negation operators λ (corresponding, by Lemma 24, to logical operator $\neg_\lambda$ equal to $\Box_{\overline{R}_{\leq}^{-1}}\neg_c$) and ρ (corresponding, by Lemma 24, to logical operator $\neg_\rho$ equal to $\Box_{\overline{R}_{\leq}}\neg_c$) which compose the closure operator $\Gamma = \rho\lambda$, for any two hereditary-subsets $S, S_1 \in H(\underline{\Upsilon})$, as follows:

1. $S^M = \rho\lambda S$;
2. $(S \cap S_1)^M = S^M \cap S_1^M$;
3. $(S \cup S_1)^M = \rho\lambda(S^M \cup S_1^M)$;
4. $(S \Rightarrow_h S_1)^M = \rho\lambda(S^M \Rightarrow_h S_1^M)$.

Lemma 25 *The extended Heyting algebra* $\mathbf{H}(\underline{\Upsilon})^+ = (H(\underline{\Upsilon}), \cap, \cup, \Rightarrow_h, \bot^0, \Diamond_\lambda, \Box_\rho)$*, where* $\Diamond_\lambda = \bot^0 \cup (\underline{\Upsilon} \backslash \lambda\_)$ *and* $\Box_\rho = \bot^0 \cup \rho(\underline{\Upsilon} \backslash \_)$*, is an algebraic dual of the S4 extended bimodal logic (by modal logical operators* $\Diamond_{\overline{R}_{\leq}^{-1}}$ *and* $\Box_{\overline{R}_{\leq}}$*, respectively), so called IM-logic (Intuitionistic Modal logic [26]).*

Proof It is easy to show that $\Diamond_\lambda$ is an additive algebraic modal operator and $\Box_\rho$ is a multiplicative algebraic modal operator. In fact, for any two $S_1, S_2 \in H(\underline{\Upsilon})$ (we recall that $\bot^0 = \{\bot\}$ is an empty database with the empty relation $\bot$), we have:

$$\begin{aligned}\Diamond_\lambda(S_1 \cup S_2) &= \{\bot\} \cup \big(\underline{\Upsilon} \backslash \lambda(S_1 \cup S_2)\big) = \{\bot\} \cup \big(\underline{\Upsilon} \backslash (\lambda S_1 \cap \lambda S_2)\big)\\ &= \{\bot\} \cup \underline{\Upsilon} \backslash (\lambda S_1) \cup \underline{\Upsilon} \backslash (\lambda S_2) = \Diamond_\lambda(S_1) \cup \Diamond_\lambda(S_2);\\ \Diamond_\lambda(\bot^0) &= \{\bot\} \cup \big(\underline{\Upsilon} \backslash \lambda(\{\bot\})\big) = \{\bot\} \cup (\underline{\Upsilon} \backslash \underline{\Upsilon}) = \{\bot\} \cup \emptyset = \bot^0;\end{aligned}$$

and hence $\Diamond_\lambda$ is an additive modal algebraic operator, while,

$$\begin{aligned}\Box_\rho(S_1 \cap S_2) &= \bot^0 \cup \rho\big(\underline{\Upsilon} \backslash (S_1 \cap S_2)\big) = \bot^0 \cup \rho\big(\underline{\Upsilon} \backslash (S_1) \cup \underline{\Upsilon} \backslash (S_2)\big)\\ &= \bot^0 \cup \big(\rho\big(\underline{\Upsilon} \backslash (S_1)\big) \cap \rho\big(\underline{\Upsilon} \backslash (S_2)\big)\big) = \Box_\rho(S_1) \cap \Box_\rho(S_1);\\ \Box_\rho(\underline{\Upsilon}) &= \bot^0 \cup \rho(\underline{\Upsilon} \backslash \underline{\Upsilon}) = \bot^0 \cup \rho(\emptyset) = \bot^0 \cup \underline{\Upsilon} = \underline{\Upsilon},\end{aligned}$$

and hence $\Box_\rho$ is a multiplicative modal algebraic operator. Thus, these two algebraic operators correspond to the logical modal operators $\Diamond_{\overline{R}_{\leq}^{-1}}$ and $\Box_{\overline{R}_{\leq}}$. □

Consequently, WMTL can be regarded just as a fragment of classical bimodal logics containing S4, whose dual is the Heyting algebra

$$L_{alg} = \mathbf{F}(\underline{\Upsilon}) = \big(\mathcal{F}(\underline{\Upsilon}), \cap, \rho\lambda \cup, \rho\lambda \Rightarrow_h, \bot^0\big),$$

where $\rho\lambda = \Gamma = T$, $\rho\lambda \cup = \sqcup$ and, from Corollary 34, $(\rho\lambda \Rightarrow_h) = \widehat{\Rightarrow}$.

Let us show that this database Heyting algebra $L_{alg} = \mathbf{F}(\underline{\Upsilon})$ can be embedded into a complete database Boolean algebra extended by three multiplicative algebraic modal operators.

Corollary 36 *The algebra* $\mathbf{BA} = (Ob_{\underline{\mathbf{DB}}_{sk}}, \cap, \cup, -, \bot^0, \underline{\Upsilon})$ *is a complete Boolean algebra with negation operator defined for every* $A = TA \in Ob_{\underline{\mathbf{DB}}_{sk}}$ *by* $-A \triangleq \bot^0 \cup (\underline{\Upsilon} \backslash A)$*, such that there is a homomorphism:*
(i) $\overrightarrow{\_} : \mathbf{BA} \to \mathbf{BA}_{\underline{\Upsilon}}$,
into the complete Boolean algebra $\mathbf{BA}_{\underline{\Upsilon}} = (\underline{\Upsilon}, \wedge_v, \vee_v, \neg_v, \bot, \overrightarrow{\underline{\Upsilon}})$ *(Proposition 70).*

Consequently, the database Heyting algebra $L_{alg} = \mathbf{F}(\underline{\Upsilon})$ *can be embedded into the N-modal algebra* $\mathbf{BA}^+ = (Ob_{\underline{\mathbf{DB}}_{sk}}, \cap, \cup, -, \bot^0, \underline{\Upsilon}, \Box_1, \Box_2, \Box_3)$*, where* $\Box_1 = \lambda_{\overline{R}_{\leq}} -$ *is the S4 algebraic universal modal operator (Example 42) and, from Lemma 25,* $\Box_2 = -\Diamond_\lambda - = -\lambda$ *and* $\Box_3 = \Box_\rho = \rho-$*, such that for every* $A = TA \in Ob_{\underline{\mathbf{DB}}_{sk}}$ *and* $B = TB \in Ob_{\underline{\mathbf{DB}}_{sk}}$*:*

(ii) $\Box_i(A \cap B) = \Box_i(A) \cap \Box_i(B)$ *and* $\Box_i(\underline{\Upsilon}) = \underline{\Upsilon}$, *for* $i = 1, 2, 3$.

Proof It is easy to verify that for each closed object $A = TA \in Ob_{\underline{\mathbf{DB}}_{sk}}$, $A \cap -A = \bot^0$ and $A \cup -A = \underline{\Upsilon}$, and hence **BA** is a Boolean algebra and complete because its lattice is complete as well. Moreover, for any two $A = TA$, $B = TB$ in $Ob_{\underline{\mathbf{DB}}_{sk}}$ we have:

(a) $\overrightarrow{A \odot B} = \overrightarrow{A} \odot \overrightarrow{B}$, for $\odot \in \{\cap, \cup, \backslash\}$,

from the fact that $\mathfrak{J}_A \subseteq A$, and $\overrightarrow{A} = \bigcup \mathfrak{J}_A$, and $\mathfrak{J}_{A \odot B} = \mathfrak{J}_A \odot \mathfrak{J}_B$ (for $\odot = \cup$ it holds directly, for $\odot = \cap$ it holds from the distributivity between $\cap$ and $\cup$, and analogously for the set difference $\odot = \backslash$). Consequently,

$$\overrightarrow{A \odot B} = \bigcup \mathfrak{J}_{A \odot B} = \bigcup (\mathfrak{J}_A \odot \mathfrak{J}_B) = \Big(\bigcup \mathfrak{J}_A\Big) \odot \Big(\bigcup \mathfrak{J}_B\Big) = \overrightarrow{A} \odot \overrightarrow{B}.$$

The mapping $\overrightarrow{\_} : Ob_{\underline{\mathbf{DB}}_{sk}} \to \underline{\Upsilon}$ is well defined and idempotent, for each $A = TA \in Ob_{\underline{\mathbf{DB}}_{sk}}$, $\overrightarrow{A} = \overrightarrow{R}_A \triangleq \bigcup_{R \in A} \overrightarrow{R} \in \underline{\Upsilon}$ with the bottom $\overrightarrow{\bot} = \bot$ and the top unary relation $\overrightarrow{\underline{\Upsilon}}$. Let us show that it satisfies the homomorphism properties for all other (non-nullary) algebraic operators:

$$\overrightarrow{A \cap B} \overset{\text{(from (a))}}{=} \overrightarrow{A} \cap \overrightarrow{B} = (\overrightarrow{\overrightarrow{A}}) \cap (\overrightarrow{\overrightarrow{B}}) = \overrightarrow{A} \wedge_v \overrightarrow{B};$$

$$\overrightarrow{A \cup B} \overset{\text{(from (a))}}{=} \overrightarrow{A} \cup \overrightarrow{B} = (\overrightarrow{\overrightarrow{A}}) \cup (\overrightarrow{\overrightarrow{B}}) = \overrightarrow{A} \vee_v \overrightarrow{B};$$

$$\begin{aligned}
\overrightarrow{(-A)} &= \overrightarrow{\bot^0 \cup (\underline{\Upsilon} \backslash A)} \overset{\text{(from (a))}}{=} \overrightarrow{\bot^0} \cup (\overrightarrow{\underline{\Upsilon}} \backslash \overrightarrow{A}) \\
&= \{\bot\} \cup \big\{\langle a\rangle \mid \langle a\rangle \in \overrightarrow{\underline{\Upsilon}}, \langle a\rangle \notin \overrightarrow{A}\big\} \\
&= \{\bot\} \cup \big\{\langle a\rangle \mid a \in \mathcal{U}, \langle a\rangle \notin (\overrightarrow{\overrightarrow{A}})\big\} = \{\bot\} \cup (\neg_v \overrightarrow{A}) \\
&= \big\{\{\langle\rangle\}\big\} \cup (\neg_v \overrightarrow{A}) = \neg_v \overrightarrow{A},
\end{aligned}$$

from the fact that an empty tuple $\langle\rangle$ is a tuple of every relation. □

It is easy to verify that the free multiplicative modal algebraic operators that extend the Boolean database algebra **BA** into the N-modal algebra $\mathbf{BA}^+$ are defined for each $A = TA \in Ob_{\underline{\mathbf{DB}}_{sk}}$ by:

$$\Box_i(A) = \big\{R \in \underline{\Upsilon} \mid \forall R' \in \underline{\Upsilon}(R, R') \in R_i \text{ implies } R' \in A\big\},$$

where $R_1 = R_{\leq}^{-1}$, $R_2 = \overline{R}_{\leq}^{-1}$ and $R_3 = \overline{R}_{\leq}$ are derived from the poset $R_{\leq}$ (or $\leq$) of the complete lattice $L_{\Upsilon} = (\underline{\Upsilon}, \leq, \wedge, \vee, \bot, R_{\Upsilon})$ in Lemma 22.

The embedding of the database Heyting algebra

$$L_{\text{alg}} = \mathbf{F}(\underline{\Upsilon}) = \big(Ob_{\underline{\mathbf{DB}}_{sk}}, \cap, \sqcup, \Rrightarrow, \bot^0\big)$$

into the N-modal algebra $\mathbf{BA}^+ = (Ob_{\underline{\mathbf{DB}}_{sk}}, \cap, \cup, -, \bot^0, \Upsilon, \Box_1, \Box_2, \Box_3)$ is given by the following equalities of these algebra operators:

$$TA \sqcup TB = \Box_3 - \Box_2 - (TA \cup TB), \quad \text{and}$$
$$TA \widehat{\Rightarrow} TB = \Box_3 - \Box_2 - \Box_1(-TA \cup TB),$$

for any $TA, TB \in Ob_{\underline{\mathbf{DB}}_{sk}}$.

9.2.4 Weak Monoidal Topos and Intuitionism

From the fact that a propositional formula ϕ is a theorem of intuitionistic logic iff it is **HA**-valid (i.e., valid in any Heyting algebra, as specified by point 8 in the introduction, Sect. 1.2), and the fact that the **DB** weak monoidal topos is based on a "closed" Heyting algebra $L_{alg} = (Ob_{\underline{\mathbf{DB}}_{sk}}, \preceq, \otimes, \oplus, \widehat{\Rightarrow}, \widehat{\neg}) = (Ob_{\underline{\mathbf{DB}}_{sk}}, \subseteq, \cap, T\cup, \widehat{\Rightarrow}, \widehat{\neg})$ (defined in Proposition 71), we conclude that each theorem of intuitionistic logic is a theorem for this weak monoidal topos.

We have seen that this subset of "closed" Heyting algebras (by a Birkhoff-polarity closure operator Γ, as demonstrated by Corollary 34) has a Kripke semantics (in Proposition 72) different from the ordinary Kripke semantics of intuitionistic logic (given in the introduction, Sect. 1.2), we can suppose that the **DB** weak monoidal topos can have more theorems than intuitionistic logic.

An intuitionistic *general* frame is a triple $\mathfrak{F} = (W, R_\leq, \mathcal{H})$, where $(W, R_\leq)$ is an intuitionistic frame (with poset $R_\leq$) and $\mathcal{H} \subseteq H(W)$ is a set of upsets (hereditary subsets of W) such that $\emptyset$ and W belong to $\mathcal{H}$, and $\mathcal{H}$ is closed under $\cap$, $\cup$ and $\Rightarrow_h$ is a relative pseudo-complement defined, for any two hereditary subsets $U, V \in \mathcal{H}$), by

$$U \Rightarrow_h V = \{a \in W \mid \forall b((a, b) \in R_\leq \text{ and } b \in U \text{ implies } b \in V)\}$$
$$= W \backslash \bigcup_{b \in U \backslash V} \{a \in X \mid (b, a) \in R_\leq\}.$$

Hence, from a general frame $\mathfrak{F}$ we derive its complete Heyting algebra

$$\mathfrak{F}_+ = (\mathcal{H}, \subseteq, \cap, \cup, \Rightarrow_h, \emptyset, W).$$

Every intuitionistic Kripke frame can be seen as a general frame, where $\mathcal{H}$ is the set $H(W)$ of *all* hereditary subsets of $\mathfrak{F}$.

- A general frame $\mathfrak{F}$ is called *refined* if for each $a, b \in W$, $(a, b) \notin R$ implies that there is a $V \in \mathcal{H}$ such that $a \in V$ and $b \notin V$;
- A general frame $\mathfrak{F}$ is called *compact* if for each $S \subseteq \mathcal{H}$ and $S' \subseteq \{X \backslash V \mid V \in \mathcal{H}\}$, if $S \cup S'$ has the finite intersection property (i.e., every intersection of finite elements of $S \cup S'$ is nonempty) then $S \cap S' \neq \emptyset$;
- A general frame $\mathfrak{F}$ is called *descriptive* if it is refined and compact;

- A general frame $\mathfrak{F} = (W, R, \mathcal{H})$ is called *rooted* if there exists $a_0 \in W$ such that $a_0 \leq b$ for all $b \in W$ and $W \backslash \{a_0\} \in \mathcal{H}$. (Notice that if W is a complete lattice then $a_0 = 0$.)

For every *complete* Heyting algebra **H**, let $\mathbf{H}_*$ denote the descriptive frame of all prime filters of **H**, i.e., $\mathbf{H}_* = (W, R, H(W))$, where W is the set of all prime filters and R is the inclusion $\subseteq$. For every descriptive frame $\mathfrak{F}$, let $\mathfrak{F}^*$ denote its complex Heyting algebra of all admissible sets of $\mathfrak{F}$. Then, the following duality (isomorphism) for Heyting algebras (Sect. 8.4 in [3]) is valid:
(*) $\mathbf{H} \simeq (\mathbf{H}_*)^*$ (and $\mathfrak{F} \simeq (\mathfrak{F}^*)_*$) (standard *duality*),
and hence for every complete Heyting algebra **H** there exists an intuitionistic descriptive frame $\mathfrak{F} = (W, R_{\leq}, \mathcal{H})$ such that **H** is isomorphic to $(\mathcal{H}, \subseteq, \cap, \cup, \Rightarrow_h, \emptyset, W)$.

An intermediate logic **L** is called *Kripke complete* if there exists a class $\mathcal{K}$ of Kripke frames such that $\mathbf{L} = Log(\mathcal{K}) = \bigcap\{Log(\mathfrak{F}) \mid \mathfrak{F} \in \mathcal{K}\}$, where $Log(\mathfrak{F}) = \{\phi \mid \mathfrak{F} \models \phi\}$. It was demonstrated that if an intermediate logic **L** is Kripke complete, then **L** is Kripke complete w.r.t. the class of its rooted frames.

An intermediate logic **L** has the *finite model property* (fmp) if for each formula which is a non-theorem of this logic there is a *finite* model (thus a frame) of it in which this formula is not true. That is, **L** has fmp if there exists a class $\mathcal{K}$ of *finite* Kripke frames such that $\mathbf{L} = Log(\mathcal{K})$ (i.e., **L** is complete w.r.t. $\mathcal{K}$). For example, a method for constructing such finite counter-models which uses the loop-checker is presented in [25], while a method not requiring loop-checker, based on contraction-free sequent calculi is presented in [20]. Moreover, it was demonstrated that every intermediate logic **L** is complete to its finitely generated rooted descriptive frames (Theorem 8.36, [3]).

The fact that intuitionistic logic **IPC** is complete with respect to its finite frames (or finite Heyting algebras) is well-known; a proof can be found in the book [3], in Corollary 2.33 (and also Theorem 2.57). In a personal correspondence, Michael Zakaryaschev wrote to me that he does not remember now who was the first to prove this result; but he guesses that it has been known since the 1940s, thus:
(**) $\mathbf{IPC} = Log(\mathcal{K}_F)$,
where $\mathcal{K}_F$ is a class of all finite rooted descriptive Kripke frames.

We defined the WMTL by the class of Heyting algebras $\mathcal{HA}_{DB}$ in Definition 64, which are the complex algebras of Kripke frames in Proposition 72, with binary accessibility relations $(W, \leq)$ equal to (also infinite) complete lattices $(\underline{\Upsilon}, \leq)$ with $\bot$ and $R_{\Upsilon} = F(\underline{\Upsilon})$ its bottom and top elements, respectively. Let as denote the class of these Kripke frames by $\mathcal{K}_{DB}$ (corresponding to the class of Heyting algebras $\mathcal{HA}_{DB}$). It is easy to verify that $\mathcal{K}_{DB}$ is a class of rooted descriptive frames, so that for each (autoreferential) frame $\mathfrak{F} \in \mathcal{K}_{DB}$, its complex algebra (see Proposition 72)

$$\mathfrak{F}^* = \mathbf{F}(\underline{\Upsilon}) = L_{\text{alg}} = \left(Ob_{\underline{\mathbf{DB}}_{sk}}, \subseteq, \cap, T\cup, \widehat{\Rightarrow}, \widehat{\neg}, \bot^0, \underline{\Upsilon}\right)$$

is such that for any two $V_1, V_2 \in Ob_{\underline{\mathbf{DB}}_{sk}} = \mathcal{F}(\underline{\Upsilon})$,

$$V_1 \sqcup V_2 = T \cup (V_1, V_2) = W \backslash \left(\bigcup_{b \in W \backslash \bigcup_{c \in V_1 \cup V_2} \{b \in W \mid c \nleq b\}} \{a \in W \mid a \nleq b\} \right);$$

$$V_1 \widehat{\Rightarrow} V_2 = W \backslash \Big(\bigcup_{b \in W \backslash \bigcup_{c \in V_3} \{b \in W | c \nleq b\}} \{a \in W \mid a \nleq b\} \Big),$$

where $V_3 = W \backslash (\bigcup_{c \in V_1 \backslash V_2} \uparrow c)$.

Dually, an autoreferential frame $\mathfrak{F}$, for any complete Heyting algebra

$$\mathbf{H} = (\underline{\Upsilon}, \leq, \wedge, \vee, \rightharpoonup, \neg, \bot, R_{\underline{\Upsilon}}) \in \mathcal{HA}_{DB},$$

is defined by Proposition 72 by reduction of this Heyting algebra to the underlying poset $\underline{\mathbf{H}} = (\underline{\Upsilon}, \leq)$. Hence, we obtain an *autoreferential duality* (in the place of the standard duality (*)):

Proposition 73 (AUTOREFERENTIAL DUALITY) *For any complete Heyting algebra $\mathbf{H} = (\underline{\Upsilon}, \leq, \wedge, \vee, \rightharpoonup, \neg, \bot, R_{\underline{\Upsilon}}) \in \mathcal{HA}_{DB}$ we obtain the following isomorphism:*
1. $\downarrow : \mathbf{H} \simeq (\underline{\mathbf{H}})^* = \mathbf{F}(\underline{\Upsilon})$.
Dually, for each autoreferential frame $\mathfrak{F} = (\underline{\Upsilon}, \leq) \in \mathcal{K}_{DB}$ we obtain the equality
2. $\mathfrak{F} = \underline{(\mathfrak{F}^*)}$.

Proof From Proposition 72, we obtain

$$(\underline{\mathbf{H}})^* = \mathbf{F}(\underline{\Upsilon}) = L_{\text{alg}} = \big(Ob_{\underline{\mathbf{DB}}_{sk}}, \subseteq, \cap, T\cup, \widehat{\Rightarrow}, \widehat{\neg}, \bot^0, \underline{\Upsilon}\big).$$

Let us show that $\downarrow$ is indeed an isomorphism between the Heyting algebra $\mathbf{H}$ and $\mathbf{F}(\underline{\Upsilon})$, that is,

$$\downarrow : (\underline{\Upsilon}, \leq, \wedge, \vee, \rightharpoonup, \neg, \bot, R_{\underline{\Upsilon}}) \simeq \big(\mathcal{F}(\underline{\Upsilon}), \subseteq, \cap, T\cup, \widehat{\Rightarrow}, \widehat{\neg}, \bot^0, \underline{\Upsilon}\big).$$

Based on the isomorphism of their underlying complete lattices (Proposition 68), it is enough to show the homomorphic property for their relative pseudo-complements. In fact, for any $a, b \in \underline{\Upsilon}$,

$$\bigvee((\downarrow a) \widehat{\Rightarrow} (\downarrow b)) = \bigvee\Big(\bigcup\{\downarrow c \mid \downarrow c \cap \downarrow a \subseteq \downarrow b\}\Big)$$

(from the homomorphic property for meets, Proposition 68)

$$= \bigvee\Big(\bigcup\{\downarrow c \mid c \wedge a \leq b\}\Big) = \bigvee(\{c \mid c \wedge a \leq b\}) = a \rightharpoonup b.$$

Thus,

$$\downarrow (a \rightharpoonup b) = \downarrow \bigvee((\downarrow a) \widehat{\Rightarrow} (\downarrow b)) = \Gamma((\downarrow a) \widehat{\Rightarrow} (\downarrow b))$$

(the closure of a closed object is equal to itself)

$$= (\downarrow a) \widehat{\Rightarrow} (\downarrow b),$$

so that the homomorphic property is satisfied.

Notice that inverse of this isomorphism $\downarrow$ is the homomorphism:

$$\bigvee : (\mathcal{F}(\underline{\Upsilon}), \subseteq, \cap, T\cup, \widehat{\Rightarrow}, \widehat{\neg}, \bot^0, \underline{\Upsilon}) \simeq (\underline{\Upsilon}, \leq, \wedge, \vee, \rightharpoonup, \neg, \bot, R_{\underline{\Upsilon}}). \quad \square$$

Let $\mathcal{K} \subset \mathcal{K}_{DB}$ be a strict subset of all finite Kripke frames, so that $\mathcal{K} \subset \mathcal{K}_F$, and consequently, $Log(K)$ can be different from **IPC** (**). In fact, we obtain:

Lemma 26 *For the subset $\mathcal{K} \subset \mathcal{K}_{DB}$ of all finite rooted descriptive frames in Proposition* 72, *based on finite complete lattices $(\underline{\Upsilon}, \leq)$, we obtain that $Log(\mathcal{K}) =$* **CPC**.

This subset $\mathcal{K}$ corresponds to the subset of Heyting algebras in $\mathcal{HA}_{DB}$ with top elements $\underline{\Upsilon}$ obtained from finite database universes $\mathcal{U}$.

Proof From the isomorphism (in Proposition 68, when $X = \underline{\Upsilon}$ with $\Gamma = T$),

$$\downarrow : (\underline{\Upsilon}, \leq, \wedge, \vee, \bot, R_{\underline{\Upsilon}}) \simeq \big(\mathcal{F}(\underline{\Upsilon}), \subseteq, \cap, \sqcup, \{\bot\}, \underline{\Upsilon}\big) = \big(Ob_{\underline{\mathbf{DB}}_{sk}}, \subseteq, \cap, T\cup, \bot^0, \underline{\Upsilon}\big),$$

the top element $\underline{\Upsilon} = \downarrow R_{\underline{\Upsilon}}$ is finite, thus defined by a finite universe $\mathcal{U}$, i.e., when $\mathfrak{I}_{\underline{\Upsilon}} = \{R = \langle d\rangle \mid d \in \mathcal{U}\}$ is finite, and hence from Lemma 23, $\underline{\Upsilon} \simeq \mathfrak{I}_{\underline{\Upsilon}}$, i.e., $\underline{\Upsilon} = T\underline{\Upsilon} = T\mathfrak{I}_{\underline{\Upsilon}})$.

From the fact that **CPC** $=$ **IPC** $+ (\phi \vee \neg\phi)$ (see the introduction, Sect. 1.2), it is enough to show that for each Heyting algebra

$$L_{\text{alg}} = \big(Ob_{\underline{\mathbf{DB}}_{sk}}, \subseteq, \cap, T\cup, \widehat{\Rightarrow}, \widehat{\neg}, \bot^0, \underline{\Upsilon}\big) \in \mathcal{HA}_{DB}$$

with finite set of atoms $\mathfrak{I}_{\underline{\Upsilon}}$, for any closed database $B = TA \in Ob_{\underline{\mathbf{DB}}_{sk}}$, $B \sqcup \widehat{\neg} B = T(B \cup \widehat{\neg} B) = \underline{\Upsilon}$. In fact, from Lemma 23, for each $C = TC \in Ob_{\underline{\mathbf{DB}}_{sk}}$, $C \simeq \mathfrak{I}_C$, that is, $C = TC = T\mathfrak{I}_C$. Thus,

$$\widehat{\neg} B = T\Big(\bigcup\{C \in Ob_{\underline{\mathbf{DB}}_{sk}} \mid C \cap B \subseteq \bot^0\}\Big) = T\Big(\bigcup\{C \in Ob_{\underline{\mathbf{DB}}_{sk}} \mid \mathfrak{I}_C \cap \mathfrak{I}_B = \emptyset\}\Big),$$

and hence $\mathfrak{I}_{\widehat{\neg} B} = \mathfrak{I}_{\underline{\Upsilon}} \backslash \mathfrak{I}_B$.

Consequently, $\mathfrak{I}_{T(B \cup \widehat{\neg} B)} = \mathfrak{I}_{B \cup \widehat{\neg} B} = \mathfrak{I}_B \cup \mathfrak{I}_{\widehat{\neg} B} = \mathfrak{I}_{\underline{\Upsilon}}$, so that, from $B \cup \widehat{\neg} B \simeq \mathfrak{I}_{B \cup \widehat{\neg} B}$ we obtain $B \sqcup \widehat{\neg} B = T(B \cup \widehat{\neg} B) = T\mathfrak{I}_{B \cup \widehat{\neg} B} = T\mathfrak{I}_{\underline{\Upsilon}} = T\underline{\Upsilon} = \underline{\Upsilon}$. $\square$

For example, the class of all database mapping systems with the universe equal to a finite domain $\mathcal{U} =$ **dom**, i.e., without Skolem constants and, consequently, with all tgds without existential quantification on the right side of material implication (Data Integration with complete information) has WMTL equal to classical propositional logic.

Consequently, from the fact that the subset of all finite Heyting algebras in $\mathcal{HA}_{DB}$ defines the CPL, the intuitionistic property of WMTL has to be determined by its subclass of *infinite* complete Heyting algebras.

Let us consider the infinite complete Heyting algebra obtained from the canonical model of **IPC**, based on theories with disjunction property.

Given a set of formulae Δ and a formula ϕ, for **IPC** (as for **CPC**) the Deduction theorem is valid: if $\Delta, \phi \vdash_{\mathrm{IL}} \psi$ then $\Delta \vdash_{\mathrm{IL}} \phi \Rightarrow \psi$. A *theory* is a set of formulae Δ that is closed under **IPC**-consequence. A set of formulae Δ has the *disjunction property* if $\phi \vee \psi \in \Delta$ implies $\phi \in \Delta$ or $\psi \in \Delta$.

The method of canonical models used in modal logics can be adapted to the case of intuitionistic logic. Instead of considering the maximal consistent sets of formulae, we consider theories with the disjunction property. The *canonical model* (or Henkin model) of **IPC** is a Kripke model $\mathcal{M}^C = (\mathfrak{F}^C, V)$ based on a frame $\mathfrak{F}^C = (W^C, R^C)$, where the set of possible worlds W^C is equal to the set of all consistent theories with the disjunction property, and the binary accessibility relation $R^C \subset W^C \times W^C$ is the inclusion. The canonical valuation $V : PR \to H(W^C)$ on $\mathfrak{F}^C$ is defined by putting for each $\Delta \in W^C$, $\Delta \in V(p)$ (i.e., $\mathcal{M}^C \models_\Delta p$) iff $p \in \Delta$.

The point of completeness proof lies in showing that, for each formula ϕ and each $\Delta \in W^C$, $\mathcal{M}^C \models_\Delta \phi$) iff $\phi \in \Delta$, or in a different notation, $\mathcal{M}^C \models_\Delta \phi$) iff $\Delta \in V(\phi)$.

The canonical frame $\mathfrak{F}^C$ is isomorphic to the frame $(\mathbf{H}_{\mathrm{IL}})_*$ obtained from the Lindenbaum–Tarski quotient Heyting algebra $\mathbf{H}_{\mathrm{IL}} = (\mathbf{IPC}_{\sim_{\mathrm{IL}}}, \sqsubseteq, \sqcap, \sqcup, \rightharpoonup, \neg)$ (introduced in Sect. 1.2), with a mapping $\sigma : W^C \to \mathbf{IPC}_{\sim_{\mathrm{IL}}}$ defined for each consistent theory (with disjunction property) $\Delta \in W^C$, $F = \sigma(\Delta) = \{[\phi] \mid \phi \in \Delta\}$. In fact, F is a prime filter:

1. If $\phi, \psi \in \Delta$ (i.e., $[\phi], [\psi] \in F$) then $\phi \wedge \psi \in \Delta$ (i.e., $[\phi] \sqcap [\psi] = [\phi \wedge \psi] \in F$);
2. If $\phi \in \Delta$ (i.e., $[\phi] \in F$) and $\vdash_{\mathrm{IL}} \phi \Rightarrow \psi$ (i.e., $[\phi] \sqsubseteq [\psi]$) then (by MP) $\psi \in \Delta$ (i.e., $[\psi] \in F$);
3. If $\phi \vee \psi \in \Delta$ (i.e., $[\phi] \sqcup [\psi] = [\phi \vee \psi] \in F$) then $\phi \in \Delta$ or $\psi \in \Delta$ (i.e. $[\phi] \in F$ or $[\psi] \in F$).

The complex Heyting algebra of this frame is

$$(\mathfrak{F}^C)^* = (H(W^C), \subseteq, \cap, \cup, \Rightarrow_h, \neg_h, \emptyset, W^C),$$

which, in the case when PR is an infinite set, is an infinite complete Heyting algebra. It is well known that a canonical model is complete for **IPC** (that is, $\vdash_{\mathrm{IL}} \phi$ iff $[\phi]$ is valid in $\mathbf{H}_{\mathrm{IL}}$).

Another example of single infinite complete Heyting algebra which is complete for **IPC** is given by well known topological interpretation of **IPC** based on the family of open sets $\mathcal{O}(W)$ (which generalize the power-set of W).

A pair $(W, \mathcal{O}(W))$ is called a *topological space* if $W \neq \emptyset$ and $\mathcal{O}(W)$ is the set of subsets of W such that $\emptyset, W \in \mathcal{O}(W)$ and:

1. If $V_1, V_2 \in \mathcal{O}(W)$ then $V_1 \cap V_2 \in \mathcal{O}(W)$; (finite intersection closure)
2. If $V_i \in \mathcal{O}(W)$, for every $i \in \mathcal{N}$, then $\bigcup_{i \in \mathcal{N}} V_i \in \mathcal{O}(W)$. (infinite union closure).

For example, when W is a set of real numbers $\mathbb{R}$, $\mathcal{O}(\mathbb{R})$ is the family of all open sets of real numbers. We denote by $\|\phi\|$ the open set of $\mathbb{R}$ assigned to a formula ϕ, so that we obtain a valuation $\|\_\| : PR \to \mathcal{O}(\mathbb{R})$ extended inductively for all formulae by:

1. $\|\phi \wedge \psi\| = \|\phi\| \cap \|\psi\|$;
2. $\|\phi \vee \psi\| = \|\phi\| \cup \|\psi\|$;

3. $\|\phi \Rightarrow \psi\| = Int(\mathbb{R}\backslash\|\phi\| \cup \|\psi\|)$;
4. $\|\neg\phi\| = Int(\mathbb{R}\backslash\|\phi\|)$,

where $Int(S)$ is the interior of the set S (i.e., the largest open subset of S). The set S is *open* if $Int(S) = S$, and *closed* if $Clo(S) \triangleq \mathbb{R}\backslash Int(\mathbb{R}\backslash S) = S$ (the operation *Int* corresponds to the universal modal operator denoted usually by '$\Box$', while *Clo* to its dual existential modal operator '$\Diamond$').

In fact, from the algebraic point of view, we obtain a Boolean algebra

$$(\mathcal{P}(\mathbb{R}), \cap, \cup, -, 0, 1)$$

with 0 equal to the empty set and 1 equal to $\mathbb{R}$, and $-$ the set-complement ($-S \triangleq \mathbb{R}\backslash S$ for any subset $S \in \mathcal{P}(\mathbb{R})$), so that $(\mathcal{P}(\mathbb{R}), \cap, \cup, -, Int, 0, 1)$ is a topo-Boolean (or S4) algebra (which satisfies normal modal properties for the multiplicative modal operator, $Int(S \cap S') = Int(S) \cap Int(S')$, $Int(1) = 1$, and "open" properties $Int(S) \subseteq S$, $Int(S) \subseteq Int(Int(S))$, for any $S, S' \in \mathcal{P}(\mathbb{R})$), that is, an N-modal algebra for $N = 1$.

Hence, for $\|\phi\| = -\{0\} = \mathbb{R}\backslash\{0\}$ we obtain $\|\neg\phi\| = Int(\{0\}) = \emptyset$, thus the excluded middle does not hold, $\|\phi \vee \neg\phi\| = (-\{0\}) \cup \emptyset \neq \mathbb{R}$. Consequently, this infinite Heyting algebra $(\mathcal{O}(\mathbb{R}), \subseteq, \cap, \cup, Int(\mathbb{R}\backslash\_ \cup \_), Int(\mathbb{R}\backslash\_), \emptyset, \mathbb{R})$ is complete for **IPC**, such that [21], $\bigvee_{i\in I} S_i = \bigcup_{i\in I} S_i$ and $\bigwedge_{i\in I} S_i = Int(\bigcap_{i\in I} S_i)$.

Notice that for the dual monotonic "closure" operator *Clo* we have the dual properties:

1. $Clo(S \cup S') = Clo(S) \cup Clo(S')$, $Clo(0) = 0$; (modal additivity property)
2. $S \subseteq Clo(S)$, $Clo(Clo(S)) \subseteq Clo(S)$; (closure property)
3. $\bigvee_{i\in I} S_i = Clo(\bigcup_{i\in I} S_i)$ and $\bigwedge_{i\in I} S_i = \bigcap_{i\in I} S_i$.

Let us consider the topological interpretation of WMTL:

Example 46 (CLOSURE OPERATION IN THE TOPOLOGICAL SPACE $(\mathbb{R}, \mathcal{O}(\mathbb{R}))$) Let us consider the topological space $(\mathbb{R}, \mathcal{O}_W(\mathbb{R}))$ where $\mathcal{O}_W(\mathbb{R})$ is the set of all open intervals $(a, b) = \{x \in \mathbb{R} \mid a < x < b\}$, $\mathcal{O}_W(\mathbb{R}) = \{(a, b) \subseteq \mathbb{R} \mid a < b\} \cup \{\emptyset\} \subset \mathcal{O}(\mathbb{R})$. So, the valuation $\|\_\| : PR \to \mathcal{O}_W(\mathbb{R})$ is extended inductively for all formulae by:

1. $\|\phi \wedge \psi\| = \|\phi\| \cap \|\psi\|$;
2.

$$\begin{aligned}\|\phi \vee \psi\| &= Clo(\|\phi\| \cup \|\psi\|) \\ &= (\min(glb(\|\phi\|), glb(\|\psi\|)), \max(lub(\|\phi\|), lub(\|\psi\|)));\end{aligned}$$

3. $\|\phi \Rightarrow \psi\| = Clo(Int(\mathbb{R}\backslash\|\phi\| \cup \|\psi\|))$;
4. $\|\neg\phi\| = Clo(Int(\mathbb{R}\backslash\|\phi\|))$,

where $Int(S)$ is the interior of the set S (i.e., the largest open subset of S), and $Clo(S)$ is the least open set (open interval) in $\mathcal{O}_W(\mathbb{R})$ such that $S \subseteq Clo(S)$ and for each $S \in \mathcal{O}_W(\mathbb{R})$, $Clo(S) = S$ (so that $Clo(\emptyset) = \emptyset$ and $Clo(\mathbb{R}) = \mathbb{R}$). It is easy to verify that *Clo* is a closure topological operator, and that $\|\neg\phi\| = \mathbb{R}$ if $\|\phi\| \neq \mathbb{R}$; $\emptyset$ otherwise.

Consequently, $\|\phi \vee \neg\phi\| = \mathbb{R}$, i.e., the excluded middle is valid and hence the closure in the topological space is a model of **CPC** and not of **IPC**.

Consequently, it is natural to consider the subclass of *infinite* Heyting algebras in $\mathcal{HA}_{DB}$ in order to examine their intuitionistic properties (because we obtained by Lemma 26 that the subset of finite Heyting algebras in $\mathcal{HA}_{DB}$ corresponds to **CPC**).

Let us show that indeed the class of Heyting algebras for databases $\mathcal{HA}_{DB}$ in Definition 64 (where for each **dom**, $L_{\text{alg}} = (Ob_{\underline{\mathbf{DB}}_{sk}}, \preceq, \otimes, \oplus, \widehat{\Rightarrow}, \widehat{\neg}) \in \mathcal{HA}_{DB}$), that is, its WMTL logic is not CPL (standard propositional) but intuitionistic (or strictly an intermediate) logic:

Theorem 19 *The WMTL does not satisfy the "excluded middle" axiom of CPL.*

Proof We have to show that the excluded middle axiom 12, $\phi \vee \neg\phi$ (see Sect. 1.2 in the introduction), does not hold in WMTL, that is, there exists a Heyting algebra in the class $\mathcal{HA}_{DB}$ with $B = TA \in Ob_{\underline{\mathbf{DB}}_{sk}}$ such that $B \oplus \widehat{\neg} B \neq \underline{\Upsilon}$.

By Definition 31 in Sect. 5.1 of Σ_R (SPJRU) algebra, for a given set of variables (relational symbols) X, the set of terms with variables $\mathcal{T}_P X$ of this algebra is defined inductively, thus finitely, by the set of its basic finitary operators. Thus, in Sect. 5.1.1, it is shown that for each database instance A, the database TA is obtained by a unique surjective mapping $\alpha_\# : \mathcal{T}_P X \twoheadrightarrow TA$, and the following initial algebra commutative diagram

$$
(\mathcal{A}, \alpha) \quad \mapsto \quad
\begin{array}{ccccc}
X = \mathcal{A} & \xrightarrow{inl_X} & \mathcal{T}_P X & \xleftarrow{inr_X} & \Sigma_P(\mathcal{T}_P X) \\
 & \searrow^{\alpha} & \downarrow{\alpha_\#} & & \downarrow{\Sigma_P(\alpha_\#)} \\
 & & TA & \xleftarrow{h_A} & \Sigma_P(TA)
\end{array}
$$

where $\alpha : \mathcal{A} \to A \subseteq TA$ defines this particular database instance A of a schema $\mathcal{A}$ with the set of relational symbols in X.

Let us consider an infinite universe of values $\mathcal{U} = \mathbf{dom} \cup SK$, with an infinite schema $\mathcal{A}$ with unary relational symbols, so that $A = \alpha^*(\mathcal{A}) = \perp^0 \cup \{R_i = \alpha(r_i) = \{\langle d_i \rangle\} \mid d_i \in \mathcal{U}\}$ = (from Lemma 23) $= \perp^0 \cup \mathfrak{J}_{\underline{\Upsilon}}$.

Let us consider the closed object $B = TA = T(\perp^0 \cup \mathfrak{J}_{\underline{\Upsilon}})$, so that TB is the set of all *finite* relations (because $\mathcal{T}_P X$ is an infinite set of all finite-length terms of SPJRU algebra, by considering that in Σ_R we have an infinite set of unary 'select' operations "_ WHERE C_i" for the conditions C_i equal to "$name = \overline{d_i}$" for each distinct $d_i \in \mathcal{U}$).

Let us show that the "excluded-middle" axiom does not hold for the Heyting algebra with this infinite universe $\mathcal{U}$,

$$L_{\text{alg}} = (Ob_{\underline{\mathbf{DB}}_{sk}}, \preceq, \otimes, \oplus, \widehat{\Rightarrow}, \widehat{\neg}) = (Ob_{\underline{\mathbf{DB}}_{sk}}, \subseteq, \cap, T\cup, \widehat{\Rightarrow}, \widehat{\neg}) \in \mathcal{HA}_{DB}.$$

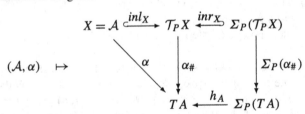

That is, $B \oplus \widehat{\neg} B \neq 1 = \underline{\Upsilon}$ where $\underline{\Upsilon}$ is the top element 1 of this complete distributive lattice defined by this universe $\mathcal{U}$ (while $\bot^0 = \{\bot\}$ is its bottom element 0). In fact,

$$\widehat{\neg} B = T\left(\bigcup\{C \in Ob_{\underline{\mathbf{DB}}_{sk}} \mid C \cap B = \bot^0\}\right) = \bot^0,$$

from the fact that $\mathfrak{J}_B = \mathfrak{J}_A = \mathfrak{J}_{\underline{\Upsilon}}$, and for each $C = TC \in Ob_{\underline{\mathbf{DB}}_{sk}}$, from Lemma 23, $\mathfrak{J}_C \subseteq C$, so that from $C \cap B = \bot^0$ it must hold that $\mathfrak{J}_C \cap \mathfrak{J}_B = \mathfrak{J}_C \cap \mathfrak{J}_{\underline{\Upsilon}} = \emptyset$, thus $\mathfrak{J}_C = \emptyset$ and, consequently, $C = \bot^0$. Hence, we obtain that $B \oplus \widehat{\neg} B = T(B \cup \widehat{\neg} B) = T(B \cup \bot^0) = T(B \cup \{\bot\}) = TB = TTA = TA \subset 1 = \underline{\Upsilon}$, because for an infinite universe $\mathcal{U}$ the top object (instance database) $\underline{\Upsilon}$ is composed of *also infinite* (but with a finite arity) relations over this universe, while TA is composed of only finite relations. □

Let us consider an example where the excluded middle does not hold:

Example 47 Let us consider a continuation of Example 44 where $A = can(\mathcal{I}, D)$ has two infinite binary relations R_1 and R_2:

$R_1 = \|r\|_{can(\mathcal{I},\mathcal{D})}$	$R_2 = \|s\|_{can(\mathcal{I},\mathcal{D})}$
a, b	b, ω_0
b, ω_1	ω_1, ω_2
ω_1, ω_3	ω_3, ω_4
ω_3, ω_5	ω_5, ω_6
...	...

Let us consider a unary but infinite relation

$$R = \{\langle\omega_{4i}\rangle \mid i = 0, 1, \dots\} = \{\langle\omega_0\rangle, \langle\omega_4\rangle, \langle\omega_8\rangle, \dots\} \in \underline{\Upsilon}.$$

Let us show that $R \notin TA \oplus \widehat{\neg} TA = T(TA \cup \widehat{\neg} TA)$. In this case, $\widehat{\neg} TA$ is a finite database with all finite relations with values in the finite set $\mathbf{dom} \backslash \{a, b\}$, and from Example 44 we obtain that $R \notin T(\{R_1, R_2\}) = TA$, and from the fact that R has no value in $\widehat{\neg} TA$, $R \notin (T\{R_1, R_2\} \cup \widehat{\neg} TA) \subseteq T(T\{R_1, R_2\} \cup \widehat{\neg} TA) = TA \oplus \widehat{\neg} TA$, so that for this instance-database $TA \in Ob_{\underline{\mathbf{DB}}_{sk}} \subseteq \underline{\Upsilon}$ the excluded middle does not hold.

A duality between finite Heyting algebras and finite posets has been known for a long time. One such description was provided by the Jongh and Troelstra in [4]. They used the join-irreducible elements to get a representation of finite Heyting algebras. So that any finite (thus complete) Heyting algebra $\mathbf{H} = (X, \leq, \wedge, \vee, \rightharpoonup, \neg, 0, 1)$ is isomorphic to the Heyting algebra of hereditary subsets

$$\mathbf{H}(X^+) = (H(X^+), \subseteq, \cap, \cup, \Rightarrow_h, \neg_h, \emptyset, X^+),$$

where X^+ is the subset of join-irreducible elements of X.

Theorem 20 *The weak excluded middle* $\neg\phi \vee \neg\neg\phi$ *is a theorem of WMTL. Thus WMTL, differently from* **IPC**, *satisfies the De Morgan laws.*

Proof We have to show that for each $A = TA \in Ob_{\underline{\mathbf{DB}}_{sk}}$, $\widehat{\neg}A \oplus \widehat{\neg}\widehat{\neg}A = \underline{\Upsilon}$, that is, for each (also infinite) relation $R \in \underline{\Upsilon}$, $R \in T(\widehat{\neg}A \cup \widehat{\neg}\widehat{\neg}A)$.

Hence $\mathfrak{J}_{\widehat{\neg}A} = \mathfrak{J}_{\underline{\Upsilon}} \backslash \mathfrak{J}_A = \mathfrak{J}_{\underline{\Upsilon}} \backslash \mathfrak{J}_{\widehat{\neg}\widehat{\neg}A}$, so that at least one of $\mathfrak{J}_{\widehat{\neg}A}$ and $\mathfrak{J}_{\widehat{\neg}\widehat{\neg}A} = \mathfrak{J}_A$ has to be finite. Let us consider the case when $\mathfrak{J}_{\widehat{\neg}\widehat{\neg}A} = \mathfrak{J}_A$ is finite. In what follows, the values in $\mathfrak{J}_A$ will be denoted by a_i while the values in $\mathfrak{J}_{\widehat{\neg}A}$ will be denoted by x_i.

Thus, $R = R_1 \cup R_2$, where R_1 in each tuple has at least one value a_i, while R_2 has only the tuples with values x_i in the infinite set $\mathfrak{J}_{\widehat{\neg}A}$, so that
1. $R_2 \in \widehat{\neg}A$ can be an infinite relation.
R_1 can be an infinite relation as well because it has not only the values a_i of the finite set $\mathfrak{J}_A$ (in every its tuple) but also the values x_i of the (possibly infinite) set $\mathfrak{J}_{\widehat{\neg}A}$.

Let us take any tuple $\mathbf{t}_I \in R_1$ composed of the finite number $n \geq 1$ of finite substrings (the arity of each relation is always finite) composed of only values in $\mathfrak{J}_A$: we denote by $i_m \geq 1$, $m = 1, \ldots, n$, the position of the mth substring of a length k_m, composed of only elements in $\mathfrak{J}_A$ (so that $i_m \leq i_l$ iff $m \leq l$), and set $i_0 = 1$, $k_0 = l_0 = 0$, and $i_{n+1} = ar(R) + 1$.

The remaining part of this tuple $\mathbf{t}_I$ is composed of substrings with the values x_i in $\mathfrak{J}_{\widehat{\neg}A}$. We can have at maximum $n+1$ such substrings, each one of the length $l_{m+1} = i_{m+1} - (i_m + k_m)$, $m = 0, 1, \ldots, n$.

The I is a complete ordered set of indexes of values a_i in this tuple $\mathbf{t}_I$, thus equal to:

$$I = \langle i_1, \ldots, i_1 + k_1 - 1, i_2, \ldots, i_2 + k_2 - 1, \ldots, i_n, \ldots, i_n + k_n - 1 \rangle.$$

The complete ordered set of indexes of values x_i in $\mathfrak{J}_{\widehat{\neg}A}$ in this tuple $\mathbf{t}_I$ is then

$$J = \langle i_m + k_m, \ldots, i_{m+1} - 1, i_{m+1} + k_{m+1}, \ldots, i_{m+2} - 1, , \ldots, i_l + k_l, \ldots, i_{l+1} - 1 \rangle,$$

where $m = 0$ if $i_1 > 1$; 1 otherwise, and $l = n$ if $i_n + k_n < i_{n+1} = ar(R) + 1$; $n - 1$ otherwise.

Let us consider, for example, the tuple

$$\mathbf{t}_I = \langle x_1 x_2 x_3 a_1 a_2 x_4 x_5 a_3 a_4 a_5 x_6 a_6 x_7 x_8 x_9 \rangle,$$

so that $n = 3$; $i_1 = 4, i_2 = 8,\ i_3 = 12, i_4 = ar(R) + 1 = 16$; $k_1 = 2,\ k_2 = 3,\ k_3 = 1$; and $l_1 = 3,\ l_2 = 2,\ l_3 = 1,\ l_4 = 3$, and hence,

$$I = \langle 4, 5, 8, 9, 10, 12 \rangle \quad \text{and} \quad J = \langle 1, 2, 3, 6, 7, 11, 13, 14, 15 \rangle.$$

Let us define a function $f : \{1, \ldots, n\} \to \mathcal{N}$ such that $f(1) = 0$ and $f(l) = \sum_{1 \leq j \leq l-1} k_j$, for $2 \leq l \leq n$. Then we define the relation:

$$R_I = R_1 \text{ WHERE } \bigwedge_{1 \leq l \leq n} \left(\bigwedge_{0 \leq m \leq k_l - 1} (nome_{i_{l+m}} = a_{f(l)+m+1}) \right),$$

such that it is exactly a subrelation of R_1 with all tuples that have at positions defined by ordered list I the values a_i in $\mathbf{t}_I$, that is, $\pi_I(R_I) = \pi_I(\{\mathbf{t}_I\})$ (equal to $\{\mathbf{t}\}$, with the tuple $\mathbf{t} = \langle a_1 a_2 a_3 a_4 a_5 a_6 \rangle$ in the example above).

Consequently,

2. $R_1 = \bigcup_{I \in S} R_I$, for a *finite* set S.

We recall that S is finite because the set $\mathfrak{J}_A$ is a finite set of values, so that R_1 is a finite union, thus a *finite-length* term of Σ_R (SRRJU) algebra (Definition 31 in Sect. 5.1), and, in order to show that $R_1 \in T(\widehat{\neg} A \cup \widehat{\neg}\widehat{\neg} A)$, it is enough to show that $R_I \in T(\widehat{\neg} A \cup \widehat{\neg}\widehat{\neg} A)$ for all $I \in S$. In fact, the projection

3. $R_I^{(1)} \triangleq \pi_J(R_I) \in \widehat{\neg} A$

is a (also infinite) relation with only values in $\mathfrak{J}_{\widehat{\neg} A} = \mathfrak{J}_{\underline{\Upsilon}} \backslash \mathfrak{J}_A$, while

4. $R_I^{(2)} \triangleq \{\mathbf{t}\} = \pi_I(R_I) \in \widehat{\neg}\widehat{\neg} A$,

is the relation with a single tuple composed of all values in the finite set $\mathfrak{J}_A$. Thus, let us show that from these two relations, $R_I^{(1)}, R_I^{(2)} \in \widehat{\neg} A \cup \widehat{\neg}\widehat{\neg} A$ we obtain that $R_I \in T(\widehat{\neg} A \cup \widehat{\neg}\widehat{\neg} A)$. First of all, we define the relation:

5. $R_I^{(3)} \triangleq (R_I^{(2)} \text{ TIMES } R_I^{(1)}) \in T(\widehat{\neg} A \cup \widehat{\neg}\widehat{\neg} A)$, and hence

6. $R_I = \pi_{M_I}(R_I^{(2)} \text{ TIMES } R_I^{(1)}) \in T(\widehat{\neg} A \cup \widehat{\neg}\widehat{\neg} A)$,

where, for $h(m) = \sum_{0 \le j \le m} k_j$ and $g(m) = h(n) + \sum_{0 \le j \le m} l_j$, for $0 \le m \le n$, we have the following list of indexes:

$$M_I = \big\langle \overbrace{1 + g(0), \ldots, g(1)}^{l_1}, \overbrace{1 + h(0), \ldots, h(1)}^{k_1}, \ldots,$$

$$\overbrace{1 + g(m), \ldots, g(m+1)}^{l_{m+1}}, \overbrace{1 + h(m), \ldots, h(m+1)}^{k_{m+1}}, , \ldots,$$

$$\overbrace{1 + g(n-1), \ldots, g(n)}^{l_n}, \overbrace{1 + h(n-1), \ldots, h(n)}^{k_n}, \overbrace{1 + g(n), \ldots, g(n+1)}^{l_{n+1}} \big\rangle,$$

where, if $l_1 = 0$, we eliminate the first sublist l_1, and, if $l_{n+1} = 0$, we eliminate the last sublist l_{n+1} from the list of indexes above.

In fact, $ar(R_I^{(3)}) = ar(R)$, and M_I is only the permutation of columns of $R_I^{(3)}$ corresponding to R_I.

Consequently, from facts 1–4, for all $R_I^{(1)}, R_2 \in \widehat{\neg} A$, $R_I^{(2)} \triangleq \{\mathbf{t}\} \in \widehat{\neg}\widehat{\neg} A$, $I \in S$ (where S is *finite*), for any $R \in \underline{\Upsilon}$,

$$R = R_2 \bigcup_{I \in S} \pi_{M_I}(R_I^{(2)} \text{ TIMES } R_I^{(1)}) \in T(\widehat{\neg} A \cup \widehat{\neg}\widehat{\neg} A).$$

Thus, $T(\widehat{\neg} A \cup \widehat{\neg}\widehat{\neg} A) = \widehat{\neg} A \oplus \widehat{\neg}\widehat{\neg} A = \underline{\Upsilon}$.

If $\mathfrak{J}_{\widehat{\neg}\widehat{\neg} A} = \mathfrak{J}_A$ is infinite (it happens only when we have the cyclic tgds with existential quantifiers on the right-hand side of implication, so that *all* infinite Skolem constants in *SK* are inserted into a subset of relations of a database A), then $\mathfrak{J}_{\widehat{\neg} A} = \mathfrak{J}_{\underline{\Upsilon}} \backslash \mathfrak{J}_A$ is finite (in this case $\widehat{\neg} A$ contains only a finite number of values in **dom**, with $\mathcal{U} = \mathbf{dom} \cup SK$) and hence we obtain analogous (dual) case to the previous one. □

Let us show that the *weak* excluded middle holds for $TA \in Ob_{\underline{\mathbf{DB}}_{sk}}$ for which instead the excluded middle does not hold:

Example 48 Let us consider a continuation of Example 47 where $A = can(\mathcal{I}, D)$ has two infinite binary relations R_1 and R_2:

$R_1 = \|r\|_{can(\mathcal{I},\mathcal{D})}$	$R_2 = \|s\|_{can(\mathcal{I},\mathcal{D})}$
a, b	b, ω_0
b, ω_1	ω_1, ω_2
ω_1, ω_3	ω_3, ω_4
ω_3, ω_5	ω_5, ω_6
...	...

Let us consider a unary but infinite relation

$$R = \{\langle\omega_{4i}\rangle \mid i = 0, 1, \dots\} = \{\langle\omega_0\rangle, \langle\omega_4\rangle, \langle\omega_8\rangle, \dots\} \in \underline{\Upsilon},$$

so that $R \notin TA$, and $R \notin TA \oplus \widehat{\neg} TA = T(TA \cup \widehat{\neg} TA)$ (the excluded middle does not hold).

Let us show that, instead, the weak excluded middle is valid, so that

$$R \in \widehat{\neg} TA \oplus \widehat{\neg}\widehat{\neg} TA = T(\widehat{\neg} TA \cup \widehat{\neg}\widehat{\neg} TA).$$

Hence, $\widehat{\neg} TA$ is a finite database with all finite relations with values in the finite set $\mathbf{dom}\backslash\{a, b\}$. However, $\widehat{\neg}\widehat{\neg} TA = T(\bigcup\{TB \in Ob_{\underline{\mathbf{DB}}_{sk}} \mid TB \cap \widehat{\neg} TA = \bot\})$. So let us consider a database $TB = T(\{R, \bot\}) \in Ob_{\underline{\mathbf{DB}}_{sk}}$ that satisfies $TB \cap \widehat{\neg} TA = \bot^0$ and $R \in B \subseteq TB$, and hence

$$R \in T\Big(\bigcup\{TB \in Ob_{\underline{\mathbf{DB}}_{sk}} \mid TB \cap \widehat{\neg} TA = \bot\}\Big) = \widehat{\neg}\widehat{\neg} TA$$
$$\subseteq (\widehat{\neg} TA \cup \widehat{\neg}\widehat{\neg} TA) \subseteq T(\widehat{\neg} TA \cup \widehat{\neg}\widehat{\neg} TA).$$

Consequently, $R \in T(\widehat{\neg} TA \cup \widehat{\neg}\widehat{\neg} TA) = \widehat{\neg} TA \oplus \widehat{\neg}\widehat{\neg} TA = \underline{\Upsilon}$.

It is well known that the logic of weak excluded middle is a strict intermediate logic **KC** (called Jankov's or De Morgan logic as well).

It is (by Jankov's Theorem [9]) the strongest intermediate logic extending **IPC** that proves the same negation-free formulae as **IPC** and can be embedded into S4.2 modal logic where the accessibility relation $R \subseteq W^2$ (a partial order) of the frame is reflexive, transitive and *directed* (convergent) as well. That is, if $(a, b) \in R$ and $(a, c) \in R$ then there exists a $d \in W$ such that $(b, d), (c, d) \in R$.

Notice that this property is satisfied if the poset R is a complete lattice, where $d = b \vee c$, for example.

With this final result we demonstrated a concrete difference between the intuitionistic logic of standard topoi and our superintuitionistic, or intermediate, (Jankov's or De Morgan) logic of our weak monoidal topos for the database mapping systems.

9.3 Review Questions

1. What are topoi and what are the most known examples for them? Why is the **DB** category not a topos (provide at least two main reasons)? What is the relationship between the database metric space and the database lattice? Why does the maximal distance between two databases mean that these two databases have no view in common? Does it mean that the intersection of the sets of values (from the universe $\mathcal{U}$) contained in these two databases is empty? Can we choose another metrics, based on the merging instead on the matching operation?
2. The main idea, in order to define the subobject classifier in the **DB** category, is to interpret the *true* and *false* arrows of the commutative diagrams for the subobject classifier by the top and bottom values, respectively, in the complete algebraic database lattice L_{DB}. Is the interpretation of this database lattice, as the set of the "logic values" of a particular many-valued propositional database logic, appropriate in order to provide the underlying propositional logic? Can the meet and join lattice operators be interpreted by the logic operators of conjunction and disjunction? Is it possible to obtain the pullback diagram for the logic dual (with "false" arrow) so that it has the same vertex $LimP = A \times B$ instead of $\Upsilon \times B$?
3. Enumerate the topos properties which are still valid in the **DB** category. Enumerate the most important features of the topos which are not satisfied in the **DB** category. What is the consequence of the non-well-pointedness property in the **DB** category w.r.t. the extensionality principle (which is valid in the **Set** category of sets and functions between them)? Explain why in a standard topos if its initial object is equal (up to isomorphism) to its terminal object then all objects are isomorphic, and hence such categories degenerate? Why doesn't this degeneration happen in the **DB** category, where from the duality property we have that the initial and terminal objects are equal to the zero object $\bot^0$? Can we then consider that **DB** is a kind of a generalization of a standard topos where the isomorphism of the initial and terminal objects does not cause the degeneration, and why is such a generalization important not only for the database mappings applications but also for the general theoretical point of view?
4. Demonstrate that the subobject classifier pullback diagram with $A = \Omega \times \Omega$,

$$\begin{array}{ccc} (\Omega \times \Omega) \times \Upsilon & \xrightarrow[p^A]{} & \Omega \times \Omega \\ \Big\downarrow p^{\Upsilon} & (\Omega) & \Big\downarrow \bigwedge \\ \Upsilon & \xrightarrow{true} & \Omega \end{array}$$

defines the "conjunction truth-function" characteristic arrow

$$\bigwedge = C_f = [id_\Omega, id_\Omega] : \Omega \times \Omega \rightarrow \Omega$$

for the subobject monomorphism $f = \langle true, true \rangle : \Upsilon \hookrightarrow \Omega \times \Omega$, with $B = \Upsilon$, and provide the exact expressions for the arrows p^A and p^B. Demonstrate that the diagram above reduces to the analogous diagram in the standard topos:

$$\begin{array}{ccc} \Upsilon & \xrightarrow[f = \langle true, true \rangle]{} & \Omega \times \Omega \\ \Big\downarrow id & (\Omega) & \Big\downarrow \bigwedge \\ \Upsilon & \xrightarrow{true} & \Omega \end{array}$$

Why is $(\Upsilon \times \Upsilon) \times \Upsilon \approx \Upsilon$ true? Show that for any two logic values (i.e., closed objects) $\widetilde{f}, \widetilde{g} \in L_{DB}$, for any two morphisms $f, g : \Upsilon \rightarrow \Omega$, their logic conjunction $\widetilde{f} \cap \widetilde{g}$ is equal to the information flux of the arrow

$$\bigwedge \circ \langle f \circ g, \bot^1 \rangle : \Upsilon \rightarrow \Omega.$$

5. Demonstrate that the subobject classifier pullback diagram with $A = \Omega \times \Omega$,

$$\begin{array}{ccc} \Upsilon \times \bot^0 & \xrightarrow[p^A]{} & \Omega \times \Omega \\ \Big\downarrow p^{\Upsilon} & & \Big\downarrow \bigvee \\ \Upsilon & \xrightarrow{false} & \Omega \end{array}$$

defines the "disjunction truth-function" characteristic arrow

$$\bigvee = 1 - C_f = \langle id_\Omega, id_\Omega \rangle : \Omega \times \Omega \rightarrow \Omega$$

for the subobject monomorphism $f = \langle \bot^1, \bot^1 \rangle : \bot^0 \hookrightarrow \Omega \times \Omega$, so that $B = \bot^0$ and the arrow $p^A = false \uplus \bot^1$. Demonstrate that the diagram above reduces to the analogous diagram in the standard topos:

$$\begin{array}{ccc} \Upsilon & \xrightarrow{\langle false, false \rangle} & \Omega \times \Omega \\ \big\downarrow id & & \big\downarrow \bigvee \\ \Upsilon & \xrightarrow{false} & \Omega \end{array}$$

Show that for any two logic values (i.e., closed objects) $\widetilde{f}, \widetilde{g} \in L_{DB}$, for any two morphisms $f, g : \Upsilon \to \Omega$, their logic disjunction $T \cup (\widetilde{f}, \widetilde{g}) = T(\widetilde{f} \cup \widetilde{g})$ is equal to the information flux of the arrow $\bigvee \circ \langle f, g \rangle : \Upsilon \to \Omega$.

6. Based on the points 4 and 5 above, we conclude that the basic and derived truth-values are expressed by the homset $\mathbf{DB}(\Upsilon, \Omega)$, with bottom arrow $false : \Upsilon \to \Omega$ and top arrow $true : \Upsilon \to \Omega$, and hence, dually, by their information fluxes $Ob_{\mathbf{DB}_{sk}} = \{\widetilde{f} \mid f \in \mathbf{DB}(\Upsilon, \Omega)\}$, with the bijection $\beta : \mathbf{DB}(\Upsilon, \Omega) \cong Ob_{\mathbf{DB}_{sk}}$ provided by Lemma 19. By reduction to simple databases, we obtain the complete sublattice of truth-values

$$L_{\underline{DB}} = (Ob_{\underline{\mathbf{DB}}_{sk}}, \preceq, \otimes, \oplus) = (Ob_{\underline{\mathbf{DB}}_{sk}}, \subseteq, \cap, T\cup) \subset L_{DB}$$

in Definition 63. Why can't we simply extend this lattice by the standard negation operator (i.e., the set difference $\underline{\Upsilon}\backslash\_$) and hence obtain a Boolean algebra for the standard propositional logic?

7. How did we define, based on the Birkhoff polarity over the complete infinitary distributive lattice $(W, \leq, \wedge, \vee, 0, 1)$, a Dedekind–McNeile closure operator and the autoreferential Kripke-like semantics for such lattices? Why is it interesting for us to consider the negation operators which can be derived by the Birkhoff polarity from the incompatibility relations equal to partial orders? We obtained a homomorphism $\Gamma : \mathbf{H}(W) \to \mathbf{F}(W)$ between the standard Heyting algebras, whose elements are the set of the hereditary subsets of W denoted by $H(W)$, and the Heyting algebras of *closed* hereditary subsets $\mathcal{F}(W) \subseteq H(W)$. Why do we need to work with this second, that is, the closed version of Heyting algebras?

8. Why can't the "vectorization" operation $\overrightarrow{\_}$ which transforms each database instance A into an unary relation $\overrightarrow{A}$ be used in order to have an observationally equivalent relation of a given database, such that $T(\{\overrightarrow{A}\}) = TA$? For which strict subclass of the database mapping systems is such an equivalence still valid? What are the fundamental properties of the mapping $F : Ob_{\underline{DB}} \to \underline{\Upsilon}$ introduced in Proposition 69, and in which way can we represent any closed object TA in the **DB** category from the relation $F(A)$?

9. The weak monoidal topos database logic WMTL is defined by a strict subclass of the Heyting algebras. Why can we consider this weak monoidal topos logic as an abstract-behavior logic? In which way do we provide an embedding of WMTL into the intuitionistic propositional bimodal logics, and how can we translate the closure operator Γ (analogous to the database power-view operator T) obtained from Dedekind–McNeile Galois connection? What is the Gödel-like translation from WMTL into the S4 extended bimodal logic? By using

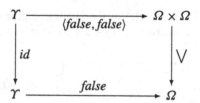

the autoreferential Kripke model of the Heyting algebra $\mathbf{F}(W)$ of the closed hereditary subsets, for any two $V_1, V_2 \in \mathcal{F}(W)$, demonstrate that

$$V_1 \sqcup V_2 = \Gamma \cup (V_1, V_2) = W \setminus \Big(\bigcup_{b \in W \setminus \bigcup_{c \in V_1 \cup V_2} \{b \in W \mid c \nleq b\}} \{a \in W \mid a \nleq b\} \Big)$$

10. How can the database Heyting algebra $L_{alg} = F(\underline{\Upsilon})$, of closed simple objects in the **DB** category, be embedded into the N-modal extension of the Boolean algebra $\mathbf{DB}_{\underline{\Upsilon}}$ obtained by "vectorization" of the instance databases? Explain what is the relationship between the weak monoidal topos **DB**, its Heyting algebra $L_{alg} = F(\underline{\Upsilon})$ and the intuitionistic logics **IPC**. What is the finite model property (fmp) for the intermediate logics, and how is it used to define **IPC** by the class $\mathcal{K}_F$ of all finite rooted descriptive Kripke frames?
11. WMTL is defined by the class of Heyting algebras $\mathcal{HA}_{DB}$ in Definition 64, which are the complex algebras of Kripke frames in Proposition 72, with the class $\mathcal{K}_{DB}$ of these Kripke frames (corresponding to the class of Heyting algebras $\mathcal{HA}_{DB}$). Show that $\mathcal{K}_{DB}$ is a class of rooted descriptive frames, so that for each (autoreferential) frame $\mathfrak{F} \in \mathcal{K}_{DB}$, its complex algebra is

$$\mathfrak{F}^* = \mathbf{F}(\underline{\Upsilon}) = L_{alg} = \big(Ob_{\underline{\mathbf{DB}}_{sk}}, \subseteq, \cap, T\cup, \widehat{\Rightarrow}, \widehat{\neg}, \bot^0, \underline{\Upsilon}\big).$$

Explain the meaning of the autoreferential Kripke semantics for any complete Heyting algebra $\mathbf{H} = (\underline{\Upsilon}, \leq, \wedge, \vee, \rightharpoonup, \neg, \bot, R_{\Upsilon}) \in \mathcal{HA}_{DB}$ by the isomorphism

$$\downarrow : \mathbf{H} \simeq (\underline{\mathbf{H}})^* = \mathbf{F}(\underline{\Upsilon}).$$

12. What is the strict subclass $\mathcal{K}$ of the class of rooted descriptive frames $\mathcal{K}_{DB}$ (corresponding to the class of Heyting algebras $\mathcal{HA}_{DB}$) such that $Log(\mathcal{K})$ corresponds to the classical propositional logic **CPC**? What is the relationship between the class of all database mapping systems without incomplete information (which are obtained by using only the FOL tgds and egds) and this class $\mathcal{K}$ of rooted descriptive frames? Why is the intuitionistic property of WMTL determined only by its subclass of *infinite* complete Heyting algebras? What are other significative examples of the infinite complete Heyting algebras? Explain why the "exclude-middle" axiom is not satisfied by WMTL, while it satisfies De Morgan laws.
13. (Open problem) It is demonstrated that WMTL is a Jankov's or De Morgan intermediate logic. Are there other theorems of the weak monoidal database topos logic (based on its infinitary distributive complete lattice of closed database instances in **DB** category) which would produce a strictly more powerful intermediate logic w.r.t. this De Morgan logic?

References

1. N.D. Belnap, A useful four-valued logic, in *Modern Uses of Multiple-Valued Logic*, ed. by J.-M. Dunn, G. Epstein (D. Reidel, Dordrecht, 1977)

2. G. Birkhoff, *Lattice Theory*. Amer. Math. Soc. Colloquium Publications, vol. XXV (1940). Reprinted 1979
3. A. Chagrov, M. Zakharyashev, *Modal Logic*. Oxford Logic Guides, vol. 35 (Oxford University Press, Oxford, 1997)
4. D. de Jongh, A.S. Troelstra, On the connection of partially ordered sets with some pseudo-Boolean algebras. Indag. Math. **28**, 317–329 (1966)
5. S. Eilenberg, G.M. Kelly, Closed categories, in *Proceedings of the Conference on Categorical Algebra*, La Jolla, 1965 (Springer, Berlin, 1966), pp. 421–562
6. M.C. Fitting, Bilattices and the semantics of logic programming. J. Log. Program. **11**, 91–116 (1991)
7. M. Ginsberg, Multivalued logics: a uniform approach to reasoning in artificial intelligence. Comput. Intell. **4**, 265–316 (1988)
8. R. Goldblat, *Topoi: The Categorial Analysis of Logic* (North-Holland, Amsterdam, 1979)
9. V.A. Jankov, Ob ischislenii slabogo zakona iskljuchennogo tret'jego. Izv. Akad. Nauk SSSR, Ser. Mat. **32**(5), 1044–1051 (1968)
10. S.A. Kripke, Semantical analysis of intuitionistic logic I, in *Formal Systems and Recursive Functions*, ed. by J.N. Crossley, M.A.E. Dummett (North Holland, Amsterdam, 1965), pp. 92–130
11. S.M. Lane, T. Moerdijk, *Sheaves in Geometry and Logic: A First Introduction to Topos Theory* (Springer, Berlin, 1991)
12. F.W. Lawvere, Introduction, in *Toposes, Algebraic Geometry and Logic*. Lecture Notes in Mathematics, vol. 274 (Springer, Berlin, 1972)
13. W. Lawvere, Metric spaces, generalized logic, and closed categories, in *Rendiconti del Seminario Matematico e Fisico di Milano, XLIII* (Tipografia Fusi, Pavia, 1973)
14. C.I. Lewis, Implication and the algebra of logic. Mind **21**, 522–531 (1912)
15. S. Maclane, Sets, topoi, and internal logic in categories, in *Logic Colloquium 1973*, ed. by H.E. Rose, J.C. Shepherdson (North-Holland, Amsterdam, 1975), pp. 119–134
16. Z. Majkić, Autoreferential semantics for many-valued modal logics. J. Appl. Non-Class. Log. **18**(1), 79–125 (2008)
17. Z. Majkić, Bilattices, intuitionism and truth-knowledge duality: concepts and foundations. J. Mult.-Valued Log. Soft Comput. **14**(6), 525–564 (2008)
18. Z. Majkić, A new representation theorem for many-valued modal logics. arXiv:1103.0248v1 (2011), 01 March, pp. 1–19
19. R. Paré, C0-limits in topoi. Bull. Am. Math. Soc. **80**, 556–561 (1974)
20. L. Pinto, R. Duckhoff, Loop-free construction of counter-models for intuitionistic propositional logic, in *Symposia Gausianna Conference*, New York (1995), pp. 225–232
21. H. Rasiowa, R. Sikorski, *The Mathematics of Metamathematics*, 3rd edn. (PWN-Polisch Scientific Publishers, Warsaw, 1970)
22. G. Servi, On modal logics with an intuitionistic base. Stud. Log. **36**, 141–149 (1977)
23. V.B. Sheltman, Kripke type semantics for propositional modal logics with intuitionistic base, in *Modal and Tense Logics*, Institute of Philosophy, USSR Academy of Sciences (1979), pp. 108–112
24. M. Tierney, Sheaf theory and the continuum hypothesis, in *Lecture Notes in Mathematics*, vol. 274 (Springer, Berlin, 1972)
25. J. Underwood, A constructive completeness proof for the intuitionistic propositional calculus, in *Proc. of Second Workshop on Theorem Proving and Analytic Tableaux and Related Methods*, Marselle, France, April (1993)
26. F. Wolter, M. Zakharyaschev, On the relation between intuitionistic and classical modal logics. Algebra Log. **62**, 73–92 (1997)
27. F. Wolter, M. Zakharyaschev, Intuitionistic modal logics as fragments of classical bimodal logics, in *Logic at Work*, ed. by E. Orlowska (Springer, Berlin, 1999), pp. 168–186

Index

Z. Majkić, *Big Data Integration Theory*, Texts in Computer Science,
DOI 10.1007/978-3-319-04156-8,

Printed in the United States
By Bookmasters